Handbook of
Industrial Chemistry

Volume 1

Unit Operations, Heavy Inorganic Chemicals, Explosives & Propellants, Catalysis, Corrosion, Electroplating & Electroless Plating, Pollution Prevention, Environment & Energy, Safety Considerations in Chemical Process

Handbook of Industrial Chemistry

Volume 1
Unit Operations, Heavy Inorganic Chemicals, Explosives & Propellants, Catalysis, Corrosion, Electroplating & Electroless Plating, Pollution Prevention, Environment & Energy, Safety Considerations in Chemical Process

K.H. Davis
M.Sc., Ph.D., F.R.I.C.

F.S. Berner
M.Sc.

Edited by

S.C. Bhatia
B.E. (Chemical), M.B.A.

CBS

CBS PUBLISHERS & DISTRIBUTORS PVT. LTD.
NEW DELHI • BENGALURU • CHENNAI • KOCHI • MUMBAI • PUNE

ISBN: 81-239-1056-8

First Edition: 2004
Reprint: 2005, 2015

Published by:
Satish Kumar Jain for CBS Publishers & Distributors Pvt. Ltd.,
4819/XI Prahlad Street, 24 Ansari Road, Daryaganj, New Delhi - 110002
delhi@cbspd.com, cbspubs@airtelmail.in • www.cbspd.com
Ph.: 23289259, 23266861, 23266867 • Fax: 011-23243014

Corporate Office: 204 FIE, Industrial Area, Patparganj, Delhi - 110 092
Ph: 49344934 • Fax: 011-49344935
E-mail: publishing@cbspd.com • publicity@cbspd.com

Branches:
• *Bengaluru:* 2975, 17th Cross, K.R. Road, Bansankari 2nd Stage, Bengaluru - 70
  Ph: +91-80-26771678/79 • Fax: +91-80-26771680
  E-mail: cbsbng@gmail.com, bangalore@cbspd.com
• *Chennai:* No. 7, Subbaraya Street, Shenoy Nagar, Chennai - 600030
  Ph: +91-44-26681266, 26680620 • Fax: +91-44-42032115
  E-mail: chennai@cbspd.com
• *Kochi:* 36/14, Kalluvilakam, Lissie Hospital Road, Kochi - 682018
  Ph: +91-484-4059061-65 • Fax: +91-484-4059065
  E-mail: cochin@cbspd.com
• *Mumbai:* 83-C, Dr. E. Moses Road, Worli, Mumbai - 400018
  Ph: +91-9833017933, 022-24902340/41 • E-mail: mumbai@cbspd.com
• *Pune:* Bhuruk Prestige, Sr. No. 52/12/2+1+3/2,
  Narhe, Haveli (Near Katraj-Dehu Road Bypass), Pune - 411041
  Ph: +91-20-64704058/59, 32342277 • E-mail: pune@cbspd.com

Representatives:
• Hyderabad: 0-9885175004
• Nagpur: 0-9021734563
• Vijayawada: 0-9000660880
• Kolkata: 0-9831437309, 0-9051152362
• Patna: 0-9334159340

Printed at:
J.S. Offset Printers, Delhi

Preface

This volume dealing with inorganic aspects is the first volume of the comprehensive reference cum text book on industrial chemistry. This useful handbook is specifically intended for practising chemical engineers, industrial chemists, research students and students undergoing graduate or postgraduate studies in chemistry.

This useful handbook endeavours to a comprehensive, accurate and most recent information on industrial chemistry. As the academic education becomes more complex it is essential for industrial chemists and research students to be familiar with the realisation of scientific principles and latest developments in technological aspects.

The encyclopaedic range of the subject of industrial chemistry has been condensed into a handy, easy to use reference book presented in a concise, compact and lucid manner, stressing the essential facts and latest methods for students–thus forming the necessary criteria for a systematic study of industrial chemistry. The reference text book clarifies and explains how chemical processing transform raw materials into usable and profitable products by chemicals.

Each chapter covers an important aspects of industrial chemistry with accurate and upto date account of each topic. Chapter 1 gives an overview of chemical process industry and includes the systematic analysis of chemical processes along with the role of 'Dot com' in chemical industry. Chapter 2 is devoted to unit operations which are particular kind of physical changes used in the industrial production of various chemicals and related materials. Chemical operations may be classified as techniques of operation, specialised operation and unit operations. Chapter 3 focuses on high pressure processes which is one of the outstanding developments of chemical industry during the past two decades.

Chapter 4 explains the latest techniques of treatment of industrial waste-water, including sludge disposal, membrane processes and their applications. Chapter 5 is devoted to various instrumental methods, acid base titrations, chemical methods of separation, chromatography and melt crystallisation. Chapter 6 discusses manufacture, properties and uses of salt and salt based products such as–sodium chloride, soda ash, chlorine, baking soda, caustic soda, along with by-products of salt industry. Chapter 7 familiarises the students with sulphur, sulphuric acid, hydrochloric acid and various other chemicals, such as hydrofluoric acid, fluorine, aluminium, boron, copper, strontium and their compounds. Chapter 8 is devoted to various industrial gases such as oxygen, carbon dioxide, hydrogen, acetylene, along with their separation from air. Chapter 9 focuses on various types of fertilisers, phosphorus, potassium salts and their compounds. Kinetics and synthetic aspects of ammonia along with manufacturing aspects are also discussed. Nitrogen, an extremely useful and versatile element with its increasingly numerous applications is the subject of chapter 10. Synthetic ammonia, urea, nitric acid – their preparation properties uses, reactions and equilibrium, rate of catalysis of the reactions are also discussed in the chapter.

Chapter 11 is devoted to chemical explosives and propellants which highlights the potential energy of explosives, properties, manufacture and composition. Electrochemical process has completely revolutionised the production of certain primary products and at such lowered cost as to permit the development of new secondary industries utilising these cheap raw materials. Keeping this in mind chapter 12 focuses on electrothermal and electrochemical processes.

Chapter 13 is devoted to insulating materials and describes briefly the electrical, thermal, mechanical and chemical properties of insulating materials. The other materials (which are used as insulation) gases, liquids, (mineral oils, chlorinated hydrocarbons, fluorochemicals), and solids (coating powders, insulating varnishes, mica insulations) are also discussed. Chapter 14 focuses on cement and lime and discusses manufacturing, uses and applications along with chemical admixtures, coloured concrete and utilisation of flyash as part replacement of cement in concrete. Chapter 15 is devoted to ceramic, refractories and potteries. The basic raw materials, manufacturing aspects and types of kilns used are briefly discussed. Chapter 16 elaborates on glass, which has many uses because of its transparency, high resistance to chemical attack etc. Chapter 17 is devoted to photographic products industry and discusses the various photographic processes including the special applications of photography. Chapter 18 explores nuclear industries based on radium, uranium and plutonium and emphasis the importance of protection from radioactivity as well as waste disposal. Chapter 19 discusses metallurgy of iron and steel along with various compounds of iron and steel and chapter 20 focuses on metallurgy of various non ferrous metals such as copper, silver, gold, zinc, cadmium, mercury, nickel, and aluminium.

Chapter 21 focuses on corrosion and its prevention. Various types of corrosion, formation of corrosion in industries and preventive methods are explained in detail. Chapter 22 deals with electroplating and electroless plating. Briefly discussed are-substrate in electroplating, cleaner production options to electroless plating process, electromigration and metal treatment aspects. Chapter 23 is devoted to mechanism of catalysts and biocatalysts. Chapter 24 focuses on pollution prevention and waste minimisation. Various environmental pollutants such as, air, water, solid wastes, radio active wastes, noise pollution their causes and control are discussed in detail. Chapter 25 deals with environment and energy. Basic concepts of energy conservation in chemical process industries and energy crisis and alternatives are also explained. Chapter 26 focuses on safety considerations in chemical process industries including storage, handling and transporation.

The text in this book is throughout supplemented with diagrams and tables. The treatment of all topics is in a cogent, lucid style aimed at enabling the reader to grasp the information quickly and easily. The references and index are also given at the end of the book.

While painstaking care has gone into producing a useful and exhaustive reference textbook, the authors would welcome any constructive criticism and creative feedbacks from the students, teachers and professionals from the industry.

We wish to thank all those colleagues who provided figures and photographs and offered important advice during its preparation.

K. H. Davis

M. S. Berner

Contents at a Glance

Contents

Chapter 9. Fertilisers, Phosphorus and Potassium Salts **261-334**

Chapter 10. Nitrogen Industry **335-355**

CHAPTER 1

Chemical Industry : An Overview

INTRODUCTION

Products of the chemical process industry are used in all areas of everyday life. The raising of food plants and animals requires chemical fertilisers, insecticides, food supplements, and disinfectants. Many building materials have been chemically processed, for example, metals, concrete, roofing materials, paints, and plastics. Clothing utilises many synthetic fibres and dyes. Transportation depends upon gasoline and other fuels. Written communication uses paper and printing ink and electronic communication requires many chemically processed insulators and conductors. The nation's health is maintained by drugs and pharmaceuticals, soaps and detergents, insecticides and disinfectants—all products of the chemical process industry. In addition, many chemicals never reach the consumer in their original form but are sold within the industry for further processing or use in the production of other chemicals for consumer use. It is often said that the chemical industry is its own best customer.

The chemical process industry is a sprawling complex of raw-material sources, manufacturing plants and distribution facilities which supplies thousands of chemical products. No clear limit can be used to define the chemical process industry and within the industry it is not possible to classify completely the various industries.

Any definition or description of the chemical process industry is bound to be incomplete. Most processes in the chemical industry involve a chemical change. The term 'chemical change' should be interpreted to include not only chemical reactions but also physico-chemical changes, such as the separation and purification of the components of a mixture. Purely mechanical changes are usually not considered part of the chemical process, unless they are essential to later chemical changes. As an example, the manufacture of the plastic polythene, using ethylene produced from petroleum or natural gas, involves a chemical process. On the other hand, the moulding and fabrication of the resulting plastic resin into final shapes for consumer products would not be considered part of the chemical process.

Some industries which depend on chemical changes are not usually considered part of the chemical process industry, because of tradition, special processes, or large volume of a particular product, such as paper or steel. Many complex chemical processes in the food industry, for example, cheese making, which involves a fermentation reaction, are not considered to be part of the chemical process industry. On the other hand, the fermentation of sugars to produce beverage and industrial alcohols is often listed as a chemical industry. The huge metallurgical industry is usually distinguished from the chemical process industry because of the special nature of the processes and the vast quantities of the products; but the processing of metals might also be considered to be merely one segment of the chemical industry.

One possible list of the important chemical process industries is given in Table 1.1. Not all industries which utilise chemical processing are included, nor are all parts of the listed industries necessarily part of the chemical process industry. An examination of the list will show the tremendous scope of the industry and will also demonstrate the problems involved in any such listing or classification. A consideration of a few of the products of the industries listed in Table 1.1 will show the wide scope of chemical processing. Only a few typical products are listed.

Table 1.1. Chemical process industries.

Industry	Typical products	End uses
Inorganic chemicals	Sulphuric acid	Fetilisers, chemicals, petroleum refining, paints, pigments, metal processing, explosives
	Nitric acid	Explosives, fertilisers
	Sodium hydroxide	Chemicals, rayon and film processing, petroleum refining, pulp and paper processing, lye, cleansers, soap, metal processing
Organic chemicals	Acetic anhydride	Rayon, resins, and plastics
	Ethylene glycol	Antifreeze, cellophane, dynamite, synthetic fibres
	Formaldehyde	Plastics
	Methanol	Formaldehyde manufacture, antifreeze, solvent
Petroleum and petrochemicals	Gasoline	Motor fuel
	Kerosene	Jet fuel
	Oils	Lubricating, heating
	Ammonia	Fertiliser, chemicals
	Ethyl alcohol	Acetaldehyde manufacture, solvent, other chemicals
	Alkyl aryl sulphonate	Detergent
	Styrene	Synthetic rubber, plastics
Pulp and paper	Paper	Books, records, newspapers, etc.
	Cardboard	Boxes
	Fibreboard, etc.	Building materials
Pigments and paint	Zinc oxide	
	Titanium dioxide	Pigments for paint, ink, plastic,
	Carbon black	rubber, ceramics, linoleum
	Lead chromate	
	Linseed oil	Drying oil
	Phenolic resins	Basic lacquer, varnish, and
	Alkyd resins	enamel constituents
Rubber	Natural rubber (isoprene)	Automobile tyres, moulded goods and sheeting, footwear, insulation
	Synthetic rubbers (GR-S, neoprene, butyl)	
Plastics	Phenol-formaldehyde	
	Polystyrene	Various uses in all areas of
	Polymethylmethacrylate	everyday life

(Contd ...)

Industry	Typical products	End uses
	Polyvinyl chloride	
	Polyethylene	
	Polyesters	
Synthetic fibres	Rayon	
	Nylon	
	Polyesters	Cloth and clothing
	Acrylics	
Minerals	Glass and ceramics	Windows, containers, bricks, pipe
	Cement	Concrete for construction of buildings, highways, etc.
	Coal	Fuel, coke and its by-products.
Cleansing agents	Soap	
	Synthetic detergents	Household and industrial cleaning
	(sodium alkyl aryl sulphonates)	
	Wetting agents	
Biochemicals	Pharmaceuticals and	Health and medicinal applications
	drugs	
	Fermentation products:	
	Penicillin	Medicinal use
	Ethyl alcohol	Solvent, beverage
	Food products	Human sustenance
Metals	Steel	
	Copper	
	Aluminium	Building material, machinery, etc.
	Zirconium	
	Uranium	Nuclear fuel

The primary petroleum products are gasoline, other fuels, lubricants, and petrochemicals. Since the Second World War petrochemicals have assumed a tremendous role in the International economy. They are usually organic chemicals made from components of petroleum or natural gas. Paradoxically, the largest tonnage petrochemical is the inorganic chemical ammonia, produced by reaction of hydrogen from natural gas or petroleum with nitrogen from air. It is used as a fertiliser and as a starting material for other products. Many plastics, such as polyethylene and polypropylene, derive their starting materials from natural gas or petroleum. Primary components of many synthetic detergents are produced in the oil refinery.

The pulp and paper industry uses many unusual processes and is often considered separate from the chemical process industry, even though it employs many chemical engineers. The manufacture of paint involves combination of inorganic pigments or organic dyes for colour with various natural or synthetic organic vehicles, extenders, and drying agents. This very old industry still relies heavily on the 'art' or 'practice' in its production of paints, varnishes, and other surface coatings; although many newer developments are based on a better understanding of the scientific principles.

The division between the rubber and plastics industries is not distinct. The various synthetic rubbers are plastics with properties similar to natural rubber. Natural rubber is made from the sap of a tree, whereas most synthetic rubbers are derived from petroleum. Synthetic fibres are usually polymeric organic compounds. Rayon made from natural cellulose was the first large-volume synthetic textile fibre, and nylon was the first purely synthetic textile fibre.

Important chemical process industries utilising mineral raw materials are glass and ceramics, cement, and coke and its by-products. The important by-products of coke are benzene and toluene compete with the same materials produced in the oil industry. Petroleum, natural gas, and sulphur are also minerals. The trend in the field of cleansing agents has been away from soap produced from natural fats toward synthetic detergents, many of which are produced using raw materials from the oil refinery.

The biochemical process industry includes pharmaceuticals, fermentation products, and food. These industries either use processes involving biological action (such as fermentation to produce penicillin) or produce products that are biologically active (such as penicillin in its medicinal effects). Pharmaceuticals are produced by controlled natural biological processes or by synthetic organic chemistry. Although much of the food industry is not generally considered part of the chemical process industry, those products which have been highly processed, such as sugar and hydrogenated shortening, may be considered to be part of the industry. Fermentation processes produce industrial and beverage alcohols, acetone, and acetic acid by biological action on various sugars.

As stated earlier, the metals industry is usually considered apart from the chemical process industry, although recovery and purification of many of the newer metals involve much chemical processing. Many chemical engineering principles are applied in the production of all metals. The production of iron and steel involves many high-temperature chemical reactions and phase equilibrium. Aluminium is produced by electro- chemical means.

Many new metals, such as uranium, zirconium, and titanium, are produced by processes involving standard chemical engineering operations. Many new metallurgical processes are being developed by chemical companies, rather than by the traditional metals companies. From the above discussion it is evident that any listing or classification of chemical process industry is arbitrary and subject to ambiguities. Perhaps a satisfactory definition of a chemical process industry is 'an industry whose principal products are manufactured by processes based upon the chemical and physical principles included in the field of chemical engineering'.

SYSTEMATIC ANALYSIS OF CHEMICAL PROCESSES

Systematic analysis of chemical processes elucidated many underlying principles which could be used in the synthesis of new processes. Modern chemical processes are often extremely complex operations involving hundred of pieces of equipment. Without a systematic approach, it would be impossible to analyse an existing process or to design a new process. Therefore, chemical processes are broken down into individual steps that recur in many other processes. The general principles of these steps have been carefully studied and are frequently well understood. If the principles are known, it is possible to design the step to do the best possible job. The typical chemical process is analysed with the following interdependent considerations:

1. Mass and energy balances.
2. Thermodynamics and kinetics.
3. Unit operations.
4. Plant equipment.
5. Ancillary equipment.
6. Process flow diagrams.
7. Instrumentation and control.
8. Economics.

Mass and Energy Balances

It relates to the fundamental principles that engineers and scientists employ in performing design calculations and predicting the performance of plant equipment. These includes the conservation laws for both mass and energy, thermochemistry, chemical reaction equilibrium, chemical kinetics, the ideal gas law, and phase equilibrium.

Thermodynamics and Kinetics

Thermodynamics deals with the transformation of energy from one form to another. Many important practical conclusions can be derived from the two fundamental laws of thermodynamics. The energy balance, is an expression of the first law of thermodynamics. The second law states that in a process involving heat transfer alone energy may be transferred only from a higher temperature to a lower one. The thermodynamic analysis of a process leads to conclusions concerning the feasibility and efficiency of the various process steps. Thermodynamics also is useful in determining the composition of phases in equilibrium and in predicting the distribution of chemical species in reaction equilibrium.

Kinetics considers the rate at which chemical compounds react. Data on rate of reaction are necessary in the design of the industrial chemical reactors.

Unit Operations

These are particular kind of physical changes used in the industrial production of various chemicals and related materials. Filtration, evaporation, distillation, fluid flow and heat transfer are some of the examples.

Plant Equipment

The plant equipment provides details on a number of commonly used process units i.e., reactors, heat exchangers, columns of various types (distillation, absorption, adsorption, evaporation, extraction), dryers, and grinders. The purpose of each unit or operation and the many configurations in which the units can be found are also discussed. The reactor is often the heart of a chemical process. It is the place in the process where raw materials are usually converted into products, and reactor design is therefore a vital step in the overall design of the process.

Heat exchangers

The transfer of heat to and from process fluids is an essential part of most chemical processes. The most commonly used type of heat transfer equipment is the shell and tube heat exchanger.

Mass transfer equipment

Distillation

Distillation is probably the most widely used separation process in the chemical industry. Its applications range from the rectification of alcohol, which has been practised since antiquity, to the fractionation of crude oil. The separation of liquid mixtures by distillation is based on differences in volatility between the components. The greater the relative volatilities, the easier the separation.

Adsorption

In the adsorption process, one or more components in a mixture are preferentially removed from the mixture by a solid (referred to as the adsorbent). Adsorption is influenced by the surface area of the adsorbent, the nature of the solvent being adsorbed, the pH of the operating system (liquid application), and the temperature of operation. These are important parameters to be aware of when designing or evaluating an adsorption process.

The adsorption process is normally performed in a column. The column is run as either a packed or fluidised-bed operation. The adsorbent, after it has reached the end of its useful life, can either be discarded or regenerated. This operation can be applied to either a gas mixture or a liquid mixture.

Absorption

The process of absorption conventionally refers to the intimate contacting of a mixture of gases with a liquid so that part of one or more of the constituents of the gas will dissolve in the liquid. The contact usually takes place in some type of packed column.

Evaporation

The processing industry has given the operations involving heat transfer to a boiling liquid the general name evaporation. The most common application is the removal of water from a processing stream. Evaporation is used in the food, chemical, and petrochemical industries, and it usually results in an increase in the concentration of a certain species. The factors that affect the evaporation process are concentration in the liquid, solubility, pressure, temperature, scaling, and materials of construction.

Extraction

Extraction (sometimes called leaching) encompasses liquid-liquid as well as liquid-solid systems. Liquid-liquid extraction involves the transfer of solutes from one liquid phase into another liquid solvent; it is normally conducted in a mixer-settler, plate and agitated-tower contracting equipment, or packed or spray towers. Liquid-solid extraction, in which a liquid solvent is passed over a solid phase to remove some solute, is carried out in fixed-bed, moving-bed, or agitated-solid columns.

Drying

Drying generally involves the removal from solids of relatively small amounts of water or organic liquids, whereas evaporation removes larger amounts. Drying removes the liquid as a vapour by warm gas (usually air) currents. Drying can be accomplished on a batch or continuous basis.

Ancillary Equipments

These are devices for transporting gases and liquids to, from, or between units of process equipment. Some of these devices are simply conduits for the moving of material (pipes, ducts, fittings, stacks); others control the flow of material (valves); still others provide the mechanical driving force for the flow (fans, pumps, and compressors). This also covers storage facilities, holding tanks, materials handling devices and techniques, and utilities (e.g., gas, steam, water) along with air, water, and solid waste control equipment.

Process Flow Diagrams

To the practising engineer, particularly the chemical engineer, the process flow sheet is the key instrument for defining, refining, and documenting a chemical process. The process flow diagram is the authorised process blueprint, the framework for specifications used in equipment designation and design; it is the single, authoritative document employed to define, construct, and operate the chemical process. This type of diagram is also used in other processes and industries.

Instrumentation and Control

In all chemical processes it is necessary to know such process data as flow rates, compositions, pressures and temperatures, so that the operator and production engineer can tell that the process is functioning properly. In the typical chemical process many instruments are used to measure, indicate and record the necessary process data.

Automatic control may be applied to almost any process variable that can be measured. Control of a variable is often maintained by the measurement of another more easily measured variable. Automatic control reduces the number of human operators. It can give faster and more accurate control than a human operator.

In other words it is collective term for sensing devices used to measure, record, and control chemical process variables such as temperature, pressure, flow rate, thickness, liquid level, pH, etc. Such instruments permit automatic correction of variables on a continuous basis. Increasingly sophisticated developments in automatic control technology have enabled many chemical processes to be carried out with a minimum of personnel, particularly petroleum refining. The ultimate in sophisticated instrumentation is utilised in nuclear reactor control.

Economics

No matter how efficiently a process operates to produce a final product of high purity, the process is a failure if the product cannot be sold at a profit. Before a chemical plant is built, a thorough market analysis is made to determine how much of the product can be sold and at what price. Often it is possible to sell more of the product if the price is lower.

Present and future competition from other producers must be carefully evaluated. As the plant is designed, many economic analysis are made to determine the least expensive design which will produce the desired quantity of product at a minimum price.

If the cost of manufacture is sufficiently below the selling price to give an attractive profit, the plant will be built. If the product is successful and profitable, a competitor may find the market attractive and enter it with perhaps a somewhat better product produced at a lower price by an improved process. It is then necessary for the older producer to improve his process and product, or he will be forced out of the market.

FLOW SHEET AS A REPRESENTATION OF A CHEMICAL PROCESS

The simplified flow sheets are prepared to illustrate the process. These are often for a special purpose and do not show all the details of a process. A common types of flow sheet shows the major unit operations and chemical reactors with their interconnecting piping and an identification of the materials being processed.

On a process flow sheet the equipment is arranged logically to show the flow of materials through the process, but a photograph show the final physical arrangement determined by structural requirements, without regard to the process flow.

All tall distillation columns may be grouped together for structural support, and large heat exchangers may be grouped for ease of maintenance. The successful chemical engineer must learn to read flow sheets and to relate them to the actual plant layout.

Flow sheet shown instrumentation and controls are often prepared. Flow sheet symbols for various operations are shown in Fig. 1.1.

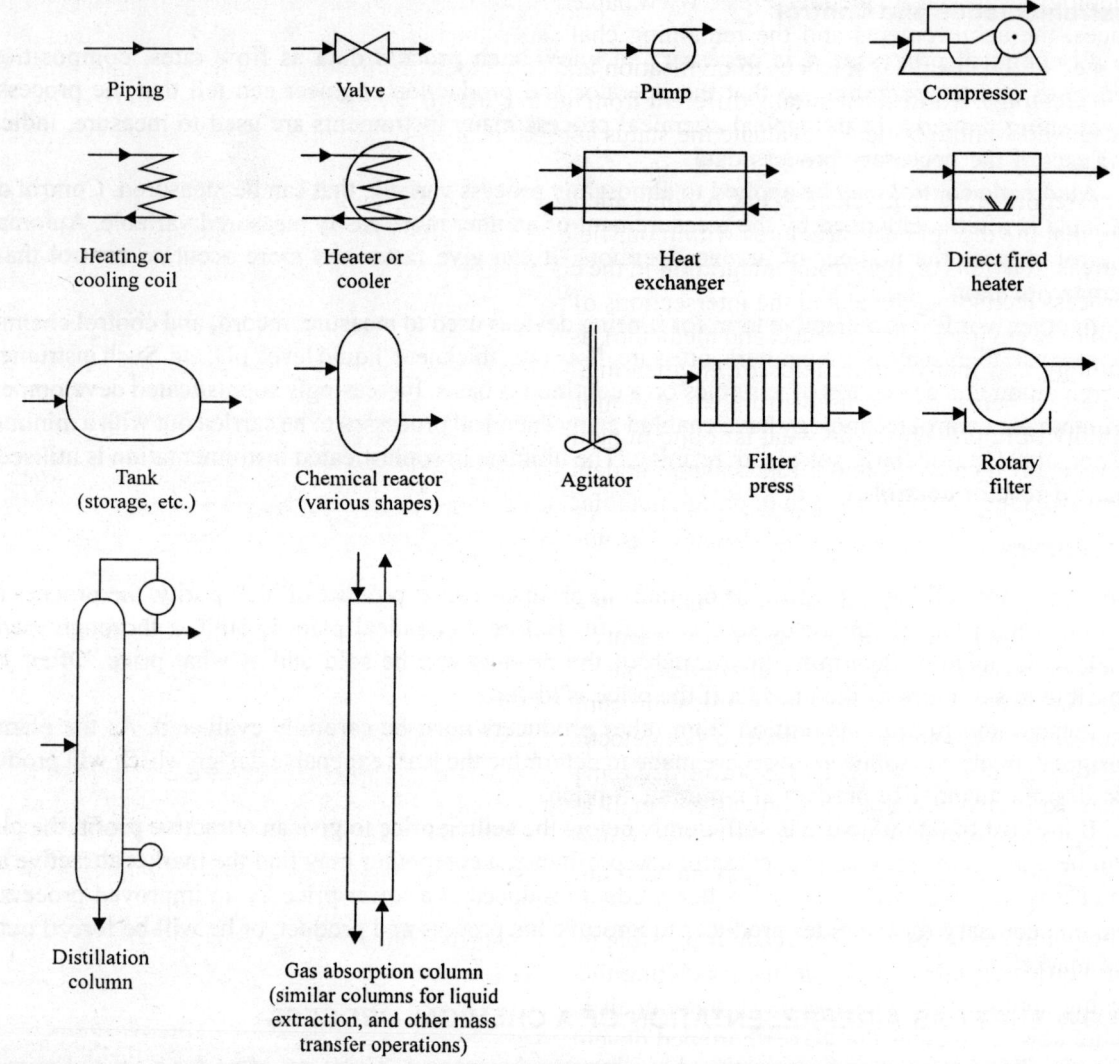

Fig. 1.1. Flow-sheet symbols for various operations.

These symbols are not standard, because equipment varies widely, but they will be used along with others in the flow sheets. There are many variations on the symbols shown, especially in the chemical reactors, which have a wide variety of sizes and shapes. Product flow sheets can be prepared for large chemical plants in which the products of certain chemical processes are the starting materials for others. In product flow sheets the chemical processes are shown as simple blocks, with the starting material and product streams flowing to and from each process block.

GRAND CHALLENGES FOR CHEMISTRY AND CHEMICAL ENGINEERING

A new report, 'Beyond the Molecular Frontier: Challenges for Chemistry and Chemical Engineering' prepared under the auspices of the US National Research Council's Board on Chemical Sciences and

Technology [National Academies Press; www.nap.edu] outlines the status of research in the chemical sciences, the achievements and the remaining challenges for the future. The report underscores the relevance of the chemical sciences to civilisation and concludes that the sciences have a major role to play in creating a world substantially different from the one in which we live. The report will be of value to those who want to critically evaluate the status and future goals of important sectors of the chemical sciences.

The report notes that sciences has become increasingly inter-disciplinary, and that it is critical to ensure that the disciplinary structures within our fields do not hinder the future of chemical sciences in new areas. It dwells on the strong integration in the chemical sciences, right from the molecular level to the process technology level and the intersections of the chemical sciences with all the natural sciences (agriculture, environmental science, and medicine), as well as with materials science, physics, information technology and many other fields of engineering. The combined efforts of chemists and chemical engineers, it notes, can be used to invent new molecules, and new ways to make them, offering the possibility of improvement on what is found in nature, to make our lives safer and more pleasant.

The report recognises that a fundamental understanding of processes, which has emerged over the last few decades, can now be used to design manufacturing methods with unprecedented reliability. The time-scale in which measurements are made have undergone phenomenal evolution, and moved frontiers to the investigation of processes that take place on the femto-second time-scale, at which molecular bonds are broken or formed. A more detailed understanding of chemical reaction pathways and the mechanisms of physical transformations represent a challenge that will increase our ability to manipulate reactions and processes for practical applications.

The dramatic improvements in chemical instrumentation has changed the frontiers of measurement in astonishing ways and moved the levels of detection 'from a mole to molecule' – an improvement by over 20 orders of magnitude. This has afforded remarkable control of processes in real time and routine analysis can now be done on samples as small as a microgram, rather than a gram or more. The challenges and opportunities herein will be to evolve techniques for detection of toxins, explosives, etc., efficiently and rapidly, and to device advanced control systems that make chemical manufacturing more precise and environ-friendly.

The computer revolution has made it possible to approach a number of important goals: predicting the properties of unknown materials, predicting the pathways of chemical and physical processes and designing optimal processes. While all these goals have not yet been met, when they are we can expect to create new substances that have shortened development times, by, by-passing substantial amount of empirical experimental work; optimally meeting out needs in areas such as medicines and advanced materials.

While modern chemists have invented and chemical engineers have learned to manufacture medicines that have let us conquer many diseases, there is much to be done in understanding the chemical pathways of biological processes and as this happens biology will increasingly become a chemical science and chemistry a life science.

The chemical sciences must remain ambitious if it is to make the maximum possible contribution to our welfare. It will be necessary for chemists and chemical engineers to produce major new discoveries, revolutionary new technologies and important new additions to the quality of our life. The astonishing developments of the 20th century, in all of the sciences, including chemicals, have made it possible to dream of goals that might have been earlier unthinkable.

The chemical sciences and the chemical industry today face many challenges that make it an exciting field to work in. If they are to deliver on what they promise, they deserve the financial support of government and the sanction and appreciation of society at large.

Opportunities and Challenges for Chemical Professionals

1. Learn to synthesise and manufacture new substances using compact synthetic schemes and processes with high selectivity for the desired product, with low energy consumption and benign environmental effects.
2. Develop new materials and measurement devices to protect against terrorism, accident, crime, disease, by identifying dangerous substances and organisms using methods with high sensitivity and selectivity.
3. Understand and control how molecules react — over all time scales and the full range of molecular size.
4. Learn to design and produce new materials and molecular devices with properties that can be predicted, tailored, and tuned before production.
5. Understand the chemistry of living systems, in detail.
6. Develop medicines and therapies that can cure currently untreatable diseases.
7. Develop self-assembly as a useful approach to synthesis and manufacture of complex systems and materials.
8. Understand the complex chemistry of the earth, so that we can maintain its livability.
9. Develop unlimited and inexpensive energy, to pave the way for a truly sustainable future.
10. Design and develop self-optimising chemical systems.
11. Revolutionise design of chemical processes to make them safe, compact, flexible, energy efficient, environmentally benign and conducive to rapid commercialisation of new products.
12. Communicate effectively to the public, the contributions of chemistry and chemical engineering to society.
13. Attract the best and the brightest young students into the chemical sciences, to meet these challenges.

'DOT COM' IN THE CHEMICAL INDUSTRY

Recently we have seen the emergence of the so-called 'dot com' companies in many 'business to business (B-B) and 'business to consumer' (B-C) segments, including ones that hope to service one or the other sector of the chemical industry. Globally there are more than hundred 'dot com' companies that have created a marketplace where chemicals can be bought, sold, auctioned, reverse-auctioned and what have you. In India too, more than thirty companies have ventured into this uncharted area, or plan to do so in the short term, and each has a different revenue and operational model that they hope will bring in the cash that presently seems a mirage.

The valuation of these companies in the stock markets has defined traditional models, and even the most successful of them all – Amazon.com, which pioneered the sales of books through the Internet – is reported to be just about breaking-even, after bleeding steadily since inception. This has not deterred the markets from viewing these companies favourably, and unbelievably high evaluations have been the engine that has propelled growth in the segment.

The deluge of press releases and announcements is a far indicator of the battle in the marketplace among the heavyweights serving the chemical industry. Many of these firms are roping in chemical companies as channel-partners who will source or sell their products through 'third party' sites. Some chemical companies, after evaluating the efficiencies that such a system could bring about, have even picked up equity stakes in these 'dot coms', as have publishing houses that have a wealth of information at their disposal (not to mention a lucrative subscription base). While some chemical companies have actually begun to trade on-line, there is a sizeable number sitting on the fence, watching and evaluating their options. Clearly these are early days for the business, but most companies while unsure whether this will be the way trading will be done in the future, are keen not be left behind least it does.

Typically how does such a market place work? For one, buyers and sellers need to register at the site, usually for a small fee (which can differ for buyers and sellers, depending on the model followed), or sometimes even for free, and post their offers. Visitors to the site, must again do the same, and can view the options available (specifications, price, quantity, locations, discounts etc.), before deciding on who to do business with. A deal is struck for the required quantity, at a negotiated price, on mutually agreed delivery and payment terms. A commission is usually payable to the company hosting the site – and this will hopefully form the bulk of revenue that the 'dot.com' hopes to generate. From this point on too, various models are being tried out. At least one company operating in the Indian scenario, and many internationally, offer to complete the transaction on behalf of his client (usually the buyer), by becoming involved in logistical operations (taking delivery of the material, shipping it across etc.), and even collecting payment – for a higher commission, of course.

What are the challenges in the business? Finance for one – most 'dot coms' do not expect to break even for some time, and expect to sustain operations through capital raised from IPOs. Indeed a few may even be in business attracted by the prospects of quick money at the bourses. A recent survey revealed that up to 60 per cent of the total cost of a project – which runs into a few million dollars (much less here in India) – comes from the marketing and advertising expenses, followed by the costs in creation and maintenance of the site. Getting visitors is a tough proposition, and it takes full-page ads screaming in full colour to make once presence heard in a crowded Internet. Revenues of 'Dot com' companies could also be impacted by chemical companies who could do their negotiations under their umbrella, only to move offline to conclude the negotiations without paying the commissions due.

Issues Worrying Chemical Companies

Some of the issues that chemical companies are worried about, but 'dot coms' claim to have addressed include.

Security: How secure are the sites from encroachers, hackers?

Legality: How does one enforce a contract on the net? This is particularly important in India, where cyber-trading laws are still to be legislated (although they are expected to be in place soon).

Responsibilities of the involved parties: What is the responsibility of the trading platform, the credit evaluation company, the shippers, etc.

Off-spec. materials: How does one pin down responsibility when materials does not meet the specifications indicated.

Bogus deals: How does one prevent bogus deals (that have been reported but not concluded) that could influence price one way or the other? (this has been reported in one instance for a petrochemical).

In the Indian context will Internet trading be a cheaper and more efficient way of doing business? The answer remains to be found, but the realities in this country are indeed very different, and the models followed elsewhere may need modification to fit in with the local scenario. In the chemical industry for one, the fragmentation level is high and Internet penetration is still low and expensive (only about in 5 companies surveyed recently had an Internet address, and even fewer used it for anything other than e-mail). Few of the Indian 'dot coms' have an authentic database that provide their customers some choice (efficiencies disappear when you have just one seller!), and creating one is an arduous task. Most activity in India is still restricted to providing a common ground for buyer and seller, and making the 'first contact'. Most of the subsequent transactions are expected to take place offline, till at least the payment gateways are in place. Large companies could be reluctant to source their raw materials requirements through third parties, fearing many of the issues raised earlier, and could prefer to do so through their own supply chain management systems that rope in just their preferred vendors and buyers (this is already happening). Logistics in India are usually a nightmare and the advantage of concluding a deal over the Internet could easily disappear if one has to chase the supplier or the transporter through phone calls and faxes to ensure delivery schedules!

So what of the future? There is no doubt that the Internet is a powerful tool, which can revolutionise the manner in which trading is done, and make geographical boundaries meaningless. It's ability to transform the market place by providing a medium to interact directly is unsurpassed and will threaten the role of the 'middle-man' in the years to come. But will it replace the handshake when it comes to finalising a transaction. Not likely. A shakeout is expected among these companies globally – and India will be no exception.

packing. The liquid is sprayed into the top of the tower and, on descending, meets a counter-current of the gases passing up the tower. The liquid containing the dissolved gas leaves the base of the tower and the undissolved gas leaves the top of the tower. The liquid may be recycled until it will dissolve no more gas, or it may be "stripped" of its dissolved gas and used again.

Bubble Plate or Sieve Plate Columns

These are sometimes used instead of a packed column and are often more economical to operate, but the presence of solids in the liquid may tend to block the fine holes in the sieve plate column.

Spray Towers

This type is a short, unpacked tower into which liquid is injected at the top as a very fine spray. They are not used frequently but are employed in air-conditioning plants and as vent condensers.

Jet Scrubbers

The absorbing liquid is forced under pressure through a nozzle and into an orifice, then dispersed in a chamber where the gas is sucked in and absorbed. Caustic soda solution recycled continuously may be used to remove acid gases: if water is used it is discharged to an effluent treatment plant.

CALCINATION

This unit operation may be defined as the heating of materials to a high temperature in rotary calciners or kilns to convert metals to their oxides, carbonates to their oxides, burn off unwanted organic substances, and for drying. In the latter case it is used particularly for drying substances which contain water of crystalisation, which, since it is chemically bonded to the substance, is more difficult to remove than from those which are simple "water-wetter".

The *rotary calciner* is similar to a rotary drier, except that it is specially constructed in stainless steel or special metals lined with refractory materials which can withstand the higher temperatures required. The direct-heating type consists of an inclined metal cylinder lined with a refractory material, down which the crushed material being treated passes continuously from screw conveyors or hoppers and is discharged at the lower end. The lower end is connected to a furnace heated by burning gases; for example, producer gas or a solid fuel may be mixed with the material being changed to the calciner itself. It is sometimes necessary to fit gas absorbers to avoid passing noxious gases to the atmosphere, or cooling chambers at the discharge end using sprays of cooling water. Indirect-heat calciners employ the same principle, but use a stainless steel cylinder rotating within a stationary cylinder lined with refractory material. The fuel gas is burned between the walls of the inner and outer cylinders, and the materials fed down the inner cylinder. The inner cylinder extends beyond the ends of the outer cylinder to accommodate the bearings, driving machinery, feeders, etc. Sometimes ores break down to a powder by the calcination process and this eliminates the need to crush the material.

CRYSTALLISATION

Solids occur in the form of crystals when the atoms and molecules within the bulk are arranged in an ordered manner to form a definite geometrical pattern or shape. This arrangement gives rise to a pure form of the solid since impurities will not fit into the spaces and are rejected. The formation of crystals by a crystallisation process is, therefore, often used to produce pure chemicals. Furthermore, crystallisation improves the appearance of chemicals, makes handling, and often drying, easier.

Crystallisation may be defined as the removal of a solid from a solution by increasing its concentration above its saturation point, that is by becoming "supersaturated". This is achieved by cooling, evaporating, or both.

The simplest method of crystallisation is the addition of a solid to a hot liquid (the solvent) until no more will dissolve. The saturated solution thus produced is allowed to stand and cool, perhaps over a few days, when crystals are deposited on the bottom of the tank, leaving a *mother liquor*. The mother liquor is separated by filtration, and the crystals freed of mother liquor wetting their surfaces by careful washing with cooled, pure solvent. The crystals thus obtained are called the *first crop*. To increase the yield, the mother liquor is sometimes evaporated to a smaller volume and the crystallisation operation repeated. In this case the liquor when separated from the crystals—the *second crop*—is often called the *grandmother* (or *granny*) liquor.

Decolourising charcoal is often added to the hot, saturated solution prior to cooling. After agitating the suspension for a short time the charcoal is removed by filtration and the crystallisation is completed in the normal manner in another vessel. If charcoal is not used, the hot solution is normally passed through a *polishing filter* to remove small quantities of undissolved impurities or dirt and obtain a clear, bright solution.

Sometimes it is necessary to add a small quantity of crystals to the hot solution to start the crystallisation process; this is known as *seeding*. The same effect may be achieved by gently scraping the inside wall of the tank, by agitation, or simply by contamination with atmospheric dust.

Agitated batch crystallisers are tanks, often with a tapered bottom, fitted with a suitable agitator and cooled by circulating cooling water or brine through an outer jacket or an immersed coil. This method (i) is quicker, by virtue of the forced cooling and the better heat transfer achieved by the agitation; (ii) produces crystals of similar size by keeping the temperature of the mother liquor more uniform and preventing the small crystals from settling to the bottom of the tank; and (iii) prevents the crystals from sticking together.

Simple evaporators used for crystallisation may be shallow, open pans fitted with a steam jacket. The crystals deposited by the evaporation of the solvent (usually water) are prevented from sticking together or to the pan walls by slow-moving rakes.

Vacuum crystallisers are closed, lagged vessels to which vacuum may be applied to a hot, concentrated solution to induce cooling by evaporation.

Continuous crystallisers employ the recirculation of cooled, supersaturated solution through a loosely packed bed or suspension of crystals when crystal growth occurs. The thick magma or slurry is continuously siphoned or pumped off and filtered. The mother liquor may be saturated again and returned to the system.

DISTILLATION

The term simple distillation generally refers to the separation of a mixture of two or more liquids by boiling, to produce a vapour of the more volatile (lower boiling point) component and the condensation of this vapour back to a liquid which is collected in a receiving vessel (Fig. 2.2)

This unit operation depends on the fact that liquids have different vapour pressures at a given temperature. When a liquid mixture is volatilised the vapour produced is richer in the more volatile component and the liquid remaining in the boiler is richer in the less volatile component. Hence a partial separation of the liquids is achieved. If the condensed vapour—the distillate—is heated again, the vapour produced will be even richer in the more volatile component. This process of redistillation can be repeated many times until the separation of the liquids is nearly complete, but this is a lengthy operation and may be achieved more conveniently by fractional distillation.

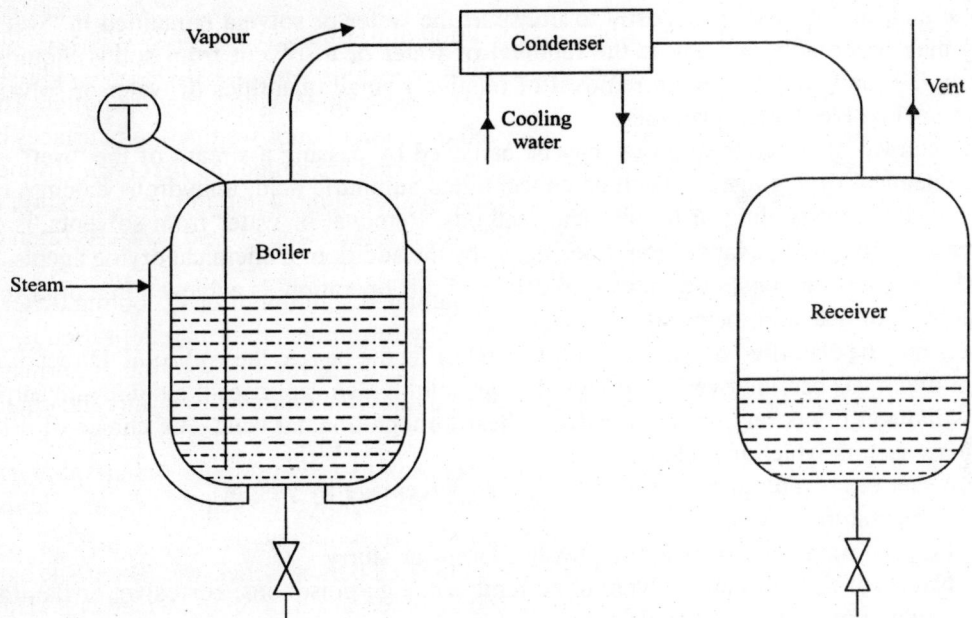

Fig. 2.2. Simple distillation plant.

Fractional distillation or rectification is a combination of redistillation and partial condensation of the vapour and employs a fractionating column. Part of the condensed vapour is returned to the boiler down the fractionating column where the liquid is brought into contact with the vapour rising up the column on its way to the condenser. This results in the less volatile components of the liquid mixture in the column becoming enriched, whilst the vapour leaving the top of the column is enriched in the more volatile components. The liquid condensate being returned to the boiler is called reflux, and the reflux ratio is the number of gallons (or pounds) of liquid sent back per gallon (or pound) of the liquid collected in the receiver. Generally, the higher the reflux ratio the greater the efficiency of separation. Columns may be filled with a packing material such as Raschig rings, Berl saddles, or small glass cylinders, or they may contain a series of perforated bubble-cap plates or sieve plates.

Continuous distillation is achieved by returning the liquid condensate to a point approximately midway up the fractionating column, where the liquid in the column is about the same composition as the returning liquid (reflux). The relatively pure vapour leaving the top of the column may be taken off continuously leaving only the unwanted "still bottoms" in the boilers. The section of the column above the inlet is called the rectifying section, and that below is called the stripping section.

Steam distillation is a special method employed for distillation liquids with a high boiling point or those that decompose when heated to their boiling point. This is carried out by heating the liquid in the presence of water or injecting live steam into it through a pipe. The operation usually involves two immiscible liquids (for example aniline and water). In this case, the vapour mixture condenses to give two liquid layers which may be separated in the receiver. Another method employed for distillation heat-sensitive compounds is *vacuum distillation*: the pressure in the system is reduced to lower the boiling point of the liquid to be distilled.

DRYING

This unit operation is used in most chemical processes as chemicals are generally required for use in the dry state and, in any case, it is costly to transport the water or solvent contained in "wet" materials. Drying may be broadly defined as the removal of water or a solvent from solids, liquids, or gases. The term is usually applied to the removal of relatively small quantities of water or solvent, whereas evaporation involves larger quantities.

The removal of water from gases may be achieved by passing a stream of the "wet" gas through special chemical drying agents (such as concentrated sulphuric acid, anhydrous calcium chloride, or silica gel), by rapid cooling or by physical methods. Removal of water from solvents, is effected by azeotropic distillation or evaporation, freezing, or by the addition of chemical drying agents. The drying of solids, which is the most important application of this operation, is achieved in a drier in two steps; the supplying of heat and the removal of vapour.

Driers may be classified by the method of transferring the heat to the wet solid. Direct driers employ a stream of hot gas passing over or through the wet solid which vaporises the liquid and carries it away, whereas indirect driers rely on the transfer of heat through a metal wall. The choice of drier depends upon many factors which include:

1. Water or solvent content of the wet solid (percentage by weight).
2. Temperature required.
3. Form of material, for example powder, lumps or slurry.
4. Nature of the solid and solvent to be removed, e.g., poisonous, corrosive, or flammable.
5. Drying time (or drying cycle).
6. Bulk density of wet and dry solid.

Many solids are sensitive to high temperatures and may discolour or decompose if overheated: drying at lower temperatures is usually achieved by the use of vacuum driers. Overheating may also cause some solids to melt and form glass-like beads, or a molten mass, which sticks to the supporting medium. Particular care must therefore be taken with solids with low melting points (these are usually organic substances) to control the drier temperature.

A special method of vacuum drying for chemicals which are very sensitive to heat is called freeze-drying. An aqueous (water) solution of the material is frozen in shallow trays and placed in a chamber which is then evacuated to a higher vacuum. If the vacuum is kept high enough to maintain the water-vapour pressure in the chamber below the vapour pressure of the ice itself, the ice will pass directly into the vapour state without melting, leaving the dry material on the trays.

Rotary and tumbler driers are most suitable for drying solids which do not conduct heat very well or are very dusty or hazardous, since this type keep physical contact and handling to a minimum. It is necessary, however, that the solids break down easily into a powder as the moisture or solvent is removed, otherwise the materials tends to form a hard cake on the walls or form into balls which will not dry easily. The load of material in this type of drier should preferably be about a half of the internal volume to attain maximum efficiency.

Tray driers are simple to operate and commonly used, but involve considerable manual handling of material. It is important that the wet solid should be spread thinly and evenly on the trays, usually to a depth of no more than half an inch. Too thick a layer of solid will result in a long drying cycle, a uneven rate of drying throughout the bulk, and the tendency to form lumps. It is desirable to check the internal temperature of the drier frequently, particularly those which do not have automatic temperature control. It is often advantageous to turn the material over and carry out regular visual inspections during the drying cycle. Any visible change in its appearance or form should be reported to the supervisor immediately. At the end of the normal drying cycle a sample of the dry solid is taken and submitted to the analytical laboratory for a moisture determination. When unloading trays from racks it is good practice to start removing them from the bottom, since this avoids the risk of dislodged dirt or dust falling on the contents of the tray below. The principle is applied in reverse when loading trays on racks (the first tray is placed on the top of the rack). The discharge of driers often gives rise to dust in the atmosphere which may be harmful. Dusts may be poisonous, corrosive, dermatitic, and, in some case, explosive. The chemical operator must keep dust to a minimum and ensure he is adequately protected by covering exposed skin and wearing a dust mask.

EVAPORATION

Evaporation is the separation of one liquid from another or from a solution or suspension of solid particles by changing the liquid (often water) to the vapour state. This operation differs from that of drying in that relatively large amounts of liquid are involved, thus evaporation often precedes drying in chemical processing. Distillation is concerned with the vapour itself and the separation of its components, whereas evaporation processes are usually concerned with the material which remains in the evaporator.

Evaporation is achieved by the application of heat, which may be supplied by the sun, fire, steam, hot water, electricity, hot oil, or special materials. This heat is used to (i) heat the compound to its boiling point; (ii) supply the latent heat of vaporisation (which is the amount of heat required to change a compound from liquid to vapour at its boiling point). Since the supply of heat is expensive, the design and operation of evaporators is concerned primarily with its efficient utilisation. For example, the heat of the vapour may be utilised to heat another vessel (multiple effect evaporators). Heat transfer depends on the temperature difference between the materials supplying and receiving the heat, the area over which it is being transferred, and the material of the heating surface.

Most evaporators employ steam-heated, tubular heating surfaces, over which the liquid to be vaporised is circulated. The efficiency of an evaporator may be reduced considerably by the formation of scale on the heating surface, which may be removed by chemical or mechanical means. Another problem is the formation of foam which may seriously restrict the capacity of an evaporator, but this may be controlled by the use of special chemical defoaming agents or gas or air jets.

FILTRATION

This unit operation covers a wide range of mechanical methods designed to separate solids from liquids or gases. In a simple filter (Fig. 2.3) the mixture of solid and liquid, or solid and gas, is pressed against one side of a porous bed called the filter medium, allowing only the liquid or gas to pass, leaving the solid in the form of a cake on the surface of the filter medium.

The liquid containing the suspension of solid particles to be separated is called the slurry, and the clear liquor obtained by the separation, the filtrate. The wet solid retained by the filter medium is known as the filter cake. This cake must be washed with pure water or solvent to remove the residual filtrate adhering to it; this is called the filter wash.

There is a resistance to flow of liquid passing through a filter medium which causes a pressure drop across the filter. This resistance increases as the solid cake builds up on the filter medium and it is necessary, therefore, to apply a force to maintain the filtration at a steady rate. The nature of this force provides a convenient means of classifying the many types of filter. Thus, we have gravity filter, vacuum filters, pressure filters, and centrifugal filters.

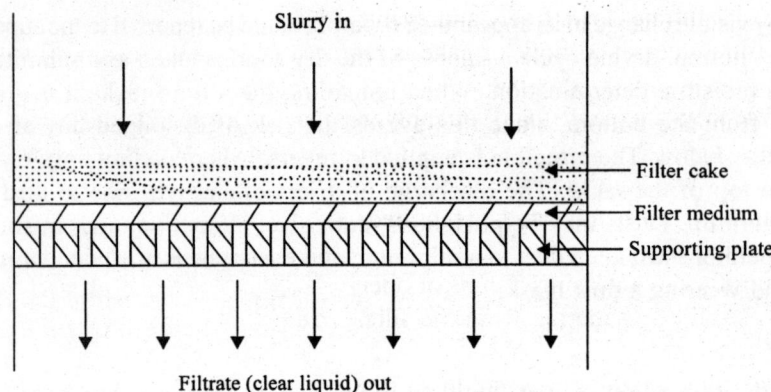

Slurry in

Filter cake

Filter medium

Supporting plate

Filtrate (clear liquid) out

Fig. 2.3. A simple filter.

Gravity filters are the least efficient since they rely only on the weight of the liquid itself to cause it to flow through the filter medium. Hence they are used only for filtering solids with a large particle size, requiring a filter medium or relatively large pores to prevent their passage through it, and therefore a low resistance to flow. On the other hand pressure filters may be operated with slurry inlet pressures of up to 60 psi to handle finely divided solids. Therefore, the choice of medium depends on the solid to be filtered and the type of filter to be used. A further consideration is the chemical nature of the liquid, for example whether it is acid or alkaline, corrosive, or capable of dissolving the filter medium. A wide variety of filter media are now available in different pore sizes, each offering resistance to particular chemical attack. They include cotton, jute, wool, paper, cellulose-asbestos, woven metal cloth, glass cloth, rubber cloth, and a range of synthetic materials, such as Terylene, Nylon, and polypropylene. When fitting a filter medium, care must be taken to place it perfectly flat on the supporting plate, frame, or wall, and in a symmetrical position so that the edges may be properly sealed to prevent solids by-passing it. Naturally, a torn or holed cloth is useless and must be replaced.

When a slurry contains finely divided solids, or is slimy or viscous, it may be necessary to use a filter-aid. Filter-aids are light, powdered materials (such as decolorising charcoal, magnesia, or diatomaceous earth) which may be added to the slurry or spread over the filter medium, as pre-coat, to prevent clogging of the pores and speed filtration.

The operating steps in a filtration procedure usually consist of the following:
1. Setting up.
    (a) Check that plates, frames, or other supporting media are clean.
    (b) Select filter medium, inspect, and fit in correct position on supports.
    (c) Check that components are assembled in the proper order.
    (d) Seal off the whole assembly.
    (e) Set all associated valves in correct positions.
    (f) Apply heat to jacket as necessary.
2. Loading
    (a) Start slurry feed to inlet side of filter.
    (b) Adjust internal pressure, vacuum, or speed of rotation to that required.
    (c) Regulate rate of slurry feed as necessary.
    (d) Check that filtrate is free from solid (take sample if necessary).

(e) Check that filtrate flows freely and evently.

(f) When filtrate ceases to flow, or cake is required depth, stop feeding slurry to filter.

3. Washing.

(a) Continue to rotate (centrifuge), or apply vacuum or compressed air or gas until filtrate flow stops.

(b) Feed pure wash liquor to filter inlet and repeat (a).

The whole operation may be repeated as required.

4. Discharge (manual).

(a) Open up filter by releasing clamps or opening lid.

(b) Remove solid by scraping from the filter medium (for plate and frame filters, clear one frame at a time).

(c) Readjust filter medium or wash down filter.

(d) Take representative sample of filter cake.

PRECIPITATION

A precipitation process involves the formation of a precipitate, which is a suspension of particles of a solid in a liquid. If a chemical reaction takes place in a solvent to form a product which will not dissolve in that solvent, then the product will be rejected as a solid precipitate. Another method of forming a precipitate is by the addition of a concentrated solution of a dissolved solid to a large quantity of a pure liquid in which the solid will not dissolve. This process is used extensively to isolate chemical products or waste materials from process liquor.

Precipitation should not be confused with crystallisation. A precipitate is produced instantaneously, its particles are amorphous (that is they have no particular shape or form) and they do not "grow" in size like crystals. Precipitated solids are often of a lesser purity than crystals since sometimes spongy aggregates of particles are formed which trap (occlude) liquid or solid impurities within the bulk. For this reason, vessels used for precipitation are usually provided with very efficient means of agitation to ensure that the particles of the precipitate are properly dispersed in the liquid. In most cases, only limited cooling is necessary, and sometimes it is not required at all.

SIZE REDUCTION AND SIZE SEPARATION

Size reduction may be described in many terms, such as crushing, grinding, cutting, shearing, comminution, breaking, disintegration, or pulverisation. The choice of term depends on the machine used to reduce the size of the particles, the method it employs, and the degree of fineness required. This is a very common unit operation since many materials cannot be used conveniently unless they are reduced in particle size, or made to a convenient shape. Materials may be available in the form of boulders, lumps, crystals, or agglomerates, but are often required for use as fine powders. For example, crystalline material may be milled to a fine powder to enable it to be compacted into a tablet for effective medicinal use. In some cases, material is required in the form of a powder to enable it to be dissolved readily in a solvent, to chemically react, or simply to flow freely in processing equipment. In all cases size reduction brings about an increase in surface area of the solid and it is this effect for which the operation is most utilised.

Size reduction may be achieved in several stages using different devices: for example large boulders or rocks may be reduced to lumps or pebbles using a jaw crusher, then reduced to a coarse powder using a hammer mill, and finally, to a very fine powder using a micropulveriser.

Size separation is a process which often follows size reduction: its purpose is to separate and classify the solid particles according to their size. The main methods are screening, magnetic separation, and mechanical classifying. Magnetic separation is limited to iron and similar metals, and is often used to separate iron impurities from other materials to eliminate damage to chemical plant or discoloration of the products.

Classifiers work on the principle that the rate at which solid particles settle in a liquid depends on the particle size. The larger particles may be drawn off at the bottom and the smaller particles as an overflow at the top. A device of this type, used for the classification of very fine particles, is called an elutriator, and the operation of a series of such devices, elutriation. A suspension of about 10 per cent solids is fed into the side of a column against an upward current of water when smaller particles flow upwards and overflow into another vessel and larger particles settle out at the bottom.

Screen are widely used for the separation of wet or dry solid particles. They consist of metal bars, perforated, holed, or slotted metal sheets, or fine wire mesh. The material to be separated passes over the screen which is shaken or vibrated, manually or mechanically, to allow all the smaller particles to fall through the mesh. The mesh is given a number according to the size of the hole or aperture, and for wire-mesh screens the number indicates the number of holes per inch of the screen surface. Standard scales, which take into account the width of the wire itself, are given by the British Standards Institution (B.S.I.).

SOLVENT EXTRACTION

There are two types of solvent extraction: solid-liquid and liquid-liquid extraction. The former, often called leaching, is the process of removing a substance from a mixture of solids by mixing with a liquid (the solvent) in which the substance it is required to separate is dissolved. Liquid-liquid extraction is based on the same principle, but in this case a solution of a dissolved substance is intimately mixed with another liquid (the solvent). If the two liquids are immiscible, two separate liquid layers are formed when the mixture is allowed to settle, the dissolved substance being concentrated in the second liquid. The liquid now containing the dissolved substance is called the rich layer or extract and the remaining spent layer is called the raffinate. In many cases one of the solvents used is water, and these are termed aqueous extractions. Solvents commonly used in conjunction with water are chloroform, carbon tetrachloride, ethylene dichloride, benzene, toluene, xylene, and petroleum ethers.

A simple batch, liquid-liquid extraction is illustrated in Fig. 2.4.

If the liquids are partially soluble (that is, part of each liquid dissolves in the other), it may be necessary to perform a second extraction when the raffinate is mixed again with fresh solvent. It is convenient if the initial solvent closen is less dense than the added pure solvent, because it forms an upper raffinate layer. In this way mixing of the solvents may be assisted, and should a second extraction be required, it may be carried out directly in the same extraction vessel after running off the first extract.

A solvent extraction process is usually followed by distillation to remove the solvent from the extract and render the solvent available for re-use.

A simple flowsheet for a continuous, single extraction process with solvent recovery is shown in Fig. 2.5.

Mixers achieve intimate contact between the two liquids by mechanical or air agitation or jet mixing (in which one liquid is sprayed through at nozzle into the other liquid) or by external recirculating pumps.

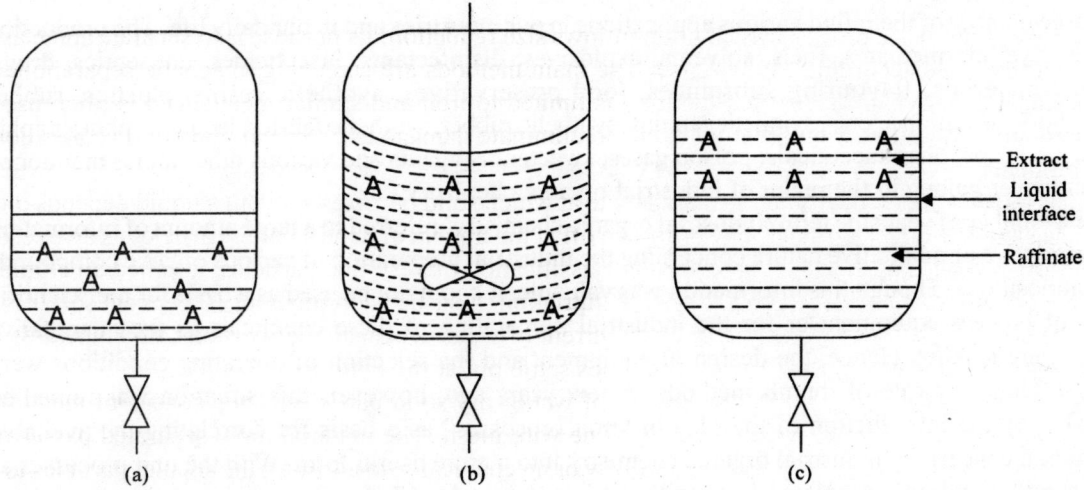

Fig. 2.4. (a) Solvent containing dissolved substance A; (b) second solvent added and mixed by agitation; (c) separation of two liquid layers on settling. Raffinate is run off.

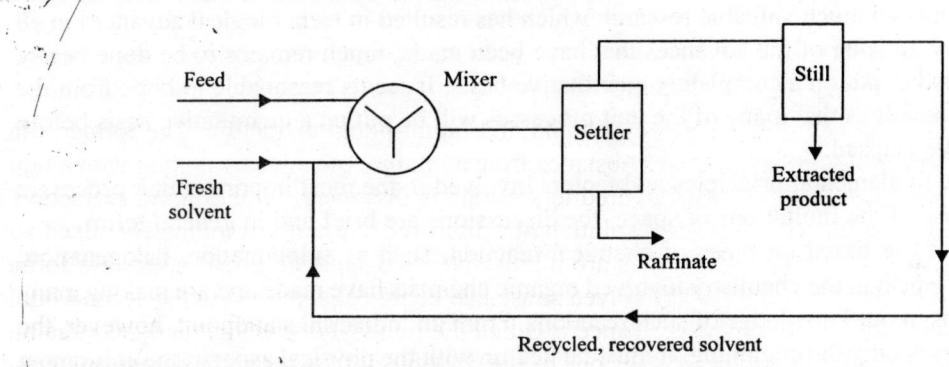

Fig. 2.5. Continuous, single extraction.

Multi-stage extraction is used to achieve a higher efficiency of separation, in which the product is almost completely removed from the raffinate. The solvent is split up into several portions and fed to a series of mixers and settlers. The disadvantage of this method is the need to use large volumes of solvent. A more complicated system, called countercurrent, multi-stage extraction uses a series of mixers and settlers arranged as before, but the feed liquid and pure solvent are passed through the system in opposite directions, that is countercurrently. Continuous countercurrent operation may be carried out by means of spray columns, packed columns (similar to those used in distillation), plate columns, or, sometimes, bubble cap columns. Similar methods are employed in solid–liquid extrcation– all designed to bring about intimate contact between the solid and pure solvent. Leaching systems, as they are often called, may employ batchwise, single and multiple, continuous, and countercurrent extraction.

ORGANIC UNIT PROCESS

Organic chemistry or the chemistry of carbon compounds deals with more than 250,000 compounds. While at the present time many of these compounds are merely scientific or laboratory curiosities, a very large number of them find various applications in our industries and in our daily life. The production of dyes, dye intermediates, fuels, solvents, explosives, disinfectants, insecticides, antiseptics, drugs, perfume materials, flavouring substances, food preservatives, synthetic resins, plastics, rubber accelerators, floatation agents, synthetic tannins, synthetic rubber, synthetic fabrics, lacquers, photographic chemicals, soaps, pharmaceuticals, poison gases, sweetening agents and various other topics mentioned in this chapter belong to the realm of industrial organic chemistry.

For a number of years the term 'industrial organic chemistry' referred to a large amount of information of a descriptive or qualitative nature concerning the industrial preparation of various organic compounds and compositions. Though this information was valuable, it could not be used as a basis for the scientific design of process equipment or for the industrial preparation of these chemicals as the quantitative aspects were lacking. Hence, the design of equipment and the selection of operating conditions were largely effected by rule-of thumb methods. A few years ago, however, this situation was remedied somewhat by the introduction of the idea of "unit processes" as a basis for correlating the available information concerning industrial organic chemistry into a more useful form. With the unit processes as a framework, hundreds of isolated facts and principles were classified under a few main headings. This classification has been of great value to manufacturers, to research workers, and to designers as it has emphasised the wide applicability and utility of the fundamental principles underlying each of the unit processes. It has stimulated much valuable research which has resulted in technological advances in all phases of the industry. In spite of the advances that have been made, much remains to be done before the unit processes can be put on a completely quantitative basis. It seems reasonable to hope from the analogy to the unit operations that many of the unit processes will be put on a quantitative basis before many more years have elapsed.

In this chapter the fundamental principles and factors involved in the most important unit processes are discussed. Because of the limitations of space, the discussions are brief and in general terms.

The unit processes are based on types of chemical reaction, such as sulphonation, halogenation, oxidation, etc. With respect to the chemistry involved organic chemists have made and are making many valuable contributions to our knowledge of such reactions. From an industrial standpoint, however, the chemistry of a process is only the beginning. It must be tied up with the physical aspects, the equipment needed and the economics involved. Hence, to design and operate equipment for carrying out processes on a commercial scale much more information is necessary than is usually obtained under the almost ideal laboratory conditions where the material is first prepared. The following information is needed over a wide enough range of conditions to permit the optimum conditions to be determined: (i) the yield that can be expected; (ii) the time necessary to carry out the reaction; (iii) the kind of apparatus suitable; and (iv) the cost per unit of product.

The first item is concerned with equilibrium conditions; the second relates to reaction rates (chemical kinetics); the third is dependent upon the materials used and the type of process; and the fourth is concerned with economics.

Equilibrium

A system is said to be in equilibrium when all forces and reactions are balanced by other forces and reactions so that no net change takes place with time. All spontaneous processes tend to approach a state of equilibrium and remain in equilibrium so long as external conditions are unchanged. Equilibrium conditions therefore represent the maximum yield that can be expected from a given reaction. In industrial processes it is necessary to know the effect of the important variables on the equilibrium so that the optimum conditions can be determined. In general, *LeChatelier's* principle can be used to make qualitative forecasts of the effect of changing one of the operating variables. According to this principle, if a stress is applied to a system in equilibrium, the equilibrium will tend to shift in such a way as to relieve the applied stress. The conclusions reached from the application of this principle are, in general, similar to those discussed below, for which quantitative calculations can be made. This principle has been a very useful generalisation to guide research work even though the quantitative aspects are lacking for many cases.

Quantitative information on the effect of changing one of the important variables on a system in equilibrium is highly desirable wherever it is possible to obtain it. For many systems it is possible to make such calculations with a reasonable degree of accuracy by means of the equilibrium constant for the reaction. The equilibrium constant is a number characteristic of a given reaction at a given temperature and is independent of pressure or concentration. The familiar mass-action law expresses the relation between the "active-masses" of materials present in a system and the equilibrium constant for the reaction. For reactions taking place under conditions such that the reactants may be considered ideal gases, partial pressures may be used as active masses, while in dilute solutions of non-ionised substances concentrations can be used. For other homogeneous systems the activity function can be used. Hence, the equilibrium constant makes it possible to calculate the quantitative yields that can be expected from a given reaction if sufficient data are available to evaluate the constant. Equilibrium constants may be evaluated from direct experimental methods or by calculation by thermodynamics using the free energy concept. The experimental method is laborious and in many cases may not be reliable unless extreme precautions are taken. The thermodynamic method is also limited by the lack of accurate thermal data but even where accurate data are lacking, approximate data are useful for indicating trends and tendencies.

Once the equilibrium constant at various temperatures are known, it is possible to calculate the effect of the principal variables on the equilibrium conversion under various conditions provided the necessary data are available. From such calculations and the Principle of *LeChatelier* certain generalisations are evident and are worthy of consideration. These are as follows:

1. The equilibrium conversion of an endothermic reaction is increased by a rise in temperature, while the equilibrium conversion is decreased by a rise in temperature in the case of an exothermic reaction.

2. Pressure affects the equilibrium conversion of a gaseous reaction in which there is a change in volume due to a change in the total number of models of gaseous components. If a reaction produces a decrease in volume, the equilibrium conversion will be increased by an increase in pressure. If there is no change in volume, pressure is without appreciable effect. If a reaction produces an increase in the number of moles of gaseous components, the equilibrium conversion will be decreased by an increase in the total pressure on the system.

3. The presence of an excess of one reactant tends to increase the equilibrium conversion of the other reactants present.

4. The addition of one of the reaction products to the initial reacting system will decrease the equilibrium conversion of the reaction. Likewise, removal of one of the reaction products as soon as formed will increase the equilibrium conversion of the reaction.

5. Dilution of a reacting system with an inert gas has the same qualitative effect as decreasing the total pressure on the system. If the reaction produces an increase in the total number of moles of gaseous components, dilution with an inert gas will increase the equilibrium conversion. If there is no change in the total number of moles of gaseous components, dilution with an inert gas will have no effect on the equilibrium conversion.

In making and interpreting calculations of the equilibrium conversion for a given system, the limitations of such calculations should always be borne in mind. Equilibrium calculations are made for a single reaction, though it is possible to calculate the equilibrium conversion for certain cases where more than one reaction takes place. Many organic reactions involve a complex mechanism and are complicated by side reactions to an extent which cannot be predicted by theoretical consideration alone. Equilibrium calculations for such system are of no value in predicting what will happen when the reaction is carried out experimentally. However, if the reaction can be carried out in a manner to give primarily one product, equilibrium calculations can be of great value. Reactions of carbon monoxide and hydrogen furnish examples of both types of reactions. By the proper selection of catalysts and operating conditions, it is possible to carry out the reaction between CO and H_2 to give primarily methanol. Under these conditions calculated equilibrium conversions agree quite well with those obtained in commercial practice, and such calculations have been of great value in the development of synthetic methanol. On the other hand, in the Fisher-Tropsch synthesis, the same materials are reacted in the presence of a different type of catalyst under operating conditions but slightly different from those employed in the methanol synthesis, and a whole series of complex liquid products are obtained. Obviously in cases of this sort where it is impossible to predict what products will be obtained, equilibrium calculations are of no value.

It should be emphasised also that the equilibrium conversion is the limiting condition toward which the reaction is directed and beyond which it cannot go. Industrial reactions, however, usually are not allowed to reach equilibrium but are stopped at some intermediate point because of economic considerationd. Hence, the actual extent to which a reaction will proceed depends upon the rate of reaction and the time allowed for it to proceed, as well as on the limiting equilibrium composition.

Reaction Rate

The rate at which a reaction proceeds is determined by a number of factors: the order mechanism of the reaction, the active masses present, the character of the system, the temperature, the presence of catalysts, the pressure, the energy of activation, side reactions, successive reactions, reverse reactions, etc., all affect the rate to a certain extent. Although no general procedure has been developed for weighing the effect of each of these factors on a given reaction for the purpose of exactly predicting the reaction rate, certain principles have been developed which are useful in a quantitative or semi-quantitative way.

Order of reactions

The relation existing between the concentration of the reactants and the rate of reaction determines the order of a reaction. If the rate of reaction is directly proportional to the first power of the concentration of a single reactant, the reaction is said to be a first order one. Second order reactions are those in which the rate of reaction is proportional to the product of the concentration of two reacting molecules, while third order reactions are those in which the rate of reaction is proportional to the product of the concentrations of three reaction molecules. The probability of reactions of an order higher than third order is so remote that they are given no further consideration. It should be pointed out at this point that it is not possible to predict the order of reaction by merely writing down the equation for the reaction. The order of reaction does not necessarily bear any relation to the ordinary equation used to represent the reaction, but it must be determined, experimentally. Reaction rates do not necessarily fit in with any "order" as defined above. On the other hand, certain reactions do approximate first order reactions even when heterogeneous.

First order reactions ordinarily involve a single reactant molecule A , which decomposes into products as follows:

$$A \longrightarrow B + C + D \ldots$$

If n_A = moles of A per unit mass of reacting system, t = time, an k is a constant (reaction velocity constant) a first order reaction can be expressed by the equation.

$$\frac{-dn_A}{dt} = kn_A$$

which on integration gives

$$ln\frac{n_{A0}}{n_A} = kt$$

where n_{A0} = moles of A present in a unit mass of system when $t = 0$.

Analogous equations for second and third order reactions may be formulated. In the cases of homogeneous gaseous reactions carried out under conditions such that the ideal gas laws apply, such equations show that the rates of second order reactions are directly proportional to pressure while third order reactions are proportional to the square of the pressure. Under the same conditions first order reactions are independent of pressure.

The majority of industrial organic processes are so complicated by the existence of side reactions, existence of more than one phase, consecutive reactions, reversible reactions, and complex compositions of the original reactants that they do not behave as first, second, or third order reactions. Here, no wide use of mathematical formulations for predicting reaction rates has been made.

Such equations that have been developed are empirical in origin and in general prove to be merely approximations applicable only over limited ranges of conditions. Complex systems, however, sometimes may be handled satisfactorily by method of approximations applicable only over limited ranges of conditions. Complex system, however, sometimes may be handled satisfactorily by methods of approximations analogous to those found satisfactory for simple cases. Even in cases where mathematical expression proves inadequate some information of a semi-quantitative nature can be predicted from the kinetics of similar reactions.

Effect of temperature on rate of reaction

It is well known that temperature has marked effect on the rate of reaction. For some time the statement that the rate of reaction is doubled by increasing the temperature 10 or 15°C has been known to be inaccurate.

In some cases it takes 50°C or more to double the speed of a reaction. The most satisfactory mathematical relationship expressing the temperature effect on reaction velocity for a homogeneous system is the empirical equation developed by *Arrhenius*:

$$\frac{d(ln\,K)}{dT} = \frac{E}{RT^2}$$

which on integration gives:

$$ln\,k = \frac{-E}{RT} + C$$

where E = "energy of activation" per mol
      k = reaction velocity constant at temperature T
      R = the gas law constant
      T = the absolute temperature
and   C = a constant of integration

In order to carry out the integration it was necessary to assume E to be constant. Since this assumption only holds for a limited temperature range, use of the *Arrhenius* equation should be restricted to temperatures in the neighborhood of those for which the values of E are known. From the integrated equation above it is evident that the limiting value of the reaction velocity constant, k, is determined by the value of C. This means that under conditions such that the quantity $\frac{E}{RT}$ is very small as compared to C, the rate of reaction is practically independent of temperature changes, which is direct contradiction to the usual effect of temperature on reaction rate. These remarks should be sufficient to emphasise the necessity of making certain that the *Arrhenius* equation holds for the conditions under consideration before using it to predict the effect of temperature on reaction rate.

For the majority of homogeneous reactions the *Arrhenius* equation is found to be relatively satisfactory over considerable changes of temperature. In our present state of knowledge the energy of activation E must be measured experimentally. Hence, if rate data are available at two different temperatures to permit E and C to be evaluated, it is possible to predict the velocity of a homogeneous reaction over the temperature range for which the equation holds.

The effect of temperature on the reaction of a homogeneous reaction is readily apparent from the integrated form of the *Arrhenius* equation. Since the reaction velocity constant, k, depends upon the difference between the values of C and $\frac{E}{RT}$, an increase in the value of T decreases the value of $\frac{E}{RT}$ which is substracted from C, and hence the rate of reaction increases because the reaction velocity constant k, increases in value.

The marked effect of temperature changes on reaction rates can be used to explain one of the difficulties encountered in carrying out exothermic reactions under adiabatic conditions. If a large quantity of heat is liberated by the reaction, the reaction mass will increase in temperature which in turn increases the reaction rate. If this continuous building up of the reaction rate is allowed to proceed unchecked the reaction will soon reach explosive violence.

However, the equilibrium conversion decreases, for this type of reaction, with an increase in temperature and this tends to act as a safety valve by decreasing the amount of heat liberated due to the lower conversion. Hence, all exothermic reaction cannot be expected to reach explosive violence if carried out under adiabatic conditions, but this explanation should be sufficient to account for the necessity of controlling the temperature in many reactions.

In this discussion of rate of reaction no consideration had been given to the effect of reverse reactions, side reactions, successive reactions, temperature gradients and the like on the speed of reaction. Many cases of industrial importance are known in which it is not possible to neglect such effects. Other reactions are so complicated that a complete and rigorous analysis would be impracticable even though the basic theory were available. However, it should be emphasised again that in certain cases satisfactory methods of approximation have been developed whereby complex systems may be treated by relatively simple methods.

Heterogeneous reactions

Heterogeneous reactions, particularly those involving catalysts, make up many of the most important industrial reactions of the present day. The rates at which such reactions proceed is determined by a number of factors, such as type and condition of catalyst if present, rate of diffusion, type of reaction, character of reacting molecules, rate of heat transfer, etc., in addition to those already mentioned. Correlations based on unit processes have brought out certain qualitative factors regarding the reaction rate of such systems but the quantitative aspects are lacking. Much work has recently been done on this type of reaction, and it is hoped that an early release of such information will permit a quantitative treatment in the near future.

All of these general principles have been known for some time, but few applications to industrial technology were made before the 'unit processes' emphasised their wide applicability and utility.

Function of the pilot plant

From a casual reading of the preceding discussion on equilibria and reaction rates one might conclude that such studies are interesting but of no practical value. Such a conclusion is far from the truth. The chemical industry spends huge amount for research in pilot plant equipment to obtain information on reaction rate and equilibrium conditions, as well as to test various arrangements of experimental equipment. In these studies, the fullest use of the available information is made in order to cut down the experimental work to a minimum. The principles just discussed are used as a basis for deciding what experimental work shall be done and in extrapolating and interpreting the data to get the most information from a minimum of experimental work. Though the equilibrium and kinetic considerations are not yet quantitative they are definitely not a waste of time as they serve as a frame-work for correlations which eventually will bring order out of chaos, just as similar work did for the unit operations.

Organic Unit Processes

Since the unit processes are based on chemical reactions, unit process equipment is essentially an apparatus in which the chemicals are brought together under the proper conditions and allowed to react. In many cases the equipment can be quite simple but in others it needs to be very elaborate with very careful control of operating conditions provided. For example, some reactions can be carried out in open vessels by simply mixing the materials together with very little heating or cooling. In other cases high-pressure equipment built of special corrosion-resistant materials and controlled by elaborate automatic devices is necessary. In this connection it should be pointed out that once the chemicals are brought together, the operator has no control over the chemical reactions that take place. All that he can do is to control over the chemical reactions so that the optimum yield of the desired product is obtained. Hence, the principal concerns of the equipment designer are to select the materials most suitable for the apparatus, and to get an arrangement of apparatus which will permit the necessary control of the principal factors affecting the reaction, such as temperature, pressure, concentration, time of contact, homogeneity of reaction mass, presence or absence of light or catalysts, etc. In the following section a brief discussion of a number of organic unit processes is given. Application of the fundamental principles already discussed are pointed out and the types of equipment used for the various unit processes are also discussed.

Halogenation

Halogenation may be defined as a process in which a halogen atom is introduced into an organic compound. Halogenation may occur by (i) addition; (ii) substitution for an element; and (iii) replacement of a group such as the hydroxyl or sulphonic acid group.

Organic halogen compounds are prepared by halogenation reactions on a large scale and are widely used industrially. Halogen derivatives are used to a considerable extent directly as cleaning fluids ($CHCl=CCl_2$, CCl_4); plant stimulants ($CH_2Cl \cdot CH_2Cl$); refrigerants (CF_2Cl_2, CH_3Cl); moth repellant and deodorant (p-Cl-C_6H_4-Cl); anesthetic ($CHCl_3$); general solvents (CCl_4); etc. The greater proportion of these compounds, however, is used as raw materials for a very large number of syntheses, e.g., in the preparation of alcohols, phenols, amines, alkylene oxides, ethers, dyes, acids, synthetic rubbers, hydrocarbons, alkaryl compound, etc. Of the halogen derivatives the chlorine compounds are the most important because they can be prepared cheaply.

Character of reactions

Halogenation reactions are carried out in both liquid and vapour phases. Both thermal and catalytic reactions are employed. In the catalytic reactions, the halides or iron, antimony, phosphorus, and sulphur as well as actinic light serve as important catalysts. In the preparation of halogenated aromatic compounds, the catalysts not only speed up the reactions, but also influence the point at which the halogen enters the aromatic compound.

Controlling homogeneous chlorinations

The reaction of chlorine with vapours of a paraffin hydrocarbon is exothermic and so vigorous in a homogeneous system that precautions must be taken to prevent it from reaching explosive violence. Hence, numerous procedures have been devised to control the reaction rate and to avoid undesirable reactions which result in diminished yields. The following such methods are used:

1. Employing an excess of the hydrocarbon so that only a limited amount can react per cycle. The excess hydrocarbon is recovered and recycled.
2. Reacting the materials in the presence of a diluent gas, such as steam, nitrogen, or HCl.
3. Reacting the hydrocarbon with a halogen gas that reacts mildly and then replacing the combined halogen with chlorine.
4. Employing a liquid solvent in which the hydrocarbon is soluble.
5. Carrying out the reaction in successive stages by mixing only part of the chlorine with the hydrocarbon and cooling between stages.
6. "Hot chlorination" of olefins; at temperatures above 500°C, the chlorine substitution" reaction predominates over the addition reaction.
7. Use of an excess of chlorine under high pressure. In this case the excess chlorine serves as the diluent gas and helps to keep the reaction temperature down.
8. Employing a moltern salt bath in contact with the reagent. The temperature of the bath can be maintained at the desired temperature by cooling and salts can be used which serve as catalysts for the desired reaction.

A study of the above methods shows that they are really applications of the mass action law and the *Le Chatelier* Principle. For example, the use of a diluent gas, a liquid solvent, a moltern salt bath, or an excess of hydrocarbon or chlorine assists in controlling the reaction temperature by increasing the heat capacity of the system and hence prevents large increases in temperature due to the exothermic reaction. In addition, the use of an excess of chlorine or hydrocarbon tends to increase the degree of completion of the reaction, while the use of a diluent decreases the reaction rate not only by making it more difficult for the reacting molecules to get together but also because the added heat capacity of the diluent tends to keep the temperature of the system down. In the case of the liquid solvent and molten salt bath, the inert material serves both as a diluent and as a medium for removing the heat of reaction.

Chlorination of aliphatic compounds

Because of the variety of products made by chlorinating aliphatic compounds, it is difficult to generalise on the conditions for chlorination of these materials. Reactions are carried out using both batch or continuous processes and reacting the materials in either liquid or vapour phase. The temperatures usually employed in liquid phase reactions range from about 30 to 200°C., and in vapour phase reactions from 200 to 500°C. Atmospheric or moderate pressures are commonly used but recent work indicates that high pressures may be desirable in certain cases. The chlorination of aliphatic compounds other than hydrocarbons usually takes place at temperatures lower than are necessary for the paraffins.

The chlorination of aromatic compounds

The chlorination of aromatic compounds is generally carried out in the liquid phase, though the more volatile compounds are sometimes chlorinated in the vapour phase. Since chlorine may enter either the side chain or the ring, conditions must be properly selected in order to obtain the maximum yield of the desired compounds. In general, hot chlorination in the absence of a catalyst favours introduction of chlorine in the side chain, while chlorination at a lower temperature in the presence of a catalyst, such as iron, favours the replacement of nuclear hydrogen.

In certain halogenations, because of the orienting influence of groups already present, it is necessary to employ two or more chlorination methods to get the desired chlorine compound. In the chlorination of aromatic materials which are normally solids at ordinary temperatures it is frequently necessary to employ solvents. Inert solvents, or solvents such as benzene and nitrobenzene, which ordinarily react readily with chlorine, can be used in cases where chlorine reacts much more readily with the solute than with the solvent.

Apparatus for chlorination

The conditions for carrying out halogenations vary so widely that no general rules for the design and construction of equipment can be formulated. For example, chlorinations are carried out in jacketed vessels, in packed towers, in molten salt baths, and in tubular equipment. Both continuous and batch processes are used. Due to the corrosive action of the reagents and the acids formed by the reactions, the materials of construction must be carefully selected. With non-aqueous media, the usual materials are iron, alloys, copper, or lead though vitreous silica, glass or glass-lined equipment may be utilised. Tantalum seems to be the best material for use when both H_2O and HCl are present, but because of cost its use is limited. Lead-lined steel reactors are used where aqueous HCl may be present and lead coils are used for temperature control.

In addition to corrosion resistance, chlorination equipment should have the following features: (i) provision must be made for dissipating the heat of reaction. In liquid phase reactions carried out in autoclaves, this is usually accomplished by circulating a cooling medium through the jacket of the chlorinator, or through coil suspended in the reaction vessel; (ii) agitation must be provided for heterogeneous systems. This may be accomplished by outside circulation with a pump or by providing a mechanical stirrer in the reaction mass; (iii) provision must be made for getting the reacting materials into the chlorinator and for removing the products of the reaction; (iv) careful control of the concentrations must be provided in order to prevent undesirable side reactions and explosions; (v) sight glasses should be provided for observation of the reaction and to detect the escape of unreacted chlorine. In case the reaction is photocatalytic, provision should be made for exposing the reaction mass to light. A variety of methods have been developed for commercial chlorinations embodying these features.

Hydrolysis (including alkali fusion)

Hydrolysis may be defined as a process in which a double decomposition reaction is carried out with water as one of the reactants. In this discussion the meaning of the term has been extended to cover also the alkali fusion reactions which are very similar to hydrolysis reactions. This process is used principally for the conversion of —X to —OH; ester to alcohol and acid or alcohol and soap; —CN to —CONH$_2$ to —COOH; —OSO$_2$H to —OH; and for cleavage of C—C linkages in large molecules, such as those of proteins and carbohydrates.

Uses of hydrolysis

Hydrolysis reactions are widely employed for splitting fats and to form glycerol and fatty acids or soaps. All carbohydrates, including sugar, cellulose, starch, and other polysaccharides can be hydrolysed. The hydrolysis of starch is the basis of a large industry. Corn sirup for confectionery trade or table use and corn sugar (glucose) are the principal products from the hydrolysis of corn starch. The hydrolysis of wood cellulose to glucose is being carried out on an industrial scale. Two processes have been reported, one employing dilute sulphuric acid (0.2 to 0.1% by weight) and the other fuming hydrochloric acid as the hydrolysis reagent. The hydrolysis of organic halides and alkyl esters of sulphuric acid (prepared from unsaturated hydrocarbons and sulphuric acid) are widely used methods for the preparation of alcohols and phenols such as amyl alcohol, ethyl alcohol, ethylene glycol, isopropyl alcohol, and phenol. The latter material is in great demand for use in plastics.

Conditions for hydrolysis

In practically all cases of industrial importance hydrolysis reactions are carried out in the presence of either acids or bases as catalysts. Dilute hydrochloric acid and sulphuric acid are the most commonly used acids and caustic soda, sodium carbonate, and sodium bicarbonate are the most common alkalies used. Enzymes are used as catalysts for hydrolysis reactions in the brewing industry.

Most hydrolysis reactions are carried out in the liquid phase and the temperatures are usually in the range of 80 to 220°C. The pressures used vary from one to 200 atmospheres. Though the use of a high temperature may affect the equilibrium point adversely, in commercial practice the highest temperature practicable is employed in order to complete the reaction in the minimum time. Several methods of increasing the rate of hydrolysis are worthy of mention. Since many of the organic compounds which are hydrolysed are insoluble in water, efficient agitation, by means of an agitator or recirculation by means of a pump, in conjuction with an emulsifying agent markedly increases the rate of hydrolysis. Other methods include pressure percolation followed by immediate cooling (wood hydrolysis), carrying out that reaction in successive stages with removal of part of the products between stages (soap making), and recirculation of an equilibrium mixture of by-product to prevent further formation of these materials (recirculation of diphenyl oxide in the preparation of phenol). Though increasing the concentration of hydrolysing reagent tends to increase the rate of reaction, in several cases it has been found that such high concentrations result in undersirable by-products and are therefore avoided in these instances.

Vapour-phase hydrolysis reactions and alkali fusions are carried out at temperatures ranging from 300 to 500°C. Only a few vapour phase hydrolytic reactions are used commercially, while alkali fusion is limited essentially to the replacement of —SO$_3$H by OH.

Hydrolysis equipment

In the matter of equipment design, alkaline processes in general have the great advantage over acidic processes since they can usually be carried out in steel or iron reaction vessels. Such vessels are simple in design, usually requiring only some method for heating and agitating the reaction mass. Even the severe conditions of caustic fusion do not seem to greatly shorten the life of cast iron pots in which they are carried out. Both batch and continuous processes are used.

Acid processes, on the other hand, present a difficult problem in equipment design. Lead lined equipment is necessary for processes in which sulphuric acid is used. In addition to the lead-lining, acid-resisting brick or Pyrex glass brick can be employed for additional protection. With hydrochloric acid, corrosion is even more severe unless the reaction is carried out with dilute acid (less than 1% HCl by weight) and at low temperatures, in which cases bronze and copper suffice. For higher concentrations, especially at higher temperatures, stoneware, or iron, lined with an acid-resisting tile joined with a special cement, is necessary. Rubber-lined iron is used to some extent and tantalum is finding favour in continuous large scale processes where HCl is used as the catalyst.

Esterification

The process of preparing an ester is called "esterification". Esters of organic acids can be represented by the general type formula RCOOR', when R and R' represent hydrocarbon or sustituted hydrocarbon radicals. Esters of inorganic acids can be represented by the formula ROA', where R represents a hydrocarbon or substituted hydrocarbon radical and —OA' represents an inorganic acid radical (such as —ONO_2, —OSO_3H, etc.). A variety of reagents can be used for the preparation of esters but the usual reagents are alcohols or phenols with acids or their anhydrides in the presence of dehydrating agents (HCl, H_2SO_4, etc.)

Esters have several important properties which make them of great value as industrial chemicals. Large quantities of acetates, such as butyl acetate, ethyl acetate, amyl acetate, are used as solvents, especially in the lacquer industry. Because of their pleasant odour and taste quite a large number of esters find use in perfumes, flavours, and perfumed soaps. Several of the esters have valuable plasticising properties, e.g., dibutyl phthalate and diethyl phthalate. Esters of nitric acid, such as glycerol trinitrate, glycol dinitrate, and cellulose nitrate, have valuable explosive properties. Cellulose nitrate (called nitrocellulose) and cellulose acetate are valuable as plastics, while the latter is also used in large quantities as synthetic fabrics.

Operation conditions

The operating conditions for industrial esterification reactions are mild in comparison with most of the other unit processes. The temperatures range from about 10 to 200°C. Atmospheric pressure or slightly above is commonly used. Though an increased temperature speeds up esterification, the use of this method for the volatile esters makes it necessary to use pressure vessels. Fortunately, by the use of catalysts, esterification reactions can be carried out in a reasonable time at relatively low temperatures, and this is the method employed industrially. Sulphuric acid and hydrochloric acid are the most common catalysts used.

Esterification reactions are reversible and the equilibrium conditions are such that a considerable proportion of the original reactants are present when equilibrium is established. Hence, it is usually necessary to employ some method to increase the degree of completion of the desired reaction. The methods used successfully are: continuous removal by distillation of water or ester as it is formed in the reaction; passing a gas slowly through the reaction mass to remove H_2O; addition of liquids, such as benzene, ethylene chloride, and carbon tetrachloride to assist in bringing over the water in a constant-boiling mixture.

Equipment for esterification

The simple esters are prepared in standard distillation apparatus, the ester being formed either in the still body or in the fractionating column itself. Copper apparatus is generally used for this type of esterification. Both batch and continuous processes are employed.

Nitration

Nitration is the process by which the union of the —NO$_2$ group directly to a carbon atom is effected. In organic nitro compounds the nitrogen in the —NO$_2$ group is united directly to a carbon atom. In organic nitro compounds the nitrogen in the —NO$_2$ group is united directly to a carbon atom. In organic nitrates the nitrogen of the —NO$_2$ group united to a carbon by means of an intermediate oxygen atom. The organic nitrates are really esters of nitric acid and they are discussed under the unit process of esterification.

In most cases nitration is carried out by the use of mixed acids; for example, HNO$_3$ admixed with a dehydrating agent, such as oleum, sulphuric acid, acetic anhydride, or acetic acid. Other nitrating agents are HNO$_3$ alone, NaNO$_3$ + H$_2$SO$_4$, N$_2$O$_4$, nitrosulphonic acid (HSO$_3$NO$_2$) and organic nitrates, such as acetyl and benzoyl nitrate.

Uses of nitro compounds

The nitro compounds prepared by nitration reactions are used principally as intermediates for organic syntheses, dye manufacture, and for explosives. Nitrobenzene, dinitrobenzene, nitronaphthalene, dinitrochlorobenzene, *p*-nitrotoluene-*o*-sulphonic acid, and *m*-nitro-*p*-toluidine are some of the important nitro compounds.

Nitration of an aromatic compounds

Nitration of an aromatic compounds is usually a liquid phase reaction in which a mixture of nitric acid and sulphuric acid is the nitrating agent. The concentration of both the nitric acid and sulphuric acid, as well as the proportion of mixed acid to compound being nitrated, are important variables to consider when nitric acid and sulphuric acid, as well as the proportion of mixed acid to compound being nitrated, are important variables to consider when nitration conditions are chosen. Since the nitration reaction is highly exothermic, some method of controlling the temperature must be provided to prevent the formation of undersirable oxidation products and to keep the reaction from reaching explosive violence. Nitration temperatures range from 0 to about 90°C., with temperatures in the neighborhood of 40 to 50°C being commonly used. Nitrators usually have internal cooling coils to remove the heat of reaction and keep the temperature at the desired level. Agitation is also provided to avoid local super-heating, to prevent the formation of more highly nitrated products, and to increase the speed of the reaction.

Nitration of paraffins

Though long thought to be impossible, methods of nitrating paraffin hydrocarbons with nitric acid at high temperatures have been discovered and already commercial developments have taken place. The nitroparaffins are excellent solvents for many of the synthetic resins such as cellulose acetate, cellulose nitrate, and vinylite. In addition, they are quite reactive chemically and may soon become important as intermediates in the preparation of chemicals. In contrast to the low temperatures and high acid concentrations used for the preparation of aromatic nitro compounds, nitrations of paraffin hydrocarbons are carried out at higher temperatures (130 to 140°C) with relatively dilute acids. Vapour phase nitrations are performed at much higher temperature (390 to 450° C). An excess of hydrocarbon is employed to prevent explosions, and the reactions are performed in stainless steel tubular apparatus.

Although continuous nitration apparatus has been developed, batch processes are used in most cases. The apparatus consists of a jacketed cylindrical type of nitrator equipped with internal acid-resisting iron cooling tubes and a marine type propeller for agitation. The cooling tubes are of the closed end type and are designed for close temperature control by circulating brine or cold water. The propeller operates in a draft tube, with vanes arranged to contract swirl, and provides passive agitation and circulation. The bottom of the nitrator slopes to a sump to permit complete discharge through the blow pipe. The outside cooling jacket is of welded steel construction. Connections for installing temperature indicating devices are provided so that the operators can keep a close check on the temperature of the reaction mass at all times.

Amination by reduction

Amination is process in which an amine is formed by the production of the linkage $-\overset{|}{\underset{|}{C}}-NH_2$. Two methods of amination are commonly used: namely: (i) amination by reduction; and (ii) amination by ammonolysis. Amination by reduction of nitro compounds is the usual method employed in the industrial preparation of aromatic amines, while ammonolysis is commonly used for the preparation of aliphatic amines and can be used for aromatic amines.

The discussion of amination by reduction is limited to the reduction of nitro compounds. Many reducing agents are resorted to under variable conditions. The following are used industrially:

Fe and HCl; tin and zinc with acids, or alkali or metal sulphides, such as Na_2S and $(NH_4)_2S$, in solution or suspension; gaseous hydrogen or CO; $FeSO_4$; $Na_2S_2O_4$; and Fe or Zn in alkaline solution.

Aromatic amines are used principally for the preparation of synthetic organic chemicals and dye intermediates but they also find some use as solvents emulsifying agents, insecticides, rubber accelerators, antiseptics, and medicinals.

Conditions for reduction

In technical aminations nascent hydrogen is commonly produced from the reaction of finely divided iron with dilute hydrochloric acid. The reactions are carried out at the boiling point of the solution at atmospheric pressure. The iron not only provides active hydrogen by reacting with the acid present but also furnishes the metal adsorption surface for the reaction. It also enters into the reaction by regeneration of ferrous chloride (catalyst) and acts as an oxygen carrier. Hence, it is necessary to use a finely divided iron and to keep it thoroughly mixed with the rest of the reaction mass. This requires the use of efficient, sturdy agitators. Sleeve and propeller or double-impeller type agitators have been found satisfactory for this purpose. As the reaction is distinctly exothermic, the heat of reaction is sufficient to maintain the material at the reaction temperature as long as iron is being added. Most reducers are steam jacketed, however, in order to permit the charge to be heated initially, and to supply the necessary heat to complete the reaction after all of the iron has been added. This procedure results in considerably saving in overall time required for the reaction.

Equipment for reduction

The cast iron jacket is integral with the side wall and covers the lower part of the body. This jacketing permits very uniform heating of the areas adjacent to the reactants and prevents dilution due to the introduction of live steam into the reducer when heating is necessary. This reducer is equipped with cast iron liner plates to protect the body against abrasive action of the iron borings in the reducer charge.

These removable liners may be taken out and replaced through the door on the side. The charge and sludge can be removed from the reducer by a quick-acting valve faced with a wooden plug. A plow type of rabble is used in the above reducer. This type of rabble is necessary to lift the heavy iron borings from the bottom of the reducer to give a homogeneous reaction mass. A simple form of lifting device is also provided which permits raising the rabble out of the sludge in an emergency or for cleaning.

Amination by ammonolysis

Amination by ammonolysis refers to those reactions wherein an amino compound is formed as a result of the action of ammonia. This process is utilised mainly for the substitution of the —Cl, —OH, and —SO₃H groups by the —NH₂ group, but some reactions involving simple addition and the replacement of oxygen are also of industrial importance.

Conditions for ammonolysis

Though the great majority of aminations are carried out in the liquid phase using aqueous ammonia, anhydrous ammonia can be used, and the ammonolysis of alcohols, oxides, and aldehydes may often be carried out most advantageously in the vapour phase. Ammonolysis reactions are usually carried out at temperatures ranging from about 165 to 210°C. With the solutions ordinarily used this requires pressures of 500 to 950 lb. per square inch, but in some cases pressures as high as 3,000 pounds per square inch are employed. If the compound being aminated is insoluble in aqueous ammonia, it is necessary to provided efficient agitation to create a homogeneous reaction mass. In agreement with *Le Chatelier's* principle, the use of a high ratio of concentrated ammonia to compound being aminated promotes ammonolysis. Because of the high vapour pressure of ammonia at the amination temperatures commonly used, it is necessary to limit the free space above the solution. If this is not done, enough ammonia will enter this vapour space to materially reduce the concentration of ammonia in the liquid, thereby decreasing the rate of reaction. Copper and copper salts are particularly effective as catalysts for ammonolysis reactions.

Apparatus for ammonolysis

Since ammonolysis reactions are carried out at elevated temperatures and pressures, either high pressure autoclaves or tubular reaction equipment must be used. Autoclaves are used for batch process and tubular equipment for continuous processes. Jacketed, steam-heated vessels are used up to temperatures of about 190°C. When higher temperatures are necessary oil or hot water under pressure is the customary heating fluid. It is usually necessary to provide for mechanical agitation if autoclaves are used for the reaction. Horizontal vessels, provided with a number of rotating splash arms, are preferred in the ammonolysis of compounds difficult to wet out and which are converted only at high temperatures.

Diazotisation and coupling

Diazotisation is the process by which a diazo or diazonium compound is prepared. Diazotisation involves the reaction between a primary aromatic amine and nitrous acid. Both diazo and diazonium compounds may be formed simultaneously. Diazo compounds have the structure R—N=N—X, while diazonium compounds have the structure

$$
\begin{array}{c}
\text{X} \\
\mid \\
\text{R—N} \equiv \text{N}
\end{array}
$$

Coupling is the process by which a diazo-or diazonium compound is united with coupling compounds such as aromatic amines and phenols. The two unit process are discussed together because of their similarity. Diazotisation is usually an intermediate step in the preparation of azo dyes in which coupling is the final step, and because of their instability the diazonium or diazo compounds are usually not separated as such from the solution in which they are prepared. Diazotisation can be employed for the preparation of organic compounds not readily prepared by other means.

Conditions for diazotization

Technical diazotisations are usually carrieds, out under mild conditions in aqueous acid solutions in which amines are soluble. With the usual concentrations employed, the reactions are so rapid, even at low temperatures, that most diazotisations are completed in from 5 to 60 minutes. Low temperatures, between 0 and 20 °C., are used to avoid undesirable reactions of the diazonium salts. Because of the mild reaction conditions, the speed of the reaction, and the quantities of material made, the equipment used for technical diazotisations is relatively simple. The reaction vessels usually consist of tubs or tanks provided with agitation. The materials of construction used must not reduce the diazo compound and must withstand the dilute acids used. Wooden tubs, tile-lined iron tanks, stoneware, rubber, enamel-lined iron, and lead are suitable materials. Lead is ordinarily the only metal used in direct contact with diazonium salts, but some diazotisations are carried out in concentrated sulphuric acid solutions in iron reaction vessels.

Coupling represents the last step in the preparation of azo dyes from diazonium salts. Coupling reactions may be used to form solid dyes, or for the formation of the dye directly on the cloth. The reaction conditions are similar to those of diazotisation with respect to temperature, time and concentrations as well as types of equipment.

The azo dyes formed by diazotisation and coupling are used not only to colour textiles but also a wide variety of other materials such as inks, paints, gasoline, and solvents.

Oxidation

Oxidation may be defined as the process whereby oxygen is introduced into, or hydrogen removed from an organic compound by means of an oxidising agent. The following oxidising agents are used: Air, O_2, $KMnO_4$, K_2CrO_7 + Acid, $KClO_3$, NaOCl, PbO_2, H_2O_2, HNO_3, Ferric salts, etc.

Oxidation reactions constitute one of the most powerful means of synthesizing organic chemicals. Complete oxidation, however, results in the formation of carbon dioxide and water, which usually are less valuable than the original substances. Complete oxidation is likely to occur unless the proper precautions are observed, but if the reaction can be controlled so that only partial oxidation occurs, valuable chemicals are frequently obtained. Formaldehyde, acetic acid, phthalic anhydride, camphor, vanillin, benzaldehyde, benzoic acid, alizarin, anthraquinone, and indigo are a few of the many important compounds prepared by oxidation reactions.

There is a marked difference in the behaviour of aromatic and aliphatic hydrocarbons with regard to oxidation.

1. *Aromatic hydrocarbons* are fairly resistant to oxidation and, in general, the reactions are carried out at elevated temperatures employing an active catalyst and an excess of oxidising agent. There seem to be points of resistance where it is relatively simple to stop the oxidation at an intermediate stage. Naphthalene to phthalic anhydride is an example.

2. *Aliphatic hydrocarbons*, on the other hand, are easily oxidised and there are apparently no real points of resistance to oxidation in these compounds. This makes it necessary to control the

temperature carefully and use only limited amounts of oxidising agent when these materials are oxidised. Even then there is a marked tendency to form complex mixtures which are difficult to separate into pure compounds.

Control of partial oxidation

Oxidation reactions of organic compounds are exothermic, and are accompanied by a decrease in free energy. Hence, equilibrium is favourable, and usually it is necessary to take steps to limit the extent of the reaction rather than force it to completion. No matter how favourable the equilibrium may be, suitable reaction rates must be obtained before useful processes are possible. The methods that have been used to provide these favourable rates and for limiting the extent of oxidation have resulted in the great variety of organic oxidation reactions now in use.

Conditions for oxidation

Both liquid-phase and vapour-phase oxidation methods are successfully utilised. Liquid-phase reactions are employed in cases where high molecular weight, complex, thermally unstable substances are dealt with, and where the oxidising agent is relatively nonvolatile. The preparation of vanillin, camphor, quinone alizarin, fatty acids from petroleum, and acetic acid are examples. This type of reaction is conducted at low or moderate temperatures, and the extent of oxidation may be really controlled by (i) limiting the time of contact with oxidising agent; (ii) controlling the temperature; (iii) limiting the amount of oxidising agent; and (iv) varying the type of oxidising agent used.

Vapour-phase oxidation may be applied to materials readily vaporised which do not decompose at the elevated temperatures employed. The preparation of methanol, maleic acid, and phthalic anhydride are examples. The products of oxidation must be thermally stable and fairly resistant to further oxidation before this type of oxidation can be used. Solid or vapour-phase catalysts can be successfully employed to assist in obtaining the desired reaction. The temperatures are usually high. A short time of contact, low oxygen concentrations, selective catalysts, and low conversion per pass are necessary to obtain reasonable yields of the products.

Since most oxidation reactions are highly exothermic, some means must be provided for removing heat fast enough to prevent the temperature from rising to a point where complete oxidation takes place. In liquid-phase reactions the temperature can be controlled by the rate of addition of oxidising agent, by removal of heat in the form of latent heat, by cooling and recirculating part of the reaction mass, or by use of a cooling medium circulated through jacketed vessels or cooling coil. Since the temperatures are usually low, and the rate of heat generation under close control, these methods usually suffice. With vapour-phase reactions, however, rather severe conditions are encountered and more elaborate methods must be used for temperature control. These reactions may be carried out in tabular equipment which has reactively high ratio of heat transfer surface to volume of reacting materials. The use of a catalyst is a good heat conductor with a tube that is relatively "black" which enables more heat to be transmitted to the tube wall by radiation seems to be of value in controlling the reaction temperature. The removal of heat from the outer surface of the tube may be accomplished by the use of boiling liquids such as mercury, sulphur, and diphenyloxide. These materials not only have high heat capacities and low resistance to heat flow, but also permit accurate control of the temperature merely by controlling the pressure.

Equipment

The equipment used for oxidation reactions may be simple or very elaborate. Liquid phase reactions may be carried out in closed jacketed kettles provided with suitable means for regulating the rate of flow of reactants and products. Apparaturs for vapour-phase oxidation must be more elaborate than for liquid-phase reactions. Most organic compounds are sensitive to heat and have a tendency to decompose even though they are not oxidised. This means that it is essential to bring the substance to the reaction temperature in the minimum of time. The difficulty of controlling the temperature at the desired level has already been mentioned. Finally, to avoid further oxidation or decomposition, the reaction products must be cooled immediately after leaving the reaction zone. Many modifications of reactors are available for the oxidation of organic compounds.

Hydrogenation

Hydrogenation may be defined as a process in which gaseous hydrogen is caused to react with an organic compound by addition, substitution, or molecular cleavage (hydrogenolysis).

The hydrogenation of petroleum products is important to the petroleum industry for the preparation of gasoline and specialty products, as well as for improving such materials as kerosene, Diesel fuel, and lubricating oils.

Conditions for hydrogenation

Hydrogenation reactions are usually carried out at temperatures varying from 150 to 250°C., though both lower and higher temperatures are sometimes used. For most cases, two opposing tendencies must be balanced in choosing the optimum reaction temperature. Increasing the temperature speeds up the reaction but also tends to affect the equilibrium adversely since the reactions are exothermic. Fortunately, the development of effective catalysts have made it possible to use lower temperatures where equilibrium conditions are favourable and still maintain satisfactory reaction rates. Since most hydrogenation reactions result in a decrease in volume as the reaction proceeds, the degree of completion is favoured by increasing the pressure. Hence, the use of pressure is common. The range of pressures used for hydrogenation of organic compounds varies from a few to over 200 atmospheres. Catalysts are essential in hydrogenation reactions, and a wide variety of them are available.

Equipment

Although high pressures are already used, the trend in commercial hydrogenations seems to be toward still higher pressures. The design of such equipment, however, is somewhat complicated , and it is necessary to use alloy steels. Equipment is available for both batch and continuous operations. Batch operations can be carried out in autoclaves provided with agitation and using a catalyst suspended in the material being hydrogenated.

These hydrogenators are operated at pressures varying from 20 to 75 pounds per square inch and temperatures varying from 325 to 425° F. In this reactor the catalyst is suspended in the oil which is sprayed into the top of the reactor. Hydrogen is bubbled into the bottom of the reactor to provide a countercurrent flow of the reactants. Internal heating coils provide for temperature control. The catalyst is kept in suspension by the action of the oil circulating pump, by the agitation produced by the hydrogen gas, and by the mechanical agitator mounted on the side of the reactor.

In continuous processes the catalyst is maintained stationary in forged steel reactors and the material being hydrogenated is passed through it continuously. Because of the thick reaction vessel walls which make heat transfer somewhat difficult, it is usually better to add all of the heat needed to the material before it gets to the reaction vessel. Since hydrogenation reactions are usually exothermic, it is sometimes necessary to remove heat once the reaction is started.

This can be accomplished by heat exchange within the reactor, or by cooling part of the material outside the reactor and returning it to maintain the desired temperature.

Sulphonation

Sulphonation is the process by which the $-\overset{\displaystyle O}{\underset{\displaystyle OH}{S}}\!\!=\!\!O$ group is united with a carbon or nitrogen of an organic compound. In sulphonation with sulphuric acid, water is formed as one of the products of the reaction. One molecule of water is formed for each sulphonic acid group introduced into the compound being sulphonated.

Sulphonic acids prepared by sulphonation reactions are used principally for the preparation of dyestuffs, though some of them are used directly and as intermediates for other chemicals. Certain sulphonated oils, such as sulphonated castor oil, olive oil, and other vegetable oils have long been used in the glue, paper, leather, and textile industries while others find special uses as wetting agents, detergents, and constituents of soaps.

Conditions for sulphonation

Sulphonation reactions are carried out under a wide variety of conditions. Except for a few types of aliphatic compounds such as vegetable oils and unsaturated compounds, sulphonation reactions are limited to aromatic compounds. The following are important factors in sulphonation.

Strength of sulphonating agent, proportion of sulphonating agent to compound sulphonated, temperature of reaction, time of reaction, agitation, and the presence of catalysts. Concentrated sulphuric acid is the most common sulphonating agent, but the use of oleum or other agents, such as $ClSO_3H$, SO_3, Na_2SO_3, is sometimes advantageous. The reactivity of H_2SO_4 as a sulphonating agent depends upon the SO_3 concentration, since it is known that sulphonation stops at a definite SO_3 concentration which is different for each compound undergoing treatment. Since water is formed in the reaction, it is necessary to add sufficient acid to prevent the SO_3 concentration from falling to this minimum value before the desired reaction is complete. In order to avoid a large excess of sulphonating agent the use of oleum to take up the water formed has been employed. Another method of accomplishing the same result is to carry out the reaction at a low temperature and distill off the water formed.

In sulphonation reactions, an increase in temperature not only accelerates the rate and degree of reaction, but also influences orientation. Most sulphonations are carried out at a temperature below 250°C. It is especially important to stop the reaction as soon as the desired conversion is reached in order to avoid polysulphonation, rearrangements, or migration where these are possible. Liquid-phase catalytic processes are the rule–the common catalysts being salts of vanadium, mercury and alkali metals. Experience has shown that agitation is one of the principal requisites of good sulphonation practice. This is necessary to secure contact between the reactants as well as to prevent local overheating with its consequent ill effects.

Equipment

In the construction of sulphonators, cast iron apparatus is suitable with concentrated acid of 88 to 98% strength, but steel equipment is desirable where oleum or a solution containing free SO_3 is utilised. The apparatus must provide for agitation and for heat transfer. Neither of these factors presents great difficulty and there are a number of commercial vessels designed for this unit process.

In cases where continuous heating or cooling is required, or where temperatures higher than those obtainable with low pressure steam are necessary, thermocoil sulphonators may be used. The chief characteristic of this type of sulphonator is a series of tubes cast into and integral with the walls of the vessel. The steel tubing and cast iron wall form a homogeneous mass insuring high efficiences in heat transfer. The steel tubes are designed for high pressures steam or some other high temperature heating medium can be used. Since a number of inlets must be provided for the heating medium, the surface temperature of the vessel can be kept practically uniform and hence hot spots are eliminated.

The Friedel-Crafts Reaction

The term condensation is generally considered to include reaction in which union is effected between two or more of the same or different molecules, with or without the elimination of water or some other inorganic compound. Hence, condensation, in the broad sense, takes in such a variety of type reactions that it is not very helpful to consider it as a unit process. However, one type of condensation, known as the Friedel-Crafts reaction, is important enough to demand inclusion as a separate unit process.

An exact definition of the Friedel-Crafts reaction is impossible because it embraces such a variety of unusual but useful reactions. In the broad sense, and reaction catalysed by a metallic halide may be considered as a Friedel-Crafts reaction. Taken in a narrow sense, however, this unit process is understood to involve the union of a comparatively active compound, such as ethyl chloride, ethylene, CO, CCl_4 or acetyl chloride, with a relatively stable hydrocarbon or substituted hydrocarbon, such as benzene or anisole, in the presence of the halides of metals such as Al, Fe, B, Sn, or Ti. Aluminum chloride is the usual condensation agent, and hydrochloric acid is usually eliminated in the reaction.

There are two principal types of Freidel-Craft reaction: (i) the strictly catalytic type in which the aluminum chloride, or other halide, is always present in small amounts to act as a catalyst, and can be recovered and re-used; and (ii) the type in which the aluminum chloride must be present in greater than catalytic amounts—at least one mole per mole of product.

Use of the reaction.

A wide variety of synthetic organic chemical, such as hydrocarbons, aldehydes, acids, ketones, halogen compounds and other derivatives, can be prepared by Friedel-Crafts reactions. Acetophenone, propiophenone, benzophenone, benzaldehyde, benzoylbenzoic acid, ethyl benzene, phenyl ethyl alcohol, and tricresyl phosphate are a few of the compounds prepared on an industrial scale by this reaction. These materials are used as intermediates in dyes and perfumes, in the preparation of synthetic resins, and as plasticisers for rubber and plastics.

Conditions for the process

A number of chemical and physical factors have an important effect on the course and extent of the Friedel-Crafts reaction. The major ones are temperature, concentration of organic reactants, molal ratio of metal halide to organic reactants, purity of reacting chemicals, and size of aluminum chloride particles. The quantity and purity of aluminum chloride must be carefully regulated to obtain the maximum yield of the desired products. Many reactions are unfavourarbly influenced by impurities in the aluminum chloride, such as $FeCl_3$, TiC_4, $MgCl_2$, excess HCl, etc. Impurities in the reacting chemicals may remove $AlCl_3$ as double compound, making it no longer available to act as catalyst. The state of aggregation of the $AlCl_3$ has considerable influence on the process. It has been found that a fairly coarse type of $AlCl_3$, in the form of granules or small lumps, should be used with efficient agitation. The removal of the hydrochloric acid liberated by the reaction is desirable in most Friedel-Crafts syntheses. This is usually accomplished by drawing a slow stream of warm dry air over the charge. Control of the temperature at the optimum for a particular reaction is essential if good yields are expected. Moderate elevations in temperature not only affect orientation but also bring about secondary condensations. With excessive heating, almost all Friedel-Crafts reaction masses can be largely converted into oily masses of complex or uncertain composition. Friedel-Crafts sysntheses are usually carried out in a liquid medium. An excess of one of the reaction material may be used as the solvent, or a non-reactive solvent such as carbon disulphide, petroleum ether, acetylene tetrachloride, or nitrobenzene can be utilised.

There is no standard design of apparatus for carrying out Friedel-Crafts reactions, but there are certain basic principles that can be applied to the design or selection of such equipment. The handling of aluminum chloride and certain other reagents must be done in a manner to avoid scrupulously the entrance of water into the system. Water not only causes hydrolysis of aluminum chloride, resulting in decreased yields, but also creates a corrosion problem due to the hydrochloric acid released by the hydrolysis. Due to the absence of moisture, most Friedel-Crafts reactions can be carried out in iron reactors. Certain compounds are adversely influenced by the presence of metallic iron at the reaction temperature, and such reactions are best carried out in enamel-lined or lead-lined reaction vessels. Some of the recently developed reactors have been made of the newer alloys which are resistant to the corrosive action of hydrochloric acid and acid chlorides.

Alkylation

Alkylation can be defined as the unit process whereby an alkyl or an aralkyl radical is introduced by addition or substitution into an organic compound. This process is used principally in one of the following five general types of reactions: (i) addition to a tertiary nitrogen; (ii) addition to a metal to form alkyl-metallic compounds; (iii) substitution for H in an —OH group of an alcohol or phenol; (iv) substitution for hydrogen attached to nitrogen; and (v) substitution for hydrogen in carbon compounds. The following are some of the important reagents used:

$(CH_3)_2SO_4$, $(C_2H_5)_2SO_4$, CH_3OH, C_2H_5OH, CH_3Cl, alkyl iodides, olefins and $C_6H_5CH_2Cl$.

The products obtained by alkylation are used in making a great variety of materials, such as medicinal preparations, dyes, explosives, intermediates, plastics, solvents, rubber accelerators, perfumes, photographic chemicals, gasoline, etc.

Conditions for alkylation

Because of the variety of materials prepared by alkylation reactions in addition to the diversity of alkylating agents which may be used, it is difficult to list the general conditions under which alkylation reactions are carried out. The concentration of alkylating agent may be high, particularly when alcohols are used as the alkylating agent. But sometimes it is advisable to dilute the alkylating agent in order to control the reaction better. Most alkylations are carried out in the liquid phase at temperatures below 200°C., but some are performed in the vapour phase at much higher temperatures around 400°C. Pressure is necessary in numerous instances in order to keep the reactions in the liquid phase. Catalysts are important in alkylation reactions, sulphuric acid being the most common catalyst. The proportions of reacting materials determine the type of products obtained. In some cases it is desirable to use an excess of alkylating agent to obtain better yields, but in others, especially where there is more than one place in the molecule where alkylation can take place, it is necessary to use only a limited amount of the agent.

Equipment

Alkylation reactions are usually carried out in autoclaves or jacketed kettles provided with agitation. The autoclaves are generally constructed of steel, with or without an anti-corrosion liner of lead, tinned iron, tinned copper, or enamel coated iron. The autoclaves can be heated directly or by the use of a heating medium circulated through a jacket or coils within the autoclave. By the use of alkylating agents with low vapour pressures, it is sometimes possible to carry out the reactions in agitated jacketed kettles of standard design.

Polymerisation

"Polymerisation can be defined as the chemical combination of a number of similar units to form a single molecule, wherein polymerisation phenomena as well as condensation reactions occur.

The unit process of polymerisation has become increasingly important . The synthetic resins are being used more as substitutes for glass, ceramic ware, wood, bone, metals and alloys, in addition to filling many needs for structural materials which could not be supplied by any other substances available at any price.

In the preparation of synthetic resins, condensation as well as polymerisation reactions may occur, especially in the early stages of the reaction. The final products obtained are exceedingly complex, high molecular weight materials which make structural determinations almost impossible. Enough is known about the reaction conditions, however, to permit many useful materials to be made.

The conditions used in the preparation of synthetic resin depend not only upon the reaction materials but also upon the type of products desired. By making suitable changes in the reaction conditions it is possible to obtain a great variety of products from essentially the same basic raw materials. The products obtained depend upon a number of factors such as heat, light, pressure, catalysts, time of reaction, and concentration, as well as the chemical character of the reactants themselves. The presence of certain unsaturated linkages in one or more of the reactants seems to be one of the important characteristics of the raw materials. Polymerisation reactions are sensitive to heat, and control of the temperature is essential in the building of large molecular-weight products. Some of the common methods used to control the temperature are: use of moderate sized reaction chambers with good mixing, dilution with solvents, carrying out the reaction under vacuum, or maintaining the reaction mass at the boiling point of one of the reactants. Catalysts are important in polymerisation reactions, many of the commonly used acids, salts, alkalies, peroxides, and oxides being employed. Pressure is another important factor in effecting polymerisation reactions, especially where secondary condensations or rearrangements are involved. The removal of the volatile products of the reaction under reduced pressure often helps govern the reactions as well as assists in controlling the reaction temperature. On the other hand, high pressures are sometimes desirable especially where volatile reacting materials are used and where high temperatures are needed. Solubility also plays an important part in polymerisation reactions. It is necessary that all of the constituents obtained be mutually soluble at all temperatures and under all conditions so that there will be no tendency for separation of one of the constituents with its consequent ill effects on the finished article.

The equipment used for resin manufacture usually consists of some type of autoclave fitted with the necessary equipment for controlling the reaction.

Thermal decomposition (Pyrolysis)

Thermal decomposition is a unit process of vast importance to our industrial civilisation. The decomposition of wood furnishes charcoal, acetic acid, methanol. etc. The pyrolysis of coal furnishes not only gas for industrial and domestic heating purposes, and coke, an essential material to the metallurgical industries, but also supplies the raw materials for the important aromatic organic chemicals industry.

The thermal decomposition of petroleum is also of great importance in converting certain petroleum fractions into more valuable products. Since the decomposition of each of these raw materials provides the basis for a large industry, the decomposition of each material will be discussed separately.

Thermal decomposition of wood

Wood consists principally of three materials: cellulose, 40 to 63%; lignin, 24 to 37%; and hemicelluloses, 6 to 20%. Most woods also contain from 0.2 to 1.0% ash, 0.1 to 0.3% nitrogen, and varying amount of gums, tannins, volatile oils, fats, waxes, resin and colouring matter.

The exact chemical nature of the principal constituents of wood is unknown, though a great deal is known, concerning their properties. Cellulose is known to have the empirical formula $(C_6H_{10}O_5)_x$ but exact value of x is still uncertain.

The destructive distillation of wood for preparation of charcoal with the recovery chemicals as by-products has been a well established industry. Synthetic methods for producing organic chemicals, normally recovered as by-products of this industry, seriously threatened its existence.

The thermal decomposition of wood is chemically complex. The raw material is made up of several complex constituents which probably have different decomposition temperatures. Moreover, these constituents seem to decompose in a manner which is dependent upon the temperature. Ordinary wood is not uniform throughout, and the products formed upon heating depend upon the size of the wood pieces heated. In addition there is the possibility that a number of the original decomposition products undergo further changes before they are removed from the reaction zone, and hence, the end products may not throw much light on the actual mechanism involved.

Sequence of reactions

When heated in the absence of air, wood begins to evolve water and carbon dioxide at comparatively low temperatures. When the temperature reaches the neighbourhood of 280°C gas evolution proceeds rapidly and the reaction changes from an endothermic to an exothermic one. Noncondensable gases such as hydrogen, carbon monoxide, carbon dioxide, methane, and illuminants are evolved, besides many aliphatic chain compounds, such as acids, alcohols, esters, and ketones. A tar is also formed, particularly at high temperatures, which contains various aromatic compounds, such as phenols. The higher the temperature employed the greater the quantities of noncondensable gases found, and the smaller the yields of condensable products. The residual charcoal contains a higher percentage of carbon as the temperature of decomposition is raised.

There is some doubt as to the source of the various important chemicals obtained from the destructive distillation of wood, though distillation of the major constituents separately gives some indication of the sources. Lignin is probably the source of methanol and higher phenols. Cellulose yields little methanol but considerable acetic acid, while cellulose tar yields principally phenol. Pentosans give acetic acid and considerable furfural, in addition to the usual gas and tar by-products. Allyl alcohol, ammonia compounds, amines and pyridine are also usually found in the products in relatively small amounts. The destructive distillation of agricultural wastes yields materials similar to those obtained from wood.

The thermal decomposition of coal is of great industrial significance since this process not only furnishes fuels but also the raw materials for aromatic chemicals. The decomposition of coal is carried out to obtain three principal types of products: namely, (i) gas for industrial and domestic heating; (ii) coke, for heating purposes and for the metallurgical industries; and (iii) aromatic chemicals.

It is believed that coal originated from the partial decomposition of vegetable matter through the action of moisture, heat, pressure, and bacteria over long periods of time. All vegetable matter has a chemical composition similar to that of wood, but the constituents are present in different proportions from those found in wood. One theory of coal formation is that decomposition of this vegetable matter to form coal occurred with the elimination of compounds containing carbon, oxygen, and hydrogen, such as CH_4, H_2O, CO, and CO_2, leaving a material remaining which had a higher percentage of carbon than the original vegetable matter. During the decomposition process the material lost its cellular structure and the cellulose, lignin, and hemicelluloses no longer retained their identity.

The destructive distillation of coal yields products similar to those obtained from wood, but in different proportions. The products obtained depend upon such factors as the: type of coal; temperature, time, and pressure of decomposition; catalysts; type of coking oven used; etc. In general, higher yields of aromatic compounds and nitrogen compounds are obtained from the decomposition of coal than from wood.

Unit processes of the petroleum industry

The petroleum industry makes use of a number of unit processes. For convenience these processes are discussed together, but it should be noted also that in actual practice they do occur simultaneously. These processes in the past have been thermal reactions almost exclusively but recently the trend has been strongly toward catalytic reactions. The unit processes included in this section are: Cracking (thermal decomposition, pyrolysis); Hydrogenation; Dehydrogenation; Alkylation; Polymerisation; Isomerisation; Aromatisation (cyclisation).

Definition of terms

Before discussing the factors involved in the various unit processes used in petroleum refining, a brief discussion of the scope and meaning of each process will be given. The terms thermal decomposition and pyrolysis refer to the breaking up of large molecules into smaller ones by the action of heat. In referring to petroleum products this is usually called 'cracking'. Originally only high boiling stocks, such as kerosene, gas oils, and residues, were used as charging stocks for cracking and it was thought that the action was solely one of decomposition.

Now almost any petroleum product, including gases, may be used as the charging stock and it is known that polymerisation and condensation reactions occur simultaneously with decomposition at the conditions usually employed for cracking.

From cracking processes a number of low molecular weight saturated and unsaturated hydrocarbons are obtained in addition to those required for gasoline.

These hydrocarbons serve as the raw materials for most of the other unit processes. By polymerisation, unsaturated hydrocarbons, particularly propylene and butylene, are united to form longer chain unsaturated compounds which can be converted into saturated hydrocarbons in the gasoline range by hydrogenation. Unsaturated hydrocarbons, such as the butylene and propylene produced by cracking, can also be made from saturated aliphatic hydrocarbons by dehydrogenation. Alkylation accomplishes the purposes of both polymerisation and hydrogenation by combining in one step a saturated iso-hydrocarbon with unsaturated hydrocarbon to produce longer chain saturated hydrocarbons into branched chain hydrocarbons which have higher anti-knock characteristics.

Aromatic hydrocarbons are usually more valuable than aliphatic or naphthionic hydrocarbons and several methods are available for synthesizing aromatics form hydrocarbon is called aromatisation or aromatic cyclization. The conversion of an aliphatic hydrocarbon into a cycioparaffin is called cyclozation.

Cracking

The oldest and most important unit process of the petroleum industry is cracking, or pyrolysis. By cracking processes, the yield of gasoline from crude oil has been more than doubled. The primary purpose of cracking is to increase the yield of high anti-knock gasoline from crude oil. For that purpose a number of cracking processes have been developed. These processes can be classified under the following headings:

1. *Viscosity breaking.* Conversion of viscous heavy crudes and residues into fuel oils of low viscosity by a short-time decomposition, usually conducted at low cracking temperatures (800 to 875° F).

2. *Mixed-phase non-catalytic cracking.* A number of important cracking processes fall in this classification. The mixed-phase processes are carried out at temperatures ranging from 820 to 1100°F, and at pressures varying from 50 to 1500 pounds per square inch. As usually practiced, the operation is carried out with moderate conversions per pass combined with recycling of a clean distillate stock. Recently, selective cracking in combination units has been finding favour. This process is based on employing the optimum conditions of time, temperature, and pressure for each fraction of the charging stock.

3. *Vapour-phase cracking.* The vapour phase processes are not as common as the mixed-phase process and are usually more expensive. The operating pressures are lower, usually less than 75 pounds per square inch, and the temperatures higher (1000 to 1200°F) than for mixed-phase processes. Only comparatively light distillates can be processed and it is important to get a homogeneous phase. If any liquid is present, it is thrown to the tube walls and causes the formation of coke. The quantity of gas produced is greater, and the yield of gasoline lower than for mixed-phase processes. The gasoline produced, however, has a high anti-knock rating and can be sold at a premium price as a blending stock.

4. *Reforming.* A cracking process for converting low octane number gasoline or naphthas into high octane number products. The process is carried out in mixed-phase or vapour phase with or without the presence of catalysts. Dehydrogenation, decomposition, and aromatisation reactions take place in the process. The naphthalenes and paraffins present in the feed stock are partially converted to olefins and aromatics, and some low molecular weight gaseous compounds are also produced. Reforming operations are carried out under fairly severe conditions. The temperatures range from 930° to 1040°F, and the pressures from 250 to 1000 pounds per square inch.

5. *Coking.* A cracking operation carried out under conditions such that coke instead of a liquid residue is obtained. This is accomplished by keeping the material at cracking temperatures long enough for complete decomposition of high molecular weight materials to take place.

6. *Catalytic cracking.* Catalytic cracking has become of commercial importance and an important adjunct to thermal processes. The temperature conditions used in catalytic cracking are very similar to those used in thermal cracking, but lower pressures, are usually employed. The catalysts used in these processes do not appreciably affect the rate of reaction but do exert a directive influence on the reactions that take place. In these processes temperature control is of paramount importance. The cost and life of the catalyst and the cost of regeneration are items which must be considered in an installation of this type.

Two types of hydrogenation are practiced commercially on oil products; namely simple and destructive hydrogenation. Simple hydrogenation consists merely of addition of hydrogenation carbon-carbon double bonds and is exemplified by the hydrogenation of iso-octene to iso-octane. Most of the commercial installations for hydrogenation, with the exception of equipment for isoocatene production, are examples of destructive hydrogenation. This type of hydrogenation involves the breaking of carbon to carbon bonds and the addition of hydrogen to the products formed. Hence, it may be considered as a combination of simple hydrogenation with polymerisation, cracking, dehydrogenation, cyclization, etc. A wide variety of charging stocks can be employed for destructive hydrogenation and the materials obtained vary from aviation gasoline to lubricating oils. The temperatures employed in destructive hydrogenations are so high (750 to 1050°F) that cracking reactions take place simultaneously. Because to these high temperatures it is necessary to employ high hydrogen pressures–usually around 200 atmospheres. Though not essential to petroleum hydrogenation reactions, catalysts are advantageous and are generally used.

Dehydrogenation is one of the more recent developments in the application of unit processes to petroleum processing. The general use of the term is so broad that it is almost useless to refinery technologists. Hence, its use is generally restricted to the dehydrogenation of materials boiling within the gasoline range or lower. Both non-destructive and destructive hydrogenation may be accomplished. The former term refers to the removal of hydrogen from hydrocarbons without cracking, while the latter term indicates that rupture of the molecule occurs in addition to the removal of hydrogen.

Non-destructive dehydrogenation is catalytic, but destructed dehydrogenation can be carried out either with or without catalysts. In the dehydrogenation of a paraffin hydrocarbon, the formation of hydrogen is favoured by increasing the temperature and by using a low pressure. Accordingly, temperatures above 950° F and atmospheric pressure is used for the process. The principal use of the olefins formed is as a charging stock for alkylation.

Alkylation

The broad use of this term has already been discussed. In the petroleum industry its use is restricted to alkylation processes employing low molecular weight materials for the production of high octane number gasoline. Both catalytic and thermal processes are in use. The catalytic process is carried out at low temperatures, 30° to 100° F., and atmospheric pressure. In this process isobutane is combined with isobutylene to give almost pure iso-octane. The thermal alkylation process represents the other extreme, operating at 920 to 960° F at about 4500 pounds per square inch. In this process ethylene is combined with isobutane to produce neohexane.

The reactor is used for the low temperature alkylation of hydrocarbon from petroleum using sulphuric acid as the catalyst. In this reaction contact time, temperature, and agitation are important variables which are dependent upon the type of equipment used. Thorough agitation is provided by a pump which circulates the oil-acid mixture through a series of baffles and over the tubes of the heat exchanger. Close temperature control is provided by the heat exchanger, and time of contact can be adjusted by regulating the amount of material recycled. Polymerisation is used to a great extent than any of the other unit processes in oil refining except cracking. Both thermal and catalytic processes are employed and low molecular weight hydrocarbons are used as the charging stock. The thermal process operates at high pressures and high temperatures but utilises part of the saturated hydrocarbons as well as the unsaturates in the charge. Thermal polymerisation processes are carried out at temperatures around 1000°F under pressures varying from 600 to 2500 pounds per square inch. Catalytic polymerisations also are carried out under high pressures, 150 to 1500 pounds per square inch, but the temperatures are much lower than in thermal polymerisations, around 175° to 400°F. In these processes only unsaturated hydrocarbons are polymerised.

Isomerisation is used to produce isobutane and isopentane from their normal isomers. The process is catalytic and is carried out at 200°F under a pressure of 200 pounds per square inch.

It is known, however, that low pressures (atmospheric to 40 pounds gauge) and very high temperatures (1100° F and higher) favour this unit processes. Toluene is being prepared from petroleum by this process. The equipment used for the unit processes of oil refining is practically all "tailor-made" for each installation. The reaction vessels or chambers are probably the most specialised pieces of apparatus employed. The high temperatures necessary in most cases are obtained by passing the material through tubes placed in gas or oil-fired furnaces. The equipment used for separation of the reaction products consists of standard type of unit operations equipment.

Miscellaneous unit processes

There are a number of miscellaneous unit processes such as neutralisation, condensation, acylation, nitrosation, carboxylation, decarboxylation, molecular rearrangement, etc., which could be included in a discussion of this kind. Acrylation and codensation reactions are included to a certain extent in the Friedel-Crafts reactions. However, it is recognised that some of these unit processes may become very important in the future.

CHAPTER 3

High Pressure Processes

INTRODUCTION

One of the outstanding developments of the chemical industry during the past two decades has been use of high pressure processes. Not only have well-known products been made by new and more economical processes operating at high pressures, but new chemical synthesis have been made possible and commercially feasible. In some cases the development of high pressure processes has opened up completely new fields of chemical development.

Products such as ammonia, methanol and urea had been made commercially for many years by other processes which have now been largely supplanted by the newer pressure processes. For example, prior to the direct synthesis process, ammonia was obtained largely as a by-product of the coking of coal and this placed a definite limit to the amount that could be produced.

Another source of ammonia was calcium cyanamide and the process for producing it in this way was developed on a large scale but, due to high power requirement and handling costs on large amounts of solid materials, it did not offer the promise of a truly cheap and abundant source of ammonia and has been almost entirely abandoned in favour of the direct synthesis.

The freeing of ammonia from its dependence on the processing of coal caused a veritable revolution in the whole economy of chemical industry. A cheap and unlimited supply of fixed nitrogen in a form readily converted to nitric acid has had an effect on the development of the chemical industry whose importance it would be difficult to exaggerate. Ammonia has become not only a cheap raw material for the chemical industry but also may hold the key to cheap fertilisers for agriculture. It is of the utmost importance to national defense as a raw material for nitric acid manufacture.

A similar story can be told with reference to methanol, which prior to its direct synthesis was available only as a by-product of hard wood distillation. As a result of the introduction of the high pressure synthesis, a relatively cheap and unlimited source of it was a great stimulus to the development of the phenol-formaldehyde plastics which depend on methanol as a raw material. High pressure operations and processes do not differ qualitatively from the operations and processes discussed but the use of high pressure presents special possibilities and problems which are worthy of special consideration.

For the present purpose, high pressure will be defined as pressures above 50 atm. (750 lbs. per sq. in.). Pressures as high as 1000 atm. are now used in chemical industry and higher ones could undoubtedly be employed, provided the temperature is not too high, if there were any good reason for their use. At the present time even 1000 atm. is rather uncommon and most high pressure processes operate under 500 atm. Very much higher pressures (of the order 3,000,000 lbs. per sq. in.) have been produced in some gears and bearings. Pressures of the order of 50,000–75,000 lbs. per sq. in. are developed in heavy artillery. Many laboratories both in industry and in the universities are investigating properties and reactions under "super" pressures and one may confidently expect that in the not far distant future, pressures higher than those now in use will find application in industry.

REASONS FOR USE OF PRESSURE

There are only two primary reasons for the use of pressure in chemical processes, viz.: (i) to shift equilibrium (Effect of pressure on equilibrium); and (ii) to maintain a liquid phase.

Effect of Pressure on Equilibrium

Qualitatively, the effect of pressure on any chemical equilibrium is to shift it in the direction of a volume decrease according to the *Le Chatelier-Braun principle*. The quantitative effect is given by well-known thermodynamical principles first worked out in rigorous and general form by Willard Gibbs. For good first approximations it can be assumed that the effect of pressure on the equilibrium of liquid phase reactions is negligible and that for gas reactions, the equilibrium constant in terms of partial pressure, K_P, is independent of the pressure.

For a rigorous treatment one must take account of the fact that the equilibrium constant as well as the equilibrium state varies with the pressure. For example, in the case of the ammonia equilibrium, the equilibrium constant at 450°C increases 3.5 fold as the pressure increases from 1 to 1000 atmospheres. In the case of the methanol equilibrium the effect may be even greater depending on the temperature.

A typical example of the effect of pressure on chemical equilibrium is given by the data on the ammonia synthesis equilibrium shown in Fig. 3.1. It is not at once evident from a study of this figure why high pressures are necessary in this process because if the reaction could be carried out at 200°C, good conversions would be obtained at low pressures. This brings out the fact that equilibrium must always be considered in conjunction with reaction rate. No practicable catalyst has been discovered which will permit the carrying out of this reaction at less than 400°C and in most cases the gases leave the catalyst at a temperature between 50 and 600°C. Thus it may be said that the reason for using pressure is to counteract the unfavourable effect of temperature on the equilibrium.

The degree of conversion continues to increase with the pressure but above about 900 atm. any gain from this increase is more than offset by the practical difficulties of working at higher pressures. In fact, some engineers with experience in this field believe that the point of diminishing returns is reached at about 350 atm. In laboratory experimentation ammonia has been synthesised at pressures as high as 5000 atm. It is of interest to note that at such a high pressure, almost complete conversion to ammonia can be achieved without the aid of any catalyst.

Maintaining a Liquid Phase

The second main reason for carrying out reactions under pressure is the very simple one that pressure is necessary to maintain a condensed phase. In such a case the reaction is carried out at the vapour pressure of the system and naturally at a temperature below the critical. In this category we may place such processes as the production of phenol by liquid-phase hydrolysis of chlorobenzene and the liquid-phase cracking of petroleum oils. It is recognised that the pressure may have other concomitant effects such as shifting the equilibria but the primary reason for its use is to "keep the liquid in the pot".

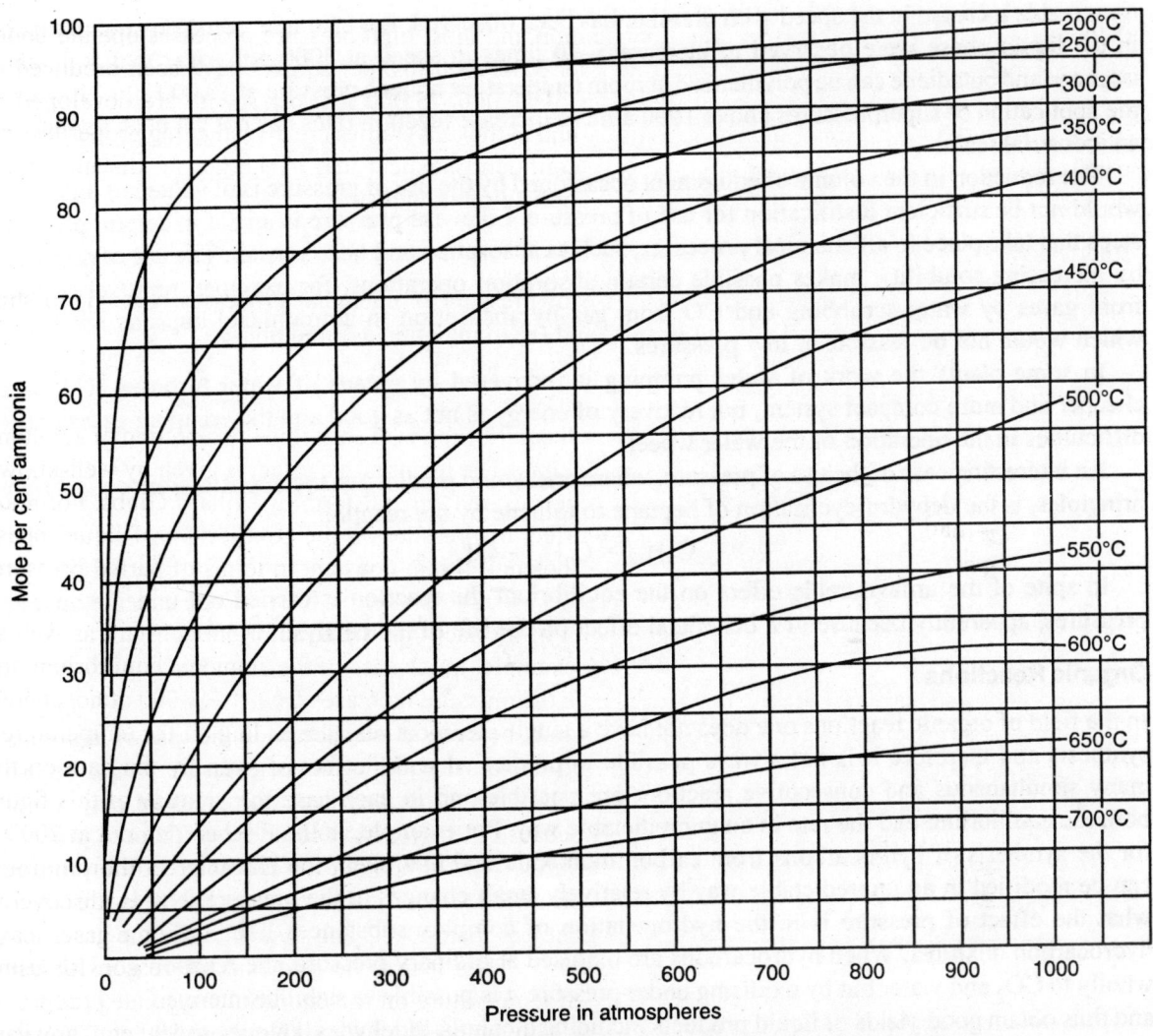

Fig. 3.1. Equilibrium per cent ammonia in a 3 to 1 hydrogen-nitrogen mixture.

The upper limit in such a case is the critical pressure of the system, the highest pressure at which a liquid phase is possible (this strictly applies only to a single component; for binary and multi-component systems, liquids are possible over limited composition ranges, at pressures considerably higher than the critical). Most organic liquids have critical pressures around 40–60 atm., whereas for water it is 218 atm.

Other Advantages of High Pressure

There are a few other gains resulting from the use of pressure which constitute additional reasons for its use but which are generally not primary reasons dictating such use. For example, rate of chemical reactions increases with the pressure due to increased concentrations and this means greater throughput, or less time for reaction. This is, of course, particularly true of gaseous reactions where the concentration is approximately proportional to the pressure but even with some liquid and solid-phase reactions a remarkable increase in the speed with pressure has been observed. For example, some organic reactions in the liquid phase were observed to increase 5–10 times in speed at 3000 atm. over that at 1 atm.; isoprene and butadiene can be polymerised at room temperature under a pressure of 10,000 atm. However, the application of superpressures above 1000 atm. to increase reaction rates has not yet been applied on an industrial scale.

The reduction in the volume of equipment occasioned by the use of pressure is of value but this alone would not be sufficient justification for use of pressure. Likewise pressure is an aid in certain physical steps that take place in all chemical processes, such as absorption and heat transfer. The use of pressure, by increasing solubility, makes possible certain absorption operations, for example, removal of CO_2 from gases by water-scrubbing and CO from gas by absorption in ammoniacal cuprous solutions, which would not be feasible at low pressures.

In some plants the work of water pumping is recovered by means of water turbines. This is a cheaper and more compact system, but recovery of energy is not as good and the escaping gases cause difficulties in the operation of the water wheel.

An interesting case of the use of pressure, where one would predict just the opposite from equilibrium principles, is the dehydro-cyclisation of heptane to toluene by the reaction:

$$C_7H_{16} = C_7H_8 + 4H_2$$

In spite of the unfavourable effect on the equilibrium the reaction is carried out under moderate pressures, apparently because of a beneficial effect on the life of the catalyst.

Organic Reactions

In the field of organic reactions one does not have a single clean-cut reaction as in the case of ammonia synthesis and therefore it is not always possible to predict what the effect of pressure would be. So many simultaneous and consecutive reactions are possible and in each case the pressure may affect both the equilibrium and the rate in an unpredictable way. For example, in the Fischer-Trcpsch process for the synthesis of hydrocarbons from carbon monoxide and hydrogen, the character of the product can be modified in an unpredictable way by relatively small changes in the pressure. Nor is it evident what the effect of pressure is in the hydrogenation of complex substances like coal or petroleum-hydrocarbon mixtures. When hydrocarbons are oxidised at ordinary pressure, the reaction goes almost wholly to CO_2 and water but by oxidising under pressure it is possible to stabilise intermediate products and thus obtain good yields of liquid products including alcohols, aldehydes, ketones and acids. Though this process has not yet found industrial application it is cited because it illustrates an important effect of pressure which is difficult to predict and which may lead to future applications.

Practically the only clean-cut organic reaction facilitated by pressure is the methanol synthesis reaction:

$$CO + 2H_2 = CH_3OH$$

By suitable choice of conditions this reaction can be made to take place practically to the exclusion of all others. The extent of this reaction as a function of pressure, temperature and proportion of reactants can be fairly accurately predicted from the thermodynamic data. The only other simple reaction between these two reactants is the formation of formaldehyde but investigation of the equilibrium in this reaction has shown that only very small yields are to be expected even at 1000 atm. Many other reactions between these two substances are possible but all attempts to produce any one pure compound have failed. Once the attempt is made to build up compounds with more than one carbon atom, there seems to be no way to control the reaction within narrow limits and a complex product results. This is the case, for example, in the synthesis of the higher alcohols from carbon monoxide and hydrogen.

CATALYSTS

Practically all high pressure synthesis are dependent upon catalysts to bring about a sufficiently rapid reaction to make them commercially feasible; hence it is appropriate to say a few things about these important agents. At a sufficiently high temperature, most reactions would be rapid without a catalyst but then other conditions would be quite unfavourable. For example, the equilibrium generally becomes less favourable as temperature increases and the problem of materials to withstand the combination of pressure and temperature becomes exceedingly difficult. It is usually essential to keep the temperature under 600°C, at least.

In spite of all the research that has been done on the theory of catalysis there is only one reliable way to determine the best catalyst to use and that is the time-honoured method of trial and error. Of course there are certain guides that aid one in the preliminary narrowing down of the field. Certain elements or compounds are generally known to favour certain types of reactions. For example nickel or copper or certain chromites are known to favour hydrogenation, whereas certain oxides like alumina or tungsten oxide promote dehydration but even these very general rules breakdown in specific instances. There are so many variables affecting not only the activity of a catalyst but also certain physical properties which may be of equal importance. Thus, a catalyst suitable for industrial use must be rugged enough to resist disintegration due to purely mechanical forces, must have long life, and be resistant to heat and to poisons. The development of a catalyst with the necessary physical and chemical properties for a specific case requires many hundreds of purely empirical tests and most companies who are operating catalytic processes have elaborate catalyst testing laboratories.

The specification of the elements or compounds that are active in a catalyst is far from a complete characterisation of the catalyst. Physical state may be just as important as chemical composition. For example zinc oxide is a methanol catalyst but the activity depends entirely on the methanol of preparation and subsequent treatment of the oxide. Some forms of zinc oxide are wholly inactive as catalysts and others will exhibit varying degrees of activity. What may seem to be minor variables in method of treating the oxide—variables involved in the precipitation, drying or heat treatment of the catalyst—may exert an important influence on the properties.

In some cases the active catalyst is a metal and in other cases it may be an oxide or a salt. There is a certain amount of ambiguity in referring to catalysts which may confuse the uninitiated. For instance, it is commonly said that the usual ammonia catalyst is a promoted iron oxide. It is true that the catalyst is charged into the reactor as an oxide but before use it is reduced to metallic iron which is the true catalyst. In another case one may speak of a methanol catalyst as consisting of 45 mole % chromium and 55 mole % zinc. This does not mean that the catalyst is a mixture of these elements, for it is actually in the form of oxides which are not reduced to the metal. It merely means that these two elements are present in this proportion.

Most catalysts do not consist of a single element or compound but of a mixture of two or more. For example, most ammonia synthesis catalysts consist largely of metallic iron as the active agent but small amounts of other substances, often called promoters, are essential for a good catalyst. In this case the promoters are usually basic oxides such as those of the alkali or alkaline earth elements or of aluminium. One common ammonia catalyst is doubly promoted with potassium and aluminium oxides. In the case of methanol, zinc oxide, chromium oxide and metallic copper are definitely known to exhibit catalytic activity but a combination of zinc and chromium oxides is much better than either one alone. In this case the two oxides are used in roughly equal proportions so that there is no clear-cut distinction between catalyst and promoter as in the ammonia case. The properties of a methanol catalyst can also be modified in an important way by the addition of relatively small amounts of other elements. For example, a catalyst containing only zinc and chromium (as oxides, of course) will give substantially pure methanol but the addition of small amount of alkalies, or of copper or manganese or other elements will lead to the formation of significant amounts of higher alcohols in addition to methanol.

Most catalysts are used in the form of a stationary bed through which the reactants pass and hence they must be in lump or granular form. This form can sometimes be produced without special mechanical means by proper control over the conditions used in manufacture. In other cases it is necessary to compress the powdered catalyst into pellets. In still other cases the catalyst is held on a supporting material which is itself in suitable granular form such as pumice or Alfrax or is impregnated on a powdered material such as kieselguhr which is then pressed into pellets. A recent interesting development in the petroleum industry is a process in which the catalyst is suspended as a very fine powder in the moving stream of the reactants, then separated by mechanical means, re-activated if necessary, and again dispersed in more reactant mixture.

SURVEY OF PRESSURE PROCESSES USED IN INDUSTRY

The number of possible processes which might advantageously be carried out under pressure is very large and this discussion will be confined to a few of the most important ones that are used industrially. These, with some of the important operating conditions, are summarised in Table 3.1.

The process in the table are believed to be in actual operation on an industrial scale. Many other reactions have been investigated in the laboratory and on a semi-commercial scale which may be the basis of future industrial processes.

In addition to the chemical reactions listed, there are a number of physical processes in which pressure is a useful tool, as for example in the recovery of helium from natural gas, the manufacture of liquid oxygen, the re-pressuring of petroleum-producing wells in which the oil production is from a saturated vapour in the critical region and maintenance of the pressure is necessary to continued production, electrolysis of water under pressure, and the use of high pressure steam to explode wood to a fibrous material from which useful products are pressed.

METHANOL SYNTHESIS

This has been chosen as a typical high pressure catalytic process about which to give more detail than was possible within the confines of Table 3.1.

The reactants, hydrogen and carbon monoxide in the proportion of 2 to 1, may be produced in a variety of ways from different raw materials though only three sources of the gases are believed to be in use at the present time. These three sources are:
1. By-product gases from the fermentation of corn.
2. By-product gas from carbide furnaces.
3. Coal and water.

In the case of (1) the gas is a mixture of CO_2 and H_2 in about equal proportions. Methanol can be synthesised from these two gases just as well as from CO and H_2 but of course water is formed at the same time so that the product is a solution of methanol in water instead of substantially pure methanol. This can be a avoided by first converting the CO_2 to CO by passing it over a bed of hot coke where the following reaction takes place

$$CO_2 + C = 2CO$$

Table 3.1. Outline of most important industrial pressure processes.

Process	Reaction	Temperature °C	Pressure Atm.	Raw materials	Catalyst	Phase	Remarks
Ammonia synthesis	$N_2 + 3H_2 = 2NH_3$	400–600	100–1000	Coke, air, water by-product H_2, natural gas	Promoted iron oxide	Gas	
Methanol synthesis	$CO + 2H_2 = CH_3OH$	300–450	100–1000	Coke, water, by-product H_2 and CO, natural gas	Zn and Cr oxides with or without promoters	Gas	
Urea synthesis	$CO_2 + 2NH_3 = CO(NH_2)_2 + H_2O$	200	375	Ammonia and by-product CO_2	Non-catalytic	Liquid	
Higher alcohol synthesis	$nCO + 2nH_2 = C_nH_{2n+1}OH + (n-1)H_2O$	300–450	200–1000	Same as methanol	Alkalised methanol catalysts.	Gas	Many catalysts claimed. One actually in use is not known. Product is a mixture of primary and secondary alcohols with from 1 to 9 carbon atoms.
Hydrolysis of chlorobenzene	[structure] Cl + Na_2CO_3 + H_2O = [structure] OH + NaCl + $NaHCO_3$	300–350	200–300	Chlorobenzene, diphenyl oxide, soda ash	Copper	Liquid	Diphenyl oxide added in equilibrium amount to prevent further conversion to this by-product.
Ammonolysis of chlorobenzene	[structure] Cl + $2NH_3$ = [structure] NH_2 + NH_4Cl	190–210	60	Aniline, ammonia	Cuprous salts	Liquid	Phenol and diphenylamine formed as by-products.
Propionic acid synthesis	$CO + C_2H_5OH = C_2H_5COOH$	150–350	100–800	By-product CO and ethyl alcohol	Mineral acids, boron trifluoride	Liquid	A similar reaction between CO and MeOH to give HAc

(Contd...)

Process	Reaction	Temperature °C	Pressure Atm.	Raw materials	Catalyst	Phase	Remarks
Methyl formate synthesis	$CO + CH_3OH = HCOOCH_3$	—	350	MeOH and by-product CO	—	—	is also believed to be in commercial use Step in production of methyl methacrylate
Higher alcohol from fatty acids	$RCH_2COOH + 2H_2 = RCH_2CH_2OH + H_2O$	350–400	200	Natural fats and oils and hydrogen.	Copper, zinc and other chromates	Liquid	Many possible variants of this general reaction.
Production of hexaline (cyclohexanol)	$\text{(benzene-OH)} + 3H_2 = \text{(cyclohexane-OH)}$	175	—	Hydrogen and phenol	Nickel	Liquid	
Tetraline and decaline production	$C_{10}H_8 + 2H_2 = C_{10}H_{12}$	—	—	Naphthalene and hydrogen	Nickel and copper	Liquid	Decahydronaphthalene produced by same reaction.
Hydrogenation of coal and tar	Complex	400–500	200–700	Coal or coal tar, water, by-product H_2	Mo, W or tin compounds	Liquid and vapour	Fuels and lubricant for internal combustion engines and fuel oil produced.
Polymer gasoline	Polymerisation of olefins	200–600	15–200	Unsaturates in cracking still gases or light saturated hydrocarbons	Phosphoric or sulphuric acids on a carrier or non-catalytic	Vapour	The purely thermal process uses higher temperatures and pressures than the catalytic one. When raw material is saturated hydrocarbons, both cracking and polymerisation occur in one operation.

(Contd....)

Process	Reaction	Temperature °C	Pressure Atm.	Raw materials	Catalyst	Phase	Remarks
Cracking petroleum oils	Complex	500–600	50–70	Various petroleum fractions	None	Liquid	—
Iso-octane production	$2C_4H_8 = C_8H_{16}$	150–200	50–100	Butanes	Phosphoric acid	—	Similar to polymer gasoline process but using substantially pure butenes as raw materials. Product is hydrogenated to iso-octanes.
Hydrogenation of petroleum	Complex	—	200	Various petroleum products and hydrogen	Metals oxides	Liquid	Can produce variety of products depending on charging stock and conditions.
Thermal alkylation	Isobutane + ethylene → Neohexane	500–525	200–300	Various light saturated hydrocarbons	None	Vapour	The saturated hydrocarbons are cracked to produce ethylene.
Petroleum reforming	Complex, including cracking polymerisation, isomerisation and cyclisation	500–600	50–100	Various petroleum fractions	None	Vapour and liquid	Production of more desirable products from less desirable ones (for example, high octane gasoline from lower octane gasoline) without great change in volatility range.
Synthesis of ethylene glycol	$CH_2O + CO + H_2O =$ $\begin{array}{c} CH_2OH \\ | \\ COOH \end{array}$ $\begin{array}{c} CH_2OH \\ | \\ COOH \end{array} + 2H_2 = \begin{array}{c} CH_2OH \\ | \\ CH_2OH + H_2O \end{array}$	—	—	Formaldehyde, CO, H_2 and water	—	—	Very recent application and no data on reaction conditions available. The formaldehyde is made from methanol by oxidation.

In the case of (2) the gas is nearly pure CO, a portion of which is converted to hydrogen by means of the water-gas reaction, namely:

$$CO + H_2O = CO_2 + H_2$$

The CO_2 may be removed in a variety of ways leaving a substantially pure hydrogen which is then mixed with the original gas in the proper proportion.

In the case of (3) the coal is coked in ovens and the resulting coke is used to produce blue water gas which is then mixed with hydrogen obtained from the coke oven gas by low temperature separation.

Other possible sources of the two reaction gases are: (i) blue water gas from coke, a part of which is converted to hydrogen through the water gas reaction; (ii) by-product CO from phosphorus manufacture; (iii) coke-oven or other coal gases; (iv) petroleum cracking gases; and (v) natural gas. In certain localities natural gas should offer a very cheap source of raw material for methanol synthesis. One of the difficulties in such a source is seen by considering the reaction which occurs when natural gas (assumed for simplicity to be methane) is treated with steam and passed over a catalyst, namely

$$CH_4 + H_2O = CO + 3H_2$$

Since a 2 to 1 gas is necessary, either hydrogen must be removed or carbon monoxide added. There are several possibilities for doing this, only two of which will be mentioned briefly. The CO and H_2 can be separated by a low temperature distillation process and the CO added to some of the original gas mixture resulting from the reaction with steam, or some of the natural gas can be burned and the CO_2 separated from the flue gas by suitable means. The CO_2 is then mixed with the natural gas and steam to bring about the net reaction

$$3CH_4 + CO_2 + 2H_2O = 4CO + 8H_2$$

A simplified flow sheet showing only the main features of one process for methanol synthesis starting from coal and water as the raw materials is given in Fig. 3.2. The coal is converted to coke and gas in ovens of standard type. The coke is then fed to water-gas generators where a blue gas containing approximately equal volumes of CO and H_2 is produced. This gas contains hydrogen sulphide which would poison the methanol catalyst and cause other troubles later in the process so it is removed at this point by any one of several well-known processes, as for example the Thylox process which uses a sodium arsenate solution as the absorbing agent. The gas is then compressed to about 30 atmospheres in 2 or 3 stage compressors and scrubbed with water under pressure in a packed tower to remove carbon dioxide (formed in the gas generators) along with some other impurities. The gas now consists of approximately equal volumes of hydrogen and carbon monoxide and hydrogen is added to adjust the composition to the desired 2 to 1 mixture. This hydrogen is produced from the coke oven gas by a low temperature process in which all the other constituents of the gas are liquefied, leaving the hydrogen which is a non-condensable gas at the temperature ($-300°F$) used. The purified and compressed gas mixture containing hydrogen, carbon monoxide and small amounts of other gases, principally nitrogen, methane and carbon dioxide, now enters a recirculation system where the synthesis takes place. The gases from the recirculating pump mix with fresh, make-up gas and enter the converter in which they first exchange heat with the hot gases leaving the catalyst bed. In this way the entering gases are heated to about 300°C. before entering the catalyst chamber. This heating is necessary to start the reaction at a sufficiently rapid rate. Since the reaction is exothermic the gases leave the catalyst at a somewhat higher temperature (approximately 400°C.) and the temperature is continually maintained at the proper level by heat exchange between ingoing and outgoing gases. Of course, when a converter is cold after a shut-down, some auxiliary means of heating has to be used to start the reaction.

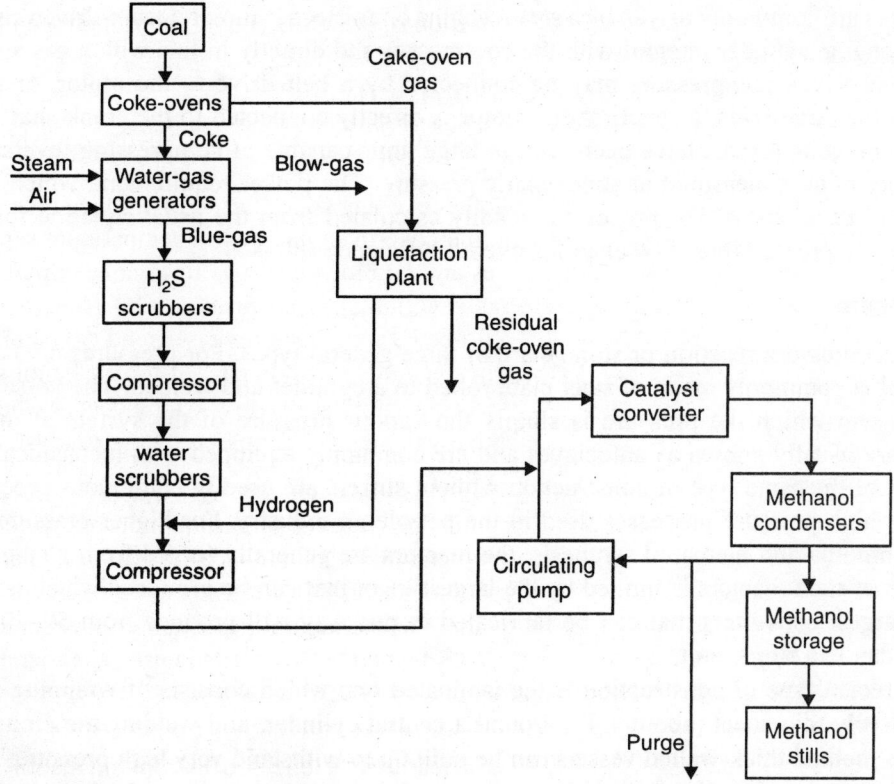

Fig. 3.2. Simplified flow sheet of process for methanol synthesis.

Depending on the catalyst, temperature, pressure, space velocity and other factors, from 10–20% of the carbon monoxide is converted to methanol in one pass through the converter. By cooling the gases with water in a tubular cooler, the majority of the methanol (and higher alcohols if they are present) is condensed to a liquid and flows to a receiver whence it is expanded to a low pressure storage system. The uncombined hydrogen and carbon monoxide are then recirculated to the converter. Since the make-up gas contains a certain amount of inert gas and since some inerts such as methane are formed in side reactions, a certain proportion of the gas must be continuously purged from the circulatory system to prevent the inerts from building up. In order to avoid too great a loss of active gas through the purge, the inerts in the system are generally maintained at a fairly high figure, of the order of 20–30%. The only disadvantage of this is that it reduces the effective pressure of the reaction.

EQUIPMENT

Compressors

Gas is compressed in reciprocating compressors, both for mechanical and thermodynamic reasons, the compression is done in more than one stage with inter-stage cooling. Compressors for pressures up to 300 atm. are fairly well standardised. Machines for pressures as high as 1000 atm. have been built but are quite special.

Compressors are commonly driven by a steam engine or an electric motor. Steam-driven compressors have a steam engine cylinder integral with the compressor and directly in line with a gas compression cylinder. Power-driven compressors may be connected by a belt drive to the motor, or as is more common to in the large sizes, a synchronous motor is directly connected to the crank shaft.

High pressure compressors have been built in large units capable of compressing several thousand cu. ft. of gas per minute measured at atmospheric pressure. The power requirement, which may be an important item in the cost of the process, is readily calculated from the usual equation for adiabatic compression with a reasonable allowance for overall efficiency (80–85%).

Pressure Vessels

Vessels for high pressure reaction or storage are of three general types. For pressures up to about 100 atm. the vessel is commonly made of steel plate rolled to a cylinder and welded. The vessels used for batch reactions in which the pressure is simply the vapour pressure of the system at the reaction temperature, are usually known as autoclaves and are commonly equipped with mechanical agitators. Larger vessels of the same type of construction without stirrers are used in continuous processes such as the various high pressure processes used in the petroleum industry. For higher pressures, such as used in the ammonia and methanol synthesis, the reactors are generally forged from a single ingot of steel. The size of such reactors is limited by the largest ingot that can be produced which is about 225 tonnes. The largest converters that can be fabricated in this way will produce from 50–70 tonnes of ammonia per day in a single unit.

The most recent type of construction is the laminated one which consists of wrapping successive layers of relatively thin sheet (about 1/4") around a central cylinder, and welding the circumferential seams. By this method thick-walled vessels can be built up to withstand very high pressures and larger vessels can be made than from forgings which makes possible larger units with a corresponding reduction in costs. Furthermore, only the inner cylinder need be of a special material to withstand the corrosive action of reacting materials.

All pressure vessels have bolted on heads and their gas-tightness is assured by the use of a gasket of some yielding material, usually copper or aluminium or soft steel.

Materials of Construction

For the lower pressures and in the absence of hydrogen, mild steel may be used in the construction of pressure vessels. Hydrogen causes decarburisation of mild steel above about 200°C, with resultant cracking along crystal boundaries. For the combination of high temperature, high pressure and hydrogen containing gases, alloy steels containing chromium, nickel, vanadium, tungsten, or molybdenum and combinations of them are used, a common one containing 1% chromium, 0.2% vanadium and 0.3% carbon. The elastic limit depends on the heat treatment but is generally in the range of 65,000–85,000 lbs. per sq. in. Where severe corrosion is encountered, special alloys high in nickel and chromium must be employed. For the very severe conditions encountered in such a process as the Claude for ammonia synthesis operating at 550–650°C and 1000 atm., the converters are made of a nickel-chromium-iron alloy containing about 25.5% iron, 12% chromium, 60% nickel and 2.5% tungsten.

The tensile properties of all steels are little affected by temperatures up to 250°C, above 300°C a fairly rapid decrease in strength sets in, and above about 400–450°C, the proportional limit becomes very low and the so-called 'crepe' phenomenon becomes important. This is a continuous elongation that occurs without increase in the stress. An important property of steels to be used at high temperatures and pressures is the crepe stress which is defined as the stress that will produce 1% elongation in a certain number of hours, usually from 25,000 to 100,000. In contrast to the ordinary short-time tensile tests, the crepe stress can only be evaluated by long-time tests of at least 1000 hours.

Liquid Pumps

Liquid pumps for high pressures have generally been of the reciprocating plunger-type though recently centrifugal pumps have been developed for large rates of flow at pressures as high as 3000 lbs. per sq. in. These are multi-stage pumps driven either by steam turbines or high speed electric motors and there appears to be no reason why still higher pressures cannot be developed if the demand arises. For low and moderate flow rates the reciprocating pump still dominates the field.

Miscellaneous Equipment

Valves, packagings, pipe and fittings, pressure gauges, flow meters, safety valves and other small items of equipment are not essentially different for high pressure processes from those employed at low and moderate pressures, but little details of design are important and may make the difference between success and failure. The construction must, of course, be heavier and frequently of special steels to secure higher strength without so great an increase in weight. For the highest pressures, seamless tubing is generally used in place of pipe with welded seams. Joints are made with flanges and soft metallic gaskets or the lens ring gasket. Fittings may be forged or they may be machined from bar stock. Sight glasses constitute a special problem which has necessitated more of a departure from the conventional designs.

CHAPTER 4

Industrial Waste-water and its Treatment

INTRODUCTION

All industrial operations produce some waste-water which must be returned to the environment. Waste-waters can be classified as: (i) domestic waste-waters; (ii) process waste-waters; and (iii) cooling waste-waters. Domestic waste-waters are produced by plant workers, shower facilities, and cafeterias. Process waste-waters result from spills, leaks, and product washing. Cooling waste-waters are the result of various cooling processes and can be once-pass systems or multiple-recycle cooling systems. Once-pass cooling systems employ large volumes of cooling waters that are used once and returned to the environment. Multiple-recycle cooling systems have various types of cooling towers to return excess heat to the environment and require periodic blow down to prevent excess buildup of salts.

Domestic waste-waters are generally handled by the normal sanitary-sewerage system to prevent the spread of pathogenic microorganisms which might cause disease. Normally, process waste-waters do not pose the potential for pathogenic microorganisms, but they do pose potential damage to the environment through either direct or indirect chemical reactions. Some process wastes are readily biodegraded and create an immediate oxygen demand. Other process wastes are toxic and represent a direct health hazard to biological life in the environment. Cooling waste-waters are the least dangerous, but they can contain process waste-waters as a result of leaks in the cooling systems. Recycle cooling systems tend to concentrate both inorganic and organic contaminants to a point at which damage can be created.

WASTE-WATER CHARACTERISTICS

Organics

Organic compounds in waste-waters have created most of the pollution problems as a result of their effect on oxygen resources in the environment. The low-molecular-weight water-soluble organics tend to be biodegraded by bacteria and fungi with the utilisation of oxygen. As the complexity of organic molecules increases, their solubility and biodegradability decrease. The total chemical oxygen demand (COD) of organic compounds in waste-waters is measured by the dichromate COD test. A 2-h reflux with concentrated sulphuric acid and potassium dichromate with silver sulphate and mercuric sulphate catalysts is adequate for complete oxidation of all but a few aromatic organic compounds. In recent years' the total organic carbon (TOC) apparatus has been developed to give a TOC measurement similar to the COD measurement. Unfortunately, the TOC data have not been correlated with COD data except for a few industrial wastes.

Inorganics

The inorganics in most industrial wastes are the direct result of inorganic compounds in the carriage water. Soft-water sources will have lower inorganics than hard-water or salt water sources. In a few instances, the industrial processes add inorganic compounds in the waste-waters. While domestic waste-waters have a balance in organics and inorganics, many process waste-waters from industry are deficient in specific inorganic compounds. Biodegradation of organic compounds requires adequate nitrogen, phosphorus, iron, and trace salts.

Ammonium salts or nitrate salts can provide the nitrogen, while phosphates supply the phosphorus. Either ferrous or ferric salts or even normal steel corrosion can supply the needed iron. Other trace elements needed for biodegradation are potassium, calcium, magnesium, cobalt, molybdenum, chloride, and sulphur. Carriage water or demineraliser waste-waters or corrosion products can supply the needed trace elements for good metabolism. Occasionally, it is necessary to add specific trace elements or nutrient elements.

pH and Alkalinity

Waste-waters should have pH values between 6 and 9 for minimum impact on the environment. Waste-waters with pH values less than 6 will tend to be corrosive as a result of the excess hydrogen ions. On the other hand, raising the pH above 9 will cause some of the metal ions to precipitate as carbonates or as hydroxides at higher pH levels. Alkalinity is important in keeping pH values at the right levels. Bicarbonate alkalinity is the primary buffer in waste-waters. It is important to have adequate alkalinity to neutralise the acid waste components as well as those formed by partial metabolism of organics. Many neutral organics such as carbohydrates, aldehydes, ketones, and alcohols are biodegraded through organic acids which must be neutralised by the available alkalinity. If alkalinity is inadequate, sodium carbonate is a better form to add than lime. Lime tends to be hard to control accurately and results in high pH levels and precipitation of the calcium which forms part of the alkalinity. In a few instances, sodium bicarbonate may be the best source of alkalinity.

Temperature

Most industrial wastes tend to be on the warm side. For the most part, temperature is not a critical issue below 37°C if waste-waters are to receive biological treatment. It is possible to operate thermophilic biological waste-water treatment systems up to 65°C with acclimated microbes. Low-temperature operations in northern climates can result in very low winter temperatures and slow reaction rates for both biological treatment systems and chemical treatment systems. Increased viscosity of waste-waters at low temperatures makes solid separation more difficult. Efforts are generally made to keep operating temperatures between 10 and 30°C if possible.

Dissolved Oxygen

Oxygen is a critical environmental resource in receiving steams and lakes. Aquatic life requires reasonable dissolved-oxygen (DO) levels. Central Pollution Control Board (CPCB) has set minimum stream DO levels at 5 mg/L during summer operations, when the rate of biological metabolism is a maximum. It is important that waste-waters have maximum DO levels when they are discharged and have a minimum of oxygen-demanding components so that DO remains above 5 mg/L. DO is a poorly soluble gas in water, having a solubility around 9.1 mg/L at 20°C and 101.3-kPa (1-atm) air pressure. As the temperature increases and the pressure decreases with higher elevations above sea level, the solubility of oxygen decreases. Thus, DO is a minimum when BOD rates are a maximum. Lowering the temperature yields higher levels of DO saturation, but the biological metabolism rate decreases. Warm-waste-water discharges tend to aggravate the DO situation in receiving waters. Summary of industrial wastes— their original character treatment is given in Table 4.1.

Table 4.1. Summary of industrial wastes: their origin, character and treatment.

Industries producing waste	Origin of major wastes	Major characteristics	Major treatment and disposal methods
1	2	3	4
Textiles	Cooking of fibres, desizing of fabrics	Highly alkaline, coloured, high BOD and temperature, high suspended solids.	Neutralisation, chemical precipitation, biological treatment aeration and trickling filtration.
Leather goods	Unhairing, soaking, deliming and bating of hides	High total solids, hardness, salts, sulphides, chromium, pH, precipitated lime and BOD.	Equalisation, sedimentation and biological treatment.
Laundry trades	Washing of fabrics	High turbidity, alkalinity and organic solids.	Screening, chemical precipitation floatation and absorption.
Canned goods	Trimming, cutting, juicing and blanching of fruits and vegetables	High in suspended solids, colloidal and dissolved organic matter.	Screening, lagooning, soaking, absorption or spray irrigation.
Dairy products	Dilutions of whole milk, separated milk, butter milk and ghee	High in dissolved organic matter, mainly protein, fat and lactose.	Biological treatment by trickling filtration, activated sludge.
Brewed and distilled beverages	Steeping and pressing grain, residue from distillation of alcohol, condensate from stillage evaporation.	High in dissolved organic solids containing nitrogen or their fermented products.	Recovery, concentration, centrifugation and evaporation, trickling filtration, reused in feeds, digestion of slops.
Meat and poultry products	Stockyards, slaughtering of animals, processing of bones and fats, residues in concentrates, grease and wash water, picking of chickens.	High in dissolved and suspended organic matter, blood, other proteins, and fats.	Screening, settling and/or floatation, trickling filtration.
Animal feedlots	Excreta from animals	High in organic suspended solids and BOD.	Land disposal and anaerobic lagoons.
Beet sugar	Transfer, screening and juicing water, drainings from lime.	High in dissolved and suspended organic matter containing sugar and protein.	Reuse of waters, coagulation and lagooning.
Pharmaceutical products	Mycelium, spent filtrate and wash waters.	High in suspended and dissolved organic matter, including vitamins.	Evaporation and drying.
Yeast	Residue from yeast filtration	High in solids (mainly organic) and BOD.	Anaerobic digestion, trickling filtration.

(Contd ...)

Industries producing waste	Origin of major wastes	Major characteristics	Major treatment and disposal methods
1	2	3	4
Pickles	Lime water, brine, alum and turmeric, syrup, seeds, and pieces of cucumber.	Variable pH, high suspended solids, colour, and organic matter.	Good house keeping, screening, equalisation.
Coffee	Pulping and fermenting of coffee bean.	High BOD and suspended solids.	Screening, settling and trickling filtration.
Fish	Rejects from centrifuge, presses evaporator and other wash water wastes.	Very high BOD, total organic solids and odour.	Evaporation of total waste; large remainder to sea.
Rice	Soaking, cooking and washing of rice.	High BOD, total and suspended solids (mainly starch)	Lime coagulation, digestion.
Soft drinks	Bottle washing; floor and equipment cleaning; syrup storage tank drains.	High pH, suspended solids and BOD	Screening, plus discharge in municipal sewer.
Bakeries	Washing and greasing of pans, floor washings.	High BOD, grease floor washings, sugar, flour, detergents.	Amenable to biological oxidation
Water production	Filter backwash, lime soda sludge, brine, alum sludge.	Minerals, and suspended solids.	Direct discharge to streams or indirectly through hold lagoons.
Pulp and paper	Cooking, refining, washing of fibres, screening of paper pulp.	High or low pH, colour, high suspended, colloidal and dissolved solids, inorganic filters.	Settling, lagooning, biological treatment, aeration, recovery of products.
Photographic	Spent solutions of developer and fixer.	Alkaline, containing various organic and inorganic reducing agents.	Recovery of silver; discharge of wastes into municipal sewer.
Steel	Coking of coal, washing of blast furnace fuel gases, pickling of steel.	Low pH, acids, cyanogen, phenol and ore, cokes, lime stone, alkali, oils, mill scale, and fine suspended solids.	Neutralisation, recovery, reuse, chemical coagulation.
Metal-plating	Stripping of oxides, cleaning and plating of metals.	Acid metals, toxic, low volume mainly mineral matter.	Alkaline chlorination of cyanide, reduction and precipitation of chromium, with lime.
Iron-foundry	Washing of used sand by hydraulic discharge.	High suspended solids, mainly sand, some clay and coal.	Selective screening, drying of reclaimed sand.
Oil fields and refineries	Drilling muds, salt, oil and some natural gas, acid sludge and miscellaneous oils from refining.	High dissolved salts, high BOD, odour, phenol and sulphur compounds from refinery.	Diversion, recovery, acidification and burning of alkaline sludge, leak and spill prevention, floatation.
Fuel oil use	Spills from fuel tank, filling waste, auto crank case oils.	High in emulsified and dissolved oils.	
Rubber	Washing of latex, coagulated rubber, exuded impurities from crude rubber.	High BOD and odour, high suspended solids, variable pH, high chlorides.	Aeration, chlorination, sulphonation, biological treatment.
Glass	Polishing and cleaning of glass.	Red colour, alkaline non settleable suspended solids.	Calcium chloride precipitation.
Naval stores	Washing of stumps, drop solution, solvent recovery, and oil recovery water.	Acid, high BOD	By product recovery, equalisation recirculation and trickling filtration.

(Contd ...)

Industries producing waste	Origin of major wastes	Major characteristics	Major treatment and disposal methods
1	*2*	*3*	*4*
Glue manufacturing	Lime wash, acid washes, extraction of non-specific proteins.	High COD, BOD, pH, chromium; periodic strong mineral acids.	Amenable to aerobic treatment, and floatation.
Wood preserving	Steam condensates	High in COD, BOD, solids phenols.	Chemical coagulation, oxidation pond and other aerobic biological treatment.
Candle manufacturing	Wax spills, stearic acid condensates.	Organic (fatty) acids.	Anaerobic digestion.
Plywood manufacturing	Glue washing	High BOD, pH, phenols, potential toxicity.	Settling ponds, incineration.
Acids	Dilute wash waters, many varied dilute acids.	Low pH, low organic content.	Upflow or straight neutralisation, burning when some organic matter is present.
Detergents	Washing and purifying soaps and detergents.	High in BOD and saponified soaps.	Floatation and skimming, precipitation with $CaCl_2$.
Corn starch	Evaporator condensate or bottoms when not reused or recovered, syrup from final washes, wastes from bottling up process.	High BOD and dissolved organic matter, mainly starch and related material.	Equalisation, biological filtration anaerobic digestion.
Explosives	Washing TNT and gun-cottage for purification, washing and pickling of cartridges.	TNT, coloured acid, odorous and contains organic acids alcohol from powder and cotton, metals, acid, oils and soaps.	Floatation, chemical precipitation biological treatment, aeration, chlorination, neutralisation, adsorption.
Pesticides	Washing and purification products such as 2,4, D and DDT	High organic matter, benzene ring structure, toxic to bacteria and fish, acid.	Dilution, storage, activated carbon absorption, alkaline chlorination.
Phosphate and phosphorus	Washing, screening, floating rock, condenser bleed off silica and fluoride.	Clays and limes, low pH, high suspended solid, phosphorus, settling of refined waste.	Lagooing, mechanical clarification, coagulation and
Formaldehyde	Residues from manufacturing synthetic resins and from dyeing synthetic fibres.	Normally high BOD and HCHO toxic to bacteria in high concentrations.	Trickling filtration, absorption on activated charcoal.
Plastics and resins	Unit operation, from polymer preparation and use, spills and equipment wash downs.	Acids, caustic dissolved organic matter such as phenols, formaldehyde etc.	Discharge to municipal sewers, reuse, controlled discharge.
Steam power	Cooling water, boiling blow down, coal drainage.	Hot, high volume, high inorganic and dissolved solid.	Cooking by aeration, store of ashes, neutralisation.
Coal processing	Cleaning and classification of coal, leaching with water.	High suspended solids, mainly coal, low pH, high H_2SO_4.	Excess acid waste settling, froth floatation, drainage control and sealing of mines.
Nuclear power and radioactive materials	Processing ores, laundering of contaminated clothes, research lab wastes, processing of fuel, power plant cooling waters.	Radioactive elements can be very acid 'hot'.	Concentration and containing of dilution and dispersion.

Pretreatment

Many industrial-waste-water streams should be pretreated prior to discharge to municipal sewerage system or even to a central industrial sewerage system. Pretreatment of individual streams should be considered whenever these streams might have an adverse effect on the total treatment system.

Equalisation

Equalisation is one of the most important pre-treatment devices. The batch discharge of concentrated waste is best suited for equalisation. It may be important to equalise waste-water flows, waste-water concentrations, or both. Periodic waste-water discharges tend to overload treatment units. Flow equalisation tends to level out the hydraulic loads on treatment units. It may or may not level out concentration variations, depending upon the extent of mixing within the equalisation basin. Mechanical mixing may be adequate if the wastes are purely chemical in their reactivity. Biodegradable wastes normally require aeration mixing so that the microbes are kept aerobic and nuisance odours are prevented. Diffused aeration systems offer better mixing under variable load conditions than mechanical surface aeration equipment. Mixing and oxygen transfer are both important with biodegradable waste-waters. Operation on regular cycles determines the size of the equalisation basin. There is no advantage in making the equalisation basin any larger than necessary to level out waste-water variations. Industrial operation on a 5-day, 40-h week will normally make a 2-day equalisation basin as large as needed for continuous operation of the waste-water treatment system under uniform conditions.

Neutralisation

Acidic or basic waste-waters must be neutralised prior to discharge. If an industry produces both acidic and basic wastes, these wastes may be mixed together at the proper rates to obtain neutral pH levels. Equalisation basins can be used as neutralisation basins. When separate chemical neutralisation is required, sodium hydroxide is the easiest base material to handle in a liquid form and can be used at various concentrations for in-line neutralisation with a minimum of equipment. Yet, lime remains the most widely used base for acid neutralisation. Limestone is used when reaction rates are slow and considerable time is available for reaction. Sulphuric acid is the primary acid used to neutralise high-pH waste-waters unless calcium sulphate might be precipitated as a result of the neutralisation reaction. Hydrochloric acid can be used for neutralisation of basic wastes if sulphuric acid is not acceptable. For very weak basic waste-waters, carbon dioxide can be adequate for neutralisation. Characteristics of waste-waters from selected organic industries is given in Table 4.2.

Grease and Oil Removal

Grease and oils tend to form insoluble layers with water as a result of their hydrophobic characteristics. These hydrophobic materials can be easily separated from the water phase by gravity and simple skimming, provided they are not too well mixed with the water prior to separation. If the oils and greases form emulsions with water as a result of turbulent mixing, the emulsions are difficult to break. Separation of oil and grease should be carried out near the point of their mixing with water. In a few instances, air bubbles can be added to the oil and grease mixtures to separate the hydrophobic materials from the water phase by flotation. Chemicals have also been added to help break the emulsions. American Petroleum Institute (API) separators have been used extensively by the petroleum industry to remove oils from waste-waters. The food industries use grease traps to collect the grease prior to its discharge.

Table 4.2. Characteristics of waste water from selected industries in India.

Industry	Flow	pH	Suspended solids mg/l	BOD mg/l	COD mg/l	BOD/COD	BOD load per kg prod.	Other important characteristics
1	2	3	4	5	6	7	8	9
Dairy	7 m³/m³ milk	8.0	690	816	1340	0.6	5.7	Oil and grease, ready putrescibility
Distillery	14 m³/m³ rectified spirit	4.0–4.5	—	40,000	80,000	0.5	560	High TDS, Cl, SO_4 and putrescibility.
Cotton textile	25 m³/1000 metre cloth	7–10	375	350	525	0.66	8.75	TDS 3000 mg/l, % sodium 90 and colour
Viscose rayon kg hide	570 m³/tonne	2.8–4.1	200	—	210	—	polysulphides	Zinc, 6–18 mg/l
Tannery	3.2 m³/1000 kg hide	8.0	1900	1700	3500	0.48	5.4	Cr^{+3} 0.35–0.45 kg/100 kg hide, Cl 4600 mg/l, tannin, colour.
Pulp and paper Craft process	310 m³/tonne	6.5–8.2	375	160	610	0.26	50	Lignin, colour, suspended fibre Solids.
Primary polymers								
Polythylene		1.44–5.8	—	200–4000	—	—		
Polypropylene		1.44–5.8	—	200–4000	—	—		
Polystyrene		1.8–3.6	—	1000–3000	—	0.05–0.025		
Polyvinyl chloride		5.5–11	50–500	1000–2000	—	0.05–.25		
Cellulose acetate		0.036–.72	500–2000	1000–5000	—	0.4–0.5		
Butyl rubber		7.2–21.6	800–2000	2500–5000	—	0.32–0.4		
Miscellaneous substances Isocyanate		18.0–36.0	1000–2500	4000–8000	—	0.25–0.31		Nitrogen
Phenyl glycerine		18.0–36.0	1000–2500	4600–8000	—	0.25–0.31		Phenol content
Parathion		10.8–78.8	1500–3500	3000–6000	—	0.5–0.59		Solids, pH
Tributyl phosphate		3.6–14.4	500–2000	1000–3000	—	0.5–0.67		Phosphorus
Coke oven	3 m³/T of coke	8.6–11.2	82	1200	2500	0.41	3.6	NH_3, phenol, CN, CNS, tar and oil
Refinery processed	1.7 m³/T oil	—	300	200	—	—	0.34	Free oil, 2500 mg/l emulsified oil, 100 mg/l mercaptane 115 mg/l

Unfortunately, grease traps are designed for regular cleaning of the trapped grease. Too often they are allowed to fill up and discharge the excess grease into the sewer or are flushed with hot water and steam to fluidise the grease for easy discharge to the sewer. A grease trap should be designed for a specific volume of grease to be collected over specific time periods. Care should be taken to design the trap so that the grease can easily be removed and properly handled. Neglected or poorly designed grease traps are worse than no grease traps at all.

Toxic Substances

Legislation has made it illegal for industries to discharge toxic materials in waste-waters. Each industry is responsible for determining if any of its waste-water components are toxic to the environment and to remove them prior to the waste-water discharge. The CPCB has identified a number of priority pollutants, which must be removed and kept under proper control from their origin to their point of ultimate disposal. Major emphasis has been placed on heavy metals and on complex organics that have been implicated in possible cancer production. Pretreatment is essential to reduce heavy metals below toxic levels and to prevent discharge of any toxic organics. Fortunately, toxic organics can ultimately be destroyed by various chemical oxidation systems. Incineration appears to be the most economical method for destroying toxic organics. To make incineration economical, the organics must be kept separated from the dilute waste-waters and treated in their concentrated form. If the heavy metals can not be reused, they must be concentrated and placed into insoluble materials which will not leach the heavy metals. Toxic substances currently pose the greatest challenge to industries since very little attention has been paid to these materials in the past. Characteristic of waste-waters from some selected chemical and allied industries is given in Table 4.3.

Table 4.3. Characteristics of waste-waters from some organic chemical industries.

Industry	Flow	BOD	COD	$\dfrac{BOD}{COD}$	Other important characteristics
	$m^3/tonne$	mg/l	mg/l		
Dyes and pigments	180–900	200–400	500–2000	0.2–0.4	Heavy metals, solids, colour, pH.
Primary petrochemical	0.18–5.4	100–1000	500–3000	0.2–0.33	Phenol, oil and pH.
Ethylene propylene	0.36–7.2	100–1000	500–3000	0.2–0.33	Phenol, pH
Primary intermediates					
Toluene	1.1–11.0	300–2500	1000–5000	0.3–0.5	—
Xylene	0.72–11	500–4000	1000–8000	0.5	—
Methanol	1.1–11	300–1000	500–2000	0.5–0.63	Oil and solids.
Butanol	0.72–7.2	500–4000	1000–8000	0.5	Heavy metals.
Ethyl benzene	1.1–11	500–3000	1000–7000	0.43–0.5	Heavy metals.
Chlorinated hydrocarbons	0.18–3.6	50–150	100–500	0.3–0.5	Oils, solids, pH.
Secondary intermediates					
Phenol	1.8–9	1200–10000	2000–15000	6.63–0.67	Phenols, solids
Acetone	1.8–5.4	1200–5000	2000–10000	0.5–0.63	—

(Contd...)

Industry	Flow m^3/tonne	BOD mg/l	COD mg/l	$\dfrac{BOD}{COD}$	Other important characteristics
Glycerine glycols	3.6–18	500–3500	1000–7000	0.5	—
Terephthalic acid	3.6–10.8	1000–3000	2000–4000	0.5–0.77	Heavy metals.
Urea	0.36–7.2	50–300	100–1050	0.29–0.5	—
Acrylates	3.6–10.8	500–5000	2000–15000	0.25–0.33	Solids, cyanide and colour.
Acetic anhydride	3.6–28.8	300–5000	500–8000	0.63	pH
Acrylonitrite	3.6–30	200–700	500–1500	0.4–0.48	CN, colour, pH.
Butadiene	0.36–7.2	25–200	100–400	0.25–0.5	Oil and solids.
Styrene	3.6–36	300–3000	1000–6000	0.3–0.5	—
Vinylchloride	0.036–0.72	200–2000	500–5000	0.4.	—

PRIMARY TREATMENT

Waste-water treatment is directed toward removal of pollutants with the least effort. Suspended solids are removed by either physical or chemical separation techniques and handled as concentrated solids.

Screens

Fine screens such as hydroscreens are used to remove moderate-size particles that are not easily compressed under fluid flow. Fine screens are normally used when the quantities of screened particles are large enough to justify the additional units. Mechanically cleaned fine screens have been used for separating large particles. A few industries have used large bar screens to catch large solids that could clog or damage pumps or equipment following the screens.

Grit Chambers

Industries with sand or hard, inert particles in their waste-waters have found aerated grit chambers useful for the rapid separation of these inert particles. Aerated grit chambers are relatively small, with total volume based on 3-min retention at maximum flow. Diffused air is normally used to create the mixing pattern shown in Fig. 4.1, with the heavy, inert particles removed by centrifugal action and friction against the tank walls. The airflow rate is adjusted for the specific particles to be removed. Floatable solids are removed in the aerated grit chamber. It is important to provide for regular removal of floatable solids from the surface of the grit chamber; otherwise, nuisance conditions will be created. The settled grit is normally removed with a continuous screw and buried in a landfill.

Gravity Sedimentation

Slowly settling particles are removed with gravity sedimentation tanks. For the most part, these tanks are designed on the basis of retention time, surface overflow rate, and minimum depth. A sedimentation tank can be rectangular or circular. The important factor affecting its removal efficiency is the hydraulic flow pattern through the tank. The energy contained in the incoming-waste-water flow must be dissipated before the solids can settle. The waste-water flow must be distributed properly through the sedimentation volume for maximum settling efficiency. After the solids have settled, the settled effluent should be

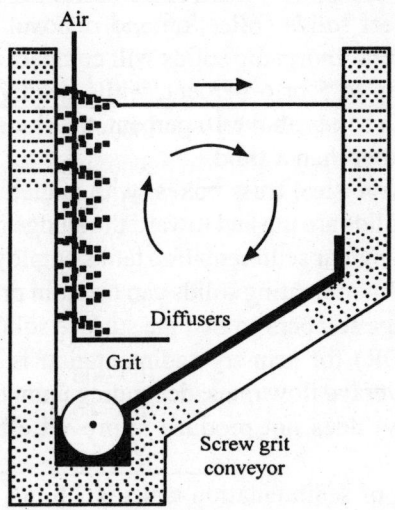

Fig. 4.1. Schematic diagram of an aerated grits chamber.

collected without creating serious hydraulic currents that could adversely affect the sedimentation process. Effluent weirs are placed at the end of rectangular sedimentation tanks and around the periphery of circular sedimentation tanks to ensure uniform flow out of the tanks. Once the solids have settled, they must be removed from the sedimentation-tank floor by scraping and hydraulic flow. Conventional sedimentation tanks have sludge hoppers to collect the concentrated sludge and to prevent removal of excess volumes of water with the settled solids. Schematic diagrame of a rectangular sedimentation tank is shown in Fig. 4.2.

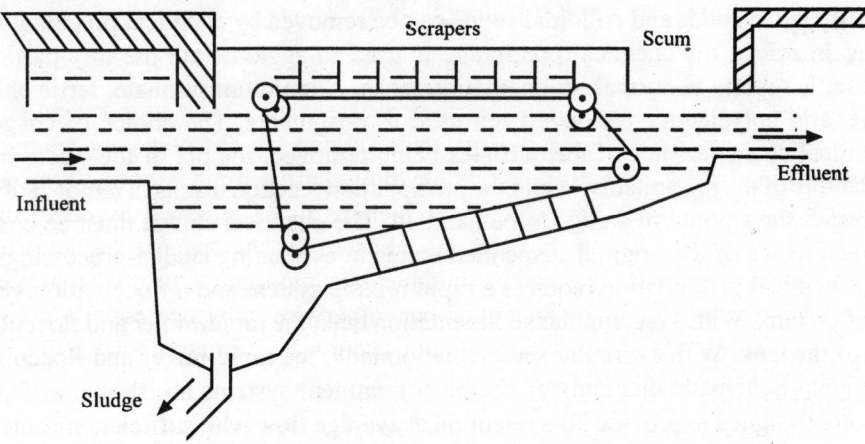

Fig. 4.2. Schematic diagram of a rectangular sedimentation tank.

Design criteria for gravity sedimentation tanks normally provide for 2-h retention based on average flow, with longer retention periods used for light solids or inert solids that do not change during their retention in the tank. Care should be taken that sedimentation time is not too long; otherwise, the solids will compact too densely and affect solids collection and removal. Organic solids generally will not compact to more than 5 to 10 percent. Inorganic solids will compact up to 20 or 30 percent. Centrifugal sludge pumps can handle solids up to 5 or 6 percent, while positive-displacement sludge pumps can handle solids up to 10 percent. With solids above 10 percent, the sludge tends to lose fluid properties and must be handled as a semisolid rather than a fluid.

Circular sedimentation tanks have steel truss boxes with angled sludge scrapers on the lower side. As the sludge scrapers rotate, the solids are pushed toward the sludge hopper for removal on a continuous or semi continuous basis. The rectangular sedimentation tanks employ chain-and-flight sludge collectors or railmounted sludge collectors. When floating solids can occur in primary sedimentation tanks, surface skimmers are mounted on the sludge scrapers so that the surface solids are removed at regular intervals.

The surface overflow rate (SOR) for primary sedimentation is normally held close to 40.74 $m^3/(m^2.day)$ [1000 gal/(ft^2.day)] for average flow rates, depending upon the solids characteristics. Lowering the SOR below 40.74 $m^3/(m^2.day)$ does not produce improved effluent quality in proportion to the reduction in SOR.

Generally, the minimum depth of sedimentation tanks is 3.0 m (10 ft), with circular sedimentation tanks having a minimum diameter of 6.0 m (20 ft) and rectangular sedimentation tanks having length-to-width ratios of 5:1. Chain-and-flight limitations generally keep the width of rectangular sedimentation tanks to increments of 6.0 m (20 ft) or less. While hydraulic overflow rates have been limited on the effluent weirs, operating experience has indicated that the recommended limit of 186 $m^3/(m.day)$ [15,000 gal/(ft.day)] is lower than necessary for good operation. A circular sedimentation tank with single-edge weir provides adequate weir length and is easier to adjust than one with a double-sided weir. More problems appear to be created from improper adjustment of the effluent weirs than from improper length.

Chemical Precipitation

Lightweight suspended solids and colloidal solids can be removed by chemical precipitation and gravity sedimentation. In effect, the chemical precipitate is used to agglomerate the tiny particles into large particles that settle rapidly in normal sedimentation tanks. Aluminium sulphate, ferric chloride, ferrous sulphate, lime, and polyelectrolytes have been used as coagulants. The choice of coagulant depends upon the chemical characteristics of the particles being removed, the pH of the waste-waters, and the cost and availability of the precipitants. While the precipitation reaction results in removal of the suspended solids, it increases the amount of sludge to be handled. The chemical sludge must be considered along with the characteristics of the original suspended solids in evaluating sludge-processing systems.

Normally, chemical precipitation requires a rapid mixing system and a flocculation system ahead of the sedimentation tank. With a rectangular sedimentation tank, the rapid-mixer and flocculation units are added ahead of the tank. With a circular sedimentation tank, the rapid-mixer and flocculation units are built into the tank. Schematic diagrams of chemical treatment systems are shown in Fig. 4.3 and 4.4. Rapid mixers are designed to provide 30-s retention at average flow with sufficient turbulence to mix the chemicals with the incoming waste-waters.

The flocculation units are designed for slow mixing at 20-min retention. These units are designed to cause the particles to collide and increase in size without excessive shearing. Care must be taken to move the flocculated mixture from the flocculation unit to the sedimentation unit without disrupting the large floc particles.

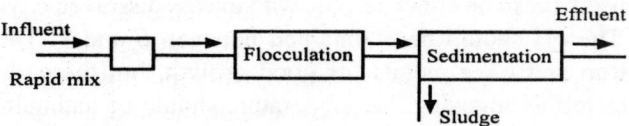

Fig. 4.3. Schematic diagram of a chemical precipitation system for rectangular sedimentation tanks.

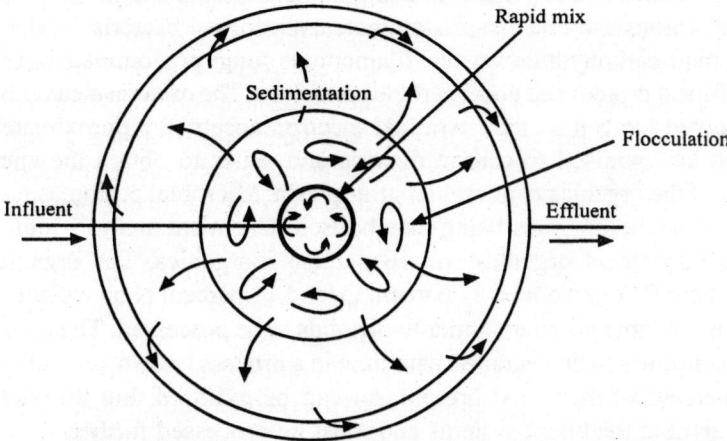

Fig. 4.4. Schematic diagram of a chemical precipitation system for circular sedimentation tanks.

SECONDARY TREATMENT

Concepts of Biological Treatment

Secondary treatment is designed to remove colloidal and soluble organics through microbial metabolism. Biological treatment systems convert the biodegradable organics in solution into suspended organics which flocculate and are removed by gravity sedimentation. Colloidal solids tend to be adsorbed onto the microbial flocs.

Biological waste-water treatment systems employ bacteria as the primary microorganisms responsible for removing excess bacteria and producing a clarified effluent. The bacteria average 0.5 to 1.0 µm in diameter and 1.0 to 2.0 µm in length. Protozoa range from 10 µm, to several hundred micrometers in size. The bacteria cannot metabolise solid organics but rather must convert solid particles to soluble organics prior to metabolism. Fortunately, the bacteria have enzymes on their cell surface which can hydrolyse the complex organics to simple organic molecules. Bacteria are widely distributed throughout the environment with different biochemical characteristics. Soil is the best source of concentrated microbes, since the bacteria in the soil are responsible for the stabilisation of waste materials in the soil. Normally, the desired bacteria are extract from the soil by growing them in the desired waste-waters until sufficient numbers have been obtained for the desired reactions. While specialised bacteria are available for purchase, it is not normally necessary to do so as soil will supply the desired microorganisms more economically.

In order to grow, bacteria must have a suitable environment with all the proper nutrients. The environment must provide good mixing for adequate contact between the microorganisms and the pollutants being metabolised. It can be either aerobic with excess dissolved oxygen or anaerobic without any dissolved oxygen. The pH should be maintained between 6 and 9. There should be sufficient nitrogen, phosphorus, iron and trace metals for good growth, but there should not be too high a concentration of heavy metals to be toxic. The temperature should be maintained between 5 and 35°C, allthough it is possible to operate thermophilic systems up to 65°C. If the environment is not balanced, improper microbial growth and inadequate treatment can occur. One of the major problems lies with filamentous microbes which adversely affect flocculation and sedimentation. In the absence of adequate oxygen and iron, filamentous bacteria can predominate over normal bacteria. Under low-pH conditions and low-nitrogen or high-carbohydrate wastes, filamentous fungi predominate over bacteria.

The best quality effluent is produced under aerobic conditions. The bacteria metabolise the biodegradable organics, using dissolved oxygen as their terminal electron acceptor. Approximately one-third of the organics metabolised are oxidised to carbon dioxide and water to obtain the energy to convert the remaining two-thirds of the organics to microbial protoplasm. Microbial protoplasm is similar regardless of the bacteria species or the organics being metabolised. Chemical analysis indicates that microbial protoplasm contains 90 percent organics and 10 percent inorganics. The organic fraction contains approximately 50 percent C, 7 percent H, 31 percent O, and 12 percent N by weight. The major problem in aerobic treatment is the large volume of microbial solids to be processed. The microbial protoplasm is not stable but rather continues to be degraded with time in a process known as "endogenous respiration." Approximately 80 percent of microbial protoplasm can be oxidised, but 20 percent of the organic fraction is stable in aerobic treatment systems and must be processed further.

In the absence of dissolved oxygen, the bacteria must find another electron acceptor. Chemically bound oxygen is the primary electron acceptor under anaerobic conditions. NO_3, NO_2, SO_4, CO_2, and organic oxygen compounds are used as electron acceptors. Nitrates are reduced to nitrites and to various intermediates before being reduced to nitrogen gas. It is important to recognise that denitrification results in the ultimate production of nitrogen gas that is insoluble in the waste-waters rather than in the production of ammonia or ammonium ions. If the nitrates are needed as a source of nitrogen for the bacteria, they will be reduced to ammonium ions in proportion to the nitrogen needs for protoplasm. A special group of bacteria use sulphates for their electron acceptor and can produce thiosulphates, free sulphur, or hydrogen sulphide. The methane bacteria can reduce carbon dioxide with the production of methane, which is relatively insoluble in water and can be used as a source of energy. Organic oxygen compounds such as carbohydrates are broken down to simple organic acids, aldehydes, ketones, and alcohols which can be metabolised by the methane bacteria to methane and carbon dioxide. Since it takes the same amount of energy to produce a unit of microbial protoplasm aerobically and anaerobically, less microbial cell mass is produced per unit of organic pollutants metabolised anaerobically than aerobically. The anaerobic reactions do not yield as much energy to the microbes as the aerobic reactions. The anaerobic end products contain considerable energy that is not available to the microbes. Current research into anaerobic waste-water treatment systems is being stimulated by the production of methane, which can be used as an energy source, and by the lower microbial solids production. Approximatley 80 to 95 percent of the organics can be converted to methane and carbon dioxide, with 5 to 20 percent being converted to microbial solids.

Time is a critical variable in sizing waste-water treatment systems. A definite time period is required to metabolise a given amount of organic matter by a unit of cell mass. By retaining the microbes in the treatment system, the treatment time per unit of organic matter is reduced. Unfortunately, the time for aerobic treatment is controlled by oxygen transfer. Under anaerobic conditions, contact between the microbes and the organic pollutants controls the total reaction time. With proper design, more organic matter can be treated anaerobically than aerobically in the same time period. Unfortunately, the anaerobic system cannot produce as high-quality effluent as the aerobic system. The net result is that anaerobic systems are often used as the first stage of biological treatment, with aerobic systems being used as the second stage.

WASTE MANAGEMENT

Lagoons

Lagoons are low-cost, easy-to-operate waste-water treatment systems capable of producing satisfactory effluents.

Facultative lagoons

These lagoons have been designed to use both aerobic and anaerobic reactions. Normally, facultative lagoons consist of two or more cells in series. The settleable solids tend to settle out in the first cell and undergo anaerobic metabolism with the production of organic acids and methane gas, which bubbles out to the atmosphere. Algae at the surface of the lagoon utilise sunlight for their energy in converting carbon dioxide, water, and ammonium ions into algal protoplasm with the release of oxygen as a waste product. Aerobic bacteria utilise the oxygen released by the algae to stabilise the soluble and colloidal organics. Thus, the bacteria and algae form a symbiotic relationship as shown in Fig. 4.5. The interesting aspect of facultative lagoons is that the organic matter in the incoming waste-waters is not stabilised but rather is converted to microbial protoplasm, which has a slower rate of oxygen demand. In fact, in some facultative lagoons, inorganic compounds in the waste-waters are converted to organic compounds with a total increase in organics within the lagoon system.

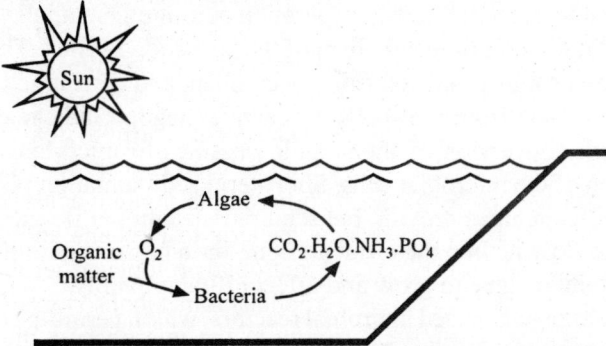

Fig. 4.5. Schematic diagram of oxidation-pond operations.

Facultative lagoons are designed on the basis of organic load in relationship to the potential sunlight availability. The facultative lagoons can be designed on the basis of 2.2g/(m^2.day) [20 lb BOD5/ (acre.day)]. In some cases the organic load can be increased to 3.4 to 4.5 g/(m^2. day) [30 to 40 lb BOD5/(acre.day)], or 6.7 g/(m^2.day)[60 lb BOD5/(acre.day)]. The depth of lagoons is normally maintained between 1.0 and 1.7 m (3 and 5 ft). A depth less than 1.0 m (3 ft) encourages the growth of aquatic weeds and permits mosquito breeding. In dry areas, the maximum depth may be increased above 1.7 m (5 ft) depending upon evaporation. Most facultative lagoons depend upon natural wind action for mixing and should not be placed in screened areas where wind action is blocked.

Effluent quality from facultative lagoons is related primarily to the suspended solids created by living and dead microbes. The long retention period in the lagoons allows the microbes to die off, leaving a small particle that settles slowly. The release of nutrients from the dead microbes permits the algae to survive by recycling the nutrients. Thus, the algae determine the ultimate effluent quality. Use of series ponds with well-designed transfer structures between ponds permits maximum retention of algae within the ponds and the best-quality effluent. Normally the soluble BOD5 is under 5 or 10 mg/L with a total effluent BOD5 under 30 mg/L. The effluent suspended solids will vary widely during the different seasons of the year, being a maximum of 70 to 100 mg/L in the summer months and a minimum of 10 to 20 mg/L in the winter months. If suspended-solids removal is essential, chemical precipitation is the best method available at the present time. Slow sand filters and rock filters have been studied for suspended-solids removal; they work well as long as the effluent suspended solids are relatively low, 40 to 70 mg/L.

Aerated lagoons

These lagoons originated from efforts to control overloaded facultative lagoons. Since the lagoons were deficient in oxygen, additional oxygen was supplied by either mechanical surface aerators or diffused aerators. Mechanical surface aerators were quickly accepted as the primary aerators because they could be quickly added to existing ponds and moved to strategic locations. Unfortunately, the high-speed, floating surface aeration units were not efficient, and large numbers were required for existing lagoons. The problem was simply one of poor mixing in a very shallow lagoon.

Eventually, diffused aeration equipment was added to relatively deep lagoons [3.0 to 6.0 m (10 to 20 ft)]. Mixing became the most significant parameter for good oxygen transfer in aerated lagoons. From an economical point of view, it was found that a completely mixed aerated lagoon with 24-h retention provided the best balance between mixing and oxygen transfer. As the organic load increased, the fluid-retention time also increased. Short-term aeration permitted metabolism of the soluble organics by the bacteria, but time did not permit metabolism of the suspended solids. The suspended solids were combined with the microbial solids produced from metabolism and discharged from the aerated lagoon to a solids-separation pond. Data from the short-term aerated lagoon indicated that 50 per cent BOD5 stabilisation occurred, with conversion of the soluble organics to microbial cells. The problem was separation and stabilisation of the microbial cells. Short-term sedimentation ponds permitted separation of the solids without significant algae growth, but required cleaning at frequent intervals to keep them from filling with solids and flowing into the effluent. Long-term lagoons permitted solids separation and stabilisation, but also permitted algae to grow and affect effluent quality.

Aerated lagoons were simply dispersed microbial reactors, which permitted conversion of the organic components in the waste-waters to microbial solids without stabilisation. The residual organics in solution were very low, less than 5 mg/L BOD5. By adding oxygen and improving mixing, the microbial metabolism reaction was speeded up, but the stabilisation of the microbial solids has remained a problem to be solved.

Anaerobic lagoons

These lagoons were developed when a major fraction of the organic contaminants consisted of suspended solids that could be removed easily by gravity sedimentation. The anaerobic lagoons relatively are deep [3.0 to 6.0 m (10 to 20 ft)], with a short fluid-retention time (3 to 5 days) and a high BOD5 loading rate, up to 3.2 kg/(m^3.day) [200 lb/1000 ft^3.day)]. Microbial metabolism in the settled-solids layer produces methane and carbon dioxide, which quickly rise to the surface, carrying some of the suspended solids. A scum layer that retards oxygen transfer and release of obnoxious gases is quickly produced in anaerobic lagoons. Mixing with a grinder pump can provide a better environment for metabolism of the suspended solids. The key for anaerobic lagoons is adequate buffer to keep the pH between 6.5 and 8.0. Protein wastes have proved to be the best pollutants to be treated by anaerobic lagoons, with the ammonium ions reacting with carbon dioxide and water to form ammonium bicarbonate as the primary buffer. High-carbohydrate wastes are poor in anaerobic lagoons since they produce organic acids without adequate buffer, making it difficult to maintain a suitable pH for good microbial growth.

Anaerobic lagoons do not produce a high-quality effluent, but are able to reduce the BOD load by 80 to 90 percent with a minimum of effort. Since anaerobic lagoons work best on strong organic wastes, their effluent must be treated by either aerated lagoons or facultative lagoons. An anaerobic lagoon is simply the first stage in the treatment of strong organic waste-waters.

Trickling Filters

For years trickling filters were the mainstay of biological waste-water treatment systems because of their simplicity of design and operation. Trickling filters were displaced as the primary biological treatment system by activated sludge because of better effluent quality. Tricking filters are simply fixed-medium biological reactors with the waste-waters being spread over the surface of a solid medium where the microbes are growing. The microbes remove the organics from the waste-waters flowing over the fixed medium. Oxygen from the air permits aerobic reactions to occur at the surface of the microbial layer, but anaerobic metabolism occurs at the bottom of the microbial layer where oxygen does not penetrate.

Originally, the medium in tricking filters was rock, but rock has largely been replaced by plastic, which provides greater void space per unit of surface area and occupies less volume within the filter. A plastic medium permitted trickling filters to be increased from a medium depth of 1.8 m (6 ft) to one of 4.2 m (14 ft) and even 6.0 m (20 ft). The waste-waters are normally applied by a rotary distributor or a fixed-spray nozzle. The spraying or discharging of waste-waters above the trickling-filter medium permits better distribution over the medium and oxygen transfer before reaching the medium. The effluent from the trickling filter medium is captured in a clay-tile underdrain system or in a tank below the plastic medium. It is important that the bottom of the trickling filter be open for air to move quickly through the filter and bring adequate oxygen for the microbial reactions.

If a high-quality effluent is required, trickling filters must be operated at a low hydraulic-loading rate and a low organic-loading rate. Low-rate trickling filters are operated at hydraulic loadings of 2.2×10^{-5} to 4.3×10^{-5} $m^3/(m^2.s)$ [2 million to 4 million gal/(acre.day)]. High-rate trickling filters are designed for 10.8×10^{-5} to 40.3×10^{-5} $m^3/(m^2.s)$ [10 million to 40 million gal/(acre.day)] hydraulic loadings and organic loadings up to $1.4/300$ $kg/(m^3.day)$ [90 lb BOD5/(1000 ft^3.day)]. Plastic-medium trickling filters have been designed to operate up to 108×10^{-5} $m^3/(m^2.s)$ [100 million gal/(acre.day)] or even higher, with organic loadings up to 4.8 $kg/(m^3.day)$ [300 lb BOD5/(1000 ft^3. day). Low-rate trickling filters will produce better than 90 percent BOD5 and suspended-solids reductions, while high-rate trickling filters will produce from 65 to 75 percent BOD5 reduction. Plastic-medium trickling filter will produce from 59 to 85 percent BOD5 reduction depending upon the organic-loading rate. It is important to recognise that concentrated industrial wastes will require considerable hydraulic recirculation around the trickling filter to obtain the proper hydraulic-loading rate without excessive organic loads. With high recirculation rates the organic load is distributed over the entire volume of the trickling filter for maximum organic removal. The short fluid-retention time within the trickling filter is the primary reason for the low treatment efficiency.

Rotating Biological Contactors (RBC)

The newest form of trickling filter is the rotating biological contactor with a series of circular plastic disks, 3.0 to 3.6 m (10 to 12 ft) in diameter, immersed to approximately 40 percent diameter in a shaped contact tank. The RBC disks rotate at 2 to 5 r/min. As the disks travel though the waste-waters, a small layer adheres to them. As the disks travel into the air, the microbes on the disk surface oxidise the organics. Thus, only a small amount of energy is required to supply the required oxygen for waste-water treatment. As the microbes build up on the plastic disks, the shearing velocity that is crated by the movement of the disks through the water causes the excess microbes to be removed from the disks and discharged to the final sedimentation tank.

Rotating biological contactors have been very popular in treating industrial wastes because of their relatively small size and their low energy requirements. Unfortunately, there have occurred a number of problems which should be recognised prior to using RBCs. Strong industrial wastes tend to create excessive microbial growths which are not easily sheared off and which create high oxygen-demand rates with the production of hydrogen sulphide and other obnoxious odours. The heavy microbial growths have damaged some of the disks and caused some shaft failures. The disks are currently being covered with plastic shells to prevent nuisance odours from occurring. Air must be forced through the covered RBC systems and be chemically treated before being discharged back into the environment. Recirculation of waste-water flow around the RBC units can distribute the load over all the units and reduce the heavy initial microbial growths. RBC units also work best under uniform organic loads, requiring surge tanks for many industrial wastes. The net result has been for the cost of RBC units to approach that of other treatment units in terms of organic matter stabilised.

RBCs should be designed on both—a hydraulic-loading rate and an organic-loading rate. Normally, hydraulic-loading rates of up to 0.16 $m^3/(m^2.day)$ [4 $gal/(ft^2.day)$] of surface area are used with organic loading rates up to 44 $kg/(m^2.day)$ [9 lb $BOD5/(ft^2.day)$]. Treatment efficiency is primarily a function of the fluid-retention time and the organic-loading rate. At low organic-loading rates, the RBC units will produce nitrification in the same way as low-rate trickling filters.

Activated Sludge

Activated sludge has been the most widespread biological waste-water treatment system because of its high quality effluent. The high energy requirements for activated sludge have caused some questions to be raised about the continued use of activated sludge systems. There is no doubt that these systems are energy intensive, but there does not appear to be any way to provide treatment without the expenditure of a definite amount of energy. Care should be taken with regard to activated sludge systems providing a high degree of treatment with a minimum expenditure of energy.

Modifications

The modifications of activated sludge systems offer considerable choice in processes. Complete-mixing of activated sludge is the most popular system for industrial wastes because of its ability to absorb shock loads better than other modifications. Contact stabilisation is a modification of activated sludge that is best suited to waste-waters having high suspended solids and low soluble organics. Contact stabilisation employs a short-term mixing tank to adsorb the suspended solids and metabolise the soluble organics, a sedimentation tank for solids separation, and a reaeration tank for stabilisation of the suspended organics. Extended aeration systems are actually long term aeration, completely mixed activated sludge systems. They employ 24 to 48h aeration periods and high mixed liquor suspended solids to provide complete stabilisation of the organics and aerobic digestion of the activated sludge in the same aeration tank. The oxidation ditch is a popular form of the extended aeration system employing mechanical aeration. Pure oxygen systems are designed to treat strong industrial waste in series, completely mixed units having relatively short contact periods. One of the latest modifications of activated sludge employs powdered activated carbon to adsorb complex organics and assist in solids separation. Another modification employs a redwood-medium trickling filter ahead of a short-term aeration tank with mixed liquor recycled over the redwood-medium tower to provide heavy microbial growth on the redwood as well as in the aeration tank. Fig. 4.6 shows schematic diagrams of the various modifications of activated sludge.

Each activated sludge process employs an aeration tank for waste stabilisation with conversion of the organics to new microbial solids and a sedimentation tank for solids separation. Solids are normally recycled from the sedimentation tank back to the aeration tank as a seed for microorganisms. Since more solids are produced than can be oxidised, there are always some solids that must be continuously wasted from the activated sludge system. Short-term aeration systems require more solids wasting than long-term aeration systems.

Aeration systems

These systems control the design of aeration tanks. Aeration equipment has two major functions: mixing and oxygen transfer. Diffused aeration equipment employs either a fixed speed positive displacement blower or a high-speed turbine blower for readily adjustable air volumes. Air diffusers can be located along one side of the aeration tank or spread over the entire bottom of the tank. They can be either fine-bubble or coarse-bubble diffusers. Fine bubble diffusers are more efficient in oxygen transfer, but require more extensive air cleaning equipment to prevent them from clogging as a result of dirty air. Mechanical surface aeration equipment is more efficient than diffused aeration equipment but is not as flexible. Economics has dictated the use of large power aerators, but tank configuration has tended to favour the use of greater numbers of lower-power aerators. Mixing is a critical problem with mechanical surface aerators since they are a point source pump of limited capacity. Experience has indicated that bearings are a serious problem with mechanical-aeration equipment. Wave action generated within the aeration tank tends to produce lateral stresses on the bearings and has resulted in failures and increased maintenance costs. Slow-speed mechanical-surface-aeration units present fewer problems than the high-speed mechanical-surface-aeration units. Deep tanks, greater than 3.0 m (10 ft), require draft tubes to ensure proper hydraulic flow through the aeration tank. Short-circuiting is one of the major problems associated with mechanical aeration equipment. Combined mechanical and diffused aeration systems have enjoyed some popularity for industrial waste systems that treat variable organic loads. The mechanical mixers provide the fluid mixing with the diffused aeration varied for different oxygen transfer rates.

Diffused aeration systems transfer from 20 to 40 mg/(L O_2.h). Combined mechanical and diffused aeration systems can transfer up to 65 mg/(L O_2.h) while mechanical-surface aerators can provide up to 90 mg/(LO_2.h). Pure oxygen systems can provide the highest oxygen transfer rate, up to 150 mg/(L O_2.h). Aeration equipment must provide sufficient oxygen to meet the peak oxygen demand; otherwise, the system will fail to provide proper treatment. For this reason, the peak oxygen demand and the rate of transfer for the desired equipment determine the size of the aeration tank in terms of retention time.

Economics dictates a balance between the size of the aeration tank and the size of the aeration equipment. As the cost of power increases, economics will favour constructing a larger aeration tank and smaller aerators. It is equally important to examine the hydraulic flow pattern around each aerator to ensure maximum efficiency of oxygen transfer. Improper spacing of aeration equipment can waste energy.

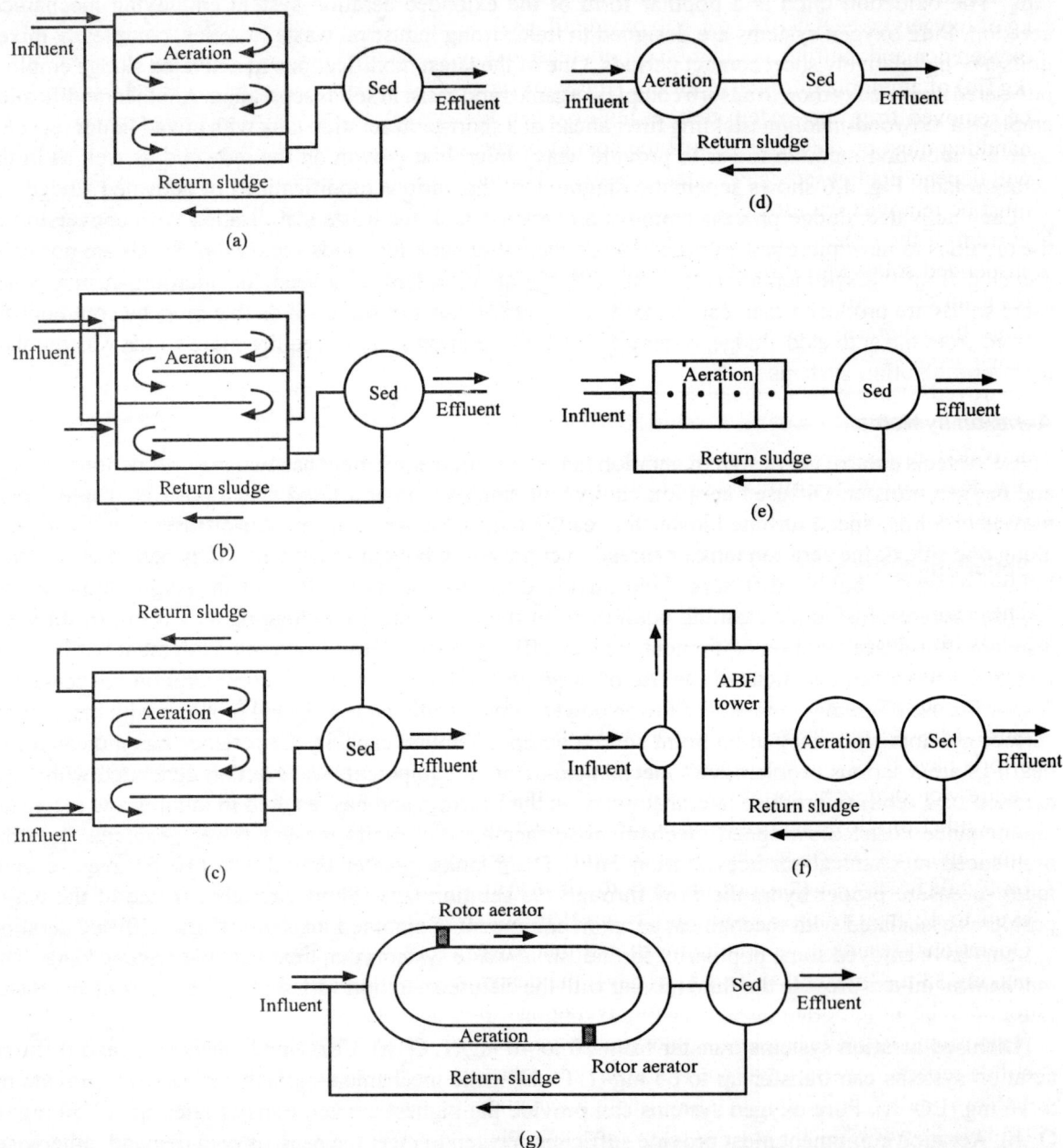

Fig. 4.6. Schematic diagrams of various modifications of the activated-sludge process. (a) Conventional activated sludge; (b) Step aeration; (c) Contrast stabilisation; (d) Complete mixing; (e) Pure oxygen; (f) Activated biofiltration (ABF); (g) Oxidation ditch.

There is no standard aeration tank shape or size. Aeration tanks can be round, square, or rectangular. Shallow aeration tanks are more difficult to mix than deeper tanks. Yet aeration tank depths have ranged from 0.6 m (2ft) to 18 m (60 ft). The oxidation ditch systems tend to be shallow, while some high rate diffused aeration systems have used very deep tanks to provide more efficient oxygen transfer.

Regardless of the aeration equipment employed, oxygen transfer rates must provide from 0.6 to 1.4 kg of oxygen/kg BOD5 (0.6 to 1.4 lb oxygen/lb BOD5) stabilised in the aeration tank for carbonaceous oxygen demand. Nitrogen oxidation can increase oxygen demand at the rate of 4.3 kg (4.3 lb) of oxygen /kg (lb) of ammonia nitrogen oxidised. At low oxygen transfer rates, more excess activated sludge must be removed from the system than at high oxygen transfer rates. Here again the economics of sludge handling must be balanced against the cost of oxygen transfer. The quantity of waste activated sludge will depend upon waste-water characteristics. The inert suspended solids entering the treatment system must be removed with the excess activated sludge. The soluble organics are stabilised by converting a portion of the organics into suspended solids, producing from 0.3 to 0.8 kg(0.3 to 0.8 lb) of volatile suspended solids/kg(lb) of BOD5 stabilised. Biodegradable suspended solids in the waste-waters will result in destruction of the original suspended solids and their conversion to a new form. Depending upon the chemical characteristics of the biodegradable suspended solids, the conversion factor will range from 0.7 to 1.2 kg (0.7 to 1.2 lb) of microbial solids produced /kg(lb) of suspended solids destroyed. If the suspended solids produced by metabolism, are not wasted from the system, they will eventually be discharged in the effluent. While considerable efforts have been directed toward developing activated sludge systems which totally consume the excess solids, no such system has proved to be practical. The concept of total oxidation of excess sludge is fundamentally unsound and should be recognised as such.

Sedimentation Tanks

These tanks are an integral part of any activated sludge system. It is essential to separate the suspended solids from the treated liquid if a high quality effluent is to be produced. Circular sedimentation tanks with various types of hydraulic sludge collectors have become the standard secondary sedimentation system. Square tanks have been used with common wall construction for compact design with multiple tanks. Most secondary sedimentation tanks use centre feed inlets and peripheral weir outlets. Recently, efforts have been made to employ peripheral inlets with submerged orifice flow controllers and either centre weir outlets or peripheral weir outlets adjacent to the peripheral inlet channel.

Aside from flow control, basic design considerations have centre on surface overflow rates, retention time, and weir overflow rate. Surface overflow rates have been slowly reduce from 33 $m^3/(m^2.day)$ [800 gal/(ft^2.day)] to 24 $m^3/(m^2.day)$ to 16 $m^3/(m^2.day)$ [600 gal/(ft^2.day) to 400 gal/ft^2.day)] and even to 12 m^3 (m^2.day) [300 gal/(ft^2.day)] in some instances, based on average raw-waste flows. Operational results have not demonstrated that lower surface overflow rates improve effluent quality, making 33 $m^3/(m^2.day)$ [800 gal/(ft^2.day)] the design choice in most systems. Retention time has been found to be an important design factor, averaging 2-h on the basis of raw-waste flows. Longer retention periods tend to produce rising sludge problems, while shorter retention periods do not provide for good solids separation with high return sludge flow rates. Effluent weir overflow rates have been limited to 186 $m^3/(m.day)$ [15,000 gal/(ft.day)] with a tendency to reduce the rate to 124 $m^3/(m.day)$ [10,000 gal/ (ft.day). Lower effluent-weir overflow rates are obtained by using dual sided effluent weirs cantilevered from the periphery of the tank. Unfortunately, proper adjustment of dual side effluent weirs has created more hydraulic problems than the weir overlfow rate. Field data have shown that effluent quality is not really affected by weir overflow rates up to 990 m^3/(m.day) [80,000 gal/(ft.day)] or even 1240 m^3/ (m.day) [100,000 gal/(ft.day)] in a properly designed sedimentation tank. A single peripheral weir, being easy to adjust and keep clean, appears to be optimal for secondary sedimentation tanks from an operational point of view.

Depth tends to be determined from the retention time and the surface overflow rate. As surface overflow rates were reduced, the depth of sedimentation tanks was reduced to keep retention time from being excessive. It was recognised that depth was a valid design parameter and was more critical in some systems than retention time. As mixed liquor suspended solids (MLSS) concentrations increase, the depth should also be increased. Minimum sedimentation tank depths for variable operations should be 3.0 m (10 ft) with depths to 4.5 m (15 ft) if 3000 mg/L MLSS concentrations are to be maintained under variable hydraulic conditions. With MLSS concentrations above 4000 mg/L, the depth of the sedimentation tank should be increased to 6.0 m (20 ft). The key is to keep a definite freeboard over the settled sludge blanket so that variable hydraulic flows do not lift the solids over the effluent weir.

Scum baffles around the periphery of the sedimentation tank and radial scum collectors are standard equipment to ensure that rising solids or other scum materials are removed as quickly as they form. Hydraulic sludge collection tubes have replaced the centre sludge well, but they have caused a new set of operational problems. These tubes were designed to remove the settled sludge at a faster rate than, conventional sludge scrapers. To obtain good hydraulic distribution in the sludge collection tubes, it was necessary to increase the rate of return sludge flow and decrease the concentration of return sludge. The higher total inflow to the sedimentation tank created increased forces that lifted the settled solids blanket at the wall, causing loss of excessive suspended solids and lower effluent quality. Operating data tend to favour conventional secondary sedimentation tanks over hydraulic sludge collection systems. Return sludge rates normally range from 25 to 50 percent for MLSS concentrations up to 3300 mg/L. Most return sludge pumps are centrifugal pumps with capacities up to 100 percent raw waste flow.

Gravity settling can concentrate activated sludge to 10,000 mg/L, but hydraulic sludge collecting tubes tend to operate best below 8,000 mg/L. The excess activated sludge can be wasted either from the return sludge or from a separate waste sludge hopper near the centre of the tank. The low solids concentrations result in large volumes of waste activated sludge in comparison with primary sludge. Unfortunately, the physical characteristics of waste activated sludge prevent significant concentration without the expenditure of considerable energy. Gravity thickening can produce 2 percent solids, while air floatation can produce 4 percent solids concentration. Centrifuges are able to concentrate activated sludge from 10 to 15 percent solids, but the capacity is limited. Vacuum filters can equal the performance of centrifuges if the sludge is chemically conditioned. Filter presses and belt-press filters can produce cakes with 15 to 25 percent solids. It is very important that the excess activated sludge formed in the aeration tanks be wasted on a regular basis; otherwise, effluent quality will deteriorate. Care should be taken to ensure that sludge thickening systems do not control activated sludge operations. Alternative sludge handling provisions should be available during maintenance on sludge thickening equipment. At no time should final sedimentation tanks be used for the storage of sludge beyond that required by daily operational variations.

PHYSICAL CHEMICAL TREATMENT (PCT)

Concept of PCT

Physical chemical treatment processes have been developed to treat waste-waters that are either toxic to or difficult to treat with biological treatment. In a few instances PCT has been used to replace biological treatment. The first step in PCT is the use of chemical precipitants to remove the suspended solids by flocculation and gravity sedimentation. The clarified effluent is then passed through an activated carbon bed to remove soluble organic compounds. If it is desired to remove soluble salts, the effluent can be passed through ion-exchange resins. PCT treatment can produce an effluent of sufficient quality for reuse. Sand media or multimedia filters have also been used after chemical precipitation and gravity sedimentation to produce a higher quality effluent to be applied to the activated carbon. Reverse osmosis has also been used as a final treatment when a very high quality of water is desired for reuse.

Limitations of PCT

Continuous chemical additions and disposal of chemical sludge have proved to be a major limitation of PCT. Chemical coagulants not only add suspended solids to the waste suspended solids but also add soluble salts to the effluent. Inadequate suspended solids separation results in carry-over of solids into the activated carbon bed, which acts as a filter as well as an adsorber of organics. Unfortunately, most activated carbon beds are not designed for easy removal of suspended solids. For this reason, multimedia filters have often been placed after chemical precipitation and sedimentation to ensure removal of the suspended solids prior to carbon adsorption. Reverse osmosis has proved to be a necessity when a high quality effluent is desired for reuse, but it simply concentrates the contaminants into a smaller volume that must be handled as concentrated waste-waters. The unit operations and their application in envionmenal pollution control are given in Table 4.4.

Table 4.4. Unit operations and thier applications in environmental pollution control.

Unit operation 1	Principles involved 2	Application 3	Efficiencies and merits of the process 4	Remarks 5
Sedimentation	Under the influence of gravitational force, suspended substances with higher specific gravity values get deposited to the bottom, while with lesser values float the surface of the medium of suspension.	To remove suspended solids	Rate of settling depends upon size and specific gravity (density) of the particles and viscosity and specific gravity of the medium of suspension	Dissolved and colloided particles are not influenced.
Filtration (a) Screening	Particles when allowed to pass through screen of lesser pore diameter, get retained on the surface.	To separate solids with comparatively larger dimensions.	This operation is faster than sedimentation and the percentage of removal varies almost directly with pore size of the screen; Solids recovered are nearly dry.	Only coarser particles are separated.
(b) Bed filtration	Physical process of solid retention in a bed filtration, is a combination of at least three actions: (i) Particles larger	Drying of sludge.	Efficiency of the process depends upon construction of bed, (absorptive property, size, packing and length of bed material), type of effluent to be filtered (pH, viscosity and surface tension), and pressure head.	Fine particles, and certain dissolved and colloidal substances are also removed.

(Contd...)

Unit operation	Principles involved	Application	Efficiencies and merits of the process	Remarks
1	*2*	*3*	*4*	*5*
	than the filter medium opening cling to the filter. (ii) Particles, smaller than the openings in the filter medium, get absorbed into the filter. (iii) Particles of different sizes cling to already filtered caked material.			
Floatation	By attachment of an air bubble to or inclusion in a suspended solid structure of liquid phase, the bulk density of the paired system may be less than the density of the parent system causing the agglomeration floated to the top.	To convert finite suspended solids and some colloidal emulsified and dissolved substances to floating matter.	Detention time and temperature two important factors which define the effectiveness of the process. Satisfaction of immediate oxygen demand, associated with partially reduced compounds of the effluent, and minimisation of odour nuisance are the merits of the process.	
Evaporation	Under the influence of heat, water and low boiling compounds of the effluent evaporate, leaving behind other dissolved and suspended solids.	To reduce effluent bulk, to concentrate toxic chemicals or their easy destruction to separate low boiling liquids and to recover costly chemical.	Rate depends upon temperature difference between the effluent bulk and the surroundings, exposed surface area and rate of vapour removed from the surface.	
Dialysis	Solutes with unequal diffusion powers, when subjected to pass through membranes, get separated.	To separate costly toxic chemicals present in a dissolved state in the effluent.		
Ion-exchange	Certain compounds (mainly resins) possess a property to exchange one of their ions from the like ions of surrounding medium	To separate costly chemicals present in a dissolved state and also certain undesirable compounds	Ease of replacement depends on the nature of the forces binding replaceable ions to the crystal, concentrations of exchanging ions in the liquid phase, charges on the ions, ion size and accessibility of the lattice ion.	Only ions or ionisable substances are removed.

(Contd ...)

Unit operation 1	Principles involved 2	Application 3	Efficiencies and merits of the process 4	Remarks 5
	due to forces like electronegativity and ionisation potential.			
Absorption	Under the influence of different surface forces, many substances (e.g., clays, activated carbon, etc.) may retain certain type of compounds on their exposed surface.	To remove and/or recover toxic, substances, colour producing compound, and refractory organic toxic metals from the effluent bulk.	Capacity of a given (solid material) absorbent to absorb or extract is a complex function of several variables: Chemical and physical properties of the solid absorbent, composition and concentration of the liquid phase, temperature of the system, type of contact and duration of the contact.	Absorbent bed in the form of percolation bed in addition serves as a filtering medium, while in the contact method, as coagulant for finely divided suspended solids. These services, however, interfere with the quantities absorbed.
pH-control (a) Neutralisation	When alkaline or acidic materials are added to a solution, the medium gets correspondingly due to neutralisation reactions.	To remove undesirable acidity or basicity of effluent and precipitation of certain unwanted compounds soluble in acidic or basic medium. To develop a medium for bio-chemical activities.		
(b) Acidification		For breaking of emulsion, precipitation of colloids, removal of compounds, volatile in acidic solutions such as cyanides and sulphides.		
(c) Custicisation		By product recovery or salvage.		
Flocculation	Precipitates formed by reaction of coagulation chemicals, however, have the property of forming large flocs of high surface area. As these flocs move	To remove colloidal and finely divided particles and to reduce the organic load of a waste.	Efficiency of the process gets considerably reduced in the presence of surface active chemicals like soaps and detergents.	Dairy wastes due to presence of excessive detergent may not be flocculated.

(Contd ...)

Unit operation	Principles involved	Application	Efficiencies and merits of the process	Remarks
1	2	3	4	5
	through the liquid in the settling tank, they remove other suspended solids by absorption or mechanical agglomeration.			
Chlorination	When chlorine gas is added to water, it hydrolyses to form hypochlorous acid. The reactions: $Cl_2 + H_2O \rightleftharpoons HOCl + HCl$. The acid produced, the alkalinity and the hypochlorite ion react with most reducing substances and oxidise them to a higher oxidation number. Electrical particles are also affected thus aiding their coagulation.	Odour abatement, disinfection, reduction of BOD or treatment of plant effluent, removal of gases to control bulking of activated sludge, to correct ponding on trickling filters are the main applications of the process.	Chlorination reaction are not instantaneous, but proceed at a rate dependent on several variables, concentration of hypochlorous acid, dissolved chlorine, and of the and temperature. An increease in temperature usually resuls in the acceleration of the chlorination reactions, although it may also such as chlorate formation.	Chlorine in solution reacts with ammonia or with organic nitrogen compounds like protein aminoacids for corresponding organic derivatives.
Electrolytic oxidation	Under certain pH, electrode potential and current density, some resistant organics are activised, which then get oxidised by the nascent oxygen produced at the electrodes during the processes. Due to presence of highly active ionic species, neutralisation of charges on colloidal particles is also aided resulting in their precipitation.	For oxidation of toxic, colour producing and refractory organics, disinfection and precipitation of colloids.		
Oxidation	Substances containing elements at a low positive oxidation potential and with a	To convert complex resistant organic to simpler non-polluting compounds.		

(Contd ...)

Unit operation 1	Principles involved 2	Application 3	Efficiencies and merits of the process 4	Remarks 5
	strong tendency to change to a higher number, when brought into contact with oxygen, get oxidised.			
(a) Aeration		To increase DO of the effluent, needed in oxidations, both chemical or biological, and its degassing.		Breaking of emulsions and removal of volatile and odour producing substances is achieved in addition.
(a) Ozonisation	Ozone when brought into contact with elements of low positive oxidation potential, also oxidises them.	For oxidation of resistant organics, disinfection, accelerated oxidation of organics and compounds producing colour and colour.	To achieve more efficient utilisation, batch treatment rather than continuous flow treatment may be preferred. Temperature is an important variable to define the efficiency of the process.	The end products and by-product of oxidation with ozone are the same as those from air oxygen oxidation and are generally not objectionable.
Aerobic treatment	In the presence of dissolved oxygen and proper environmental conditions, certain microorganisms utilise organic waste as food, converting it into simpler non-polluting compounds such as carbondioxide, nitrates, sulphates and water. Complex cell tissues and protein materials are also synthesised which are then agglomerated and removed from the waste by settling.	To remove organics.	Aerobic biological treatment operations for industrial wastes require for their successful performance, a community of proper microorganisms, food material, an appropriate environment and a supply of dissolved oxygen. Almost any organic matter with the possible exception of hydrocarbons and ethers may be oxidised by the process. Temperature influences the metabolic activities of the microbiological population (i.e., biological reaction rate constants), gas transfer rates and the settling characteristics of the biological solids, and thus plays a significant role in determining the efficiency of the process.	Germicidal and resistant organics like cyanides and phenols can be destroyed by special microorganism but after a prolonged period of acclimatisation
(a) Activated sludge process	In this process aerobic oxidation of waste is performed	to remove dissolved, colloidals and coarse solid organic matter.	For optimum activity, the kinetics of activated sludge require a young flocculant sludge in the logarithmic	Process is very susceptible to interference by

(Contd ...)

Unit operation	Principles involved	Application	Efficiencies and merits of the process	Remarks
1	*2*	*3*	*4*	*5*
	by active growths created previously and an aerobic environment is achieved by the use diffused or mechanical aeration. Organic matters from the wastes get absorbed on the gelled structure; by collision, electrical attractions, chemical reactions of surface atoms and finally converted to simple compounds by oxidation enzyme systems.		growth state by controlled sludge wastage, continuous loading of the organisms and elimination of anaerobic conditions of any point with oxidation treatment.	toxic materials. However, tolerance may be built up over a period of time if concentrations are slowly and uniformly increased. Various degrees of efficiencies are obtained by controlling the contact period of the concentration of the active floc.
(b) Trickling filtration	In this process bacteria and other microorganisms are cultivated in a porous bed of some material with a large surface area and the entire system is maintained in aerobic condition by free passage of oxygen in the unit. The waste water is trickle through in during which the organics are oxidised. Gases produced are removed continuously in the effluent and solid residue of cell tissue that cling to the porous media for a time are finally Sloughed off with the filter effluent.	Dissolved and colloidal organic matter.	Efficiency is affected both by waste and filter variables. Waste variables are the composition of the waste, strength of hydraulic loading, pH and temperature while filter variables include depth of filter, size and uniformity of filter medium, uniformity of waste distribution over the filter, and air availability. The trickling filter has more resistance to toxic waste than the activated sludge and also recuperates more promptly from an over dose.	A reasonable degree of tolerance to more toxic materials can be developed in the filter microorganics by gradual build up of toxic concentrations but shock load or sudden surges of concentrated poison may impair the efficiency of the filter temporarily or permanently.
(c) Lagooning	The process is essentially the same as the activated	Dissolved colloidal and suspended organic matters.	There is little or no decrease in the organic content of waste passing through an oxidation period.	It is not suitable for use in

(Contd ...)

Unit operation 1	Principles involved 2	Application 3	Efficiencies and merits of the process 4	Remarks 5
	sludge process excluding the recycling of active growths from the treated effluent. Oxygen is supplied either through compressed air, mechanical surface aeration or by photo synthetic aeration.		But putrescible raw wastes are converted into relatively stable algae cells. Temperature, pH, dissolved oxygen concentrations, sun light and presence of sufficient nutrient minerals are important variables to decide the efficiency of the process.	populated areas because of its nuisance value from odour and insects.
Anaerobic treatment	Under anaerobic conditions, i.e., in the absence of gaseous or dissolved oxygen, certain groups of micro-organism may carry the digestion of complex organic wastes. These are mainly hydrolyte and organisms. by their action convert complex organic compounds to a low molecular weight organic acids and alcohol. These are then converted to carbon dioxide and methane by methane bacteria.	To remove complex organics and volatile solids.	A major advantage of the anaerobic treatment process is that reactions can be carried out in depth, without need for large surfaces or interfacial areas. It is less expensive but it has the limitation that its final effluent is less satisfactory than that from aerobic treatment because it is dark in colour odorous and with a high residual BOD. It takes place in mixed or enriched cultures and hence may be maintained easily on a large scale. It is applicable to any type of substrate except lignin and mineral oil, and converts the entire substrate quantitatively to CO_2 and CH_4. It produces lesser quantities of in-offensive sludge which may also dry quickly. Efficiency of the process depends on temperature, pH, waste loadings, presence of toxic materials and absence of oxygen.	The ultimate products of anaerobic decomposition are less oxygenated than products obtained under aerobic conditions.

BIOREACTOR SYSTEMS FOR INDUSTRIAL WASTE-WATER TREATMENT (AEROBIC METHODS)

Although biological treatment is one of the steps in treatment of industrial waste-water, it is a crucial step and a lot depends on its efficiency. Unfortunately this is a neglected area, and little attention is paid while choosing the bioreactor systems. This forces a management of the situation by alternatives, which is neither advisable nor attractive. Full advantage of the activity of micro-organisms can only be obtained if one has the right kind of bioreactor.

There are a large number of bioreactor systems developed and available in the market, and it is important to make a proper assessment of them. This section will discuss bioreactor systems in two parts – the first will deal with aerobic systems, and the second with anaerobic systems and special designs of bioreactors. The bioreactors discussed here (both parts) are intended for reduction of COD/BOD. If N and/or P removal or bio remediation is to be achieved, special designs of reactors using different micro-organisms are to be used and they are not covered in these discussions. Typically, waste-water treatment is carried out in three stages:

1. Primary stage essentially comprises of screening, equalisation tank, flow balancing, sedimentation etc., steps.
2. Secondary stage comprises of biological treatment using a bioreactor to promote biological oxidation in one or few tanks.
3. Tertiary stage, which is a polishing stage.

The objectives of biological treatment of waste-water can be removal of carbonaceous organic matter (which is measured as BOD, COD, TOC), removal of nutrients (such as phosphorus and nitrogen), removal of fats and grease, heavy metal removal or degradation of other specific pollutants. The knowledge of sewage treatment, which has accumulated over last 130 years or so, has become the basis and guideline for treatment of industrial waste-water. But, for reasons mentioned below, this approach can not work and industrial effluent treatment needs more careful handlings. Industrial waste-waters are difficult to treat for the following reasons:

1. Variation in load with the size of industry and the process used.
2. Variation in concentration of the pollutants.
3. Variation in toxic materials.
4. Seasonal and daily changes in amount of organic load, depending on the product produced.
5. They are often nutritionally unbalanced.
6. They are complex and differ in composition and nature of pollutants.

Biological treatment performs best in a narrow range of pH (6 to 10), organic load (4:1), low oil and grease content (<50 ppm), inhibitory pollutants (<10 ppm), sufficient N and P and the presence of inorganic trace elements. As a result, there cannot be any uniform way of dealing with these effluents. Although physical and chemical methods are also used for effluent treatment, they have their own limitations and disadvantages. Thus, biodegradation is often a must, at least as part of the whole treatment system. Thus:

1. Agricultural wastes, animal farm wastes, waste from food processing industries and breweries will satisfy the needs of biodegradation.
2. Wastes from the pharmaceutical, chemical and pesticide industries will have little biodegradable organic matter and could contain toxic chemicals, and/or inhibitors of biodegradation.
3. Oil refineries, petrochemical industries will have difficulty to degrade hydrocarbons with no supportive nutrients.
4. Pulp and paper mills will have biodegradable organics plus toxic chemicals.
5. Tanning industry will have toxic chemicals and no biodegradables.
6. Foundries, metal processing, and mining industries will have more inorganics.

The type of treatment plant to be used will depend on the complexity and nature of pollutants. Non-biodegradables need to be tackled by physical/chemical methods, but biodegradable pollutants differ in the degree of attack possible. The general classification of pollutants can be:

1. Readily degradable organics (e.g., from food industry).
2. Complex organics (e.g., from organo-chemical industry).
3. Reactable inorganics (e.g., heavy metal industry, electroplating).
4. Inert inorganics (e.g., coal mining, quarrying).

Biological treatment of industrial wastes is a relatively neglected area, less properly understood and controlled. Many do not give priority to choosing the right kind of biological treatment method, often choosing the activated sludge process or its variant. Many handle the biological treatment stage in the most crude manner: add some sewage sludge and expect BOD/COD reduction. Ignorance about the composition of effluent just cannot be excused. Appointing a microbiologist or a bio-technologist or regular consultancy with experts in this field is often felt as a burden by many. The result is less than satisfactory functioning of the treatment plant.

Choice of the Bioreactor System

The bioreactors discussed in both the parts still deal only with the COD/BOD or specific pollutant degradation and if N, P removal or bio-remediation are to be achieved, there are some special designs of reactors for different kind of micro-organisms which are to be used. Moreover, large number of variations exist in the basic design meant for the particular purpose, e.g., biogas production from waste-water. Immobilisation of micro-organisms, or enzymes has resulted in smaller but effective reactors and their use is increasing. The choice of bio-treatment plant should take the following criteria into consideration:

1. Process should be eco-friendly.
2. Process should be flexible to take varying loads and characteristics.
3. Process should be operator-friendly, requiring little personal attention.
4. Capital costs and operating costs should be low.
5. Process should not be energy-intensive, and should generate energy, if possible.
6. Continuous operation should be possible.
7. Automation should be easy.
8. It should achieve the targeted degradation.
9. Space requirements should be minimum.
10. It should avoid dilution, as water is a costly resource.
11. It should produce less sludge, as sludge disposal involves considerable costs.

Weightage to the above mentioned points may differ for different industries as per their size, location, resources, etc. Waste-water systems broadly fall into two categories: aerobic and anaerobic. A comparison of aerobic and anaerobic systems is shown in Table 4.5.

Table 4.5. Comparison of aerobic and anaerobic systems.

Aerobic Treatment	*Anaerobic Treatment*
Range of waste-waters can be used	Limitations in treatment applications
Process stability and control	Less process stability and control
Require more power input	Require less power input
Produce more sludge	Produce less sludge
Percentage BOD removed is more	Percentage BOD removed is less
Nitrogen removal is better	Nitrogen removal is poor
Phosphorus removal is better	Phosphorus removal is poor
Can cope up with low substrate level wastes	Advantageous for high substrate level wastes
Works in mesophilic range of temperature	Advantageous at higher temperatures
Reductive dechlorination occurs	Can operate at high organic load
Nitrification, denitrification, ligninase activity occurs	Micro-organisms remain dormant for several months

Suspended Growth Systems

Activated sludge process

It operates as homogeneous continuous culture and is a well-aerated system. Biosorption and flocculation remove the organic matter rapidly, while oxidation and biosynthesis proceed at a lower rate. Subsequently flocs settle into the next stage of secondary sedimentation tank, and a portion of the floc may be returned as inoculum. BOD and suspended solids are reduced by 85–95%. Organisms in activated sludge are similar to that in percolating filters.

In activated sludge, conditions are not suitable for micro invertebrate grazer population. Consequently, plants do not suffer from nuisance of flies, and fungi are less dominant, resulting in less sludge bulking. Protozoans are abundant in the activated sludge, but nematodes and rotifers are small in number. The contents of the reactor are referred to as Mixed Liquor Suspended Solids (MLSS) or Mixed Liquor Volatile Suspended Solids (MLVS) and consist of micro-organisms and inert and non-biodegradable suspended matter.

The design of the process gives controlled environment for micro-organisms. Design considerations are :

1. Nutrient requirements: The BOD: nitrogen: phosphorus ratio must not be more than 10:5:1. Dosing of inorganic nitrogen may be done if required.
2. Working is satisfactory over pH range of 6–9.
3. Ambient temperature 30–40°C is suitable.
4. Fatty matter 0.1–0.25 kg FOG (fats, oils and grease) kg^{-1} MLSS per day is maximum allowable. If FOG is more, it should be chemically pretreated.
5. Loading rate 0.1–0.3 kg BOD kg^{-1} MLVSS per day or 0.2–1.2 kg BOD m^{-3} reactor/day.
6. Aerator efficiency 2 kg oxygen kwh^{-1} and BOD removal rate 0.5–0.6 kg BOD kwh^{-1} are normally observed.
7. Excess biological sludge 0.3–0.5 kg, ss/kg BOD applied. Treatment and disposal of excess sludge can cost up to 50% of overall treatment costs.
8. BOD removal is more than 95% and with dilute waste-waters, i.e., less than 1000 mg/l BOD, effluent with BOD 20 mg.l^{-1} quality is produced.

The original activated sludge process had drawbacks like, high operating costs, difficulty in operation and maintenance, and used to produce large surplus of biomass. Other disadvantages of the process are:

1. It is a sensitive process.
2. It can not withstand shock loads in the flow rates, pH, and concentrations.
3. It requires long acclimatisation periods for organisms to metabolise tough organic compounds.
4. Higher concentration of oil, hydrocarbons is a problem.

Useful organisms to degrade tough pollutants are always small in number and any nutrients added favour ordinary micro-organisms rather than the special ones. Thus, activated sludge process should not be considered the ideal one for any waste-water treatment. To combat some of these disadvantages, various modifications of the activated sludge process have been developed and are discussed in the following section.

Tapered aeration

Here aeration capacity is related to demand and it is less at the outlet than at the inlet.

Step aeration

Here feeding as well as aeration is done in steps in the system throughout the length of the tank.

Contact stabilisation

Here returned sludge is aerated to encourage organisms to utilise any stored nutrients, and more wastes are digested. Sludge volume is reduced through aerobic digester stage. It is similar in principle to the extended aeration treatment, where aeration and mixing of sludge and waste-water is in the same unit. Powdered Activated Carbon (PAC) may be added as an additive to the activated sludge process. This absorbs organic compounds, which can not be degraded, reduces colour and metal, besides lowering the effective level of inhibitors.

Carrier activated sludge process (CASP)

Activated sludge systems modified by the addition of inert suspended solid particles (carriers) are widely used. These CASP systems combine the characteristics of fluidised bed reactor and conventional activated sludge process.

Here biomass concentration may vary from 8,000–30,000 mg/l of MLVSS, and the disadvantages experienced in the fluidised bed system—like ineffective suspended solids removal and operational difficulties—are overcome. With minor modifications in the existing process installations, many conventional activated sludge systems can be converted into a CASP. Some of the advantages of carrying out such a modification are:

1. Surface area is considerably increased.
2. Biomass concentration is much more than in conventional activated sludge process.
3. Biomass is attached as well as in the flocs form.

Carriers may include sand, plastic, glass, powdered activated carbon, clay, etc. Micro-carriers that are used help floc formation in addition to providing of attached growth support. Biological flocs capture particulate matter and hence effectively remove fine suspended solids. CASP systems commercially available include:

1. INKA system (US).
2. Biofix system (France) having 20–40% of aeration tank volume as medium.
3. Bio-2 sludge process (Germany).

All of them show improved substrate removal and increased (20–40% more) organic loading, more stable operation, reduced sludge yield, and increased air transfer efficiency. The Ring Lace Submerged filter system marketed by Kajima Corporation, Japan is used in 110 installations in Japan. Loosely woven polyvinylidene chloride fiber strings in the form of rope are used. In China, synthetic fibre media are used.

Advanced activated sludge process

Most of them operate with pure oxygen, and can handle higher biomass concentration, reducing residence time. Bulking (i.e., excessive growth of filamentous bacteria and fungi, which inhibit sludge settling) is also inhibited. These systems are having very wide applications to different category of effluents. Efficiency of COD removal is 85–95%.

Examples of closed tank systems of this type are UNOX, OASES, F3O and Marox, but they are difficult to monitor and maintain. However, there is no loss of unutilised oxygen and no nuisance of odour or aerosol formation. As combustible materials like oil, fats, grease burn vigorously, hydrocarbons should not be allowed to accumulate. Open tank systems include the Vitox, Megox, Primox and Simplox systems, which have the advantages of ease of access for maintenance and monitoring, higher treatment capacity, with no nuisance of odour and aerosol formation. However, loss of unutilised oxygen can occur.

Deep shaft process

This is analogous to the Activated Sludge Process (ASP). 20 Deep Shaft Reactors are currently in operation all over the world (mainly in Japan and Canada). The reactor is 50–150 metres below the ground, and its diameter is 0.5–10 metre. Oxygen transfer is the key feature. Major advantage of this reactor is that it requires less land. In this reactor, circulation of liquid occurs by airlift principle. Liquid velocity is 1–2 $m.s^{-1}$, and air is carried down the shaft. Location of air injection is important. There are two problems in this reactor: gas disengagement and micro bubble formation in the outlet material, which causes difficulty in settlement and hence turbidity. BOD reduction achieved is from 1060 $mg.l^{-1}$ to 60 $mg.l^{-1}$.

High contact reactor (Kvaerner HCR)

The HCR system is a biological waste water treatment process which works according to the "activated sludge" principle. The process combines a loop reactor and a special patented two-phase jet aeration system that operates in conjunction with a sludge separation unit (usually a standard clarifier). The bio-culture is vigorously mixed with waste-water in the presence of finely dispersed air bubbles to achieve the maximum possible mass transfer-oxygen to the biomass. Degradable organic matter is there by quickly converted to carbon dioxide, water and new bio-culture. The HCR system show excellent degradation and sludge reduction compared to traditional technologies.

The process requires less land, less investment and operating costs, show good flexibility to pollution fluctuations, easy installation and has capacity to treat formaldehyde, phenol, furfural and resin acids efficiently.

The process has been employed on the full scale for the treatment of high strength effluent in a number of European pulp and paper plants, as well as in a municipal treatment facility in China. HCR systems are also operating in a variety of other industries, such as food processing, brewery, dairy, and landfill leachate. This process, commercialised by the well known Norwegian environmental company, Kvaerner Water, is now available in the US exclusively from Biothane Corporation.

Granular aerobic sludge system (Aerothane GASS)

The Aerothane Process was developed for post treatment of anaerobically pretreated waste-waters. It is an aerobic polishing stage for odour control and requires small place for its tower type reactor, which are aerated by means of bubbles. Internal baffle plates create controlled gas lift loops, which ensure perfect mixing in the system. An internal settler retains the biomass, maximising the amount of active biomass in the reactor.

Due to the excellent sludge retention, the system is high loaded and has liquid residence time of 1–4 hours. Aerothane process reduces sulphide (which causes sewer corrosion) from anaerobic effluent prior to discharge, causes partial nitrification to further reduce oxygen demand of the waste-water, and removes other odours of anaerobic activity. Comparison of different treatment system is given in Table 4.6.

Table 4. 6. Comparison of performance of different treatment systems.

	Trickling filter	Activated sludge (conventional)	RBC	FBBR	Activated sludge (pure oxygen)
Biomass conc. (MLVSS mg/l)	2000–7000	700–2500	10000–20000	10000–50000	3000–5000
Surface area (m²/m³ reactor vol)	12–30		40–50	800–1200	
Process loading rate*		0.5–1.2		8–16	1.2–2.4

* As BOD removal per m³ of reactor volume per day.

Fixed film systems

Biological filters

Micro-organisms are attached to an inert support medium which is packed into a tower or a tank. In the tank, while waste-water percolates down the reactor with proper distribution, the air is introduced from the bottom. Microbial slime develops on media support by using the organic matter and oxygen. When the thickness of the slime increases, extra biomass sloughs off. This sludge is collected by gravity in a sedimentation tank. Sludge is later on treated and disposed off the same way as for activated sludge process. Loading rates of aerobic filters are 3–10 kg BOD m^{-3}d^{-1} and BOD removal is up to 70%. There is no recycling of microbial sludge done. Excess sludge produced is 0.5–1 kg SS kg^{-1} BOD. Innovations in tank designs and support media have occurred over the years. Development of slime, metabolic activities of micro-organisms in slime, increase in thickness of slime, detachment of slime, and formation of new slime is a continuous process. Facultative bacteria like *Achromobacter, Flavobacterium, Pseudomonas, Alcaligenes,* filamentous forms like *Sphaerotilus natans, Baggiatoa,* nitrifying bacteria like *Nitrosomonas, Nitrobacter* and *algae, fungi, protozoa* are commonly present. On the basis of hydraulic and organic loading rates, filters are of two classes: low rate and high rate.

Rotating biological contactors (RBC)

This is one of the principal types of fixed-film moving-medium systems used as digesters. RBC was first used for dairy waste-water in the US. Cheese waste-water, poultry wastes, bakery wastes, winery wastes, paper unit wastes, wastes of meat packaging units, etc. have been treated afterwards on pilot plant and commercial scale. RBCs are very popular and useful systems now. The RBC systems find applications in:
1. Secondary treatment.
2. Nitrification.
3. Landfill leachate and runoff.
4. Pretreatment of water supplies.
5. Denitrification.
6. Phosphate removal, etc.

RBC has discs which are 2–3 metre in diameter, 10–20 mm wide and mounted on a horizontal shaft. The distance between the adjacent discs is 20 mm. Discs are partly submerged (40% area in the medium), and are rotated at 1–7 revolutions per minute. Air is sparged to avoid the risk of anaerobiosis. Biomass on the discs is 200 g dry weight per sq. metre of disc surface which is equivalent to

40–60 kg/m^3 MLVSS in the activated sludge system. Biological growth is 2–3 mm thick. Standard loading allowed is 6–20 g BOD per sq. metre of disk-day and 80–90% BOD reduction occurs. RBCs are simple to operate, have low maintenance, require less space. It accommodates shock loading, and foaming/aerosol/airstripping are reduced. Fast start-up, efficient mixing, easier oxygenation, little sloughing off of biomass are some of the other advantages. Problems of clogging, ponding or filter fliers which are present in trickling filters are absent here. Less experience of the process, less operational control, odour problem, undesirable growth of Baggiotoa, Thithrix and more sludge production, etc. are some of the disadvantages.

Fluidised bed reactor (FBR) or fluidised bed biofilm reactor (FBBR)

It is a combination of attached growth (trickling filter) and suspended growth (activated sludge) systems. Biological film is developed and maintained on a solid support medium consisting of small particles, which are maintained in suspension by the upward flow of liquid being treated. Support medium particles neither sink nor overflow. Reactors are generally cylindrical with perforated distribution plates and tapered or conical entry sections.

Fluidisation of support particles is allowed but clumping is prevented. Support media used may be sand, carbon, fly ash, anthracite, glass, calcinated clay, etc. with particle size 0.2 to 0.3 mm. Smaller the size, larger is the surface for biofilm formation and lower is the settling rate in the reactor. Also, fabricated porous media may be used to allow biomass development within the porous structure. Spheres of knitted S.S. wire (6 mm diameter), knitted polypropylene toroids (20 mm diameterorous forming particles of 53 mm diameter), reticulated polyester foams cut into cuboids of 25 × 25 × 10 mm, polypropylene squares of 25 mm, etc. are the examples of fabricated support media. FBRs are now used for waste-water treatment of soya processing units, denitrification, canning industry, dairy, distilleries, food processing units, paper and pulp manufacturers, oil refining units. FBR has advantages like saving of space, easy to install, modules can be added, suitable for high strength industrial wastes, low biomass sludge production, and is more efficient. There is continuous research going on in the FBR systems to reduce energy consumption, to reduce start-up time, to improve process stability and to make them suitable for hazardous wastes also.

Inverse fluidised bed biofilm bioreactor (IFBBR)

Fluidised bed reactors used today are mostly operating with upflow systems, consisting of gas-solid, liquid-solid or gas liquid-solid phases. Upflow fluidisation is inconvenient for certain applications carrying aerobic processes. Collision between the bioparticles and shear stress affects the biofilm formation and then biofilm thickness. Inverse fluidisation removes these drawbacks and gives higher performance. IFBBR shows 17 times higher efficiency in COD/BOD reduction when compared with mechanical surface aeration systems.

BIOREACTOR SYSTEMS FOR INDUSTRIAL WASTE-WATER TREATMENT (ANAEROBIC METHODS AND SPECIAL DESIGNS)

Anaerobic digestion is microbial fermentation wherein absence of high levels of nitrates and sulphates, organic matter is converted into CO_2 and methane. Earlier, anaerobic treatment was used for sludge digestion and stabilisation. It was once considered to be sensitive to toxicity and less reliable in performance (i.e., COD reduction). These doubsts have been removed and anaerobic treatment is now preferred due to a number of advantages:

1. Very low sludge production.
2. Lower energy consumption.
3. Production of methane, which has high calorific value.
4. Process can operate at high organic loading rate.
5. No environmental nuisance of odour and aerosol formation, as in aerobic methods.
6. Micro-organisms can remain dormant for several months and can become operational within a week of start-up. This is suitable for seasonally produced waste-waters.

Contact Digesters

This is anaerobic equivalent of activated sludge process. It consists of a stirred tank and a tank under anaerobic conditions. The output of completely stirred tank (digester) is settled under anaerobic conditions and a part of settled sludge is returned to the digester. This results in concentration of sludge and longer retention time. This enables retaining of mathanogenic organisms over a wide range of loading. Separation of the bacteria is hindered by gassing of the effluent, hence the effluent is degassed before settling the biomass. Contact digesters are currently used for treating wastes from sugar processing, distilleries, citiric acid and yeast production, food processing factories, farm slurries, etc. Contact digesters are not much affected by suspended solids in feed as it happens in retained biomass type of digesters. 95–97% COD removal is achieved. One of the problems with contact digesters is the poor settlement of solids because of attachment of product gas to solid particles. This is overcome by vacuum degassing which is very common. Dorr Oliver's Clarigester, is a modification of a contact digester with improved solid retention.

Packed Bed Reactors—Packed Column Reactors (PCR)

They are simple in design, easy to construct and operate, and reported to have advantages over suspended growth reactors like Continuously Stirred Tank Reactors (CSTRs). Organisms are contained within the packing medium in an enclosed vessel and liquid wastes pass upwards. Systems wherein the liquid waste-water flows downward are also available, but they have less biomass concentration. In these systems, organisms do not form slime on packing, and regular back-washing prevents clogging. This will also prevent high concentration of suspended solids sloughing continuously into the effluent. The treatment rate is directly related to surface area of packing. PCR is 15–30 times efficient than CSTRs.

Anaerobic filters (columns) packed with anthracite and coal particles have been used for denitrification of uranium recovery waste-waters and found effective for effluents with concentrations of 500–5000 mg NO_3-N per litre. Polystyrene spheres in anaerobic filter or Neptune-Microfloc filter containing sand, silica, anthracite coal are also used. In upflow anaerobic fixed film filters, hydraulic retention time is 70 hours and biogas produced is 2.6 $m^3/m^3/d$. In downflow anaerobic fixed film filters, hydraulic retention time is 10–20 hours and biogas produced is 2.6 $m^3/m^3/d$. These systems have achieved 60–99% COD reduction.

Expanded Bed Reactor (EBR)

The basic mode of operation is similar to the packed filters and fluidised bed processes. The expanded bed system uses the operating mode of fluidised bed reactor. In this reactor, velocities are maintained to preserve delicate attached living film. Separation, retardation and biocoagulation of fine suspended solids occurs. Besides, maximum biomass concentration is achieved. However it is a better bioreactor as shearing of microbial biofilm is minimum and maximum biomass concentration is achieved.

Anaerobic Continuously Stirred Tank Reactor

Biobulk (from Biothane Corporation) is a conventional anaerobic contact process that is applicable for waste streams containing high strength COD/BOD concentrations and fats, oils and grease (FOG) concentrations greater than 150 mg/l. Biobulk is a "medium loaded" system with volumetric loadings of 2–5 kg. COD/m^3/day. The system is economical in operation and of proven reliability.

The Biobulk Continuously Stirred Tank Reactor (CTSR) has specially designed internal mixing capability to ensure that the waste-water is in constant contact with the biomass. CTSRs are cost effective when flows are low, i.e., 150,000 gpd or less. With these low flows, longer hydraulic retention times are possible allowing for degradation of the solids. The Biobulk Process is an excellent treatment choice for ice cream plants and other food processing facilities which discharge effluents high in biodegradable fats and oils. Removal efficiencies with this technology have been found to consistently average above 90% with respect to COD, FOG and BOD and close to 75% with respect to the organic fraction of TSS.

Anaerobic Baffled Digesters

These reactors have walls across the tank built from the top to bottom, so that waste-water flowing along the tank has to go alternately under and over the baffles. The baffles tend to keep the bacteria in the tank and also help to prevent problems with floating solids.

Upflow Anaerobic Sludge Blanket Reactors (UASB)

UASB digester is one of the most effective and economic methods of anaerobic digestion of wastes. UASB digesters possess all the advantages of anaerobic digestion mentioned earlier like less sludge production, less energy consumption, biogas production etc. Other advantages are compact reactor, easy start-up, prolonged maintenance-free, noiseless, aesthetically satisfactory working.

UASB digesters are very popular systems today, with about 200 operating plants world-wide. Research efforts of Lettinga *et al.*, and two Dutch companies marketing the process are behind the success of UASB. Contact digesters have difficulty of loading rate while filters in anaerobic filters are likely to clog. To overcome these problems, flocculation of bacteria is sought without use of inert carrier material in UASB. Active bacteria are present as high density granular sludge, which is retained in digester tank despite gassing and upflow velocity of waste-water.

In an UASB Reactor, waste-water is fed from the bottom. As it flows upward through the reactor, organic matter in the waste is degraded anaerobically by micro-organisms resident in the sludge blanket. Besides converting organics to cell mass, biogas, rich in methane, is produced as a by-product. A Gas Liquid Solid Separator (GLSS) provided near the top of the reactor, enables sludge to settle into the blanket, biogas to escape into the dome at the top of the reactor, and treated supernatant to flow out of the reactor. High sludge concentration in sludge blanket and low concentration of suspended solids in the reactor overflow are characteristic features of a good UASB digester. Sludge granulation is complex and is not fully understood. Initially 10–15% inocula of granular sludge is required. Netherlands has carried out a lot of studies in sludge granulation. Hydrodynamics of the digester created by feed distribution and the shape of the digester are important in retaining the correct granular form. Well adapted sludge may be sufficient in 1% volume only. The type of waste-water being treated has an important role in the formation of granules. Presence of calcium, magnesium, aluminum, silicon, ammonia is important for successful granule formation. A large proportion of filamentous micro-organisms (e.g., Methanothrix spp) is also essential. Baffles in the digester, promote gas solid separation along with the shape of digester and upward liquid velocity.

The biomass responsible for biodegradation forms two discrete zones in reactor vessel. The key to the success is its ability to retain biomass solids in the reactor. 10–15 kg COD/m^3 reactor volume/day loading can be handled. Efficiency of UASB is 70–95% COD removal. Methane produced is about 0.15 to 0.35 N m^3/kg COD destroyed. UASBs have been extensively used to treat food processing wastes. UASB has been used in experiments for dentirification in Netherlands and Japan. 70% removal of nitrates and 63% reduction in COD has been achieved. A full scale version of 200 m^3 of UASB has successfully treated sugar beet waste at loading rate of 16 kg COD m^{-1}d^{-1} and hydraulic retention time of 4 hours, achieving 90% COD removal. UASB has also been used for purely methanogenic process with acidogenic phase being carried out in a separate reactor. Biothane reactors are used for food, chemical, and fermentation industry wastes. The UASB reactor contains no mechanical or moving part involving wear and tear. Thus it is virtually maintenance free and involves few operational problems. When properly designed and made, a UASB provides trouble-free service for many years.

Expanded Granular Sludge Bed (EGSB)

As an outgrowth of successful operating experience with UASB technology, an Expanded Granular Sludge Bed (EGSB) process that incorporates the sludge granulation concept of the UASB has been developed by Biothane Corporation. Biobed—the trademark of the EGSB system—uses no carrier material as a mechanism for biomass retention within the reactor as in other types of anaerobic fluidised or expanded bed technologies. It has highly settleable granular biomass. This process can be perceived either as an ultra high rate UASB or a modified conventional fluidised bed. Applications for Biobed include brewery, chemical, fermentation products and pharmaceutical waste-waters. The Biobed process offers several inherent advantages like simplicity, flexibility over other types of anaerobic treatment systems. Because this system is designed to operate at high COD loadings (15–30 kg COD/m^3 reactor volume), it is very space efficient, requiring a smaller size (slender and vertical construction) than a UASB system. The process (although extremely effective), consists of only two major components within the reactor: those being the settlers in the top of the tank, and the feed distribution at the bottom of the tank. High recalculation ratios create inherent hydraulic balancing capacity, and allow for treatment of waste-water containing inhibitory, but biodegradable substances.

Special Designs of Biotreatment Systems

Periodic biological reactors or sequencing batch reactors (SBR)

One of the most recent advances in biological treatment systems is the use of Periodic Biological Reactors: Activated sludge systems are used in various forms today in developed and developing countries. Mostly they operate at 50% efficiency. There are various reasons for this. High investment and maintenance costs are problems in developing countries. Requirement of skilled personnel and difficulties at operator level, variable flow of waste-water causing system disturbance, ineffective recycling of sludge are some of the reasons for poor performance. SBR is a fill and draw reactor containing activated sludge. Conceptually it works in the following stages:

1. Fill period.
2. React period.
3. Settle period.
4. Draw period.
5. Idle period.

All these stages are carried out in a single tank and result is equalisation, organic removal and sedimentation. There are various modifications of SBRs which are in use today. These are :

1. SBRs — Suspended growth.
2. SBBR — Sequencing Batch Biofilm Reactor.
3. GAC — SBBRs (Granular Activated Carbon SBBR).
4. SS — SBRs (Soil Slurry SBRs).

Applications of SBRs are for nitrification and denitrification of secondary effluents, landfill leachates, liquid from land remediation sites. Efficiency of SBRs comes from the following facts:

1. Frequent alternation of factors such as availability of nutrients, electron donors and electron acceptors allow simultaneous enrichment of a variety of microbial strains in multispecies biocommunity.
2. Exploitation of metabolic capabilities of various strains is achieved.
3. Bacteria maintain high metabolic rate even with low substrate concentrations.

The patented Intermittent Cycle Extended Aeration System (ICEAS) process is a cost effective and economical means of providing biological treatment for both municipal and industrial waste-water. The process aerates, settles, and decants in a time sequenced, continuous operation in a single basin. The process combines continuous inflow with intermittent decant. This allows for minimum basin size and eliminates the need for multiple basins. It can handle variety of flows. Plant expansions can be accomplished by installing additional basins in parallel. Since the introduction of this unique waste-water treatment system in the 1970s, ABJ designs have been incorporated into hundreds of facilities world-wide. The world's largest SBR type waste-water treatment plant, located in Kunming, People's Republic of China, began operation in November 1997. The process is designed (a total of 14 basins) to process 300,000 m^3/d (80 mgd) of domestic and light industrial waste-water at the Kunming No. 3 plant.

Membrane Bioreactors

The Membrane bioreactor is an innovative system for the treatment of waste-water, and is used in a compact activated sludge process, which incorporates a membrane separation process for liquid-solid separation instead of the usual settling process. Suspended solids can be removed completely and bacterial free treated water is produced. The sludge concentration (10–40 kg MLSS m^3) and reactor capacity (10–50 kg BOD/m^3/d) are very high in membrane biotreatment (up to 40 times higher than in conventional treatment). Instead of gravity settlement, biomass retention is achieved by a cross-flow filtration process, and even slow growing micro-organisms can be enriched efficiently. The membrane bioreactor offers an improved degradation capacity for persistent chemicals, and xenobiotic compounds, which is often done by bacteria having long generation times. The membrane bioreactor has proved its suitability as an efficient system for degradation of recalcitrant compounds like phenanthrene. Significantly higher biomass concentrations (5 gl^{-1}) and utilisation rates (400 ppm.d^{-1}) than in corresponding batch experiments were achieved. It is not economical to run a membrane bioreactor at low (conventional) treatment capacity. Two types of membrane modules are in use:

1. Submerged membrane module (usually hollow fibre).
2. Cross-flow membrane module.

These modules are incorporated in a plug-flow membrane bioreactor or in a distributed flow reactor. Industrial effluents are often complex in nature. Many a time organic material may be accompanied toxic chemicals, high salt concentration or acidic pH. Inhibitory effects on biodegradation of otherwise degradable matter can be removed by the use of membrane bioreactors. When acid and salt are present along with phenol, the membrane bioreactor can be used where phenol is separated from inorganic components by the use of silicon rubber membrane. Then phenol is transferred to a biological growth medium and degradation is achieved. Similarly, the use of suspended and immobilised cultures to degrade s-triazines in pesticide manufacturing waste-water has been reported. Here waste-water acted as source of nitrogen and glucose or glycerol is added as a source of carbon. Here also 3.5% salts which is otherwise inhibitory to biodegradation are eliminated by membrane separation. The inhibitory effect of inorganic salts is also reported while degrading 3,4 dichloroaniline (3,4 DCA). Effluents from the production of aminonaphthalene and aminonaphthol sulfonic acids also have 8–10% inorganic salts which are inhibitory. Waste-water containing 3-chloronitrobenzene also has pH <1and salts>4% w/w and hinders the biodegradation process. In all these examples membrane separation is useful to remove the inhibitory components. Chemical industry effluents often may have (priority) target pollutant which needs to be eliminated to avoid its adverse effects on the environment. Specialised cultures having capacity to degrade such priority pollutants can be used. But if easily metabolisable substrate is present along with such target specialised pollutant in the effluent, then these specialised cultures divert their metabolism from target pollutant to easily degradable substrate. In such circumstances, specialised cultures in conjuction with membrane bioreactors dedicated to treatment of 'target pollutant' are advocated.

Membrane Anaerobic Reactor System (MARS)

Dorr Oliver had developed and patented a Membrane Sewage Treatment System (MSTS) containing activated sludge process plus ultrafiltration step. In this reactor, membrane rejects solutes of 1–1000 nm. Pressure driven membrane separation takes place. High biomass concentration is achieved. The use of ultrafilter or microporous membrane can be made to retain and maintain high concentration of biomass in the reactor. In contact process, the settling tank is used for the purpose, but settling is not efficient above 10000 mg/l MLSS. Membrane will be successful. Growth of micro-organisms in anaerobic conditions is slow, but this can be overcome by concentration of biomass by a membrane.

SLUDGE PROCESSING

Objectives

Sludges from primary and secondary treatment systems pose a major processing problem in waste-treatment systems. These sludges consist of concentrated unstabilised organics together with inert organics and inert inorganics. The inert sludges must be collected, concentrated, dewatered, and returned to the environment. The biodegradable organics in the sludges create problems and necessitate further processing. Normal processing of sludges consists of thickening to minimise the total volume to be handled and biological treatment or heat treatment for stabilisation.

Thickening

Gravity thickening has been used to concentrate solids. It is possible to thicken primary sludges to 6 to 8 percent and secondary sludges to 2 percent. Since these concentrations can be reached in a well-designed and operated sedimentation tank, there is little value in gravity thickening and some negative benefits. Microbial activity creates odour nuisances along with hydrolysing some of the organics. The return flow returns some solids and soluble organics to the treatment system for removal a second time.

Floatation

Air floatation has proved to be successful in concentrating secondary sludges to about 4 percent concentration. The incoming solids are normally saturated with air at 275 to 350 kPa (40 to 50 psig) prior to being released in the floatation tank. As the air comes out of solution, the fine bubbles collect under the suspended solids and carry them to the surface of the tank. The air bubbles compact the floating solids to the maximum extent. Normally, the air-to-solids ratio is about 0.05 by weight. The thickened solids are scraped off the surface, while the effluent is drawn off the middle of the tank and returned to the treatment system. In large flotation tanks with high flows, the effluent rather than the incoming solids is pressurised and recycled to the influent The size of the floatation tanks is determined primarily by the solids loading rate, directly or indirectly. A solids loading of 25 to 97 kg/(m^2.day) [5 to 20 lb/(ft^2.day)] has been found to be adequate. On a flow basis, this translates into 0.14 to 2.7 L/(m^2.day) [0.2 to 4 gal/(ft^2.min)] surface area. For the most part, air floatation equipment has been developed on a trial and error basis with various equipment manufacturing companies supplying their own units.

Centrifugation

Both basket and solid bowl centrifuges have been used to concentrate waste sludges. Field data have shown that it is possible to obtain 10 to 15 percent solids with waste activated sludge, 15 to 25 percent solids with a mixture of primary and waste activated sludge, and up to 30 to 35 percent solids with primary sludge alone. Centrifuges result in 85 to 90 per cent solids capture with good operation. The problem is that the concentrate contains the fine solids not easily removed. The concentrate is normally returned to the treatment process, where it may or may not be removed. Economics do not favour cetrifuges unless the sludge cake produced is at least 20 to 25 percent solids. For the most part, centrifuges are designed by equipment manufacturers from field experience. With varying sludge characteristics, centrifuge characteristics will also vary widely.

Anaerobic Digestion

Since the organics in both primary and secondary sludges contain biodegradable compounds, concentrated sludges can be treated by anaerobic digestion. Anaerobic digesters are large covered tanks with detention times of 20 to 30 days, based on the volume of sludge added daily. Currently, two anaerobic digesters are constructed of equal size and operated in series so that each unit has a 10- to 15-day retention time. The first digester is heated with an external heat exchange to 35 to 37°C to speed the rate of reaction. Mixing provides good contact between the microbes and the incoming organic solids. Gas mixing and mechanical mixers have been used to provide mixing in the anaerobic digester. The second digester is basically a solids-separation unit and is not normally equipped for either heating or mixing. The supernatant is recycled back to the treatment plant, while the settled sludge is allowed to concentrate to form 6 to 8 percent solids before being returned to the environment.

Anaerobic digestion results in the conversion of the biodegradable organics to methane, carbon dioxide, and microbial cells. Because of the energy in the methane, the production of microbial mass is quite low, less then 0.1 kg/[0.1 lb volatile suspended solids (VSS)/lb] BCOD metabolised except for carbohydrate wastes. The production of methane is 0.35 m^3/kg (5.6 ft^3/lb) BCOD destroyed. Digester gases range from 50 to 80 percent methane and 20 to 50 percent carbon dioxide, depending on the chemical characteristics of the waste organics being digested.

There are two major groups of bacteria in anaerobic digesters: acid forming and methane forming. Acid forming bacteria break down the complex organics to organic acids, which are metabolised by the methane bacteria. It is important that the acid forming bacteria do not produce excess acid; otherwise the pH will fall to the toxic level for methane bacteria. Acid production is best controlled by the addition of organics to the anaerobic digester. A continuous, uniform addition of fresh solids will keep the system in good equilibrium. If continuous addition of solids is not possible, additions should be made at as short intervals as possible. Alkalinity levels are normally maintained at about 3000 to 5000 mg/L to keep the pH above 6.5 as a buffer against variable organic acid production with varying organic loads. Proteins will produce an adequate buffer, but carbohydrates will require the addition of alkalinity to provide a sufficient buffer.

Anaerobic Filter

Anaerobic digestion is limited by the ability of the system to retain the methane bacteria at high levels. The anaerobic filter was developed to retain the methane bacteria on a fixed medium so that high volumes of relatively dilute organic wastes could be treated by anaerobic digestion with the production of methane gas. Considerable research has been done on anaerobic treatment of various types of soluble organic waste-waters. With a fixed medium, fluid retention time is reduced to a few hours or a few days, depending upon the strength of the waste-waters. Waste-waters containing up to 2000 mg/L BCOD can be treated in a few hours, while waste-waters containing up to 10,000 mg/L BCOD require several days' retention. Approximately 90 percent of the BOD is metabolised with proper contact between the microbes and the waste-waters. Recirculation of waste-waters around the filter help provide increased organic reduction by furnishing optimum contact between the microbes and the wastes. Normal recycle rates are 6 to 10 times raw waste-water flow rates.

Aerobic Digestion

Waste activated sludge can be treated more easily in aerobic treatment system than in anaerobic systems. The sludge has already been partially aerobically digested in the aeration tank. For the most part, only about 25 to 35 percent of the waste activated sludge can be digested. An additional aeration period of 15 to 20 days should be adequate to reduce the residual biodegradable mass to a satisfactory level for dewatering and return to the environment. One of the problems in aerobic digestion is the inability to concentrate the solids to levels greater than 2 percent. A second problem is nitrification. The high protein concentration in the biodegradable solids results in the release of ammonia, which can be oxidised during the long retention period in the aerobic digester. Limiting oxygen transfer to the aerobic digester appears to be the best method to handle nitrification and the resulting low pH. High power costs for aerobic digestion help keep the oxygen supply close to the amount required.

Chemical Conditioning

Lime, alum, and various ferric salts have been used to condition sludge prior to dewatering. Lime reacts to form calcium carbonate crystals, which act as a solid matrix to hold the sludge particles apart and allow the water to escape during dewatering. Alum and iron salts help displace some of the bound water from hydrophilic organics and form part of the inorganic matrix. Chemical conditioning increases the mass of sludge to be ultimately handled from 10 to 25 percent, depending upon the characteristics of the individual sludge. Chemical conditioning can also help remove some of the fine particles by incorporating them into insoluble chemical precipitates.

Thermal Conditioning

Heat conditioning of waste sludges was first applied toward the total oxidation of organics, but operating problems resulted in shifting heat treatment from a total oxidation process to a heat conditioning process.

By raising the temperature to 180 to 230°C for 15 to 60 min, it is possible to dewater the remaining solids without the addition of chemicals. To keep the system fluid, the heat treatment system must be operated at a pressure of 1380 to 2070 kPa (200 to 300 lbf/in^2). After the heat reaction, the solids are normally separated in a covered sedimentation tank. The gases over the tank are odorous and generally passed through an incinerator for complete combustion. The liquid supernatant is returned to the treatment process and retreated. Approximately 40 to 50 percent of the organics are returned to the treatment process, creating a very heavy load on the biological units. Economics does not favour thermal conditioning of sludges at the present time.

Vacuum Filtration

Vacuum filtration has been the most common method employed in dewatering sludges. Vacuum filters consist of a rotary drum covered with a cloth filter medium. Various plastic fibres as well as wool have been used for the filter cloth. The filter operates by drawing a vacuum as the drum rotates into chemically condition sludge. The vacuum holds a thin layer of sludge, which is dewatered as the drum leaves the sludge vat, carrying the attached sludge into the air. As the drum rotates the cloth to the opposite side, air pressure jets replace the vacuum, causing the sludge cake to separate from the cloth medium as the cloth moves away from the drum. The cloth travels over a series of rollers, with the sludge being separated by a knife edge and dropping onto a conveyor belt by gravity. The dewatered sludge is moved on the conveyor belt to the next concentration point, while the filter cloth is spray-washed and returned to the drum prior to entering the sludge vat. Vacuum filters yield the poorest results on waste activated sludge and the best results on primary sludge. Waste activated sludge will concentrate to between 12 and 18 percent solids at a rate of 4.9 to 9.8 kg dry cake/ (m^2.h) [1 to 2 lb/(ft^2.h)]. Primary sludge can be dewatered to 25 to 30 percent solids at a rate of 49 kg dry cake/ (m^2.h) [10 lb/ft^2.h)].

Pressure Filtration

Pressure filtration has been used increasingly since the early 1970s because of its ability to produce a drier sludge cake. The pressure filters consist of a series of plates and frames separated by a cloth medium. Sludge is forced into the filter under pressure, while the filtrate is drawn off. When maximum pressure is reached, the influent sludge flow is stopped and the pressure filter is allowed to discharge the residual filtrate prior to opening the filter and allowing the filter cake to drop by gravity to a conveyor belt below the filter press. The pressure filter operates at a pressure between 689 and 1380 kPa (100 and 200 psig) and takes 1.5 to 4 h for the pressure cycle. Normally, 20 to 30 min is required to remove the filter cake. The sludge cakes will vary from 20 to 25 percent for waste activated sludge to 50 percent for primary sludge. Chemical conditioning is necessary to obtain good dewatering of the sludges.

Belt Press Filters

The newest filter for handling waste activated sludge is the belt press filter. The belt press utilises a continuous cloth filter belt. Waste activated sludge is spread over the filter medium, and water is removed initially by gravity. The open belt with the sludge moves into contact with a second moving belt, which squeezes the sludge layer between rollers with ever increasing pressure. The sludge cake is removed at the end of the filter press by a knife blade, with the sludge dropping by gravity to a conveyor belt. Belt press filters can produce up to 20 percent solids.

Sand Beds

Sand filter beds can be used to dewater either anaerobically or aerobically digested sludges. They work best on relatively small treatment systems located in relatively dry areas. The sand bed consists of coarse gravel graded to fine sand in a series of layers to a depth of 0.45 to 0.6 m (1.5 to 2 ft). The digested sludge is placed over the entire filter surface to a depth of 0.3 m (12 in) and allowed to sit until dry. Free water will drain through the sand bed to an open pipe underdrain system and be removed from the filter. Air drying will slowly remove the remaining water. The sludge must be cleaned from the bed by hand prior to adding a second layer of sludge. The sludge layer will drop from an initial thickness of 0.3 m (12 in) to about 0.006 m (1/4 in). An open sand bed can generally handle 49 to 122 kg dry solids/ $(m^2.year)$ [10 to 25 $lb/(ft^2.year)$]. Covered sand beds have been used in wet climates as well as in cold climates, but economics does not favour their use.

SLUDGE DISPOSAL

Incineration

Incineration has been used to reduce the volume of sludge after dewatering. The organic fractions in sludges lend themselves to incineration if they do not have too much water. Multiple hearth and fluid bed incinerators have been extensively used for sludge combustion.

A multiple hearth incinerator consists of several hearths in a vertical cylindrical furnace. The dewatered sludge is added to the top hearth and is slowly pushed through the incinerator, dropping by gravity to the next lower layer until it finally reaches the bottom layer. The top layer is used for drying the sludge with the hot gases from the lower layers. As the temperature of the furnace increases, the organics begin to degrade and undergo combustion. Air is used to add the necessary oxygen and to control the temperature during combustion. It is very important to keep temperatures above 600°C to ensure complete oxidation of the volatile organics. One of the problems with the multiple hearth incinerator is volatilisation of odorous organics during the drying phase before the temperature reaches combustion levels. Even afterburners on the exhaust gas line may not be adequate for complete oxidation. Air pollution control devices are required on all incinerators to remove fly ash and corrosive gases. The ash from the incinerator must be cooled, collected and conveyed back to the environment, normally to a sanitary landfill for burial. The residual ash will weigh from 10 to 30 percent of the original dry weight of the sludge. Supplemental fuels are needed to start the incinerator and to ensure adequate temperatures with sludges containing excessive moisture, such as activated sludge. Heat recovery from wastes is being given more consideration. It is possible to combine the sludges with other wastes to provide a better fuel for the incinerator. A fluid bed incinerator uses hot sand as a heat reservoir for dewatering the sludge and combusting the organics. The turbulence created by the incoming air and the sand suspension requires the effluent gases to be treated in a wet scrubber prior to final discharge. The ash is removed for the scrubber water by a cyclone separator. The scrubber water is normally returned to the treatment process and diluted with the total plant effluent. The ash is normally buried.

Sanitary Landfills

Dewatered sludge, either raw or digested, is often buried in a sanitary landfill to minimise the environmental impact. Increased concern over sanitary landfills has made it more difficult simply to bury dewatered sludge. Sanitary landfills must be made secure from leachate and be monitored regularly to ensure that no environmental damage occurs. The moisture content of most sludges makes them a problem at sanitary landfills designed for solid wastes, requiring separate burial even at the same landfill.

Land Spreading

The nutrient content of most sludges makes them useful as fertilisers or as soil conditioners if properly mixed with the surface soil. Land spreading has gained in popularity in agricultural areas. Normally the rate of application of sludge to land is controlled by the nitrogen content of the sludge. Since nitrogen uptake varies with different crops, nitrogen application is limited to approximately twice the annual uptake of nitrogen by the proposed crop. Approximately one half of the nitrogen is readily available in sludge. Nutrient release with sludge is slower than with chemical fertilisers, allowing the nutrients to become available as the crop needs it. Activated sludge appears to be an excellent soil conditioner because the humus material in the sludge provides a good matrix for root growth, while the nutrient elements are released in approximately the right combination for optimal plant growth. There is a growing concern over heavy metals in some sludges, and care should be taken to minimise heavy metal concentration in sludges placed on the land. Since heavy metals cannot be easily removed from sludge it is important to prevent them from entering the waste-water treatment system. Greater concern will be placed on other potentially toxic or hazardous materials, including some organic compounds such as pesticides and PCBs. Land spreading of sludge requires careful application of the sludge at the surface and its mixing with the soil. Soil microbes will assist in further stabilisation of any biodegradable organics remaining. Land spreading of sludge will become more popular as energy and nutrients become scarcer.

RECYCLING AND REUSE

Water Conservation

Water shortages are increasing in some areas and will increase further as the population grows. Industries will be faced with increased water conservation measures. Less water will be wasted, and waste concentrations will generally increase. In some areas, dual water systems will be employed to permit use of treated waste-water effluents for lawn sprinkling and other non potable uses. Potable water will be carefully regulated to ensure that it is safe for use. Periodic droughts have provided data on methods to conserve limited water resources while minimising the adverse impact on the industry involved. More and more industries are setting up and maintaining regular water conservation programmes.

Waste Segregation

Greater efforts will be made to segregate toxic and hazardous wastes. The statutory requirements of the Central Pollution Control Board are designed to encourage minimum loss of toxic and hazardous chemicals. Many industries will find that it is economical to segregate certain wastes and to reuse them in the process or to keep them segregated as future raw materials. Waste segregation may make it easier to handle certain wastes prior to mixing with carriage waters. There is no single approach to waste segregation. Each industry must examine its own processes and determine exactly where wastes are being generated and how each waste stream can be controlled. Unit operations and their applications in environmental pollution control are already given in Table 4.4

MEMBRANE PROCESSES AND THEIR APPLICATIONS

Membrane technology is well situated by the virtue of its special features to play a major role in industrial and biomedical development. New membrane separation technologies are overcoming the commercialisation barrier. Research and development in membrane technologies are expanding all over the world. Due to continuing interest in the development of novel membranes for various uses, newer membranes are being synthesised as well as explored.

Here we will provide brief overview of such membrane technologies as they have evolved over last five to six years. The emphasis will be on processes, technologies/techniques, membranes and their applications in many fields such as biomedical, industry, separation and in management of environment. Although membrane processes have been studied for more than half a century, they have only been recognised in recent past as important industrial operations. The interest in membrane processes is due from the basic advantage that the solute separation is affected from the solvent without a phase change as compared to conventional methods of evaporation and crystallisation.

Membrane Processes

Membrane processes can be broadly classified on the basis of driving forces namely:

1. Pressure driven processes which work by the application of hydrostatic pressure in reverse osmosis (RO), micro filtration (MF), ultrafiltration (UF), gas permeation, (GP), pervapouration (PV).
2. Voltage driven processes that work by the application of electric potential across the membrane interface in electrodialysis (ED).
3. Concentration driven processes that work by the transmembrane difference in dialysis and liquid membrane processes.

The important membrane processes are listed in Table 4.7.

Table 4.7. Important membrane processes.

Membrane process Preferable permeating component	Separation potential	Driving force released by
Reverse osmosis Solvent	Aqueous low molecular mass solutions, Aqueous solutions	Pressure difference (100 bar)
Ultrafilteration Solvent	Macromolecular solutions, emulsions, Suspensions	Pressure difference (90 bar)
Osmosis Solvent	Aqueous solutions	Concentration difference
Dialysis Solute (ions)	Aqueous solutions	Concentration difference
Electrodialysis Solute (ions)	Aqueous solutions	Electric field
Liquid membrane technique	Aqueous low molecular mass solutions	Concentration difference

Reverse osmosis (RO)

When a hydrostatic pressure exceeding the osmotic pressure is applied to a system that is separated by a semi permeable membrane, the solvent is forced to flow from the higher concentration side to the lower concentration side. Since the flow direction is reverse of osmosis, it is called reverse osmosis.

For solute diffusion equation is

$$N_A = D_A K \frac{C_A{}' - C_A{}''}{l}$$

Where K is Henry's constant and $C_A{}'$ and $C_A{}''$ are concentrations of solute in upstream and downstream respectively. For the solvent, a permeation equation holds:

$$N_B = \frac{Q}{l}[\Delta P - \sum_{i \neq B} \pi i]$$

where i is the osmotic pressure due to solute i, and $\hat{E}$ is the applied hydrostatic pressure difference.

Ultrafiltration (UF)

Ultrafiltration is a separation process in which large molecules or colloidal particles are filtered from the solution by means of a suitable membrane. Ultrafiltration is frequently used as a synonym for reverse osmosis. Nevertheless it is possible and convenient to distinguish these two processes. Michaels (scientist) proposed to use reverse osmosis for membrane separation involving solutes whose molecular dimensions are within one order of magnitude of those of the solvent and ultrafiltration to describe separations involving solutes of molecular dimensions greater than 10 solvent molecular diameters and below the limit of resolution of the optical microscope (0.5). For both the processes hydrostatic pressure forces the solvent to permeate a membrane keeping solute on one side of the membrane. Both the processes, require external application of energy as hydrostatic pressure to accomplish separation of components, whereas dialysis or a regular diffusion process does not require such force, since separation takes place spontaneously.

Osmosis

The term was originally used to represent the transport of solute as well as solvent across membrances. Dutrochet (scientist) distinguished between two different transport phenomena as end osmosis and exosmosis. Today it is called dialysis or diffusion. Osmosis is diffusion process where only the solvent permeates through the membrane from the less concentrated side to more concentrated solution. If the solute is transferred then it is called dialysis. Hydrostatic pressure is the driving force in any osmosis process. In 1885 Vant Hoff (scientist) showed that the osmotic pressure π_A in dilute solution is related to the concentration of the solute C_A by

$$\pi_A = RTC_A$$

For solutions of higher concentration of the solutes, the following equation is used to evaluate the osmotic pressure

$$\pi_A = \frac{RT}{\vartheta_B} \int_0^{C_B} \left(\frac{\partial^{ln} Q_B}{\partial C_B} \right)_{P,T} dC_A$$

where Q_B and C_B are partial molar volume and the activity of the solvent respectively. If the observed osmosis is either greater or lesser than that predicted by Vant Hoff's law, it is called anomalous osmosis.

Dialysis

Dialysis is basically a different process while diffusion refers to phenomenon itself. Dialysis ascribes the separation of substance in solution by means of the unequal diffusion rates through the membrane. Dialysis is the first membrane process of separation. Graham (scientist) was successful in separating a colloidal solution using a device called dialyser consisting of a membrane containing the open end of a cylinder immersed in a container of water. The cylinder contained a solution to be dialysed, the dialysing cell was kept in a place until the desired separation was achieved. Later on, equilibrium dialysis cell consisting of two interchangeable compartments separated by a membrane was developed. The governing equation in dialysis is the diffusion gas equation.

$$W = D_0 S \frac{C_1 - C_2}{l}$$

Electrodialysis (ED)

Electrodialysis is a process in which solute ions move across membrane by application of an electric field. The electrodialysis requires an external energy to maintain the separation process as electric potential. The application of electric energy will drive electrolytes from a dilute to a more concentrated solution. Ion selective membranes are employed in electrodialysis. Electrodialysis involves two separate driving forces that are concentration difference and electrical potential difference. The net transport of an ion can be expressed as the sum of two flow rates.

$$\frac{V}{St} = \frac{b_1 \, \Delta E}{R'} + \frac{b_2 D \, \Delta C}{l}$$

Where b_1 and b_2 are the proportionality constants, ΔE is the electric cell potential, R is the electric resistance of cell, V is the volume flow and St is the area at time, 't'. The first term represents the ion flux driven by the electric potential that is directly proportional to the electric current across the membrane. The second term is the flow rate due to simple diffusion under a concentration gradient. If the applied potential gradient is in the direction of concentration, positive sign (+) is used, if opposite, (–) sign is used.

Liquid membrane technique

Ionic separations by liquid membrane technique have appeared recently in separation process scenario. Various types of liquid membranes that can be used as effectively are :
1. Bulk liquid membrane.
2. Thin sheet supported liquid membrane.
3. Hollow fibre supported liquid membrane.
4. Emulsion liquid membrane.

The techniques offer certain distinct advantages e.g., lower operating cost and space requirements, low energy consumption and low solvent inventory that eventually provides opportunities for economic use of even exotic and expensive reagents. It proved to be superior than conventional methods. Recently unified liquid membrane has attracted increasing attention and acclaimed to be a novel separation technique. These are based on the principles of liquid membranes, solvent extraction and dispersion technique.

Hybrid Processes

Recently a variety of hybrid processes are emerging where a traditional separation process and membrane separation process is used in an integrated fashion to achieve what cannot be attained economically by either separation technique alone. Alternatively, a hybrid process is needed since the individual technique may have a severe deficiency. Separation concepts have also emerged wherein two separation mechanisms are simultaneously operative within one device, such a hybrid process may be called a composite separation process.

Membrane Materials

There are over 130 different materials in the patent and scientific literature from which membranes have been made; many more are being added every year. Some of the membrane materials and processes are listed in Table 4.8.

Table 4.8. Some membrane materials and processes.

Material	Processes
Cellulose acetate	ED, UF, RO
Cellulose triacetate	UF, RO
Polyacrylonitrile	UF
Polyacrylonitrile-polyvinylchloride copolymer	UF
Polyamide (Aromatic)	UF, RO
Polybenzimidazole	RO
Polyimide	UF, RO
Polyvinylidene fluoride	UF

Pore and pore size distribution play an important role in determining the transport phenomenon and ultimate quality of the product. RO membranes have a pore size range of 0.0001 to 0.001 m whereas microfiltration membranes have pore sizes 0.02 μ to 10 μ.

Applications of Membrane Processes

Membrane processes have enjoyed increasing popularity because of its applications in various industrial and biomedical fields.

Desalination of brackish water

With increasing industrial activity, water requirement for industrial use as well as drinking purpose is rising day by day. The salinity of ground water has also increased due to various reasons. Nuchem Weir has put up Asia's largest desalination plant using RO for production of process water from the treated sewage water at the rate of 360 m^2/h. Thus the potential to develop the alternative source of water for industrial use as well as for drinking requirement is enormous and membrane processes can be put to use to conserve this invaluable resource on this planet.

Natural colour concentrates

Synthetic colours are being banned for use in food and beverage preparations. Attempts are being made to develop natural colorants. The natural colorants used in food are anthocyanins, betamines and carthamines. Natural colour extracts contain these colorants from 0.1 to 0.2% and 6–8% of dissolved solids. This demands higher quality use of these natural colour extracts. UF and RO processes can be used to prepare high concentration of natural colouring material for more practical industrial applications.

Pollution control

RO, UF and ED find wide application in pollution control. Requirement of fresh water by industries is increasing and estimated to be about 30×10^6 m^2/day. With the increased requirement, effluent discharge will also increase in volume and contaminants level. Membrane processes can play a vital role in reducting discharge by providing recyclable water. Some of the processes that can be converted to zero effluent discharge process through membrane processes are: (i) pulp and paper; (ii) sugar and distilleries; (iii) electroplating; (iv) petrochemical effluents; (v) chlor alkali industry; (vi) cooling tower and blow down tower.

Food technology

India is an agricultural land and traditional methods of food preparation are still being followed in processing of food products. A look around the world shows a mixed approach and change from traditional processing. Though the applications have been few, yet the potential is very large and will provide energy saving, processing without phase change and higher recovery of quality products. Some of the potential areas in food technology are: (i) UF and RO in concentration of fruit juice and milk; (ii) RO in manufacture of khoa; (iii) UF, RO and ED in concentration of cheese.

Membrane reactors

Interest is now growing in high performance fermentation processes that are continuous, operate at high dilution rate, maintain high cell densities (about 100 gm/lit) and permit unwanted products to be removed from the system. In a membrane bioreactor, the main reaction vessel is coupled to a membrane module. The membrane module separates the cells from the product and recycles it back to the fermenter. The process has a great promise.

Ion selective electrodes

Compact membranes or liquid membranes are the basic components of analytical devices, namely, ion selective electrodes. Glass electrodes with a suitable glass composition can be used for pH determination in alkaline media. Devices based on the glass electrode can be used to determine certain gases present in gases or liquid phase. Properties analogous to those of glass electrode are exhibited by a number of ion selective electrodes, with fixed ion exchange sites. The membrane in these electrodes is a layer of solid electrolyte. The most important of this group is fluoride ion selective electrode. The electrode has many applications—from fluoride determination in water to analysis of saliva in tooth enamel.

Artificial kidneys

The most widely known use of synthetic membranes today is that of artificial kidneys. The human kidney has a number of important functions such as excretion of products from the blood, regulation of water, maintenance of acid base balance, regulation of blood pressure and formation of red blood cells. The first three functions can be closely imitated by artificial kidneys. Such kidney consists of a haemodialyser and its supporting equipment. Haemodialyser is a membrane containing apparatus to dialyse the blood. It is a one step operation, whereas natural kidney works in a two step operation to eliminate waste materials and keep the balance of electrolytes and water.

Drug encapsulation

The technique of drug encapsulation has received much attention in recent years. The more conventional way of drug encapsulation, administration, injection and oral tablets frequently causes undesirable side effects or the concentration of drug in the body increases sharply yielding a peak that is responsible for side effects. In other situations, it is desired to have a large quantity of drugs in a body and be released slowly over a long period of time. Examples are contraceptive devices containing progesterone and hormone implants to make up the deficiencies caused by the endocrine failure. Numerous polymeric materials have been studied for implants. The basic requirements for the material are biocompatibility and high permeability for drugs. It is well known that silicon rubber is highly bio compatable and also possesses very high permeabilities for many drugs and hormones. A recent study of testosterone encapsulation by silicon rubber membrane shows that the release rate can be effectively controlled by changing the thickness membrane and also modifying the membrane structure.

Dialysis by semipermeable microcapsules

The micro-encapsulation technique is used in micro-dialysis. A procedure has been developed in which semipermeable microcapsules are prepared with an ultra thin semipermeable membrane to permit rapid mass transfer. In ordinary haemodialysis, an important concern is to attain a large membrane area in a small bulk volume. This is easily achieved by micro-encapsulation.

The microdialysis can also be readily combined with other separation processes such as adsorption or catalysis, or both. When some adsorbents i.e., activated charcoal, ion exchange resins, activated zirconium charcoal and zirconium phosphates are coated with a thin membrane, it permits only the permeation of toxic materials without interacting with plasma proteins and platelets. The removal of creatinine and uric acid was very efficient with micro-capsules of activated charcoal, combined action of micro-dialysis and catalysis is employed to remove urea.

This was accomplished by adding selective adsorbents such as ion exchange resins or zirconium phosphate. Besides haemodialysis, the micro-encapsulation technique also demonstrates its usefulness and flexibility in other separation processes. By hoosing appropriate membranes, adsorbents, and catalysis, one can achieve various degrees of separation for many different systems. In conclusion, membrane technology and science has tremendous potential and bright future to serve the needs of humankind and help to preserve out environment.

CHAPTER 5

Instrumental Methods of Chemical Analysis

INTRODUCTION

The sub-division of chemistry concerned with identification of materials (qualitative analysis) and with determination of the percentage composition of mixtures or the constituents of a pure compound (quantitative analysis). The gravimetric and volumetric (or "wet") methods (precipitation, titration, and solvent extraction) are still used for routine work; indeed, new titration methods have been introduced, e.g., cryoscopic, pressure-metric (for reactions that produce a gaseous product), redox methods, and use of a fluoride-sensitive electrode. However, faster and more accurate techniques (collectively called instrumental) have been developed in the last few decades. Among these are—infra-red, ultra-violet, and x-ray spectroscopy, where the presence and amount of a metallic element is indicated by lines in its emission or absorption spectrum; colorimetry, by which the percentage of a substance in solution is determined by the intensity of its colour; chromatography of various types by which the components of a liquid or gaseous mixture are determined by passing it through a column of porous material, or on thin layers of finely divided solids; separation of mixtures in ion-exchange columns; and radioactive tracer analysis. Optical and electron microscopy, mass spectrometry, microanalysis, nuclear magnetic resonance (NMR), and nuclear quadrupole resonance (NQR) spectroscopy all fall within the area of analytical chemistry.

VOLUMETRIC METHODS

Titrimetry

Qualitative analysis deals with the determination of the nature of the constituents of a given material. On the other hand, *quantitative* analysis deals with the determination of the proportions of the various constituents of a given material. One of the most important branches of quantitative analysis is volumetric or *titrimetric* analysis which depends on the accurate measurements of volumes of solutions undergoing a chemical change. In short, a volumetric experiment consists in dissolving a known weight of a substance in water and making up the solution to an exact volume. A measured volume of this solution is completely reacted with another solution in presence of a substance called indicator. The volume of the latter solution is determined. The whole of this process is known as titration. Knowing the volume of the reacting solutions, the results are calculated. *Titrimetric* analysis is a quick process and thus has the major advantage over the other types of analysis such as gravimetric analysis.

It may be pointed out that *titrimetric* analysis can be carried out only if a suitable reaction is possible between the reactants, e.g., an acid and a base. A volumetric experiment between the reactants must satisfy the following conditions :

1. A suitable reaction should be possible between the substance to be taken in the burette and the titration flask, e.g., a reaction between an acid and a base; an oxidising agent and a reducing agent.

2. There should be only one reaction occurring between the solutions taken in the burette and the titration flask. There should be no side reaction which may complicate the process of calculations.

3. The reaction between the solution taken in the burette and the titration flask should be complete within a reasonable time, i.e., reaction should be quite rapid.

4. The reaction between the solution taken in the burette and the titration flask should be simple so that a definite weight relationship exists between the reactants.

5. The reaction between the solution taken in the burette and the titration flask should be possible at about room temperature, i.e., at not very high temperature; in dilute solutions and not any very special conditions.

6. A suitable indicator must be available which can help to locate the exact end point.

Keeping the above points in view, chemists have developed the following types of volumetric analysis:

1. Acid base titrations (acidimetry and alkalimetry).

2. Oxidation-reduction titrations (redox titrations).

3. Precipitation titrations.

4. Complexometric titrations.

Let us discuss the theory of the above types of titrations.

Acid-Base Titrations

It is a well known fact that an acidic solution has a pH value of less than 7 while an alkaline solution has a pH value greater than 7. Therefore, when an acidic solution is reacted with an alkaline solution, pH value changes. These changes in pH value can be determined by any one of the following methods:

1. *Visual method:* Organic compounds such as methyl orange, phenolphthalein, methyl red and thymol blue are sensitive to pH value and have different colours at different pH values. Using any one of these compounds as an indicator, it is possible to determine the changes in pH value and thus the end point.

2. *Electrical conductivity measurements:* We know that electrical conductivity depends upon the number and mobility of the ions. As the acid and base undergo neutralisation, electrical conductivity changes which helps to locate the end points. Titrations based on electrical conductivity measurements are called conductometric titrations.

3. *E.M.F. measurements:* Since the electrical potential of hydrogen electrode varies with the concentration of H^+ ions (i.e., pH value) with which it is in contact, it is possible to determine the changes in pH when an acid is neutralised by a base by measuring e.m.f. Such type of titrations which are based on the measurement of electrical potential are called potentiometric titrations.

Quite commonly acid-base reactions are studied in the laboratory using organic compounds known as indicators. Examples of commonly used indicators in acid-base titrations are methyl orange, phenolphthalein or methyl red.

Oxidation-Reduction (or Redox) Titrations

Oxidation involves loss of electrons while reduction involves gain of electrons. Therefore, it is possible to titrate an oxidising agent (which will accept electrons) against a reducing agent (which will lose electrons). Such type of reactions in which oxidation and reduction occur at the same time are known as redox reactions and titrations involving redox reactions are known as redox titrations. Some of the commonly used oxidising and reducing agents in the redox titrations are as follows:

1. Oxidising agents
   (a) $KMnO_4$ and dil. H_2SO_4
   (b) $K_2Cr_2O_7$ and dil. H_2SO_4
   (c) Ceric sulphate $[Ce(SO_4)_2]$ and dil. H_2SO_4
   (d) Iodine solution
   (e) Potassium iodate (KIO_3) in HCl solution
2. Reducing agents

   (a) Ferrous sulphate, $FeSO_4.7H_2O$
   (b) Ferrous ammonium sulphate (or *Mohr's salt*) $FeSO_4.(NH_4)_2SO_4.6H_2O$

   (c) Oxalic acid $\begin{vmatrix} COOH \\ \\ COOH \end{vmatrix} .2H_2O$

   (d) Oxalates such as potassium oxalate
   (e) Sodium thiosulphate
   (f) Potassium iodide
   (g) Arsenious oxide, As_2O_3
   (h) Antimonius compounds such as tartar emetic.

There is no universal oxidising agent which can be titrated against every reducing agent and vice-versa. Hence, the choice of an oxidising agent to be used against a particular reducing agent depends upon the reaction conditions and standard reduction potential of the oxidising agent. Therefore, it will be useful to give a summary of the various type of redox titrations.

Permanganometry

Permanganate ion in acidic solution is a strong oxidising agent and its action can be represented by the following reaction:

$$MnO_4^- + 8H^+ + 5e^- \rightleftharpoons Mn^{2+} + 4H_2O; \; E^\circ = 1.15 \text{ V}$$

[*Note:* In permanganometry, potassium permanganate is used as an oxidising agent as it is easily available in good purity].

It is evident from the above equation that the equivalent weight of $KMnO_4$ is one-fifth of its formula weight, i.e.

$$\frac{39 + 55 + 64}{5} = 31.6$$

Permanganate titrations are carried out in presence of dil. H_2SO_4 as it has no action on permanganate ion. Hydrochloric acid is not used in permanganate titrations particularly because HCl is liable to be oxidised to chlorine by $KMnO_4$. Then more of $KMnO_4$ will be consumed in the redox titration. It is also not advisable to use dil. HNO_3 in these titrations because HNO_3 is itself an oxidising agent and will interfere with the oxidation action of $KMnO_4$.

Since potassium permanganate has dark pink colour, it itself acts as an indicator and the end point is the appearance of a permanent pink colour in the solution.

The reducing agents which are generally titrated against acidified $KMnO_4$ are ferrous salt such as ferrous sulphate, ferrous ammonium sulphate known as *Mohr's salt*, oxalic acid, oxalates such as potassium oxalate and ferrous oxalates.

It may be mentioned that ferrous salts, are titrated against acidified $KMnO_4$ at room temperature. However, oxalic acid and oxalates are titrated while keeping the oxalic acid, oxalate solutions quite warm (around 60–70°C). This is due to the fact that oxalic acid and oxalates on oxidation with acidified $KMnO_4$ produce carbon dioxide according to the equation:

$$\begin{array}{c} COOH \\ | \\ COOH \end{array} + [O] \longrightarrow 2CO_2 + H_2O$$

In order to expel the CO_2 gas produced in the above reaction, the reaction mixture has to be kept hot so that the reaction proceeds in the forward direction.

It is also possible to estimate ferric salts by carrying out titration with acidified $KMnO_4$ after reduction of ferric salts with zinc and sulphuric acid.

Moreover, non-reducing cations like Ca^{2+}, Zn^{2+} which form precipitated oxalates can also be estimated by dissolving the washed precipitate in dilute sulphuric acid and then titrating the resulting solution against acidified $KMnO_4$.

Dichrometry

Acidified potassium dichromate is also a powerful oxidising agent though not as powerful as a potassium permanganate. However, potassium dichromate has the following advantages over potassium permanganate as an oxidising agent.

1. It is obtainable in very pure state so that it serves as an excellent primary standard.
2. Its aqueous solutions are stable for a long time as its solution is not decomposed by light.
3. Its solution has no action on organic matter such as rubber so that its solution can be used even in burettes having rubber taps.
4. Its titrations can be carried out even in presence of hydrochloric acid provided the concentration of the acid is not more than 20%.

Acidified dichromate acts as an oxidising agent according to the following equation:

$$Cr_2O_7^{2-} + 14H^+ + 6e^- \rightleftharpoons 2Cr^{3+} + 7H_2O$$

It is evident from the above equation that the equivalent weight of $K_2Cr_2O_7$ is one-sixth of its formula weight, i.e.,

$$\frac{2 \times 39 + 2 \times 52 + 7 \times 16}{6} = 49$$

The end point in potassium dichromate titrations is determined by any one of the following methods.

By the use of external indicator

Potassium dichromate cannot act as its own indicator because it forms green coloured chromic salts during redox reaction. Therefore, potassium ferricyanide is used as external indicator during potassium dichromate titrations. On a white tile, drops of the potassium ferricyanide solutions are placed with a glass rod. During titration a drop of the solution of ferrous ions taken in the titration flask is touched with the drop of the indicator on the tile.

The development of blue colour indicates the presence of ferrous ions according to the reaction.

$$2K_3[Fe(CN)_6] + 3Fe^{2+} \xrightarrow[\text{Blue}]{} Fe_3[Fe(CN)_6]_2 + 6K^+$$

At the end point, there will be no more ferrous ions so that no blue colour is produced rather a brownish yellow colour is formed.

By the use of internal indicator

Since the use of external indicator is quite cumbersome, analytical chemists have developed internal indicators for dichromate titrations. These are:

1. 1% solution of diphenylamine in conc. H_2SO_4; or
2. 1% solution of diphenylbenzidine in conc. H_2SO_4; or
3. 0.2% aqueous solution of sodium diphenylamine sulphonate.

With the above indicators, phosphoric acid is always used which lowers the oxidation potential of ferrous-ferric system by forming a complex $[Fe(HPO_4)]^+$. The above indicators impart green colour to the ferrous ions which changes to intense-purple or-violet at the end point.

However, now-a-days *N*-phenylanthranilic acid is used as indicator for dichromate titrations as with this indicator there is no need to add phosphoric acid. In this case at the end point, the colour changes from green to violet red.

Redox reactions involving iodine

An aqueous solution of iodine is a mild oxidising agent according to the following equation.

$$I_2 + 2e^- \rightleftharpoons 2I^-$$

Therefore, an aqueous solution of iodine is used to estimate reducing agents such as sodium thiosulphate, sodium sulphite, arsenites and antimonites according to the reaction:

$$2S_2O_3^- + I_2 \rightleftharpoons S_4O_6^- + 2I^-$$

$$SO_3^{2-} + I_2 + H_2O \rightleftharpoons SO_4^{2-} + 2H^+ + 2I^-$$

$$AsO_3^{2-} + I_2 + H_2O \rightleftharpoons AsO_4^{3-} + 2H^+ + 2I^-$$

$$SbO_2^- + I_2 + H_2O \rightleftharpoons SbO_3^- + 2H^+ + 2I^-$$

All these titrations which involve the use of standard iodine solution for the estimation of a reducing agent are called iodiometric titrations.

On the other hand, it is possible to estimate active oxidising agents like permanganates, dichromates, hydrogen peroxide and cupric ions by treating with excess of potassium iodide to liberate iodine. The liberated iodine is then estimated by titration against standard sodium thiosulphate solution. The following reactions are involved in these titrations:

With acidified $KMnO_4$ solution

$$2KMnO_4 + 3H_2SO_4 \longrightarrow K_2SO_4 + 2MnSO_4 + 3H_2O + 5(O)$$

$$2KI + Of + H_2SO_4 \longrightarrow K_2SO_4 + H_2O + I_2$$

With acidified $K_2Cr_2O_7$ solution

$$K_2Cr_2O_7 + 4H_2SO_4 \longrightarrow K_2SO_4 + Cr_2(SO_4)_3 + 4H_2O + 3(O)$$

$$2KI + OI + H_2SO_4 \longrightarrow K_2SO_4 + H_2O + I_2$$

With copper sulphate solution

$$CuSO_4 + 2KI \longrightarrow CuI_2 + K_2SO_4$$
$$2CuI_2 \longrightarrow Cu_2I_2 + I_2$$

All those titrations in which an oxidising agent is estimated by first reacting with potassium iodide and estimating the liberated iodine with standard sodium thiosulphate solution are known as iodiometric titrations.

In the titrations involving iodine solution, freshly prepared starch solution is used as an indicator which forms blue coloured complex with iodine. It may be pointed out that in iodometry, starch solution indicators may be added at any time while in iodiometric titrations, starch solution is added only when the solution is faintly yellow indicating the close approach of the end point.

Precipitation Titrations

There is another class of reactions which involve the formation of a precipitate when a solution taken in the burette reacts with a solution taken in the titration flask. Such type of titrations are known as precipitation titrations. The best example of such type of titrations is the reaction between silver nitrate solution and a halide solution such as sodium chloride solution.

$$AgNO_3 + NaCl \longrightarrow AgCl \leftarrow + NaNO_3$$

Titrations involving the use of silver nitrate to estimate chloride, bromide, iodide or thiocyanate content in a solution are called argentometric titrations.

Complexometric Titrations

There are many reactions which involve the formation of a soluble undissociated stoichiometric complex at the end point when a substance taken in burette reacts with a substance taken in the titration flask. Such type of titrations are called complexometric titrations. For example, a reaction between silver nitrate solution and sodium cyanide solution results in the formation of a very stable and soluble complex Na $[Ag(CN)_2]$ at the end point.

$$AgNO_3 + 2NaCN \longrightarrow Na[Ag(CN)_2] + NaNO_3$$
$$\text{Complex}$$

In the complexion $[Ag(CN)_2]^-$, silver is the central atom while cyanide ions are attached to it are called ligands. Since two cyanide ions are attached to silver, it means co-ordination number of silver in this complex is two. In general, ligands contain highly electronegative atoms such as nitrogen, oxygen or halogens. Such ligands contain lone pairs of electrons which act as donor sites. Some metal ions such as aluminium, bismuth and lead easily form complexes with ligands having oxygen donor sites. However, metal ions like iron, cobalt, nickel, copper, zinc, cadmium and mercury form stable complexes with ligands containing nitrogen donor atoms. In some cases, the central metal ion links with donor atoms of the ligand in such a way that a ring is formed. Such a ligand is called chelating agent. Complexes formed by chelating agents are more stable than those formed by non-chelating agents. Commonly used chelating agents in analytical chemistry are EDTA (Ethylenediamenetetracetic acid), DMG (Dimethylglyoxime) and 8-hydroxyquinoline (known as oxine). In volumetric analysis, EDTA finds numerous applications as it forms very stable complexes with metal ions such as magnesium and calcium. Titrations involving chelating agents are called chelometric titrations. However, titrations involving the use of EDTA are known as EDTA titrations.

EDTA is commonly known as complexone III, Sequesterene, Versene, Nullapon etc. It is a polyprotic acid and so its abbreviated formula is H_4Y. It dissolves readily in water and is obtainable in high degree of purity. It can co-ordinate with metal ions through its two nitrogen atoms and four oxygen atoms of the carboxyl groups.

Indicator in EDTA titrations

In these titrations, Eriochroma Black T (Erio T) is used as indicator. Erio-T has the interesting property that it can also form complexes with metal ions through its oxygen and nitrogen donor sites but the complexes thus formed are less stable than metal-EDTA complexes. Consequently, the indicator releases metal ions at the end point and then undergoes change in the colour.

Most of the EDTA titrations are performed in presence of buffers having pH 8 to 10 wherein the indicator has blue colour. Therefore, the titration of a solution containing magnesium ions in presence of a buffer (such as NH_4Cl in NH_4OH having pH 10) against EDTA using Erio Black T changes its colour from red to blue at the end point.

Advantages of EDTA titrations

1. EDTA forms stable and soluble complexes which have definite composition as compared to other titrants.
2. EDTA is commonly used in the form of its disodium salt having the abbreviated formula Na_2H_2Y which is itself a primary standard.
3. Since EDTA forms soluble complexes, the end point is reached quite readily.

Indicator Theories

Indicator is a substance which indicates by its colour change the completion of a reaction. For example, in the neutralisation reaction between sodium hydroxide and hydrochloric acid (taken in the burette), phenolphthalein turns pink to colourless. Now let us try to find out as to why do the indicators change colour at the end point.

This can be best explained with reference to acid-base titrations. It has been found out that the indicators used in neutralisation titrations are complex organic compounds which are themselves weak acids or bases. Therefore, like other weak acids and bases, they associate or dissociate according to pH of the solution. Hence, they change their colour with change in pH as shown in Table 5.1.

Table 5.1. Colour changes of indicators with pH.

Indicator	pH range	Colour	
		Acid solution	*Alkaline solution*
Thymol blue	1.2–2.8	Red	Yellow
Methyl orange	3.1–4.4	Red	Yellow
Methyl red	4.2–6.3	Red	Yellow
Litmus	5.5–7.4	Red	Blue
Phenolphthalein	8.3–10.0	Colourless	Red

The following two theories have been put forward to explain the colour change of indicators at the point in neutralisation reactions.

Ostwald theory

The main points of this theory are as follows:

1. An indicator is a weak acid or a weak base.
2. Ionisation of the indicator causes change in its colour. In other words, the unionised molecules of the indicator possess different colour than the ions.
3. Being weak electrolytes, indicators are feebly ionised in solution. However, the addition of a strong acid or strong base increases their ionisation considerably which causes change in the number of coloured ions.
4. An acidic indicator such as phenolphthalein must possess a coloured anion while a basic indicator such as methyl orange must possess a coloured cation.

In the light of this theory, let us explain the behaviour of phenolphthalein as well as methyl orange in neutralisation reactions.

Phenolphthalein as an indicator

Phenolphthalein is a weak organic acid having the formula HPh. Being a weak acid, it exists mostly as unionised molecules. These unionised molecules are colourless while on ionisation it produces H^+ ions (colourless) and Ph^- ions (pink) according to the reaction:

$$HPh \rightleftharpoons H^+ + Ph^- \qquad \qquad ...(5.1)$$
$$\text{(Colourless)} \quad \text{(Colourless)} \quad \text{(Pink)}$$

Since the reaction (5.1) is in equilibrium, the addition of an acid (or H^+ ions) increases the H^+ ion concentration and shifts the equilibrium in the backward direction making the solution colourless. Thus in acidic solution phenolphthalein remains colourless.

On the other hand, addition of a base like NaOH solution, produces OH^- ions which react with H^+ ions to produce feebly ionised water molecules as shown below:

$$HPh \rightleftharpoons H^+ + Ph^-$$
$$NaOH \rightleftharpoons OH^- + Na^+$$
$$\Updownarrow$$
$$H_2O \text{ (Freely ionised)}$$

Consequently, the equilibrium of reaction (5.1) is shifted in the forward direction resulting in the formation of greater number of Ph^- ions which are pink coloured. Therefore, in basic solution phenolphthalein gives pink colour.

This theory could explain as to why a weak base like NH_4OH cannot be titrated against a strong acid like HCl using phenolphthalein as indicator. This is because a weak base like NH_4OH is feebly ionised and hence produces very small number of OH^- which are insufficient to shift the equilibrium of reaction (5.1) towards the right until a large excess of NH_4OH is added. Therefore, solution remains colourless, i.e., phenolphthalein fails to work as an indicator for such titrations which involve weak bases.

Methyl orange as an indicator

Methyl orange is a weak organic base having the formula MeOH. Being a weak base, it exists mostly as unionised molecules. These unionised molecules have yellow colour while on ionisation. It produces Me^+ ions (red) and OH^- ions (colourless) according to the reaction:

$$MeOH \rightleftharpoons Me^+ + OH^-$$
$$\text{(Yellow)} \quad \text{(Red)} \quad \text{(Colourless)}$$

Since the reaction (5.2) is an equilibrium, the addition of a base like NaOH increases the concentration of OH⁻ ions and thus pushes the equilibrium in the backward direction making the solution yellow in colour. Therefore, in basic solution methyl orange gives yellow colour.

On the other hand, addition of an acid like HCl solution, produces H^+ ions which react with OH^+ (already present in solution due to reaction 5.2) to produce feebly ionised water molecules as shown below:

$$MeOH \rightleftharpoons Me^+ + \quad OH^-$$
$$HCl \rightleftharpoons Cl^- + \quad H^+ \qquad \qquad ...(5.3)$$

$$\Updownarrow$$

$$H_2O \text{ (freebly ionised)}$$

Consequently, the equilibrium of the reaction (5.2) is shifted in the forward direction resulting in the formation of greater number of Me^+ ions which are red coloured. Therefore, in acidic solution, methyl orange gives red colour.

This theory could explain as to why methyl orange cannot be used as an indicator in neutralisation reactions involving a weak acid like acetic acid (CH_3COOH). This is because a weak acid like acetic acid produces a very small number of H^+ ions which are insufficient to shift the equilibrium of reaction (5.2) to the forward direction until a very large excess of the acetic acid is added. Therefore, methyl orange fails to work as an indicator in such titrations which involve weak acids.

Quinonoid theory

The main points of this theory are as follows:

1. The acid-base indicators are organic aromatic compounds which can exist in at least two tautomeric forms.
2. The tautomeric forms of the indicator exist in equilibrium with one another.
3. One of the tautomeric form of the indicator can exist in an acid solution while the other form can exist in an alkaline solution.
4. The two tautomeric forms possess different colours depending upon the pH value of the solution. Therefore, as the pH of the solution changes, the colour of the solution changes due to conversion of one form into the other.
5. The two tautomeric forms possess different structures known as *benzenoid* (benzene-like) form and *quinonoid* (quinone-like) form.
6. The *quinonoid* form is usually deeper in colour than the *benzenoid* form.

On the basis of this theory, let us explain the colour changes with change in pH using phenolphthalein or methyl orange as indicator.

Phenolphthalein as indicator

Phenolphthalein is a diprotic acid which is colourless in acidic medium as it possesses benzenoid structure. It gives pink to deep red colour in basic medium while it is colourless in acidic medium.

Methyl orange as indicator

Methyl orange is an organic base which has a yellow colour in the benzenoid form. On the addition of an acid, it forms a cation which exists in quinonoid form wherein it has pink colour.

Advantage of quinonoid theory

This theory has the advantage that it could explain not only the behaviour of indicators in acid-base titrations but even in redox titrations. For example, diphenyl amine is used as an indicator in dichromate titrations. In presence of reducing agent diphenylamine is colourless having benzenoid structure. On oxidation, it changes into diphenyl-benzidine.

Errors in Titrimetry and their Rectifications

In titrimetrical analysis, the following types of errors can crepe in:
1. *Operational errors:* They arise due to the individual analyst who may not be following proper analytical techniques. Examples are:
   (a) Improper washing of the apparatus such as burette, pipette and titration flask.
   (b) Use of improper quantity of the indicator in each titration.
   (c) Improper weighing of the reagents.
2. *Personal errors:* They arise from constitutional inability of an individual analyst to make correct observations. Examples are:
   (a) Due to colour blindness, a person may not be able to judge the correct end point in a visual titration.
   (b) An analyst having small height may not be able to correctly read the burette.
3. *Instrumental errors:* They arise from faulty construction of the balances, uncalibrated weights, uncalibrated burettes and pipettes.
4. *Reagent errors:* They arise from use of reagents which are substandard, i.e., impure acids, bases and indicators.
5. *Errors of method:* These errors arise due to faulty choice of method of analysis. Examples are:
   (a) Incompleteness of a reaction. For example, incomplete neutralisation between an acid and a base.
   (b) Occurrence of side reactions. For example, in titration between liberated iodine from $CuSO_4$ solution and sodium thiosulphate solution, atmospheric oxygen may take part in the reaction and cause a side reaction.
   (c) Reaction of substances other than the constituent being determined may take place. For example, water containing oxygen may oxidise ferrous sulphate solution rather than being oxidised by acidified $KMnO_4$.

Rectification of errors in titrimetry

1. Operational errors can be minimised by carefully washing the apparatus; weighing the reagents correctly; and using the same quantity of the indicator for each titration.
2. Personal errors can be minimised by taking the help of a person who does not suffer from constitutional inability.
3. Instrumental errors can be minimised by using good quality balance, calibrated weights and properly calibrated glass apparatus.
4. Reagent errors can be rectified by using reagents of analar grade chemicals of E. Merck or B.D.H.
5. Errors of methods can be rectified by carefully carrying out the experiment in such a way that external factors do not harm the main reaction.

CHEMICAL METHODS OF SEPARATION

Recently, new methods have been developed to separate inorganic compounds at the micro-level. Some of these methods are ion exchange, solvent extraction and chromatography. They are quite simple, rapid and do not require any sophisticated apparatus. In this unit, we shall explain briefly the underlying principles of these methods.

Ion Exchange Method

Originally, this method was used for softening (or deionising) water. But now this technique has been modified to such an extent that it can be used for separation of anions, cations (particularly lanthanides) which are otherwise difficult to separate. It is possible to separate a mixture of amino acids by ionexchange method.

The early ion exchangers were complex inorganic compounds such as silicates, fuller's earth, and synthetic aluminosilicates (zeolites). But now organic synthetic resins have replaced inorganic ion exchangers. Organic ion-exchange resins should have the following special properties:

1. They should have reactive —OH, —COOH or —SO_3H group as exchange sites which are reversible.
2. They should be almost insoluble in water as well as organic solvents.
3. They should have an open, permeable molecular structure so that ions and solvent molecules can move freely in and out of the molecular structure.

Types of ion exchangers

There are two types of ion exchange resins as.

Cation exchangers

They are acidic resins as they contain sulphonic acid groups ($RSO_3^-H^+$) or carboxylic acid groups (RCOOH). The resins containing sulphonic acid groups are known as strong acid resins and find numerous applications in chemistry. The resins containing carboxylic acid groups are known as weak acid resins. These type of resins are made by first copolymerising styrene and divinyl benzene, the polymer is then sulphonated to get the requisite type of cation exchange resin.

Anion exchangers

They are basic resins as they are either quarternary or tertiary amines or contain hydroxyl groups. The resins which are quarternary or tertiary amines are known as strongly basic anion exchangers. On the other hand, resins which contain —OH groups are known as weakly basic anion exchangers. Structurally in anion exchangers, —SO_3H group of the cation resin is replaced with —CH_2—NH_3^+ or —OH^- group.

Selectivity of ion exchange process depends upon the following factors:

1. At the same concentration, the greater the charge on the ions, the strongly they are held by the resin.
2. At the same concentration and ions having the same charge the larger the ion (less hydrated), the more strongly it is held by the resin.
3. Selectivities widen with increase in cross-linkage of the resin.

Mechanism of ion exchange

Different views have been put forth to explain the mechanism of ionexchange. These are: (i) crystal lattice exchange; (ii) double layer theory; and (iii) donnan membrane theory.

Crystal lattice exchange

According to this view, the ion exchange in resins is quite similar to exchange of crystal lattice ions. The exchange of ions on the resins occurs throughout the entire gel structure of the resin and not on the exposed surface only.

Double layer theory

According to this theory, as in colloids, a double layer exists with inner fixed layer and outer mobile layer of charges. The charged layers are due to absorbed ions and the concentrations of the ions on the outer layer depends upon the concentration and pH of the external solution. As soon as the foreign ions present in the external solution come in contact with the outer layer new ions enter and replace the old ions in the outer layer and consequently, new equilibrium is established.

Donnan membrane theory

First as Donnan membrane theory, there is unequal distribution of ions on two sides of a membrane, in the same way, in ion exchange equilibria, the interface between solid and liquid phases is considered as membrane. There is greater activity of ions on one side of the membrane (which is free of non-diffusible ions, i.e., colloidal micelle to which exchangeable ions are attached) as compared to the other. The exchange of ions occurs so that the activity ratio become equal on both sides.

Applications of ion exchange method

In recent years, ion exchange methods have become quite popular and have numerous applications in analytical chemistry as well as in industry. Some of their applications are listed below :

1. This method was first of all used for softening or deionising water and is still used quite extensively.
2. Using ammonium citrate-citric acid, buffered at pH–5.5 cation exchange resins are used to separate lanthanide ions.
   Similarly, zirconium hafnium, niobium and tantalum can be separated by cation exchange resins.
3. Using EDTA solutions, it has been possible to separate quantitatively a mixture of alkali and alkaline earth metals.

Solvent Extraction Method

This method has become quite popular in recent years as it has the following advantages: (i) this method is quite simple and clean; (ii) it is quite rapid and convenient; (iii) it does not require any sophisticated apparatus or instrumentation; (iv) it can be carried out at macro-level as well as micro-level; and (v) it can also be used to purify an inorganic compound.

Solvent extraction is defined as the process by which a substance may be extracted from dilute solution (usually aqueous or inorganic solutions) into an immiscible solvent (usually organic solvent) with or without the use of a complexing agent. The organic solvents used for extraction may be polar like chloroform, nitrobenzene, tetrahydrofuran, butyl acetate, ethers, methyl isobutyl ketone etc., or non-polar like benzene, carbontetrachloride, petroleum ether, kerosene, xylene, *n*-hexane, etc.

Nerst distribution law governs the partition of a solute or substance say X between two immiscible solvents expressed by the equilibrium:

$$X_{aq} \rightleftharpoons H_{org}$$

The ratio of activities of X in organic (*org*) or aqueous (*aq*) phases will be constant and independent of the total concentration of X at a given temperature.

$$K = \frac{[X_{org}]}{[X_{aq}]}$$

K = partition coefficient which is independent on the total solute concentration in either of the phases

$[X_{org}]$ = activity of X in organic phase

$[X_{aq}]$ = activity of X in aqueous phase.

At constant temperature, the value of X is constant. This is applicable theoretically to ideal solutions only but in practice many substances follow the expression and activities may be replaced by molar concentration. Quite often, K is approximately equal to the ratio of the solubility of X in each solvent.

Generally, ionic substances are insoluble in non-polar solvents but soluble in polar solvents with high dieiectric constant due to the solvation of ions. On the other hand, covalent compounds are mostly soluble in non-polar solvents. Therefore, the extraction of an ion from aqueous solution into organic phase can take place only if the ion forms a species showing preference for organic phase either by: (i) chelate formation; (ii) solvate formation; and (iii) ion-pair formation.

Chelate formation

Chelating agents form an important class of organic bases which can interact with a metal atom/ion from more than one position in the ligand molecule in such a way that the metal atom/ion is bound in a stable 4 or 5 or 6 membered ring. Such complexes are called chelates and organic bases forming these complexes are called chelating agents. A widely used chelating agent is 8-hydroxyquinoline (C_9H_7NO). Most of its chelates are soluble in organic solvents.

A wide variety of chelating agents are in use for the separation of metal ions at micro as well as macro levels. For example, extraction of uranium with 8-hydroxyquinoline is done in $CHCl_2$ and that of Fe with cupferron is done in CCl_4 solvent.

Solvate formation

Extraction by this method is very clean and can be performed at microgram concentration of the metal ions as well as macrogram or pilot plant level. It involves solvation of the extracted species into the organic phase. There are several kinds of solvating solvents, for example, ether, ethylacetate, methyl-isobutyl ketone (MIBK), tributylphosphate $(C_4H_9O)_3PO$(TBP), tributylphosphine oxide $(C_4H_9)_3PO$ trioctylphosphine oxide $(C_8H_{17})_3PO$, etc. This method is frequently used for the extraction of transition metals and inner transition metals. For example, uranium can be conveniently separated from lead and thorium if the nitrate in water is extracted with etherial phase as a solvate $[UO_2(NO_3)_2.2H_2O.2Et_2O]$. Lead and thorium are left in aqueous phase. Uranium can be extracted on industrial scale using TBP as an extractant (for solvate formation) in kerosene phase. This method is based on the principle of solvation of the metallic species into the organic phase.

A number of covalent compounds like $ZnCl_2$, $HgCl_2$, $CeCl4$, OsO_4, RuO_4, etc., undergo easy extraction into organic phase of hydrocarbons, halohydrocarbons or ethers. The mechanism involves solvation of these substances into the organic phase.

Ion-pair formation

Another category of extractants involve extraction by non-pair formation. The extractants are mostly high molecular weight amines. As for example, tertiary amines like trioctylamine TOA $(C_6H_{17})_3N$, triiso-octylamine [OA (i-C_8H_{17})$_3$N trioctylamine oxide (C_8H_{17})$_3$NO, etc. are best extractants for anionic complexes with mineral acids. A typical example showing mechanism of extraction of a solution of $FeCl_3$ in concentrated hydrochloric acid, $HFeCl_{l4}$ is as follows:

$$Fe_{aq}^{3+} + 4Cl^-_{aq} \rightleftharpoons FeCl_4^-{}_{aq}$$

$$R_3NO_{org} + H_{aq}^+ + Cl^-_{aq} \rightleftharpoons R_3NH_{org}^+ Cl_{org}^-$$

$$R_3NH_{org}^+Cl_{org}^- + FeCl_4^-{}_{aq} \rightleftharpoons [R_3NH^+FeCl_4^-]_{org} + Cl_{org}^-$$

Ion-pairs involving big metallic cations or anions with ions of the opposite charges, such as, tetraphenylarsonium perrhenate $[C_6H_5)_4As]^+$ ReO_4^- or tetraphenylarsonium permanganate $[(C_6H_5)_4As]^+$ MnO_4^- undergo easy extraction with organic phase of chloroform.

Choice of a solvent

The following aspects must be looked into while considering the choice of a solvent or an extractant :
1. The solvent should be practically insoluble in the aqueous phase.
2. There should be significant difference in their densities and organic phase should preferably form the upper layer.
3. The organic solvent should have low toxicity and if possible low inflammability.
4. The dissolved metal ion in the organic phase should be easily recoverable.
5. The distribution ratio of the extractable metal ions should be high and that of impurities low.

Applications of solvent extraction method

Some examples involving the use of solvent extraction procedures for the separation of metal ions are given below :
1. Uranium can be conveniently separated from impurities of lead and thorium by the etherial extraction of aqueous solution of uranium (VI) (uranyl UO_2^{2+}) in saturated solution of NH_4NO_3 with 1.5 M concentration of HNO_3.
2. Extraction of plutonium and uranium and their separation from a solution of uranium fuel elements in a nuclear reactor is based on the process of extraction.

Using oxidation-reduction procedures and extraction or precipitation, bulk of unwanted fission products are removed. The aqueous solution containing UO_2^{2+} and PuO_2^{2+} is then treated with an organic phase containing methylisobutyl ketone (MIBK). Both UO_2^{2+} and PuO_2^{2+} get extracted into organic phase which on washing with aqueous SO_2 affects their separation. Plutonium (VI) is reduced by SO_2 to Pu^{4+} which goes into aqueous phase, whereas UO_2^{2+} remains unaffected in the organic phase.

In another method, UO_2^{2+} and Pu^{4+} are extracted into an organic phase of kerosene containing 30% TBP (tributylphosphate) as complexing agent from an aqueous solution of $6NHNO_3$. The organic phase containing UO_2^{2+} and PU^{4+} is then reduced with aqueous SO_2, when PU^{4+} gets changed to Pu^{3+} and goes into aqueous phase and UO_2^{2+} remain in organic phase.
3. Extraction and separation of a mixture of Zr (IV) and Hf (IV) or a mixture of Nb (V) and Ta (V) is based on the extraction by solvation of their halide from the acidic aqueous medium into an organic phase using MIBK as extractant.
4. Absolute alcohol is used for the separation of calcium from a mixture of Ca, Ba and Sr nitrates. Calcium nitrate, $Ca(NO_3)_2$ forms a solvated complex with alcohol and goes into solution whereas others (Sr and Ba nitrates) remain insoluble.

ADSORPTION

Just as the particles on the surface of a liquid experience inward pull resulting in surface tension, in the same way, particles on the surface of a solid experience inward pull because of unbalanced attractive interaction with other particles which surround them only on one side and not all sides. Moreover, when a solid is broken, some inter-atomic bonds break. As a result, valencies of the atoms at the surface remain unsatisfied. Consequently, the surface of a solid has a tendency to attract and hold molecules of a gas, a liquid or a dissolved solute. In other words, there is greater concentration of a substance (which comes in its contact) on the surface of solid than in the bulk. This is known as adsorption. Therefore, adsorption is defined as the phenomenon of higher concentration of any molecular species at the surface than in the bulk of a solid (or liquid). For example, if a piece of coconut charcoal is introduced into a gas jar containing ammonia, after a few minutes it is found that, there is no smell of ammonia in the jar indicating that the whole of ammonia has been taken up by charcoal.

The substance on whose surface absorption takes place is called adsorbent. In the above example, charcoal is the adsorbent. The substance which gets adsorbed on the adsorbent is called adsorbate. In the above example, ammonia is the adsorbate. It has been established that non-polar substances like nitrogen get adsorbed on the surface of a solid through van der Waals' forces (known as physical adsorption) while polar substances like carbon monoxide may get adsorbed on the surface of a solid through chemical interaction known as chemical absorption or chemisorption. The phenomenon of adsorption differs from absorption in the following respects :

1. In adsortion, there is increased concentration of the particles of a substance on the surface while in adsorption the particles actually penetrate into the entire bulk of the absorbing substance.
2. Adsorption is rapid in the beginning but slows down with the passage of time while absorption occurs at a uniform rate.

Some of the important characteristics of adsorption are as follows:

1. The process of adsorption is selective and specific depending upon the nature of adsorbate and adsorbent.
2. Adsorption of gases on the surface of solids is a rapid and reversible process.
3. Adsorption of gases on solid increases with increase in the pressure of the gas. Similarly, in case of solutions, the adsorption on a solid increases with increase in the concentration of the solute in the solution.
4. Since adsorption is an exothermic process, according to *Le Chateliers' Principle,* adsorption decreases with increase in temperature.

Mechanism of Adsorption

Langmuir put forth certain views about the mechanism of adsorption. On the basis of his views, he also derived a mathematical equation known as *Langmuir's adsorption isotherm* which could explain the phenomenon of adsorption of gases on solids in a fairly satisfactory manner. To derive Langmuir adsorption isotherm, Langmuir made the following assumptions :

1. A gas molecule on striking the surface of a solid gets condensed there known as *inelastic collision.* After some time, the molecule evaporates off; the time lag between condensation and evaporation causes adsorption.
2. An equilibrium always exists between the adsorbed and the unadsorbed gas.
3. The adsorbed layer of the gaseous molecules on the surface of the adsorbent is normally one molecule thick.

Langmuir derived the well-known equation known as *Langmuir adsorption isotherm* which is as follows:

$$\frac{x}{m} = \frac{k_3 p}{1 + k_1 p}$$

where x is amount of the gas adsorbed, m is the mass of the adsorbent, k_1 and k_3 are constants, characteristics of the system, and p is the pressure of the gas. Langmuir's Adsortion Isotherm has been verified at various pressures and has been found to hold good.

Langmuir's adsortion equation for solution is obtained by substituting concentration C for pressure. Then we have,

$$\frac{x}{m} = \frac{k_3 C}{1 + k_1 C}$$

Applications of Adsorption

The phenomenon of adsorption is extensively employed in the industrial and laboratory processes. Some important applications of adsorption are as follows:

Ion exchange resins

These are either inorganic polymers containing silicate groups or organic polymers containing —COOH, —SO_3H, —NH_2 or NH groups which have the property of selective adsorption of ions from a solution. They are used in (i) softening of water on an industrial scale, and (ii) separating a mixture of rare-earths into individual components.

Adsorption indicators

In titrations involving precipitation reactions, dyes such as eosin, fluorescein are used as indicators and are known as adsorption indicators. The working of these indicators is based on their preferential adsorption by the precipitates.

Chromatography

All chromatographic techniques are based on the selective adsorption of different substances by an adsorbent.

Miscellaneous applications

The phenomenon of adsorption finds applications in decolourisation of oils; sugar juice; gas masks; dyeing of cloth; in detergents and dehumidisers.

Chromatography Method

Tswettt described a new technique known as chromatography to separate a mixture of coloured substances into individual components. The original technique has been modified and extended so that it has been used to separate a mixture of coloured as well as colourless substances into individual components. Moreover, this technique can also be used to test the purity of a substance.

Principle of chromatography

Chromatography is based upon the selective removal of the components of one phase from it as it flows through another phase which remains stationary. The selective removal itself may be due to either of the following two processes:

1. *Adsorption:* In this process, the components of a mixture in the liquid or gas phase undergo selective adsorption to the surface solid phase.
2. *Partitioning:* In this process, the components of a mixture undergo selective dissolution in immiscible solvents in which they have different relative solubilities.

Types of chromatography

The main types of chromatography are as follows: (i) column (or adsorption) chromatography; (ii) paper chromatography; (iii) thin layer chromatography; (iv) ion exchange chromatography; (v) gas chromatography; and (vi) high performance liquid chromatography (HPLC). Let us discuss briefly the various types of chromatographic techniques.

Column (or adsorption) chromatography

This technique is used to separate non-polar substances and constituents of low volatility.

Principle of adsorption chromatography

The various constituents of the mixture get adsorbed to different extents on the fixed solid phase such as alumina or silica gel which is packed as a long column. The constituents get separated into distinct bands which are removed with the help of a solvent. The extracted component along with the solvent is known as *elute*. From the *elute*, the solvent is removed to get the individual component.

It is believed that there is always a competition between solute and solvent molecules for the adsorption sites. Consequently, a dynamic equilibrium gets established at the interface where the solute and solvent molecules get attracted and settled at the solid surface for a short while and then leave the solid surface to re-enter the mobile liquid phase. However, the desorbed molecules which have greater affinity for the solid surface flow into the mobile layer slowly while those desorbed molecules which have lesser affinity for the solid surface flow into the mobile layer quickly.

Characteristics of good adsorbent

1. It should possess high and selective adsorption power.
2. It should be uniform in size.
3. It should be finely divided to offer greater surface area for adsorption.
4. It should be pure as impurities cause irreversible adsorption.
5. It should be chemically inert.

Commonly used adsorbents are activated alumina, activated charcoal, magnesium oxide, silica gel, starch and fuller's earth.

As a result of experimental investigations, it has been found out that greater the number of double bonds and hydroxyl groups present in a molecule, greater is the adsorption. Thus, we have adsorption series as:

Acids, bases > alcohols > ketones, aldehydes > unsaturated hydrocarbons > saturated hydrocarbons.

Adsorption chromatography is mainly used to separate a mixture of organic compounds such as a mixture of *ortho* and *para*-nitroanilines; a mixture of blue and red dyes.

Paper chromatography

Mechanism of paper chromatography

This technique is based upon a mechanism which is partly adsorption and partly partition. A mixture of the solute is placed on a strip of chromatographic paper such as Whatmann No. 1 or 3 and a solvent is allowed to move along the paper strip. The solvent extracts the solute because of distribution of the solute between the two solvents.

Types of paper chromatography

There are three main types of paper chromatgraphic techniques as
1. Ascending paper chromatography
2. Descending paper chromatography
3. Radical paper chromatography

Advantages of paper chromatography

Its main advantages are as follows:
1. The procedure is very simple and precise.
2. It is a reasonably rapid process.
3. Only small quantities of the material are required.
4. No costly apparatus as well as reagents are required.

Thin layer chromatography (TLC)

Although paper chromatography is a good technique for separation of a mixture yet it suffers from the following defects:
1. Quite often, the sample spots tend to spread out on the filter paper as the fibres of the filter-paper are quite coarse.
2. It is not possible to remove the spot from the filter-paper either for identification or quantitative analysis.
3. It works only for those substances which are separated on cellulose.

To overcome the above defects, a new technique known as thin layer chromatography has been developed. In this technique, a solid adsorbent such as alumina or silica gel is mixed with a little binder such as hydrated calcium sulphate or starch. It is then spread as a thin layer on a glass plate or plastic sheet. The plate thus formed is known as chromatoplate. By means of a fine capillary, solution of the sample to be analysed is applied on chromatoplate and then dried. Then it is allowed to stand in jar containing solvent covered with lid. After some time, spots are located either through naked eye or by applying suitable reagent.

Mechanism of thin layer chromatography

Its mechanism is the same as that of paper chromatography.

Advantages of thin layer chromatography

1. It is a very rapid process as compared to paper chromatography.
2. It is used for large number of sample separations.
3. In this method sharp spots are obtained while in paper chromatography diffused spots are formed.
4. In this method, even a solution containing acid or alkali can be separated whereas paper chromatograph cannot be used for such solutions.

Applications of thin layer chromatography

Some important applications of thin layer chromatography are as follows:

1. Quantitative and qualitative analysis of organic and inorganic compounds.
2. Checking impurities in a solvent.
3. Separation of plasticisers, inks, dyes and anti-oxidants.

Ion Exchange Chromatography

Essentially, this technique of chromatography is similar to column or thin layer chromatography. However, the adsorbent is either a cellulose derivative or a synthetic ion-exchange resin. For example, if a sample solution containing electrolyte and non-electrolyte is allowed to pass through a column of ion-exchange resin, the electrolyte flows down the column while non-electrolyte particles diffuse into the resin particles. Therefore, electrolyte will appear first in the effluent. Thereafter, the water is made to flow down the column of resin when non-electrolyte also gets detected in the effluent. In this way, it is possible to separate ionic from non-ionic substances. For example, NaCl and ethanol; NaCl and ethylene glycol; HCl and CH_3COOH can be separated by this technique.

Gas Chromatography

Essentially, this technique is similar to liquid-liquid chromatography except that a mobile liquid phase is replaced by a moving gas phase.

Types of gas chromatography

There are two types of gas chromatographic techniques as:

1. *Gas-liquid chromatography:* In this type, the separation is effected by partitioning the sample between a mobile gas phase and a thin layer of non-volatile liquid coated on a solid support.
2. *Gas-solid chromatography:* In this type, the separation is effected by partitioning the sample between a mobile gas phase and solid of large surface area. For example, separation of CO_3 from H_2, O_2, N_2, CO or C_2H_2 using silica gel column.

In this method, the vapours of the sample are allowed to enter the column inlet. The solutes get adsorbed at the head of the column. The separated components are detected by special detectors such as thermal conductivity, flame ionisation or electron capture detectors.

High Performance Liquid Chromatography (HPLC)

So far we have described adsorption, partition and ion exchange chromatography which are examples of liquid column chromatography. The classical liquid column chromatography involved the use of a long (0.5 to 5 m) glass tube with diameter $12–50 \times 10^{-3}$ m containing large sized particles of the solid having diameter $150–200 \times 10^{-6}$ m which acts as a stationary phase. The mobile liquid phase moves down the column on account of its own pressure leading to separation of components. But the rate of flow was very low and so the process of separation took a very long time. The use of vacuum and by pumping did not help to increase significantly the rate of flow. Later experimental studies revealed that the rate of flow did increase by: (i) increasing the height of the column; and (ii) decreasing particle size of the packings. In view of the above facts, an improved chromatographic method has been developed which is known as high performance liquid chromatography (HPLC). The diagram indicating the layout of the equipment employed in HPLC is shown in Fig. 5.1. Let us briefly discuss each component of the apparatus.

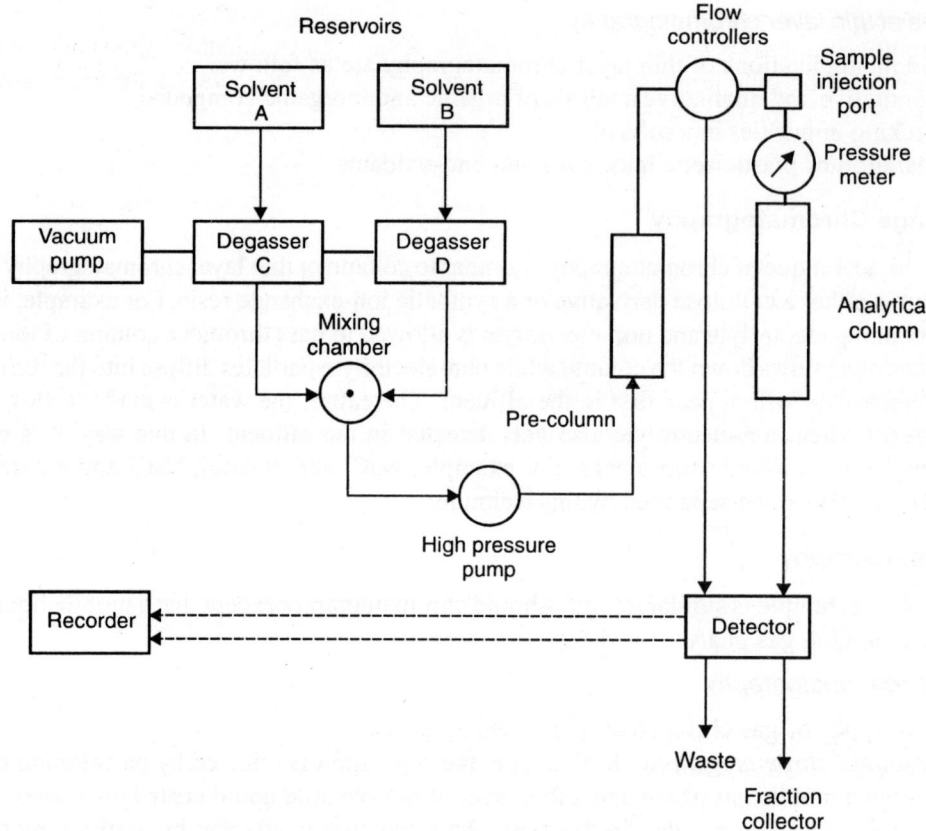

Fig. 5.1. A typical high performance liquid chromatography unit.

Solvent reservoir and degassing system

There are one or more reservoirs (A and B) which can hold about 2 litres of a solvent. A separation which is carried out using a simple solvent is called isocratic elution and is not very efficient. Therefore, two or more solvents having different polarities are used instead of a simple solvent for more efficient separation. The pressure of dissolved gases such as nitrogen or oxygen in the solvents form bubbles in the column and the detector. These bubbles cause band spreading and interfere with the working of the detector. Therefore, the solvents have to be degassed (i.e., gases have to be removed). Degassers (C and D) are either vacuum pumping system or a distillation system. The degassed solvents from reservoirs are led into a mixing chamber at varying rates.

Pumps

In HPLC, screw-driven, reciprocating or pneumatic pumps are used which have output of 4000 to 6000 psi with a flow delivery rate of at least 3 ml min^{-1}. Although high pressures are generated by these pumping devices, yet they are not dangerous because liquids are not very easily compressible.

Precolumns

A precolumn contains a packing which has larger particle size but is chemically identical with that in the analytical column.

The main functions of the precolumn are:

1. To remove impurities from the solvent and thus prevent contamination of the analytical column.
2. To saturate the mobile phase and thus prevent the stripping of the stationary phase from the packing of the analytical column.

Sample injection system

Sample injection is effected by means of a syringe and self-sealing septum of silicone or teflon.

Analytical column

The analytical column is made of either glass or stainless steel. Glass column can be used for pressure below 600 psi only. The column is 0.15 to 1.5 m long with internal diameter of 2 to 3×10^{-3} m.

The column is packed with finely divided silica gel, alumina or celite (diatomaceous earth). Recently, a new packing material known as *pellicular particles* has been developed. It consists of small glass beads which are coated with about 2×10^{-6} layer of porous material such as silica gel, alumina or ion exchange resin. Pellicular packings have the advantage that the rate at which equilibrium is established between the phases is high which results in better efficiency. However, pellicular packings have the disadvantage of limited sample capacity.

Temperature control

Most HPLC separations are carried out at room temperature. However, if temperature control is desired, water-jacketed columns are used.

Detectors

Depending upon the nature of the sample, a detector is used. Most commonly used detectors are ultra-violet, visible, infra-red absorption detectors. Even mass spectrometry has been employed as a sensitive detector.

Recorders

The chromatogram is recorded with potentiometric recorders in conjunction with the particular detector.

Applications of HPLC

1. It is used in separation of components in pharmaceutical and pesticide industries.
2. In conjunction with ion exchange, HPLC has been used to separate nucleic acids and vitamins.
3. It has been used to determine the molecular masses of polymers and biochemicals.
4. HPLC is particularly suitable for non-volatile substances (including inorganic ions) and thermally unstable materials.

ELECTROPHORESIS

It is a well established fact that the particles of colloids bear either positive or negative charge due to the presence of adsorbed ions. Therefore, on passing electric current through a colloidal solution, the charged colloidal particles migrate to the oppositely charged electrodes which is known as electrophoresis. Hence, electrophoresis was defined as the migration of colloidal particles through the solution under influence of an electrical field. Later on, it was found that the process of electrophoresis can be applied to colloidal aggregates as well as monodispersed ions. Methods based on electrophoresis provide powerful tools in the hands of analytical chemists to separate components of a wide variety of biological materials such as proteins, gastric juices, nucleic acids and vitamins.

Types of Electrophoretic Methods

Methods employed to separate a mixture into its components based on the property electrophoresis are called electrophoretic methods. These methods are of two types as described below:

Free Solution Method

In this method, the sample solution is introduced as a band at the bottom of a U-tube having electrodes near the ends of the tube which is filled with a buffered solution. On passing electric current, the charged particles start moving towards one or the other electrode. However, the rates of migration of different particles of the mixture are different:

1. Different particles have different charge to mass ratios.
2. Different particles have different inherent mobilities in the medium.

Consequently, different species get deposited at the electrodes at different intervals of time and at different places leading to separation of the mixtures.

Drawbacks of free solution method

Although this method played an important role in the development of biochemistry yet it suffers from the following defects:

1. The separated components have a tendency to get mixed by convection currents.
2. The bands of separated species can be detected only by using elaborate optical systems.

Stabilising Medium Method

The various drawbacks associated with free solution method get removed if the process of electrophoresis is carried out in a stabilising medium such as paper, a layer of finely divided solid or a column packed with a suitable solid. Consequently, components of a mixture get separated by electrophoresis in combination with adsorption or ion exchange method. Depending upon the type of stabilising medium, different types of electrophoretic methods are known such as electrochromatography, zone electrophoresis, electromigration and ionophoresis. Of all these methods, the most common method is known as electrophoresis or electrochromatography which involves the migration of changed solute particles through the influence of electrical field. It has been found that the migration of charged particles in paper depends on the following factors :

1. Nature and magnitude of charge on the solute particles.
2. Concentration of the electrolyte.
3. pH value of the medium.
4. Temperature of the medium.
5. Viscosity of the medium.
6. Adsorption capacity of solute particles.
7. Voltage applied.
8. Mobility of the ions in the opposite directions.

The distance which a charged particle travels under the influence of electric field is given by the relationship:

$$D = \frac{Ute}{qk}$$

where   D is the distance travelled;

U is the mobility of ion at time t;

e is current applied;

q is the cross-sectional area; and

k is the conductivity.

Experimental Methods of Electrochromatography

There are numerous methods for carrying electrochromatography but we shall discuss here only three simple and common methods of electrochromatography. These are as follows:

Horizontal strip method

A strip of filter paper is stretched horizontally between two containers A and B which are filled with a buffer solution. The filter paper is well soaked with the buffer solution. To check evaporation of the buffer solution, the apparatus is housed in an air-tight container (Fig. 5.2).

The sample is placed at the centre of the strip and a direct current of 100 to 300 volts is applied across the two electrodes. After some time, the filter paper is removed, dried and bands of components are detected by colorimetric reagents.

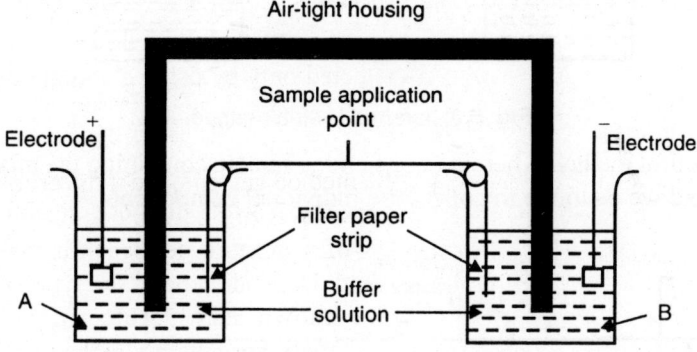

Fig. 5.2. Horizontal strip method.

Inverted V-Strip Method

This method is quite similar to the horizontal strip method. However, the filter paper strip is in the form of an inverted 'V' (Fig. 5.3).

The sample under test is placed at the apex of the 'V'. After electrolysis, the cationic species move down one arm of 'V' while anionic species move down the other arm of 'V'. Thereafter, each species is separately detected.

Curtain electrochromatography

In this method, a curtain of filter paper immersed in a buffer solution is held vertically (Fig. 5.4). The sample is continuously applied on the filter-paper from the sample reservoir. One end of the filter paper is in fluted form.

On passing electric current, separation of components occurs and the separated components can be detected by special reagents. Alternatively the separated components are allowed to fall in different sample collection tubes.

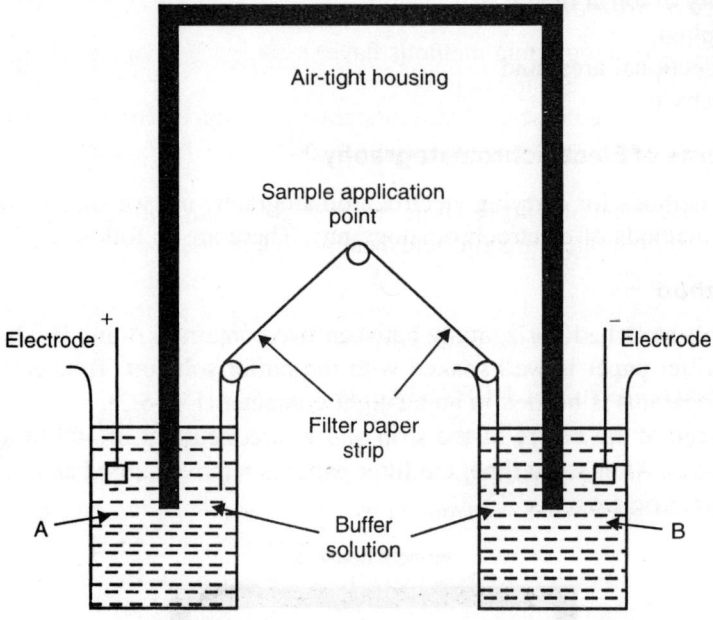

Fig. 5.3. Inverted V-strip method.

Thus, this is a beautiful method wherein at one end, a sample containing the mixture is continuously fed and at the other end we continue to collect the individual components.

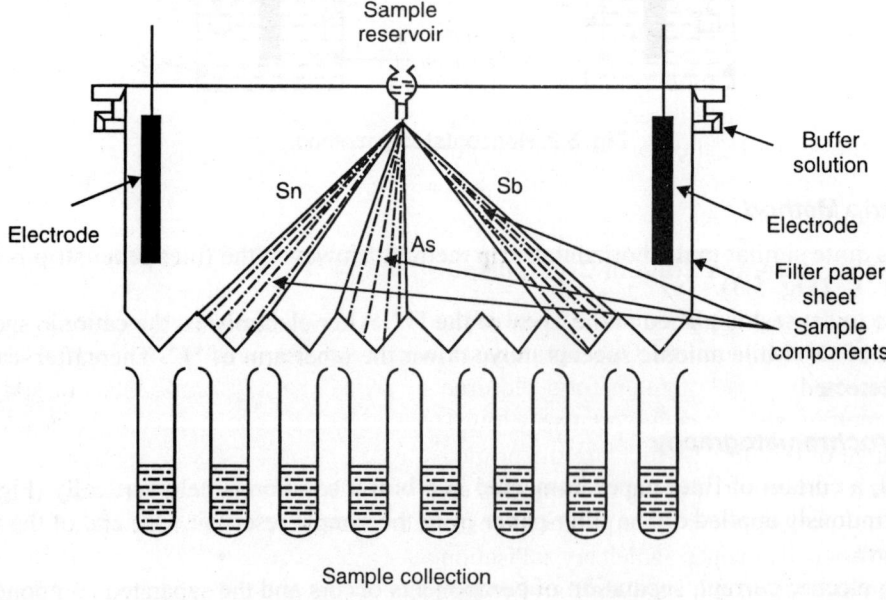

Fig. 5.4. Curtain electrochromatography.

Applications of electrochromatography

In recent years, electrochromatographic methods have become very popular as analytical tools. Some of their important applications are listed below:

1. For clinical diagnosis, a clinical chemist separates through electrochromatography proteins and other large molecules contained in serum, urine, spinal fluid, gastric juices and other body fluids.
2. Biochemists have been able to separate through electrochromatography alkaloids, antibiotics, amino acids, carbohydrates, organic acids, natural pigments, nucleic acids, steroids and vitamins from natural sources.
3. Electrochromatography has been successfully employed to separate a mixture containing a number of metallic ions from a complexing medium.

SEPARATION TECHNOLOGY: MELT CRYSTALLISATION

The chemical industries are beginning to focus on manufacturing processes that address an enhanced environmental awareness on the part of the public and more stringent Governmental regulations for chemical disposal and emissions. As a result, new separation technologies are being developed and implemented.

Unit operations that minimise solvents and reduce operator exposure to chemicals are needed in today's chemical industries. An example of the currently available technology is the melt crystallisation that offers unique separative potential for problematic and high-purity applications in addition to environmental and health benefits.

Over the past few years, more stress is being given to the separation of organics by using melt crystallisation. Two factors, which are responsible for this growth are escalating requirements for purity and increasing environmental concerns.

Principles of Melt Crystallisation

Melt crystallisation is very simple technique to understand. If an impure molten material is cooled to its freezing point and further heat is removed, then some of the material will solidify.

In most of the systems, the solid will be a pure component. Impurities will concentrate in the remaining melt, which is known as the residue. Separating the solid from the residue and remelting it then recovers purified product.

Melt Crystallisation Verses Solvent Crystallisation

When crystallising non-polar organics, solvent crystallisation requires organic solvents, which are, in general, neither cheap nor harmless. On laboratory scale their use is manageable, but large scale processing becomes expensive because of the efforts required to prevent solvent emissions. In addition, most organics have melting points that are relatively low. Over 70 per cent of these substances have a melting point in the range of 0–200°C, which are the prime candidates for melt crystallisation.

The volume of the material being processed is considerably less in melt crystallisation as compared to solvent crystallisation. Thus the equipment costs and energy consumption are much lower in melt crystallisation. Other advantages of melt crystallisation over solvent crystallisation are no solvent recovery; no product contamination with solvent; and reduced environmental hazard. But the growth rate is moderate in melt crystallisation, as compared to solvent crystallisation. Also the operating temperature is higher in melt crystallisation.

Fundamentals

The study of the phase equilibria that drive separation is very important in melt crystallisation. *Wynn* and *Perry* have discussed the following two types of systems.

Eutectic systems

For binary systems, three phases can co-exist only at a single point on a composition/temperature plot. At all other compositions, a liquid phase can co-exist with only a single solid phase. A single solid phase consist of only one component, we must have a pure solid phase in equilibrium with a liquid mixture. A phase diagram is shown in Fig. 5.5.

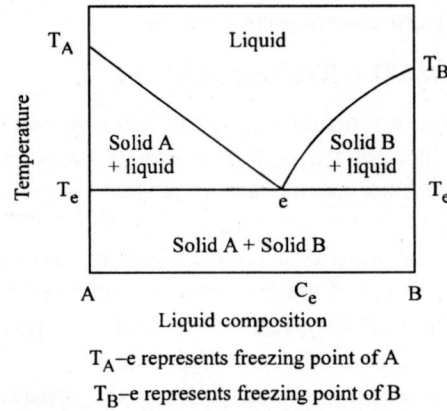

T_A–e represents freezing point of A

T_B–e represents freezing point of B

Fig. 5.5. Phase diagram of binary eutectic system.

Solid solution systems

In some cases one component can substitute for other in the crystal lattice and there can be two solid phases, a solid and liquid. A phase diagram for this system is shown in Fig. 5.6.

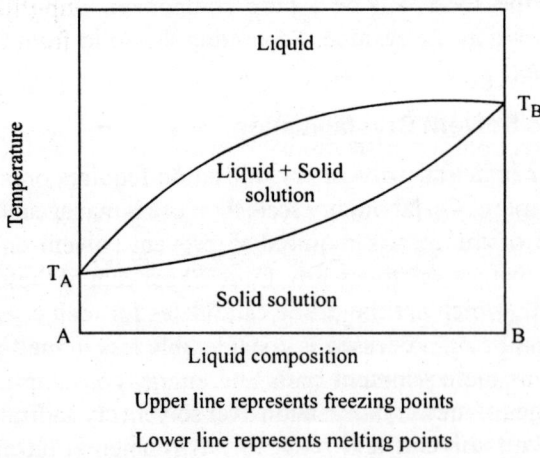

Upper line represents freezing points

Lower line represents melting points

Fig. 5.6. Phase diagram of binary solid-solution system.

Melt Crystallisation Verses Distillation

Both the separation techniques depend on following three elements which has been described by *Wynn*.

1. Phase equilibria that provide the driving force for separation.
2. Mass-transfer rates that allows the phase to equilibrate.
3. Phase separability.

In melt crystallisation, liquid phases are totally miscible; and solid phases are not miscible. Solid phase is pure, except at eutectic point. Hence ultra-high purity is easy to achieve. But the recovery is limited by eutectic composition and phase separation is slow. This contrasts with better-known distillation in which both liquid and vapour phases are totally miscible. As neither phase is pure, ultra-high purity is difficult to achieve. There is no limit on recovery and also phase separation is rapid and complete. If a material is prone to decomposition at distillation temperature, it will likely be stable at its freezing point. Compared to distillation, melt crystallisation may be slower, but it is also gentler and kinder.

Need of Melt Crystallisation

Though distillation is much faster than crystallisation, melt crystallisation is a cheaper process to install or operate. Difficulties with distillation arise in major two areas: low relative volatility and thermal instability. In case of mixture of dichlorobenzene isomers, as the boiling points of the *para-* and *ortho-*isomers differ by only 2°C, separation by distillation requires a large number of stages and a high reflux ratio. Even then, top purities cannot be reached. In contrast, the *para*-isomer crystallises particularly well from the melt, and end purities of 99.99 per cent can be easily obtained. Thermally unstable compounds actually are quite often distilled, but decomposition is the main problem because the decomposition rate increases with temperature. The colour or odour of a product can be spoiled by decomposition products that are very difficult to remove. Dimerisation or polymerisation can cause real operating problems in distillation columns, which lead to column blockage. To avoid these problems stabilisers can be used. But stabilisers are only effective if they remain in the liquid. If local temperature gradients separate the stabiliser, then polymerisation may start. Hence melt crystallisation is now the preferred method, e.g., production of the high purity glacial acrylic acid.

Column Crystallisation (Suspension Crystallisation)

Conducting crystallisation inside a column with a countercurrent flow of crystals and liquid can produce higher product purity than conventional crystallisation and distillation. Schematic representation of the components of a column crystalliser is shown in Fig. 5.7.

It consists of a freezing section, a purification section and a melting section. In freezing section, primary crystallisation takes place. The crystals generated are then mechanically transported through a purification section, usually adiabatic, in which crystals are in differential countercurrent contact with melt generated in melting section. In melting section crystals from purification section are melted, part of the melt is withdrawn as product and part is returned to purification section.

If the crystals are denser then the melt, they may be separated by gravity settling. In many column crystalliser mechanical transport devices are used in purification section in order to move crystals. The transport device may be an auger or a helix rotating in an annular space, about a stationary central column. The separation attainable in a column crystalliser depends on the phase relationships of the system.

With a simple eutectic system, the initial crystallisation should produce pure crystals of the major component in a melt that becomes progressively richer in the minor component, until the eutectic is reached. In practice, impurity is retained within crystal defects and crystals retain impure melt on their external surfaces. Adhering melt may be removed by countercurrent contact with pure melt.

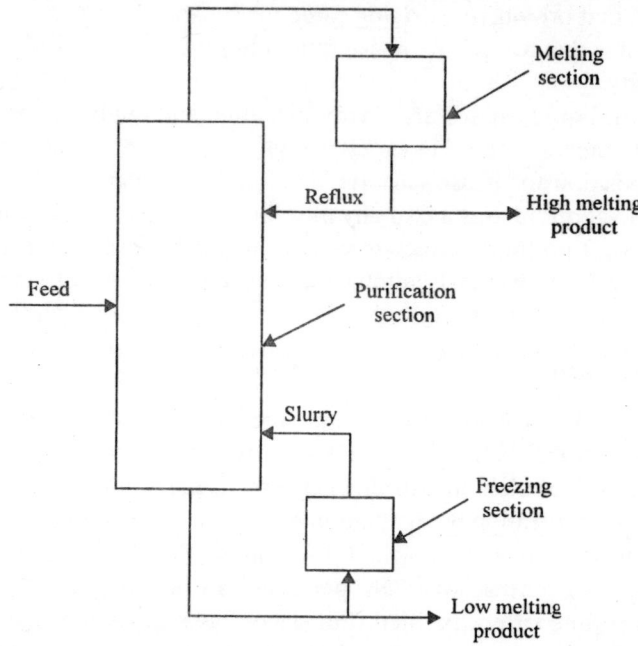

Fig. 5.7. Representing schematic representation of the components of a column crystallisers.

In case of a solid solution system, it is not possible to obtain pure product in a single crystallisation stage and several stages of partial melting and refreezing will be required to attain high purity. Such multistage crystallisation can be carried out in a column if suitable temperature gradient is provided along the length of the column. Crystals are transported to regions of progressively higher temperature, where they melt.

Crystallisation conducted inside a column is categorised as either end-fed or centre-fed depending on whether the feed location is upstream or down-stream of the crystal forming section.

In case of an end-feed column crystalliser as shown in Fig. 5.8, first the crystals are formed by indirect cooling of the melt in scraped surface heat exchangers and then the resultant slurry is introduced into the column at the top. This type of column has no mechanical internals to transport solids and instead relies upon an imposed hydraulic gradient to force the solids through the column into the melting zone. Residue liquid is removed through a filter directly above the melter. A pulse piston in the product discharge improves washing efficiency and column reliability.

In case of a centre-feed column crystalliser, liquid feed enters the column between the hot purifying section and the cold freezing or recovery zone as shown in Fig. 5.7. Crystals are formed internally by indirect cooling of the melt through the walls of the refinery and recovery zones. Positions of the column may be vertical or horizontal, depending on the requirements.

In some cases, melt crystallisation will reach higher purities than distillation, but at the expense of recovery. In such case, both techniques can be combined. The feed is first pre-distilled to remove the bulk of the impurities and then melt crystallised to polish the final product.

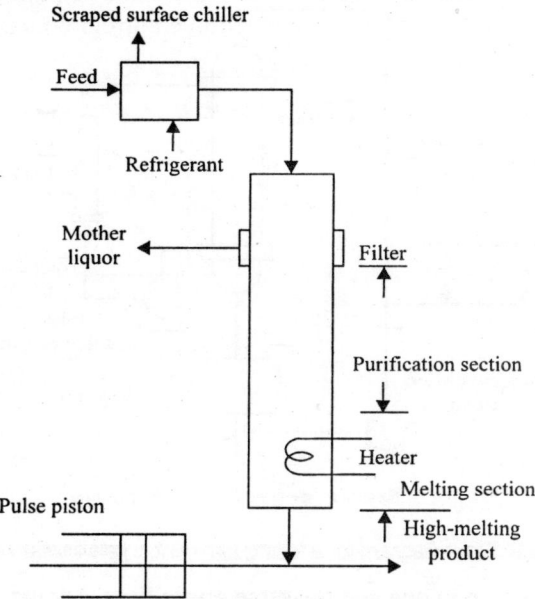

Fig. 5.8. End-fed column crystalliser.

Falling Film Crystallisation

Falling film crystallisation utilises progressive freezing principle to purify melts and solutions. The technique established to practice the process is inherently cyclic. Instead of flooding the crystalliser, a crystalline layer is formed by sub-cooling a liquid film on a vertical surface inside a tube. This coating is then grown by extracting heat from a falling film of melt (or solution) through a heat transfer surface. Impure liquid is then drained from the crystal layer and the product is then reclaimed by melting. Variants of this technique have been perfected and are used commercially for many types of organic materials.

Fig. 5.9 shows the process schematically. The crystalliser contains vertical tubes in which the crystal layers grow as cylinders. The collecting tank beneath the tubes is built integrally with the crystalliser. At the start of a stage, it is filled with a batch of material. The circulating pump is then started to irrigate the tubes, while coolant temperature ramping is begun on the shell side of the crystalliser. The melt circulation rate is high compared to the rate of crystal deposition so that conditions of temperature and composition are uniform down the length of tubes.

Shell-side temperature is ramped down at a constant rate until the collecting-tank level drops to a preset value, indicating the desired amount of product is suspended in the tubes. At this point, melt circulation ceases and the crystals drain free of residue. Shell-side temperature is then ramped up to a value just below the product's melting point to sweat out including impurities and assist the draining process.

In some cases even with sweating, the product from a single stage may not be at the specified product purity, so multistaging may be required. Impurities can be included inside the crystal structure or the crystal surface may simply wetted with residue. In this case the product is reprocessed by repeating the crystallisation, sweating and melting steps.

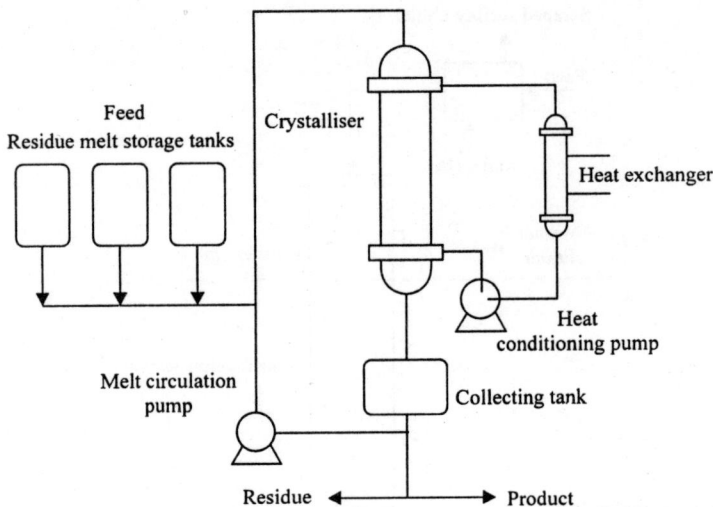

Fig. 5.9. Flow diagram for a falling-film melt crystallisation plant.

All the stages are normally run in one and the same crystalliser. Adding a purification stage does not mean adding a second crystalliser, but reprocessing material in the one unit. At the end of the first stage, molten product ends up in the collecting tank. This material is then re-crystallised using the same equipment.

Applications

In the chemical industry, melt crystallisation finds primarily applications in the separation of materials with close boiling points; or of azeotropic mixtures i.e., for separation problems that are difficult to solve by rectification as well as for purification of temperature-sensitive products.

It is one of the separation technique applied in the separation of organics (such as close boiling hydrocarbons), isomers, heat sensitive materials, and so on. The technique is now routinely used to purify many chemicals without the need for organic solvents. Some chemicals which can be purified by using this technique are given Table 5.2.

Table 5.2. Chemicals purified by melt crystallisation technique.

Acetic acid	Acrylic acid
Adipic acid	Benzene
Biphenyl	Bisphenol-A
Caprolactam	Chloroacetic acid
p-Chloro toluene	p-Cresol
Combat (proprietary)	Dibutyl hydroxy toluene (BHT)
p-Dichlorobenzene	2,5-Dichlorophenol
Dicumyl peroxide	Diene
Heliotropin	Hexachloro cyclobutene
Hexamethylene diamine	Isophthaloyl chloride

(Contd...)

Isopregol	Lutidine
Maleic anhydride	Naphthalene
p-Nitrochloro benzene	p-Nitro tolune
Phenol	Picolin
Pyridine	Stilbene
Terephthaloyl chloride	Tertiary butyl phenol
Toluene diisocyanate	Trioxane
p-xylene	3,4-xylidine

CHAPTER 6

Salt and Salt Based Products

INTRODUCTION

The salt industry is as old as mankind. Salt has long been an essential part of the human diet. It has served as an object of worship and as a medium of exchange, lumps of salt being used in Tibet and Mongolia for money. Its distribution was employed as a political weapon by ancient governments, and in Oriental countries high taxes were placed on salt. Salt is a vital basic commodity for life but is also a source of many of the chemicals which are now the mainstay of our complex industrial civilisation.

Many sodium salts are of definite industrial necessity. Most of them are derived directly or indirectly from ordinary salt so far as their sodium content is concerned. In one sense, the sodium may be viewed simply as a carrier for the more active anion to which the compound owes its industrial importance. For instance, in sodium sulphide, it is the sulphide part that is the more active. Similarly, this is the case with sodium thiosulphate and sodium silicate. The corresponding potassium salt could be used in most cases; however, sodium salts can be manufactured more cheaply and in sufficient purity to meet industrial demands. Sodium chloride is the basic raw material of a great many chemical compounds, such as sodium hydroxide, sodium carbonate, sodium sulphate, hydrochloric acid, sodium phosphates, sodium chlorate and chlorite, and it is the source of many other compounds through its derivatives. Practically all the chlorine produced in the world is manufactured by the electrolysis of sodium chloride. Salt is used in the regeneration of sodium zeolite water softeners and has many applications in the manufacture of organic chemicals. Salt is obtained in three different ways, namely, solar evaporation of sea water (Fig. 6.1) on the Pacific Coast or from western salt-lake brines, from mining of rock salt, and from well brines. The purity of the salt obtained from the evaporation of salt water is usually about 98 to 99%. Mined salt varies very widely in composition, depending on the locality. Some rock salt, however, runs as high as 99.5% pure. The solution obtained from wells is usually about 98% pure and depends to a great extent on the purity of the water forced down into the well to dissolve the salt from the rock bed. For many purposes, the salt obtained from mines and by direct evaporation of salt solutions is pure enough for use; however, a large portion must be purified to remove such materials as calcium and magnesium chlorides. Solar evaporation is used extensively in dry climates, where the rate of evaporation depends on the humidity of the air, wind velocity, and amount of solar energy absorbed. The mining of rock salt uses methods similar to coal mining. As the salt is removed, large rooms are formed, supported by pillars of salt.

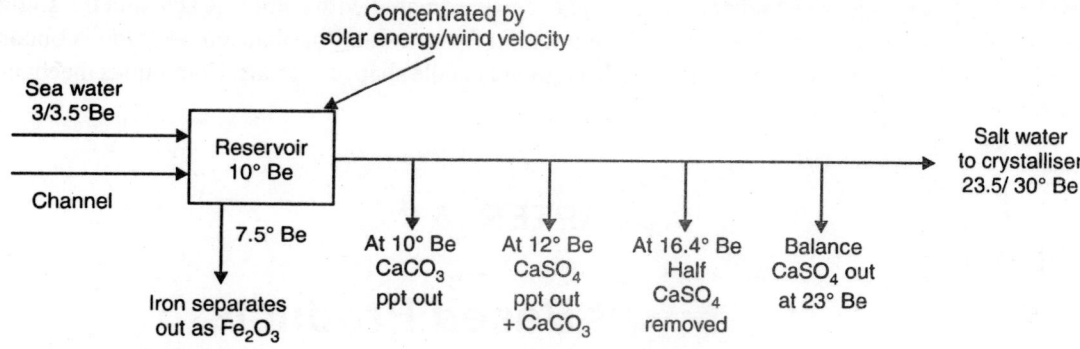

Fig. 6.1. Solar evaporation.

BY-PRODUCTS OF SALT INDUSTRY

The important by-products of salt industry are: (i) calcium carbonate; (ii) calcium sulphate; (iii) bromine; (iv) magnesium sulphate; (v) magnesium chloride; (vi) potassium chloride; and (vii) sodium sulphate.

1. *Calcium carbonate:* As already explained before, when brine is concentrated by evaporation to 3.5° Be the calcium carbonate starts separating out and continues till the concentration reaches 10°Be. At this stage most of the calcium carbonate is crystallised out.

2. *Calcium sulphate:* As the evaporation of brine is continued beyond 10°Be about half the calcium sulphate initially present is crystallised. This continues till the concentration reaches to 25°Be.

3. *Bromine:* Bromine is recovered from the bitterns, that is mother liquor of 31°Be after crystallisation of salt. First the pH is adjusted with sulphuric acid to about 3.5 and castor oil is also added to reduce foaming.

   Equi valent amount of chlorine is added to replace bromine from its salt.

   $$2NaBr + Cl_2 = Br_2 + 2NaCl$$

   The operation is carried out in tower packed with rasching rings in nine sections.

   Steam at 4.5 kg/cm^2 is passed at the bottom section and chlorine at the second section from bottom. Fig. 6.2. shows the flow chart of the bromine manufacturing unit using bitterns as raw material.

   The bittern solution is sprayed countercurrent to the flow of chlorine and steam. Bromine vapour is condensed by cooling with water in tantalum condenser. The impure bromine contains chlorine and water.

   It is passed through U-tube the other end of which is connected to the ninth section of the tower. From here it goes to the gravity separator. Here most of the water is separated and water layer is recycled to the fourth section of the tower to recover bromine present in it. Liquid bromine is purified by distillation with 2.8 kg/cm^2 closed steam.

4. *Magnesium salt:* There are different methods for the recovery of magnesium salts. One method consists in treating the residual brine after removal of bromine, with milk of lime, when magnesium hydroxide is precipitated.

   $$MgCl_2 + Ca(OH)_2 = CaCl_2 + Mg(OH)_2$$
   $$MgSO_4 + Ca(OH)_2 = CaSO_4 + Mg(OH)_2$$

Magnesium hydroxide is settled and filtered. The slurry is treated with sulphuric acid and the solution is crystallised in vacuum crystallisers to give commercial Epsom salt. Pharmaceutical grade is obtained by re-crystallising the product in vacuum crystalliser giving needle shaped crystals. Continuous mechanical crystalliser gives granular product

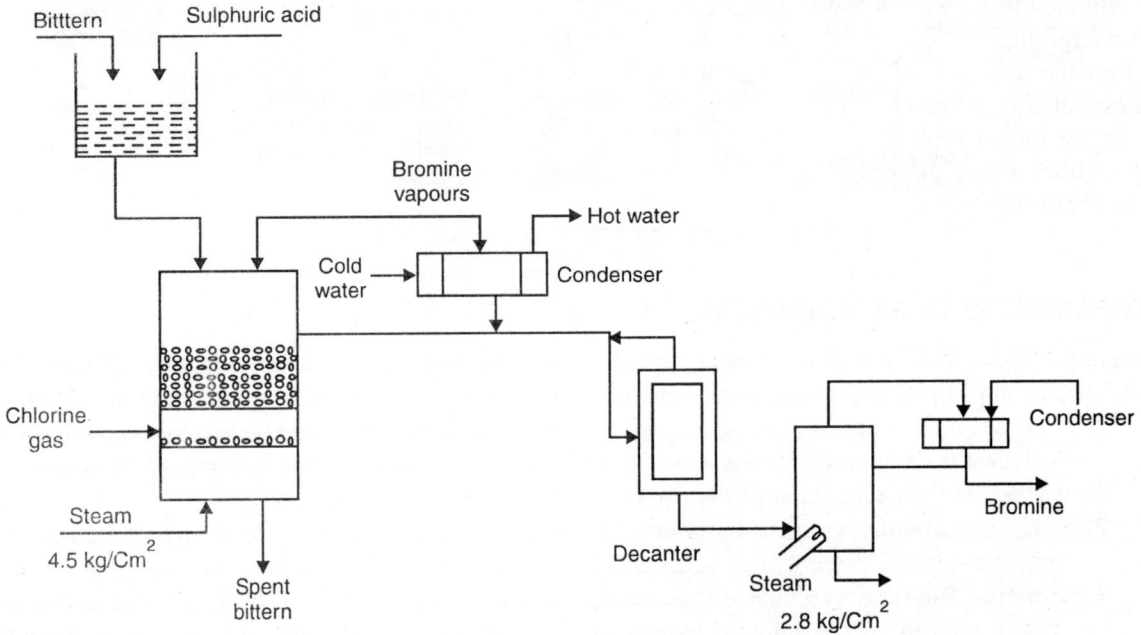

Fig. 6.2. Manufacture of bromine from bittern.

Epsom salt is also obtained by chilling the spent bittern at 34°Be to 10°C by refrigeration, when about 50 per cent of the magnesium sulphate present separates out.

5. *Magnesium chloride:* The above bittern is further concentrated to 38–39.5° Be when $MgCl_2 6H_2O$ crystal separates out on cooling. In case the sulphate content of brine is high the purity of magnesium chloride is also low. The inland salt works produce sulphate free bittern.

6. *Mixed salts:* Several methods are used for the production of mixed salts. All the methods are based on the principle of heating, cooling by refrigeration or by adding chemicals from outside. Central Salt and Marine Chemical Research Institute, Bhavnagar have developed a process based on addition of excess of calcium chloride to remove sulphates as calcium sulphate. The treated bittern is evaporated and cooled when a mixture of magnesium sulphate and potassium chloride separate out. The mother liquor is further concentrated and cooled to crystallise out a mixture of potassium and magnesium chlorides known as carnallite. Potassium chloride is separated from carnallite by dissolving in water and evaporating in copper vessel till a boiling point of 160°C. It is gradually cooled to 115°C when potassium chloride separates out.

There are no potash deposits in India and most of the potassic fertiliser is at present imported.

Recovery of valuable by-products of national importance has not been given proper attention by salt works in the country. Effects should be made to recover all the by-products of this industry which in turn would add to the economy.

SODIUM CHLORIDE (COMMON SALT)

Process of Manufacture

From saturated brine by multiple-effect evaporation process

Saturated brine for the production of "evaporated" salt is usually obtained by pumping water into an underground salt deposit and removing the saturated solution from an adjacent interconnected well, or from the same well by means of an annular pipe. Besides sodium chloride, the brine will contain more or less calcium sulphate, calcium chloride, and magnesium chloride, plus traces of hydrogen sulphide and ferrous ions (Fig. 6.3). Such a solution may have the following approximate composition:

Other minor impurities account for the remainder. Dissolved hydrogen sulphide (H_2S) may amount to about 0.015%.

Ingredients	Per cent
Water	73.5
Sodium chloride	26.3
Calcium sulphate	0.12
Calcium chloride	0.003
Magnesium chloride	0.007
	99.93

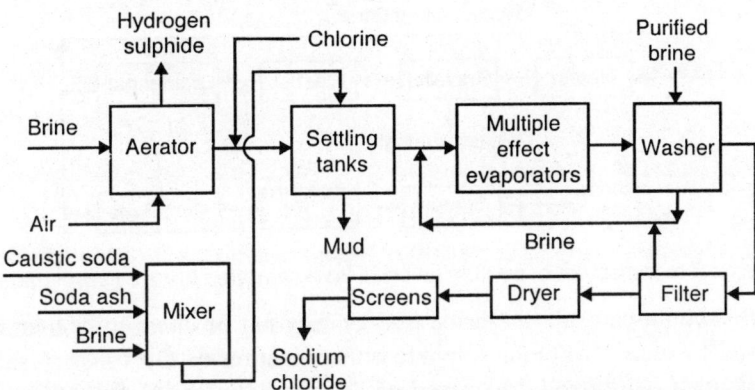

Fig. 6.3. Flow diagram for manufacture of sodium chloride from saturated brine by multiple-effect evaporation.

The chemical treatment the brine receives varies from one plant to another, and in a few cases, where very pure rock salt deposits are available, it may require none at all. Typically, however, the brine is first aerated to remove most of the hydrogen sulphide. Addition of a small amount of chlorine will complete hydrogen sulphide removal by oxidation, and also oxidise ferrous ions to ferric. The brine is then pumped to settling tanks, where it may be treated with a dilute solution of caustic soda and soda ash to remove most of the calcium, magnesium, and ferric ions. After clarification the treated brine is pumped to multiple-effect evaporators. As water is removed, salt crystals form and are removed as a slurry. After screening to remove lumps, the slurry is sent to a conical washer, where the salt crystals are washed counter-currently with fresh brine. By this washing, actually hydraulic classification, fine, light crystals of calcium sulphate are removed from the mother liquor of the slurry and returned to the evaporator. Eventually, the calcium sulphate concentration in the evaporation system builds up to the point where it must be removed by "boiling-out" the evaporators.

The washed slurry is filtered, the mother liquor is returned to the evaporators, and the salt crystals from the filter are dried and screened. Salt thus produced from the typical brine described is of 99.8% purity or greater. Many salt manufacturers do not treat the raw brine, but control calcium and magnesium impurities by watching the concentrations in the evaporators and bleeding off sufficient brine to maintain a predetermined level. By such methods, salt of better than 99.5% purity can be made consistently. In either case, the final screening of the dried salt yields various grades, depending on particle size. The finest grade (sometimes made by grinding) is a flour salt; the next coarsest is table salt; the coarsest is industrial salt. These products may be packed as such, or blended with small amounts of other materials as required. Free-flowing table salts are made by blending 0.5 to 2% magnesium carbonate, hydrated calcium silicate, or tricalcium phosphate with the salt. Iodised salt after blending contains 0.01% potassium iodide, a stabiliser such as 0.1% sodium carbonate, and 0.1% sodium thiosulphate.

From saturated brine by open pan (Grainer) process

Salt in the form of a hopper-like crystal (grainer salt) is made by causing the salt crystal to form on the surface of hot, quiescent brine held in an open pan or grainer. The grainer is a flat, open pan, 4.5 to 6 m wide, 45 to 60 m long, and about 60 cm deep. Beneath the submerged steam coils is a system of reciprocating rakes for salt removal (Fig. 6.4).

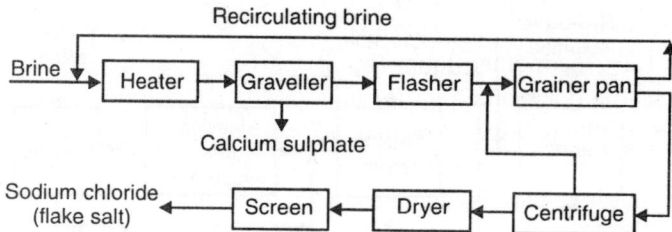

Fig. 6.4. Flow diagram for manufacture of sodium chloride from saturated brine by open pan–Grainer process.

As in the case of vacuum-pan salt, the brine may or may not be chemically treated. Also, different methods may be used to reduce calcium sulphate contamination of the finished salt. In the process shown in the flow diagram, saturated brine mixed with recirculating brine from the grainer is heated to 110°C, at which temperature calcium sulphate is much less soluble than at lower temperatures. The precipitated calcium sulphate is then removed in a graveller or bed of stones. The purified brine is then flash-cooled to such a temperature that sodium chloride is still retained in solution, but any remaining calcium sulphate crystals re-dissolve. The slurry is then pumped to the grainer, where evaporation takes place slowly at 96°C. Flat, hopper-shaped crystals formed on the surface and then fall to the bottom of the grainer, where they grow further before being removed by the rake system. The wet crystals are centrifuged, dried, and screened. When the incoming brine has been treated, salt of 99.98% sodium chloride can be obtained. Even without chemical treatment, salt of 99.8% purity is not unusual.

Because of its high surface-volume relationship, grainer salt dissolves rapidly even though it is coarse. It is preferred in the butter and cheese industries. The grainer process is very wasteful of heat, and there are only a few plants operating to satisfy the demand for this special type of salt.

From rock salt by mining

About 35% of the salt produced comes from mines. The salt deposits vary in colour from light reddish-brown to a sort of off-gray. Purity is usually about 98.5%. After chunks of salt are blasted loose, they are preliminarily crushed in the mine and then crushed again at the surface. The remaining processing consists of a series of grinding and screening operations to produce salt crystals of various sizes. The product is cheaper than evaporated salt and sold for those uses in which small amounts of impurities are allowable, such as ice cream manufacture and the salting of hides. Some mines have as much as 20% waste in the form of fines. An evaporation process called the "recrystallisation process" has been developed to recover these fines as high-purity evaporated salt without resorting to chemical treatment. A slurry of rock salt fines is heated and sent to a saturator, where all the salt dissolves and part of the calcium sulphate separates. Final separation of salt and the remaining calcium sulphate is accomplished in a double-effect crystallising evaporator. Salt of 99.99% purity is produced.

From sea water by solar evaporation

At points along the West Coast of the Great Salt Lake, salt is produced from sea water by solar evaporation. In these areas, there are no rock salt deposits, and annual evaporation exceeds precipitation. For example, during the 7-month period from April to October, there is less than 125 mm of rain, but evaporation is equivalent to about 840 mm. Here sea water (containing about 3.7% solids of which about 78% is sodium chloride) is caught in ponds at high tide. After the water evaporates to about sp. gr. 1.21, at which concentration most of the calcium sulphate precipitates, it is pumped to another pond from which it flows to the rest of the system by gravity. When salt crystals start to form, the brine is run into crystallising ponds. Here about 75% of the salt separates as the evaporation continues. When the specific gravity reaches 1.25 to 1.29, the mother liquor (bittern) contains 300 to 400 g/litre of solids and is run off. It may either be discarded or worked up for other soluble salts.

At the end of the season, the ponds are drained, and the salt is harvested mechanically. It is crushed, washed with salt brine, and dried to produce an industrial grade (about 95% pure). If table salt or dairy salt is desired, the solar salt is re-dissolved and evaporated.

Other processes

A process has been proposed whereby salt will be produced as a by-product of the production of fresh water by salt-water desalination. Multiple-effect evaporation of sea water would convert 75% of the water to fresh, potable water. The remaining brine would be sent to solar evaporation ponds, and the harvested salt would be purified by the recrystallisation process previously described for upgrading rock salt. A similar effect would occur if other means for desalination, such as reverse osmosis or dialysis, are employed.

Sylvinite is a mixed crystal of sylvite (KCl) and halite (NaCl). It is usually worked for the potassium chloride content with the halite going to waste, but in a few cases the sodium chloride is recovered.

Uses

Chlor-alkali industry, soda ash, food processing, other chemicals, feed dealers and mixers.

Properties

Colourless or white hygroscopic cubic crystals. Flake salt has flat, hopper-shaped crystals. Formula wt 58.44, mp 801°C, sp. gr. 2.165, bp 1413°C (also reported as 1473°C). Soluble in water (35.7 g/100 g at 0°C, 39.8 g at 100°C). Slightly soluble in ethanol. pH of aqueous solution 6.7 to 7.3.

SODA ASH (SODIUM CARBONATE)

Soda ash, chemically known as sodium carbonate (Na_2CO_3), is a versatile inorganic chemical with a wide variety of applications in the industry and domestic sector. Its major application is in the soaps and detergents industry. Glass industry is also one of the large users of soda ash and ranks second. Apart from these two industries, it is used in a number of other industries such as sheets/hollow-wares/sleeves/tubes, paper, textile, dyes and dyestuffs, sodium silicate, sodium bicarbonate, sodium dichromate, etc. Large quantities of soda ash is used in the country for washing of clothes by dhobis and laundries, purification of industrial water and as an auxiliary chemical in other industries like resins, vanaspati, petroleum etc.

Caustic soda and chlorine are two basic products widely used in the chemical industry in India, either as raw material or as auxiliary chemical. Caustic soda is mainly used in the manufacture of pulp and paper, newsprint, viscose yarn, staple fibre, aluminium, cotton, textiles, toilet and laundry soaps, detergents, dyestuffs, drugs and pharmaceuticals, vanaspati, petroleum refining etc. Chlorine is used in the manufacture of PVC, pulp and paper, bleaching powder, textiles and a host of other inorganic and organic chlorinated compounds like metallic chlorides, refrigerants, chlorinated solvents etc. Large quantities of chlorine are also used for water purification.

Sodium carbonate Na_2CO_3, better known as soda ash, is the neutral salt of carbonic acid. It is a white crystalline, hygroscopic solid and is one of the most important raw materials used in the chemical industry. Presently it ranks eleventh in the list of commodity chemicals by volume. The normal article of commerce is a highly purified material of over 99% purity, and differences in bulk density are the only major distinction between the various grades. Soda ash competes with sodium hydroxide (caustic soda) in the production of some sodium compounds, and in certain applications some switching between the two is seen depending on economic conditions, importantly availability and price.

Production of the alkaline materials (sodium carbonate and potassium carbonate) was carried out from ancient times until the 1800s by the combustion of marine and land vegetation, followed by calcination at red heat and leaching of the ash. The term 'soda ash' can be traced to this process. The soda ash obtained in this way was a low-purity material, was very expensive and the processes too primitive for mass production, especially as they consumed vast quantities of vegetation.

Significant developments have taken place since, and the technologies for manufacture based on readily available raw materials have seen significant improvements, which have made soda ash a readily available, high volume, low value inorganic commodity chemical of significant commercial importance to several industries.

Properties

Physical properties

Sodium carbonate i.e. soda ash exists as various hydrates – anhydrous, monohydrate, heptahydrate and decahydrate. These have differing number of water molecules associated, as indicated by their names, and show differing physical properties (Table 6.1).

Most of the commercial product is anhydrous sodium carbonate. Minor quantities of sodium carbonate monohydrate and sodium carbonate decahydrate are sold and used in speciality applications.

Anhydrous sodium carbonate is an odourless, hygroscopic powder of alkaline taste. On exposure to air it gradually absorbs one mol. of water–about 15%. In air at 96% relative humidity its weight can increase 1.5% within 30 minutes. One part of soda ash dissolves in 3.5 parts water at room temperature, in 2.2 parts water at 35°C. It combines with water with evolution of heat. Its aqueous solution is strongly alkaline with a pH of 11.6. Soda ash containers should be kept well closed. Soda ash is soluble in glycerol but insoluble in alcohol.

Table. 6.1. Physical properties of sodium carbonate and its hydrates.

	Anhydrous	*Monohydrate*	*Heptahydrate*	*Decahydrate*
Formula	Na_2CO_3	$Na_2CO_3 H_2O$	$Na_2CO_3 7H_2O$	$Na_2CO_3 10H_2O$
M 1	105.99	124.00	232.10	286.14
Density at 20°C, g/cm^3	2.533	2.25	1.51	1.469
Melting point, °C	851	105[a]	35.37[b]	32.0
Heat of fusion, J/g	316			
Specific heat capacity at 25°C, Jg^{-1} K^{-1}	1.043	1.265	1.864	1.877
Heat of formation, Jg	10.676			
Heat of hydration, J/g		133.14	646.02	858.3
Crystal structure	monoclinic	rhombic bipyramidal	rhombic pseudohexagonal	monoclinic
Refractive indices, $\eta_\alpha, \eta_\beta, \eta_\gamma$	1.410, 1.537, 1.544	1.420, 1.506, 1.524	—	1.405, 1.425 1.440
Heat of solution[c], J/g	−222	−79.6	197	243

[a] in its own water of crystallisation.

[b] Incongruent.

c integrated enthalpy of solution at infinite dilution.

Sodium carbonate monohydrate is an odourless, crystalline powder with alkaline taste. It is stable at ordinary temperature, but dries out somewhat in warm, dry air or above 50°C and becomes anhydrous at 100°C. One part is soluble in 3 parts of water, 1.8 parts of boiling water, and 7 parts of glycerol, but is insoluble in alcohol.

Sodium carbonate decahydrate is also known as Nevite or Soda. The technical product is called as Sal soda or washing soda. It has transparent crystals and it effloresces on exposure to air. One part is soluble in 2 parts of cold water, in 0.25 parts boiling water, and in glycerol. It is also insoluble in alcohol.

Chemical properties

Even though the melting point of anhydrous sodium carbonate is 851°C, water vapour reacts with sodium carbonate above 400°C to form sodium hydroxide and carbon dioxide. If sodium carbonate is stored under moist conditions, its alkalinity decreases due to absorption of moisture and carbon dioxide from the atmosphere. Soda ash is decomposed by acids with effervescene.

The thermal decomposition of sodium carbonate to sodium oxide and carbon dioxide in a vacuum in the absence of chemically activating substances such as water vapour begins at about 1000°C. The dissociation pressure is about 200 Pa and increases to 10.3 kPa at 1450°C.

Sodium carbonate reacts exothermically with chlorine above about 150°C to form NaCl, CO_2, O_2 and $NaClO_4$. The elements platinum, gold, vanadium, titanium, zirconium, aluminium, molybdenum, tungsten and iron (1200°C) are attacked by fused sodium carbonate, with liberation of carbon dioxide and formation of complex metal oxide—sodium oxide compounds. At higher temperature, the Na_2O

formed reacts with excess metal to form sodium metal vapour. Sodium carbonate reacts very slowly with copper and nickel at high temperature (more than 1500°C).

Health and safety

Due to its alkaline reaction, sodium carbonate has irritating effect on the skin and mucous membranes. However, exposure to soda ash is ordinarily not hazardous but soda ash dust may produce temporary irritation of the nose and throat. Although some become accustomed to working in the dust and suffer relatively little discomfort, others are allergic to alkaline materials and develop a condition of dermatitis. The skin irritations experienced by workmen exposed to soda ash dust in hot weather are usually more severe because soda ash is likely to dissolve in perspiration.

Soda ash is corrosive to the eyes. It produces several corneal, irital an conjunctival effects (tissue destruction). If skin is attacked, the affected part should be thoroughly washed with large quantities of water, and a non-irritant dressing applied, if necessary. If eyes are splashed with sodium carbonate, the eyelids should be held open while the eye is treated for several minutes with running water or saline solution. Soda ash is harmful if ingested and may be corrosive to the linings of the stomach.

The lowest known lethal dose of sodium carbonate is 4000 mg/kg (rat, oral). For humans, oral ingestion of more than 15 gm is potentially lethal. In case of ingestion, vomiting should not be induced, but large amounts of water and dilute lemon juice or vinegar (two tablespoons per glass of water) should be drunk. The stomach can be carefully irrigated for 15 minutes at most with the usual precautions, but use of this treatment for longer periods is strongly contraindicated (danger of perforation). Medical help should be sought. Simultaneous exposure to the dusts of lime and soda ash should be avoided. The two react to form caustic soda in the presence of moisture or perspiration.

MANUFACTURE OF SODIUM CARBONATE

Process of Manufacture

Some of the processes used for manufacturing sodium carbonate are discussed as under :

Leblanc process

In the Leblanc process, common salt was treated with sulphuric acid to make sodium sulphate and hydrochloric acid. The sodium sulphate was treated with limestone and coal to produce black ash, which contained sodium carbonate, calcium sulphide, and some unreacted coal. The sodium carbonate was leached with water from the black ash. The chemical reactions of the Leblanc process are:

$$2NaCl + H_2SO_4 \rightarrow Na_2SO_4 + 2HCl$$
$$Na_2SO_4 + 2C \rightarrow Na_2S + 2CO_2$$
$$Na_2S + CaCO_3 \rightarrow Na_2CO_3 + CaS$$

A major disadvantage of the Leblanc process compared to the Solvay process is that it involves mainly solid-phase reactions and consumes large amounts of energy (at red heat or just below). The waste products (calcium sulphide and hydrochloric acid) are another disadvantage. Calcium sulphide causes both atmospheric and water pollution.

Modern processes

The modern processes for the manufacture of soda ash may be classified into two categories: Natural Processes and Synthetic Processes. The natural processes utilise the natural deposits of sodium carbonate for extracting soda ash; while the synthetic or chemical processes use various chemicals to produce soda ash. In the last two decades the importance of natural processes has significantly increased.

Natural soda ash processes

From ores

Natural evaporate deposits containing sodium carbonate in some form are found throughout the world. Some brines also contain sodium carbonate. The proportion of sodium carbonate in the salt mixture in these natural deposits seldom exceeds 40%. The accompanying compounds include sodium hydrogen carbonate, borax, sodium sulphate, sodium chloride and potassium chloride. The composition of sodium carbonate containing minerals is given below:

Trona	$Na_2CO_3.NaHCO_3.2H_2O$
Natron	$Na_2CO_3.10H_2O$
Thermonantrite	$Na_2CO.H_2O$
Hanksite	$2Na_2CO_3.9Na_2SO_4.KCl$
Pirssonite	$Na_2CO_3.CaCO_3.2H_2O$
Gaylussite	$Na_2CO_3.CaCO_3.5H_2O$

Only trona is of commercial importance. Each deposit throughout the world has its own distinctive characteristics and each requires different processing techniques.

A sodium carbonate deposit is normally, a massive, brown matrix of trona containing other alkaline minerals, as well as shales. The bed material which is mined varies slightly in quality but is usually 90% sodium sesquicarbonate. The insoluble materials are primarily dolomite shales and clays. The typical trona analysis on a water-free basis is given in Table 6.2.

Table 6.2. Typical trona analysis.

Major components	Weight %
Na_2CO_3, $NaHCO_3$	90.1
NaCl	0.08
Na_2SO_4	0.03
SiO_2	0.002
Organic	Trace
Water insoluble	9.8

Mining

Operations for each mine are generally concentrated in one thick bed of trona. The seams above and below the major bed are not recovered. Access is gained by circular, vertical concrete-lined shafts. The mines are worked by mining techniques that are similar to coal mining: room and pillar, long wall, and short wall. However, the equipment is modified to withstand the heavier and more abrasive trona ore. The extraction rates of in-place ore range from 50 to 62%. Longwall and shortwall mining techniques which involve hydraulically supporting the overburden while undercutting the trona seam have recently been introduced. The raw techniques allow over 70% recovery of the ore compared to 50% recovery for conventional room and pillar methods.

Solution mining of trona is made difficult by its low solubility in cold water and the 'blinding' of dissolving face with insoluble impurities. This method avoids expensive mine development costs and allows ore recovery in beds that are too deep for conventional mining. Pairs of wells are drilled into the ore bodies and frac-conneted using high-pressure pumps. The solvent is then circulated through the wells and sent to the processing plant.

Another mining process involves the recovery of sodium carbonate decahydrate from alkaline ponds. FMC mines this material from its solar evaporation pond using a bucket wheel dredge. The decahydrate slurry is dewatered, melted, and processed to soda ash.

Purification of trona

Several processes are used to refine trona ore and are simpler than the Solvay process. Generally, two processes, named the monohydrate and sesquicarbonate according to the crystalline intermediates, are used to produce refined soda ash from trona. Both the processes involve the same unit operations but in different sequences.

In the monohydrate process, the trona ore is calcined to impure soda ash, which is then purified. The sesquicarbonate process produces soda ash by calcination of purified sodium sesquicarbonate obtained from the trona. The simplified flow diagrams of the monohydrate process and the sesquicarbonate process are shown in Fig. 6.5. Most of the soda ash is made using the monohydrate process. However, earlier, the sesquicarbonate process was used.

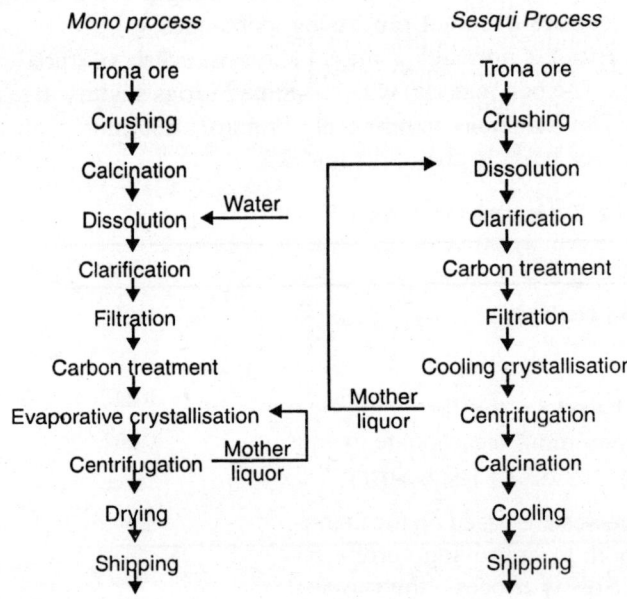

Fig. 6.5. Simplified flow diagrams for soda ash from trona.

Waste disposal

The new natural soda ash plants operate without discharge of liquid wastes to local waterways. Solar evaporation ponds are used for liquid waste disposal. Solid wastes, which for trona based processes are predominantly the insoluble material in the ore, are stockpiled. Although, environmental regulations increase construction and operating costs, their impact on natural soda ash plants is less severe than on ammonia-soda ash plants.

Production from brine

Soda ash is also produced from natural brines available from lakes. More complicated processes are required than for trona because of the complex nature of brines. The soda ash is obtained by carbonating and cooling the brine to crystallise sodium bicarbonate, the bicarbonate is filtered and calcined to light soda ash which is densified by conversion to the monohydrate followed by calcining. The procedure results in a dense ash with properties equivalent to Wyoming trona derived ash. Other salts, most notably borax, are also recovered from the lake.

In another process, brine extracted from brine lenses in the lake is evaporated to give sodium carbonate in the form of the crystalline double salt, burkeite ($Na_2CO_3.2Na_2SO_4$). Lithium compounds may also precipitate and are eventually separated from the sodium sulphate. The brine is further processed for other by-products. The burkeite is redissolved and cooled to precipitate the sodium sulphate, which is removed. The remaining solution is further treated to recover sodium carbonate. By-products include sodium sulphate, potassium chloride and borax.

Oil shale deposits

Consideration has also been given to mining nahcolite and dawsonite deposits associated with the oil shale deposits. These minerals could be co-produced with oil shale petroleum to reduce manufacturing costs of both petroleum and mineral products. Commercial practicability of such an operation has yet to be proved. However, there is an increasing interest in the use of nahcolite for scrubbing sulphur dioxide from the stack gases of power plants.

Synthetic Processes

Synthetic processes to manufacture soda ash are mainly the ammonia-soda process, usually called Solvay process. Today, the Solvay process, or its variations, is the dominant technology used throughout the world. Its continued success is based on the raw materials (salt and limestone) being more readily available than natural alkali.

Solvay process

The Solvay process is based on the formation of sparingly soluble sodium bicarbonate by reacting sodium chloride with ammonium bicarbonate in aqueous solution.

$$NaCl + NH_4HCO_3 \longrightarrow NaHCO_3 + NH_4Cl$$

All the soda ash processes are based on the manipulation of saline phase chemistry, an understanding of which is important both to improving current processes and to the economic development of new alkali resources. In the Solvay process, the raw materials are common salt and limestone. Ammonia enters into the process but is not consumed, and only a very small amount is lost. Consequently, it is not a raw material in the usual sense.

The overall chemical reaction for the entire process is:

$$CaCO_3 + 2NaCl \longrightarrow Na_2CO_3 + CaCl_2$$

However, this does not take place directly, but in a number of steps. These reactions and the required apparatus for the process are shown in Fig. 6.6.

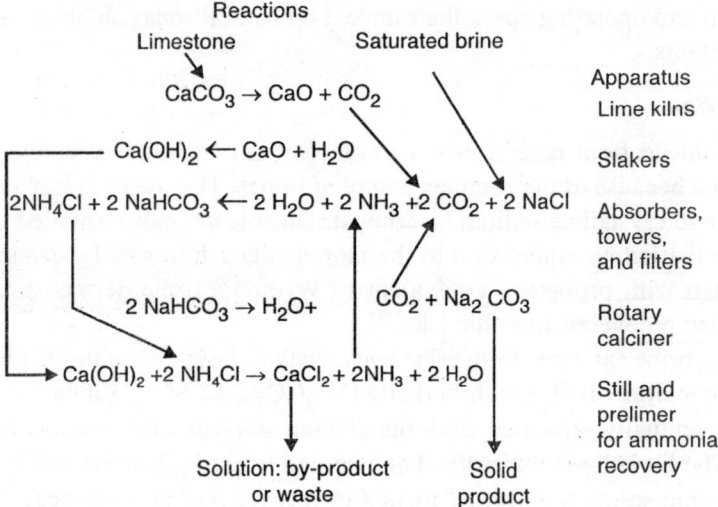

Fig. 6.6. Reaction scheme of the Solvay ammonia-soda process.

The Solvay process includes the following stages:

1. Burning of limestone or chalk to produce carbon dioxide and lime, and production of milk of lime, $Ca(OH)_2$.
2. Production of a saturated salt solution (Brine).
3. Saturation of the salt solution with ammonia (Ammonia absorption).
4. Precipitation of bicarbonate by the introduction of carbon dioxide (from stages 1 and 6).
5. Filtering and washing of precipitated bicarbonate.
6. Thermal decomposition of bicarbonate to sodium carbonate (Calcination).
7. Recovery of ammonia by distillation of the mother liquor from stage 4 (bicarbonate precipitation) with milk of lime (the ammonia liberated is recycled to stage 3).

It is convenient to consider the calcination of limestone with fuel in a kiln to produce carbon dioxide and lime (CaO), as the first step. The lime is discharged from the kiln and may be slaked with excess water to form a thick milk of lime. Common salt, in the form of near saturated purified brine, is treated with ammonia and carbon dioxide producing ammonium chloride and precipitating sodium bicarbonate, which is filtered and thermally decomposed, at about 200°C, to sodium carbonate. The filtrate contains ammonium chloride, some unreacted sodium chloride and the excess of both ammonia and carbon dioxide; the latter probably exists as bicarbonate ions, although the ammonia is not fully bicarbonated. All of the comparatively expensive ammonia is recovered in two steps. Heating alone suffices to drive off the 'free' ammonia corresponding to the bicarbonate and hydroxide ions. The hot solution, containing only ammonium chloride and unreacted salt is then treated with the milk of lime to recover 'fixed' ammonia. The liquid remaining after distillation is usually discarded in its entirety, because only a small fraction can be used for calcium chloride production, depending on demand. The reactions in Fig. 6.6 are reversible. The bicarbonation of the sodium salt is never complete and the reaction is carried only as far as economically feasible. The Solvay ammonia-soda process for the manufacture of soda ash is shown schematically in Fig. 6.7. The process employs rugged large-scale equipment, the design of which has evolved over the years.

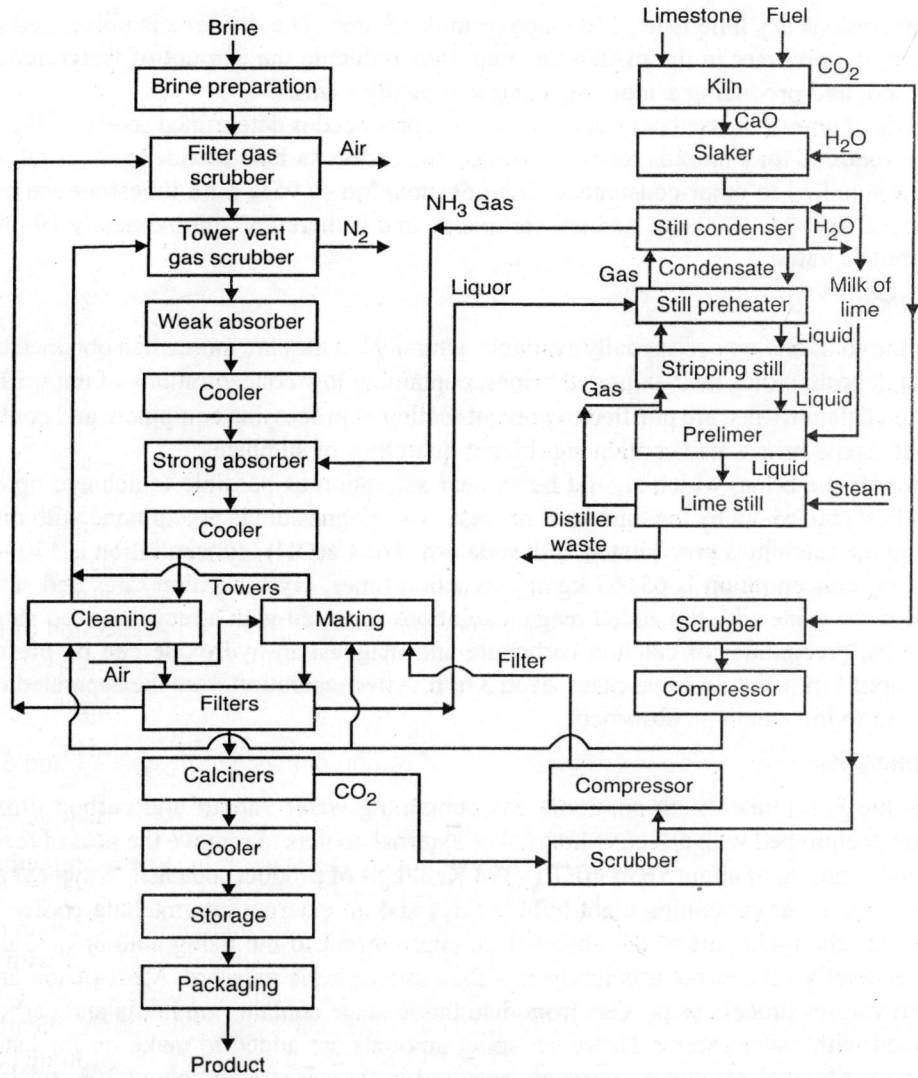

Fig. 6.7. Flow diagram for the production of ammonia-soda by the Solvay process.

Limestone burning and lime slaking

The limestone (of high purity, SiO_2 less than 3%, Fe_2O_3–Al_2O_3 less than 1.5%) is graded to a reasonably uniform coarse size. Although other fuels may be used, the limestone is usually mixed with about 7% metallurgical-grade coke or anthracite and then burned in vertical shaft kilns. Furnace gas CO_2 contents of 37–42 vol. % can be obtained. If liquid or gaseous fuels are used, the CO_2 contents of the gas is lower. The hot, dust-laden gas is cooled and washed by direct contact with water, and is compressed before sending it to the carbonation stage. Quicklime from the kiln is reacted with used cooling water (50–65°C) in horizontal rotating drums to produce a suspension of $Ca(OH)_2$ in water (milk of lime). Water addition is controlled to give a free calcium oxide content of 230–300 g/lit. In favourable circumstances, the CaO content of the milk of lime can reach 5.5 mol/lit.

In some operations dry lime is used in place of milk of lime. The dry lime is pulverised and added continuously to the prelimer in the distillation step, thus reducing the amount of water added and the steam consumed, and producing a more concentrated distiller waste.

The quantity of limestone used per tonne of soda ash produced is determined solely by the amount of calcium oxide required for ammonia recovery. Hence, the excess carbon dioxide is either released to the atmosphere or supplied to other consumers. The consumption of 95% pure limestone varies between about 1100–1200 kgs. per 1000 kgs. sodium carbonate, and is therefore approximately 10–20% higher than the theoretical value.

Brine preparation

Sodium chloride solutions are occasionally available naturally but they are more often obtained by solution mining of salt deposits. Raw, near-saturated brines, containing low concentrations of impurities such as magnesium and calcium salts, are purified to prevent sealing of processing equipment and contamination of the product. Some brines also contain significant quantities of sulphates.

Purification of the brine, which should be as near saturation as possible to achieve optimum Na^+ utilisation, is best carried out by the lime soda process. The magnesium is precipitated with milk of lime $[Ca(OH)_2]$ and the calcium is precipitated with soda ash. The $Ca(OH)_2$ concentration is 170–185 kg/m^3 and the Na_2CO_3 concentration is 65–68 kg/m^3. Reaction times, crystallisation rates and settling rates are improved if the brine with the added reagent solutions is mixed with a recycled seed slurry. Under these conditions, precipitates of calcium carbonate and magnesium hydroxide can be produced with settling rates upto 1 m/h and, in some cases, even 3 m/h. After separation from the separated impurities, the brine is sent to the ammonia absorbers.

Ammonia absorption

The strong brine is saturated with ammonia gas containing water vapour and carbon dioxide in an absorption tower equipped with effective internal or external coolers to remove the heat of reaction. The mixture requires cooling of about 1650 mJ/T (=394 Kcal/kg.) of product soda ash. A typical absorption plant consists of a tower containing eight bubble trays and an external intermediate cooler. The brine descends through the main part of the absorber counter-current to the rising ammoniacal gases. This operation is generally carried out at slightly less than atmospheric pressure. Most of the ammonia is obtained from various process steps. Gas from distillation stage contains ammonia and carbon dioxide and is saturated with water vapour. However, small amounts are added to make up for losses. In the absorption stage, the total amount of ammonia required in the process and about 20% of the required carbon dioxide are added. The ammoniacal brine obtained contains about 85–90 kg/m^3 NH_3 and 40–50 kg/m^3 CO_2. The NaCl content is reduced from about 300–260 g/lit due to the increase in the specific volume of the liquor caused by the addition of ammonia and by dilution with water. For favourable yields, more than the theoretical equivalent of NH_3 must be present and in practice about half a tonne of NH_3 is recycled per tonne of Na_2CO_3 made.

Precipitation of bicarbonate (making)

The ammoniated brine from the absorber coolers is pumped to the top of one column in a block of columns used to precipitate bicarbonate. It is an exothermic process in which the ammoniacal brine is reacted with carbon dioxide from the lime kiln or from the calcination stage. Lime kiln gas, compressed to about 414 Kpa (60 psi), enters the bottom of the cleaning column and bubbles up through the solution absorbing most of the carbon dioxide. This process is accompanied by the evolution of considerable heat, 1450 MJ/T (347 Kcal/kg), that must be removed to maintain yield. In most ammonia-soda plants, cast iron columns are used. These bubble-tray columns are equipped with tubular coolers in the lower part. Bicarbonate crystals gradually foul the heat exchange surfaces and crystallising columns must be alternately used as the cleaning column (every 3–4 days). Gases, mostly nitrogen but some carbon dioxide and ammonia, are vented from the crystallising tower columns and collected before being recycled to the absorber.

A few soda plants still use the Honigmann process in which carbonation is carried out in three to five vessels in series. The temp. profile in both column and vessel processes exhibits a maximum of 50–60°C. The suspension of bicarbonate in mother liquor leaves the carbonation stage at 30°C.

Filtration of bicarbonate

Sodium bicarbonate is generally separated from the mother liquor in continuous rotary vacuum filters or band filters, or in centrifuges. Air drawn through the vacuum filter (or the vent gas from the centrifuge operation) is returned to the ammonia absorber. Mother liquor adhering to the bicarbonate crystals is washed off with condensate produced during the production process or with softened water. The quantity of washing water required depends on the particle size of the precipitated product and varies between 0.3–1 m^3/T of bicarbonate. The filter cake is carefully washed to control residual chloride and to meet customer specifications. The filter cake often called crude bicarbonate or ammonia soda contains 9–10% moisture as residual diluted filter liquor. The cake is made up of a sodium bicarbonate and small amounts (5 mol% on a dry basis) of ammonia primarily as ammonium bicarbonate. Crude bicarbonate has the following approximate composition:

$NaHCO_3$ 75.6%, Na_2CO_3 6.9%,

NH_4HCO_3 3.4%, $NaCl + NH_4Cl$ 0.4%, H_2O 13.7%

Centrifuged products contain only 7–9% water. Calcination of 1 MT of crude bicarbonate yields about 520–560 kgs. of soda ash.

Calcination of bicarbonate

Crude filtered bicarbonate is calcined continuously by indirect heating. Various techniques are used to heat [2430 KJ/(581kcal/kg)] this material to 175–225°C in the calciners. The decomposition temperature of $NaHCO_3$ is 87.7°C at a total CO_2–H_2O pressure of 0.1 MPa. Thermal decomposition of crude bicarbonate to carbonate liberates carbon dioxide, ammonia and water vapour. The product is technical grade soda ash, which contains sodium chloride. The hot soda ash discharged from the calciner is cooled, screened, and packaged, or shipped in bulk. This product called light ash has a bulk density of around 590 kg/m^3. A certain amount is sold in this form; the majority, however is converted to dense ash. Dense soda ash is manufactured by hydrating light ash to produce the larger sodium carbonate monohydrate crystals and then dehydrating them. Hydration can be carried out in two ways: by feeding light ash and water to mixers or blenders, or by adding light ash to a saturated soda ash solution containing a slurry of monohydrate crystals. The monohydrate crystals are then fed to a continuous dryer and the dehydrated product is screened before packing and shipping. Bulk densities typically run between 960–1040 kg/m^3. Wet calcination can also be used for this purpose. This is carried out in a tower with steam in counter-current flow. In this thermal decomposition, a sodium carbonate-sodium bicarbonate equilibrium is established, which limits the yield to 85–88%.

Today, calcination is almost always carried out in rotary calciners. These are heated either externally with oil, gas or coal, or internally with steam, which passes through heating tubes in the calciner counter current to the flow of solid material. Condensing steam gives much better heat transfer than combustion gas, so that the output from steam-heated rotary calciners is considerably higher than that of calciners heated by combustion processes. The advantages of steam heated calciners compared with those heated by combustion include the absence of thermal damage to the calciner, smaller space requirements and hence lower equipment costs, and about 15% less fuel combustion.

Gases extracted from the calciners contain CO_2, NH_3, H_2O and some leakage air, together with large amounts of dust (sodium carbonate and bicarbonate), which is removed in cyclones or electrostatic filters. The gas is then cooled and washed with brine and water until it is free of ammonia. Carbon dioxide, produced at 95% or higher purity, is compressed and recycled to the carbonating tower in order to enrich the make up kiln-gas feed. The condensate produced by gas-washing is fed to the ammonia recovery stage.

Recovery of ammonia

The filter liquor contains unreacted sodium chloride and substantially all the ammonia with which the brine was originally saturated. The ammonia may be fixed or free. Fixed ammonia (ammonia chloride) corresponds stoichiometrically to the precipitated sodium bicarbonate. Free ammonia includes salts such as ammonium hydroxide, bicarbonate, and carbonate and the several possible carbon-ammonia compounds that decompose at moderate temperatures. A sulphide solution may be added to the filter liquor or corrosion protection. The sulphide is distilled for eventual absorption by the brine in the absorber. As the filter liquor enters the distiller, it is preheated by indirect contact with departing gases. The warmed liquor enters the main cake, tile or bubble-cap-filled sections of the distiller where heat decomposes the free ammonium compounds and steam strips the ammonia and carbon dioxide from the solution.

The carbon dioxide-free solution is usually treated in an external, well-agitated lining tank called a 'prelimer'. Then the ammonium chloride reacts with milk of lime and the resultant ammonia gas is vented back to the distiller. Hot calcium chloride solution, containing residual ammonia in the form of ammonium hydroxide, flows back to a lower section of the distiller. Low pressure steam sweeps practically all of the ammonia out of the lined solution. The final solution known as 'distiller waste' contains calcium chloride, unreacted sodium chloride, and excess lime. It is diluted by the condensed steam and the water in which the lime was conveyed to the reaction. Distiller waste also contains inert solids brought in with the lime.

The waste liquors are usually pumped to settling basins where the suspended solids are deposited. In some plants, calcium chloride, $CaCl_2$, is recovered from part of this solution, close control of the distillation process is required in order to thoroughly strip carbon dioxide, avoid waste of lime, and achieve nearly complete ammonia recovery. The hot (56°C) mixture of wet ammonia and carbon dioxide leaving the top of the distiller is cooled to remove water vapour before being sent back to the ammonia absorber.

Waste disposal

Large volumes of liquid waste containing suspended and dissolved solids are produced in an ammonia–soda plant. The largest quantity comes from the distiller operations where, for every tonne of product soda ash, nearly 10 cubic metres of liquid wastes are produced. This waste contains about one tonne of calcium chloride, one-half tonne of sodium chloride, and other soluble and suspended impurities. Traditionally, after settling the suspended solids in large basins, the waste was discharged into local water-ways, a practice no longer acceptable in many countries. In some countries, waste is sent to the ocean rivers or underground wells. The cost to comply with environmental regulations and increasing operating costs relative to natural ash production have forced shutdown of all synthetic soda ash plants in USA. However, several processes have been proposed to reduce or eliminate the waste streams.

Modified ammonia-soda processes

Many variants of the ammonia-soda process exist, and are used depending on regional requirements.

One such process is the ammonium chloride process (known as the dual process as it combines soda ash production with ammonium chloride production). The importance of the process is due to the high cost of imported rock salt and use of the ammonium chloride as a fertiliser. The dual process requires the use of solid sodium chloride to form NH_4Cl, but gives higher sodium conversion (+ 90%) than that obtained in the Solvay process.

Important differences exist between equipment used for the conventional Solvay process and that used for the dual process. For example, in the dual process, ammonia is not recovered, hence no distillation equipment is required. Also, lime kilns are not required if other sources of CO_2 are available.

This process has two variations named the co-production process and the mono-production process. In the co-production process, solid ammonium chloride is produced as a by-product by direct contact cooling crystallisation. In the mono-production process, the crystalline ammonium chloride reacts with a lime slurry to release the ammonia for recycle, sending the resulting calcium chloride solution to waste. Most Japanese plants have adopted some aspects of the new Asahi process which it is claimed requires less energy than the basic Solvay process and, by using direct contact cooling instead of conventional heat exchangers, reduces fouling.

Other Synthetic Routes of Manufacturing Soda-ash

Production from nepheline

In the former Soviet Union (USSR), pottery clay is produced from nepheline. The process also yields some sodium carbonate. Finely ground nepheline on sinterine with very pure limestone and after grinding the clinker with water gives the aluminate solution, which is treated with carbon dioxide to precipitate aluminium oxide hydrate and give a solution of alkali-metal carbonates. This solution is evaporated and fractionated to give sodium and potassium carbonate. In the process for converting nepheline to aluminium oxide, 4 MT of nepheline gives about 1 MT of sodium carbonate.

Huls process

Chemische Werke Huls AG has developed a process to produce soda ash and hydrochloric acid from salt via an amine-solvent system. A potential advantage of the Huls process is that, under some market conditions hydrochloric acid may be more easily sold than either ammonium or calcium chloride.

Akzo process

Akzo Zout Chemie has developed a process to produce vinyl chloride and soda ash from salt using an amine-solvent system catalysed by a copper-iodide mixture. This procedure theoretically requires half the energy of the conventional Solvay process.

Natural vs Synthetic

The dense soda ash produced from the ammonia-soda process is chemically and physically equivalent to that produced from trona (natural).

It contains slightly more impurities (Table 6.3) than the trona-based soda ash, but these slightly higher impurities cause no serious problems for the consumers.

There is wide pricing disparity between natural and synthetic soda ash. The basic reason for this is the difference in manufacturing cost between natural and synthetic ash. Local cost differentials, e.g., in fuel, may also be important (extremely site specific).

The natural plant requires approximately one-third less energy than synthetic plants. The operating labour requirements for the natural soda ash plant are less than half the synthetic. In general terms, the synthetic ash plant is a sophisticated chemical process made up of many different steps, all of which require careful control reflected in labour requirements.

Maintenance labour and supplies for the natural ash plant are about one-half those of the synthetic, showing that much less equipment is required to manufacture natural soda ash compared with synthetic soda ash. Also, the waste disposal problems in the natural process is far less serious than that in the synthetic process. The natural soda ash processes produce no large volumes of associated wastes, as against large volumes from Solvay processes.

Analyses and quality specifications

Anhydrous, technical grade soda ash is produced either by ammonia-soda or natural process. The synthetic ash is commercially available in two forms: light ash and dense ash; while the natural ash is available as medium and dense ash. Typical analysis of these forms and their granulometry are given in the Table 6.3.

Table 6.3. Soda ash analyses, % and granulometry.

Constituent	Synthetic ash		Natural ash	
	Light	Dense	Medium	Dense
Na_2O	58.3	58.3	58.4	58.4
Na_2CO_3	99.7	99.7	99.8	·99.8
Na_2SO_4	0.015	0.015	0.04	0.04
NaCl	0.18	0.18	0.03	0.03
Fe	0.0010	0.0010	0.0006	0.0006
H_2O	0.10	0.10	0.12	0.12

Light soda ash has a bulk density of 0.5–0.6 kg/lit, while the dense soda ash has a bulk density of 1.05–1.15 kg/lit.

In addition to particle size and bulk density, soda ash quality is also assessed by several chemical tests. The different physical properties of the various grades are dependent both on the crystalline product calcined (bicarbonate, monohydrate or sesquicarbonate) and on calcination conditions.

Typical properties of commercial soda ash are listed in the Table 6.4. Dense ash is preferred by the glass industry because its size and bulk density minimise segregation when mixed with other glass batch materials. Light and intermediate (medium) grades of soda ash are preferred for some detergent applications where surfactant carrying capacity and dissolution are important.

The above grades of soda ash are suitable for most applications and usually adequate for the food industry, other, special grades include dense granulated soda and briquette soda. So-called washing soda crystals, the decahydrate is now of minor importance.

Table 6.4. Typical properties of commercial soda ash.

Property	Synthetic ash		Natural ash	
	Light	Dense	Intermediate	Dense
Bulk density, g/ml.	0.59	1.04	0.80	1.04
NaCl, ppm	8000-1000	5000-300	300	300
Na_2SO_4, ppm	200	200	300	300
FeO_3, ppm	80-20	80-20	5	3
Size distribution, cumulative % on				
Sieve no. Size, μm				
+40 420	3	11	90	11
+100 149	17	90	89	90
+200 74	63	99	99	99
+325 44	86	99	99	99
Pore volume, mL/g	0.4	0.1	0.3 ·	0.1
Shape[a]	A	M	N	M

[a] Particle shapes are classified as acicular. A, Monoclinic, and blocky, M, or monoclinic and needle, N.

Indian standards

Indian Standards for technical grade of soda ash is given by IS-251 (1982). The requirements are listed in the Table 6.5.

Table 6.5. Requirements for soda ash, technical.

Characteristic	Value (on dry basis)
Total alkalinity as Na_2CO_3, % by mass, min	98.5
Matter insoluble in water % by mass, max	0.15
Sulphates as Na_2SO_4, % by mass, max	0.08
Chlorides as NaCl, % by mass, max	1.00
	(0.7 for sodium dichromate industry)
Iron as Fe_2O_3, % by mass max	0.007

Storage and transport

Calcined soda ash is hygroscopic, gradually taking up moisture and carbon dioxide during transportation and storage. This converts the surface of sodium carbonate to the bicarbonate and leads to a weight increase of up to 17%. Soda ash is therefore best stored in closed, dry areas.

Soda ash is supplied either bulk in silo wagons that can be discharged pneumatically or in paper sacks. Hopper cars, pneumatic trucks, super-sacks, and multi-wall kraft bags with polyethylene liners are the usual shipping containers.

Soda crystals (washing soda) can effloresce or dissolve in the water of crystallisation, depending on the temperature and relative humidity.

They melt at 32–33°C. At lower temperature, they are stable, provided the relative humidity is within the limits listed below:

T,°C	0	10	15	25
R.H., %	59–98	64–96	68–94	76–87

At lower humidity, soda crystals effloresce, and at higher humidity the surface becomes wet.

Applications

Sodium carbonate or soda ash has a wide variety of applications in different sectors of the chemical industry. Its sodium content results in fluxing properties that make it important in the glass and silicate industries. The biggest single use of soda ash is in glass manufacture, which has steadily increased in the past. Large quantities are used for producing other chemicals. Pulp & paper and soaps & detergent also consume significant quantities of soda ash.

It is used to neutralise inorganic and organic acids or acidic salts, and to maintain a constant pH in processes where acids are liberated. It is also used in the production of sodium salts, e.g., phosphates, nitrates, chromates, citrates, tartarates, salts of fatty acids. Sodium silicates and sodium phosphates account for the largest share of the use of soda ash for chemicals.

Soda ash can be used in aqueous solution to remove sulphur dioxide from process gas or off-gases, forming sodium sulphite and sodium bicarbonate.

These reactions are also important in the production of paper pulp by the sulphite process. Other applications of sodium carbonate are summarised here:

1. *Chemical industry:* Production of bleaching agents, borax, chromates and dichromates, fertilisers, dyes, fillers, tanning agents, industrial cleaning agents, catalysts, cryolite, adhesives, metal carbonates, sodium nitrate, perborates, phosphates, silicates, ultramarine pigments, soluble silicates, etc.
2. *Soaps and detergents:* Manufacture of detergents and saponification of fats.
3. *Petrochemicals:* Neutralisation.
4. *Pulp and paper:* Cooking wood, neutralisation, cleaning, bleaching, and treatment of recycled paper.
5. *Artificial silk industry:* Deacidification of artificial silk.
6. *Textile industry:* Dyeing, bleaching, and finishing of wool and cotton.
7. *Coke ovens, gas works and hydrogenation plants:* Gas purification (desulphurisation).
8. *Iron and steel industry:* Removal of sulphur and phosphorus from pig iron, cast iron and steel, ore benefication, flotation agents, and fluxing agents.
9. *Heavy metal industry:* Digestion and benefication of ores, e.g. of antimony, lead, chrome, cobalt, nickel, bismuth and tin.
10. *Glass industry:* Raw material for the glass melt and for reacting with sand.
11. *Aluminium industry:* Reaction with bauxite.
12. *Ceramics industry:* Production of refractory materials and glazes.
13. *Enamel industry:* As a flux.
14. *Food industry:* Hydrolysis of proteins, production of margarine and starch, and softening of sugar beet juice.
15. *Various branches of industry:* Water treatment, metal degreasing.

16. *Environmental protection:* Purification of flue gases by injecting sodium carbonate or sodium hydrogen carbonates (dry process); Regeneration of acidic lakes by the introduction of briquettes of soda ash, so that organic sediments exhibit an alkaline reaction over a long period.

17. *Miscellaneous:* General cleanser, in photography, as reagent in analytical chemistry, pharmaceutics aid (alkaliser). In veterinary treatment, soda ash has been used as an emetic. In solution it is used to cleanse skin, and in eczema, to soften scabs of ringworm.

SODIUM BICARBONATE (BAKING SODA)

Process of Manufacture

From sodium carbonate (soda ash)

$$Na_2CO_3 + H_2O + CO_2 \rightarrow 2NaHCO_3$$

Close to 100% yield

Sodium bicarbonate, also known as bicarbonate of soda and baking soda, is produced by treating a saturated solution of sodium carbonate with carbon dioxide (Fig. 6.8). Sodium carbonate (soda ash) is dissolved in the return liquor (mainly filtrate) in a dissolver to yield an essentially saturated solution; this is filtered and pumped to the top of a carbonating tower, where it comes in intimate contact with carbon dioxide gas. Sodium bicarbonate crystals are precipitated, forming a slurry.

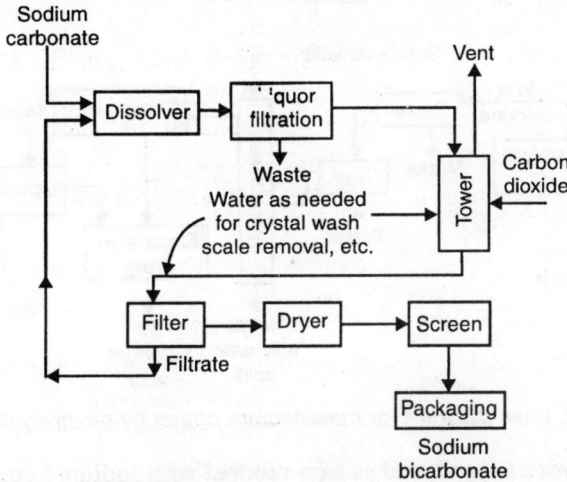

Fig. 6.8. Flow diagram for manufacture of sodium bicarbonate from sodium carbonate (soda ash).

The suspension of sodium bicarbonate thus formed is withdrawn from the bottom of the tower and filtered; the filter cake is washed. Several types of filters and centrifuges may be used. The filtrate is recycled to the dissolver.

The washed filter cake, which contains about 8% moisture, is then dried on either a continuous-belt conveyor (enclosed in a chamber) at 70°C or in a vertical-tube dryer, in which the temperature of the air is maintained at 70 to 90°C.

Following the dryer the product goes to a screening plant, where it is screened to various grades, and then to storage bins for packaging. Sodium bicarbonate of about 99.9% purity is obtained.

Uses

Food; industrial and household; pharmaceutical; fire extinguishers; soap and detergents; textile, paper, and leather; rubber and plastics etc.

Properties

White, monoclinic crystals or powder with a slightly alkaline taste. Formula wt; 84.01; Sp gr 2.20; Bulk density 850 to 1000 kg/m^3. Loses carbon dioxide progressively as the temperature is raised; the loss is significant at 100°C, and the rate increases with temperature. Soluble in water (6.9 g/100 g at 0°C, 16.4 g/100 g at 60°C). Insoluble in ethanol.

CHLORINE

Process of Manufacture

From salt by electrolysis

In the electrolysis of salt, chlorine is produced as a co-product with sodium hydroxide. Hydrogen is also obtained as a by-product (Fig.6.9).

$$2NaCl + H_2O \xrightarrow[\text{current}]{\text{Direct}} Cl_2 + 2NaOH + H_2$$

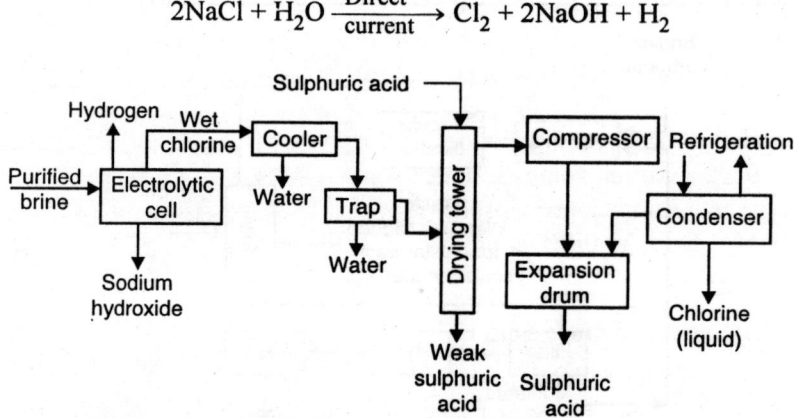

Fig. 6.9. Flow diagram for manufacture of salt by electrolysis.

In the electrolysis of salt, chlorine is produced as a co-product with sodium hydroxide. Hydrogen is also obtained as a by-product. For process details on cell operation, see Sodium Hydroxide.

From the anodes of the electrolytic cells, hot chlorine gas containing considerable water vapour is evolved. It is sent to specially constructed coolers, where the temperature of the gas is reduced to 12 to 14°C, just high enough to avoid the formation of solid chlorine hydrate, thereby circumventing the possible plugging of pipes. Most of the water vapour condenses in these coolers, which are usually of the following types: chemical stoneware chlorine lines submerged in water in trenches, water-cooled glass pipes and rubber-lined steel pipes, polyvinyl chloride (PVC) pipes, or stoneware disk coolers. Refrigeration is sometimes used, depending on the available water temperature.

The partially dried chlorine gas is then passed to drying towers, where the remaining water is removed by scrubbing with sulphuric acid, Sp. gr. 1.84. After drying, the chlorine can be handled in iron or steel equipment, whereas the extremely corrosive moist chlorine necessitates chemical stoneware, glass, rubber-lined, or PVC pipes. (In handling bulk chlorine, it is of paramount importance to exclude moisture). The dried gas goes to compressors, which almost invariably are the means of creating a partial vacuum which withdraws the chlorine from the electrolytic cells. Most plants use carbon-ring unlubricated compressors operating at about 40 psig (275 kPa). Centrifugal compressors are also used successfully. The heat of compression is removed progressively with water and refrigeration. A temperature between −30 and −45°C is attained, using single-stage or multiple-stage refrigeration, depending on the amount of compression. Most commonly employed refrigerants are of the Freon and Genetron types. Gaseous chlorine liquefies rapidly and is piped to steel storage tanks, tank cars of 27 to 50 metric tonnes capacity, or cylinders.

Some residual (less compressible) gas, called blow gas, an equilibrium mixture of air, carbon dioxide, hydrogen, and chlorine, is formed and vented for use in making such derivatives as blenching powder. Commercial liquid chlorine as produced assays higher than 99.5%. For chlorine that is to be relatively free of nonvolatile materials (chlorine content better then 99.9%), a purification procedure may be used. The chlorine to be liquefied may be passed through a bubble-cap column countercurrent to liquid chlorine. Chlorine from cells using graphite anodes may contain small amounts of chlorinated hydrocarbons arising from anode reactions; the increasingly popular metallic anodes—usually platinum-piated or -clad titanium, known as dimensionally stable anodes or DSA —do not cause such impurities in the product.

In the manufactured of caustic potash by the electrolysis of potassium chloride, chlorine is produced, as well as hydrogen, by reactions analogous to those already discussed; recovery processes are similar. Chlorine is also obtained commercially as a co-product in the manufacture of sodium from fused salt. Although chlorine is also a co-product in the electrolysis of magnesium chloride for magnesium production, this chlorine is recycled in the process and does not enter the market.

Approximately 95% of total domestic chlorine is produced as a co-product of caustic soda manufacture, as described. Another 4.5% results from other electrolytic processes: production of caustic potash and of sodium. Less than 0.5% is produced by non-electrolytic methods, as described briefly below.

From potassium chloride and nitric acid

$$3KCl + 4HNO_3 \rightarrow Cl_2 + 3KNO_3 + NOCl + 2H_2O$$

$$2NOCl + O_2 \rightleftarrows Cl_2 + 2NO_2$$

$$3NO_2 + H_2O \rightleftarrows 2HNO_3 + NO$$

$$2NO + Cl_2 \rightleftarrows 2NOCl$$

$$2KCl + 4NO_2 \rightarrow 2KNO_3 + 2NOCl$$

In the nitrosyl chloride process, chlorine and potassium nitrate are produced by treating potassium chloride with nitric acid. Potassium chloride, mixed with 63 to 66% nitric acid in a mole ratio of 1:1.7 to 1.9, is charged into a reaction tower. The reaction takes place at 75 to 125°C and atmospheric pressure to form chlorine gas along with nitrosyl chloride in a ratio of about 2:1. The nitrosyl chloride-chlorine gas mixture goes to a drying column, where it rises counter-current to 63 to 66% nitric acid at −10 to 0°C. The gases leaving the scrubber are dry enough to be handled in corrosion-resistant equipment. The gas is liquefied by cooling with brine at −15 to −20°C in a digester-condenser made of chrome–alloy steel (Fig. 6.10).

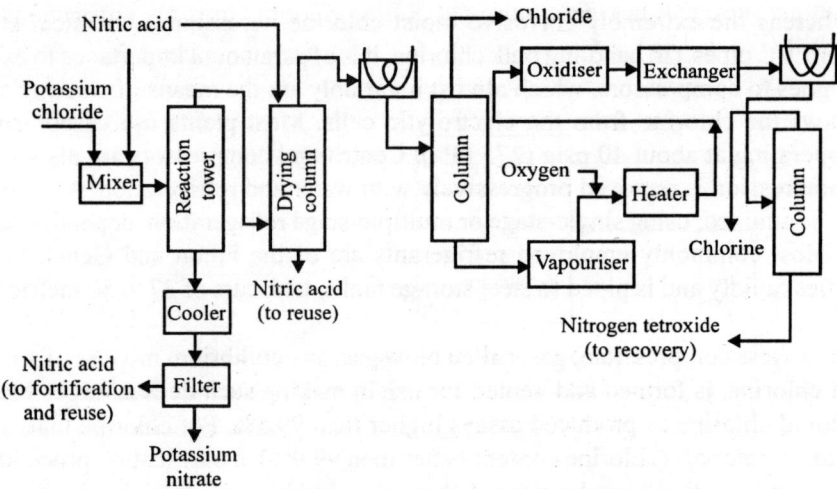

Fig. 6.10. Flow diagram for manufacture of chlorine from potassium chloride and nitric acid.

The condensed liquid is fed to a separation column from which nitrosyl chloride is removed as bottoms and chlorine gas goes overhead. The liquid nitrosyl chloride is then vapourised and heated along with added oxygen to 100 to 200°C, and sent to an oxidiser where it is converted to nitrogen dioxide and chlorine at 200 to 400°C. After cooling the gaseous effluent to 25 to 35°C in an interchanger, the mixture is liquefied with brine at –10 to –20°C. The liquid is fractionated at 15 to 60°C to send chlorine gas overhead and nitrogen tetroxide out the bottom. The nitrogen tetroxide may be converted to nitric acid by conventional methods.

This process was originally operated with salt (sodium chloride) as starting material, the coproduct being sodium nitrate. This proved uneconomical; but the analogous process using potassium salts has the attraction that the coproduct is potassium nitrate which is a superior fertiliser, especially for tobacco and other crops that cannot tolerate the chloride content of potassium chloride as a source of potash.

From hydrogen chloride and air (kel-chlor process)

Many organic chlorinations release large amounts of hydrogen chloride, and much effort has been expended on means of recovering chlorine from what is in effect a waste product. One of the more promising is the oxidation of hydrogen chloride by a series of steps not unlike the production of chlorine from potassium chloride discussed above.

$$HCl + NOHSO_4 \rightleftharpoons NOCl + H_2SO_4$$
$$2NOCl \rightleftharpoons 2NO + Cl_2$$
$$2NO + O_2 \rightleftharpoons 2NO_2$$
$$NO_2 + 2HCl \rightarrow NO + Cl_2 + H_2O$$

In the first step a solution of nitrosyl sulphuric acid in dilute sulphuric acid is treated countercurrently with a stream of hydrogen chloride and oxygen in a stripping tower. The sulphuric acid produced is flashed to remove some of the water and to cool it, and is recycled to the top of an absorber tower. The gaseous mixture of hydrogen chloride, oxygen, and nitrosyl chloride goes to a first oxidiser stage; the temperature is raised enough to dissociate nitrosyl chloride, and the oxygen reacts with the nitric oxide to form nitrogen dioxide; this reaction is carried out under 10 to 15 atm (1 to 1.5 MPa) pressure. The resulting nitrogen dioxide oxidises hydrogen chloride; thus the nitrogenous compounds function as a catalyst (Fig. 6.11).

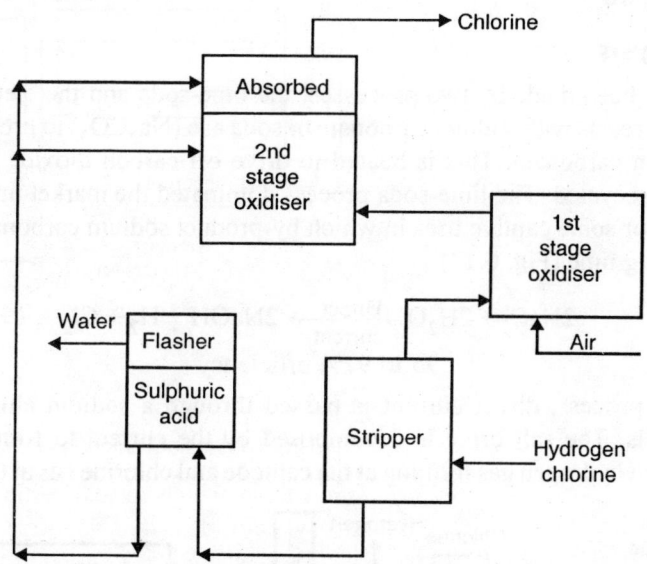

Fig. 6.11. Flow diagram for manufacture of chlorine from hydrogen chloride and Air (Kel-Chlor process).

The gases leaving the first-stage oxidiser are passed to a second-stage oxidiser, where they come in contact with a liquid obtained from the previously mentioned flashing zone. Several reactions occur here, one of which is the reverse of that occurring in the stripper. The hydrogen chloride released is largely oxidised to chlorine in this stage. The gases from the second-stage oxidiser are passed to an absorption zone, and sulphuric acid is introduced, which also removes steam from the reaction products.

The chlorine leaving the absorber tower contains a little hydrogen chloride and traces of nitrogen compounds, but is suitable for many uses. It may be further purified if necessary. Electrolysis of hydrochloric acid, to produce chlorine and hydrogen without coproduct caustic soda, has been extensively researched and even practiced, but so far has not achieved significant acceptance.

Uses

Chlorinated hydrocarbons; pulp and paper; inorganic chemicals; water and sewage; etc.

Properties

Heavy, toxic, greenish-yellow gas with a pungent, irritating odour, which may be compressed to a yellowish liquid at 15°C under a pressure of 5.7 atm or –33.6°C under atmospheric pressure. mol. wt 70.91; Sp gr. (Gas) 2.49 (0°C–Air = 1), (Liquid) 1.56 (–33.6°C); mp –101.6°C; bp –34.6°C.

Liquefaction point 6.8 atm (680 kPa) room temperature or –40°C (1 atm or 0.1 MPa). Critical temperature 146°C; Critical pressure 93.5 atm (9.35 MPa); Weight per cubic metre 3.219 kg (gas), 1468 kg (liquid); Threshold limit value 1 ppm (by volume), 3 mg/m^3.

Liquid chlorine is soluble in water (1.46% at 0°C, 0.57% at 30°C) and alkalies. Chlorine gas is soluble in water to the extent of 310 ml/100 g at 10°C and 177 ml/100 g at 30°C.

SODIUM HYDROXIDE (CAUSTIC SODA)

Process of Manufacture

From salt by electrolysis

Sodium hydroxide has been made by two processes: the lime-soda and the electrolytic. In the former, slaked lime $[Ca(OH)_2]$ reacts with sodium carbonate or soda ash (Na_2CO_3) to produce sodium hydroxide and precipitate calcium carbonate. This is heated to drive off carbon dioxide, and the resulting lime (CaO) is reslaked and recycled. The lime-soda process dominated the market until about 1940; but it is now obsolete, except for some captive uses in which by-product sodium carbonate is "recausticised" to the hydroxide by adding lime (Fig. 6.12).

$$2NaCl + 2H_2O \xrightarrow[\text{current}]{\text{Direct}} 2NaOH + H_2 + Cl_2$$
$$\text{95 to 97\% efficiency}$$

In the electrolytic process, direct current is passed through a sodium chloride (salt) solution in specially designed cells. The salt brine is decomposed by the current to form a 10 to 12% sodium hydroxide solution, with hydrogen gas forming at the cathode and chlorine gas at the anode as coproducts.

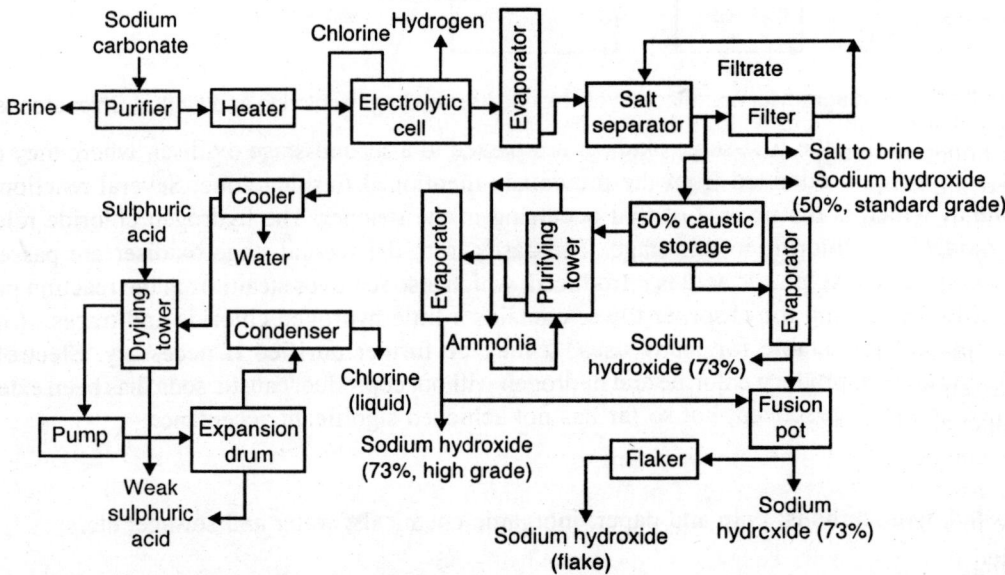

Fig. 6.12. Flow diagram for manufacture of sodium hydroxide from salt by electrolysis.

Two types of cells, the mercury cathode and the diaphragm, are used; new installations use the latter because of the danger of mercury pollution in the effluent, as well as somewhat more economical operation, although the relative economics of the two types has been a matter of debate.

In the mercury-cell process, solid salt is required for make-up of the electrolyte; the diaphragm cell can use solid salt, natural brines, or sea water concentrated by evaporation or, more recently, by dialysis. In any case, the electrolyte is a saturated (about 25%) solution of salt, which is heated and passed into purification tanks, where it is treated with sodium carbonate and some caustic soda to remove calcium and magnesium compounds. If the brine is high in sulphate, barium chloride may be added. This purification procedure is necessary both to produce a high-grade product and to reduce clogging of the diaphragm, if used, in the electrolytic cell. The purified brine is neutralised with hydrochloric acid, reheated, and fed to the cells for electrolysis.

The mercury cell, of which the Mathieson E-8 cell is typical, is more expensive to operate because of the high initial investment in mercury (ca. 1400 kg/30,000-A cell) and because of the higher voltage of operation. In this cell, mercury is the cathode; metallic sodium, not ordinarily electro-depositable from aqueous solution, deposits into the mercury as an amalgam; this is made possible by the high over voltage of hydrogen on mercury. The amalgam is decomposed in another compartment of the cell by water to produce hydrogen and a very pure (chloride-free) sodium hydroxide of 50% concentration. Chlorine is formed at the graphite anodes which operate at a current density of 4800 A/m^2 and a voltage drop of 4.5 v/cell. The current efficiency of these cells is about 95%. The cells operate at a temperature of 60 to 70°C.

The brine fed to the mercury cell, 26% sodium chloride, is reduced only to 22% sodium chloride per pass through the cell. Solid salt is used to saturate the recycle brine; this is necessary because, unlike diaphragm-cell operation, there is no water purge in the system. The brine must also be dechlorinated before recycling. The spent brine is acidified with hydrochloric acid to pH 3, and then degassed under vacuum and by blowing air counter-current to it in a spray tower. The last traces of chlorine and chlorate ion are destroyed by adding sodium bisulphite. The brine is then neutralised with caustic soda and resaturated with salt. Magnesium, calcium, heavy-metal ions, and sulphate ions are removed by adding barium or sodium carbonate to the dissolving vessel. The resulting precipitate is removed by filtration. The outstanding feature of mercury cells is the high grade and concentration of the resulting caustic liquor which may be used in the rayon industry without further purification.

There are many types and configurations of the other general type of cell, the diaphragm cell. These cells contain a diaphragm of asbestos to separate the graphite or metallic anode from the steel screen or wire cathode. Brine flows into the anode compartment, where chlorine is evolved, and then through the porous diaphragm which allows the ions to pass but reduces diffusion of the products from the cathode compartment where sodium hydroxide and hydrogen are formed.

The Hooker S-3C cell is nearly cubical and consists of three sections, one above the other. The anode assembly is in the concrete bottom of the cell, in which there is a lead casting which contains two flat copper anode-connector bars. In this base, there are 128 vertically projecting graphite anode blades.

The cathode consists of a steel frame flanged at the top and bottom. A steel screen structure is welded inside the steel frame to give an integral cathode unit. The diaphragm is applied by immersing the cathode in a bath of asbestos slurried in cell liquor and applying a vacuum to allow the asbestos to be deposited on the screen. Then the cathode with the deposited diaphragm is fitted over the graphite anodes; the concentric cell top is put into place and sealed with putty. Brine (322 g/litre and 65 to 75°C) flows into the top of the cell through a tantalum orifice, then into the anode compartment, where chlorine is formed, and then through the porous diaphragm. The cell operates with a 99.5°C catholyte, at 30,000 A or more, and with an original voltage drop of 3.95 V; this increases as the cell ages. Other characteristics of the cell are: 2500 KWh (9050 MJ)/metric tonne of chlorine; 914 kg of chlorine and 1030 kg of sodium hydroxide/day per cell; 1.4 kg of NaCl/kg NaOH; anode life 280 days, during which two or three diaphragms have been used.

The caustic soda brine solution from the cathode compartment flows out through a percolation pipe; hydrogen is removed from the back of the cell. Chlorine bubbles through the brine and accumulates at the top of the cell, whence it flows through a pipe to a header. Piping for the brine, hydrogen, and caustic liquor is usually of ceramic or polyester. Hydrogen is produced at a rate of 26.8 kg/day per cell (300 m^3/cell at standard conditions).

A variety of diaphragm cells is in use, each with slightly different design and operating characteristics but all embodying the same general principles. In particular, new installations tend to dispense with graphite anodes, which slowly disintegrate, thereby changing the geometry of the cell and introducing side products into the chlorine. In their place metallic anodes are increasingly used; these are of platinum-coated titanium, have a much longer life, and are dimensionally stable; in fact, they are known as dimensionally stable anodes (DSA).

The diaphragm cell decomposes about 50% of the brine to give a 10 to 12% caustic soda solution containing undecomposed sodium chloride. The weak caustic is charged into multieffect evaporators lined with nickel or Inconel (to avoid iron contamination), where a 50% sodium hydroxide solution is produced. Most of the salt precipitates out of the concentrated caustic and is recovered in salt separators. After filtering and washing, the salt is returned and used again to make up the charging brine. The 50% caustic soda liquor, which contains about 1% sodium chloride, may be solid as is, as standard-grade caustic liquor, purified, or concentrated further.

Several purification processes may be used and, depending on the procedure, either high-grade 50% liquor or a high-quality concentrated product is obtained. Some of the troublesome impurities in 50% caustic soda solution produced by diaphragm-cell electrolysis are sodium chloride (ca. 2%, dry basis), sodium chlorate (0.1 to 0.5%, dry basis), and colloidal iron. The last may be removed by treating the liquor with 1% by weight of calcium carbonate and filtering the resulting mixture, or by treating with chlorine. The sodium chloride and chlorates may be removed by a countercurrent treatment of the liquor with 70 to 90% ammonia in a liquid-liquid diffusion tower. The resulting 50% high-grade caustic liquor is as pure as that produced in mercury cells, and it is either sold as such or concentrated.

Another method of reducing the salt content consists of cooling the liquor to 20°C and filtering off the crystallised sodium chloride. Other processes involve both purification and concentration. For example, ammonia may be added to the 50% caustic soda solution to precipitate sodium hydroxide hydrates which contain less water than the original liquor. Alternatively, free-flowing anhydrous crystals of sodium hydroxide may be formed by treating the 50% caustic with anhydrous ammonia in counter-current pressure equipment.

The 50% standard-grade or high-grade caustic liquor may be run into single-effect evaporators, operating at 75 to 100 psi (570 to 690 kPa) steam pressure to produce either 73% standard or high-grade caustic soda liquor, depending on the purity of the dilute liquor used. These 73% solutions are handled in steam-jacketed pipes to prevent solidification, which takes place at 65°C, and are either charged into insulated tank cars equipped with steam coils or further dehydrated to yield solid sodium hydroxide.

The 70 to 75% caustic soda solution from the final evaporator is run to fusion pots constructed of special fine-grained cast iron. Here, at a final temperature of 500 to 600°C, essentially all the water is boiled off (residual H$_2$O less than 1 %). The molten caustic is generally treated with sulphur to precipitate iron and, after settling, it is discharged into thin steel drums for sale as solid caustic, or the hot anhydrous sodium hydroxide is run to a flaker to produce flaked caustic. Depending on the purity of the 73% caustic liquor charged, governed by purification or manufacturing processes, either standard-grade anhydrous caustic soda, containing 2 to 3% impurities, or high-grade anhydrous sodium hydroxide with less than 1% impurities is obtained. In addition to liquid, solid, and flake forms, caustic soda is produced in ground, powdered, and pelleted grades.

The hot, wet chlorine gas evolved at the anodes of the cells is recovered after cooling. Hydrogen is combined with chlorine to produce a high grade of hydrogen chloride, compressed for sale, used on site for chemical processes requiring it, or burned for fuel.

Uses

Chemicals; pulp and paper; aluminium; petroleum; textiles; rayon; soap and detergents; cellophane, etc.

Properties

White; deliquescent pieces; lumps or sticks. Absorbs both carbon dioxide and water from the air. Mol wt 40.00; Sp gr 2.130; mp 318.4°C; bp 1390°C. Soluble in water (42 g/100 at 0°C, 347 g/100 g at 100°C), ethanol, ether, and glycerine. Insoluble in acetone. Threshold limit value (mg/m^3) 2.

BLEACHING POWDER

When chlorine reacts with calcium hydroxide below 35°C, the product is a white, amorphous solid called bleaching powder. Its exact chemical composition is not known, but it behaves chemically as if it contained calcium hypochlorite, $Ca(OCl)_2$, and basic calcium chloride, $CaCl_2.Ca(OH)_2.H_2O$. The manufacture is carried out in towers designed to ensure that the upflowing chlorine mixes intimately with the finely-powdered slaked lime. The reaction is exothermic, and cooling is necessary to prevent the formation of calcium chlorate.

Bleaching powder reacts with carbon dioxide from the atmosphere producing hypochlorous acid, which accounts for its oxidising and bleaching actions. With dilute acids it gives chlorine, even in the cold, and this 'available chlorine', as it is called, is usually expressed as a percentage by weight of the solid, reaching about 36% in freshly made material. It can be estimated by adding a known weight of the bleaching powder to a solution containing excess of acidified potassium iodide, and titrating the liberated iodine with a standard solution of sodium thiosulphate using starch as indicator.

When some bleaching powder is warmed with a dilute solution of a cobalt salt, oxygen is evolved rapidly. This is a catalytic reaction, even traces of the cobalt compound bringing about the decomposition.

Bleaching powder is not completely soluble in water, and an aqueous suspension is normally used for bleaching wood pulp, linen and cotton, and also as a disinfectant ('chloride of lime'). Its main advantage over gaseous chlorine arises from its easier storage and safer transport.

SODIUM HYPOCHLORITE

Sodium hypochlorite is employed as a disinfectant and deodorant in dairies, creameries, water supplies, sewage disposal, and for household purposes. It is also used as a bleach in laundries. As a bleaching agent, it is very useful for cotton, linen, jute, artificial silk, paper pulp, and oranges. Indeed, much of the chlorine bought for bleaching cellulose products is converted to sodium hypochlorite before use. The most common method for making it is the treatment of sodium hydroxide solution with gaseous chlorine.

$$Cl_2 + 2NaOH \longrightarrow NaCl + H_2O + NaOCl$$

The other once widely used method was the electrolysis of a concentrated salt solution whereby the same product was made. These electrolytic cells do not have a diaphragm and are operated at high current density in a nearly neutral solution. The cells are designed to function at a low temperature and to bring the cathode caustic soda solution in contact with the chlorine given off at the anode.

SODIUM CHLORITE

Sodium chlorite ($NaClO_2$), the 80% commercial material has about 125% available chlorine.

It is manufactured from chlorine through calcium chlorate to chlorine dioxide, ending with the reaction:

$$4NaOH + Ca(OH)_2 + C + 4ClO_2 \longrightarrow 4NaClO_2 + CaCO_3 + 3H_2O$$

After filtering off the calcium carbonate, the solution of $NaClO_2$ is evaporated and drum-dried. $NaClO_2$ is a powerful but stable oxidising agent. It is capable of bleaching much of the coloration in cellulosic materials without tendering the cellulose. Hence it finds use in the pulp and textile industries, particularly in the final whitening of kraft paper. Besides being employed as an oxidiser, $NaClO_2$ is also the source of another chlorine compound, chlorine dioxide, through the reaction:

$$NaClO_2 + 1/2Cl_2 \longrightarrow NaCl + ClO_2$$

Chlorine dioxide has 2 1/2 times the oxidising power of chlorine and is important in water purification, for odour control, and for pulp bleaching.

SODIUM SULPHATE (SALT CAKE)

Process of Manufacture

From Natural Brines

Natural brines are now the principal sources of sodium sulphate. From the Searles Lake brines, sodium sulphate may be considered either a co-product or a by-product of the production of sodium carbonate, and the process is considered under that heading (Fig. 6.13).

$$Na_2SO_4 + 10H_2O \rightarrow Na_2SO_4 . 10H_2O$$
$$Na_2SO_4 . 10H_2O \rightarrow Na_2SO_4 + 10H_2O$$

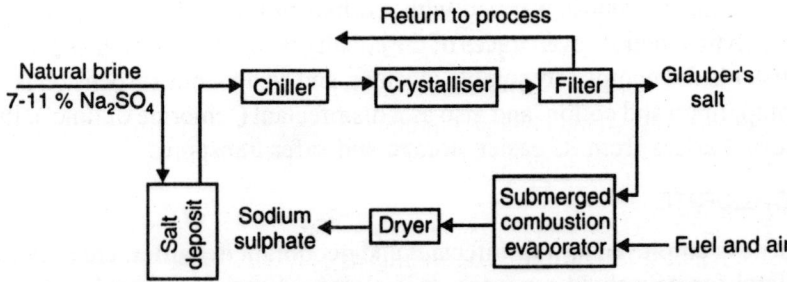

Fig. 6.13. Flow diagram for manufacture of sodium sulphate from natural brines.

The specific process used for recovery of natural sodium sulphate depends of course on the composition of the brine, and to some extent on its location. The natural brines available there run from 7 to 11% sodium sulphate, plus some sodium chloride and magnesium sulphate. To lower the sodium sulphate solubility of the brine it is saturated with sodium chloride by pumping it into a salt deposit. The salt-enriched brine leaving the well is chilled to –10 to –6°C in ammonia-cooled coils and sent to a crystalliser. The resulting Glauber's salt crystals are separated from the mother liquor by filtration. The mother liquor is returned to the process. The Glauber's salt crystals are charged to a submerged combustion evaporator where they are melted, and most of the resulting water is removed by evaporation. The wet salt product is then dried in a rotary kiln.

From salt and sulphuric acid

Still a major source of salt cake, but of diminishing importance, is the salt–sulphuric acid process for the production of hydrochloric acid. Salt cake is a by-product. Salt and sulphuric acid, sp. gr. 1.72 (slight excess) are charged to a furnace equipped with a rake agitator (Mannheim furnace), where the reacting mass is slowly heated to a temperature just below fusion (843°C). Hydrogen chloride is evolved and led through a cooling and condensing system to the absorbers. Salt cake (crude sodium sulphate) is continuously discharged from the periphery of the furnace (Fig. 6.14).

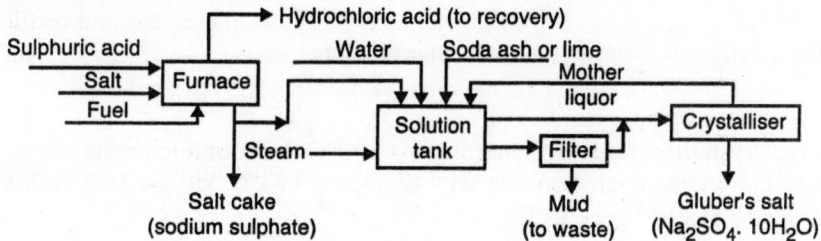

Fig. 6.14. Flow diagram for manufacture of sodium sulphate from salt and sulphuric acid.

When Glauber's salt is desired, the salt cake is dissolved in hot water to form a solution of sp. gr. 1.29. Either soda ash or lime is then added to neutralise excess sulphuric acid and to precipitate iron and alumina. The precipitate is allowed to settle, and the clear supernatant liquor is pumped to the crystalliser. The muddy bottom layer is filtered and also sent to the crystalliser. The filtered mud cake is discarded. After crystallisation the Glauber's salt is stored in closed bins to prevent desiccation. The mother liquor is returned to the solution tank for use in subsequent batches. In order for the crystals to be colourless, the crystalliser liquor must be maintained on the acid side of neutral. Free acid in the finished product will be of the order of 0.01%.

$$2NaCl + H_2SO_4 \rightarrow 2HCl + Na_2SO_4$$
$$98\% \text{ yield}$$

$$Na_2SO_4 + 10H_2O \rightarrow Na_2SO_4 . 10H_2O$$
$$95\% \text{ yield}$$

Niter cake (sodium bisulphate) may be used in the Mannheim process in place of sulphuric acid; it is mixed with the salt and fed to the furnace.

From rayon spin bath

In the manufacture of viscose rayon the coagulating or spin bath is a solution containing 9 to 11% sulphuric acid and 20% sodium sulphate, plus minor amounts of other materials. During the spinning process, 1.1 lb sodium sulphate is produced for each pound of rayon spun. The spent spin bath is recovered by further concentration of the liquor, followed by cooling in a vacuum crystalliser. Glauber's salt crystallises and may be sold as such or converted to anhydrous sodium sulphate. The mother liquor from the crystalliser is fortified with sulphuric acid and returned to the spin bath.

Other Processes

A process of some importance is Hargreaves-Robinson process. In this process sulphur dioxide, air, and steam are passed over specially prepared porous salt granules. The equation expressing the reaction is:

$$2NaCl + SO_2 + 1/2O_2 + H_2O \rightarrow Na_2SO_4 + 2HCl$$

The yield is 93 to 98%.

Two other forms of by-product salt cake of commercial importance are *chrome cake* and *phenol cake*. The former is a green form of salt cake containing a small amount of chromium. It is formed in the manufacture of chrome salts. Yellow *phenol cake* is a by-product of the manufacture of phenol by the sulphonation process.

These by-product sources are now larger than Mannheim furnace hydrochloric acid production; viscose by-product is second to natural brines, and bichromate and other by-products are third; the salt-sulphuric acid process is fourth and last.

Uses

Kraft pulp manufacture, synthetic detergents, glass manufacture.

Properties

White solid or crystals. Crystalline form is thenardite. At 100°C the rhombic crystal form changes to monoclinic, and at 500°C to hexagonal. Formula wt 142.04, mp 884°C, sp. gr. 2.70. Soluble in water (5 g/100 g water at 0°C; 40.8 g/100 g at 30°C).

SODIUM THIOSULPHATE (HYPO)

Process of Manufacture

From soda ash and sulphur dioxide

Soda ash and sulphur dioxide are simultaneously added to water to form a slurry of sodium sulphite. The sulphur dioxide is sparged well below the surface to insure complete absorption. The batch is transferred to the digestion vessel, where sulphur is added and the mixture is heated, forming sodium thiosulphate. The liquor is decanted from unreacted sulphur and concentrated to sp. gr. 1.6, causing precipitation of sodium sulphate which is filtered off and discarded.

The sodium thiosulphate pentahydrate formed in a vacuum crystalliser is centrifuged, dried, and packaged in conventional equipment. To ensure product quality, most equipment in the process is of stainless steel (Fig. 6.15).

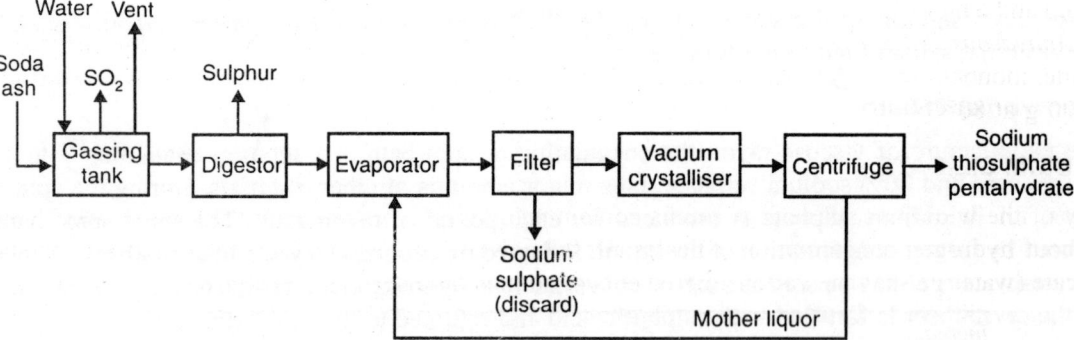

Fig. 6.15. Flow diagram for manufacture of sodium thiosulphate from soda ash and sulphur dioxide.

$$Na_2CO_3 + SO_2 \rightarrow Na_2SO_3 + CO_2$$
$$Na_2SO_3 + S \rightarrow Na_2S_2O_3$$

From sodium sulphite and sulphur

In many plants the starting material for sodium thiosulphate manufacture is sodium sulphite. In such cases the plants are essentially the same as that described beginning with the cast-iron reaction kettle.

From by-product sodium sulphide

In the manufacture of sodium sulphide, a by-product sulphide-carbonate liquor (8% Na_2S, 6% Na_2CO_3) is obtained, which may be converted to sodium thiosulphate by reaction with sulphur dioxide.

The reaction taking place is give below:

$$2Na_2S + Na_2CO_3 + 4SO_2 \rightarrow 3Na_2S_2O_3 + CO_2$$

Any excess sodium sulphide reacts with sulphur dioxide to yield sodium sulphite :

$$Na_2S + SO_2 + H_2O \rightarrow Na_2SO_3 + H_2S$$

In such a case the sulphite may be converted to the thiosulphate by treatment with sulphur. Accordingly, the plant used to produce sodium thiosulphate from by-product sodium sulphide is almost identical with that described using soda ash, sulphur dioxide, and sulphur.

From by-product of sulphur-dye manufacture

In the manufacture of sulphur dyes by fusion of sulphur, caustic soda, and certain organic compounds, the fusion product is leached with water, and the leach liquor is filtered to remove the insoluble dye. The filtrate resulting from this process contains most of the original sulphur and caustic soda in the form of sodium thiosulphate. This filtrate may be concentrated and crystallised to yield sodium thiosulphate of good quality.

Uses

Photography, Chrome tanning leather, other chemical, and miscellaneous.

Properties

White, translucent, monoclinic crystals having a cooling taste but leaving a bitter aftertaste. Formula wt 248.17, mp decomposes at 48°C, sp gr 1.685. Soluble in water (74.7 g/100 at 0°C, 301.8 g/100 g at 60°C) and ammonia. Very slightly soluble in ethanol.

Anhydrous Sodium Thiosulphate ($Na_2S_2O_3$) is the less common form and has the following properties: White, monoclinic crystals. Formula wt 158.10, Sp gr 1.667. Soluble in water (50 g/100 g at 0°C, 231 g/100 g at 80°C).

SODIUM SILICATE

Any of the widely occuring compounds containing silicone, oxygen and one or more metals, with or without hydrogen are known as silicates. Best known of the synthetic (soluble) silicates is sodium silicate (water galss). The various ingredients of sodium silicate are given below:

Ingredints	Formula	Ratio
Sodium tetrasilicate (water glass)	$Na_2Si_4O_9$	4
Sodium metasilicate	Na_2SiO_3	1
Sodium sesquisilicate	$Na_3HSiO_4 \cdot 5H_2O$	0.67
Sodium orthosilicate	Na_4SiO_4	0.5

Process of Manufacture

From sodium carbonate and silica (sand)

A variety of compounds, ranging in chemical composition from $Na_2O \cdot 4SiO_2$ to $2Na_2O \cdot SiO_2$, is produced by the fusion of silica (sand) and sodium carbonate (soda ash). By properly proportioning the reactants, the ratio of the constituent parts (Na_2O and SiO_2) may be varied to obtain a number of desired properties. Sodium silicates, varying in ratio from $Na_2O \cdot 1.6SiO_2$ to $Na_2O \cdot 4SiO_2$, are known as colloidal silicates. These are generally sold as 20 to 50% aqueous solutions called water glass. They are so named because they solidify to a glass that is water-soluble (Fig. 6.16). The expression "ratio" signifies the molar ratio of SiO_2 to Na_2O when the formulas are written in the mixed oxide form; sodium metasilicate has a ratio of 1, and water glass has ratios between 2.4 and 3.25.

Sodium metasilicate (Na_2SiO_3) is a definite crystalline compound which forms various hydrates. Substances having higher sodium oxide content are sodium sesquisilicate ($1\frac{1}{2} Na_2O \cdot SiO_2$) and sodium orthosilicate (Na_4SiO_4) ($2Na_2O \cdot SiO_2$).

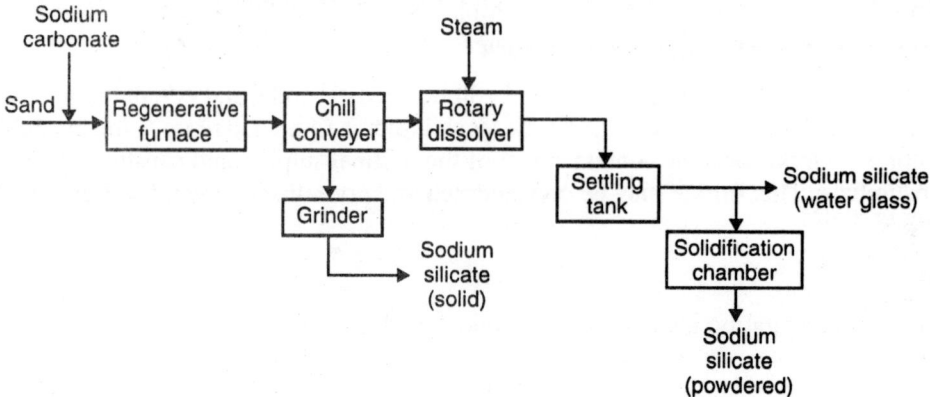

Fig. 6.16. Flow diagram for manufacture of sodium silicate from sodium carbonate and silica (Sand).

$$Na_2CO_3 + SiO_2 \rightarrow Na_2O \, SiO_2 + CO_2$$
$$(Na_2SiO_3)$$
Sodium metasillicate

The more siliceous sodium silicates are glasses, typical non-crystalline solid solutions, which are important mostly for their adhesive and binding properties. The more alkaline silicates (including sodium metasilicate) are crystalline materials with definite structures and characteristic properties. These are used chiefly as cleaners and detergents.

Sand and sodium carbonate, in selected proportions, are charged batchwise into a regenerative tank furnace resembling that used for the manufacture of glass. Fuel gas (producer, coke-oven, or natural) and air are mixed, preheated, and burned to maintain oven temperatures of 1200 to 1425°C in the hot zone. The melted materials gradually flow through the furnace evolving carbon dioxide. There is a normal shrinkage (approximately 10%) in the weight of the charge, due to the loss of gases and volatilisation of alkali oxides.

The fused melt is drawn from the furnace continuously or periodically as a thin stream. This is solidified by passage onto a moving chilled conveyer of steel moulds, in which the melt cools to a semi-transparent solid. If the hot melt is sprayed with a stream of cold water, it is shattered into fragments. The fragments are either charged into grinding and screening equipment to yield solid sodium silicates (granular) or are passed into a rotary dissolver. Here the solid material is dissolved by superheated (80-psi, 560-kPa) steam. In some plants the hot melt from the furnace is passed directly into water, in which it is dissolved with steam. The resulting solution is clarified by settling in a tank and is adjusted to the desired specific gravity. Gravities range from 1.18 to 1.91 (22 to 69°C Bé) with 1.39 and 1.56 (40 and 52° Bé) the most common. The adjusted solution is sent to storage, whence it is charged into drums and tank cars.

Dry, powdered sodium silicate may be produced by taking liquors of proper specific gravities and forcing them through a very fine opening into a solidification chamber. The chamber is swept by a rapid current of cold air which carries off the moisture. The desired sodium silicate is obtained by varying the ratio of the raw materials charged, as well as by working up the proper solution. Sodium silicate solutions may be treated with a mineral acid (H_2SO_4 and HCl) to yield hydrous silica and silicic acid. After the product is washed with water and dehydrated, silica gel results.

Sodium sulphate and carbon may be used in place of soda ash; and caustic soda solutions are used to dissolve silica in autoclaves. Potassium silicates are made by procedures analogous to that described; in much smaller quantities, lithium silicates and quaternary ammonium silicates are also produced.

Uses

Silica gel and catalysts, soaps and detergents, pigments, boxboard adhesive, water, paper and ore treatment, and roofing granules.

Properties

Sodium tetrasilicate

Water glass, Amorphous, white, deliquescent powder and lumps or clear to cloudy, colourless, aqueous solutions of varying density and viscosity. Formula wt 302.32, mp 1100°C (liquid temperature).

Soluble in hot and cold water. Insoluble in ethanol. The colloidal silicates or water glasses vary in chemical composition from $Na_2O \cdot 1.6SiO_2$ to $Na_2O \cdot 4SiO_2$. A solution containing 32% $Na_2O \cdot 1.69SiO_2$ has a specific gravity of 1.365 20°C/4 (38.8°Bé) and a viscosity of 29 centipoises (20°C°); a 32% $Na_2O \cdot 3.9SiO_2$ solution has a specific gravity of 1.298 20°C/4 (33.3° Bé) and a viscosity of 180 centipoises (20°C).

Sodium disilicate

Rhombic, pearly lustre, crystals. Formula wt 182.15, m.p. 874°C, sp. gr. 2.5. Soluble in hot and cold water.

Sodium metasilicate

Monoclinic, colourless crystals or white powder containing approximately 51.7% Na_2O and 46.2% SiO_2. Formula wt 122.06, mp 1.088°C, Sp gr 2.4. Soluble in cold and hot water, decomposing in the latter. Insoluble in ethanol. Crystallises from water as the pentahydrate and nonahydrate.

Sodium metasilicate pentahydrate

White, granular or crystalline material containing about 29.2% Na_2O and 28.7% SiO_2. Formula wt 212.14, mp 71.8°C. Very much soluble in cold and hot water.

Sodium metasilicate nonahydrate

White, efflorescent rhombic crystals. Formula wt 284.20, m.p. 40–48°C, bp loses $6H_2O$, 100°C. Very much soluble in cold and hot water; soluble (29%) in *N/2* sodium hydroxide at 18°C. Insoluble in ethanol and acids.

Sodium sesquisilicate

White, granular powder containing approximately 36% Na_2O and 24% SiO_2. m.p. 90–95°C. pH—11.6 (concentration of 0.1% by weight). Soluble in water.

Sodium orthosilicate

White, hexagonal crystals or powder. Formula wt 184.04, m.p. 1,018°C. Soluble in water.

SODIUM SULPHATE DECAHYDRATE

Properties

Large transparent crystals, small needles, or granular powder; sp. gr. 1.464 (crystals); m.p. 33°C (liquefies); loses water of hydration at 100°C. Energy storage capacity is over seven times that of water. Soluble in water and glycerol; insoluble in alcohol; solutions neutral to litmus. Non-toxic; non-flammable.

Preparation

It is prepared by crystallisation of sodium sulphate from water solution. (Glauber's salt); also occurs in nature as mirabilite.

Uses

It is used in solar heat storage; air conditioning.

SODIUM BISULPHATE

Properties

It is in the form of colourless crystals or white, fused lumps; aqueous solutions is strongly acid. Soluble in water. Sp. gr. 2.103 (13°C); m.p. 58.5°C. Non-combustible.

Preparation

A by-product in the manufacture of hydrochloric and nitric acids. Strong irritant to tissue. It can be purified by recrystallisation.

Uses

Flux for decomposing minerals; substitute for sulphuric acid in dyeing; disinfectant; manufacture of sodium hydrosulphide, sodium sulphate and soda alum; liberating CO_2 in carbonic acid baths; in thermophores; carbonising wool; manufacture of magnesia cements, paper, soap, perfumes; foods; industrial cleaners; metal pickling compounds; laboratory reagent.

SODIUM SULPHITE

Properties

It is in the form of white crystals or powder; saline, sulphurous taste. Soluble in water; sparingly soluble in alcohol. Sp. gr.: (a) 2.633; (b) mp.1.539. (a) decomposes; (b) loses $7H_2O$ at 150°C.

Preparation

1. Sulphur dioxide is reacted with soda ash and water, and a solution of the resulting sodium bisulphite is treated with additional soda ash.
2. By-product of the caustic fusion process for phenol.

Its use is prohibited in meats and other sources of Vitamin B_1.

Uses

It is used in paper industry (semichemical pulp); reducing agent (dyes); water treatment; photographic developer; food preservative and antioxidant; textile bleaching (antichlor).

SODIUM SULPHIDE

Properties

It is in the form of yellow or brick red lumps or flakes or deliquescent crystals; (a) sp. gr. 1.856 (14°C); m.p. 1180°C; (b) sp. gr. 1.427 (16°C); decomposes at 920°C. Soluble in water; slightly soluble in alcohol; insoluble in ether; largely hydrolysed to sodium acid sulphide and sodium hydroxide.

Preparation

It can be prepared by heating sodium acid sulphate with salt and coal to above 950°C, extraction with water, and crystallisation. It is flammable, dangerous fire and explosion risk. Strong irritant to skin and tissue. Liberates toxic hydrogen sulphide on contact with acids.

Uses

Organic chemicals; dyes (sulphur); intermediates; viscose rayon (sulphur removal); leather (depilatory); paper pulp; hydrometallurgy of gold ores; sulphiding oxidised lead and copper ores preparatory to flotation; sheep dips; photographic reagent; engraving and lithography; analytical reagent.

SODIUM HYDROSULPHIDE

Properties

It is in the form of colourless needles to lemon-coloured flakes. Soluble in water, alcohol, and ether. 70–72% NaSH; m.p. 55°C; water of crystallisation 26–28%.

Preparation

It can be prepared from calcium sulphide by treating it in the cold with sodium bisulphate. Contact with acids causes evolution of toxic gases.

Uses

It is used in paper pulping; dyestuffs processing; rayon and cellophane desulphurising; unhairing hides; bleaching reagent.

SODIUM NITRITE

Properties

Ii is slightly yellowish or in the form of white crystals, pellets, sticks or powder. Oxidises on exposure to air. Soluble in water; slightly soluble in alcohol and ether. Sp. gr. 2.157; m.p. 271°C; explodes at 1000°F (537°C); decomposes at 320°C. Dangerous fire and explosion risk when heated to 537°C (1000°F) or in contact with reducing materials. A strong oxidising agent. Sodium nitrite has been found to cause cancer in test animals. Its use in curing fish and meat products is restricted to 100 ppm.

Uses

It is used as diazotisation (by reaction with HCl to form nitrous acid); rubber accelerators; colour fixative and preservative in cured meats, meat products, fish; pharmaceuticals; photographic and analytical reagent; dye manufacture; antidote for cyanide poisoning.

SODIUM PEROXIDE

Properties

It is in the form of yellowish white powder, turning yellow when heated. Absorbs water and carbon dioxide from air. Active oxygen content approximately 20% by weight; sp. gr. 2.805; m.p. 460°C; b.p. 657°C. Soluble in cold water with evolution of heat.

Preparation

It is prepared when metallic sodium is heated at 300°C in aluminium trays in a retort in a current of dry air, from which the carbon dioxide has been removed. It is dangerous fire and explosion risk in contact with water, alcohols, acids, powdered metals, and organic materials. Keep dry. Toxic and irritant. Strong oxidising agent.

Uses

It can be used as an oxidising agent; bleaching of miscellaneous materials including paper and textiles; deodorant; antiseptic; organic chemicals; water purification; pharmaceuticals; oxygen generation for diving bells, submarines, etc.; textile dyeing and printing; ore processing; analytical reagent; calorimetry; germicidal soaps.

SODIUM PERBORATE

Properties

It is in the form of white odourless crystals or powder; salty taste. M.p. 63°C; loses H_2O at 130–150°C. Stable in cool, dry air but decomposes with evolution of oxygen in warm or moist air. Moderately soluble in water (with decomposition) and glycerol. pH of aqueous solutions 10.0 to 10.3. Active oxygen content 10% min.

Preparation

1. Electrolysis of a solution of borax and soda ash.
2. Crystallisation from solution of borax or boric acid, sodium peroxide, and hydrogen peroxide.

It is moderately toxic by ingestion. Fire risk in contact with organic materials; Strong oxidising agent.

Uses

It can be used in developing vat dyes; textile bleaching; synthetic detergents; neutralising cold wave preparations; dental compositions; electroplating; laboratory reagent; germicide; deodorant; mouthwash.

SODIUM AMIDE

Properties

It is in the form of white crystalline powder with ammonia odour. Decomposes in water and hot alcohol. m.p. 210°C; b.p. 400°C.

Preparation

It can be prepared when dry ammonia gas is passed over metallic sodium at 350°C. It is flammable, dangerous fire risk.

Uses

It is used in the manufacture of sodium cyanide; organic synthesis; laboratory reagent; dehydrating agent.

SODIUM CYANIDE

Properties

It is in the form of white deliquescent, crystalline powder. Soluble in water; slightly soluble in alcohol; m.p. 563°C; b.p. 149°C. The aqueous solution is strongly akaline and decomposes rapidly on standing.

Preparation

It can be prepared by absorption of hydrocyanic acid in a solution of sodium hydroxide, with subsequent vacuum evaporation. It is toxic by ingestion and inhalation. Tolerance (as CN), 5 mg per cubic metre of air.

Uses

It is used in the extraction of gold and silver from ores; electroplating; heat treatment of metals (casehardening); making hydrocyanic acid; insecticide; cleaning metals; fumigation; manufacture of dyes and pigments; nylon intermediates; chelating compounds; ore flotation.

SODIUM FERROCYANIDE

Properties

It is in the form of yellow, semi-transparent crystals. Partially soluble in water; insoluble in organic solvents. Sp. gr. 1.458. Low toxicity.

Uses

It is used in the manufacture of sodium ferricyanide; blue pigments; blueprint paper; anticaking agent for salt; ore flotation; pickling metals; polymerisation catalyst; photographic fixing agent.

CHAPTER 7

Sulphur, Sulphuric and Hydrochloric Acid

INTRODUCTION

Sulphur is one of the most important basic raw materials in the chemical process industries. It exists in nature both in the free state and combined in ores such as pyrite (FeS_2). It is also an important constituent of petroleum and natural gas (as H_2S). Various industrial and academic research groups are developing new uses for sulphur. Among some of the new uses being studied are:

1. As an additive for asphalt
2. Sulphur concretes and mortars
3. Plant and soil treatment
4. Sulphur-alkali metal batteries
5. Foamed sulphur insulation

Sulphur is an active element which combines directly with most of the known elements. It can exist in both positive and negative oxidation states and can form ionic as well as covalent and co-ordinate covalent compounds. The uses of sulphur are limited primarily to the manufacture of sulphur compounds. However, large quantities of elemental sulphur are used in the vulcanisation of rubber, in lime-sulphur sprays to destroy plant parasites, in the manufacture of artificial fertiliser and certain types of cements and electric insulators, in certain ointments and medicinals and in the manufacture of gunpowder and matches.

Sulphur compounds are used in the manufacture of chemicals, textiles, soaps, fertilisers, leather, plastics, refrigerants, bleaching agents, drugs, dyes, paints, paper and other products.

The oxides of sulphur which have been characterised have the formulae SO, SO_2, SO_3, SO_4, S_2O_3 and S_2O_7. Sulphur dioxide, SO_2 and sulphur trioxide, SO_3, are of far greater importance than the others. Sulphur dioxide can act as an oxidising agent and as a reducing agent. It reacts with water giving an acidic solution (often called sulphurous acid) and bisulphite (HSO_3) and sulphite (SO_3^{2-}) ions. The dioxide is used as a refrigerant gas, as a disinfectant and preservative, as a bleaching agent and in the refining of petroleum products.

Its major use, however, is in the manufacture of sulphur trioxide and sulphuric acid. Sulphur trioxide is used primarily in the preparation of sulphuric and sulphonic acids.

Although salts (or esters) of all the oxy acids are known, in many cases the free acids themselves have not been isolated because of their instability. Sulphurous acid is not actually known as a pure substance. Sulphuric acid (H_2SO_4) is a colourless, viscous liquid with the melting point 10.31°C (50.56°F).

It is a strong acid in water and reacts with most metals in either the dilute or concentrated form. The concentrated acid is a strong oxidising agent, especially at elevated temperatures. Pyrosulphuric acid ($H_2S_2O_7$) is an excellent sulphonating agent and loses sulphur trioxide on being heated. It also reacts vigorously with water, liberating a considerable quantity of heat. The persulphuric acids (monoperoxy-sulphuric acid, H_2SO_5, called Caro's acid and diperoxysulphuric acid, $H_2S_2O_8$, called Marshall's acid) are known as the acids and salts. Sulphonic acids are known as the esters and halides.

Sulphuric acid, H_2SO_4, is a bulk chemical of immense value. The consumption of sulphuric acid has many times been cited as an indicator of the general state of a nation's economy, and although many other indicators (such as energy consumption) might today be regarded as more important, sulphuric acid consumption still follows general economic trends and is, by far, the largest volume chemical.

Sulphuric acid, once called 'oil of vitriol', has been an important item of commerce for at least 250 years, and has been known and used since the Middle Ages, when small quantities were obtained in glass vessels when sulphur was burned with saltpeter in a moist atmosphere.

Hydrochloric acid, although not manufactured in such large quantities as sulphuric acid, is an important heavy chemical. Manufacturing techniques have changed and improved in recent years, and new procedures are now employed, such as the burning of chlorine in hydrogen.

Hydrogen chloride (HCl) is a gas at ordinary temperature and pressure. Aqueous solutions of it are known as hydrochloric acid or, if the HCl in solution is of the commercial grade, as *muriatic acid*. The common acids of commerce are 18°Bé (1.142 sp. gr.) or 27.9% HCl, 20°Bé (1.160 sp. gr.) or 31.5% HCl, and 22°Bé (1.179 sp. gr.) or 35.2% HCl. Anhydrous HCl is available in steel cylinders at a very considerable increase in cost, because of the cylinder expense involved.

The largest users of hydrochloric acid are the metal, chemical, food, and petroleum industries. The major use of hydrochloric acid is in steel pickling (surface treatment to remove mill scale). Hydrochloric acid has taken over this market because it reacts faster than sulphuric with mill scale, less base metal is attacked by it, the pickled steel has a better surface for subsequent coating or plating operations, and much smaller quantities of waste pickle liquor are produced.

SULPHUR

Process of Manufacture

From hydrogen sulphide by oxidation

The hydrogen sulphide present in "sour" natural gas or in various petroleum-refinery streams may be separated from the base fuel by standard extraction or saponification methods. In the most common process, the hydrogen sulphide is stripped from the gas with an ethanolamine solution and later liberated from the solvent.

The concentrated hydrogen sulphide may then be converted to sulphur by oxidation. Details of the process differ from plant to plant, depending on the source of the hydrogen sulphide and other factors. The following description refers to a process commonly used by petroleum refineries, in which sulphur recovery plants are almost always necessary owing to pollution control laws.

Reactions

$$2H_2S + 3O_2 \rightarrow 2SO_2 + 2H_2O$$
$$2H_2S + SO_2 \rightarrow 3S + 2H_2O$$
$$\text{Overall: } 2H_2S + O_2 \rightarrow 2S + 2H_2O$$

In the Claus process, (Fig. 7.1) one-third of the hydrogen sulphide feed to the plant is burned at 1000°C in a pressurised boiler, where 80% of the total heat of reaction is removed by generating steam. The hydrogen sulphide is converted to sulphur dioxide by the first reaction shown above; the second reaction (Claus reaction) produces sulphur.

After the removal of the molten sulphur in a condenser, the remaining gases, now 2 parts hydrogen sulphide and 1 part sulphur dioxide, are passed through a series of two or three bauxite or alumina catalyst beds at 200-260°C, with sulphur removal between each converter stage. Overhead conversion is 92–95% with two catalytic stages, 95-96% with three and 96-97% with four.

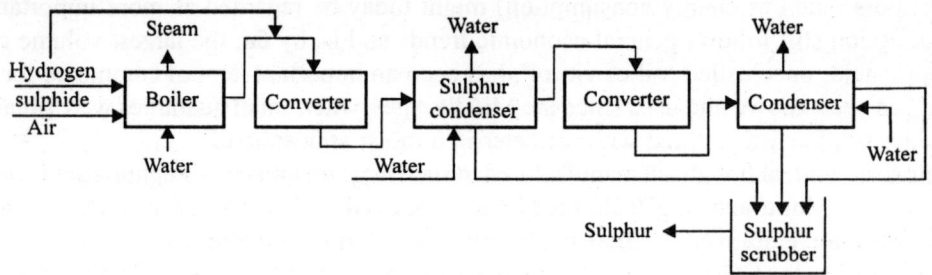

Fig. 7.1. Flow diagram for manufacture of sulphur from hydrogen sulphide.

Hydrogen or carbon dioxide in the feed gas to the plant reduces the sulphur yield by 0.25–2.5%, by allowing the formation of carbonyl sulphide (COS) and carbon disulphide (CS_2). The composition of the tail gas from the final converter is about one-third sulphur dioxide and hydrogen sulphide, one-third carbonyl sulphide and carbon disulphide, and one-third sulphur. Although this tail gas may be burned in an incinerator to yield sulphur dioxide, which is then vented to the atmosphere, pollution regulations normally do not permit this, and the tail gases must be treated. In one treatment process (Beavon process) all sulphur compounds in the Claus tail gases are reconverted to hydrogen sulphide by hydrogenation in a converter containing a cobalt molybdate catalyst. No hydrogen needs to be added, since the Claus tail gas already contains sufficient, ca. 2.5%. The gas containing hydrogen sulphide from the condenser that follows the converter may then be treated by any appropriate process to remove the hydrogen sulphide. Another treatment (Stratford process) uses sodium carbonate solution to react with the hydrogen sulphide, forming sodium hydrosulphide (NaHS) which in turn is oxidised to sulphur by sodium vanadate in solution. The finely divided sulphur froth is skimmed off, washed and dried. The effluent gas is now low enough in hydrogen sulphide to be vented.

Although what is described above refers to off-gases from petroleum refineries, the hydrogen sulphide content of other gaseous sources, such as sour natural gas, can be treated by similar methods. As a rule, differences between processes are in heat-recovery methods, or in the concentration of hydrogen sulphide in the raw material.

Properties

Sulphur (brimstone). Pale-yellow, odourless, brittle solid which commonly occurs in two crystalline forms. There is also an allotropic form known as plastic sulphur, which reverts to the crystalline form on standing. Atomic wt. 32.06; mp (rhombic) 112.8°C; mp (monoclinic) 119°C; Specific gravity (solid rhomic) 207°C; Specific gravity (monoclinic) 1.96; Liquid 1.803. Flash point (closed up) 207°C, Explosive limits Lower 30 mg/litre. Insoluble in water, soluble in various degrees in carbon disulphide, benzene, toluene, warm aniline, warm carbon tetrachloride and liquid ammonia.

Uses

sulphuric acid; pulp and paper; carbon disulphide; agriculture; rubber; sulphur dioxide; phosphorus pentasulphide etc.

SULPHUR DIOXIDE

Sulphur dioxide, SO_2, is a colourless, non-flammable gas with strong suffocating odour. It condenses at $-10°C$ at ordinary pressure to a colourless liquid. When mixed with oxygen and passed over red hot platinum, it is converted to sulphur trioxide. With water it forms sulphurous acid, H_2SO_3. Sulphur dioxide bleaches vegetable colours and is intensely irritating to the eyes and respiratory tract. Sulphur dioxide is supplied compressed in cylinders and is used in preserving fruits, vegetables, etc. Sulphur dioxide is also used as a disinfectant in breweries and food factories, bleaching textile fibres, straw, wicker ware, gelatine, glue and beet sugars. Liquid sulphur dioxide is used as a solvent.

Preparation

1. By roasting pyrites in special furnaces. The gas is readily liquefied by cooling with ice and salt, or at a pressure of 3 atm.
2. By purifying and compressing sulphur dioxide gas from smelting operations.
3. By burning sulphur.

Toxic by inhalation; strong irritant to eyes and mucous membranes, especially under pressure. Dangerous air contaminant and constituent of smog. Tolerance, 2 ppm in air. U.S. atmospheric standard 0.140 ppm. Not permitted in meats and other sources of vitamin B_1.

Properties

Colourless gas or liquid with sharp pungent odour. Soluble in water, alcohol, and ether. Forms sulphurous acid H_2SO_3. Sp. gr. 1.4337, liquid at 0°C; fp. $-76.1°C$; bp. $-10°C$; vapour pressure 3.2 atmosphere at 20°C; refractive index (liquid) 1.410 (n 24/D). An outstanding oxidising and reducing agent. Non-combustible.

Uses

Chemicals (sulphuric acid, salt cake, sulphites, hydrosulphites of potassium and sodium, thiosulphates, alum from shale, recovery of volatile substances); sulphite paper pulp; ore and metal refining; soybean protein; intermediates; solvent extraction of lubricating oils; bleaching agent for oils and starch; sulphonation of oils; disinfecting and fumigating; food additive (inhibition of browning, of enzyme-catalysed reactions, bacterial growth); reducing agent; anti-oxidant.

SULPHUR TRIOXIDE

Sulphur trioxide, SO_3 (mol. wt. 80.07) is a colourless liquid (at room temperature and atmospheric pressure) that fumes in air. Trace amounts of water or sulphuric acid can catalyse the formation of polymers. The three trimorphic phases—*alpha, beta* and *gamma*—have melting points of 62.3, 32.5 and 16.8°C, respectively. The *alpha-* form is the stable modification, while the *beta-* and *gamma-* forms are metastable. The alpha and beta forms melt to give liquid gamma SO_3. Sulphur trioxide may be prepared in the laboratory by heating fuming sulphuric acid and collecting the sublimate in a cooled receiver. If the vapour is condensed above 27°C, the gamma form is obtained as a liquid. Below 27°C and in the presence of a trace of moisture, a mixture of all three forms is obtained, which can be separated by fractional distillation.

Absolutely dry SO_3 is not corrosive to metals and shows no acid reaction. On exposure to air, it absorbs moisture rapidly, emitting dense white fumes. It combines with water with explosive violence (heat of dilution 504 cal/gm) forming sulphuric acid. Due to this acidity for water, SO_3 chars many organic substances. As a general rule sulphur trioxide reacts very briskly with all organic compounds and the organic compound may be sulphonated, oxidised (decomposed) with the release of water, or dehydrated.

Sulphur trioxide is a strong oxidising agent and at the same time it is one of the strongest known Lewis acids. The bifunctional activity is apparent, for example, in the reaction with hydrogen halides. With hydrogen fluoride, sulphur trioxide reacts as a Lewis acid to form fluorosulphuric acid. Hydrogen bromide and hydrogen iodide, on the other hand, are oxidised under the same conditions to the respective free halogens, while the sulphur trioxide is reduced to give sulphur dioxide or even hydrogen sulphide.

Sulphur trioxide reacts at 50–150°C with elemental sulphur to give sulphur dioxide. This reaction is used industrially for the production of pure sulphur dioxide (using coke or pulverised coal at elevated temperatures). Under absolutely dry conditions, sulphur trioxide is unreactive toward most metals. With metal oxides it reacts at moderately high temperature to form the corresponding metal sulphates. Due to sulphur trioxide's importance as an industrial intermediate in the production of sulphuric acid, the thermodynamics and kinetics of its generation by the oxidation of sulphur dioxide have been extensively studied, as has been the reverse reaction. Pure sulphur trioxide is extremely resistant to thermal decomposition because of kinetic inhibition, even at elevated temperatures where thermodynamic equilibrium is shifted heavily towards $SO_2 + O_2$. However, certain catalytically active substances are able to increase the rate of equilibration substantially. In the presence of metals such as platinum or of metal oxides and sulphates (e.g., of iron, copper and vanadium) the decomposition approaches equilibrium at temperatures above 700°C.

In principle, sulphur trioxide has the same toxic effects as sulphuric acid. Inhalation of the gas itself or the sulphuric acid mist formed when it comes in contact with humid air causes irritation, burning and degeneration of the tissue of moist skin, eyes and mucous membranes, especially in the respiratory tract. This happens relatively rapidly at concentrations more than 10 ppm by volume.

When storing and handling liquid sulphur trioxide, it is extremely important to prevent polymerisation, because any solid formed is very inconvenient to remove. Sulphur trioxide reacts explosively with water, so the vessels must never be cleaned with water. Concentrated sulphuric acid is required for dissolving solid deposits of sulphur trioxide. In order to maintain sulphur trioxide in the liquid state during storage, tanks and piping must be kept at about 30°C by supplementary heating. Addition of a stabiliser also effectively inhibits the formation of solid polymers at temperatures above the natural solidification point of sulphur trioxide. A number of inorganic and organic compounds are used or have been recommended as stabilisers.

Liquid sulphur trioxide is among the most dangerous of all industrial chemicals, and strict safety regulations govern its handling. In Germany, for example, every act of transporting liquid sulphur trioxide over public roads requires a special permit, which can be withdrawn at any time.

Pure sulphur trioxide is used in organic synthesis for sulphonation reactions, including the manufacture of chlorosulphonic acid, thionyl chloride, aminosulphonic acid, dimethyl sulphate and sulphamide. The reaction is difficult to control if the sulphur trioxide is introduced in liquid form, so it is preferentially supplied as a gas (pure or diluted with an inert gas), as a solution in liquid sulphur dioxide or some other solvent that is inert under the reaction conditions or as an addition compound with an organic base. Sulphur trioxide is also used in the manufacture of explosives.

SULPHURIC ACID

Properties

Physical properties

Pure sulphuric acid is a colourless, water white, odourless, slightly viscous (dense), oily, very corrosive liquid. It has a melting point of 10.4°C, a boiling pointing of 279.6°C and a density of about 1.84. It decomposes at 340°C into sulphur trioxide and water. It can be mixed with water in any ratio. Aqueous sulphuric acid solutions are defined by their H_2SO_4 content in weight-per cent terms. Anhydrous (100%) sulphuric acid is even today sometimes referred to as 'monohydrate', which simply means that it is the monohydrate of sulphur trioxide. Sulphuric acid has a great affinity for water, abstracting it from the air and also from many organic substances, and hence it chars sugar, wood, etc. It is miscible with water and alcohol with the generation of much heat and with contraction in volume. When diluting, the acid should be added to the diluent. Sulphuric acid will dissolve any quantity of sulphur trioxide, forming 'oleum' (fuming sulphuric acid). The concentration of oleum is expressed in weight-per cent of dissolved SO_3 ('free SO_3') in 100% H_2SO_4.

The physical properties of sulphuric acid and oleum are dependent on the H_2SO_4 and SO_3 concentrations, the temperature and the pressure. At constant temperature, the density of sulphuric acid increases steeply with rising H_2SO_4 concentration, reaching a maximum of about 98%. From there upto a concentration of 100% the density decreases slightly, but it rises again in the oleum range up to a concentration of about 60% free SO_3. On account of the clear relationship between density and concentration at defined temperatures in the lower concentration range, density measurement provides a quick method for determining concentrations upto about 93% H_2SO_4. Hydrometers used for this purpose were formerly calibrated in 'degrees Baume' (Be) and for that reason sulphuric acid concentration was often and sometimes still is, expressed in Be. In USA, the Baume scale is calculated utilising the formula: Degree Be = 145 − (145/sp.gr). In Europe, the Baume scale is calculated using 144.3 as the constant.

Higher concentrations are not covered because of the great difficulty in differentiating between acid concentrations in the range of 93–100% H_2SO_4 by specific gravity (density) measurements.

Sulphuric acid with free SO_3, is designated in commerce as 'oleum' (fuming H_2SO_4). Available grades contain upto about 80% free SO_3. It is colourless or slightly coloured, viscous liquid, emitting choking fumes of sulphur trioxide. It is extremely corrosive and should be handled with great care.

Chemical properties

Sulphuric acid is a strong acid with characteristic hydroscopic and oxidising properties. Like the sulphate ion, sulphuric acid is chemically and thermally very stable. The dehydrating effect of concentrated sulphuric acid is due to the formation of hydrates. Several hydrates have been identified in the solid state and these explain the irregular variation of some of the physical properties of sulphuric acid with concentration, such as its freezing temperature.

Pure sulphuric acid is ionised to only a small extent. Due to this the electrical conductivity of sulphuric acid solution has its lowest value at 100% H_2SO_4. When pure sulphuric acid is diluted with water, dissociation occurs increasingly and the conductivity rises accordingly. At higher water content the second stage of dissociation becomes increasingly important. However, on account of the diminishing total concentration of sulphuric acid, the conductivity reaches a maximum at about 30% H_2SO_4 (the exact value depends on the temperature) and decreases steeply down to 0 wt % H_2SO_4. Dilute sulphuric acid is the preferred electrolyte for industrial metal electrowinning and electroplating plants on account of its high conductivity and the chemical stability of the sulphate ion. To take advantage of the electrical conductivity maximum sulphuric acid of about 33% concentration is used in lead storage batteries.

Dilute sulphuric acid is a strong dibasic acid and dissolves all base metals forming respective metal sulphates and bisulphates (hydrogen sulphates) with the release of hydrogen. Hot, concentrated sulphuric acid has an oxidising effect, reacting with precious metals and with carbon, phosphorus and sulphur by which it is reduced to sulphur dioxide. A very important property of sulphuric acid is, its ability to decompose the salts of most other acids. Industrially important examples are given below:

1. Production of sodium sulphate and hydrogen chloride from sodium chloride.
2. Decomposition of sulphites to sulphur dioxide.
3. Decomposition of phosphate rock (natural calcium phosphates) to phosphoric acid and calcium sulphate.

The reactions of concentrated sulphuric acid with organic compounds are frequently dominated by its oxidising and hygroscopic properties, e.g., carbohydrates are decomposed to the point of carbonisation. Organic condensation reactions in which water is eliminated are promoted by sulphuric acid because it effectively removes the water as soon as it is formed. Sulphuric acid is, therefore, frequently used in industry for this purpose. It also exercises a catalytic effect on certain reactions involving organic compounds.

Sulphuric acid is oxidised both by hydrogen peroxide and anodically to diperoxysulphuric acid $H_2S_2O_8$, and the unstable monoperoxysulphuric acid (Caro's acid), H_2SO_5. Since it is a strong oxidant, Caro's acid can oxidise sulphur dioxide to sulphuric acid, a property that has been exploited in pollution control for sulphuric acid plants. Nitrogen oxides ($NO + NO_2$) react with sulphuric acid at concentrations above 70 wt% H_2SO_4 to give nitrosyl hydrogen sulphate, $NOHSO_4$.

Processes of Manufacture

Sulphuric acid is manufactured by chamber process and contact process. However, the chamber process, i.e., nitrogen oxide process is disadvantaged due to lower product concentration of around 78% compared to 98% concentration possible by the contact process.

Initially, platinum was used as catalyst in the contact process and remained the predominant catalyst until the 1930s. However, vanadium pentoxide eventually succeeded in replacing the platinum catalyst because of its insensitivity to catalyst poisons and its considerably lower cost.

With the development of vanadium pentoxide catalyst together with increasing demand for concentrated sulphuric acid, the proportionate share of world sulphuric acid output produced by chamber process declined steadily. Nonetheless, the nitrogen oxide process has continued to be the object of interest and a certain amount of development work, especially for the processing of gases with extremely low sulphur dioxide content.

In 1936–37, Lurgi introduced the wet contact process for converting moist sulphur dioxide-containing gases over a vanadium catalyst. This made it possible to process hot gases from the combustion of hydrogen sulphide in coking plants directly to sulphuric acid. In succeeding years, a number of factors affected the development of the contact process. First, the raw material basis of the industry changed progressively from mainly roaster gases to sulphur combustion gases containing higher concentrations of sulphur dioxide. Second, plant capacities increased as a result of rapid rise in the consumption of sulphuric acid by the fertiliser industry. These and other factors provided a stimulus for the introduction of improvements in the individual process steps and in the design of associated equipment (e.g., the shift to tray converters from tube converters).

In 1960, Bayer patented the so-called double-catalysis process, and the first plant using this process, built by Lurgi, started up in 1964. By incorporating a preliminary SO_3 absorption step ahead of the final catalytic stages, this improved contact process permitted a decisive increase in overall SO_2 conversion, thus reducing SO_2 emissions substantially. Because the essential difference between this process and the ordinary contact process is in the number of absorption stages, it is referred to as the 'double-absorption' process. However, this process did not substantially change the nature of the process or the process equipment. In the 1970s the principal industrial countries introduced more stringent regulations for environmental protection, which made the use of the double-absorption process more or less mandatory in new plants. Nevertheless, the conventional contact process continues to be used in countries where environmental regulations are less exacting.

In the 1970s and 1980s, the increased value of energy and production of sulphuric acid from a variety of waste products, including off-gases and spent sulphuric acid, led to a number of process and equipment modifications. In recent years also, the main thrust of development in the contact process is toward increasing the recovery and utilisation of the very substantial amount of process heat. Indeed, a large, modern sulphuric acid plant may be looked upon not just as a chemical plant but also as a thermal power plant.

Today, the contact process is the most widely used process for the manufacture of sulphuric acid throughout the world. Sulphuric acid may be produced by the contact process from a wide range of sulphur-bearing raw materials by several different process variants, depending largely on the raw material used. In some cases, sulphuric acid is made as a by-product of other operations, primarily as an economical or convenient means of minimising air pollution or disposing of unwanted by-products.

The raw materials used to make sulphuric acid are elemental sulphur, spent (diluted and contaminated) sulphuric acid, and hydrogen sulphide. Elemental sulphur is by far the most widely used. Till 1970s iron pyrite and related compounds were the predominant raw materials. A large amount of sulphuric acid is also produced as a by-product of non-ferrous metal smelting, i.e., roasting sulphide ores of copper, lead, molybdenum, nickel, zinc or others. Moderately concentrated sulphuric acid is also produced by concentration and purification of spent or waste sulphuric acid. Re-processing of sulphuric acids generated in large quantities in many processes and recycling such regenerated acid to the user is becoming increasingly important from an environmental production standpoint.

The sulphuric acid manufacturing process may be divided into the following distinctive steps:

1. Generation of sulphur dioxide gas.
2. Catalytic oxidation of sulphur dioxide to sulphur trioxide.
3. Absorbing of sulphur trioxide to form sulphuric acid.

The chemical reactions may be represented as follows:

1. $S + O_2 \longrightarrow SO_2$
2. $SO_2 + \frac{1}{2} O_2 \longrightarrow SO_3 \qquad \Delta H = -99.0 \text{ KJ}$
3. $SO_3 + H_2O \xrightarrow{\text{Catalyst}} H_2SO_4 \qquad \Delta H = -132.5 \text{ KJ}$

All the three reactions are exothermic. In addition to the above mentioned process steps, drying of sulphur dioxide gas and cooling of sulphuric acid (after SO_3 absorption) are important. The gas drying stage is not applicable to a plant of wet catalysis type which are relatively uncommon. Almost without exception, the contact process plants operate under essentially atmospheric pressure; compression is required only for driving the gases through the plant.

Where elemental sulphur or hydrogen sulphide is used as raw material, considerable heat is evolved during the initial combustion. Additional heat is generated in the next two steps. In such plants, most of the heat is typically used to produce steam utilised either for heating requirements in other processes or to generate power via turbines. In many cases, large plants of this type are essentially co-producers of steam or power and sulphuric acid; products which have significant commercial value. Where spent acid is used as raw material, it usually is decomposed in furnaces fired by gas, oil, or other fuels (sometimes H_2S or sulphur) and the high temperature gas from such furnaces can also generate steam or power.

Generation of sulphur dioxide gas

In all types of contact plants, the first step is to produce a reasonably continuous, contaminant-free gas stream containing appreciable sulphur dioxide and some oxygen. This gas is preferably dry, but plants can be designed to handle wet gas directly from H_2S combustion. This requires careful design of equipment to minimise mist formation in the condensation —absorption section of the plant. If the initial oxygen concentration of the process gas is low, additional air or oxygen must be added prior to or during catalytic oxidation to ensure that there is an excess over stoichiometric needs for conversion of SO_2 to SO_3.

In general, SO_2 gas derived from metallic sulphides, spent acids or gypsum anhydride is purified before drying by cold, i.e., wet, gas purification. Various equipment combinations including humidification towers, reverse jet scrubbers, packed gas cooling towers, impingement tray columns and electrostatic precipitators are used to clean the gas. Plants that produce good quality elemental sulphur or H_2S gas generally have no facilities for purifying SO_2.

Before the advent of relatively pure Frasch or recovered sulphur, however, hot gas purification was frequently used in which the SO_2 gas stream was passed through beds of granular solids to filter out fine dust particles just prior to its entering the converter. Sulphur shipped as a solid frequently becomes contaminated with dirt and scale during shipping and handling. In places where solid sulphur is still handled, molten sulphur is frequently filtered prior to use as an alternative to, or in combination with, hot gas purification. The air used for combustion must be dried. In the usual arrangement, filtered air from the atmosphere is drawn through a drying tower by the main blower. Since most of the plants use sulphur burning to generate SO_2 gas, it is described in detail here. Spent acid or H_2S burning and ore roasting, sintering or smelting are also used to some extent and require specialised equipments.

For sulphur burning, there is a trend toward very large single-train plants. Due to this the usual practice is to use horizontal, brick-lined combustion chambers with dried air and atomised molten sulphur introduced at one end. Atomisation typically is accomplished either by pressure spray nozzles or by mechanically driven spinning cups. Because the degree of atomisation is a key factor in producing efficient combustion, sulphur nozzle pressures are typically 2.76 Mpa (150 psi) or higher. Sulphur burners are typically designed as proprietary items by companies specialising in acid plant design and construction. Sulphur burners are normally operated at moderate pressures, in the other temperatures and concentrations are in similar proportion. SO_2 concentrations in the gas stream range from 4–14 vol%.

The gases are dried, generally in counter-current with fairly concentrated sulphuric acid in irrigated packed towers. The residual water content of the gases after drying corresponds theoretically to the partial pressure of water vapour above the drying-tower acid at the prevailing temperature and concentration. To achieve high drying efficiency, the temperature of the irrigation acid is normally maintained at 50–60°C (by cooling).

At high flame temperatures, small amounts of nitrogen react with oxygen to form nitrogen oxides, NO_x, primarily nitric oxide, NO. Ultimately, some of them form nitrosylsulphuric acid, which ends up either as trace amounts in product acids or, in considerably higher concentrations, as condensed acid collected at mist eliminators.

Conversion of SO_2 to SO_3 and its absorption

The catalytic oxidation of SO_2 to SO_3 is highly exothermic and, equilibrium becomes increasingly unfavourable for SO_2 formation as temperature increases above 410–430°C. Unfortunately, this is about the minimum temperature level required for typical commercial catalysts to function. Consequently, catalytic reactors (called as converters) are typically designed as multistage adiabatic units and have gas cooling between each stage.

To improve equilibrium or driving force for the conversion, the sulphuric acid industry has attempted one or a combination of the following methods—design modifications—increasing concentration of SO_2 in the process gas stream, increasing concentration of oxygen in the process gas stream by air dilution or oxygen enrichment, increasing the number of catalyst beds, removing the SO_3 product by interpass absorption—known as the double absorption process, lowering catalytic converter inlet operating temperature, i.e., using better catalysts, and increasing the catalytic converter operating pressure (pressure plants). In early years, the contact process frequently employed only two or three catalyst stages (passes) to obtain overall SO_3 conversions of approximately 95–96%. Later four-pass converters were used to obtain conversions from 97% to slightly better than 98%.

For sulphur-burning plants, this typically resulted in sulphur dioxide stack emissions of 1500–2000 ppm. The air pollution requirements led to the adoption of the double contact or double absorption process, which provides overall conversions better than 99.7%. Most industrialised nations have emission standards that cannot be achieved without utilising double absorption or tail-gas scrubbers.

Single absorption sulphur-burning plants

These plants were standard in the industry for many years. These single absorption plants used either relatively low strength (about 8 vol. %) SO_2 gas without air dilution, or air dilution designs and higher (about 10 vol. %) inlet gas strength. Air dilution was a common design option using additional dry air, instead of heat exchangers, to cool the process gas entering the last one or two converter passes.

The additional air improved conversion at the final converter pass by increasing oxygen concentration and reducing equivalent sulphur dioxide concentration of the process gas. Its chief advantage was reduced investment over designs with heat exchangers for inter-pass cooling.

Fig. 7.2 shows a typical four-pass, single absorption sulphur-burning plants without air dilution. Plants of this design burn sulphur to generate a process gas stream of about 7.5–9.0% SO_2. Typical converter operating conditions are listed in Table 7.1.

Table 7.1. Converter conditions for a single absorption plant.*

Converter pass temperature, °C	1	2	3	4
Inlet	410-445	430-450	430-435	425-430
Outlet	595	500	450	430-435
ΔT	150-185	50-70	15-20	5

* Using 8% SO_2, inlet gas.

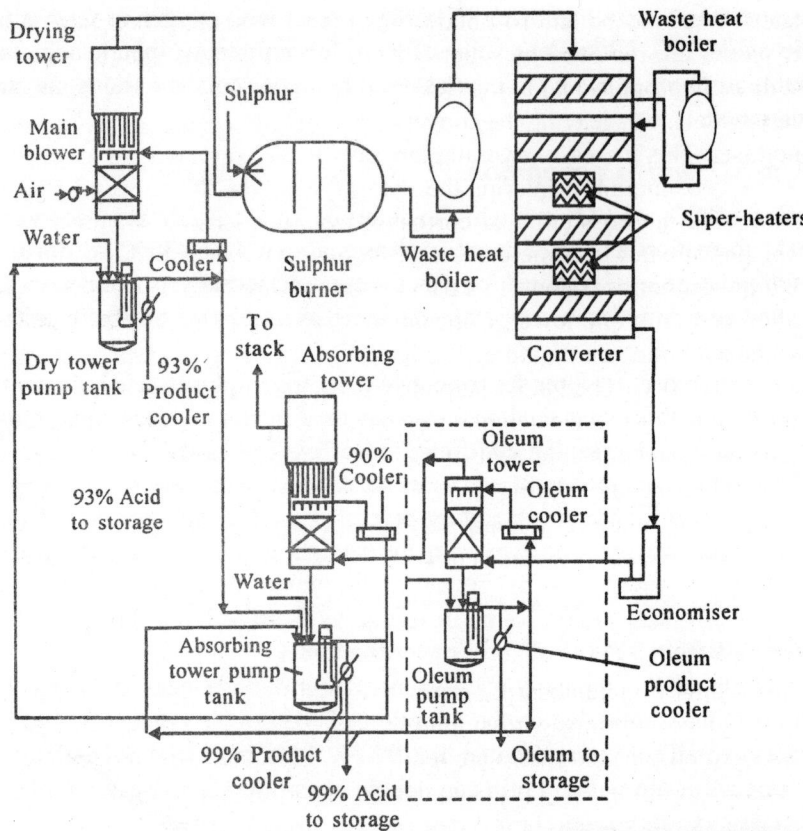

Fig. 7.2. Flow sheet for a single absorption sulphur-burning plant.

The scheme shown produces a full range of products including 66 deg. Be and 98.5% acid and oleum. If 66 deg. Be acid product is not needed, a common pump tank is usually used for both the drying and absorbing towers. If oleum production is not needed, the equipment within the dashed box is omitted and process gas enters the absorbing tower directly from the economiser.

Most of the heat of combustion from the sulphur is removed in a boiler, which reduces the process gas temperature to the desired converter inlet temperature. Typically, the inlet temperature (Table 7.1) to the first converter pass is dictated by catalyst performance, catalyst bed depth and process gas strength. Standard, i.e., sodium- or potassium-promoted, vanadium pentoxide catalysts do not have sustained catalytic activity at temperatures of more than 400–410°C, although fresh catalysts may have an initial reaction ignition temperature as low as 385°C. Such low ignition temperatures cannot be sustained by conventional catalysts. The catalyst ignition temperature is the temperature below which substantial catalytic conversion (approaching equilibrium) cannot be sustained in any given bed or pass. Newer catalysts promoted with cesium, have a considerably lower sustainable ignition temperature (about 375°C) and have proved useful in special situations.

Sulphur dioxide gas is catalytically oxidised to sulphur trioxide in a fixed bed reactor (converter) which operates adiabatically in each catalyst pass. The heat of reaction raises the process gas temperature in the first pass to approximately 600°C (Table 7.1). The temperature of hot gas exiting the first pass is then lowered to the desired second pass inlet temperature (430–450°C) by removing the heat of reaction in a steam super-heater or second boiler. To obtain optimum conversion, the heats of reaction from succeeding converter passes are removed by super-heaters or air dilution. The temperature rise of the process gas is almost directly proportional to the SO_2 converted in each pass, even though SO_2 and O_2 concentrations can vary widely. Gas leaving the converter is normally cooled to 180–250°C using boiler feedwater in an economiser. This increases overall plant energy recovery and improves SO_3 absorption by lowering the process gas temperature entering the absorption tower. The process gas is not cooled to a lower temperature to avoid the possibility of corrosion from condensing sulphuric acid originating from trace water in the gas stream. In some case, a gas cooler is used instead of an economiser.

Process gas leaving the economiser flows to a packed tower where SO_3 is absorbed. Most plants do not produce oleum and need only one tower. Concentrated sulphuric acid circulates in the tower and cools the gas to about the acid inlet temperature. The typical acid inlet temperature for 98.5% sulphuric acid absorption towers is 70–80°C. The 98.5% sulphuric acid exits the absorption tower at 100–125°C, depending on acid circulation rate. Acid temperature rise within the tower comes from the heat of hydration of sulphur trioxide and sensible heat of the process gas. The hot product acid leaving the tower is cooled in heat exchangers before being recirculated or pumped into storage tanks. Acid circulated over SO_3 absorbing towers is maintained at about 98.5% to minimise its vapour pressure. Where lower concentration product acid is desired, it is made either in separate dilution facilities, or in drying towers operated at 93–96% H_2SO_4.

Double absorption plants

Currently, the sulphur-burning double-absorption process is considered to be the standard sulphuric acid production process for conforming with sulphur dioxide emission limits, now in force in most countries. In USA, newer sulphuric acid plants are required to limit SO_2 stack emissions to 2 kg of SO_2 per metric tonne of 100% acid produced. This is equivalent to a sulphur dioxide conversion efficiency of 99.7%. Acid plants used as pollution control devices, for example those associated with smelters, have different regulations. This high conversion efficiency is not achievable by single absorption plants using available catalysts, but it can be attained in double absorption plants when the catalyst is not seriously degraded.

A typical double absorption plant design uses intermediate SO_3 absorption after the second or, more commonly, the third converter pass. This is called the 3 + 1 configuration. As of the mid-1990s, newer double absorption plants usually contained a total of four catalyst passes in a 3 + 1 configuration. Plants having five passes in a 3 + 2 configuration have also been built. The Fig. 7.3 presents a typical flow diagram for a double absorption sulphur-burning plant. Typical converter temperatures in a double absorption converter are given in the Table 7.2.

Table 7.2. Converter conditions for a 3 + 1 double absorption plant*.

Converter pass temperature, °C	1	2	3	4
Inlet	415-420	430-445	430-445	425-430
Outlet	600-610	530	470	450
ΔT	185	90	30	20

* Using 11.5% SO_2 inlet gas.

Approximately 90–95% of total sulphur trioxide produced by the double absorption process is absorbed in the inter-pass absorption tower. The sulphur trioxide produced in subsequent converter passes is absorbed in the final absorbing tower. Inter-pass absorbing tower operation is similar to an absorbing tower in a single absorption plant. In both cases, acid irrigation rates are designed so that acid temperature exiting the tower is 100–125°C. The smaller amount of sulphur trioxide absorbed in the final absorption tower of double absorption plants typically raises its acid temperature to only upto 105°C. Inter-pass and final absorbing towers of double absorption plants are very similar in size because tower diameter is dependent on total gas throughput, not sulphur trioxide concentration.

Another major difference between single and double absorption processes is that, after inter-pass absorption, the process gas is re-heated from approximately 80°C to approximately 425°C before re-entering the converter. This re-heating is accomplished in gas-to-gas heat exchangers (Fig. 7.3), using some of the heat from the initial converter passes. The gas-to-gas heat exchangers are a primary cost item. All other plant operations are very similar to the corresponding single absorption processes.

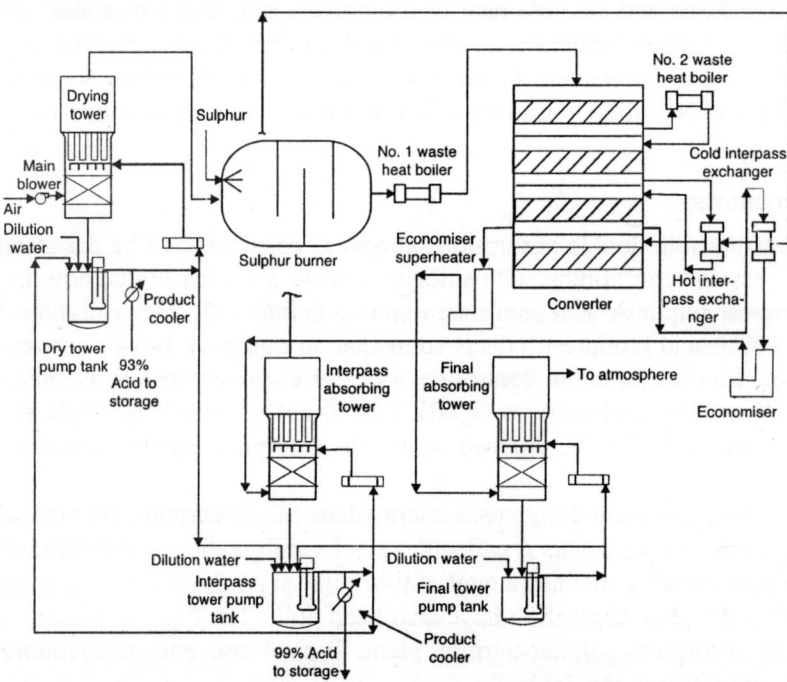

Fig. 7.3. Flow sheet for a double absorption sulphur-burning plant.

Assuming the feed gas contains 10 vol. % SO_2, the 99.7% conversion efficiency of double absorption plant corresponds to a sulphur dioxide concentration in the tail gas of about 400 ppm SO_2. Sulphur burning double-absorption plants are today designed for production capacities of 3000 T/day or more of 100% sulphuric acid in a single stream.

Wet-catalysis processes

Wet-catalysis processes differ from other contact sulphuric acid processes in that the feed gas still contains moisture when it comes into contact with the catalyst.

Sulphur trioxide formed by catalytic oxidation of sulphur dioxide reacts instantly with the moisture, to produce sulphuric acid, in the vapour phase, to an extent determined by the temperature. Liquid acid is subsequently formed by condensation of the H_2SO_4 vapours and not by the absorption of SO_2 in concentrated sulphuric acid, as in a contact process based on dry gases. The concentration of the product acid depends on the H_2O/SO_3 ratio in the catalytically converted gases as well as on the condensation temperature. The wet-catalysis process is especially suitable for processing the wet, dust-free gases obtained in the combustion of hydrogen sulphide-containing off-gases, which need only be cooled to the converter inlet temperature of about 440°C. Also their capacities are small—the largest one is of 250 TPD. A sulphur dioxide conversion of about 97.5% is achieved in the wet-catalysis process with a fourth converter.

These plants are usually designed for making 78% H_2SO_4, which has limited marketing opportunities. Therefore, the wet-catalysis method has been developed further to produce sulphuric acid of more than 90% concentration. Examples of such processes are the Concat process (Lurgi) and the WSA process (Haldor Topsoe). Another special development in this area is the wet-dry contact process with intermediate condensation.

Chamber process

Fig. 7.4 depicts the flow diagram of chamber process for the manufacture of sulphuric acid. Hot sulphur dioxide gases (425–600 °C) from the burners (7 to 9% sulphur dioxide and 9 to 12% O_2) are introduced into the base of a Glover tower. The gases pass counter-current to a cool stream of nitrous vitriol (72.8% H_2SO_4), which has been diluted from 60°Bé (77.7% H_2SO_4) acid obtained from the Gay-Lussac towers. The hot (100 to 140°C) Glover or "tower" acid (60° Bé) is cooled; part is used again in the Gay-Lussac tower and the rest is sent to storage.

Reactions

$$2NO + O_2 \longrightarrow 2NO_2$$
$$NO_2 + SO_2 + H_2O \longrightarrow H_2SO_4 + NO$$

98–99% conversion

92–96% yield (sulphur)

Partial oxidation (about 10%) of the sulphur dioxide occurs near the top of the Glover tower, where nitric oxide gas, generated by an ammonia oxidation unit, is mixed with the sulphur dioxide gases; the resulting mixture (SO_2, SO_3, N_2, O_2, NO, NO_2, N_2O_3 and steam) leaves the tower at a temperature of 70-110°C and is sucked into the lead chambers by means of a fan. Here the greater proportion of sulphur dioxide is oxidised to the trioxide and hydrated to sulphuric acid. Extra water is introduced as a spray and the sulphuric acid formed condenses on the lead walls.

This chamber acid (62-68% H_2SO_4), as it is called, is pumped to the Glover tower for concentration and nitrogen oxide removal. The gases leaving the chambers contain little sulphur dioxide but are led into the bottom of the Gay-Lussac tower counter-current to cold (35–40°C) 60°Bé (77.7%) sulphuric acid which strips nitrogen oxides from the gases, forming the nitrous vitriol containing 1–2.5% N_2O_3 by weight. The nitrous vitriol is pumped to the top of the Glover tower. The stripped gases are led to a stack and discharged to the atmosphere.

The number of lead chambers varies from 3–12 and the size depends on their design. Box-like chambers require 0.4 to 0.75 m³ space/kg sulphur burned per 24 hour, whereas externally water-cooled "ncated cones (Mill Packard type) require only 0.2 to 0.3 m³. With good operation the average chamber

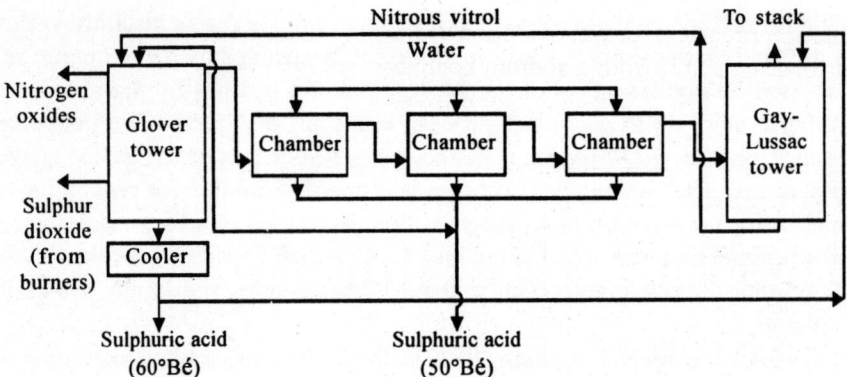

Fig. 7.4. Flowsheet for manufacture of sulphuric acid by chamber process.

The chamber process produces directly a weaker acid of sp. gr. 1.53 to 1.71 (50 to 60 Bé). Up to sp. gr. 1.84 (66 Bé) acid may be obtained by subsequent concentration. The chamber process is practically obsolete. It is described here for its historical interest and because many of the terms used in the industry (tower acid, etc.) and standard concentrations, still used for pricing, are derived from it.

MANUFACTURE OF OLEUM

To produce oleum, i.e., fuming sulphuric acid, sulphur trioxide is absorbed in one or more special absorption towers irrigated by recirculated oleum. Because of oleum vapour pressure limitations the amount of SO_3 absorbed from the process gas is typically limited to less than 70%. Since the absorption of SO_3 is incomplete, the gas leaving the oleum tower must be processed in a non-fuming absorption tower. The absorption of SO_3 for oleum production is carried out over a relatively narrow temperature range. The upper temperature is set to provide a reasonably partial pressure driving force for the oleum concentration used. The lower practical temperature limit is the freezing point of oleums, which is high enough to be a problem in shipping and handling as well. For some oleum uses it is practical to add small amounts of nitric acid as an anti-freeze.

Normally, oleum up to about 35 wt % free SO_3 content can be made in a single tower; two towers are used for 40 wt % SO_3. Liquid SO_3 is produced by heating oleum in a boiler to generate SO_3 gas, which is then condensed. Oleums containing more than 40 wt % SO_3 are usually produced by mixing SO_3 with low concentration oleum. Even the piping and storage tanks in the plant has beenwell studied. The details are available in the literature.

Equipment

Due to the highly corrosive nature of the gases and sulphuric acid product, a lot of attention has been given to various equipments used in their manufacture and also their materials of construction. The important equipment include—absorption and drying towers, acid coolers, catalytic converters, gas-gas heat exchangers, mist eliminators, oleum equipment.

Tail gas scrubbers are sometimes used on single absorption plants to meet SO_2 emission requirements, most frequently as an add-on to an existing plant, rather than on a new plant. Ammonia scrubbing is most popular, but to achieve good economics the ammonia value must be recovered as a usable product, typically ammonium sulphate for fertiliser use. A number of other tail gas scrubbing processes are available, including use of hydrogen peroxide, sodium hydroxide, lime and soda ash. Other tail gas processes include active carbon for wet oxidation of SO_2, molecular sieve adsorbents, and the adsorption and subsequent release of SO_2 from a sodium bisulphite solution (Fig. 7.4).

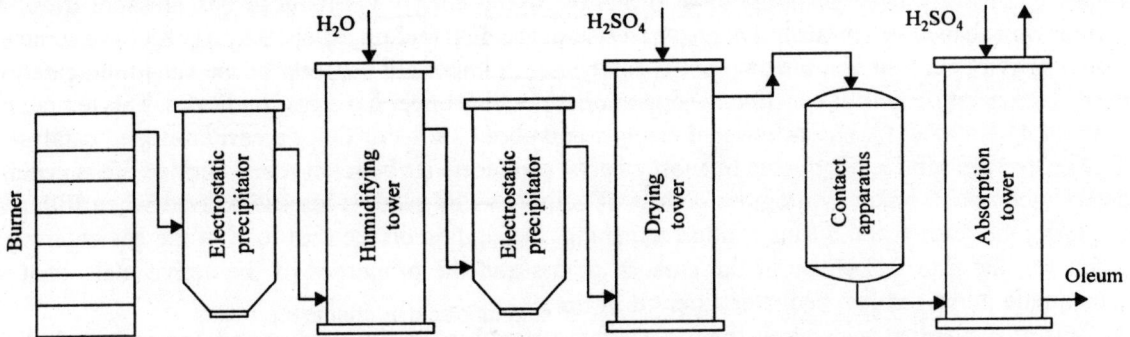

Fig. 7.4. Flow diagram for manufacture of sulphuric acid by contact process.

Small amounts of sulphuric acid mist or aerosol are always formed in sulphuric acid plants whenever gas streams are cooled, or SO_3 and H_2O react, below the sulphuric acid dew point. The dew point varies with gas composition and pressure but typically is 80-170°C. Higher and lower dew point temperatures are possible depending on the SO_3 concentration and moisture content of the gas.

Such mists are objectionable because of both corrosion in the process and stack emissions. More recently sulphuric acid mists have been satisfactorily controlled by passing streams through equipment containing beds or mats of small diameter glass or teflon fibres. Such units are called mist eliminators. Use of this type of equipment has been a significant factor in making the double absorption process economical and in reducing stack emissions of acid mist to tolerably low levels. Coalescing demister pads have been used in some single absorption plants instead of packed fibre beds to remove mist from the stack gas. For sub-micrometer particle collection these devices are not as efficient as packed fibre beds. Nevertheless, they have been used in some plants to obtain nearly invisible emissions. Successful use of a coalescing demister pad requires careful control of plant operating conditions to minimise mist formation.

Catalysts

Of all substances tested for catalytic activity toward sulphur dioxide oxidation, only vanadium compounds, platinum and iron oxide have proven to be technically satisfactory. Today, vanadium pentoxide is used almost exclusively. Commercial sulphuric acid catalysts typically consist of vanadium and potassium salts supported on silica, usually diatomaceous earth. Catalyst pellets are available in various formulations, shapes and sizes depending on the manufacturer and the particular converter pass in which they are to be used.

Commercial catalysts contain 4-9 wt% vanadium pentoxide, V_2O_5, as the active component, together with alkali-metal (usually potassium) sulphate promoters. Under operating conditions these form the liquid melt in which the oxidation of sulphur dioxide is thought actually to take place. Some catalysts also contain sodium sulphate to reduce the melting point. The carrier material is silica in the form of diatomaceous earth, silica gel or zeolites. In recent years, cesium-doped catalysts have also been developed and installed in various facilities. Cesium sulphate as a promoter reduces the melting point of the active components, resulting in significantly lower temperature limits for sustainable stable activity.

Solid cylindrical extrudates or pellets ranging from 4–10 mm diameter were used in the past. But now, ring-shaped catalysts (or star-ring) or variations, including rings having longitudinal ribs, are almost exclusively used primarily as a means of saving energy via reduced gas pressure drop. The various ring-shaped pellets also have greater resistance to dust fouling. Ring catalysts also have somewhat higher activity per unit volume than pellet catalysts. An important property of the vanadium catalyst is the low temperature limit at which stable operation is possible under fixed gas conditions. This temperature is about 410–430°C for a conventional catalyst and about 380–390°C for a cesium-doped catalyst.

The average service life quoted by most catalyst producers is about ten years. Service life is generally determined not so much by progressive loss of activity as by catalyst losses incurred when filling and emptying the reactor and during routine screening. Depending on the dust load of the gas entering the converter, the size and shape of the catalyst grains and the properties of the active melt, dust will accumulate in the catalyst bed over a period of time.

This dust eventually increases the gas-pressure drop through the catalyst bed and reduces both gas throughput and SO_2 conversion efficiency. Therefore, the catalyst must be screened from time to time. The first converter-pass catalyst pellets are screened at every significant turnaround, typically every 12–24 months. Second-pass catalyst pellets need screening less frequently because the first converter-pass catalyst bed acts as a fitter for the rest of the converter. Typical screening losses range from 10–15% of the catalyst bed per screening. Screening losses depend on screen mesh size and catalyst hardness, as well as on screening rate. In contrast to platinum, vanadium catalyst is largely insensitive to catalyst poisons. Catalyst ageing is a combination of a loss of catalytically active material from catalyst pellets and irreversible changes within the pellets. Catalyst ageing is accelerated by increasing temperatures and temperature cycling. Exposure to moisture above the dew point is not detrimental to the catalyst. But exposure to moisture at temperatures below the dew point produces irreversible damage. Prolonged exposure to moisture reduces the vanadium to the + 3 oxidation state, which is very difficult to re-oxidise under converter conditions. Moisture also damages the binders that hold together the silica support. This reduces catalyst hardness resulting in higher than normal screening losses.

Catalytic converters of different engineering firms vary in design. Stainless steel converters are frequently preferred, although carbon steel converters are also used. Stainless steel designs generally use all-welded interior construction, including stainless screens for catalyst supports. Traditional carbon steel converter designs use steel shells, sometimes partly or fully brick-lined, with cast iron and alloy internals. Essentially all designs use horizontal catalyst beds arranged one over another with gas flowing down through the catalyst. A relatively new innovation in stainless steel converter design uses structurally shaped support grids and division plates, giving improved resistance to temperature differentials during start-up and a higher strength design requiring less metal.

The design of sulphuric acid manufacturing plants has evolved to a great extent. In addition to different variations of the contact process, several special plant designs have also been installed—for example, energy-efficient plants, cement plants (cement and sulphuric acid from calcium sulphate), oxygen enriched processes. Sulphuric acid concentrators are also designed to concentrate the large quantities of dilute H_2SO_4 produced as a by-product in many chemical processes.

Storage and handling

Sulphuric acid, oleums and liquid SO_3 are very corrosive and dangerous chemicals. Carbon steel is used in concentrated sulphuric acid storage tanks because in quiescent and low temperature conditions its corrosion rate is acceptable. Carbon steel is not suitable for handling sulphuric acid in concentrations between 80–90% or less than 68% even under quiescent conditions, unless passivating agents are present. Steel tank cars, often lined to minimise iron contamination are usually employed for high concentrations of sulphuric acid. Bottom outlets or valves are not allowed, nor are internal steam coils. Tank contents must be unloaded via standpipe. Using air pressure to unload is not recommended for safety reasons, but if air pressure is used, gauge pressure should be held at less than 0.21 Mpa (30 psi). General handling precautions should be observed and equipment appropriate for exposure conditions should be worn. The acid should be handled only in areas having sufficient ventilation to prevent irritation. Acid containers must be kept closed, water must not enter containers. An emptied container retains vapour and product residue. Thus all labeled safeguards must be observed until the container is cleaned, re-conditioned or destroyed. Drums, if not safe-venting, should be periodically vented to prevent accumulations of hydrogen in dismantling lines and equipment, it should always be assumed that a spray of acid may occur and suitable precautions should be taken. Tightening flange bolts on pipe filled with acid is dangerous because of the possibility of mechanical failures. In case of a spill or leak, keep people away and up wind of the spill. If it is necessary to enter the spill area, self-contained breathing apparatus and full protective clothing must be worn. The area should be diked using sand or earth to contain the spill, the acid removed by vacuum truck and the spill area flushed with water. Washings should be neutralised with lime or soda ash and pollution control authorities notified of any runoff into streams or sewers and of any air pollution incidents. Safety showers with deluge heads, protected against freezing, should be readily available at appropriate locations in any plant producing or using sulphuric acid.

Specifications and Standards

Similar limits are generally used for other sulphuric acid concentrations, with the exception of turbidity values for high strength acids (and oleum) and SO_2 and nitrate values in oleums. Because iron sulphate is relatively insoluble in concentrated acids, the turbidities of 98–99% H_2SO_4 and oleum may be higher than shown, even at acceptable total iron concentrations. Sulphur dioxide concentrations in oleum are rarely specified or measured, but typical values are considerably higher than in acids of less than or equal to 99 wt% concentrations.

A number of different grades of sulphuric acid are produced for specialised uses, such as reagent grade, food grade and electrolyte grade. In addition some producers offer special premium priced grades that contain little or no turbidity and colour, or in some cases a maximum iron concentrations of 10 ppm. Certain objectionable elements such as arsenic, lead, mercury and selenium are not commonly specified for technical grade acid, but some producers attempt to hold each of these at less than one ppm, or less than 2–5 ppm in the case of lead, to minimise possible problems. Selenium is not usually present except at a few metallurgical-type plants or at plants using volcanic sulphur as raw material.

Typical specifications for several common types or grades of sulphuric acid are shown in Table 7.3. Indian specifications for various grades of sulphuric acid are given by IS-266. They are listed in the Tables 7.4, 7.5 and 7.6. The density composition tables for aqueous solutions of sulphuric acid are given in IS-4048. The safety code is given by IS-4262, while the limits for gaseous emissions in sulphuric acid industry are given by IS-8635. Metallurgical (smelter) plants and spent acid decomposition plants usually produce acid of good (low) colour because the SO_3 feed gases are extensively purified prior to use. In some cases, however, and particularly at lead smelters, sufficient amount of organic flotation agents are volatilised from sulphide ores to form brown or black acid. Such acid can be used in many applications, particularly for fertiliser production, without significant problems.

Table 7.3. Typical sulphuric acid specifications.

Property	Electrolyte[b,c] class I[f]	Technical[d] class I[g]	Food chemicals codex	Technical[c] 66Be (93%)
	Acid type			
H_2SO_4, wt %	93.2	93.0	—	93.2
Sp[2]. gr[4]., 15.5/15.5	1.8354	1.8347	—	1.835–1.837
Non-volatiles[i], %	0.03	0.025	—	0.02–0.03
As[i], ppm	1.0	2.5	3	
SO_{2i}[i], ppm	40		40[i]	40–80
Iron[2i], ppm	50		200	50–100
Heavy metals[i], ppm	—		20	—
Nitrate[i], ppm	5.0		10	5–20
Colour[j]	Per test		—	100–200 APHA

[b]—Limits are also specified for platinum, organics, copper, zinc, antimony, selenium, nickel, manganese, ammonium and chloride.
[c]— Fed. Spec. O-S-801E.
[d]— Fed. Spec. O-S-809E.
[c]—Typical industry sulphuric acid.
[f]—The other classes of lower strength acids are included.
[g]—One other class of lower strength acid is included.
[i]—Value given is maximum.
[j]—Value is for reducing substances as SO.

Table 7.4. Requirements for technical grade sulphuric acid.

Property	Value
Specific gravity at 25°C, min.	1.828
Sulphuric acid, % by wt., min.	93.0
Residue on ignition, % by wt., max.	0.2
Iron (as Fe), % by wt., max.	0.05
Heavy metals (as Pb), % by wt., max.	0.005

Table 7.5. Requirements for battery grade sulphuric acid (concentrated and dilute).

Property	Value	
	Concentrated	*Dilute*
Specific gravity at 25°C	1.8340	1.21
Sulphuric acid, % by wt, min.	95.0	29.7
Residue on ignition, % by wt, max.	0.06	0.02
Iron (as Fe), % by wt, max.	0.002	0.0006
Chlorides (as Cl), % by wt, max.	0.001	0.0003
Arsenic (as AS_2O_3), % by wt, max	0.0003	0.0001
Oxidisable impurities, as SO_2 and organic matter	To satisfy permanganate and charring test	To satisfy permanganate test

(Contd ...)

Property	Value	
	Concentrated	Dilute
Selenium, % by wt, max.	0.002	0.0006
Manganese, % by wt, max.	0.0001	0.00003
Copper, % by wt, max.	0.003	0.001
Zinc, % by wt., max.	0.003	0.001
Nitrate, nitrite and ammonia as nitrogen (N), % by wt, max.	0.003	0.001

Table 7.6. Requirements for pure grade and analytical reagent grade sulphuric acid.

Property	Value	
	Pure	Analytical reagent
Specific gravity at 25°C	1.8340	1.8360
Sulphuric acid, % by wt, min.	95.0	96.0
Residue on ignition, % by wt, max.	0.01	0.0025
Iron (as Fe), % by wt, max.	0.001	0.0001
Chloride (as Cl), % by wt, max.	0.0035	0.0003
Heavy metals (as Pb), % by wt, max.	0.002	0.0002
Arsenic (as AS_2O_3), % by wt, max.	0.0005	0.00001
Oxidisable impurities as SO_2 % by wt, max.	0.004	0.0005
Nitrate, nitrite and ammonia as nitrogen (N), % by wt, max.	0.003	—
Nitrate (as NO_3), % by wt, max.	—	0.00002
Ammonia (NH_3), % by wt., max.	—	0.0005

Applications

Sulphuric acid is one of the most widely used industrial chemicals. But most of its uses can be considered as indirect, because it functions as a reagent rather than as an ingredient. Surprisingly little of it appears in end products, and most ends up as spent acid or some type of sulphate waste. A number of products incorporate the sulphur of sulphuric acid, but nearly all of them are low-volume, speciality items.

Sulphuric acid is used in the manufacture of fertilisers, explosives, dyestuffs, other acids, parchment paper, glue, in purification of petroleum and pickling of metal. Formerly, dilute sulphuric acid was used in the treatment of gastric hypoacidity and the concentrated acid was used as a topical caustic.

Indirect uses

The largest single consumer of sulphuric acid by far is the fertiliser industry. Most goes into the production of phosphoric acid, which in turn is used to manufacture such fertiliser materials as triple superphosphate and mono- and diammonium phosphates. Lesser amounts are used for producing superphosphate and ammonium sulphate. About 60% of the sulphuric acid produced is utilised in fertiliser manufacture.

Substantial quantities of sulphuric acid are used as an acidic dehydrating reaction medium in organic chemical and petrochemical processes involving such reactions as nitration, condensation and dehydration, as well as in oil refining, in which it is used for refining, alkylation, and purification of crude oil distillates.

In the inorganic chemical industry sulphuric acid is used notably in the production of TiO_2 pigments, hydrochloric acid, and hydrofluoric acid. In the metal processing industry, sulphuric acid is used for pickling and descaling steel, for leaching copper, uranium and vanadium ores in hydrometallurgical ore processing and in the preparation of electrolytic baths for non-ferrous metal purification and plating. Certain wood pulping processes in the paper industry require sulphuric acid, as do some textile and chemical fibre processes and leather tanning.

Direct uses

Under certain conditions, sulphuric acid is occasionally used directly in agriculture for rehabilitating extremely alkaline soils. However, this is not a very important use in volume terms. Probably the largest use of sulphuric acid in which the sulphur becomes incorporated in the final product is organic sulphonation, particularly for the production of detergents. Other minor organic chemicals and pharmaceuticals are also made by sulphonation. One of the most familiar consumer products containing sulphuric acid—the lead-acid battery—accounts for only a tiny fraction of total sulphuric acid consumption.

Sulphuric acid has several desirable properties that lead to its use in a wide variety of applications. It typically is less costly than any other acid, it can be readily handled in steel or common alloys at normal commercial concentrations, and it is available and readily handled at concentrations more than 100 wt % (oleum).

Control of Sulphur Dioxide Pollution for the Manufacture of Sulphuric Acid in India

In India, sulphuric acid capacity during the past five years has increased from about 5 million to about 6 million tonne per year. When pollution control of sulphur dioxide is considered, the sulphuric acid plants were the first target. Rightly so, since at six million tonnes output, the traditional contact process would emit 94,000 tonnes of sulphur dioxide which result in about 144,000 tonnes of acid rain. This has been reduced to less than one-fourth of sulphur dioxide emission by the well established Double Catalysis, Double Absorption (DCDA) process.

With co-generation of power, sulphuric acid will reduce SO_2 emissions resulting from the corresponding generation of electric power, the sulphuric acid manufacture can no longer be viewed as a polluter of environment with sulphur dioxide emissions as in the past.

Modern sulphuric acid plants are viewed as power generating plants with sulphuric acid as by-product. For example, if the total capacity of six million tonnes of sulphuric acid in India was with best power generation capability, it would provide additional 180 MW to the grid. If credit is given to the saving of corresponding SO_2 emission for the power generated, the co-generation of power from sulphuric acid plants is attractive not only from economies of cheap power source, but also from lower SO_2 pollution due to corresponding power generation. For example, 180 MW would emit 80 MT/Day of SO_2 in air as permissible by Pollution Control Board. The corresponding emission from acid plants will be only 30 MT/Day of SO_2 generating the same amount of power.

Hence it has become imperative that all future sulphuric acid plants be equipped with maximum co-generation of power. This would make sulphuric acid production with no sulphur dioxide pollution and offer a sizeable reduction in SO_2 pollution for the generation of electric power.

Latest trends and achievements in emissions of SO_2 per tonne of sulphuric acid

Pollution Control Board is to become more stringent in SO_2 emissions from sulphuric acid plants. The MINAS (Minimum National Standard) provides for 4 kg of SO_2 emission per tonne of acid produced, the State Pollution Control Board insists on a figure of 3 kg of SO_2 per tonne of acid produced. Well operated DCDA plants give an average of 2 kg SO_2 per tonne of acid. However, if alkali scrubbing system is kept in operation continuously, this can be brought down to 0.5 kg SO_2 per tonne of acid produced.

A well designed alkali scrubber has to take into account the high load on the system during the unsteady state of start-up operation after short stoppage due to maintenance problems or power failure or after a prolonged annual stoppage for tower inspection and routine maintenance.

At present there are two types of scrubbing systems which are found most suitable. One is the TCA (Turbulent Contact Absorber) offered by Nissan of Japan and the second is the Ventury-cum-Fixed Bed Absorber offered by Lurgi of Germany. Both are indigenously marketed and can be supplied on turn-key basis. One of the main drawbacks in this system is the additional running cost of providing power and alkali. Since alkali scrubber cannot be economically viable on a continuous basis, a modified DCDA System using five passes instead of conventional four passes is found more attractive. In the conventional process, the fourth pass which operates at 410-430°C is found inadequate to convert the last traces of SO_2 to SO_3 due to thermodynamic and equilibrium restrictions. The new modified DCDA system employs one more, namely, the fifth bed. Allowing a cooling between 4th and 5th passes and introducing additional air to cool gases to 360°C, the approach to chemical equilibrium is favoured. In addition, use of low kindling catalysts will guarantee the stock analysis below 100 ppm without the use of alkali scrubber. This claim is yet to be proved on a plant in India.

HYDROCHLORIC ACID

Process of Manufacture

From organic chlorinations

Mostly hydrochloric acid production results as a by-product of various organic chlorination reactions, such as the chlorination of benzene to chlorobenzene. Other similar sources include the cracking of ethylene dichloride to yield vinyl chloride, the catalytic cracking of 1,1,2,2-tetrachloroethane to trichloroethylene, and the production of linear alkylbenzene. The recovery of magnesium from seawater yields some by-product hydrochloric acid. (Fig. 7.5)

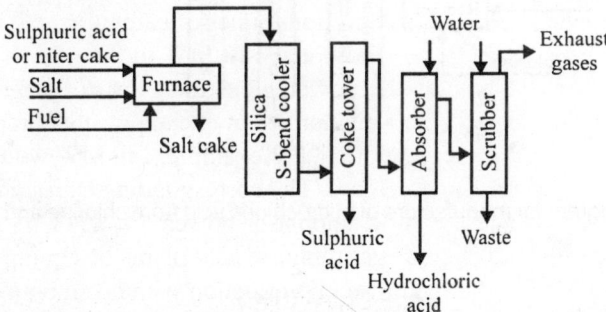

Fig. 7.5. Flow diagram for manufacture of hydrochloric acid from organic chlorinations.

Salt and sulphuric acid, sp gr 1.7 (slight excess), or salt and an equivalent amount of niter cake are charged to a furnace equipped with a rake agitator (Mannheim furnace), where the reacting mass is slowly heated to a temperature just below fusion (843°C). Hydrogen chloride is evolved and led through a cooling and condensing system to the absorbers. Salt cake (crude sodium sulphate) is continuously discharged from the periphery of the furnace.

Reactions

$$NaCl + H_2SO_4 \rightarrow HCl + NaHSO_4$$
$$NaCl + NaHSO_4 \rightarrow HCl + Na_2SO_4$$

98% yield

The combustion gases, containing hydrogen chloride (approximately 30% HCl), leave the furnace at about 840°C, pass into silica bends (coolers) externally cooled with water, and leave the coolers at about 38°C. The cooled gases then pass through a coke-packed tower, in which sulphuric acid mist and any solid particles present are removed, and thence to another series of silicas bends, where the hydrogen chloride is absorbed by water to produce hydrochloric acid. Exhaust gases from the absorber are scrubbed with water and discharged to the atmosphere. The more recently designed plants have made use of impervious graphite and structural plastics for absorbers and coolers.

A few plants use the Laury furnace, a horizontal rotary kiln, in which the reacting mass comes into contact with fuel combustion gases. These gases, containing as little as 5% hydrogen chloride, leave the furnace at 300°C, and are cooled and then recovered in the manner described previously.

In making potassium sulphate from potassium chloride, hydrochloric acid results as co-product. The reactions are analogous, with potassium chloride in place of sodium chloride.

From Chlorine and Hydrogen

Chlorine is burned in a slight excess of hydrogen to produce hydrogen chloride. Several types of burners are used: for example, the silica burner, the ceramic-lined burner, the graphite burner, and the water-jacketed steel burner. The last-mentioned burner cannot of course be used with wet gases. (Fig. 7.6).

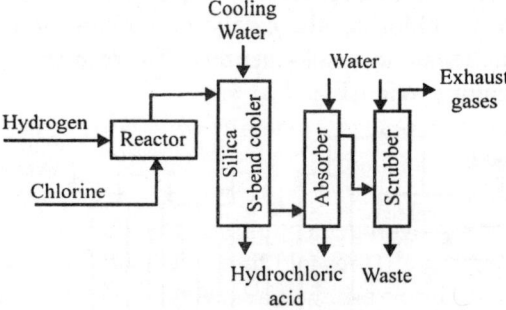

Fig. 7.6. Flow diagram for manufacture of hydrochloric acid from chlorine and hydrogen.

Reaction

$$H_2 + Cl_2 \rightarrow 2HCl$$

90–99% yield

A submerged combustion burner, operating beneath a surface of muriatic acid, has been developed. The burner gases, practically pure hydrogen chloride, are cooled, absorbed, and scrubbed in a system essentially the same as that used in the salt process. The purifying coke tower, however, is omitted, and in some systems (where the gas concentration approaches 100% HCl) the scrubber may be omitted. Strong hydrochloric acid (35%) is removed directly from the bottom of the cooler by means of a trap, and weak acid (28%) leaves the bottom of the absorber. Recently installed absorbers have been built of tantalum and structural carbon, as well as silica. This process readily gives a water-white product.

The concentration of hydrogen chloride in burner gases generally depends on the degree of chlorine utilisation. With 0.03% or less chlorine in the product gases, hydrogen chloride concentration may be as high as 98.5%. Where 0.1% chlorine is allowable, hydrogen chloride concentration may reach 99 to 99.5%. Manufacturers often insist on 100% chlorine utilisation, in which case exit gases may be only 90% hydrogen chloride because of the 5 to 10% excess hydrogen used.

Hydrochloric acid is also made by burning chlorine in methane or water gas, according to the equation:

$$2Cl_2 + CH_4 + air \longrightarrow 4HCl + CO_2$$

The recovery system is the same as that used in the chlorine-hydrogen process.

Anhydrous Hydrochloric Acid

Anhydrous hydrogen chloride (containing 99.8% HCl) is available in steel cylinders. It may be prepared by passing hot (260°C) burner gases over cold anhydrous calcium chloride. The dry hydrogen chloride is then compressed and charged into cylinders. A yield of 99% of the hydrogen chloride supplied by the burners is realised.

Another method of producing anhydrous hydrogen chloride involves distillation of 22°Bé (36% HCl) acid in a graphite stripping column under 20 psig (136 kPa). The overhead gases (97% HCl and 3% water) are cooled in a brine condenser to –12°C. A liquid containing 50% hydrogen chloride condenses; the anhydrous hydrogen chloride passes through a mist trap and is ready for use. The condensate and stripping column bottoms are recycled to appropriate places in the acid recovery plant.

Properties

Hydrochloric acid is a solution of hydrogen chloride in water. Pure hydrogen chloride is a colourless, pungent poisonous gas at ordinary temperatures and pressures.

Mol. wt. 36.46, sp. gr. (0°C) 1.268 (air = 1), density (0°C) 1.639 g/litre, mp –112°C, bp –35°C.

It is liquefiable under a pressure of 82 atm (8.2 MPa) at 51°C. At atmospheric pressure, the liquefied gas boils at –85°C and has a specific gravity of 1.194. Easily soluble in ethanol and ether, as well as in water. A constant-boiling mixture of hydrogen chloride and water contains 20.24% hydrogen chloride and boils at 110°C at 760 mm pressure. Threshold limit value 5 ppm, 7 mg/m³. Other possibilities for alleviating the hydrochloric acid surfeit are the use of hydrogen chloride in place of chlorine as a chlorinating agent, and replacement of sulphuric acid in steel pickling.

The production of hydrochloric acid is considerably less than that of sulphuric, phosphoric, and nitric acids. It is commonly used for its acid characteristics where its volatility, non-oxidising characteristics, high acidity per unit weight, and ease of neutralisation are advantageous. Its chief drawbacks are cost and extreme corrosiveness to metals.

Uses

Metal treating, chemicals, oil well acidising, food processing etc. In metal treating the largest use is in the pickling of steel. It is superior to sulphuric acid for this use; it leaves a cleaner surface, reacts more slowly with the basis metal and, perhaps most important, the spent pickle liquor can be recovered and recycled, whereas spent pickle liquor from sulphuric acid pickling is an annoying waste product. In chemical processing, techniques have been developed for using hydrochloric acid in place of chlorine for organic chlorinations (oxyhydrochlorination).

In oil-well acidising, it increases the permeability of wells by dissolving part of the limestone and dolomite in the formations. In food processing, it is used as a modifier for food starch, as a hydrolytic agent in the manufacture of sodium glutamate and gelatine, for conversion of corn starch to syrup, and for adjusting pH in brewing.

HYDROFLUORIC ACID

Process of Manufacture

From fluorspar and sulphuric acid

By treating fluorspar (calcium fluoride) with concentrated sulphuric acid in a furnace, hydrogen fluoride gas is evolved, leaving a residue of calcium sulphate. After being cleaned of dust and sulphuric acid content, the gas is condensed as 98 to 98.5% hydrogen fluoride. Distillation increases its strength to 99.9 to 99.95%. Aqueous grades may be made by dilution with water (Fig. 7.7).

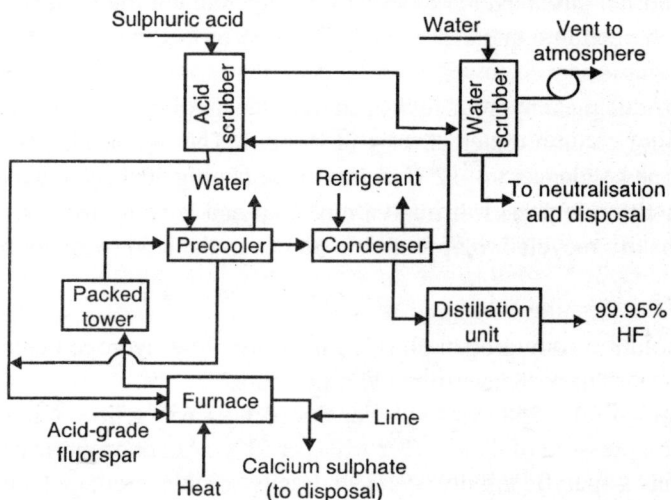

Fig. 7.7. Flow diagram for manufacture of hydrofluoric acid from fluorspar and sulphuric acid.

Reaction

$$CaF_2 + H_2SO_4 \rightarrow 2HF + CaSO_4$$
85–95% yield

Crude fluorspar, fluorite, as it comes from domestic mines, varies in calcium fluoride content from about 50 to 90%. To be suitable for acid production, the ore must be upgraded. This is usually done by flotation and results in an acid-grade fluorspar containing about 98% calcium fluoride, 1% silica, 0.03% sulphur and less than 0.1% moisture.

The finely powdered fluorspar is withdrawn from storage silos and transferred by air conveyors to steel conical-bottom hoppers. From here it is fed continuously along with concentrated (93–99%) sulphuric acid into a rotating furnace in a refractory-lined shroud. The space between the shroud and the furnace is fired with oil or gas to provide heat for the endothermic reaction. The cylindrical steel furnaces have been constructed in a variety of dimensions, up to 3.7 m in diameter and 25 m long, having a metal thickness of 5 cm. The production capacity of these furnaces ranges up to 55 metric tonnes/day.

The raw materials are fed into the furnace at a ratio of 1 mole of fluorspar to 1.1 to 1.3 moles of acid. Several methods have been proposed to keep the reacting mass of fluorspar and sulphuric acid flowing freely enough to allow complete reaction. They include two-stage reaction system with kneading, fluidised beds, and stationary and rotary kilns with fixed flights or moving hardware inside. Temperatures of 250 to 300°C are commonly reached. Reaction is accomplished using oleum, sulphuric acid, or sulphuric acid-sulphur trioxide-steam mixtures with the spar. In the last case, the heat of reaction of sulphur trioxide and steam is used to maintain the temperature.

Calcium sulphate residue, containing 1% or less unreacted fluorspar, is discharged continuously in amounts about 1.75 times the weight of the spar charge. From the feed end of the furnace, gaseous hydrogen fluoride (70–75% HF) is withdrawn at a temperature of 120–175°C. The crude gases pass through a packed tower which provides surface area for the reflux of sulphuric acid and captures dust in the gas from the furnace. A precooler condenses a mixture of hydrogen fluoride, sulphuric acid, and water, which is combined with the feed acid and returned to the furnace. The cleaned gas is condensed in a heat exchanger using a refrigerant; hydrogen fluoride boils at 19°C at atmospheric pressure. Non-condensables pass through a packed tower, using the feed acid to absorb any uncondensed hydrogen fluoride. This gas passes through another tower in which water is the absorbent to remove silicon tetrafluoride (SiF_4) and any remaining hydrogen fluoride. Motive power for gas movement is usually provided by an induced draft blower.

The 98.5% acid from the condenser is distilled to raise its assay to 99.9–99.95%. The main impurities removed by distillation are sulphuric acid, sulphur dioxide, and water. Variations are used in several plants. For instance, the furnace or retort may be a stationary, horizontal steel shell in which an alloy-steel conveyor propels the reacting fluorspar and product calcium sulphate through the unit. In one plant the fluorspar and 99% sulphuric acid are mixed in a premixer or kneader, where the initial and highly corrosive stage of the reaction is carried out. The partially reacted mass is then discharged to the rotary kiln. It is claimed that the premixing is so thorough that excess acid is not required and capital costs are reduced.

Solid and liquid particulates may be removed by such techniques as cyclones, scrubbing with sulphuric acid, coke boxes, and demisters. The gases are cooled to 20 to 30°C by sulphuric acid scrubbing and indirect heat exchangers. The resultant acid can be 98% pure, containing as impurities sulphuric acid, water, sulphur dioxide, and silicon tetrafluoride; this acid is suitable for cryolite manufacture without refining. The low boilers are readily fractionated as described. The vent gas from the last cooler will contain by-product silicon tetrafluoride and hydrogen fluoride in a mole ratio of 1:2. On absorption by water in a packed tower, fluosilicic acid is formed according to:

$$SiF_4 + 2HF \rightarrow H_2SiF_6$$

This by-product may be converted to the sodium salt for use in the fluoridation of water supplies; more often it is neutralised with lime and discarded with the calcium sulphate. Despite many efforts to utilise the calcium sulphate, it is normally dumped as landfill.

Uses

Fluorocarbons; aluminium industry; atomic energy; alkylation; stainless steel etc.

Properties

Anhydrous hydrofluoric acid is a colourless, fuming, corrosive liquid, which causes extremely serious burns on skin contact. Vapours are very irritating to the eyes and mucous membranes.

Mol. wt 20.01, sp. gr. 0.988 (13.6°C), mp –83°C, bp 19.4°C, threshold limit value 3 ppm, 2mg/m³. Although the formula of hydrogen fluoride is sometimes written H_2F_2, there seems to be no real justification for the doubled formula. Various polymerised species, up to H_6F_6, are in equilibrium.

Safety is an important item around a hydrofluoric acid plant because of the great hazards to personnel involved in handling it. Waste disposal also poses a special problem.

FLUORINE

The element is far too reactive to occur naturally, but it is found as fluorides in certain minerals, e.g. fluorspar or fluorite, CaF_2, and cryolite, Na_3AlF_6, and fluo-apatite, $CaF_2.3Ca_2(PO_4)_2$. Traces of fluorides are found in teeth and bones and in the ash of plants.

Preparation

An electrolytic method is employed, but considerable difficulty is encountered because of the very great reactivity of fluorine. The apparatus (Fig. 7.8) must be kept completely free of moisture, oil, and grease, because fluorine attacks these materials vigorously. In the modern process an electrolyte of composition KF.2HF (i.e. a mixture of KHF_2 and HF) is used at about 100°C.

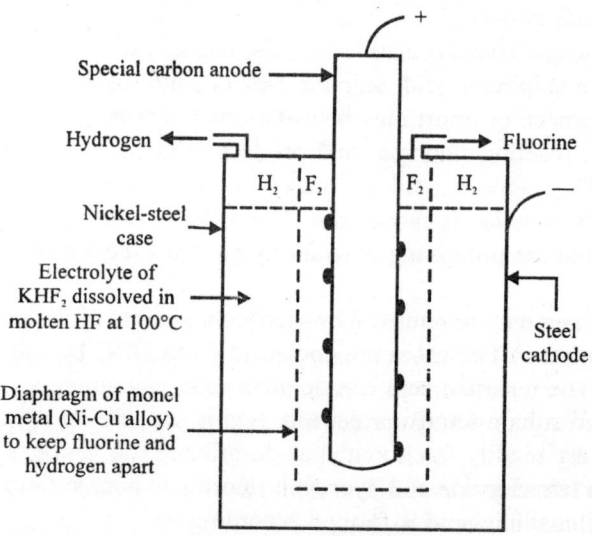

Fig. 7.8. Flow diagram for manufacture of fluorine.

The anode is made of special non-graphitic carbon impregnated with copper so that it is not readily attacked by fluorine, and the cathode is of steel. The cell itself is made of nickel steel, which is fairly resistant to attack, and a diaphragm is fitted to ensure complete separation of the products, since fluorine and hydrogen combine explosively. The electrolyte may be regarded as a solution of potassium fluoride in anhydrous hydrogen fluoride; it contains, therefore, a high concentration of potassium and fluoride ions and a much lower concentration of hydrogen ions.

The changes at the electrodes are:

$$\text{At the anode:} \quad F^- \rightarrow F + e \text{ and } 2F \rightarrow F_2 \uparrow$$
$$\text{At the cathode:} \quad H^+ + e \rightarrow H \text{ and } 2H \rightarrow H_2 \uparrow$$

Thus the overall effect of the electrolysis is to decompose hydrogen fluoride into its elements, and the cell has to be topped up with more anhydrous hydrogen fluoride from time to time. The fluorine liberated at the anode can be freed from traces of hydrogen fluoride by passing it over anhydrous sodium fluoride:

$$NaF + HF = NaHF_2$$

Properties

It is a greenish-yellow diatomic gas, slightly paler in colour than chlorine, which it resembles in smell. It melts at $-220°C$ and boils at $-188°C$. Its low critical temperature, $-129°C$, makes it impossible to store liquid fluorine at ordinary temperatures, but it can be transported in bulk as a liquid by using liquid nitrogen as a coolant.

It is the most reactive element of all, combining directly with every other element except oxygen and some of the noble gases. It is also the most electronegative element.

It explodes with hydrogen, even in the dark at very low temperatures, forming hydrogen fluoride:

$$H_2 + F_2 = 2HF$$

So great is its affinity for hydrogen that fluorine also reacts vigorously with water, hydrogen sulphide, ammonia, and hydrogen chloride and with organic compounds, many of which inflame spontaneously when brought into contact with fluorine:

Reactions

$$2F_2 + 2H_2O = 4HF + O_2$$
$$F_2 + 2HCl = 2HF + Cl_2$$
$$4F_2 + CH_4 = 4HF + CF_4$$

Fluorine combines readily at room temperature with silicon, phosphorus, and sulphur, forming fluorides in which these elements show their highest covalency, e.g., SiF_4, PF_5 and SF_6. It also reacts with carbon, particularly when heated, giving carbon tetrafluoride, and with boron giving boron trifluoride. The alkali and alkaline earth metals ignite spontaneously in fluorine at room temperature and most other metals, including platinum, react with it when heated, but some metals such as copper and nickel show some resistance to attack probably because they form impervious fluoride layers on their surfaces which hinder further action.

When fluorine is passed through a very dilute solution of sodium hydroxide some fluorine monoxide is formed as a colourless gas (pale yellow), b.p. $-146°C$; it is the only oxide of fluorine stable at room temperature:

$$2F_2 + 2NaOH = 2NaF + F_2O \uparrow + H_2O$$

Uses of fluorine and Its compounds

Fluorine is required in great quantities for making uranium hexafluoride, UF_6, which is the only volatile compound of uranium. Fluorine compounds such as CCl_2F_2 and CCl_3F are being increasingly used as refrigerants because of their inertness, stability, non-toxicity, and thermodynamic efficiency, and also as propellents in aerosols. Boron trifluoride and hydrogen fluoride are used as catalysts in the petroleum industry, and cobalt fluoride and bromine trifluoride as fluorinating agents. Other fluorides are used as disinfectants in the brewing industry and for preventing rot in timber.

The fluorocarbon, tetrafluoroethylene, C_2F_4, can be polymerised into a plastic, which is renowned for its chemical inertness; it is already important as an electrical insulator and as a lubricant. Sulphur hexafluoride is used as an electrical insulator in gas-filled high voltage cables. Molten cryolite is used as a solvent for alumina in the manufacture of aluminium. The mineral was the original source, but now-a-days a synthetic product is used.

Fluorosilicic acid, H_2SiF_6, is used for fluoridation of drinking water. By adding fluorides artificially in this way so that their concentration is only a few parts per million it has been found possible to reduce dental decay, particularly in children, presumably by strengthening the fluoride-containing enamel which protects the teeth. Toothpastes containing fluorides (e.g. stannous fluoride) are claimed to have the same beneficial effect.

ALUMINA (ALUMINIUM OXIDE)

The mineral corundum is natural aluminium oxide, and emery, ruby, and sapphire are impure crystalline varieties. The mixed mineral bauxite is a hydrated aluminium oxide.

Manufacture of Alumina

The Bayer process, the most economical method of manufacture, takes advantage of the reaction of aluminium trihydroxide and aluminium oxide hydroxide with aqueous caustic soda to form sodium aluminate:

$$Al\,(OH)_3 + NaOH \rightleftharpoons NaAlO_2 + 2\,H_2O$$
$$AlO(OH) + NaOH \rightleftharpoons NaAlO_2 + H_2O$$

The reaction equilibria move to the right with increases in caustic soda concentration and temperature. The following operations are performed:
1. Dissolution of the alumina at high temperature.
2. Separation and washing of the insoluble impurities of bauxite (red muds) to recover the soluble alumina and caustic soda.
3. Partial hydrolysis of sodium aluminate at a lower temperature to precipitate aluminium trihydroxide.
4. Regeneration of the solutions for recycle to step (1) by evaporation of the water introduced by the washings.
5. Transformation of the trihydroxide to anhydrous alumina by calcination at 1450 K. Fig. 7.9 shows the flow sheet of the Bayer process.

Variations of lime-soda sintering process are used in treatment of high-silica bauxites. Numerous sintering and acid-extraction processes have been investigated for kaolin and other clays, as well as from dawsonite and coal ash.

Electrolysis of alumina

Nearly all aluminium is produced by the electrolysis of alumina (Al_2O_3) dissolved in a molten cryolite-based bath, the Hall and Héroult process. The aluminium is deposited onto a carbon cathode which also serves as a melt container. Simultaneously, oxygen is deposited on and consumes the cell's carbon anode(s).

Cryolite ionises to form sodium (Na^+) and hexafluoroaluminate (AlF_6^{3-}) ions. The latter dissociates to form tetrafluoroaluminate (AlF_4) and fluoride (F^-) ions. Alumina dissolves at low concentrations by forming oxyfluoride ions with a 2:1 ratio of aluminium to oxygen ($Al_2OF_{2n}^{(4-2n)}$); at higher alumina concentrations, oxyfluoride ions with a 1:1 ratio of aluminium to oxygen ($Al_2OF_{2n}^{(4-2n)}$) are formed. Cells are generally operated with a 2–6 wt % Al_2O_3 in the electrolyte. Saturation ranges between 7–12% Al_2O_3, depending upon electroyte composition. Ion transport measurements indicate that Na^+ ions carry most of the current; however, aluminium is deposited. Most probably, a charge transfer occurs at the cathode interface and hexafluoroaluminate ions are discharged, forming aluminium and F^- ions to neutralise the charge of the current carrying Na^+ ions:

$$12Na^+ + 4AlF_6^{3-} + 12e \rightarrow 12(Na^+ + F^-) + 4Al + 12F^-$$

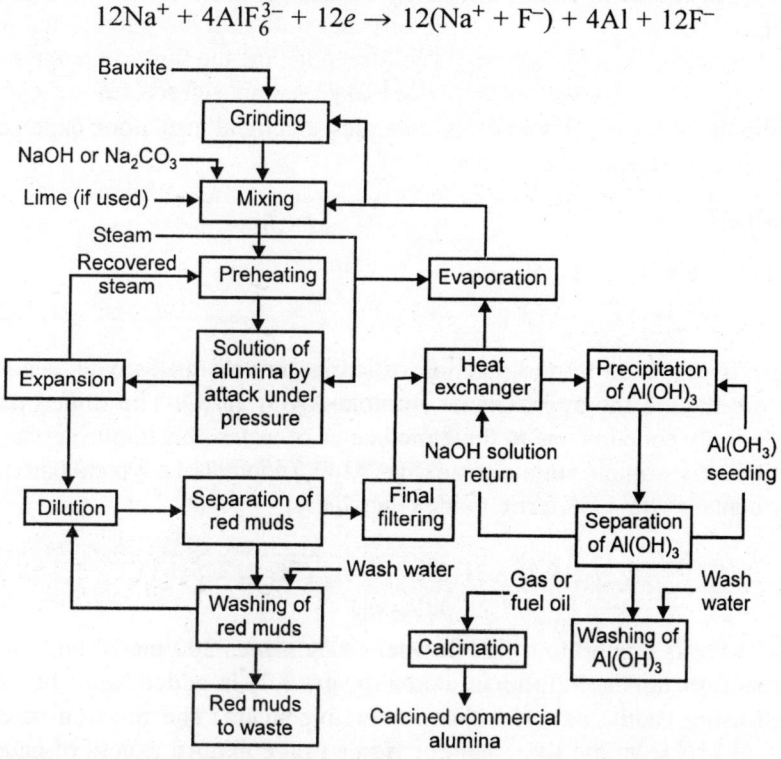

Fig. 7.9. Flow diagram for manufacture of alumina oxide by Bayer process.

Oxyfluoride ions discharge on the anode, forming carbon dioxide and AlF_6^{3-} ions:

$$3 Al_2O_2 F_4^2 + 3C + 24 F^- \rightarrow 3 CO_2 + 6AlF_6^{3-} + 12 e$$

The addition of anode and cathode equations, plus solution of alumina gives the overall reaction:

$$2 Al_2O_3 + 3C \rightarrow 4 Al + 3 CO_2$$

According to Faraday's law, one Faraday (268.A·h) theoretically should deposit one gram-equivalent to aluminium (8.994 g). In practice, only 85–95% of this amount is obtained. A modern alumina-smelting cell consists of a rectangular steel sheet lined with refractory insulation surrounding an inner lining of baked carbon which can withstand the combined corrosive action of molten fluorides and molten aluminium. Steel collector bars are joined to the carbon cathode at the bottom to conduct electric current from the cell. Current enters through pre-baked carbon anodes or through a continuous Soderberg anode.

In spite of its industrial dominance, the Hall-Héroult process has several inherent disadvantages; it requires a large capital investment; it requires expensive electric power rather than cheap thermal power; most producing countries must import alumina or bauxite; and petroleum coke supplies for anodes are limited.

Properties

Vary according to the method of preparation. White powder, balls or lumps of various mesh. Sp. gr. 3.4–4.0; m.p. 2030°C; insoluble in water; difficultly soluble in mineral acids and strong alkali. Non-combustible; non-toxic.

Uses

Production of aluminium, abrasives, refractories, ceramics, electrical insulators, paper, chromatographic analysis, fluxes, food additive and as a catalyst.

ALUMINIUM SULPHATE

Process of Manufacture

From bauxite

Aluminium sulphate is produced by the reaction of sulphuric acid on bauxite, a naturally occurring hydrated alumina ore having the approximate formula $Al_2O_3.2H_2O$. The mined material varies in composition and generally contains one to three molecules of water plus impurities such as iron, silica, titanium, and selenium. The soluble aluminium oxide (Al_2O_3) content of a typical bauxite varies from 52 to 57% and it may contain 1 to 10% ferric oxide (Fig. 7.10).

Reaction

$$Al_2O_3 \cdot 2H_2O + 3H_2SO_4 \rightarrow Al_2(SO_4)_3 + 5H_2O$$
$$92\% \text{ yield}$$

The crude bauxite ore is ground to a fine powder (80% passes 200 mesh) and charged into open lead-lined steel reaction tanks. Sulphuric acid (sp. gr. 1.7) is added, and the raw materials are thoroughly agitated using paddle agitators, hot air, or live steam. The reaction mixture is kept at a temperature of 105 to 110°C by the live steam or lead steam coils. An excess of bauxite is fed to the reactor, so that there is an excess of 0.1 to 0.2% of soluble aluminium oxide. From 15 to 20 hours are required to complete the reaction. At the end of this time, a reducing material is added to the reaction mixture to reduce the iron (ferric sulphate) to a colourless ferrous condition. Barium sulphide in the form of black ash is commonly utilised, although sodium sulphide, hydrogen sulphide, sodium bisulphite, or sulphur dioxide may also be used. If the operation is performed in batches, the charge is allowed to settle in settling tanks. Flaked glue or some similar coagulable substance is generally added to remove the finely divided suspended material remaining in the supernatant liquid. This liquid is drawn off, and the residue is washed several times. The washings are combined with the decanted liquor, which is then sent to concentrators.

The process is generally operated in a continuous manner by using a battery of combined reaction and settling tanks. A common variation of this process is the Dorr procedure, which utilises reaction agitators in series. The reactants are thoroughly mixed and heated, using mechanical agitators, air, and live steam. Black ash (barium sulphide) is added to the last reactor to reduce the ferric sulphate. The reaction mixture is sent through a series of thickeners, operating counter-currently, which remove the undissolved material. At the same time the waste is washed thoroughly, so that it contains practically no aluminium sulphate when discarded. Glue is generally added to the first thickener as a coagulant.

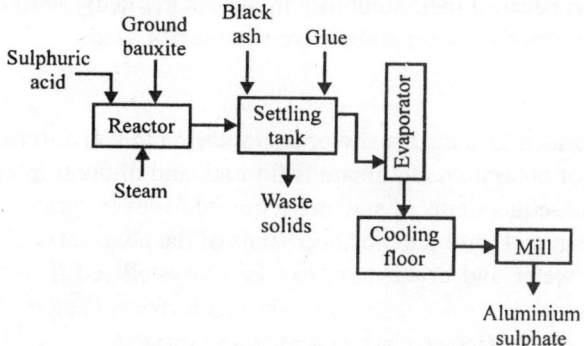

Fig. 7.10. Flow diagram for manufacture of aluminium sulphate from bauxite.

The clarified aluminium sulphate solution, from the counter-current decantation system, is concentrated in open steam-coil-heated, lead-lined evaporators. Here, the specific gravity is increased from about 1.3 to about 1.7. The concentrated solution is run into flat iron pans or onto a cooling table. The liquid quickly and completely solidifies, and when cool is broken up and ground to a uniform powder for shipment. Commercial aluminium sulphate generally contains only about 13 or 14 moles of water instead of the theoretical 18 moles. Also, it is usually in the basic form containing excess alumina. Anhydrous aluminium sulphate may be obtained by dehydration. The yield of aluminium sulphate based on the amount of aluminium oxide in both the finished product and raw material is 90 to 95%.

Properties

Colourless or white monoclinic crystals. Formula wt. 666.40; sp gr 1.69 (17°C); mp decomposes at 86.5°C. Soluble in water (86.9 g/100 ml at 0°C, 1104 g/100 ml at 100°C. Insoluble in ethanol.

Anhydrous aluminium sulphate [$Al_2(SO_4)_3$] is a white crystalline compound with the following properties:

Formula wt 342.13; sp gr 2.71; mp decomposes at 770°C. Soluble in water (31.3 g/100 ml at 0°C, 89 g/100 ml at 100°C). Insoluble in ethanol. pH of a 1% solution is 3.4.

Uses

It is used in pulp and paper; water purification; etc.

ALUMS

This series of double sulphates has the general formula:

$$M_2SO_4.N_2(SO_4)_3.24H_2O,$$

where M is univalent ion such as Na^+, K^+, or NH_4^+, and N is a trivalent ion such as Fe^{3+}, Al^{3+}, or Cr^{3+}.

e.g., $K_2SO_4.Al_2(SO_4)_3.24H_2O$ potash alum

$(NH_4)_2SO_4.Al_2(SO_4)_3.24H_2O$ ammonium alum

$K_2SO_4.Cr_2(SO_4)_3.24H_2O$ chrome alum

$(NH_4)_2SO_4.Fe_2(SO_4)_3.24H_2O$ ferric alum.

They are formed by mixing hot concentrated solutions of their component sulphates and cooling, when the alum crystals are deposited because of their low solubility in cold water. They are easily purified by recrystallisation because their solubility increases markedly with rising temperature.

The following special methods of preparation are sometimes used.

Potash Alum

Enough aluminium foil is added to some caustic potash solution to leave some undissolved at the end of the reaction. The solution of potassium aluminate is filtered, and dilute sulphuric acid is added until the solution is acidic (four gram-equivalents of acid are required for every gram-equivalent of alkali taken). The solution is then concentrated until it deposits crystals of the alum on cooling. These are filtered off, washed with a little cold water and dried; they can be recrystallised if necessary. Equations for the reactions are:

$$2Al + 2KOH + 2H_2O = 2KAlO_2 + 3H_2 \uparrow$$
$$2KAlO_2 + 4H_2SO_4 = K_2SO_4 + Al_2(SO_4)_3 + 4H_2O$$
$$\text{Alum}$$

Chrome Alum

A steady stream of sulphur dioxide is passed through a solution of potassium dichromate, acidified with sulphuric acid, until the solution turns green. Deep purple crystals of chrome alum separate from the concentrated solution on cooling. The equation for the reduction is:

$$K_2Cr_2O_7 + 3SO_2 + H_2SO_4 = K_2SO_4 + Cr_2(SO_4)_3 + H_2O$$
$$\text{Alum}$$

The crystals are filtered off, washed, and dried. The alums are isomorphous with each other, readily forming overgrowths; their crystals are usually octahedral in shape. It is important to realise that the alums are double salts and not complex salts. In solution they behave simply as a mixture of their component sulphates and give the reactions of their individual cations.

ALUMINIUM CHLORIDE

Process of Manufacture

From aluminium metal and chlorine

Scrap aluminium, or a mixture of scrap and pig aluminium, is charged to a refractory crucible furnace in which it is melted (660°C). Dry chlorine is passed into the molten charge and forms aluminium chloride which vapourises and leaves the furnace through a vapour duct in the top. The vapours are then passed into air-cooled iron condensers, where aluminium chloride sublimes below 178°C. The aluminium chloride is removed periodically (Fig. 7.11).

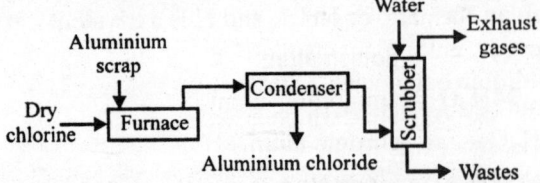

Fig. 7.11. Flow diagram for manufacture of aluminium chloride from aluminium metal and chlorine.

Reaction

$$2Al + 3Cl_2 \rightarrow 2AlCl_3$$

By proper regulation of the chlorine rate, no free chlorine remains in the exit gases from the furnace, but a protective absorber is usually placed beyond the aluminium chloride condenser to absorb unreacted chlorine and any aluminium chloride that may pass through the condenser.

Purity of the aluminium chloride depends on the quality of the aluminium used, as well as on the tightness of the system; moisture in the chlorine introduces corrosion products of the materials of construction. Usually high-purity aluminium chloride is obtained.

Special high-purity grades of crystalline aluminium chloride ($AlCl_3.6H_2O$) and solutions, are made by reaction of purified aluminium hydrate with hydrochloric acid.

Properties

Anhydrous—white deliquescent hexagonal crystals, which sublime readily. Formula wt 133.34; sp gr 2.44 25°C/4; mp 194°C (520 kPa); bp 182.7°C (100 kPa); sublimes at 178°C. Soluble in water (69.87 g/100 ml at 15°C, decomposes in hot water), ether, chloroform, and carbon tetrachloride. Insoluble in benzene. Aluminium chloride crystallises from hydrochloric acid as the hexahydrate ($AlCl_3.6H_2O$) in the form of colourless deliquescent trigonal crystals.

Uses

It is used as catalyst in: ethyl benzene; dyestuff intermediates; detergent alkylate; hydrocarbon resins; ethyl chloride; and pharmaceutical and cosmetic; pigments; roofing granules; special papers; etc.

ALUMINIUM CARBOXYLATES

The aluminium salts of carboxylic acids are derived from aluminium hydroxide, $Al(OH)_3$, by successive replacement of hydroxyl groups by carboxylate anions. For example, with formic acid the aluminium formate family would include dibasic aluminium formate, $(HO)_2Al(OOCH)$; monobasic aluminium formate, $(HO)Al(OOCH)_2$; and normal aluminium formate, $Al(OOCH)_3$. General methods of preparation for the mono- and dibasic aluminium carboxylates include direct reaction of aluminium as the catalytic electrode with an aqueous solution of the acid in an electrolytic process; reaction of the metal with the acid in the presence of catalytic amounts of mercury or mercuric chloride; and reaction by double replacement between aluminium alkoxide and the sodium salt of the organic acid. Normal aluminium carboxylates, e.g., tris(formato)aluminium, are prepared by the direct action of the acid with aluminium chloride in an organic solvent or by a double displacement reaction using a soluble aluminium salt and sodium salt of the organic acid. (Aluminium distearate is the largest selling aluminium carboxylate).

Uses

About 15–20 of the aluminium carboxylates are industrially important compounds with commercial applications. Commercial applications fall into three general areas: pharmaceutical preparations, particularly aluminium acetate and aluminium formate, based on the very low toxicity of the salts and upon either their basic or astringent properties; gelling agents, particularly aluminium mono-, di-, and tristearates, aluminium palmitate, and aluminium octanoate (2-ethylhexanoate); manufacture of cosmetics, coatings, and rocket fuels; and aluminium carboxylates, which are used in the textile industry as finishing agents for water-proofing fabric and as mordants in the dyeing process.

ALUMINIUM CHLORIDE HEXAHYDRATE

Aluminium chloride hexahydrate is usually made by dissolving aluminium hydroxide, $Al(OH)_3$, in concentrated hydrochloric acid. It is used primarily in deodorant, antiperspirant, and fungicidal preparations; for water treatment; in the manufacture of alumina and alumina–silica refractories; and in textile finishing. It may be toxic on ingestion.

ALUMINIUM BROMIDE AND ALUMINIUM IODIDE

Aluminium bromide and aluminium iodide are produced commercially only in small quantities. Anhydrous aluminium bromide, AlB_3, (density (25°C) 3.01 g/cm^3; melting point 97°C; boiling point 256°C) is used as a catalyst in Friedel-Crafts reactions. Anhydrous aluminium iodide, AlI_3, (density (25°C) 3.98 g/cm^3; melting point = 191°C) has limited use as a catalyst in the laboratory.

ALUMINIUM NITRATE

Aluminium nitrate, $Al(NO_3)_3.9H_2O$, is a white crystalline material, mp 73.5°C, soluble in cold water, alcohols, and acetone; decomposes to nitric acid and basic aluminium nitrates, $Al(OH)_x(NO_3)_y$ $(x + y = 3)$; and dissociates to aluminium oxide and oxides of nitrogen above 500°C. It is commercially produced from aluminous materials such as bauxite and nitric acid. Hydrated aluminium nitrate is used primarily as a salting-out agent in the extraction of actinides.

ALUMINIUM HYDROXIDES

Alumina monohydrates and hydrated alumina do not contain water of hydration but only hydroxide and oxide groups. Preparations of aluminium hydroxide obtained from solution of aluminium salts, alkaline aluminates, or aluminium alcoholates that are mainly amorphous are called gels. They are amphoteric, with low solubility in aqueous solutions of intermediate pH. Colloidal aluminium hydroxide is hydrophilic. Depending on the method of preparation, not all gelatinous aluminas are amorphous.

α-Alumina trihydrate is obtained by the Bayer process from crystallised alumina trihydroxide (known as gibbsite or hydrargillite, the principal constituent of tropical bauxites). Grains of α-trihydrate precipitated by the Bayer process are aggregates of relatively spherical shape, 50–100 μm in size. Solubility of α-trihydrate and heat of solution in various bases and acids is important in connection with the manufacture of aluminium chemicals. It contains about 35% chemically bound water. Typical Bayer hydrate contains 0.4 wt% of sodium ions in the crystals lattice; along with other alumina hydrates, it chemisorbs various anions effectively.

β-Alumina trihydrate is not a product of the Bayer process but is prepared by ageing of gels, gassing of sodium aluminate liquor with carbon dioxide, and calcination of aluminium oxychloride. Nordstandite, a new trihydrate is not yet commercially produced.

α-Alumina oxide hydroxide, a well-crystallised modification of AlOOH, also known as α-alumina monohydrate or boehmite, is a principal constituent of many bauxites of the Mediterranean type. Crystallised boehmite is prepared by hydrothermal digestion of aqueous slurries of trihydroxides or by solid-state reaction when gibbsite is heated in air at 383–573 K. Boehmite usually contains water in excess of the theoretical 15%. Another well-crystallised modification of AlOOH, β-alumina oxide hydroxide or diaspore, has been produced in the laboratory.

ACTIVATED ALUMINAS

These chemicals are obtained from various hydrated aluminas by controlled heating which eliminates most of the water of constitution. Their crystal structure is that of the transition aluminas: γ, η, χ or ρ alumina. Structural properties of the transition aluminas are given in Table 7.7. These transition aluminas are known as activated alumina or active alumina.

Table 7.7. Structural properties of transition aluminas.

Form	Crystal system	Molecules per unit cell	Unit axis length, nm			Density, g/cm³
			a	b	c	
γ	Tetragonal		0.562	0.780		3.2^a
δ	Orthorhombic	12	0.425	1.275	1.021	3.2^a
	Tetragonal		0.796		2.34	
η	Cubic (spinel)^b	10	0.790			2.5–3.6
θ	Monoclinic^c	4	1.124	0.572	1.174	3.56
χ	Cubic	10	0.795			3.0^a
	Hexagonal		0.556		1.344	
	Hexagonal		0.557		0.864	
κ	Hexagonal	28	0.971		1.786	3.1–3.3
	Hexagonal		0.970		1.786	
	Hexagonal		1.678		1.786	
ε	Orthohombic	4	0.773	0.778	0.292	3.71^a
	Orthorhombic^d	3	0.759	0.767	0.287	3.0

^{a} Estimated.
^{b} Space group O_h^7
^{c} Space group C_{2h}^3; angle of crystal, 103°20'
^{d} Space group D_{2h}^9 or C_{2v}^8

Preparation

Type 1, the oldest commercial form still widely used, is made from Bayer α-trihydrate. Processes produce various activated aluminas in cylindrical or spherical forms, but the pore distribution is not broad and few pores are greater than several tens of nanometers in diameter. Surface areas of these products are 300–500 m²/g. Type 2, activated bauxites with properties similar to Type 1, is obtained by thermal activation of bauxite containing alumina in the form of gibbsite. Type 3, another activated alumina, is obtained by very rapid activation of Bayer hydrate at 673–1073 K and has a pore surface area of 300–350 m²/g. Type 4 is derived from alumina gels, principally those prepared from solutions of $Al_2(SO_4)_3$ and NH_3, from $NaAlO_2$ and an acid, or from $NaAlO_2$ and $Al_2(SO_4)_3$, and produce corresponding by-product salts. Surface areas of pores are 300–600 m²/g.

CALCINED ALUMINA

These are generally obtained from Bayer hydroxide. Commercial calcined aluminas are heated to obtain α-Al_2O_3, the stable form of anhydrous alumina known also as the naturally occurring substance corundum. α-Al_2O_3 also results from solidification of molten alumina or by sintering processes. Although α-Al_2O_3 has a mp of 2326 K, the relatively fine hexagonal crystal plates obtained during calcination permit sintering to occur at much lower temperatures for ceramic-product manufacture. Higher sintering temperatures improve thermal stability for refractory applications, and sintered and fused α-Al_2O_3 products made from calcined aluminas exhibit extreme hardness, resistance to wear and abrasion, chemical inertness, outstanding electrical and electronic properties, good thermal shock resistance and dimensional stability, and high mechanical strength at elevated temperatures.

Preparation

Calcination is performed in rotary kilns and/or fluid calciners similar to those used for preparing alumina for the manufacture of aluminium by electrolysis. The particle size of the unground calcined Bayer aluminas is controlled, and the resulting porous agglomerates of alumina crystals (< 1–10 µm) are nominally 44–149 µm 100–325 mesh. Physical properties begin to exhibit the effect of crystal size above ca 2–3 µm. Bulk density decreases, angle of repose increases, attrition rate and dustiness increase, and bulk handling characteristics worsen with increasing crystal size. Special Bayer calcined aluminas can be broadly categorised according to sodium content and crystal size.

Dense sintered alumina containing large (typically 50–300 µm) tablet like crystals of α-Al_2O_3 i.e. commonly known as tabular alumina. Fused alumina is available as larger tonnage brown alumina and higher quality white alumina. Calcined sodium β-Al_2O_3, in the forms of $Na_2O.11Al_2O_3$ and $Na_2O.5Al_2O_3$, is also available.

Uses

About 90% of alumina is used in the production of aluminium metal. Tne remainder, including Bayer-source alumina, chemical-grade bauxite, and calcined abrasive and refractory-grade bauxite, is consumed in other applications such as flame-retardant fillers, preparation of aluminium compounds, pigments, adsorbents, catalysts, ceramics, refractories, and abrasives.

COPPER SULPHATE

Process of Manufacture

From copper and sulphuric acid

Copper sulphate is made by the reaction of scrap or shot copper with sulphuric acid and oxygen (air). The copper is fed to an oxidising tower; and a sulphuric acid-water mixture (20% sulphuric acid) is circulated through the bed while air is blown through the bottom of the tower. Copper sulphate pentahydrate is recovered from the resulting solution by crystallisation; the liquor is recycled to the tower along with makeup sulphuric acid (Fig. 7.12).

Reaction

$$2Cu + 2H_2SO_4 + O_2 \text{ (air)} + 8H_2O \rightarrow 2CuSO_4 \cdot 5H_2O$$

Much of the copper sulphate on the market is a by-product in copper refining. Copper is often leached from its ores with sulphuric acid; the resulting solution is treated in an electrolytic cell using an insoluble anode.

Copper is plated at the cathode, regenerating sulphuric acid. The spent electrolyte may then be used on a fresh batch of ore. The cell solution after re-use may be pumped to lead-lined tanks for evaporation and crystallisation of copper sulphate pentahydrate.

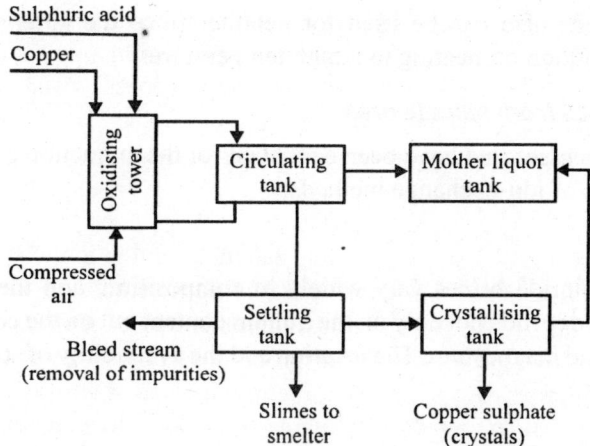

Fig. 7.12. Flow diagram for manufacturing of copper sulphate from copper and sulphuric acid.

Properties

$CuSO_4.5H_2O$ crystallises from water as the pentahydrate in the form of blue triclinic crystals which slowly effloresce in air. Formula wt 249.67; sp gr 2.286 (15.6°C/4); mp: loses $4H_2O$ at 110°C; bp: loses $5H_2O$ at 250°C. Soluble in water (24.3 g/100 g at 0°C, 205 g/100 g at 100°C). The anhydrous salt ($CuSO_4$) is obtained by dehydration as greenish-white rhombic crystals. Formula wt 159.60; sp gr 3.606 (15°C); mp 200°C; bp forms CuO at 650°C.

Uses

It is used as fungicides; algaecides; feed supplement; soil nutrient; intermediates; flotation; electroplating; petroleum etc.

LITHIUM AND LITHIUM COMPOUNDS

The applications of lithium metal are mainly in metallurgy and batteries and in the manufacture of lithium compounds, such as the amide, hydride, and nitride, and organolithium compounds.

Many of the properties of lithium are similar to those of magnesium and of the alkaline-earth metals. The resemblance to magnesium includes the high solubility of the halides (except the fluoride) in both water and polar organic solvents and the high solubility of the alkyls in hydrocarbons; the low aqueous solubility of the carbonate, phosphate, fluoride, and oxalate; the thermal instability of the carbonate and nitrate; the formation of the carbide and nitride by direct combination; and the reaction with oxygen to form the normal oxide. Lithium is widely distributed in nature; trace amounts are present in many minerals, in most rocks and soils, and in many natural waters. The lithium content of the earth's crust is estimated to be about 20 ppm. One spodumene, petalite, and lepidolite are important lithium sources from minerals. They occur mainly in granite pegmatites, which are coarse-grained igneous rocks composed largely of quartz, feldspar, and mica.

Production

Recovery from ores and brines

Spodumene is the chief lithium source for the two domestic producers of lithium products. The preferred method of extraction of lithium from spodumene is the sulphuric acid process. Methods suitable for extraction from spodumene also can be used for petalite, since the latter mineral converts to a β-spodumene-SiO_2 solid solution on heating to a high temperature.

Other recovery processes from silicate ores

Most of the numerous processes that have been described for the extraction of lithium can be classified as either alkaline methods or ion-exchange methods.

Recovery from brines

Natural predominately chloride brines vary widely in composition, and the economical recovery of lithium from such sources depends not only on the lithium content but on the concentration of interfering ions, especially calcium and magnesium. The location and the availability of solar evaporation capability also are important factors.

Lithium Metal

Process of manufacture

An electrolytic process consists of molten-salt electrolysis from a lithium chloride-potassium chloride mixture using a graphite rod as an anode. Modern industrial installations employ a 55 wt% LiCl–45 wt% KCl electrolyte at about 460°C using high purity lithium chloride as a feed. The current efficiency is about 80% and the lithium recovery is better than 98%, based on the charged chloride. The purity of the metal is 99.8% or better and the metallic impurities are less than 0.1%.

Properties

Lithium is an alkali metal with a silvery lustre and an atomic weight of 6.941. It is the first member of Group IA in the periodic system. Two stable isotopes are present in natural lithium (^7Li has an abundance of 92.6 wt% and ^6Li 7.4 wt%). Lithium has a density of 0.531 g/cm^3 at 20°C, and is the lightest of all solid elements. In general, the properties of lithium are similar to those of the other alkali metals. Physical properties of lithium are listed in Table 7.8.

Table 7.8. Physical properties of lithium.

Property	Value
At wt	6.941
At vol, cm^3	13.0
Mp, °C	180.5
Bp, °C	1336
Electronic configuration	$1s^2 \, 2s^1$
First ionisation potential, kJ/mol^a	519
Electron affinity, kJ/mol^a	52.3
Crystal structure	bcc
Lattice constant, pm	350
Metallic radius, pm	122.5
Ionic radius, pm	60

(Contd ...)

Property	Value
d_{20}, g/cm^3	0.531
Specific heat at 25°C, J/g^a	3.55
Specific heat of liquid at mp, J/g^a	4.39
Heat of fusion, J/g^a	431.8
Heat of vapourisation, kJ/g^a	ca 21.3
Electrical resistivity at 20°C, $\mu\Omega.cm$	9.446
Characteristic spectrum lines, nm	
red	670.8
orange	610.4
Vapour pressure at 702°C, kPa^b	0.065

[a] To convert J to cal, divide by 4.184.

[b] To convert kPa to mm Hg. multiply by 7.5.

The reaction of hydrogen and lithium readily gives a hydride, LiH, which is stable at temperatures from the melting point up to 800°C. Lithium reacts with nitrogen, even at ordinary temperatures, to form the reddish-brown nitride, Li_3N. Lithium burns when heated in oxygen to form the white oxide, Li_2O.

Inorganic Lithium Compounds

Lithium acetate

Lithium acetate, $CH_3COOLi.2H_2O$, is used as an alcoholysis catalyst for alkyd resin manufacture.

Lithium amide

Lithium amide, $LiNH_2$, is used in the pharmaceutical industry in the synthesis of antihistamines and analgesics. It can be prepared by the reaction of lithium hydride with ammonia. It is flammable. White crystalline solid; ammonia-like odour; sp. gr. 1.18; melts 380–400°C; decomposes in water, to form ammonia.

Lithium borates

Lithium metaborate, $LiBO_2.2H_2O$, mp 849°C, is used in special glass and enamel formulations. Lithium tetraborate, $Li_2B_4O_7$, mp 917°C, is used in the ceramic industry and as a flux in emission x-ray spectroscopy.

Lithium carbonate

Lithium carbonate is one of the most important lithium salts because it may be prepared from most of its water-soluble compounds. It is prepared from finely ground ore roasted with sulphuric acid at 250°C. Lithium sulphate is leached from the mass and converted to the carbonate by precipitation with soda ash. It is also prepared by reaction of lithium oxide with carbon dioxide or ammonium carbonate solution. Water solution is strong irritant.

Properties

It is in the form of white powder; sp. gr. 2.111; m.p. 735°C; b.p., decomposes at 1200°C. Slightly soluble in water; insoluble in alcohol; soluble in dilute acid.

Uses

Ceramics and porcelain glazes; pharmaceuticals; catalyst; other lithium compounds; coating of arc-welding electrodes; nucleonics; luminescent paints, varnishes and dyes; glass ceramics; aluminium production.

MOLYBDENUM COMPOUNDS

Halides and Oxyhalides

Molybdenum forms halogen compounds of widely different degrees of stability. The hexafluoride and pentachloride deposit molybdenum metal in the vapour phase as result of reaction with hydrogen.

The highest member of each series (MoF_6, $MoCl_5$, $MoBr_4$, and MoI_3) can be made by direct halogenation of molybdenum metal. The hexafluoride, pentafluoride, pentachloride, oxytetrafluoride, oxytetrachloride, and dioxydichloride are volatile at moderate temperatures. The entire series exhibits a wide range of stability, colour, and volatility. Few of the compounds are monomeric in their normal state. Generally, the lower halides are prepared by reduction of the highest member of the series with molybdenum metal, hydrogen, or a hydrocarbon.

Oxides and Hydroxy Compounds

Molybdenum trioxide

MoO_3, is a white crystalline powder from which most molybdenum compounds are prepared directly or indirectly. Technical grade MoO_3 is prepared commercially by oxidising molybdenite, MoS_2, in a multiple-hearth roaster. Air flow and temperature are carefully controlled to provide a low sulphur product with minimum MoO_2 content. High purity chemical grades are prepared either by sublimation of the technical grade or calcination of crystallised ammonium dimolybdate (ADM). Halogens and hydrohalides react with MoO_3 to form oxyhalides. Reaction with thionyl chloride, $SOCl_2$, gives molybdenum oxytetrachloride or pentachloride, depending on the reaction time.

Hydroxides

The structures of molybdenum blues and the series of hydrated molybdenum oxides are intimately related to those of the oxides and are characterised as genotypic. Hydrated oxides are prepared by the carefully controlled reduction of MoO_3 with either atomic hydrogen, zinc and hydrochloric acid, molybdenum metal powder and water, or lithium aluminium hydride.

Molybdates

Isopolymolybdates, a large class of inorganic compounds, consist of a cation and a condensed molybdate anion. The naming of isopolymolybdates is straightforward: they are called simply di, tri, etc., molybdates depending upon the degree of condensation.

Normal or orthomolybdates

$M_2O.MoO_3.xH_2O$ or $M_2MoO_4.xH_2O$, are prepared by direct combination of the oxides, neutralisation of slurries of MoO_3 with MOH or M_2CO_3, or by precipitation from the molybdate solution by salts of the desired metals. Acidification of molybdate ions in alkaline solution in which they exist as $(MoO_4)^{2-}$ results in the formation of polynuclear species.

Heteropolymolybdates

These form a large family of salts and free acids with each member containing a complex and high molecular weight anion.

Sulphides, selenides, and tellurides

Molybdenum forms a series of homologous compounds with sulphur, selenium, and tellurium. The disulphide, MoS_2, diselenide, $MoSe_2$, and ditelluride, $MoTe_2$, are isomorphous. These chalcogenides occur as shiny gray plates and may be prepared by the direct combination of the elements or by heating MoO_3 and the appropriate element in potassium carbonate to a high temperature.

Cyanide, thiocyanates, and carbonyls

Cyanides and thiocyanates form complex anions with molybdenum in a lower valence state. In most of these, molybdenum atoms have a co-ordination number of eight. These compounds are highly coloured. The cyanides and the thiocyanates are light sensitive, but are generally more stable in air than the complex halides. The red thiocyanate complex is formed by adding a solution of a soluble thiocyanate to a reduced acid solution of molybdenum. It is utilised in the colorimetric determination of molybdenum. The compound contains a thiocyanate to molybdenum ratio of three; the Mo is pentavalent.

Organomolybdenum compounds

In addition to the substituted molybdenum hexacarbonyl compounds, organomolybdenum sulphur compounds containing phosphorus or nitrogen have been prepared. Many of these compounds are oil-soluble and, thus, the low friction benefits of the molybdenum-sulphur interaction can be imparted to engine and gear oils.

Uses of molybdenum of compounds

The chemical applications of molybdenum have increased substantially. New applications have been developed in catalysis, lubrication, corrosion inhibition, protective coatings, inhibitive pigments, and flame and smoke retardants. Molybdenum compounds are used in the transportation and petroleum and petrochemical industries; molybdenum catalysts are prime candidates in the conversion of coal to liquid fuels.

Health and safety

Non-toxic in humans. In some areas where soil molybdenum is high, and consequently the molybdenum content of forage crops is high, animals develop a toxic reaction called molybdenosis, which can be overcome by injection or feed implementations of copper salts.

STRONTIUM AND STRONTIUM COMPOUNDS

Strontium is in Group IIA of the periodic table between calcium and barium. Strontium is present at 0.02–0.03% concentration in the earth's crust and is the fifth most abundant metallic ion in sea water. Strontium rarely forms independent minerals in igneous rocks but usually occurs as a minor constituent of rock-forming minerals.

Strontium is a hard, white metal with a melting range of 768–791°C, a boiling range of 1350–1387°C, and a density of 2.6 g/cm³, crystallising in the fcc system. Strontium-90 is a radioisotope produced by nuclear fission; four stable isotopes of strontium are known.

Strontium is produced by reduction of the oxide with aluminium *in vacuo*. There are no commercial uses of strontium metal. It has been used to remove traces of gas from vacuum tubes.

Strontium Compounds

Although many strontium compounds are known, only a few have commercial importance; the carbonate and nitrate are made in large quantities. The principal strontium mineral is celestite, $SrSO_4$. The carbonate is found as the mineral strontianite; no economically workable deposits are known. Almost all celestite is used to make strontium carbonate.

Strontium carbonate

Strontium carbonate is white impalable powder. Soluble in acids, carbonated water and solutions of ammonium salts, slightly soluble in water. Sp. gr. 3.62; loses CO_2 at 1340°C. Low toxocity. It is prepared when celestite is boiled with a solution of ammonium carbonate or is fused with sodium carbonate. It is used as a catalyst, in radiation-resistant glass for colour television tubes; ceramic ferrites; pyrotechnics.

Strontium chlorate

It is white crystalline powder, soluble in water, slightly soluble in alcohol. Sp. gr. 3.152; mp, decomposes at 120°C. It is prepared when strontium hydroxide solution is warmed and chlorine is passed in, with subsequent crystallisation. It is used in the manufacture of red-fire and other pyrotechnics; tracer bullets.

Strontium chloride

It is white in the form of crystalline needles, odourless, sharp, bitter taste, soluble in water and alcohol. Sp. gr. 3.054; m.p. 872°C; b.p. 1250°C. Sp. gr. 1.964 mp loses $6H_2O$ at 150°C.

It is prepared when strontium carbonate is fused with calcium chloride, the melt extracted with water, the solution is then concentrated and crystallised. It is used in strontium salts; pyrotechnics; electron tubes.

BORON

Boron, the fifth element in the periodic table, is composed of two stable isotopes with mass numbers of 10 and 11. Although widespread in nature, it has been estimated to constitute only 0.0001% of the earth's crust, usually occurring as alkali or alkaline-earth borates, or as boric acid. Pure forms of the element are difficult to prepare; the most common technique involves vapour deposition from a boron halide, usually in admixtures with hydrogen.

Research continues for methods of obtaining commercial quantities of the pure element, particularly as filamentary reinforcement for advanced composites.

In nuclear technology, thin films of boron are used for neutron counters, and dispersions of powdered boron in poly(vinyl chloride) or polyethylene castings are effective for shielding against thermal neutrons. Table 7.9 lists physical properties of boron.

Table 7.9. Physical properties of boron.

Property	Value
Atomic weight	10.811 ± 0.003
Melting point, °C	2190 ± 20
Boiling point, °C	3660
Coefficient of thermal expansion per °C from 25 to 1050°C	$(5–7) \times 10^{-6}$
Hardness	
Knoop, HK	2110–2580
Mohs, modified scale[a]	11
Vickers, HV	5000
Density	
Liquid[b]	2.08
α-rhombohedral crystals	2.46
Filamentary boron	
Tensile strength, MPa (psi)	3450–4830 (500,000–700,000)
Young's modulus, MPa (psi)	3040–3330 (440,000–480,000)
Structural modifications[c]	
Preparation temp, °C	
amorphous	800
α-rhombohedral	800–1100
α-tetragonal	1100–1300
β-rhombohedral	1300
Density, g/cm^3	
amorphous	2.3
α-rhombohedral	2.46
α-tetragonal	2.31
β-rhombohedral	2.35

[a] Diamond = 15.

[b] Just above melting point.

[c] The crystalline forms not listed here are designated tetragonal II (β-tetragonal), tetragonal III, and hexagonal.

Boron Compounds

At present, borax (tincal), colemanite, probertite, ulexite, and szaibelyite are the only borate minerals of commercial importance. Borax and colemanite are the most important.

Boron oxides

Boric oxide, B_2O_3, is the only commercially important oxide; however, one high oxide and several sub-oxides have been reported. Boric oxide, also known as diboron trioxide, boric anhydride, or anhydrous boric acid, normally is encountered in the vitreous state. This colourless, glassy solid usually is prepared by dehydration of boric acid at elevated temperatures. It is quite hygroscopic at room temperature, and the commercially available (Table 7.10).

A high purity grade of vitreous boric oxide (99% B_2O_3) is produced by fusing refined, granular boric acid in a glass furnace fired by oil or gas. Principal uses of boric oxide relate to its behaviour as a flux, an acid catalyst, or a chemical intermediate.

Table 7.10. Physical properties of vitreous boric oxide.

Property	Value
Vapour pressure[a], 1331–1808 K	$\log P_{kPa} = 5.849 - \dfrac{16960}{T}$
Heat of vapourisation[b], $\triangle H$, kJ/mol, 1500 K	390.4
298 K	431.4
Boiling point, °C extrapolated	2316
Viscosity, $\log \eta$, mPa.s (=cP) 1000°C	4.00
Density, g/cm³, 0°C	1.8766
18–25°C,	1.844
well-annealed 1000°C	1.528
Index of refraction, 14.4°C	1.463
Heat capacity (specific)[b], J/(kg.K) 1000 K	131.38
Heat of formation, $\triangle H_f$, kJ, 298.15 K[b] for 2B(s) + 3/2 O₂(g) = B₂O₃ (glass)	−1252.2 ± 1.7

[a] To convert kPa to torr, multiply by 7.5.

[b] To convert J to cal, divide by 4.184.

Boric acid

The name boric acid usually is associated with orthoboric acid, which is the only commercially important compound. Three forms of metaboric acid also exist. Orthoboric acid, H_3BO_3, crystallises from aqueous solutions as white, waxy plates, mp 170.9°C. When heated slowly, it loses water to form metaboric acid, HBO_2.

Most boric acid is produced by the reaction of inorganic borates with sulphuric acid in an aqueous medium. Sodium borates are the principal raw material and boric acid serves as a source of B_2O_3 in many fused products such as textile fibreglass and other borosilicate glasses. An important new use of boric acid is as a fire retardant in cellulosic materials. It also serves as a component of fluxes for welding and brazing.

A number of boron chemicals are prepared directly from boric acid. It also catalyses the air oxidation of hydrocarbons and increases the yield of alcohol by forming esters that prevent further oxidation of hydroxyl groups to ketones and carboxylic acids. The bacteriostatic and fungicidal properties of boric acid, although weak, have led to its use as a preservative in natural products. NF-grade boric acid serves as a mild, non-irritating antiseptic. Boric acid also is quite poisonous to many insects and has been used to control cockroaches and to protect wood against insect damage. A special quality grade of boric acid is added to nuclear-reactor cooling water.

Solutions of boric acid and borates

Boric acid is essentially monomeric in dilute aqueous solutions, but polymeric species may form at concentrations > 0.1 M. The conjugate base of boric acid in water is the tetrahydroxyborate anion, $B(OH)_4^-$. This species also is the principal anion in solutions of alkali-metal (1:1) borates such as $Na_2O.B_2O_3.4H_2O$. Mixtures of $B(OH)_3$ and $B(OH)_4^-$ would appear to form classical buffer systems where the solution pH is governed primarily by the acid-salt ratio. This relationship is nearly correct for solutions of sodium or potassium (1:2) borates, where the mole ratio $B(OH)_3 : B(OH)_4^- \cong 1$, and the pH remains near 9.0 over a wide range of concentrations. However, for solutions that have pH values much greater or less than 9.0, the pH changes greatly on dilution.

This anomalous pH behaviour is due to the presence of polyborates which dissociate into $B(OH)_3$ and $B(OH)_4^-$ as the solutions are diluted. Formation of polyborates also greatly enhances the mutual solubilities of boric acid and alkali borates. From a series of very accurate pH studies, Ingri calculated equilibrium constants involving the species $B(OH)_3$ and $B(OH)_4^-$, and the polyions $B_3O_3(OH)_5^{2-}$, $B_3O_3(OH)_4^-$, $B_5O_5(OH)_4^-$, and $B_4O_5(OH)_4^{2-}$. The ratio between the total anionic charge and the number of borons per ion increases with increasing pH.

Sodium borates

Disodium tetraborate decahydrate (borax decahydrate), $Na_2O.2B_2O_3.10H_2O$ [formula wt 381.43, monoclinic, sp gr 1.71, specific heat 1.611 kJ/(kg.K) (0.385 cal/g.°C)) at 25–50°C, heat of formation –6.2643 MJ/mol (–1497.2 kcal/mol)] exists in nature as the mineral borax. Disodium tetraborate pentahydrate (borax pentahydrate), $Na_2O. 2B_2O_3.5H_2O$ [formula wt 291.35, trigonal, sp gr 1.88, specific heat 1.32 kJ/(kg.K) (0.316 cal/(g.°C)), heat of formation –4.788 MJ/mol (–1143.5 kcal/mol)] is found in nature as a fine-grained deposit formed by dehydration of borax. Disodium tetraborate tetrahydrate, $Na_2O.2B_2O_3.4H_2O$ [formula wt 273.34, monoclinic, sp gr 1.91, specific heat ca 1.2 kJ/(kg.K) (0.287 cal/(g.°C)), heat of formation –4.4890 MJ/mol (–1072.9 kcal/mol)] exists in nature as the mineral kernite. Disodium tetraborate (anhydrous borax), $Na_2O:2B_2O_3$ [formula wt 201.27, sp gr (glass) 2.367, heat of formation (glass) –3.2566 MJ/mol (–778.4 kcal/mol)] exists in several crystalline forms as well as a glassy form. Sodium pentaborate decahydrate, $Na_2O:5B_2O_3.10H_2O$ [formula wt 590.34, monoclinic, sp gr 1.71] exists in nature as the mineral sborgite. Sodium metaborate octahydrate, $Na_2O.B_2O_3.8H_2O$ [formula wt 295.76, triclinic, sp gr 1.74] and sodium metaborate tetrahydrate, $Na_2O.B_2O_3.4H_2O$ [formula wt 203.68, triclinic, sp gr 191] can also be prepared.

Borax decahydrate and pentahydrate are produced from sodium borate ores, from dry lake brines, from colemanite, and from magnesium borate ores. In general, the production of fused materials is much more energy intensive than that of hydrated products, and this difference is reflected in their prices. The bulk borate products, borax decahydrate and pentahydrate, anhydrous borax, boric acid and oxide, and upgraded colemanite and ulexite, account in both tonnage and monetary terms for > 99% of sales of the boron primary products industry. A large increase in demand has been related to the use of borates in energy-conserving products (i.e., insulation).

Poisonings by boric acid have been reported following its use over larger areas of burned or denuded skin. The handling of borax or boric acid is generally not considered dangerous, however.

Borates are used mainly for glass manufacture. They also are used as fluxing agents for porcelain enamels and ceramic glazes, in soap and cleaning compositions, in agriculture as a fertiliser and herbicide, as a catalyst in the air oxidation of hydrocarbons, in the manufacture of alloys and refractories, as flux in metallurgy, as fire retardants, and as neutron absorbers in nuclear reactors.

Alkali-metal and ammonium borates

These include dipotassium tetraborate tetrahydrate, $K_2O.2B_2O_3.4H_2O$ [formula wt 305.51 orthorhombic, sp gr 1.92]; potassium pentaborate octahydrate, $K_2O.5B_2O_3.8H_2O$ [formula wt 586.42, orthorhombic, sp gr 1.74, heat capacity 329.0 J/(mol.K) (78.6 cal/(mol.K)) at 296.6 K); and diammonium tetraborate tetrahydrate, $(NH_4)_2O.2B_2O_3.4H_2O$ [formula wt 263.38, tetragonal, sp gr 1.58]. Ammonium pentaborate octahydrate, $(NH_4)_2.5B_2O_3.8H_2O$ [formula wt 544.4, sp gr 1.58, heat capacity 359.4 J/(mol.K) (85.9 cal/(mol.K)) at 301.2 K] exists in two crystalline forms, orthorhombic (α) and monoclinic (β).

Potassium tetraborate tetrahydrate may be prepared from an aqueous solution of KOH and boric acid with a B_2O_3:K_2O mole ratio of ca 2.0, or by separation from KCl–borax solution. Potassium pentaborate is prepared in a manner analogous to that used for the tetraborate, but the strong liquor has a B_2O_3:K_2O mole ratio near 5. Ammonium tetraborate tetrahydrate is prepared with a B_2O_3:$(NH_4)_2O$ mole ratio of 1.8–2.1. Ammonium pentaborate is similarly produced from an aqueous solution of boric acid and ammonia with a B_2O_3:$(NH_4)_2O$ mole ratio of 5.

The potassium and ammonium borates are low volume products with production figures of hundreds of metric tons per year for the tetra- and pentaborates. Dipotassium tetraborate tetrahydrate is used to replace borax in applications where an alkali-metal borate is needed but sodium salts cannot be used, or where a more soluble form is required.

The potassium compound is used as a component in lubricants, as a solvent for casein, as a constituent in welding fluxes, and as a component in diazo-type developer solutions. Potassium pentaborate octahydrate is used in fluxes for welding and brazing of stainless steels and non-ferrous metals. Diammonium tetraborate tetrahydrate is used when a highly soluble borate is desired, but alkali metals cannot be tolerated. Ammonium pentaborate octahydrate is used as a component in electrolytes for electrolytic capacitors, as an ingredient in flame-proofing formulations, and in paper coating.

Calcium-containing borates

Dicalcium hexaborate pentahydrate, $2CaO.3B_2O_3.5H_2O$ [formula wt 411.16, monoclinic, sp gr 2.42, heat of formulation – 3469 kJ/mol (–0.38 kcal/mol)] exists in nature as the mineral colemanite. Sodium calcium pentaborate hexadecahydrate, $Na_2O.2CaO.5B_2O_3.16H_2O$ [formula wt 810.60, triclinic, sp gr 1.95] exists in nature as the mineral ulexite.

Sodium calcium pentaborate decahydrate, $Na_2O.2CaO.5B_2O_3.10H_2O$ [formula wt 702.50, monoclinic, sp gr 2.15] exists in nature as the mineral probertite. Colemanite is used in the production of boric acid and borax, as well as for the manufacture of the "E" glass used in textile glass fibres and plastic reinforcements (where sodium cannot be tolerated). It also has limited application as a slagging material in steel manufacture and as a precursor to some boron alloys. Ulexite and proberite are used in the production of insulation fibre-glass and borosilicate glass as well as in the manufacture of other borates.

Borate melts and glasses

Most of the interest in metal borate glasses has centred on reports that indicated the existence of maxima and minima in some of the physical properties of the glasses with increasing metal oxide content. This phenomenon has been called the boron oxide anomaly. Modern theory on borate glass structure, however, indicates that the changes in the properties with alkali content are not anomalous, but are the result of well-defined structural changes in the glass at the molecular level. The borate glass compounds of commercial importance are boric oxide and disodium tetraborate pentahydrate.

Other metal borates

Borate salts or complexes of virtually every metal have been prepared. For most metals, a series of hydrated and anhydrous compounds may be obtained by varying the starting materials and/or reactions conditions. In general, hydrated borates of heavy metals are prepared by mixing aqueous solution or suspensions of the metal oxides, sulphates, or halides with boric acid or alkali-metal borates (e.g., borax). Anhydrous metal borates may be prepared by heating the hydrated salts to 300–500°C of by direct fusion of the metal oxide with boric acid or B_2O_3. Barium metaborate, $BaO.B_2O_3$, is used as an additive to impart fire-retardant and mildew-resistant properties to latex paints, plastics, textiles, and paper products, and as a preservative in protein-based glues described as technical grade cobalt tetraborate, $CoO.2B_2O_3.4H_2O$, is actually a mixture of metaborates and hexaborates. Hydrated copper metaborate, $CuO.B_2O_3.2H_2O$, has been used as a fungicide for treatment of lumber and other cellulose materials. The anhydrous salt, $CuO.B_2O_3$, is used as an oil pigment.

Zinc borates

A series of hydrated zinc borates has been developed for use as fire-retardant additives in coatings and polymers. A substantial portion of this is used in vinyl plastics where zinc borates are added alone or in combination with other fire retardants such as antimony oxide. The most commonly encountered zinc borate is $2ZnO.3B_2O_3.3.5H_2O$, which is formed when boric acid is added to solutions of soluble zinc salts.

Boron phosphate

BPO_4 is a white, infusible solid that vapourises slowly above 1450°C, without apparent decomposition. It is normally prepared by dehydrating mixtures of boric acid and phosphoric acid at temperatures up to 1200°C. The principal application of boron phosphate is as a heterogeneous acid catalyst, and a high degree of chemical purity is often unnecessary.

Boric acid esters

The term boric acid esters refers to compounds with the general formula $B(OR)_3$. Much of the chemistry of these compounds is related to the electrophilic nature of boron. A series of related compounds can be formed when electron-deficient boron accepts a fourth nucleophilic substituent leading to tetrahedral boron structures such as $NaB(OR)_4$.

Preparation

The most common preparative method for trialkoxy- and triaryloxyboranes is the reaction of the appropriate alcohol or phenol with boric acid, an inexpensive and readily available boron source. The boric acid ester prepared in the largest quantities is methyl borate, and most of it is used captively in the production of sodium borohydride. Other borate esters that have been offered commercially include ethyl borate, *n*-propyl borate, isopropyl borate, *n*-butyl borate, cresyl borate (from a mixture of *meta* and *para* cresols), and tri(hexylene glycol) diborate.

Properties

Trialkoxy (or aryloxy) borones range from colourless low boiling liquids such as trimethoxyborane (commonly referred to as methyl borate) to high melting solids. Trialkoxy (or aryloxy) boranes are typically monomeric, soluble in most organic solvents, and dissolve in water with accompanying hydrolysis to boric acid and the corresponding alcohol or phenol. This hydrolysis occurs rapidly except in cases where the boron atom is protected by bulky alkyl or aryl substituent groups. The boron atom in trialkoxy (or aryloxy) boranes is in a trigonal coplanar state with sp^2 bond hybridisation, and has a vacant *p* orbital along the threefold symmetry axis perpendicular to the BO_3 plane. This vacant orbital is readily available for acceptance of nucleophiles, such as water or alcohols. Structural studies of trialkoxy- and triaryloxyboranes have confirmed this structure having the angle between C—O bonds at 120°C except in those cases where the angles are slightly distorted by bulky substituent groups. The susceptibility of the boron atom to attack by nucleophiles is similar to that of the carbonyl carbon in carboxylic acid esters and leads to the analogy of referring to trialkoxy (or aryloxy) boranes as boric acid esters.

Trialkoxyboranes from straight-chain alcohols and triaryloxyboranes are stable to relatively high temperatures. Attempts to use this potentially useful property in borate ester-based high temperature lubricants and heat-transfer media have met with limited success because of the reactivity of these compounds to water and oxygen.

Uses

The principal application of methyl borate is as an intermediate in the commercial production of sodium borohydride. Another commercial outlet is the use of methyl borate azeotrope as gaseous flux for welding and brazing. Smaller quantities of borate esters are finding use as epoxy resin curing agents, and specific types of borate esters are used as gasoline additives to reduce engine knocking and engine deposits. Borate esters are used in hydraulic fluids and lubricants and as a variety of polymer additives. Glycol borate esters also constitute the active ingredients in biocides used in jet and diesel fuels.

Boron halides

The important physical and thermochemical properties of the boron trihalides are given in Table 7.11. The thermal conductivity of BBr_3 at 20°C is 0.112 W/(m.K). The boron trihalides are planar (sp^2) molecules with X–B–X angles of 120°C. Orbital energy assignments have been made for the trihalides based on their photoelectron spectra. Boron trihalides are highly toxic and contact should be avoided. They react vigorously (sometimes explosively) with water to yield hydrogen halides. At high temperatures they decompose to yield toxic halogen-containing fumes.

Table 7.11. Physical properties of the boron trihalides.

Property	BCl_3	BBr_3	BI_3
Melting point, °C	−107	−46	−49.9
Boiling point , °C	12.5	91.3	210
Density ρ^a (liq), g/mL	1.434_4^0	2.643_4^{18}	3.35
	1.349_4^{11}		
Critical temperature °C	178.8	300	
Critical pressure, kPa^b	3901.0		
$\triangle H_t^0$, kJ/mol gas^c	−403	−206	+18
$\triangle H_{vap}$, kJ/mol^c	23.8	34.3	
$C\rho$, J/(mol. °C), for gas at 25°C^c	62.8	67.78	
$C\rho$, J/(mol. °C), for liquid at 25°C^c	121	128	
$\triangle H_{hydrol}$ kJ/mol liquid at 25°C^c	−289	−351	
$\triangle H_{fusion}$, J/g at mp^c	18		
B—X bond energy, kJ/mol^c	443.9	368.2	266.5
B—X distance, nm	0.173	0.187	0.210

^{a} For BCl_3: ρ (g/mL) = 1.3730–2.159 × 10^{-3}°C; 8.377 × 10^{-7}°C; −44 to 5°C.
  For BBr_3: ρ (g/mL) = 2.698–2.996 × 10^{-3}°C; −20 to 90°C.
^{b} To convert kPa to mm Hg, multiply by 7.50.
^{c} To convert J to cal, divide by 4.184.

RARE-EARTH ELEMENTS

The rare earths comprise a group of 17 elements in the periodic table and have similar properties in aqueous solutions. All are metals in the elemental state and all form salts that are strong electrolytes when dissolved in water. They ionise in this medium to give triply charged ions and, because of the high charge on these ions, react strongly with water dipoles to form a tight sheath of water molecules about them. Other ions in aqueous solutions only contact this sheath, giving rise to the similar properties of rare-earth cations in water.

The group consists of the following elements: scandium (^{21}Sc), yttrium (^{39}Y), lanthanum (^{57}La), all of which appear in the group IIIB of the periodic table, and cerium (^{58}Ce), praseodymium (^{59}Pr), neodymium (^{60}Nd), promethium (^{61}Pm), samarium (^{62}Sm), europium (^{63}Eu), gadolinium (^{64}Gd), terbium (^{65}Tb), dysprosium (^{66}Dy), holmium (^{6}Ho), erbium (^{68}Er), thulium (^{69}Tm), ytterbium (^{70}Yb), and lutetium (^{71}Lu).

The rare earths are widely distributed in low concentrations throughout the earth's crust. They occur as mixtures in many massive rock formations, e.g., basalts, granites, gneisses, shales, and silicate rocks, in which they are present in amounts of 10–300 ppm. They also occur in ca. 160 discrete minerals, most of which are rare, but in which the rare-earth content, expressed as R_2O_3 or Ln_2O_3 (REO), can be as high as 60% (REO is an abbreviation used in industry for rare-earth oxide content). Ln_2O_3 is used when Y_2O_3 is not present in the ore or mineral.

Approximately ten of these occur in sufficient quantities that they may furnish some REO to commerce, but more than 95% of the REO occurs in three minerals: monazite and bastnasite for the light rare earths, and xenotime for yttrium and the heavy rare earths. Xenotime occurs mixed with monazite in alluvial deposits.

Properties

Physical

The rare earth metals alloy with most metals to form intermetallic compounds and occasionally solid solution. In rare-earth-rich alloys, other elements can change the properties of the pure metal by drastically lowering (or, more rarely, raising) the melting point by 200–300°C in some cases. Alloying with other elements can make the rare earth either pyrophoric or corrosion-resistant. Some properties of the rare earths are listed in Table 7.12.

Table 7.12. Some properties of the rare earths.

Properties	Scandium	Yttrium	Lanthanum	Cerium	Praseodymium	Neodymium
Atomic number	21	39	57	58	59	60
Atomic weight	44.9559	88.9059	138.9055	140.12	14.9077	144.24
Colour	silvery	silvery	silvery	silvery	silvery	silvery
Melting point, °C	1541	1522	918	798	931	1021
Boiling point, °C	2836	3338	3464	3433	3520	3074
Density, g/cm^3	2.989	4.469	6.146	6.770	6.773	7.008
Heat of fusion, kJ/mol^a	14.10	11.40	6.20	5.46	6.89	7.14

(Contd...)

Properties	Promethium	Samarium	Europium	Gadolinium	Terbium
Atomic number	61	62	63	64	65
Atomic weight	145	150.36	151.96	157.25	158.9254
Colour	silvery	silvery	silvery	silvery	silvery
Melting point, °C	1042	1074	822	1313	1356
Boiling point, °C	3000 (estd)	1794	1527	3274	3230
Density, g/cm^3	7.264	7.520	5.244	7.901	8.230
Heat of fusion, kJ/mol^a	7.6 (estd)	8.62	9.21	10.05	10.79

Properties	Dysprosium	Holmium	Erbium	Thulium	Ytterbium	Lutetium
Atomic number	66	67	68	69	70	71
Atomic weight	162.50	164.9304	167.26	168.9342	173.04	174.967
Colour	silvery	silvery	silvery	silvery	silvery	silvery
Melting point, °C	1412	1474	1529	1545	819	1663
Boiling point, °C	2567	2700	2868	1950	1196	3402
Density, g/cm^3	8.551	8.795	9.066	9.321	6.966	9.841
Heat of fusion, kJ/mol^a	11.06	16.87	19.90	16.84	7.66	18.65

^{a} To convert J to cal, divide by 4.184.

Chemical

The chlorides, bromides, nitrates, bromates, and perchlorate salts are soluble in water and, when their aqueous solutions evaporate, they precipitate as hydrated crystalline salts. The acetates, iodates, and iodides are somewhat less soluble. The sulphates are sparingly soluble and are unique in that they become less soluble with increasing temperature. The oxides, sulphides, fluorides, carbonates, oxalates, and phosphates are insoluble in water. The oxalate, which is important in the recovery of highly pure Ln, can be calcined directly to the oxide. Anhydrous rare-earth salts usually cannot be prepared by evaporating the water of crystallisation.

The lanthanides can form hydrides of any composition up to LnH_3. Small amounts of hydrogen dissolve interstitially but, with increasing amounts of hydrogen, a second phase (LnH_2) appears. These alloys are metallic in their properties and lose hydrogen at relatively high temperatures.

The rare earths form many compounds with organic ligands; some of which are water-soluble. Chelating agents, which are organic molecules that engulf rare-earth ions and displace some of the water of hydration, form complexes with rare-earth ions that have a much wider range in their formation constants than do those of ordinary mineral-acid salts.

Production and Processing

Monazite and bastnasite are the minerals used commercially to supply most of the rare-earth chemicals. Monazite is a brown, dense phosphate mineral. There are extensive deposits of enriched monazite sands on beaches in many parts of the world, e.g., along the south-west coast of India and the east coast of Brazil. Such deposits are dredged, pulverised if necessary, and further enriched by flotation methods. Sometimes they are also subjected to cross-belt magnetic separation since they are weakly magnetic.

Liquid-liquid extraction

As the industrial demand for individual rare earths increased, some of the rare-earth companies developed liquid-liquid extraction systems. An organic stream immiscible with water is flowed counter-current to the aqueous stream containing the rare-earth mixture. The organic phase may absorb neutral rare-earth molecules by complexing with them, e.g., as does tributyl phosphate, or it may contain a complexing molecule or anion added to it for the same purpose. Also, complexing ligands may be added to the aqueous phase so as to effect a synergistic equilibrium between the rare-earth ions and the complexant. If the various equilibria between the organic and aqueous phases have overall equilibrium constants sufficiently different for various rare earths, separation occurs.

Uses

The rare earths are used in gasoline-cracking catalysts; in carbon arcs; as additives to steel and cast irons; as polishing compounds; to make optical glass; as both colourants and decolourants for glass, depending upon the rare earth element used; and in magnets. In the electronic area, the most important industrial rare-earth compounds are garnet-based materials, e.g., yttrium iron garnet and gadolium gallium garnet, used in microwave devices, memory storage, and oxygen sensors (Y_2O_3-stabilised ZrO_2). In nuclear reactors, the rare earths are used in the form of oxides as absorbers of neutrons. They are also used as additions to superalloys, as hydrogen-storing materials, and as synthetic gems.

Health and Safety

The rare earths are considered only slightly toxic and can be handled safely with ordinary care.

SODIUM CHROMATE

Process of Manufacture

From chromite ore

Sodium dichromate is produced by the reaction of sulphuric acid on sodium chromate, which in turn is obtained by calcining a mixture of chromite ore, limestone, and soda ash (Fig. 7.13).
Reactions

$$4FeCr_2O_4 + 8Na_2CO_3 + 7O_2 \rightarrow 8Na_2CrO_4 + 2Fe_2O_3 + 8CO_2$$
$$2Na_2CrO_4 + H_2SO_4 \rightarrow Na_2Cr_2O_7 + H_2O + Na_2SO_4$$

Chromite, or chrome ironstone, is a chromium iron oxide ore containing principally ferrous chromite ($FeCr_2O_4$ or $FeO.Cr_2O_3$) plus small amounts of alumina, silica, and magnesia. High-grade chromite contains about 50% chromium (III) oxide (Cr_2O_3).

Chromite is crushed and ground to a fine powder (90 to 98% through 200 mesh) in ball mills. It is then mixed with at least 90% of the theoretical soda ash, a varying amount of lime, and sufficient filter residue (from farther along in the process) to adjust the chromic oxide content to 15 to 20%. The mixture is then blended in a rotary mixer and calcined in a rotary kiln, similar to those used in the cement industry, at 1100 to 1150°C. After a 4 hour retention time the roast from the kiln is sent directly to leaching tanks or thickeners, where the soluble chromate is dissolved in hot water and concentrated. The underflow from the last tray in the thickener is filtered. The leached residue is dried in rotary driers. A portion is recycled to the kiln feed, where it serves to keep the reacting mass in a relatively dry condition. The mother liquor from the filter is returned to the leaching tank.

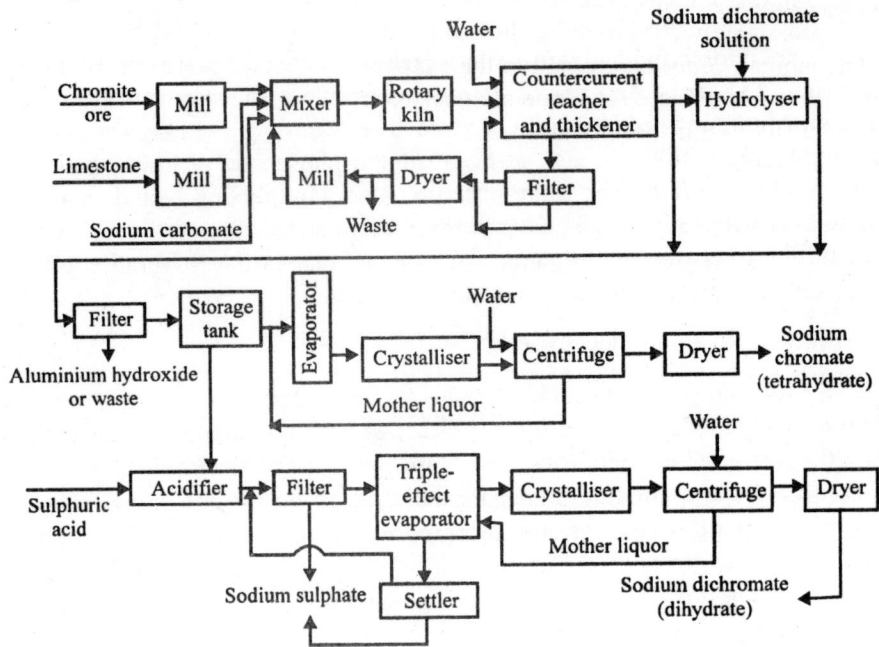

Fig. 7.13. Flow diagram for manufacture of sodium chromate from chromite ore.

The overflow from the leacher is also filtered and sent to hydrolysing tanks. Here any sodium aluminate is hydrolysed to insoluble aluminium hydrate by the slow addition of sodium dichromate solution, according to the reaction:

$$2NaAlO_2 + Na_2Cr_2O_7 + 3H_2O \rightarrow Al_2O_3.3H_2O + 2Na_2CrO_4$$

The addition of sodium dichromate solution must be closely controlled to obtain a filterable alumina hydrate instead of a gelatinous mass. After alumina removal, handling the resulting sodium chromate solution depends on the product desired.

For sodium chromate production, the liquor from the tanks is concentrated in steel evaporators. The concentrate is discharged into a crystalliser, where sodium chromate forms. The crystals are continuously centrifuged (the mother liquor is returned to the evaporators), washed, and dried in a rotary-drum dryer. The yellow crystals of sodium chromate decahydrate are packaged for shipment.

These crystals usually contain some sodium sulphate. Very pure grades of sodium chromate are therefore made from sodium dichromate by reaction with soda ash. A moist equimolar mixture of the two salts is heated in a tray dryer with steam and blown dry with air; the resulting cake is ground to yield 99.7% sodium chromate.

For sodium dichromate production, the chromate solution from the storage tank is charged into an acidifier, where sulphuric acid is added until the pH of the liquor is reduced from 8.04 to 4.7. The chromate is converted to dichromate, with the resulting formation of sodium sulphate. Most of the latter crystallises out of the boiling solution as anhydrous sodium sulphate. The crystals are removed by filtration, and the filtrate is charged into triple-effect evaporators, where the remainder of the sulphate crystallises.

The clear, hot, saturated dichromate solution (sp gr 1.7) is fed into a water-cooled crystalliser. Here, at about 20°C, approximately 70% of the sodium dichromate crystallises and is removed by continuous centrifuging. The mother liquor is returned to the evaporators. After drying in a rotary-drum dryer, the red crystals of sodium dichromate dihydrate are packaged. Both sodium chromate and dichromate may be purified by re-crystallisation from water.

Properties

Sodium chromate (Na_2CrO_4)

Yellow, rhombic crystals. Formula wt 162.00; sp gr 2.723; mp 792°C. Soluble in cold water and hot water (126 g/100 g at 100°C).

Crystallises from water as the decahydrate ($Na_2CrO_4.10H_2O$) in the form of yellow deliquescent monoclinic crystals. Formula wt 342.16; sp gr 1.483; mp 19.9°C. Very much soluble in cold water. Soluble in all proportions in hot water. Slightly soluble in ethanol.

Sodium dichromate ($Na_2Cr_2O_7.2H_2O$)

Bright-orange, deliquescent, monoclinic crystals. Formula wt 298.05; sp gr 2.348 (25°C); mp loses $2H_2O$ at 84.6°C; anhydrous at 356.7°C bp decomposes at 400°C. Soluble in water (70.6% at 0°C, 88.4% at 80°C).

Uses

It is used in pigments; chromic acid; leather tanning; metal treatment; textile and dyes.

HYDROGEN PEROXIDE

Process of Manufacture

By hydrolysis

Hydrogen peroxide is manufactured by the hydrolysis of a solution containing the persulphate ($S_2O_8^{2-}$) ion. The particular solution used (persulphuric acid, ammonium persulphate, or potassium persulphate) is prepared in an electrolytic cell in which the corresponding sulphate is oxidised anodically. Hydrolysis of the persulphate (outside the cell) is carried out under conditions that regenerate the sulphate and vapourise a mixture of water and hydrogen peroxide. The latter are separated and the peroxide concentrated by vacuum distillation.

In the embodiment of the process as shown in the flowsheet, a regenerated solution of ammonium sulphate and sulphuric acid (equivalent to ammonium bisulphate) is adjusted to proper concentration and fed to a series of electrolytic cells. In one plant the porcelain cells are 70 cm wide and 95 cm deep. The platinum anodes and graphite cathodes are so arranged that water in cooling tubes may keep the electrolyte temperature below 35°C.

The potential drop across each cell, operated at 5400 A, is 5.7 V. Ammonium thiocyanate (0.1 g/litre) is usually added to the electrolyte to increase oxygen overvoltage. Anode current density is 4000 to 11000 A/m². A current of air is swept continuously through the cell to keep the hydrogen concentration below 5%, so as to avoid explosion.

The persulphate solution leaving the cells goes to an evaporator, where it is heated by both live steam and by means of steam coils. Hydrogen peroxide and water vapours go overhead. The ammonium bisulphate solution is cooled, filtered to remove any sediment, adjusted in concentration, and returned to the cells. The cathode liquor is similarly cooled, filtered, and re-used. The hydrogen peroxide-water vapour mixture is sent to stoneware distillation towers which are operated under vacuum (40 mm Hg or 5.3 kPa). A 30 to 40% hydrogen peroxide solution is removed from the bottom of the tower. This solution may be concentrated to 70% or 90 to 98% H_2O_2 in two stages.

The 30 to 40% hydrogen peroxide solution recovered from the cell liquors is adjusted with sulphuric acid (about 0.5 g/litre) and fed to a stoneware or porcelain evaporator, where the solution is heated to 65°C at 40 to 50 mm Hg (5.3 to 6.6 kPa).

The vapours evolved are scrubbed in a packed tower with distilled or demineralised water in such ratio to the peroxide vapours that a 70% hydrogen peroxide solution leaves the bottom of the tower. This product may be fed to a second evaporator (at 75°C and the same pressure) and a second packed tower to produce an 80 to 85% solution of hydrogen peroxide. The product is cooled, stabilised with acids or an organic oxidation inhibitor, and sent to storage.

This concentration process may be run continuously by fractionating and recycling the vapours from the first evaporator. The resulting 70% hydrogen peroxide solution may be concentrated to a 90 to 98% solution in a similar second stage. From time to time, liquid bottoms from the first evaporator must be drawn off to remove impurities. The ammonium persulphate process described differs only in details from the persulphuric acid process and the potassium persulphate process. All three have been used commercially. The pure-acid process has a lower current efficiency (70 to 75%) but has the advantage that crystallisation problems within the system are eliminated.

In the potassium persulphate process, an ammonium bisulphate solution is electrolysed as before. After electrolysis potassium acid sulphate is added to the electrolyte:

$$(NH_4)_2S_2O_8 + 2KHSO_4 \rightarrow K_2S_2O_8 + 2(NH_4)HSO_4$$

The potassium salt is crystallised and separated by filtration. The solid potassium salt is then added to a sulphuric acid solution to produce a slurry. On heating with live steam, the persulphate hydrolyses to produce hydrogen peroxide vapours. As in the other processes, the potassium salt may be recovered and recycled.

By oxidation of alkylhydroanthraquinones

The modern chemical method for manufacturing hydrogen peroxide is an organic autoxidation process in which a suitable organic compound (usually a hydroquinone type) in solution is oxidised to the quinone form with the concurrent formation of hydrogen peroxide. The peroxide is extracted and concentrated; the oxidised organic compound is reduced catalytically and reused (Fig. 7.14).

In one adaptation of this process, the raw material is a 10% solution of mixed alkylated anthraquinones and their tetrahydro derivatives. The solvent is a mixture of primary and secondary nonyl alcohols with methyl or dimethylnaphthalene. This mixture is reduced with hydrogen in the presence of a catalyst of activated alumina coated with 0.7% palladium.

The reduction may be carried out at 40°C and under hydrogen pressures of 1 to 3 atm (100 to 300 kPa). After removal of the catalyst granules by filtration, the resulting hydroanthraquinone still in solution is blown with oxygen or an oxygen-nitrogen mixture at 30 to 60°C to produce hydrogen peroxide (5 to 6 g/litre) and regenerate the original anthraquinones.

The hydrogen peroxide is extracted with a sufficient quantity of water to yield a 20 to 25% hydrogen peroxide solution. The anthraquinones are recycled. The hydrogen peroxide may be concentrated as described in the previous process.

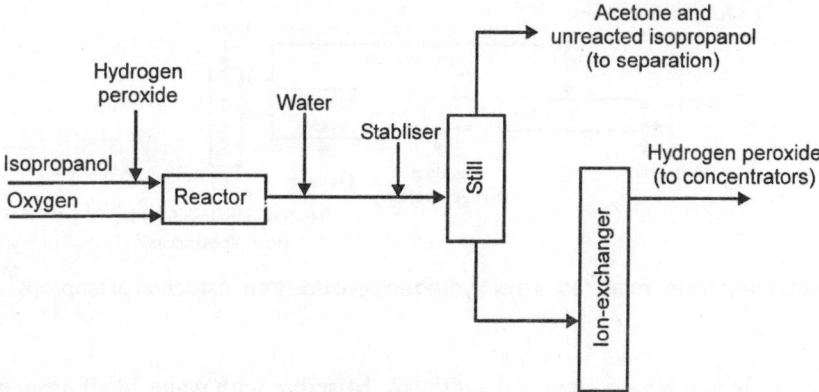

Fig. 7.14. Flow diagram for manufacture of hydrogen peroxide from oxidation of alkylhydroanthraquinones.

This process is similar to a process developed in Germany and used there during World War II. The German process made use of ethylanthraquinone in a benzene-cyclohexanol solvent and used Raney nickel as a hydrogenation catalyst. High losses of Raney nickel and the partial solubility of the solvent in water led to the economic modifications of the process described.

By oxidation of isopropyl alcohol

Hydrogen peroxide and acetone are jointly produced by the liquid-phase oxidation of isopropyl alcohol with molecular oxygen. Isopropyl alcohol and hydrogen peroxide (1 wt. %) (to eliminate the prolonged incubation period) are fed to a reactor along with 80 to 95% oxygen. The reactor is lined with glass, enamel, aluminium, or some other material that will not decompose the peroxide as it forms. Oxidation of the isopropyl alcohol takes place at 90 to 140°C and 200 to 300 psi (1.4 to 2.1 MPa) to form hydrogen peroxide and acetone.

The product (15 to 25% hydrogen peroxide) is continuously withdrawn, diluted with water to 6 to 10% hydrogen peroxide (to avoid explosion) and distilled (usually in the presence of a stabiliser). Acetone and isopropyl alcohol go overhead and thence to a further distillation for separation. Still bottoms (a weak hydrogen peroxide solution) are passed through an ion-exchange column to remove heavy metals, and then concentrated by conventional means.

Reaction

$$CH_3CH(OH)CH_3 + O_2 \rightarrow CH_3COCH_3 + H_2O_2$$
87% yield

At a conversion of 15% per pass, the yield of hydrogen peroxide is 87% based on isopropyl alcohol, and the yield of acetone is 93% (Fig. 7.15).

Properties

100% hydrogen peroxide is a water-white sirupy liquid with oxidising properties. Formula wt 34.02; Sp gr 1.438 20°C/4; mp –0.461°C; bp 155.5°C.

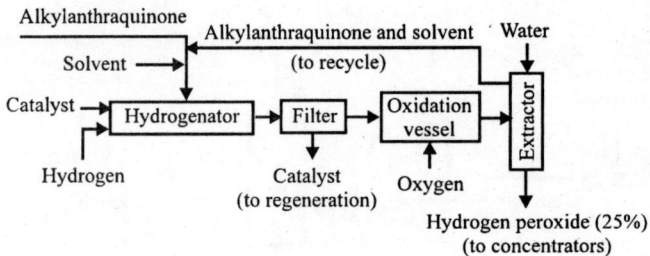

Fig. 7.15. Flow diagram for manufacture of hydrogen peroxide from oxidation of isopropyl alcohol.

Uses

It is used in textiles; pulp and paper; chemical synthesis. Miscible with water in all proportions. Soluble in ether. Insoluble in hydrocarbons. In pure form, either as is or in solution, hydrogen peroxide is quite stable. It decomposes, sometimes violently, in the presence of even traces of metallic impurities. In contact with readily oxidisable organic substances it may cause spontaneous combustion.

CHAPTER 8

Industrial Gases

INTRODUCTION

Air constitutes a large bulk of the raw materials used in the chemical and metallurgical industries. Air is used for oxidation, combustion and gasification processes. Its constituents like oxygen, nitrogen and argon are used individually for various processes. The main constituents of air are oxygen, nitrogen, argon and other rare gases.

Those of the gases which are required in large quantities for industrial purposes are termed as industrial gases. The most important and largely used industrial gases are: (i) hydrogen; (ii) oxygen; (iii) nitrogen; (iv) carbon dioxide; (v) acetylene; (vi) ethylene; and (vii) helium.

Industrial gases have a large variety of essential functions: some are used as raw materials for the manufacture of other chemicals; some are essential medicaments like oxygen and helium. Some of the gases in the liquid and solid states are used for creating cold, largely by absorbing heat via evaporation, or by performing work, or by melting. The most modern application of industrial gases is in the cryogenic techniques.

Oxygen is not only used for oxidation, combustion and gasification processes in the field of chemical and metallurgical industries, it is also used for medical application. Oxygen can be transported as compressed gas in cylinders, in liquid form in tankers and also through pipeline directly for process use.

Nitrogen is used in the fertiliser industry for production of ammonia, calcium cyanamide, as a refrigerant for frozen food industry and also for purging, inerting, blanketing applications in various industries. Argon—a rare gas—has its use in degassing of molten steel for alloy steel making. Argon is also used as a shielding gas in welding (TIG) and for electric bulb manufacture. Carbon dioxide is used as a gas or as a liquid and as solid for food preservation, beverage making, welding and other process use.

SEPARATION OF GASES FROM AIR

The gases that constitute air (except CO_2) are known as permanent gases. Permanent gases can be liquefied only below their critical temperature and under pressure, while other gases—e.g. ammonia, SO_2, CO_2, Cl_2, propane, CFC—are liquefiable gases or gases which can be liquefied at ordinary temperature under pressure. Hence, liquefaction and separation of permanent gases has been possible after the development of low temperature technology and the interpretation of thermodynamic properties and the laws applicable to permanent gases.

Thermodynamic Concept behind Liquefaction of Gases

There are three ways a gas can be liquefied: (i) external cooling; (ii) isentropic expansion through JT effect; and (iii) adiabatic expansion with external work. The last two methods are mostly used, and basic refrigeration cycles use both JT and adiabatic depending on the liquefaction cycle to be adopted. There are mainly four different commercial processes for operation of gas liquefiers. These are: (i) high pressure, Linde cycle with precooling; (ii) medium pressure claude cycle; and (iii) high pressure Heyland cycle; and (iv) low pressure Kapitza's cycle. For laboratory liquefiers up to a capacity of 20 lits/hr., VSA N_2 liquefier plant is suitable.

HYDROGEN (MANUFACTURE, PROPERTIES AND USES)

From Hydrocarbons and Steam (Steam Reformer Process)

The largest quantities of hydrogen are manufactured by catalytically reacting hydrocarbons and steam, to yield hydrogen and carbon oxides, followed by the water-gas shift reaction. The most commonly used raw material is natural gas, although other natural and refinery hydrocarbons, including naphtha cuts may be used. The following description is based on propane as a raw material.

$$CH_3CH_2CH_3 + 3H_2O \rightarrow 3CO + 7H_2$$
$$3CO + 3H_2O \rightarrow 3CO_2 + 3H_2$$
$$CH_3CH_2CH_3 + 6H_2O \rightarrow 3CO_2 + 10H_2$$

Commercial propane, obtained from either natural-gasoline plants or oil refineries, contains small amounts of organic sulphur compounds which are removed before processing. The propane from storage passes in the form of vapours through a heater at a temperature of about 370°C. The hot gases then pass over a bauxite or metallic oxide catalyst which converts the sulphur compounds (mercaptans, organic sulphides, and carbonyl sulphide) to hydrogen sulphide. After cooling, the gases are scrubbed with aqueous sodium hydroxide and water to remove the soluble sulphides. When natural gas is used as the process feed, sulphur is removed by passing the gas through drums containing activated carbon (Fig. 8.1).

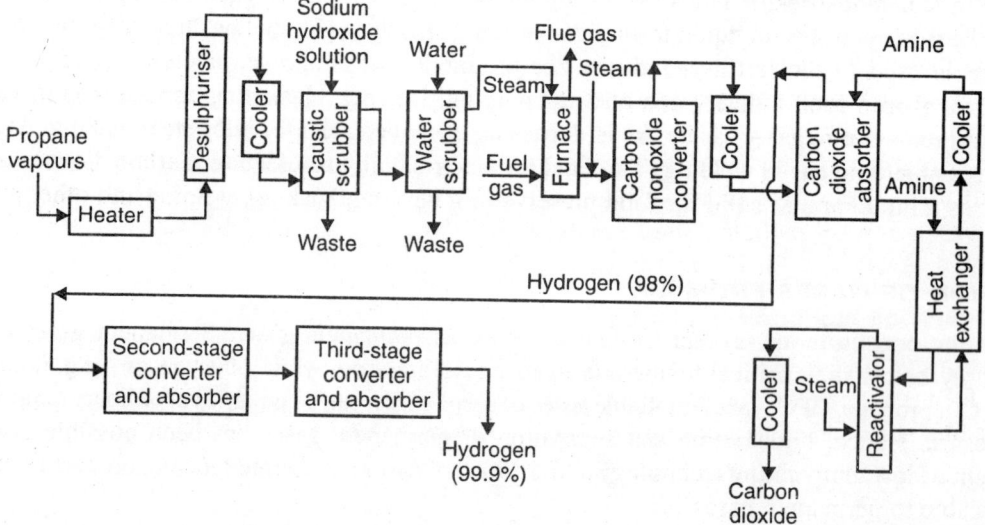

Fig. 8.1. Flow diagram for manufacture of hydrogen from hydrocarbons and steam (steam reformer process).

The sulphur-free propane vapours are mixed with steam and passed into the top of a reforming furnace. One type of furnace consists of a number of vertical nickel–chromium–iron alloy tubes, 7.5 to 20 cm in diameter and about 7.5 m long, mounted in a refractory furnace. Heat for the endothermic reaction is supplied by multiple horizontal burners located at various levels, with the flue gases passing upward counter-current to the process gas. At temperatures of 760 to 980°C, the propane gas passes down the tubes over a supported nickel catalyst at a space velocity of about 600 volumes/hour per volume of catalyst. The propane is converted to hydrogen, carbon monoxide, and carbon dioxide, with a trace of a methane remaining in the mixture.

Operating pressure in the reformer may be as high as 600 psi (4.2 MPa). Although the steam-hydrocarbon reaction is favoured by low pressure, there are compensating economic advantages in operating at high pressure. Accordingly, new plants usually are designed for high-pressure operation.

In any event, the reformed gases are cooled to about 370°C by mixing with steam, and then passed into the first-stage carbon monoxide converter containing an iron oxide catalyst promoted with chromium oxide. Here the exothermic conversion reaction (water-gas shift) takes place at a temperature of about 425°C and a space velocity of 100 (or greater) volumes of gas/volume of catalyst per hour. Both this catalyst and the nickel reforming catalyst are rugged and have a normal life of 1 year or more.

From the converter, the gases containing a small amount of carbon monoxide are cooled to about 38°C and passed into a packed (or bubble-tray) tower. Here aqueous 15 to 20% monoethanolamine is circulated down through the counter-currently blowing gas (Girbotol process). The amine solution absorbs the carbon dioxide and, after passing through heat exchangers, is run to the top of a reactivating tower. Here the carbon dioxide is desorbed by steam generated by heating the solution in a reboiler at the bottom of the tower. The carbon dioxide removed amounts to about 30 volumes/100 volumes of hydrogen and, since it is recovered at a purity of 99.8%, it is available as a useful by-product. The regenerated amine solution is then returned to the system. At atmospheric pressure and 38°C hydrogen gas containing 20% carbon dioxide may be purified to 0.1% carbon dioxide by scrubbing, with the monoethanolamine absorbing 15 to 30 m^3 carbon monoxide/m^3 of solution circulated. Approximately 0.12 kg steam/litre of solution is required for regeneration. The carbon dioxide-free hydrogen coming from the absorber still contains about 1% carbon monoxide. This is removed by passing through two more stages of carbon monoxide conversion, followed by carbon dioxide removal. From the last absorber, the purified hydrogen analysis better than 99.9% pure. The gas may be compressed to about 150 psi (1 MPa) and charged into storage tanks. Many new plants require only one stage of shift conversion. A highly active low-temperature catalyst is used.

Shift converter gas may also be purified by absorption of carbon dioxide in hot carbonate solutions. In one modification (Vetrocoke process), the potassium carbonate solution is activated with arsenic trioxide. Small amounts of carbon oxides may be removed by conversion to methane by passing the mixture over a nickel catalyst at 315 to 425°C.

From Hydrocarbons by Partial Oxidation

Synthesis gas, a mixture of hydrogen and carbon monoxide, may be made by partially burning any hydrocarbon feed (natural gas through fuel oil) with oxygen. The carbon monoxide is then reacted with water vapour in a shift converter to form more hydrogen (Fig. 8.2).

$$C_nH_{2n-2} + mO_2 \rightarrow xCO_2 + n - xCO + n - 1 - yH_2 + yH_2O$$
Fuel oil

$$CO + H_2O \rightarrow CO_2 + H_2$$

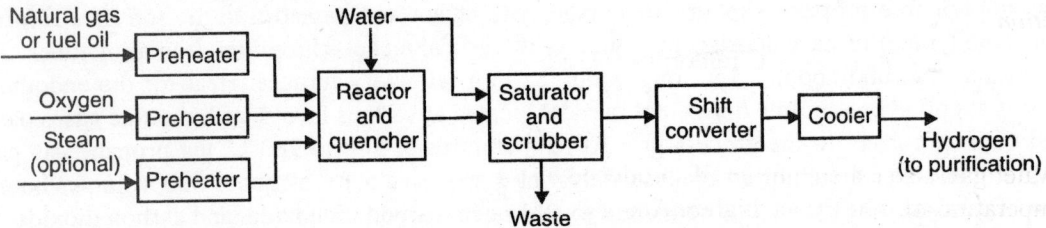

Fig. 8.2. Flow diagram for manufacture of hydrogen from hydrocarbons by partial oxidation.

Natural gas, for example, is preheated to 650°C and sent to a refractory-lined pressure vessel where it reacts with preheated oxygen. Hydrocarbon feed rate varies from 1000 to 3000 volumes/unit volume of reactor space. The oxygen/carbon ratio is held between 1.0 and 1.2 to hold carbon formation below 14 kg/million ft^3 (0.5 kg/1000 m^3) of gas. Customary operating pressure is 200 to 500 psi (1.4 to 3.5 MPa). The reaction products are removed continuously from the reaction zone. In addition to the major products, carbon monoxide and hydrogen, the burner effluent contains 5 mole % water, less than 2 mole % carbon dioxide, less than 0.1 mole % oxygen and about 0.5% unreacted methane. The effluent gases are quenched rapidly, in less than 1 sec., from 1425°C to 530 or 815°C. In a somewhat similar process used largely in Europe, the partial combustion is moderated by the presence of nitrogen, steam, or carbon dioxide; the quench is carried out in a waste-heat boiler.

In either case the gases may be scrubbed (to remove carbon particles) and saturated, and then passed to a shift converter, where the carbon monoxide reacts with steam to produce carbon dioxide and hydrogen. The hydrogen may be purified by conventional means.

From Water Gas and Steam

Hydrogen is produced along with carbon dioxide by catalytically reacting water gas (40% carbon monoxide–50% hydrogen) with steam at elevated temperatures. The carbon dioxide is removed by scrubbing the gas; relatively pure hydrogen remains. The process is particularly adaptable to low-purity hydrogen containing nitrogen and carbon monoxide for the synthesis of ammonia and methanol, respectively (Fig. 8.3).

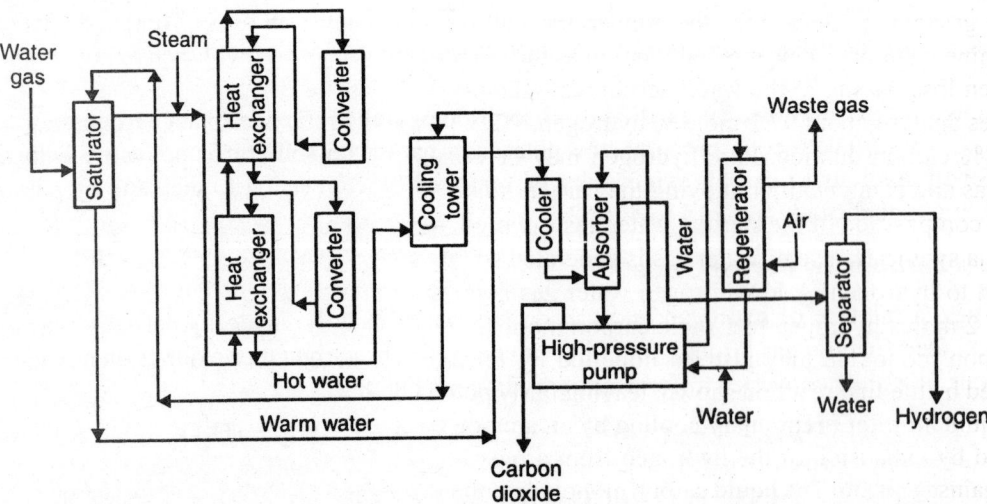

Fig. 8.3. Flow diagram for manufacture of hydrogen from water gas and steam.

Reaction

$$C \text{ (amorph)} + H_2O \rightarrow CO + H_2 \text{ (water gas)}$$

$$CO + H_2O \rightarrow CO_2 + H_2$$

94–96% (conversion of CO)

Water gas, also called blue gas, is produced by the reaction of steam on incandescent coke or coal at a temperature of 1000°C or higher. Analysis of the gas approximates 40% carbon monoxide, 50% hydrogen, 5% carbon dioxide, and 5% nitrogen and methane. The gas from a holder is run into a saturator, where it contacts hot water and is heated to 75 to 85°C. The saturated water gas (1 volume) is then mixed with steam (3 volumes) and passed into a two-stage catalytic converter, where carbon monoxide is reacted with water vapour. The first stage operates at a temperature of 425 to 480°C, and the second at 370 to 400°C, with heat exchangers between the stages. Two stages are employed because of the exothermic character of the reaction and the decreased conversion at higher temperatures. By using this two-stage procedure on the so-called water-gas shift reaction, the major part of the conversion takes place with a relatively small amount of catalyst, whereas the balance is effected at a lower temperature conducive to high overall yield. The catalyst commonly used is iron oxide promoted with chromium oxide, and is required in amounts of 180 to 270 kg/1000 ft³ (6425 to 9600 kg/1000 m³) of water gas converted per hour. The life of the catalyst is long; it is ordinarily not necessary to remove sulphur compounds from the water gas, because the catalyst is sulphur-resistant and converts sulphur compounds into hydrogen sulphide which is removed with the carbon dioxide.

The exit gases, containing about 64% hydrogen, 31% carbon dioxide, 4% nitrogen and methane, and 1% carbon monoxide, are cooled in water towers and passed to purification units.

The method of purifying depends on the ultimate use of the hydrogen gas. For uses at high pressures, the reaction gases may be compressed to 200 to 400 psi (1.4 to 2.8 MPa) and passed through water scrubbers, where most of the carbon dioxide is removed. About 3 to 5% loss of hydrogen takes place because of its solubility at these pressures in water. The resulting gas contains 0.5 to 1.5% residual carbon dioxide and 1.5 to 2.0% carbon monoxide. These impurities may be removed by compressing the gas to 2000 to 3000 psi (14 to 21 MPa) and scrubbing with an ammoniacal cuprous formate or cuprous chloride solution.

The gaseous mixture from the converters may be purified at atmospheric pressure by scrubbing with amine solutions. This method, known as the Girbotol process, saves compression costs and avoids hydrogen loss, owing to the water-scrubbing operation. After carbon dioxide removal by this method, the gases analyse about 97.2 mole % hydrogen, 1.2% carbon monoxide, 1.2% nitrogen, 0.3% methane, and 0.1% carbon dioxide. Pure hydrogen may be obtained by starting with a lower-nitrogen-content water gas and catalytically removing the carbon monoxide.

The composition of the starting water gas is varied to suit the end use destined for the hydrogen. For ammonia synthesis, a semi-water gas is generated so that after conversion it contains the proper ratio of nitrogen to hydrogen. A low-nitrogen water gas is produced for methanol synthesis, which requires a ratio of 2 moles hydrogen to 1 mole carbon monoxide in the starting gases.

Carbon monoxide may also be liquefied by refrigeration and thus separated from the water gas produced by the first reaction shown, leaving fairly pure hydrogen. In one such process, the refrigeration is by liquid air after preliminary cooling by an ammonia refrigeration system; in another, the cooling is obtained by expansion of the hydrogen from a pressure of 20 atm (2.1 MPa) to a lower one by doing work against a piston. The liquid carbon monoxide is then easily separated from the still gaseous hydrogen.

The steam-water gas process is practically obsolete, but is described here for its historical interest, since it was the dominant process for hydrogen production at one time.

Steam-Iron Process

This process produces hydrogen by the reaction of steam at high temperature on reduced iron oxide to yield hydrogen, and then re-reducing the iron oxide with a reducing gas such as water gas or producer gas in a cyclic operation, this process is obsolete.

From Water by Electrolysis

Direct current is passed between iron or nickel-plated iron electrodes in a solution consisting of 10 to 25% caustic soda or potash. Only distilled water is added to the electrolyte, since the solute is not consumed. Operating voltage is about 2 V; current density about 650 A/m^2. Electricity required is 140 or more kWh/1000 ft^3 of hydrogen (4950 kWh or 18 GJ/1000 m^3) plus one-half that volume of oxygen (if the oxygen is recovered). Where low-cost electricity is available this process is attractive, and it is one of the ways of producing heavy water.

Hydrogen is also obtained as a by-product in other electrochemical processes, although it is not always considered economical to recover it. Typical examples are the electrolysis of sodium chloride or potassium chloride solutions for the manufacture of chlorine, sodium or potassium hydroxide, sodium chlorate, and sodium perchlorate.

Thermal Decomposition of Hydrocarbons

Hydrogen is obtained as a by-product in the manufacture of carbon black by the thermal decomposition of natural gas. A brick checker-work chamber is heated to 1090 to 1370°C, and natural gas is passed through. After the carbon black is separated, the gases contain 70 to 85% hydrogen. By scrubbing to remove carbon oxides, hydrogen pure enough for ammonia synthesis is obtained. This process is one of the cheapest sources of low purity hydrogen.

Catalytic Reforming of Petroleum Stocks

Considerable hydrogen, 400 to 800 ft^3 (11 to 22 m^3)/barrel of feedstock, is produced by petroleum refiners in catalytic reforming operations. Hydrogen-rich gas from the reforming operation contains hydrogen sulphide and low-molecular-weight hydrocarbons. Purification involves only a caustic wash (to remove hydrogen sulphide) followed by refrigeration to condense most of the water and higher hydrocarbons. The resulting gas is usually further dried by passing it over a solid adsorbent (e.g., activated alumina) and may be washed with liquid nitrogen to remove more hydrocarbons. Although the product (95% hydrogen) is suitable for ammonia synthesis and was formerly so used, present refining processes have such large requirements of hydrogen (for hydro-desulphurisation and other uses) that all or practically all by-product hydrogen is used on-site and does not reach the market.

From Methanol and Steam

Hydrogen of higher than 98% purity may be manufactured by the catalytic reaction of methanol and steam at 260°C, according to the equation:

$$CH_3OH + H_2O \rightarrow CO_2 + 3H_2$$

The exothermic reaction at moderately low temperatures may be conducted at up to 300 psi (2.1 MPa), so that the hydrogen can be delivered at elevated pressures. For 1000 standard ft^3 (28 m^3)

of hydrogen plus 18 kg of pure by-product carbon dioxide, approximately 20 litres of methanol and 15 litres of fuel are required. The ease and simplicity of fabricating the plant, handling raw materials, and purifying the hydrogen make this process particularly practical for portable plants. The present cost of methanol precludes the widespread or large-scale use of this process.

From Ammonia by Dissociation

For metal treating and hydrogen welding, a source of mixed gas (75% hydrogen–25% nitrogen) is the catalytic cracking of ammonia. In the process liquid ammonia is vapourised, preheated by hot gases from the reaction chamber, and then passed into an electrically heated reactor. In the presence of a supported nickel oxide catalyst, the ammonia may be cracked at pressures up to 20 psig (350 kPa) and temperatures ranging from 900 to 980°C. The product gases may be scrubbed with water to remove traces to unconverted ammonia, and then dehydrated to yield a dry gas.

Since each kilogram of the liquid ammonia yield 2.8 m^3 of dissociated gas containing approximately 2.1 m^3 of hydrogen and 0.7 m^3 of nitrogen, it is claimed that a single 150-lb (68-kg) cylinder of liquid ammonia can replace 33 cylinders of hydrogen. Pure nitrogen may also be obtained by burning out the hydrogen at a cost comparable to or perhaps lower than cylinder nitrogen. Although it is not economically feasible to produce pure hydrogen by this process, the cracked gas is quite satisfactory for many uses, particularly in the steel industry in the annealing process.

High-purity hydrogen

Extremely pure hydrogen (99.9%) can be produced by diffusing less pure gas through palladium films. At about 315°C, molecular hydrogen dissociates to atoms on palladium; the atoms then diffuse through the metal and recombine on the other side. The liquefaction process, described in outline below, also yields extremely pure hydrogen.

Liquid hydrogen

The steam reformer process is usually the source of hydrogen in liquefied hydrogen plants. After the removal of impurities as described, further purification is obtained by refrigeration and adsorption; feed to the liquefier must be quite pure to prevent blocking of the lines and valves by condensable impurities.

The final liquefier section depends on the *ortho-para* shift of the hydrogen molecule. Hydrogen can exist in two forms, depending on the rotation of the two atoms in the molecule. In orthohydrogen, they spin in the same direction, in parahydrogen in opposite directions. The conversion temperature is also very low, –70°C, above which Joule-Thompson expansion raises rather than lowers the temperature.

Normal hydrogen is only 25% *para* at atmospheric temperature; at liquid hydrogen temperature it is 99.7% *para*, and the heat generated by the conversion is about 11 + % of the total heat removed during liquefaction. If normal liquid hydrogen were stored, as much as two-thirds of it would be vapourised as conversion took place. Therefore the liquefier cycle includes rapid conversion over a ferric hydroxide catalyst. This takes place in two steps, at liquid nitrogen and liquid hydrogen temperatures, in the presence of liquid refrigeration. Thus liquid hydrogen is over 95% *para*. Purity may exceed 99.99%. The manufacture of hydrogen is largely an integral part of other chemical manufacturing processes, the production of ammonia and methanol. In these operations it is manufactured and used as a mixture with other gases, nitrogen in the case of ammonia synthesis and carbon monoxide in the case of methanol synthesis. Nevertheless, the manufacturing processes are essentially identical, except that the gas mixtures must be purified to produce pure (95 to 100%) hydrogen.

The steam reforming process is by far the most important; next in importance is the hydrocarbon partial oxidation process. Natural gas, liquefied petroleum gases, and propane may all be used as feedstocks, according to availability and cost. An important advantage of the partial oxidation process lies in its versatility with respect to feedstocks, particularly its adaptability to fuel oil. Of the minor manufacturing processes, ammonia dissociation (cracking) is most widely used. Several companies furnish push-button plants for producing the nitrogen-hydrogen mixture that results; the nitrogen content is not detrimental in such applications as annealing of steel, powder metallurgy, and atomic-hydrogen welding.

Large quantities of hydrogen are produced as a by-product of catalytic reforming of petroleum stocks. Some of this is used in the refinery for hydrotreating, some for hydrocracking processes, and some for ammonia manufacture. Considerable amounts are burned for fuel or flared.

Properties

Colourless, highly flammable gas; the lightest known susbstance.

Mol. wt. 2.016; mp. $-259.1°C$; Sp. gr. gas 0.06948 (air =1); bp. $-252.8°C$; Sp. gr. liquid 0.0709 ($-252.7°C$).

Soluble in water ($2.1\ cm^3/100\ g$ at $0°C$, $0.85\ cm^3/100\ g$ at $80°C$). Slightly soluble in iron, palladium, and platinum.

Uses

It is used in production of ammonia, methanol and aniline; hydrocracking; hydrogenation of edible oils to produce shortenings and margarines, many of which are low in cholesterol. Hydrogen is used in large electric generators to reduce windage losses and heat. Liquefied hydrogen is used as a rocket fuel; it has become the commercial source of the ultrapure hydrogen required in some segments of the electronics industry.

CARBON DIOXIDE (MANUFACTURE PROPERTIES AND USES)

From Steam and Natural Gas

Most (more than 60%) of merchant carbon dioxide is a by-product of the production of ammonia, which in turn is produced by the reaction of nitrogen and hydrogen. Hydrogen for this reaction is produced in several ways, but primarily by the steam reforming process summarised in the above equation. Most newer producers of carbon dioxide make use of this by-product as their source of product.

$$CH_4 + 2H_2O \rightarrow 4H_2 + CO_2$$

From Flue or Kiln Gases

Burning of fuel:
$$C + O_2 \rightarrow CO_2$$

Burning of limestone:
$$CaCO_3 + heat \rightarrow CaO + CO_2$$

All carbonaceous fuels, in burning, produce carbon dioxide along with oxides of other ingredients of the fuel, such as sulphur dioxide. When limestone is calcined to make lime, carbon dioxide is evolved and may be recovered. These sources, once important, are now relatively minor, at least so far as carbon dioxide reaching the market is concerned.

From Fermentation

The fermentation of grain to produce ethyl alcohol also produces a stoichiometric quantity of carbon dioxide. Since the advent of purely chemical methods of producing ethanol, this source has become relatively unimportant. Similar fermentations to produce beer, wine, and liquor also produce carbon dioxide, but the carbon dioxide is either usually not recovered or is used by the breweries themselves.

$$C_6H_{12}O_6 \xrightarrow{\text{Yeast}} 2C_2H_5OH + 2CO_2$$

From Sodium Carbonate and Phosphoric Acid

In plants manufacturing sodium phosphates by the reaction of phosphoric acid with sodium carbonate, pure carbon dioxide is available as a by-product, and a small amount is thus recovered for sale.

$$Na_2CO_3 + H_3PO_4 \rightarrow Na_2HPO_4 + CO_2 + H_2O$$

Recovery of carbon dioxide

Methods for recovering carbon dioxide in pure form depend primarily on whether it is in dilute and impure form, as from combustion or lime kilns, or in practically pure form, as in certain types of natural gas or as a by-product of hydrogen production, in which case it needs only to be liquefied or solidified. Recovery of carbon dioxide from flue gases will illustrate the former.

From Flue or Kiln Gases by Absorption

Coke is burned under specially designed boilers in a Carborundum-lined furnace. Combustion is controlled by drafts (about 0.6 m^3/kg of coke), so that the flue gases contain 17 to 18% carbon dioxide. At the same time steam (125 to 150 psi) is generated in the boilers to furnish power for the pumps and compressors and heat for the lye boiler. The hot (232°C) flue gases, containing oxygen, carbon monoxide, nitrogen, dust, and some sulphur and organic compounds, in addition to 18% carbon dioxide, are passed through a heat exchanger and economiser. Here the temperature of the gases is reduced to about 120°C, and the excess heat is taken up by counter-currently flowing strong lye. The gases are scrubbed in limestone-packed towers with water from the coolers to remove sulphur dioxide and dust and to reduce the temperature to about 38°C. The dilute carbon dioxide gas is then fed into the bottom of coke-filled absorption towers, where it passes counter-current to an aqueous solution of sodium carbonate (sp gr. 1.12) called weak lye. After absorbing the carbon dioxide, the solution of bicarbonate, known as strong lye, is pumped through heat exchangers operating at 104 to 120°C, where the solution is heated before entry into the lye boiler. The absorbers, operating at about 30°C, remove all but about 9% of the carbon dioxide in flue gas, which is then released to the atmosphere (Fig. 8.4).

The preheated solution of sodium bicarbonate is heated with exhaust steam (from the compressors) in the lye boiler. Here, at a temperature of about 118°C, the bicarbonate is decomposed into sodium carbonate (weak lye), which is returned through the heat exchanger to the absorbers for reuse. The liberated gas, consisting of 99.8% carbon dioxide, escapes through the rectifying section of the lye boiler, where it warms the entering strong-lye solution. The gas passes through a water cooler, where the temperature is further lowered, and the moisture in the saturated gas is condensed and returned to the weak lye. The cooled gas is collected in a gas holder. In place of sodium carbonate, other absorbents, such as aqueous potassium carbonate, monoethanolamine (10 to 20%), and triethanolamine may be used in a similar manner. Before or during liquefaction, the concentrated gas is purified by removal of organic impurities that would affect its taste and odour, and the gas is dried. Various methods or so-called cycles of liquefaction are employed, such as the precooling, binary, ternary, bleeder, or flush-cooling cycles. These operate on the raw gas or on a combination of raw and revert gases (return or back-blow gas from the solid carbon dioxide press), using two- or three-stage compression and ammonia refrigeration, carbon dioxide flash cooling, or a combination of both.

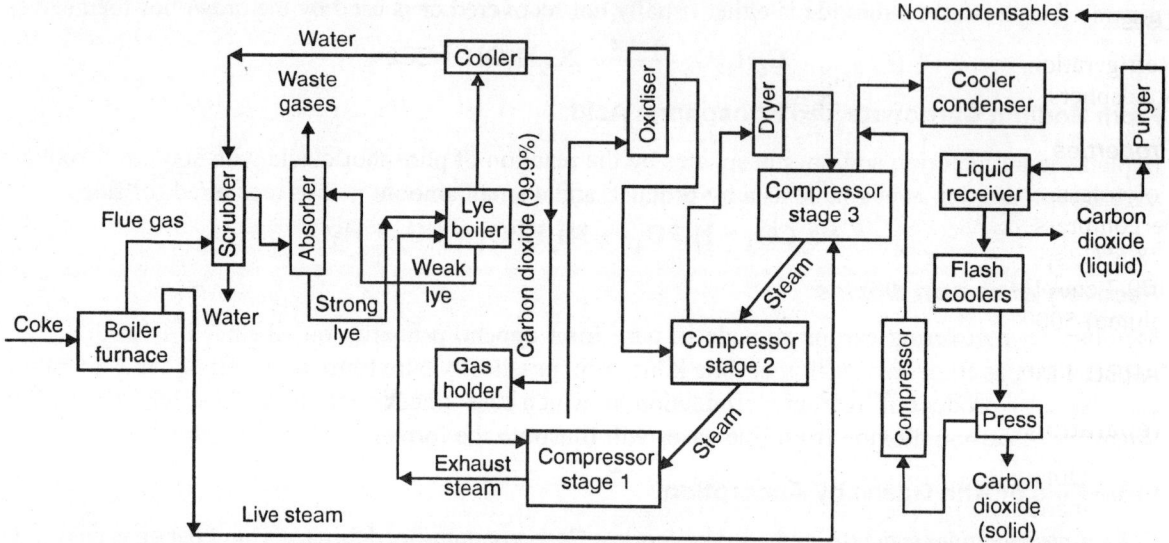

Fig. 8.4. Flow diagram for manufacture of carbon dioxide from flue or kiln gases by absorption.

In the ternary cycle, the carbon dioxide from the gas holder is cooled to 4°C and raised to a pressure of 75 psi absolute in the first stage of the primary compressor. The compressors, usually two sets, are driven by live steam from the coke boiler, and the exhaust steam from them is used in the lye boiler.

From the first stage, the gas is passed through a purification system consisting of an oil separator and a scrubber, containing potassium permanganate or dichromate solution, which oxidises organic impurities. The gas from the scrubbers, raised in temperature to about 10°C, enters the second compression stage and is discharged at a pressure of about 350 psi absolute. The temperature is lowered by water coolers to 4°C, at which point some of the water and vapourised lubricating oil (such as glycerine) are condensed.

The gas passes through a desiccant drying tower (calcium chloride), where sufficient water is removed to prevent freezing of the valves. The gas is compressed to 970 psi absolute in the third stage, passed through a cooler, and liquefied in a condenser at 26.7°C (critical temperature is 31.35°C). Non-condensable gas is purged, and the 99.9% liquid carbon dioxide, containing less than 0.1% moisture and free from organic impurities, is either fed to flash coolers or charged into cylinders.

In two step-by-step flash coolers refrigerated by expanding carbon dioxide, the temperature is lowered first to −7°C and then to −40°C. The liquid carbon dioxide is fed to a metering tank and then to the solid carbon dioxide press.

The liquid is allowed to expand through a nozzle, forming "snow" in the chamber of the press. The vapours given off (revert gas) are utilised for precooling the liquid carbon dioxide going to the flash coolers. The gas is then recompressed in the secondary compressors and combined with the raw carbon dioxide at the same pressure.

The 'snow' formed in the press is converted to a 100-kg, 50 × 50 × 25 cm block by lowering the upper ram under 2500 psi (17.5 MPa) until the block reaches a 25-cm thickness. Then the lower ram is dropped, and the block is ejected. After quartering by band saws, 25 kg, 25 cm cubical blocks are wrapped in kraft paper and either stored or shipped in insulated cars.

Uses

Refrigeration, carbonated beverages, aerosol propellant chemical intermediate, fire extinguishing, inert atmosphers, medicines, hardening of foundry moulds and cores, blowing agent.

Properties

Colourless, odourless, non-combustible gas liquefiable to a heavy, volatile, colourless liquid, which may be compressed into a white snowlike solid. mol. wt. 44.01, mp -56.6 (5.2 atm)°C, bp sublimes at -78.5°C, sp. gr., Gas 1.53 (air = 1), Liquid 1.101 (-37°C), Solid 1.56 (-79°C). Soluble in water (179.7 cm^3/100 g at 0°C, 90.1 cm^3/100 g at 20°C), acids, and alkalies. Threshold limit value (ppm by volume) 5000.

CARBON MONOXIDE

Preparation

1. Obtained almost pure by placing a mixture of oxygen and carbon dioxide in contact with inandescent graphite, coke or anthracite.
2. Action of steam on hot coke or coal (water gas) or on natural gas (synthesis gas). In the latter case, carbon dioxide is removed by absorption in amine solutions, and the hydrogen and carbon monoxide separated in a low-temperature unit.
3. By-product in chemical reactions.
4. Combustion of organic compounds with limited amount of oxygen as in automobile cylinders.
5. Dehydration of formic acid.

Highly toxic by inhalation; highly flammable, dangerous fire and explosion risk; flammable limits in air, 12 to 75% by volume. Tolerance 50 ppm (industrial workrooms); U.S. standard, 35 ppm.

Note: CO has an affinity for blood haemoglobin over 200 times that of oxygen. A major air pollutant.

Properties

Colourless gas or liquid; practically odourless. Burns with a violet flame. Slightly soluble in water; soluble in alcohol and benzene. Sp. gr. 0.96716 (air = 1.0); b.p. -190°C; solidification point -207°C; specific volume 13.8 cu ft/lb (21.1°C). Autoignition temp. (liquid) 1128°F (609°C). Classed as an inorganic compound.

Uses

Organic synthesis (methanol, ethylene, isocyanates, aldehydes, acrylates; phosgene); fuels (gaseous); metallurgy (special steels, reducing oxides, nickel refining); zinc white pigments.

OXYGEN (MANUFACTURE, PROPERTIES AND USES)

From Air by Liquefaction (Modified Linde-Frankl Cycle)

Commercial oxygen is produced in two grades: high purity, 99.5%, and low purity, 90 to 98%, although the latter is becoming less and less important since it is practically as cheap with modern technology to produce the better grades. The process consists of the liquefaction and subsequent fractionation of air. Several cycles have been proposed. These vary in methods of air compression, purification, and refrigeration, as well as in the design of heat-exchange, rectifying, evaporating, and condensing equipment. The pressures of the various cycles range from about 60 to 3000 psi (0.4 to 21 MPa).

Air consists of an invariable mixture of gases, in per cent by volume: nitrogen 78, oxygen 21, argon 0.94, hydrogen 0.01; and smaller amounts of neon, helium, krypton, and xenon. It also contains variable concentrations of carbon dioxide (0.03 to 0.07), water vapour (0.01 to 0.02), and hydrocarbon gases such as acetylene and methane, as well as local pollutants.

In the modified Linde-Frankl low-pressure cycle, reversing exchangers are used to lower the temperature of the incoming air which flows counter current to the outgoing gas. Critical cooling is obtained by the Joule-Thomson effect, that is, the cooling effect produced by the expansion of compressed air with no external work [1 kg of air at 3000 psi (21 MPa) and 10°C, when expanded to 75 psi (520 kPa), produces a cooling effect of 10 kcal (42 kJ)]. Air is filtered and compressed to 77 psi (530 kPa) in a two-stage, steam-driven centrifugal compressor, and then passed through a catalytic oxidation chamber. Here traces of hydrocarbons are converted to carbon dioxide and water. After cooling, the hydrocarbon-free air is passed through a water separator and then to automatic three-way reversing valves which select the proper air channels through the exchangers. The exchangers, which function both to cool and to purify the air, have a concentric, triple-tube construction in which the oxygen product follows the same path through the centre tube, while waste nitrogen and incoming air switch channels in the two outer annuli. This periodic switching (about every 3 min) permits the carbon dioxide and water vapour, which condense from the incoming high-pressure air, to be evaporated and purged out of the system by the low-pressure waste nitrogen (Fig. 8.5).

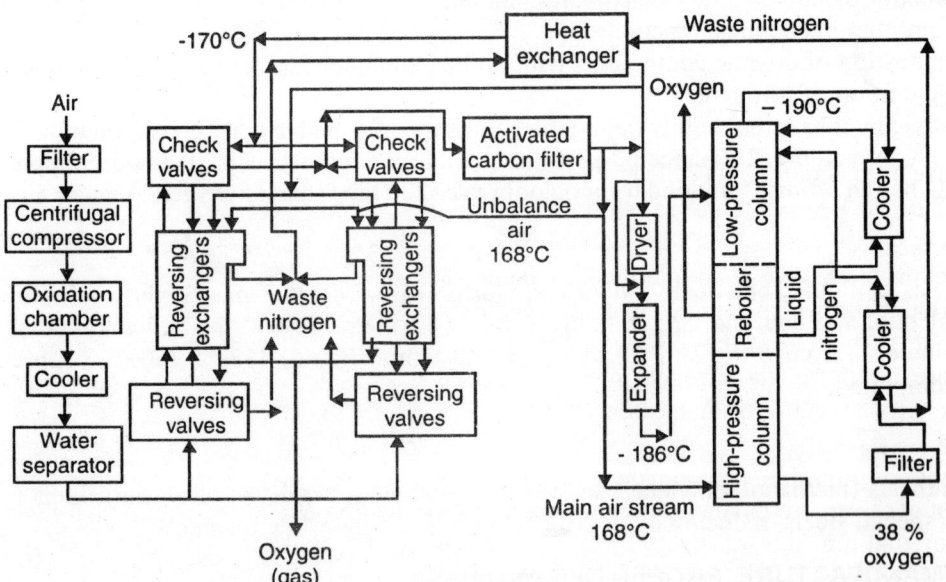

Fig. 8.5. Flow diagram for manufacture of oxygen from air by liquefaction (modified Linde-Frankl cycle).

The high-pressure air entering the exchangers is saturated with water at about 32°C and 77 psi (530 kPa). It also contains approximately 0.03 mole % carbon dioxide. As the air is cooled, it deposits first water, then ice, and finally carbon dioxide snow. Leaving the top of the last exchangers through automatic check valves and at a temperature of $-168°C$, the cold, purified air passes through activated carbon filters which remove any remaining carbonaceous impurities.

The air stream is then divided into three portions; part of the air, called the "unbalanced" stream, is returned to an extra jacket at the top of the exchangers, where it is warmed to about $-84°C$, passed through a heat exchanger against waste nitrogen, and then blended with a second portion of the air stream to give a mix temperature of $-153°C$. The air stream may be passed through silica-gel dryers and then fed into a turbo-expander.

The temperature unbalance system of diverting part of the air stream introduces a small amount of refrigeration into the reversing exchangers to compensate for heat losses. It also gives the necessary cold-end temperature difference ($2°C$) to permit constant low exchanger-outlet temperatures and insures complete purging of impurities, thereby giving excellent air purification. The silica-gel dryer is used, when starting up the cycle, to remove water contained in as yet unpurified air. The turbo expander supplies the refrigeration for the whole unit. In it the air at $-153°C$ and 77 psi (530 kPa) is expanded through nozzles against a high-speed turbine connected to an electric generator. The air leaves the turbine at its dew point, $-187°C$, and 7.5 psi (52 kPa), and is fed into the upper section of a double-fractionating column.

The third and main portion of the air stream, at a temperature of $-168°C$, enters the bottom of the double column. In the high-pressure section, operating at 71.5 psi (500 kPa), the air is condensed, and preliminary fractionation takes place. The more volatile nitrogen (bp $-195.8°C$ or 77.3K) rises to the top of the high-pressure section, where it is condensed in a condenser-reboiler by heat exchange with boiling oxygen in the low-pressure section of the column.

The nitrogen, about 94% pure, serves as reflux to the trays in the high-pressure section and is then withdrawn from the column. After passing through a subcooler, where the temperature is reduced by heat exchange with the overhead waste-nitrogen stream from the low-pressure section, the liquid nitrogen at about 71 psi (500 kPa) is expanded through a valve near the top of the low-pressure column to 7 psi (49 kPa). Here it serves as reflux to the top tray of this section, giving up some of its oxygen and forming the waste-nitrogen exhaust.

At the base of the column, so-called rich liquid air, containing about 38% of the less volatile oxygen (bp $-183°C$ or 90.1 K), is removed and passed through a glass-cloth filter to remove any traces of precipitated carbon dioxide. It is then cooled in a subcooler against the waste nitrogen gas and fed through an expansion valve at an intermediate point in the low-pressure (7-psi, 49-kPa) section of the column. Here it is further fractionated to give oxygen of desired purity and waste nitrogen. Reboiler heat is furnished at the bottom of the top section by the condensation of nitrogen at the top of the bottom (high-pressure) section. The nitrogen in the top section vapourises and passes upward through the liquid reflux, which scrubs, to some extent, the oxygen out of the feed stream to give an oxygen yield as high as 90%.

The number of trays in the column depends on the purity and yield of oxygen desired. For 90 to 95% oxygen, most of the argon remains in the product, and fractionation takes place between the oxygen and nitrogen. For making high-purity (99.5%) oxygen, fractionation must take place between argon (bp $-185.7°C$ or 87.2 K) and oxygen; this is more difficult and requires more high-capacity, high-efficiency trays. Part of the oxygen product may be withdrawn as liquid, although in this type of low-pressure cycle most of it is withdrawn as vapour at $-180°C$ from the vapour space below the bottom tray in the top section of the column. It is returned through the reversing exchangers, where it is warmed to substantially inlet temperature ($32°C$).

The waste nitrogen leaves the top of the column at $-191°C$ and is used to subcool both the nitrogen reflux and the rich air feed. It is then returned through the heat exchanger, where it cools the unbalance air and passes to the main reversing exchangers. It enters these at a temperature of $-170°C$ and a pressure at 7 psi (49 kPa) and is reheated to within $4°C$ of the entering air. After the impurities of carbon dioxide and water released by the incoming air are removed, the waste nitrogen is exhausted into the atmosphere.

The process part of the plant is enclosed by a steel section surrounding the various pieces of equipment. All the void spaces are filled with mineral wool, thus affording the required insulation. The units are of necessity particularly tight and leak-free and are usually constructed of non-ferrous metals such as copper, aluminium, and high-chrome-nickel alloys.

In some modern plants, the air is not purified before liquefaction. Rather, most of the impurities are frozen out during liquefaction and are rejected from the exchangers when the flow is reversed. Any hydrocarbons remaining in the liquid are removed by passing the liquid product from the high-pressure column through a bed of silica gel before it is fed to the low-pressure column.

Conventional Linde-Frankl Cycle

The other low-pressure process for making oxygen, the conventional Linde-Frankl cycle, differs essentially from the foregoing in its use of "cold accumulators" for purification and heat transfer. These are pressure vessels (operating at about 70 psi, 480 kPa) packed with aluminium spirals in which both waste-nitrogen and oxygen product cool the incoming air by a periodic reversal of the flow. Since the oxygen product purges the impurities, this process produces only low-purity oxygen.

A second difference is in the use of a small amount (4 to 5%) of high-pressure (3000-psi, 21-MPa) supplementary air to make up refrigeration deficiencies. The supplementary air also aids in purging the accumulators, being used in place of the unbalance air stream employed in the previously described process.

Other Cycles

Other methods for producing oxygen from air operate at moderate to high pressure and, in general, follow the basic Linde or Claude cycles. Most plants rely on chemical methods for removing carbon dioxide, and some employ chemicals also for water removal. Alkalies such as sodium or potassium hydroxide may be used for both, or sodium hydroxide for carbon dioxide removal and silica gel for water. Water is often separated from the air stream, however, by freezing in an exchanger and removing with waste nitrogen or warm refrigerant gas.

In the Linde cycle with auxiliary refrigeration, the air is compressed as high as 3000 psi (21 MPa), scrubbed with a sodium hydroxide solution, and cooled in alternating heat exchangers against product oxygen and nitrogen. At $-15°C$ most of the water is frozen out, and the air is further cooled to $-46°C$ in a second pair of alternating exchangers against refrigerated ammonia. The cold purified air is then fractionated in a double column to yield oxygen and nitrogen.

Claude Cycle

The basic Claude cycle operates at a lower pressure (about 400 psi, 2800 kPa) than the Linde cycle, but is similar in the method of air purification. The novel feature is passing up to 80% of the main air stream through an expansion machine. By this adiabatic expansion, the power consumption per unit of air is reduced by about 35%. After the air has expanded to do work, it is combined with the rest of the air, which has been allowed to expand for self-refrigeration, and passed into a fractionator.

One modification of the Claude cycle used in large plants consists of compressing filtered air, scrubbing with caustic, and drying with activated alumina or silica gel. While the air is cooling in exchangers against outgoing oxygen and nitrogen, part of it is taken off and put through a reciprocating expansion engine. The other part of the air, after completing the heat-exchange cooling, is expanded from 750 psi (5175 kPa) operating pressure to column pressure. It is combined with the air from the expander and fractionated in a double column to produce liquid oxygen, which is discharged at cylinder-filling pressure.

A further modification is the Heylandt cycle, which claims a still lower power consumption (30% less) by using high pressure (3000 psi, 21 MPa) and introducing about 60% of the air to the expander at room temperature (rather than at 88°C as in the Claude cycle). The remaining 40% is throttled.

By using any of these cycles or various modifications, it is possible to produce liquid or gaseous oxygen ranging in purity from about 90 to 99.5%. The liquid is withdrawn directly from the column, or, if gaseous oxygen is desired, it may be stored in a gas holder and then compressed prior to cylinder filling. The nitrogen, produced as a co-product, may be recovered or vented to the atmosphere. The rare gases, especially argon, may also be separated by additional fractionation. Plants separating these gases obtain an argon fraction (boiling between oxygen and nitrogen) and may process the nitrogen gas to obtain a neon-helium fraction (lower boiling than nitrogen). By special processing of liquid oxygen, a krypton-xenon fraction (higher boiling than oxygen) may be obtained.

Oxygen is also produced as a co-product with hydrogen in the electrolysis of water; only about 0.3% or less of the oxygen produced is obtained in this way.

Properties

Colourless, odourless, tasteless gas liquefiable at –190°C into a slightly bluish liquid. mol. wt. 32.00, mp –218.8°C, bp –183°C, sp. gr., Gas 1.1053 (air = 1) Liquid 1.14 (–183°C), Solid 1.426 (–252.5°C).

Most plants manufacturing oxygen also recover nitrogen, used in liquid form as a refrigerant and as a gas for blanketing sensitive chemical reactions and similar uses. Many plants also recover the rare gases argon, krypton, and xenon. The purified gases are widely used for fluorescent and display lighting. Argon in particular has gained a mass market, chiefly in gas-shielded, arc-welding processes. Although nitrogen is often considered an inert gas, many metals react with it at elevated temperatures, including especially the so-called refractory metals such as titanium and zirconium. In these cases argon must be used in all stages of their production and finishing.

Uses

Steel, welding, acetylene, missiles, medicine, chemical process industries, pulp and paper, water treatment etc.

ACETYLENE (MANUFACTURE, PROPERTIES AND USES)

From Calcium Carbide

Calcium carbide is formed when lime (essentially free of phosphates and magnesium carbonate) and coke (with low ash content), mixed in a ratio of 60:40, are heated to a temperature of 2000 to 2100°C in an electric furnace. The reaction is: $CaO + 3C \rightarrow CaC_2 + CO$. The carbon monoxide may be recovered and used as a fuel in lime kilns or as a raw material for chemical synthesis. The liquid calcium carbide is allowed to cool and solidify for 24 to 48 hours and is then crushed and screened for size. The grinding operations are carried out in an atmosphere of nitrogen to prevent explosion from acetylene generated by moisture in the air. The approximate requirements per metric tonne of carbide are: lime (991 kg), coke (683 kg), electrode paste (17 to 20 kg), electricity (3200 kWh), and labour (5 man-h). The carbide so produced is about 80% calcium carbide; 1 kg of carbide yields about 0.29 m^3 of acetylene (Fig. 8.6).

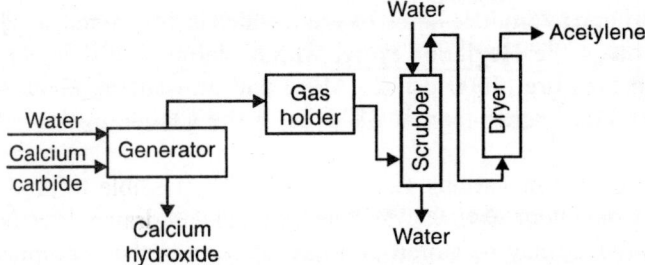

Fig. 8.6. Flow diagram for manufacture of acetylene from calcium carbide.

There are two standard methods for the production of acetylene from calcium carbide and water: the wet process and the dry process. The wet process consists of adding the carbide to a relatively large quantity of water, releasing acetylene gas while the calcium hydrate residue is discharged in the form of a lime slurry containing approximately 90% water.

$$CaC_2 + 2H_2O \rightarrow CH\equiv CH + Ca(OH)_2$$
93–95% yield

In order to eliminate the waste of calcium hydrate, the dry process was originated in which a limited amount of water (1:1 by weight) is added to calcium carbide in a generator. The heat of the reaction (1475 kcal/m^3 acetylene generated) is employed to vapourise the excess water over the chemical equivalent, leaving a substantially dry calcium hydrate suitable for reuse as a lime source. The temperature must be carefully controlled, since acetylene polymerises to form benzene at 600°C and violently decomposes at 780°C, and air mixtures may explode at 480°C. Generators of both types are usually designed to operate below 150°C and 15 psi (100 kPa). The crude gas (containing traces of hydrogen sulphide, ammonia, and phosphine) from the generator is either scrubbed with water and caustic soda solution or led to a purifier where the impurities are absorbed by the use of iron oxide or active chlorine compounds. The dry gas is fed to cylinders or manufacturing units.

From Paraffin Hydrocarbons by Pyrolysis (Wulff Process)

Acetylene may be produced by the thermal decomposition of certain hydrocarbons such as methane, ethane, propane, butane, ethylene, and natural gas. In the Wulff process the pyrolysis is carried out in a furnace containing hot brick checkerwork. The bricks in the furnace are first heated by burning a gaseous fuel in the furnace. After pyrolysis the acetylene-containing gases are quenched, and then scrubbed and purified to recover acetylene (Fig. 8.7).

ReactionS (with Butane Feed):

$$C_4H_{10} \xrightarrow{\text{Steam}} CH\equiv CH + C_2H_4 + CO + H_2$$

$$C_2H_4 \longrightarrow CH\equiv CH + H_2$$

$$2CH_4 \longrightarrow CH\equiv CH + 3H_2$$
25-50% yield

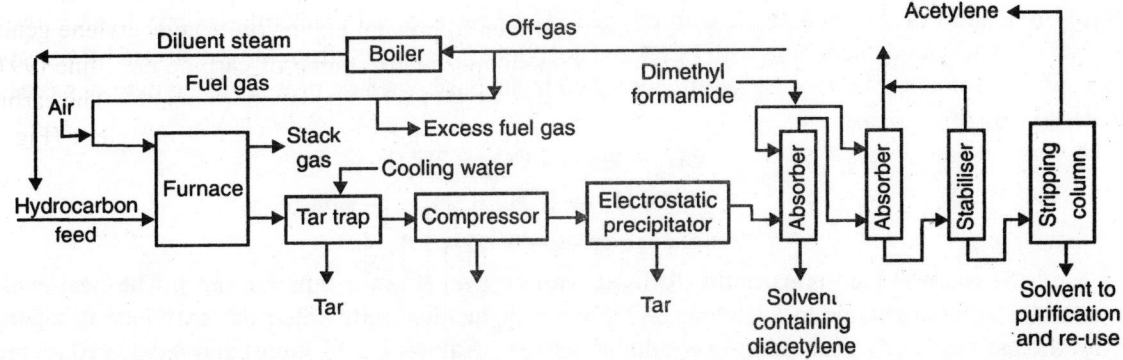

Fig. 8.7. Flow diagram for manufacture of acetylene from paraffin hydrocarbons by pyrolysis (Wulff process).

The Wulff regenerative furnace is essentially a rectangular steel box filled with refractory brick checkerwork. The furnace is operated on a 4-min cycle in which the checkerwork is heated for 1 min and the feed gas pyrolysed for 1 min. The same sequence of operation is then done in the reverse direction through the furnace. To facilitate reversal of the gas flow, fuel gas burners and hydrocarbon feed pipes are located on each side of the combustion chamber. To allow continuous flow of cracked gas to the purification train, two furnaces are usually operated on staggered cycles.

In usual operation off-gas from the acetylene recovery system is used as fuel for heating the combustion chamber. The volume of off-gas is much more than is required for fuel, so it may either be recycled to the furnace (recycle process) or used as a raw material for some other operation (once-through process).

Before the hydrocarbon feed is sent to the chamber, it is diluted with steam (up to 1:8 ratio). The feed is carried through the chamber at sub-atmospheric pressure by virtue of a large vacuum pump (see flow diagram 8.7). In this way residence time may be reduced to as little as 0.03 sec. Cracked gas leaves the chamber at about 370°C (maximum temperature in the furnace just after the heating cycle approaches 1315°C). It is then further quenched in a tar trap, where steam and various tars are removed. The gas is compressed to atmospheric pressure, passed through a knock-out drum and electrostatic precipitator, and sent to the recovery system. Usually diacetylene and acetylene are separated by absorption in dimethylformamide. By proper adjustment of solvent ratio and temperature, diacetylene may be removed in the first scrubbing column. In the acetylene absorber small quantities of ethylene, carbon dioxide, and higher acetylenes are also absorbed, but most of the acetylene-free off-gases go overhead to be used as fuel for the steam boilers, for combustion chamber heating, and whatever other good use can be made of them at the particular location.

The rich solvent containing acetylene is sent to a stabiliser, where less soluble components are removed by stripping. Acetylene is then removed from the solvent in a second stripping column. The solvent is readied for reuse by stripping out high boilers by blowing with off-gas from the acetylene absorber followed by rectification.

Yield of acetylene (98.5 to 99.3% purity) varies with the hydrocarbon feed stock used. Average yields for the once-through process are 22.5 kg/100 kg methane, 38.8 kg/100 kg ethane, and 35.5 kg/100 propane. The off-gas is principally ethylene, carbon monoxide, hydrogen, and methane.

From Natural Gas by Partial Oxidation (Sachsse Process)

Acetylene may be produced from a wide variety of hydrocarbon feed stocks (natural gas, LPG, naphtha, fuel oil, even crude oil) by high-temperature cracking. Heat for the cracking operation is developed by partial oxidation of the feed stock with oxygen. The heat evolved cracks the excess hydrocarbon to acetylene. After rapid quenching with water, the acetylene is separated from the gas stream by absorption in a suitable solvent. The process herein described is the basic Sachsse process using natural gas as raw material (Fig. 8.8).

$$CH_4 + 2O_2 \rightarrow CO_2 + 2H_2O$$
$$2CH_4 \rightarrow C_2H_2 + 3H_2$$

Methane conversion—90–95%, 30% yield

Methane (natural gas) is partially oxidised with oxygen (from a separate unit). The heat evolved cracks the excess methane to acetylene. After rapid quenching with water, the acetylene is separated from the gas stream by absorption in a suitable solvent. Natural gas (1 mole) and oxygen (0.65 moles 95% O_2) are preheated separately to 510°C and fed to a specially designed burner.

The converter is a vertical cylindrical unit built in three sections: mixing chamber, flame room, and quench chamber. After rapid and thorough mixing of the oxygen and methane in the mixing chamber, the gases are fed to the flame room through ports in a burner block designed to prevent back-travel or blowoff. The heat of combustion heats the gases to 1550°C to allow cracking of the excess methane to acetylene. Residence time is 0.001 to 0.01 sec. By rapid quenching of the resulting gases with water to 40°C, decomposition of the acetylene is prevented. The cooled effluent gases contain (on the dry basis) 8% acetylene, 54% hydrogen, 26% carbon monoxide, 5% methane, 4% carbon dioxide, and 3% nitrogen and higher acetylenes. These gases are run to a filter where carbon black (26 kg/metric tonne of acetylene produced) is removed on a moving bed of coke. Clean gas leaving the filter is compressed to 150 psi (1 MPa) and fed to an absorption unit, where acetylene is separated and purified in a manner similar to that described for the Wulff process. Acetylene of 99.5% or higher purity is produced. A variation of the Sachsse process in wide use is the BASF process.

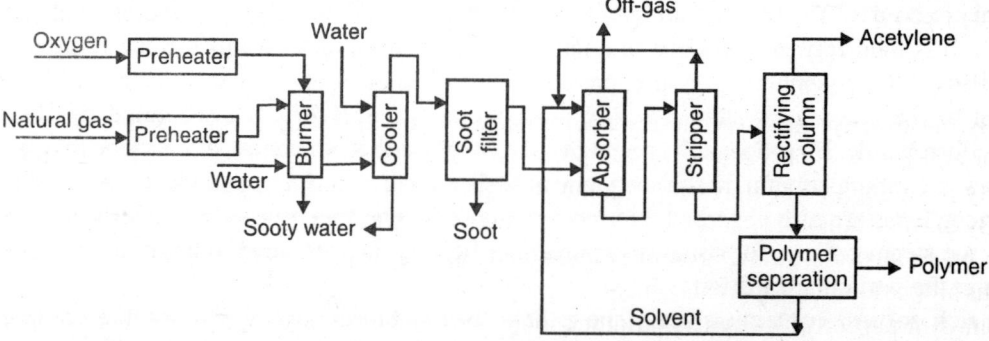

Fig. 8.8. Flow diagram for acetylene from manufacture of natural gas by partial oxidation (Sachsse Process).

Several other partial oxidation processes have been developed and are available for commercial use. The chief differences among processes are in burner design, hydrocarbon feed, recovery solvent, and number of reaction stages. Two-step processes use stoichiometric fuel-oxygen mixtures in the first step and add the raw material to be cracked to acetylene in a second step.

From Hydrocarbons by the Arc Process

Several processes have been developed in which hydrocarbon cracking is carried out in an electric arc. In these processes the electric arc provides energy at very high flux density, so that reaction time can be kept at a minimum. The design of the arc furnace appears to be a factor of major importance. In one design of an arc furnace (Huels process), gaseous feedstock enters the furnace tangentially through a turbulence chamber and then passes with a rotary motion through a pipe in which an arc is passed between a bell-shaped cathode and the anode pipe. The rotary motion of the gas causes the arc to rotate and thus reduces fouling. The arc is operated at 8000 kW direct current at 7000 V and 1150 A. Cathodes are said to last 800 hours, anodes only 150 hours. Feed to the arc is fresh hydrocarbon and recycle gas. The reaction effluent gases are quenched and purified. In one pass through the furnace, a weight yield of 35% purified acetylene may be obtained, along with 17% weight yield of ethylene and 10% weight yield of carbon black, plus hydrogen and minor products.

Uses

Vinyl chloride monomer, acrylates, acetylenic chemicals, vinyl acetate, chlorinated solvents, welding and cutting metals, carbon black.

Properties

Colourless, flammable gas; odourless, when pure, but ordinarily has a garlick-like odour because of impurities:

Mol. wt. 26.04, fp –80.8°C, Sp gr 0.618 (32°C/4) bp –84°C.

One pound of acetylene is equivalent to 14.5 ft^3 (1 kg = 0.9 m^3). Soluble in acetone (2500 ml gas/ 100 g, 30,000 ml gas / 100 g at 12 atm), ethanol (600 ml/100 g), water (100 ml/100 g), and liquid ammonia at room temperature. Ignition temperature 644°C (pure), vapour density, (air = 1), 0.91, explosive limits (% by volume in air), (lower 2.5), (upper 82), threshold limit value (ppm) 1000. Acetylene is produced on the spot where needed by means of portable generators utilising calcium carbide. Miner's lamps are examples of small generation.

SULPHUR DIOXIDE

Preparation

1. By roasting pyrites in special furnaces. The gas is readily liquefied by cooling with ice and salt, or at a pressure of 3 atm.
2. By purifying and compressing sulphur dioxide gas from smelting operations.
3. By burning sulphur.

Toxic by inhalation; strong irritant to eyes and mucous membranes, especially under pressure. Dangerous air contaminant and constituent of smog. Tolerance, 2 ppm in air. U.S. atmospheric standard 0.140 ppm. Not permitted in meats and other sources of vitamin B$_1$.

Properties

Colourless gas or liquid with sharp pungent odour. Soluble in water, alcohol, and ether. Forms sulphurous acid H_2SO_3. Sp. gr. 1.4337, liquid at 0°C; fp. –76.1°C; bp. –10°C; vapour pressure 3.2 atmosphere at 20°C; refractive index (liquid) 1.410 (n 24/D). An outstanding oxidising and reducing agent. Non-combustible.

Uses

Chemicals (sulphuric acid, salt cake, sulphites, hydrosulphites of potassium and sodium, thiosulphates, alum from shale, recovery of volatile substances); sulphite paper pulp; ore and metal refining; soybean protein; intermediates; solvent extraction of lubricating oils; bleaching agent for oils and starch; sulphonation of oils; disinfecting and fumigating; food additive (inhibition of browning, of enzyme-catalysed reactions, bacterial growth); reducing agent; anti-oxidant.

NITROUS OXIDE

Preparation

Thermal decomposition of ammonium nitrate; controlled reduction of nitrites or nitrates. Supports combustion; can form explosive mixtures with air. Narcotic in high concentrations.

Properties

Colourless, sweet-tasting gas; non-combustible; sp. gr. (gas), 1.52 referred to air; (liquid), 1.22 (−89°C); fp. −90.8°C; bp. −88.5°C; soluble in alcohol, ether, and concentrated sulphuric acid; slightly soluble in water. An asphyxiant gas.

Uses

Anaesthetic in density and surgery; propellant gas in food aerosols; leak detection.

CHAPTER 9

Fertilisers, Phosphorus and Potassium Salts

INTRODUCTION

Fertiliser is defined as any material—organic or inorganic, natural and synthetic which supplies one or more of the chemical elements required for plant growth. Sixteen elements are identified as essential elements for plant growth, some of which are required in macro-quantities e.g., N, P and K, and some even in micro-quantities. Table 9.1 indicates essential elements for plant growth.

Table 9.1. Essential elements for plant.

Primary elements	Secondary elements	Micro-elements
Carbon	Calcium	Boron
Oxygen	Magnesium	Chlorine
Hydrogen	Sulphur	Copper
Nitrogen	—	Iron
Phosphorus	—	Manganese
Potassium	—	Molybdenum
—	—	Zinc

Of the elements listed in Table 9.1, carbon, oxygen and hydrogen are supplied by air and water, and are, therefore, not treated as nutrients for the fertiliser industry. The main aim of the fertiliser industry is to provide the primary and secondary nutrients which are required in macro-quantities. The primary nutrients are normally supplied through chemical fertilisers. Whatever be the chemical, its most important ingredient for plant growth is the nutrient content.

Fertilisers are extensively used in (i) agriculture; (ii) horticulture; (iii) forest-based industry; and (iv) other industries, e.g., floriculture, aquaculture, animal husbandry etc. The fertilisers are added to the soil to supply elements identified in Table 9.1 as essential elements for plant growth. At present, the entire K_2O requirement in India is met by import and most of P_2O_5 is made from imported rock phosphates. Of the entire fertiliser requirement, about 2/3rd in the form of nitrogen (N) and, as such, nitrogen-based macro-nutrients are most important in the fertiliser industry. The minor elements or micro-nutrients are provided by speciality chemicals containing micro-elements.

Phosphorus occurs in nature in phosphate rock in pure $[Ca_3 \{PO_4\}]_2$, in apatite $[Ca_5\{PO_4\}_3F]$, in bones, teeth, and in organic compounds of living tissue. Also as phosphorite anodules on ocean floor. Phosphorus ignites spontaneously in air at 86°F (30°C). Store under water and away from heat. Dangerous fire risk. Highly toxic by ingestion and inhalation. Skin contact for the burns. Tolerance, 0.1 mg per cubic metre of air.

Potassium is far too reactive metal, but it is widely distributed in minerals, the most important deposits of which occur in places where island seas have undergone natural evaporation. The main commercial source is carnalite, but kainite and chonite and sylvinite are sometimes used. Potassium is present in most clays and silicate rocks, and also in sea water (about 0.05%) and wood ash (about 30% of which is potassium carbonate).

Potassium salts must be present in the soil in order to have normal plant growth, hence they are essential components of the fertilisers used in food production throughout the world. Various potassium salts are also basically important as raw materials for many commodities such as soaps, detergents, glass, dyes, gunpowder, and pyrotechnics.

The sylvinite is mined and treated to yield high-grade potassium chloride, and langbeinite is processed to make potassium sulphate.

FERTILISERS

There are various types of fertilisers. Some of the important fertilisers along with their applications are discussed below.

Nitrogenous Fertilisers

Nitrogenous fertilisers may be classified into different groups e.g., ammonia, ammonium sulphate, ammonium nitrate, ammonium phosphate, urea, sodium nitrate, calcium nitrate, calcium ammonium nitrate (CAN), ammonium chloride, etc. Urea, because of its high 'N' content, is the cheapest nitrogenous fertiliser. It is made from ammonia and carbon dioxide which can be produced from different feedstocks, e.g., naphtha, natural gas, fuel oil, coal etc. Fig. 9.1 indicates raw materials used throughout the world to manufacture ammonia and different nitrogenous fertilisers. The 'N' content of each of the fertiliser is also shown in Fig. 9.1. Of the different feedstocks, the most economical feedstock for manufacture of ammonia is natural gas, in view of the lowest capital cost of the plant and low energy requirement per tonne of ammonia production.

Sources of nitrogen and hydrogen for ammonia

1. *Nitrogen:* From air directly or by air separation to O_2 and N_2 and use of nitrogen.
2. *Hydrogen:* From feedstock as indicated in Fig. 9.1.

Ammonia is an intermediate product in the manufacture of nitrogenous fertilisers. It is also used for direct application to the soil and in aqueous condition with solutions of other nitrogenous fertilisers like ammonium nitrate and/or urea.

Ammonia also finds applications in the production of nitric acid, soda ash, cleaning agents, leather tanning, petroleum refining, paper and pulp industry, textiles, refrigeration, rubber and synthetic resins industry, explosives, and wood and beverage industry. Ammonia is mainly produced as an intermediate product in fertiliser plants in India. In addition to above some ammonium sulphates are available for ammonia liquors of coke oven and by-product plants in integrated steel plants.

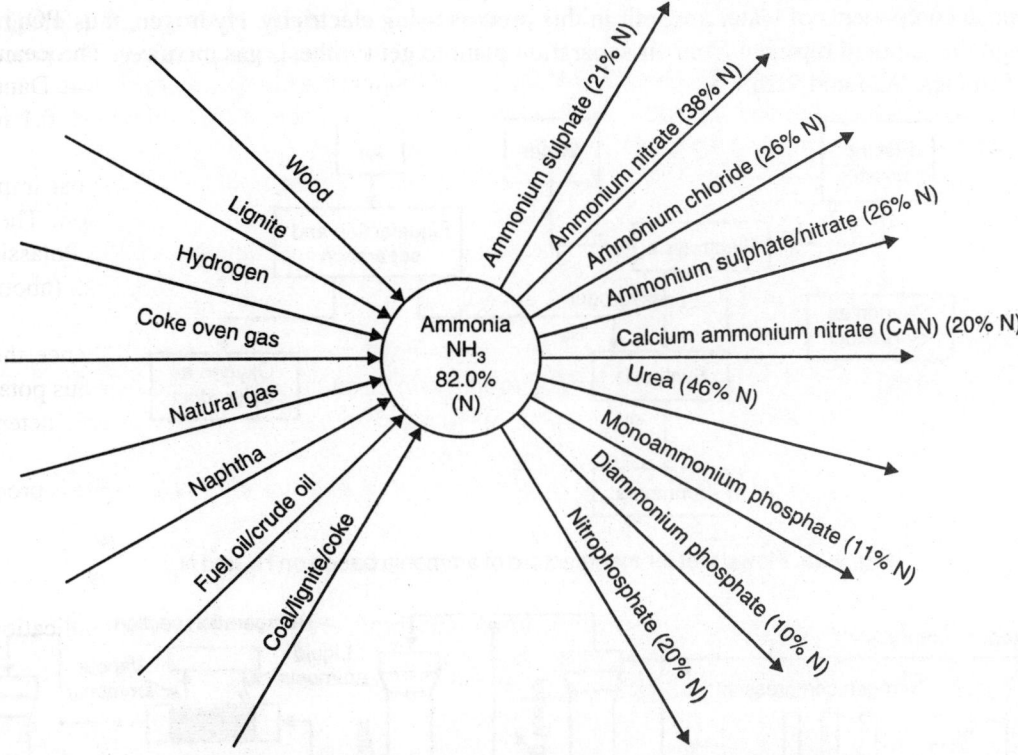

Fig. 9.1. Feed stock for ammonic and 'N' content of each fertiliser.

Feedstock for ammonia manufacture

A resume of different feedstock for ammonia manufacture is given in Table 9.2.

Table 9.2. Different feed-stock of ammonia manufacture.

Feestock	Process
Water/air	Electrolysis of water for H_2. Separation of air for N_2.
H_2 rich refinery gas	Steam reforming or partial oxidation
Natural gas	Steam reforming
Naphtha	Steam reforming or partial oxidation
Crude or residual oil	Partial oxidation
Coke-oven gas	Reforming or partial oxidation and low temperature separation
Coal and carbonaceous materials e.g., lignite	Water gas reaction or coal/lignite gasification

Methods of manufacture of synthesis gas ($3H_2 + N_2$) for ammonia manufacture

Electrolysis and air separation

The raw material used for electrolysis is water and for air separation it is air. These processes are adopted where adequate power at cheap rate is available. Heavy water production was once a determining factor in going for this process.

Chemical components of water are split in this process using electricity. Hydrogen, thus obtained is mixed with the required nitrogen from air separation plant to get synthesis gas mixtures. The process is indicated in Figs. 9.2a and 9.2b.

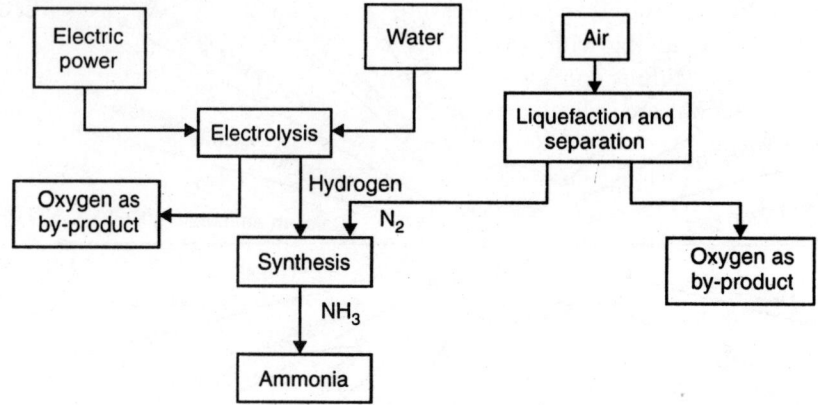

Fig. 9.2a. Flowsheet for manufacture of ammonia based on H_2 and N_2.

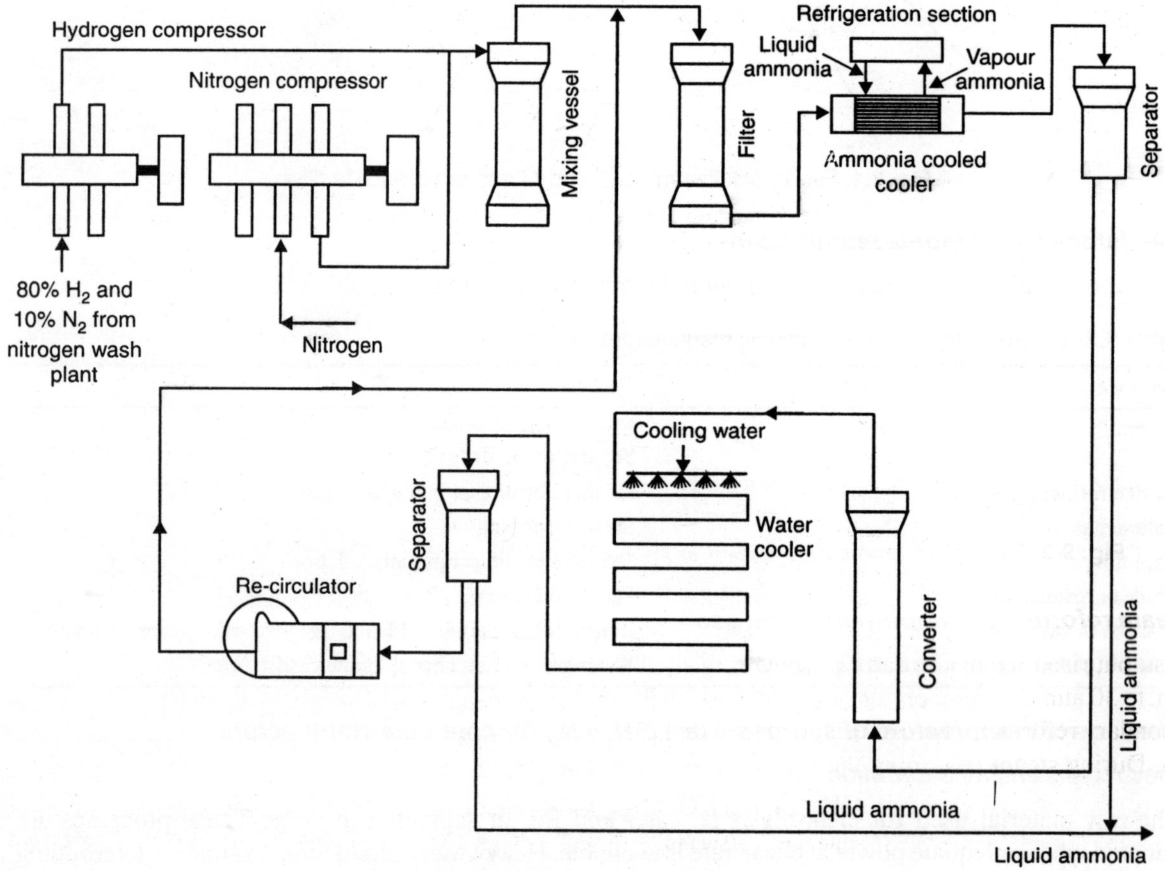

Fig. 9.2b. Ammonia plant flow diagram.

Partial oxidation of hydrocarbon feedstocks

Any hydrocarbon feedstock, namely naphtha or heavy oil and oxygen or oxygen enriched air are pre-heated separately and injected into a refractory lined chamber. The partial oxidation is carried out at temperature between 1100°C to 1500°C and at a pressure up to 80 atm. The raw gas is freed of carbon formed during oxidation by scrubbing with water. The raw gas after desulphurisation is sent to shift conversion where carbon monoxide is converted into carbon dioxide and hydrogen through steam.

The partial oxidation reactions are indicated below:

$$C_nH_{2n+2} + \left(\frac{n}{2}\right)O_2 \longrightarrow nCO + (n+1)H_2$$

Hydrogen   Oxygen   Carbon   Hydrogen
monoxide

The process flowsheet of partial oxidation process is given in Fig. 9.3.

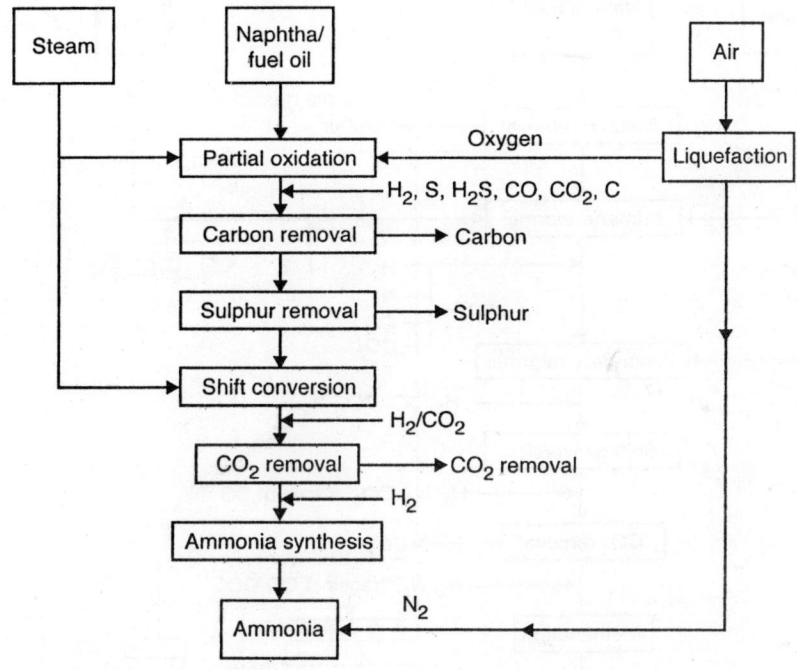

Fig. 9.3. Flowsheet for manufacture of ammonia by partial oxidation process of naptha/fuel oil.

Steam reforming of naphtha/natural gas

Desulphurised naptha or natural gas is subjected to thermal reforming over nickel catalyst at about 28 atm. to 30 atm. and temperature around 800°C in the presence of steam. This is followed by an autothermal secondary reformer where air is added to give the desired 3 : 1 hydrogen to nitrogen ratio in the synthesis gas. During steam reforming the following reaction takes place:

$$CH_4 + H_2O \longrightarrow CO + 3H_2$$
(Methane)   (steam)   (Carbon   (Hydrogen)
monoxide)

The steam reforming of naphtha or natural gas is followed by secondary reformer produces carbon monoxide. Carbon monoxide content of the 'raw gas' is converted to carbon dioxide and hydrogen by passing over activated iron oxide catalyst in the presence of steam. This supplements further hydrogen production in the raw gas as follows:

$$CO + H_2O \longrightarrow H_2 + CO_2$$

Thus, the overall reaction is:

$$CH_4 + H_2O \longrightarrow CO + 3H_2$$

$$CO + H_2O \longrightarrow CO_2 + H_2$$

$$CH_4 + 2H_2O \longrightarrow CO_2 + 4H_2$$

A typical flowsheet for manufacture of ammonia by steam reforming of naphtha/natural gas is shown in Fig. 9.4.

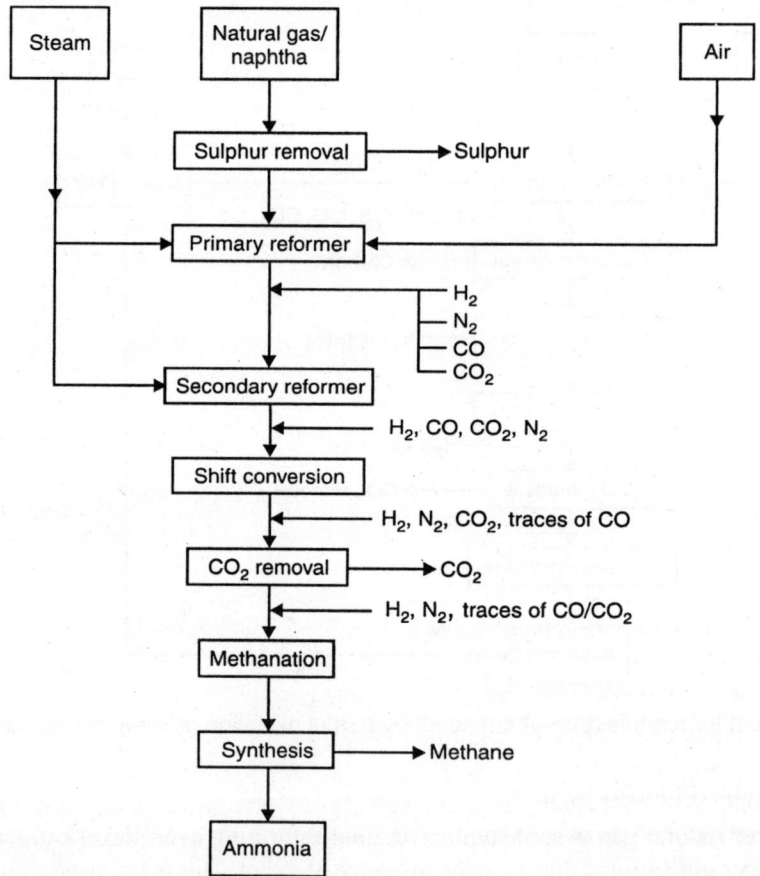

Fig. 9.4. Flowsheet for manufacture of ammonia based on naphtha/natural gas steam reforming process.

Gasification of coal or carbonaceous materials

The solid fuel gasification process of coal or other carbonaceous materials e.g., lignite can be at atmospheric pressure or under pressure. The process can be fixed bed, entrained bed or fluidised bed. Gasification or partial oxidation of carbonaceous material is achieved with oxygen and steam. Carbon monoxide content of the 'raw gas' in this case is also converted to carbon dioxide and hydrogen by passing over activated iron oxide catalyst in the presence of steam. This supplements further hydrogen concentration in the raw gas:

$$CO + H_2O \longrightarrow H_2 + CO_2$$

In the case of gas obtained from partial oxidation of hydrocarbon feedstock and carbonaceous materials, pure hydrogen obtained after purification is mixed with nitrogen from air separation plant in required quantities (3 : 1) to form synthesis gas mixtures. A typical flowsheet for manufacture of ammonia by coal/lignite gasification is shown in Fig. 9.5.

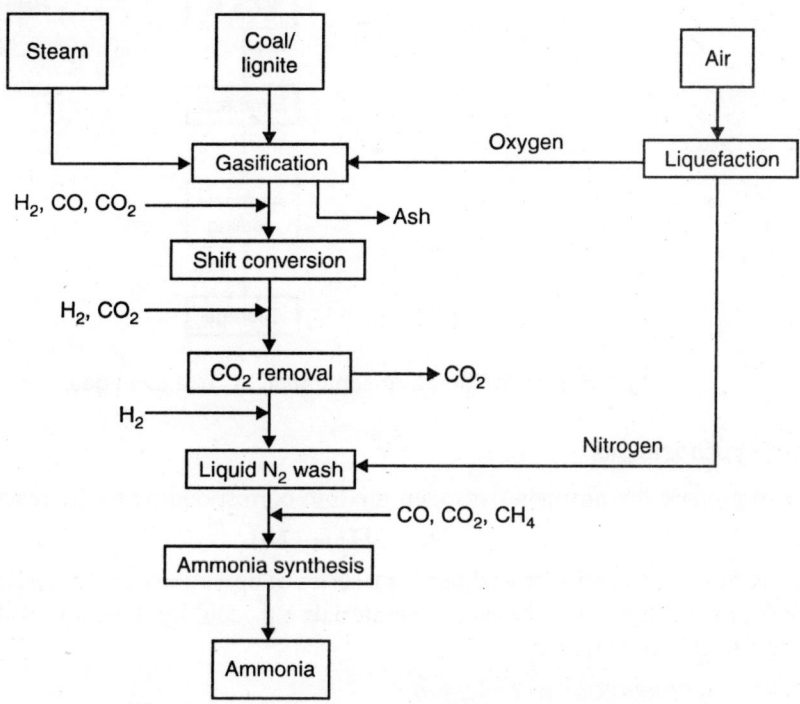

Fig. 9.5. Flowsheet for manufacture of ammonia by coal/lignite gasification.

Low temperature separation of coke oven gas

There are basically two types of process for manufacture of ammonia from coke oven gas. These are:
1. Separation of H_2 from coke oven gas by using liquid N_2 wash.
2. The methane content of coke over gas is cracked and the gas is then fractionated in a N_2 wash plant for separation of hydrogen. A typical scheme followed in the process is given below (Fig. 9.6).

Purification of coke oven gas

Carbon dioxide thus obtained in mixed state with hydrogen is removed by chemical or physical absorption process. Carbon dioxide is later recovered and can be used where necessary. Carbon monoxide that is left as an impurity is either converted to methane (an inert in synthesis loop) or absorbed by using liquid nitrogen when pure nitrogen is available from air separation plant.

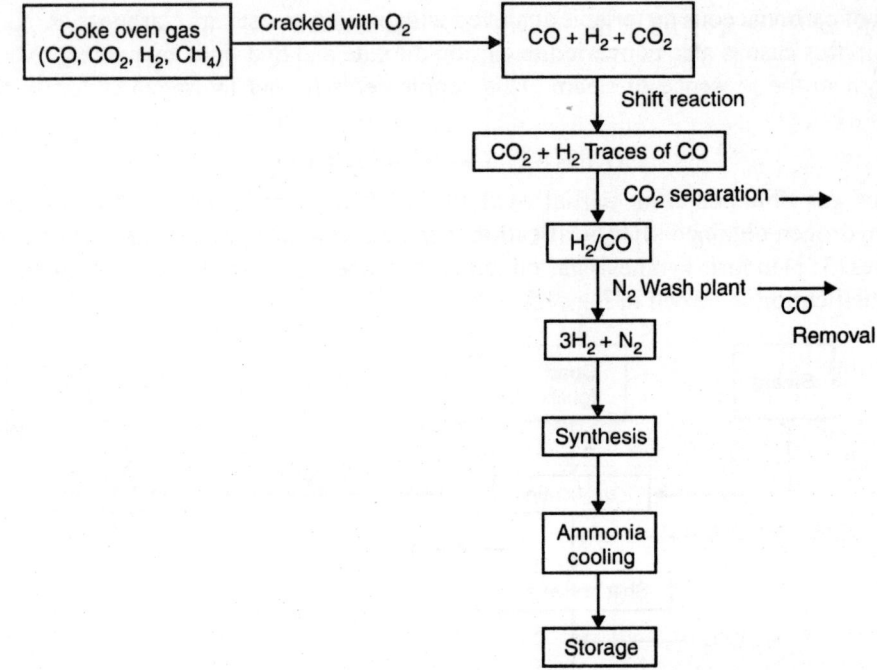

Fig. 9.6. Low temperature separation of coke oven gas.

Preparation of synthesis gas

It is necessary to prepare the nitrogen/hydrogen mixture corresponding to the reaction:

$$N_2 + 3H_2 \rightleftharpoons 2NH_3$$

This can be achieved as mentioned earlier after a series of operation employing (i) partial oxidation of hydrocarbons; (ii) gasification of carbonaceous materials e.g., coal/lignite; and (iii) the steam reforming of methane (natural gas) or Naphtha.

Partial oxidation of hydrocarbons with oxygen

The sequence of operations concerned in this case are: (i) air distillation; (ii) partial oxidation of the hydrocarbon by oxygen; (iii) removal of carbon and recovery of heat; (iv) possible removal of H_2S and conversion to sulphur; (v) catalytic conversion of 'CO' by steam (shift conversion); (vi) CO_2 removal; and (vii) CO removal by liquid nitrogen wash which also introduces the nitrogen required in the synthesis gas to form the mixture $N_2 + 3H_2$.

The treatments required for processing the synthesis gas are not different from those used to produce hydrogen. However, since excessive amount of inert gases (methane, argon, helium) cannot be tolerated in the synthesis gas, the operating conditions of partial oxidation are set so that the methane content is low. Moreover, the scrubbing with liquid N_2 must be carried out at a temperature at which the vapour pressure of nitrogen is such that the scrubbed gas removes the necessary amount of gaseous nitrogen so that:

This operation, which is specific to the use of hydrogen produced for ammonia synthesis, is similar in principle to the one used to separate carbon monoxide by scrubbing with liquid methane. Its value is, therefore, twofold:

1. To avoid the presence of residual hydrocarbons which act as diluents in the hydrogen obtained by replacing them by nitrogen which is a reactant. The effluent produced thus contains less than 1 ppm volume of CO and CH_4. On the other hand, the nitrogen content is at least 2 to 8 per cent volume, which normally precludes any other application.

2. To adjust the composition of ammonia synthesis gas according to the needs of the reaction.

Among the adaptations made to the liquid methane scrubbing scheme is the replacement of the carbon monoxide refrigeration cycle by a similar system operating with nitrogen. Fig. 9.7 indicates process details of liquid N_2 wash plant for removal of impurities.

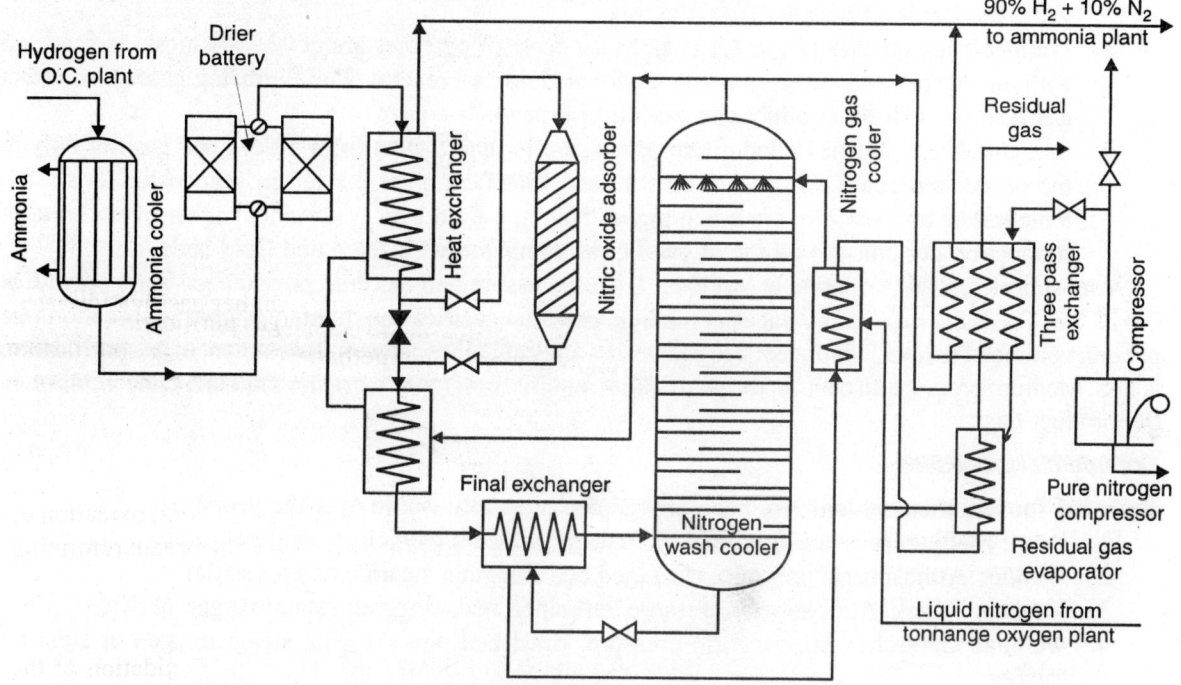

Fig. 9.7. Liquid nitrogen wash plant.

The object of the nitrogen wash plant is to remove the carbon monoxide and methane present in the hydrogen produced by the oil gasification plant and to render it suitable for ammonia synthesis. This unit can form an integral or non-integral part of the tonnage oxygen plant and has a capacity for purifying sufficient hydrogen as required by the ammonia plant.

Gasification of coal and properties required

The important properties of coal required for gasification are: (i) moisture content; (ii) ash content; (iii) ash properties; (iv) caking characteristics; (v) friability; and (iv) reactivity.

A high moisture can pose a problem in certain types of gasifiers. In such cases, pre-drying of coal becomes necessary. Coals with high ash content present problems in the gasifier. The important ash properties of coal for gasification are the melting point and melt viscosity.

The alkali metal content of a coal ash can also be a problem in coal gasification. Caking coals present problems in fixed bed and fluidised bed gasifiers, whereas non-caking coal poses no problems. The more friable the coal, the greater the quantity of fines produced during mixing and crushing. The reactivity of the coal sample determines the rate at which it will react with steam to produce hydrogen and carbon monoxide. In general, the higher the rank of coal the lower is its reactivity.

Types of gasifiers

Coal gasifiers are basically of three types: (i) fixed bed; (ii) fluidised bed; and (iii) entrained bed.
1. *Fixed bed gasifier:* Here the coal is supported by a grate and is maintained within certain upper and lower limits. Within the bed, the fuel moves slowly down through gasification zone. Fresh coal is continuously or intermittently fed at the top of the bed and the residue, principally ash removed from the bottom. Steam and air/oxygen are introduced at the bottom of the fuel bed while the gas is withdrawn from the top.
2. *Fluidised bed gasifier:* Here the coal, in the form of particles about 10–200 mesh, is fluidised with air/oxygen and steam in a conventional fluid bed reactor. The fluidising gases also effect gasification with the product gas passing overhead.
3. *Entrained bed:* This is called dilute-phase gasifier and differs from a fluid bed gasifier only in the particulate coal (100 to 200 mesh) used. Gasification is carried in suspension. Also, the temperature and velocity are much higher than in a fluid bed. It provides higher overall rate of production per unit of volume of gasifier as compared with fixed and fluid beds.

Atmospheric gasifiers operate at less then 1·1 bar pressure and medium pressure gasifiers operate at 15–25 bar. There are many advantages at medium pressure gasification, namely, higher gasification rate per unit volume of reactor, which reduces heat losses and capital investment. Investment in gas purification is less. Medium pressure also shifts the gasification equilibrium toward greater methane concentration in the product gas.

Commercial processes

There are four gasifiers which have been proved commercially viable over the years:
1. Lurgi: Medium pressure fluid bed, non-slagging steam/oxygen or steam/air type.
2. Winkler: Atmospheric pressure, entrained bed slagging, steam/oxygen gasifier.
3. Koppers Totzek: Atmospheric pressure, entrained bed, slagging steam/oxygen gasifier.
4. Wellman Galvsche: Atmospheric pressure, fixed bed non-slagging steam oxygen or stem air gasifier.

A new type of entrained bed coal gasification plant under pressure has been developed by Texaco—USA which also has future.

Process based on hydrocarbon steam reforming

In the hydrocarbon steam reforming process (Fig. 9.8) the series of operating sequences are as follows: (i) steam treatment (primary reforming); (ii) conversion of residual methane by air, which contributed the nitrogen required (secondary reforming or post-combustion); (iii) catalytic conversion of CO by steam (shift conversion); (iv) CO_2 removal; and (v) CO removal by treatment with cuprous derivatives (CO-sorb process, for example) or with methanol. Methanation can also be resorted to if one can tolerate the amount of CH_4, corresponding to that of carbon monoxide in the gas intended for ammonia synthesis.

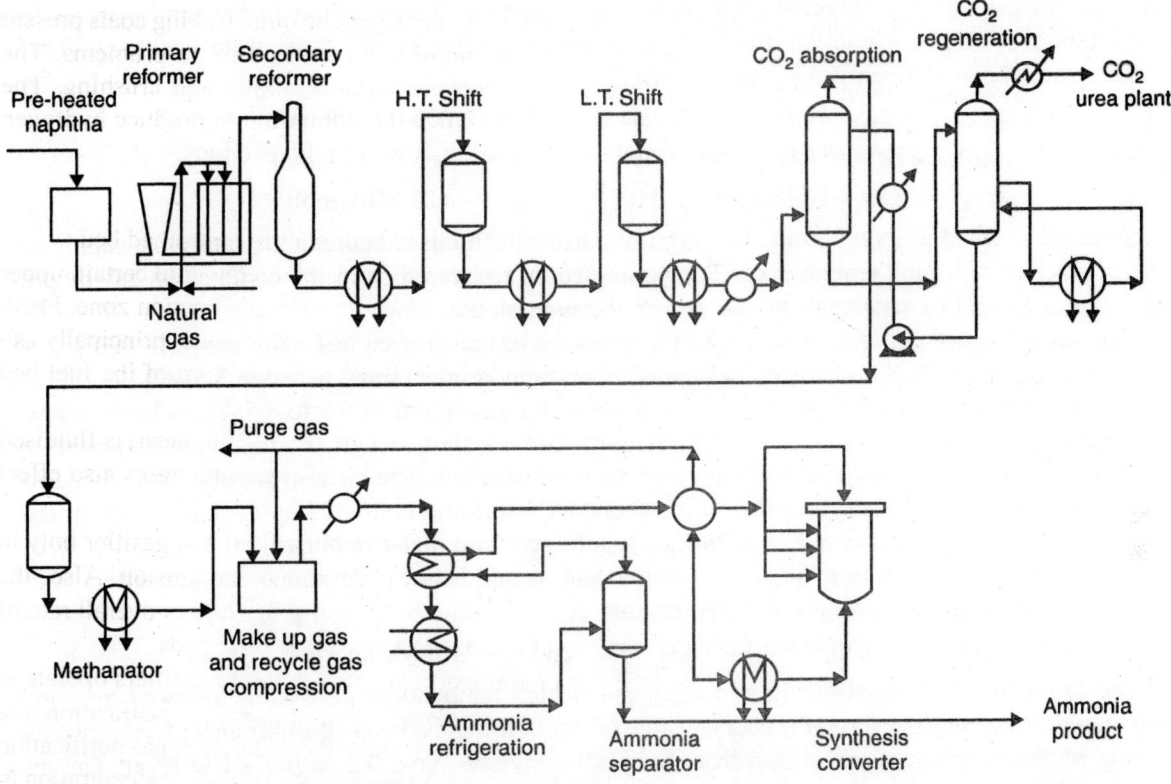

Fig. 9.8. Synthesis gas production based on naphtha/natural gas.

In this scheme the first steam reforming is so regulated that some methane substance in the gas is produced so that the next operation—conducted in the presence of air—provides the volume of nitrogen required. If the latter conversion—called secondary reforming or post combustion—were a simple selective conversion of methane yielding carbon monoxide, carbon dioxide and steam, each residual hydrocarbon molecule would contribute seven to eight molecules of nitrogen. At the outlet of the primary reforming stage, the gas composition should satisfy the following equation:

$$(H_2 + CO)/CH_4 \rightleftharpoons 25 \text{ or } 24 \text{ volumes}$$

Since 10 per cent carbon dioxide is formed at the operating temperature of the reforming reactors it is easy to calculate that the effluent must contain about 3.5 to 4.2 per cent residual methane. This value can only be reached at high temperature, low pressure and high steam ratio.

It is thus preferable to convert methane by air, so as to introduce the nitrogen required. This operation is performed at a comparable temperature in order to maintain the required thermal levels of the successive operating sequences and to avoid excessively disturbing the steam compositions. This is done in the presence of nickel based catalysts similar to those employed in the primary reforming reactor, to guarantee the conversion of low hydrocarbon contents in a dilute medium. Post-combustion is thus carried out adiabatically, between 850 and 1,000°C at a pressure that is close to that of the initial steam reforming.

In addition to avoiding an excessively severe initial steam reforming, post-combustion offers the advantage of improving the total heat recovery at high thermal load. The catalystic conversion of the CO contained in gas mixture obtained after secondary reforming is carried out in the same way as for production of hydrogen, namely in two reactors or only one. The VHSV varies from 1,500–3,000 h⁻¹.

The reactions in different stages are shown:

$$CH_4 + H_2O \rightleftharpoons CO + 3H_2 \qquad \Delta H = +206.0 \text{ MJ/k.mol.}$$
$$CH_4 + 2H_2O \rightleftharpoons CO_2 + 4H_2 \qquad \Delta H = +165.0 \text{ MJ/k.mol.}$$
$$CO + H_2O \rightleftharpoons CO_2 + H_2 \qquad \Delta H = -410 \text{ MJ/k.mol.}$$

Since the reaction is endothermic, fuel is burnt outside the tubes containing the catalyst to supply the heat of reaction at a high temperature. The waste heat is recovered from the combustion gas by pre-heating gas, air and by raising steam for power and process use.

The normal operating temperature and the pressure in the primary reformer are 750–800°C and 29.4–34.3 bar, respectively. The normal steam-carbon ratio is maintained between 3.5–4.0.

Gas from the primary reformer—with a residual methane content of 0.2 to 0.5% (dry basis) passes to the secondary reformer, a vessel filled with bulk nickel catalyst and air is added to supply heat (by combustion of hydrogen) for completing steam-methane reaction. The air also supplies N_2 needed for ammonia synthesis. However, the amount of secondary reforming is limited by the amount of air that can normally be added to give a 3 : 1 hydrogen to nitrogen ratio in the final gas.

The gas is then cooled in a waste heat boiler and passed into carbon monoxide converters (shift converters) where the following reactions occurs:

$$CO + H_2O \rightleftharpoons CO_2 + H_2 \quad \Delta H = -41.0 \text{ MJ/kg. mol}$$

The conversion is carried out in two reactors—a high temperature converter, followed by a low temperature converter. The former favours the rate of reaction and the latter equilibrium with intermediate cooling of gases, the gases emerging from the shift converter have 0.2 to 0.3% CO level. The gas mixtures then passes into an absorber where a solvent removes the carbon dioxide. Since the gas mixture still contains small amounts of CO and CO_2—which may poison the ammonia catalyst—it is passed over a nickel catalyst in the 'methaniser' to make oxides of carbon react with H_2 to form methane and water. The water is removed from the gas before it is passed to the catalyst and methane is allowed to pass since it does not harm the catalyst.

$$CO + 3H_2 \rightleftharpoons CH_4 + H_2O$$

The gas mixtures leaving the methanator contains carbon oxides less then 10 ppm and is compressed to synthesis gas pressure and mixed with recycle gas from an ammonia convertor.

Ammonia Synthesis

Haber studied the process and developed the catalyst for ammonia synthesis. The reaction is shown below:

$$N_2 + 3H_2 \rightleftharpoons 2NH_3 + Q$$

There is decrease in volume and generation of heat. The equilibrium can be shifted towards right by increasing the pressure and temperature. The heat of reaction of ammonia synthesis, depending on temperature and pressure can be computed from the following:

$$Q = 915.7 + \left(0.545 + \frac{840.6}{T} + \frac{4597 \times 10^5}{T^3} \right) P + 5.35T + 2.52 \times 10^{-4} T - 1.69 \times 10^{-6} T^3 \text{cal / mol.}$$

Thermodynamic aspect of ammonia synthesis

The reaction: $N_2 + 3H_2 \rightleftharpoons 2NH_3$, $H°_{298} = -92$ KJ/mol of N_2 is exothermic and endentropic and $\Delta H°_T = -77,294 - 54.24T + 0.01919\ T^2$ (in joules).

Thus, $\Delta H° = -107.8$ KJ/mol at 500°C, and starting with an approximate expression of the value of the equilibrium constant (Kp) as follows:

$$\log K_p = \frac{2940}{T} - 6.178$$

The production of ammonia is favoured by high pressure and low temperature.

These thermodynamic considerations imply that, in practice:

1. Once through, conversion of the feed gas is limited and recycling of the unconverted fractions result in the use of a 'synthesis loop' operating at high pressure.
2. Obtaining the high pressures associated with partial conversion of the reactants incurs large mechanical energy expenditure.
3. The use of low temperature, which partly offsets these drawbacks, tends to reduce the reaction rate.

Kinetic aspects of ammonia synthesis

To accelerate the approach to equilibrium, the oxide catalysts employed are based on Group 7 metals, exclusively iron in practice (Fe_3O_4). A number of promoters help to improve performance, including Al_2O_3, which increases the active surface area of the particles and $K_2O.SiO_2$, MgO, CaO etc. which improve stability and increase activity and resistance to poisoning. Systems currently under development make use of ruthenium derivatives to replace or to be used together with those of iron, modified by rubidium, titanium and cerium compounds.

The catalyst may be supplied to the users:

1. In its original oxide form which must be reduced in the unit itself by the mixture of $N_2 + 3H_2$. This means a conditioning interval lasting from 4–10 days.
2. In a pre-reduced, non-pyrophoric form which is immediately operational but more expensive.

The optimum process conditions will vary with the local situation and particularly with the relationship between cost of plant, capital and fuel. With the recent drastic changes in these ratios and with foreseen future changes, the optimum design for a new plant will have to be established in each individual case. This is true for the synthesis unit dealt with above, and it is equally true for the complete ammonia plant.

Ammonia production processes

The main ammonia production processes can thus be classified into three types as mentioned earlier. These are:

1. The first is built around the partial oxidation of hydrocarbons.
2. The second is built on gasification of carbonaceous matters.
3. The third is based on hydrocarbon steam reforming.

The latest improvement in ammonia production by steam reforming of natural gas includes the following:

1. ICI process characterised by the introduction of excess air in the secondary reforming step, which cuts total energy consumption drastically.
2. The Fertimont process (A Montiedison subsidiary process).

3. Technology proposed by Humphreys and Glasgow for revamping the existing reforming units with the direct introduction of part of the feed in the secondary reforming step.

4. KTI-PARC technique—which is ideal for low capacity installation and particularly relevant for India for cutting down fertiliser costs.

KTI/PARC process—ammonia manufacture

During the past decade ammonia production mainly directed to large size plants with capacity in the range of 1,000–1,500 tonnes/day. However, in many locations in the world, due to market requirements infrastructure, raw materials availability or national policy small ammonia plants in the range of 400–600 tonnes/day have better prospect.

This however can be justified if the production cost is comparable to that of a large size ammonia plant. KTI-PARC process has demonstrated that a small capacity ammonia plant is as attractive as modern large scale unit not only in production costs but also in terms of investment cost per tonne of production capacity.

KTI process: The KTI process comprises a combination of the following units:

1. Air separation plant to produce pure nitrogen.
2. Synthesis gas production by hydrocarbon steam reforming, HT shift conversion, and hydrogen purification in a modified pressure swing adsorption (PSA) system.
3. Synthesis gas compression and ammonia synthesis to produce liquid and vapour ammonia product.

The process arrangement is similar to ammonia manufacturing route based on coke oven gas as established in Europe in early sixties. At that time hydrogen rich gas was purified by a nitrogen wash plant to provide a steam of hydrogen and then reacted with pure nitrogen gas from air separation.

The flowsheet for the KTI-PARC process is given in Fig. 9.9.

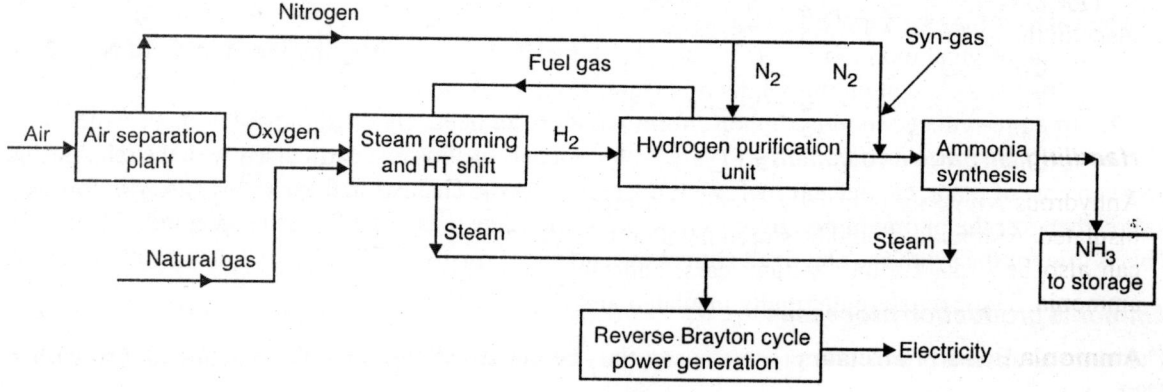

Fig. 9.9. KTI-PARC process.

Reactions: steam methane reforming and HT shift:

1. $CH_4 + H_2O = CO + 3H_2$
2. $C_2H_6 + 2H_2O = 2CO + 5H_2$
3. $C_nH_{2n+2} + nH_2O = nCO + (2n+1) H_2$
4. $CO + H_2O = CO_2 + H_2$

KTI-PARC process of ammonia manufacture comprise:

1. Pure Nitrogen Plant—The nitrogen generator is a standard process based on cryogenic separation of air at low temperature. The very pure nitrogen produced is 99.99% and used for ammonia synthesis and syn-gas purification. The oxygen enriched waste gas (40% O_2) can be used as combustion air in the reformer plant.

2. The syn-gas production section consists of the following units:
   (a) Feed desulphurisation
   (b) Steam hydrocarbon reforming.
   (c) H.T. shift conversion.
   (d) Modified PSA unit for syn-gas purification.

In this system compared to conventional ammonia gas production units, fewer processing steps are required. Thus the secondary reformer, the low temperature shift, the carbon dioxide removal and methanation steps are replaced by one PSA purification unit. The process simplification not only increases plant reliability but improves thermal efficiency.

Ammonia synthesis and make up synthesis gas compression

The make up synthesis gas from PSA unit is compressed from 23–28 bar to the 250 bar pressure. This is carried out by two multistage service compressors operating in parallel, both directly driven by electric motors. The multiservice compressor is also utilised for compressing pure nitrogen.

The process design from KTI has certain special features and they are :

1. Radial flow converter to minimise recycle compression power.
2. An efficient loop boiler which recovers up to 1.4 tonnes of saturated steam per tonne of ammonia.
3. Improved efficiency by operation without purge due to the high purity of the make up synthesis gas.

Due to rapid increase in the demand of fertiliser a number of technological improvements have taken place in the fertiliser technology. These are:

1. Construction of large size centrifugal compressors which have become increasingly reliable.
2. The use of active catalyst at low pressure, i.e., $20–25 \times 10^6$ pa abs.

Handling, storage and packing

Anhydrous ammonia is usually stored at a pressure of about 2–7 atmosphere and suitably thermally insulated. Ammonia should be stored in an unrefrigerated tank. Refrigerated storage at ambient pressure can also be used. Cylinders, tank cars, and tank trucks are used as shipping containers for liquid ammonia. These cars are thermally insulated and operate at pressure of about 15 atmosphere.

Ammonia Based Fertilisers

Ammonium sulphate

Ammonium sulphate is one of the important fertilisers produced in India. It contains 21% nitrogen and 24% sulphur and has traditionally been very popular in various parts of the country.

Raw materials and sources

The raw materials are:

1. Sulphuric acid for its recovery from coke oven plants.
2. Ammonia and sulphuric acid for neutralisation process.

3. Gypsum (natural or by-product from phosphoric acid plants), CO_2 and ammonia for process using gypsum route.

Methods of manufacture

1. *Recovery from coke ovens:* Coke oven gas contains about 1% ammonia by volume. The gas is cooled and passed into saturators containing weak sulphuric acid. Ammonium sulphate crystals formed in the saturator are recovered, centrifuged, washed and dried. This process is used in all the integrated steel plants where large coke oven batteries are in operation.

2. *Direct neutralisation:* Gaseous ammonia is directly neutralised with sulphuric acid to produce ammonium sulphate.

$$2NH_3 + H_2SO_4 \rightleftharpoons (NH_4)_2SO_4$$
Ammonium sulphate

The neutraliser reactor and the crystallisers are interconnected so that heat released during neutralisation is used to evaporate water in the slurry. Crystalliser is designed to produce uniformly sized crystals. Amorphous ammonium sulphate is prepared by reacting gaseous ammonia and sulphuric acid in spray towers. The heat of reaction removes all the water present and the dry, fine product is continuously removed from the base of the tower. This product is suitable for making dry mixed and granular fertilisers.

In the manufacture of some granular 'NPK' fertilisers ammonium sulphate is formed directly with other ammonium salts.

3. *Gypsum process:* In this process ammonia gas is absorbed in water and then converted to ammonium carbonate by absorbing carbon dioxide. Ammonium carbonate is reacted with gypsum ($CaSO_4$) to produce ammonium sulphate and calcium carbonate.

$$CaSO_4.2H_2O + (NH_4)_2CO_3 \rightleftharpoons (NH_4)_2SO_4 + CaCO_3 + 2H_2O$$

| Gypsum | Ammonium carbonate | Ammonium sulphate | Calcium, carbonate |

Calcium carbonate is removed by filtration. Ammonium sulphate solutionis evaporated, crystallised, centrifuged and dried (Fig. 9.10.)

4. *By-product ammonium sulphate from caprolactum plant:* Ammonium sulphate solution is formed during the manufacture of caprolactum (the starting material for Nylon 6). The solution is concentrated and ammonium sulphate is recovered by crystallisation, centrifuging and drying.

Properties

Nitrogen content	21%
Colour	White
Sp. gravity	1.769
pH	5.0
Bulk density kg/cum	720–1,040 (45–65 lbs/cft)
Angle of repose	32°–33°
Critical relative humidity at	
25°C	81.8
30°C	79.2
40°C	78.2

Specifications

1. Moisture per cent by wt. (max) : 1%
2. Ammoniacal nitrogen % by wt. (min) : 20.6%.
3. Free acidity (as H_2SO_4) % by wt. (max): 0.025
4. For material obtained from by-product ammonia and by-product gypsum: 0.04.
5. Arsenic (as As_2O_3)% by wt. (max): 0.01.

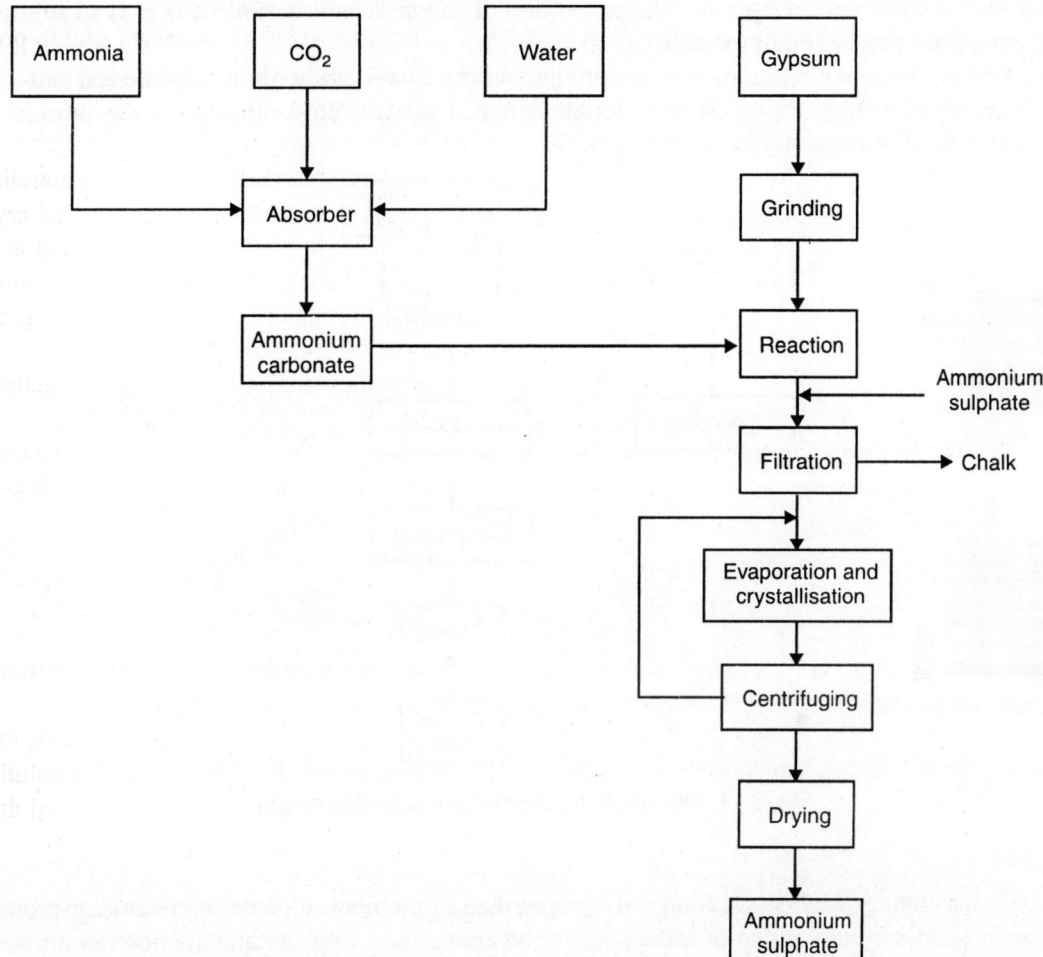

Fig. 9.10. Flowsheet for manufacture of ammonium sulphate based on gypsum process.

Handling, storage and packing

Crystalline ammonium sulphate is free flowing and does not pose any problem in handling and storage. However, it generally contains some powder material which causes linking, especially under high humidity.

Due to its susceptibility to caking and slight acidity, ammonium sulphate is normally bagged in polyethylene-lined gunny bag or high density polyethylene (HDPE) sacks.

Ammonium sulphate nitrate

This is also known as 'Montana saltpetre'. It is a straight nitrogenous fertiliser product. The product contains equimolar properties of ammonium sulphate and ammonium nitrate and contains 26% nitrogen.

Three raw materials required for the process are ammonia, nitric acid and sulphuric acid.

Method of manufacture

A mixture of sulphuric acid and nitric acid is ammoniated go get a mixture of ammonium sulphate and nitrate which is known as double salt. The proportion of sulphuric acid and nitric acid is so adjusted as to give equimolar proportion of the acids.

When such a mixture is ammoniated, the product will also have ammonium sulphate and ammonium nitrate in equimolar proportions. Such a double salt will contain 26% nitrogen in the product. The flowsheet of the process is indicated in Fig. 9.11.

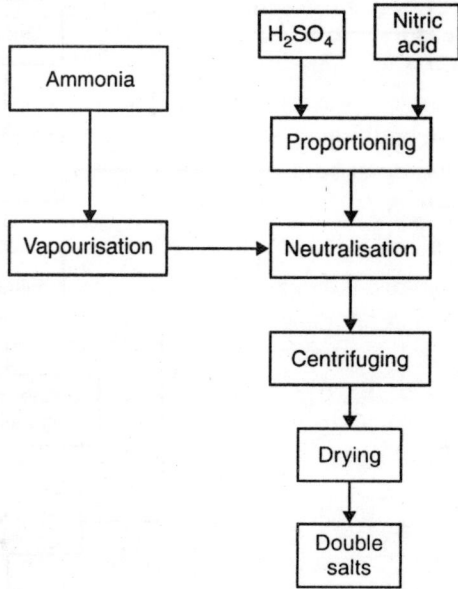

Fig. 9.11. Flowsheet for ammonium sulphate nitrate.

Properties

The double salt contains higher percentage of nitrogen than ammonium sulphate and its storage properties are superior to ammonium nitrate or mixtures of solid ammonium sulphate and ammonium nitrate.

Specifications

1. Moisture per cent by wt. (max).   1.0
2. Total ammoniacal and nitrate nitrogen per cent by wt. (min)   26.0
3. Ammoniacal nitrogen % by wt. (min)   19.25
4. Free acidity (as HNO_3) per cent by wt. (max)   0.015
5. Particle size in the range of $-4 + 2$ mm

Handling, storage and packing

Storage properties of the 'double salt' are superior to those of ammonium nitrate, but the necessary precautions have to be taken to protect the product from moisture absorption during handling and storage. The product is normally packed in gunny bags lined with polyethylene liner or HDPE open sacks.

Ammonium nitrate

Ammonium nitrate contains 35% nitrogen, is not used as fertiliser in India. It is manufactured in India at FCI/Sindri and SAIL/Rourkela for use in explosives and for making anesthetic grade nitrous oxide.

To produce ammonium nitrate, the raw materials required are nitric acid and liquid ammonia.

Process of manufacture

1. *Prilling process:* Ammonia vapour is reacted with aqueous nitric acid in a stainless steel reactor with agitation. The heat of reaction causes water to boil off and the final salt solution is of 75 per cent concentration at 140°C. It is pumped to vacuum evaporator to concentrate it further to 95.0% solids. Prills are formed by spraying hot liquor from the top of the prilling tower (60–75 m height) countercurrent to conditioned air flow. The solidified prills are about 1.5 mm in diameter and are screened and dried before coating with clay.

$$\underset{\text{Nitric acid}}{HNO_3} \quad + \quad \underset{\text{Ammonia}}{NH_2} \quad \rightleftharpoons \quad \underset{\text{Ammonium nitrate}}{NH_4NO_3}$$

2. *Crystallisation process:* The process is similar to the prilling process up to the point of evaporation. The liquor from the vacuum evaporator containing 80–85% solids, is fed to a vacuum crystalliser where crystal growth is controlled at about 40°C to yield large grains for fertiliser use. The crystal slurry with 40% crystals by wt. is centrifuged; the mother liquor is returned to the evaporator and the crystals are dried in a rotary drier at about 85°C. The process flowsheets are given in Figs. 9.12. and 9.13 for the prilling process.

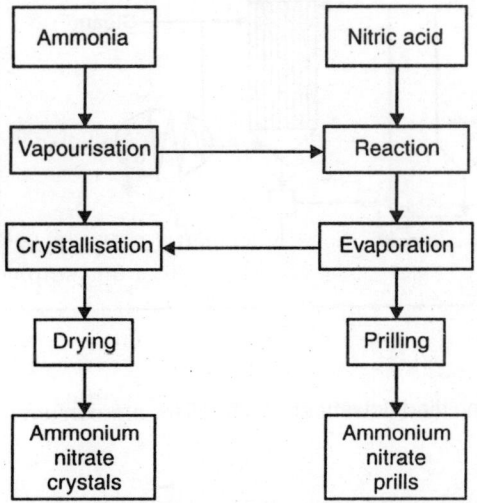

Fig. 9.12. Flowsheet for manufacture of ammonium nitrate.

3. *Stengel process:* The process patented by L.A. Stengel and assigned to a Commercial Solvents Corporation involves vapour phase reaction in a packed stainless steel reactor.

Properties

Colour	White hygroscopic crystals or granules
Nitrogen content %	35.0
Mol wt.	80.05
Specific gravity (25°–4°C)	1.66
Refractive Index	1.611
Melting point °C	170
Boiling point °C	Decomposes at 200°C
Solubility in water at 20°C gms/litre	900
Bulk density, kg per cum	998.7 (62 lbs/ft3)
Angle of repose	35°–37°
Critical humidity	as follows:

Temperature °C	10	15	20	25	30	40
Critical relative humidity	75.3	69.8	66.9	62.7	59.4	52.5

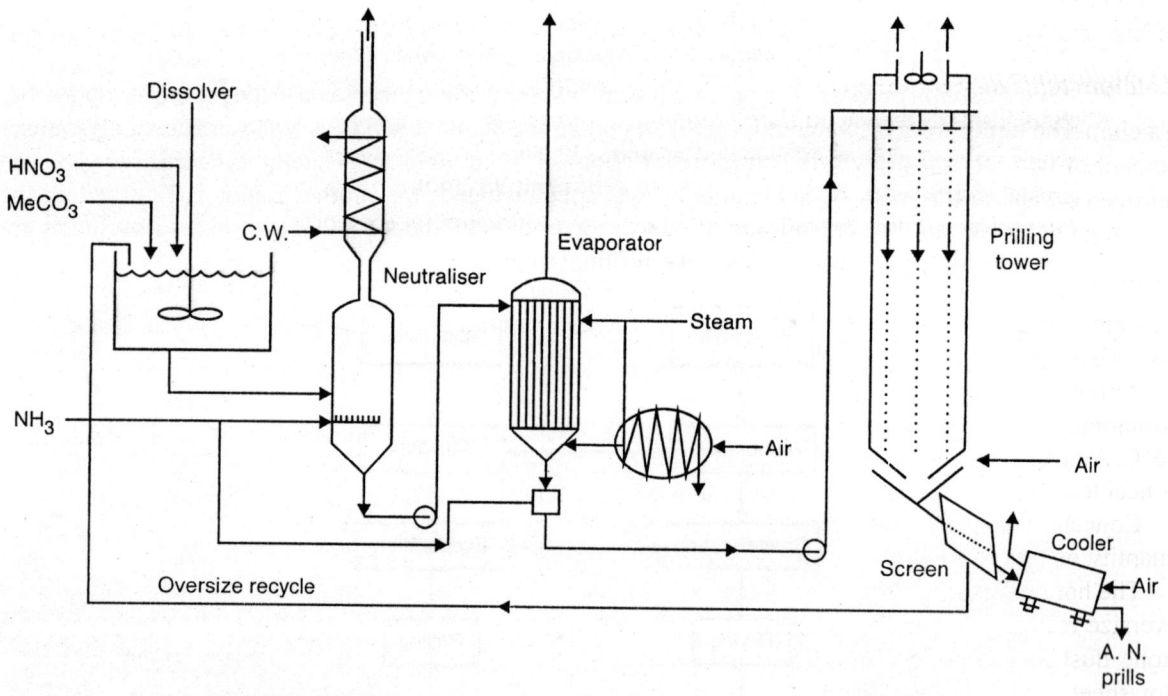

Fig. 9.13. Simplified flowsheet of a fertiliser ammonium nitrate plant.

Handling, storage and packing

Granules of ammonium nitrate are free flowing and do not normally pose any probability in handling and storage. But it should be handled very carefully because it is explosive when mixed with combustible materials or exposed to high temperature.

Ammonium nitrate is hygroscopic and so it is stored in airconditioned silos. It is bagged in polyethylene-lined jute bags with crepe paper as inside liner.

Applications of ammonium nitrate

1. *Mining industry:* For blasting in open cast mines, over burden in coal mines, in underground metalliferrous mines, tunnels, and inclines.
2. *Construction industry:* For blasting, in Hydel works, foundation of dams and buildings, irrigation canals, and underwater blasting for harbour development.
3. *Explosives industry:* For manufacture of ammonium nitrate, dynamites, ammonia gelignite dynamites, and slurry explosives etc.
4. *Drugs industry:* For the manufacture of nitrous oxide gas and as nutrient for micro-organism
5. *Dyestuffs and textile industry:* In manufacture of phthalogen blue dye, phthalocyanine blue pigment, preparing sizing solution for textiles.

Ammonium nitrate is used in bulk quantities today in many countries as an ingredient for field mixed explosive, commonly, known as ANFO.

Ammonium nitrate when mixed with combustible organic materials such as paraffin oils and wax acquires explosive properties. After many trials for field-mixed compounds, fuel oil has been found to be most suitable. Ammonium nitrate with 5.5% by wt. fuel oil (ANFO) is an oxygen balanced mixture. ANFO cannot be detonated by blasting-cap when unconfined, but can be detonated by a booster charge of high explosives. Its explosive strength is 120% of TNT as judged by the ballistic pendulum test.

Calcium ammonium nitrate

Calcium ammonium nitrate (CAN) is one of the straight nitrogenous fertiliser produced in India. Nitrogen content of CAN varies from 25 to 28 per cent. Calcium ammonium nitrate as produced in India has a nitrogen content of 25 per cent. The raw materials required to produce 'CAN' are ammonia, nitric acid, limestone or dolomite and soap stone.

Method of manufacture

Calcium ammonium nitrate is produced by granulating concentrated ammonium nitrate solution with pulverised lime stone or dolomite in a granulator.

Ammonium nitrate solution is prepared by reacting pre heated ammonia with nitric acid in a neutraliser. Ammonia is pre-heated to 85°C by vapours from neutraliser which also pre-heat nitric acid to about 65°C. Ammonium nitrate liquor of 82–83 per cent concentration, which is produced in the neutraliser, is concentrated to 92–94 per cent in a vacuum concentrator heated with steam and stored in tank.

Concentrated ammonium nitrate is pumped and sprayed into the granulator which is fed with weighed quantity of limestone powder and recycle fines from the screens.

The hot granules are dried in a rotary drier by hot air. Dried hot granules are screened and fines and oversize recycled. Granules of proper size are cooled in a rotary cooled by air and coated with soap stone dust in a coating drum. The final product is sent to storage. The process is indicated in the flowsheet in Fig. 9.14.

Properties

Nitrogen content per cent	25
Moisture content	Not more than 0.3%
Granule size	2 to 4 mm size

Specifications

Particulars		CAN of	
	25% N	*26% N*	*28% N*
Moisture % by wt. (max)	1.0	1.0	1.0
Total ammoniacal and nitrate nitrogen content % by wt. (min)	25.0	26.0	28.0
Ammoniacal nitrogen % by wt. (min)	12.5	13.0	14.0
Calcium nitrate % by wt. (max)	0.5	0.5	0.5

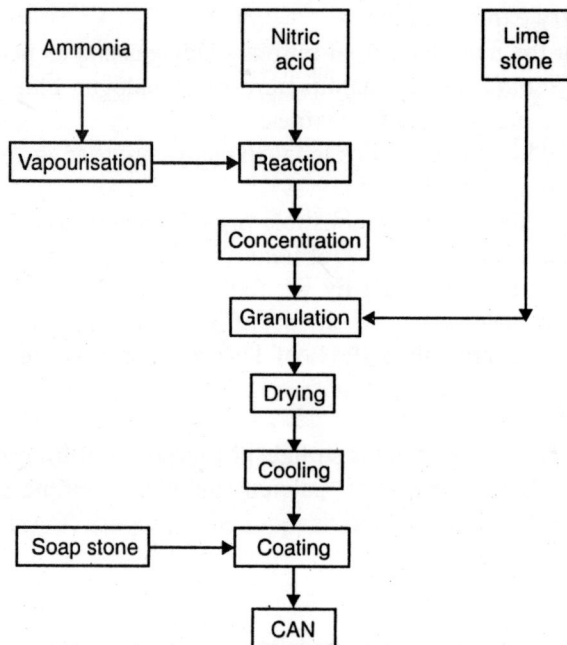

Fig. 9.14. Flowsheet for manufacture of CAN.

Particle size

The particle size of the material shall be such that material shall be retained completely on 1 mm IS sieve and not less then 80 per cent by wt. of it shall pass through 4 mm IS sieve.

Handling, storage, and packing

Granules of 'CAN' are free-flowing and do not normally pose any problem in handling and storage. Under the prevailing climatic conditions in most parts of India, storage of 'CAN' presents considerable problems. It is stored in air-conditioned silos, and the temperature in the silo is kept within 30°C. "CAN' is normally bagged in polyethylene lined jute bags.

Ammonium chloride

Ammonium chloride is used as a fertiliser for rice and some field crops. However, it may increase the residual chloride content of some soils. Besides its use as fertiliser pure ammonium chloride (technical grade) is also used in the manufacture of dry cells (battery) and as a flux in galvanising and tinning industry.

Raw materials required for the manufacture of ammonium chloride are : (i) for the direct neutralisation process—hydrogen chloride and ammonia; and (ii) for the method based on Solvay process—common salt, ammonia and carbon dioxide. In this process, ammonium chloride and soda ash are co-products.

Method of manufacture

1. *Direct reaction process:* Ammonia and hydrogen chloride are directly reacted to produce ammonium chloride:

$$NH_3 + HCl \rightleftharpoons NH_4Cl$$

Ammonia   Hydrogen chloride   (Ammon chloride)

Gaseous ammonia is bubbled into hydrochloric acid solution in a reactor. The reaction is controlled by addition of water. The slurry from the saturator is centrifuged and the crystals are washed with water and dried with warm air. This process is not widely used now a days.

2. *Modified solvay process:* The most widely used process for producing ammonium chloride is the salting out process for soda ash manufacture. This is a modification of the Solvay ammonia soda process. In the process, a 30 per cent solution of ammonia is treated with carbon dioxide in a carbonating tower to form ammonium carbonate:

$$2NH_3 + H_2O + CO_2 \rightleftharpoons (NH_4)_2CO_3$$

(Ammonium carbonate)

Further treatment with carbon dioxide yields ammonium bicarbonate:

$$(NH_4)_2CO_3 + H_2O + CO_2 \rightleftharpoons 2NH_4HCO_3$$

(Ammonium bicarbonate)

The ammonium bicarbonate reacts, as it is formed with sodium chloride, to give sodium bicarbonate and ammonium chloride:

$$NH_4HCO_3 + NaCl \rightleftharpoons NaHCO_3 + NH_4Cl$$

Ammonium bicarbonate   Sodium bicarbonate

The bicarbonate is separated by filtration, washed and calcined to produce sodium carbonate (soda ash). In the modified Solvay process, the ammonium chloride in the solution after separation of the sodium bicarbonate is salted and by ammoniating the solution, cooling below 15°C and adding washed sodium chloride. Crystals of ammonium chloride formed are centrifuged and dried in rotary dries. The process is indicated in the flowsheet in Fig. 9.15.

Properties

Nitrogen content, per cent	26.1
Colour	White, crystalline
Specific gravity	1.53
pH	4.5–5.0
Bulk density kg per cum	835
Critical humidity	
at 25°C	76.0
at 30°C	77.0

Specifications

Moisture, per cent by wt. (max)	2.0
Ammoniacal nitrogen per cent by wt. (min)	25.0
Chloride other than ammonium chloride (as NaCl) per cent by wt. (max)	2.0

Handling, storage and packing

Crystalline ammonium chloride is free-flowing and non-abrasive and does not, normally, pose any problem in handling and storage.

As it is susceptible to caking at high humidity and has slightly acidic reaction, ammonium chloride has to be bagged in multiwall paper bags or jute bags lined with polyethylene film.

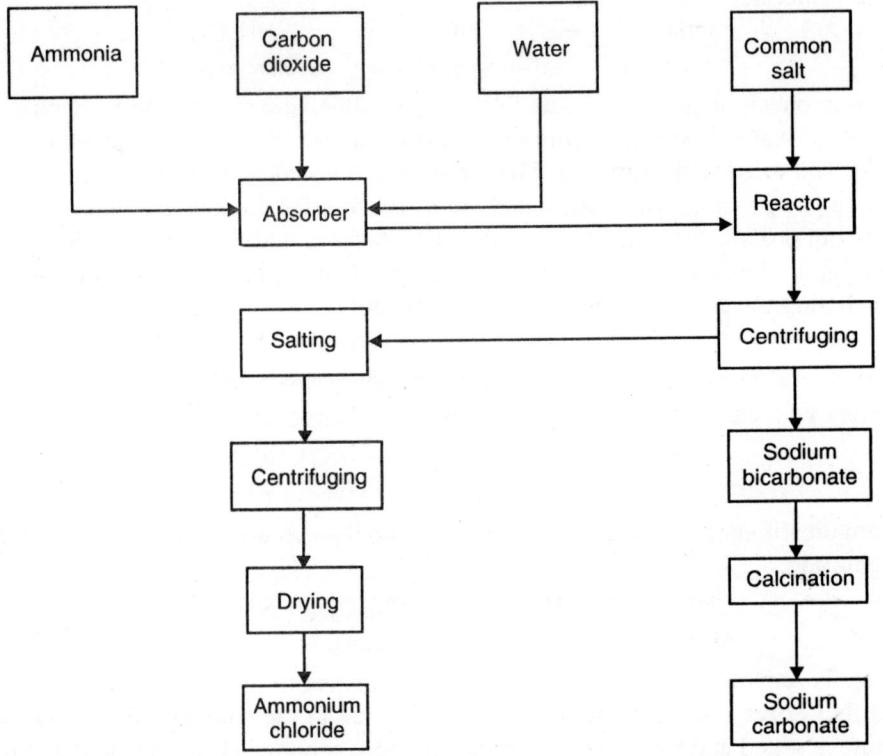

Fig. 9.15. Flowsheet for manufacture of ammonium chloride and sodium carbonate.

Urea (fertiliser grade)

Urea is now the most important nitogenous fertiliser in the country as it has a high nutrient content (46.0% N). Principal raw materials required to produce urea are carbon dioxide and ammonia.

Methods of manufacture

Ammonia and carbon dioxide are compressed and reacted at 160–220 atm and at 170–190°C in an autoclave to form ammonium carbamate. The ammonium carbamate is then decomposed to produce urea:

$$CO_2 + 2NH_3 \longrightarrow NH_4COONH_2$$
$$\text{(Ammonium carbonate)}$$

$$\underset{\text{Ammonium carbamate}}{NH_4COONH_2} \longrightarrow \underset{\text{Urea}}{NH_2CONH_2} + H_2O$$

The solution is evaporated to 99.7–99.8 per cent and the melt is prilled in tall towers countercurrent of air. A flowsheet of the process is given in Fig. 9.16.

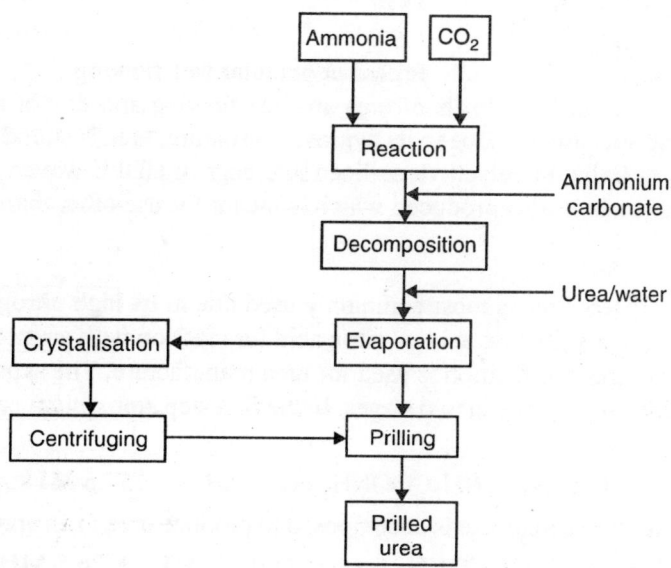

Fig. 9.16. Flowsheet for manufacture of urea.

Properties

Nitrogen content	46
Colour	White
Mol. wt.	60.05
Melting point °C	132.70
Boiling point	Decomposes
Sp. gr. 20°–4°C	1.335
Solubility in water at 20°C gm/100 gm at 100°C	100
Bulk density, kg. per cum	673–721
Angle of repose	30°

Critical humidity

Temperature °C	10	15	20	25	30	40	50
Critical relative humidity	81.8	80.0	79.9	75.8	72.5	68.0	62.5

Specifications

Moisture per cent by wt. (max)	1.0
Total nitrogen per cent by wt. (min) on dry basis	46.0
Biuret per cent by wt. (max)	1.5

Particle size: In the form of granules the material shall pass IS sieve 320 and not less then 80 per cent by wt. of it shall be retained on IS sieve 100. If in the form of prills, the material shall pass IS sieve 200 and not less then 80 per cent by weight of it shall be retained on IS sieve 100.

(i) Urea coated : 45 per cent nitrogen

Moisture per cent by wt. (max)	0.5
Total nitrogen, per cent by wt. (min)	45.0
Biuret per cent by wt. (max)	1.5
Physical condition	Prilled or granular free flowing

(ii) *Handling, storage and packing:* Prills of urea are free-flowing and do not normally pose any problem in handling and storage. Due to its hygroscopic nature, urea is stored in air-conditioned silos. It is bagged, in India, in polyethylene-lined jute bags or HDPE woven sacks.

Note: Urea of technical grade is also produced which is meant for use other than as fertiliser.

Urea process design

Among the nitrogenous fertilisers urea is most commonly used due to its high nitrogen content (46%); also the process does not require sulphuric acid or nitric acid for reaction with ammonia and the carbon dioxide recovered during syn-gas purification is used for urea manufacture. The synthesis of urea from ammonia and carbon dioxide takes place in two stages. In the first step ammonium carbamate is formed in an exothermic reaction:

$$2NH_3 \text{ (g)} + CO_2 \text{ (s)} \rightleftharpoons NH_2COONH_2 \text{ (s)}, \quad \Delta H = -157.6 \text{ MJ/k.mol}$$

In the second step, ammonium carbamate is decomposed to produce urea in an endothermic reaction:

$$NH_4COONH_2 \text{ (s)} \rightleftharpoons NH_2CONH_2 \text{ (s)} + H_2O \text{ (l)}, \quad \Delta H = +26.5 \text{ MJ/k.mol}$$

The first reaction is very fast, while the second is very slow. The effluent from the urea reactor is a highly corrosive mixture of urea, ammonium carbonate, ammonia, carbon dioxide and water. The subsequent steps in urea manufacturing process are decomposition of ammonium carbamate, recovery of unconverted CO_2 and NH_3 and concentration of urea solution to produce prilled urea.

The conversion in the reactor increases with pressure, temperature, NH_3 to CO_2 molar ratio and residence time and decreases with H_2O to CO_2 ratio. Higher pressure requires more energy whereas temperature exceeding 200°C lead to serious corrosion problem. On the other hand, the decomposition of carbamate is favoured at low pressure but this requires higher energy for recycling or unreacted components. Also, at low pressure, the evaporation of water during carbamate decomposition is greater, which retards the conversion efficiency in the reactor. Based on the mode of recycling unreacted carbamate, urea process can be classified as: (i) partial recycle process; (ii) total recycle process; (iii) slurry recycle process; and (iv) stripping process.

The process flowsheet in Fig. 9.17 gives details of production of urea solution which after concentration yield high nutrient urea (46% N).

PHOSPHORIC ACID AND PHOSPHATIC FERTILISERS

Phosphoric acid

The increasing use of balanced multinutrient fertilisers in recent years has resulted in increased importance of phosphoric acid. A small quantity of phosphoric acid is directly used as fertiliser in the soil but majority is directly used for the production of triple superphosphate, ammonium phosphates, complex fertilisers and liquid fertilisers. Pure grade phosphoric acid is also used for manufacture of detergents.

Rock phosphate and sulphuric acid form the raw materials for the production of wet phosphoric acid. For furnace acid phosphate rock, silica pebble, coke and electric power are needed.

Methods of manufacture

There are two basic methods in commercial use for the production of phosphoric acid—based on the wet process and the furnace process.

Wet process

The production of phosphoric acid by wet process follows the following steps:

1. The reaction with finely ground rock phosphate and sulphuric acid to form phosphoric acid and gypsum.

$$Ca_{10}(PO_4)_6F_2 + 10H_2SO_4 + 10H_2O \longrightarrow 10\ [CaSO_4.2H_2O] + 6H_3PO_4 + 2HF$$

Rock phosphate   Gypsum   Phosphoric   Hydrogen
                    acid   fluoride

2. Separation of gypsum from the acid by filtration.
3. Washing gypsum to remove adhering phosphoric acid.
4. Concentration of acid by evaporation to the desired concentration.

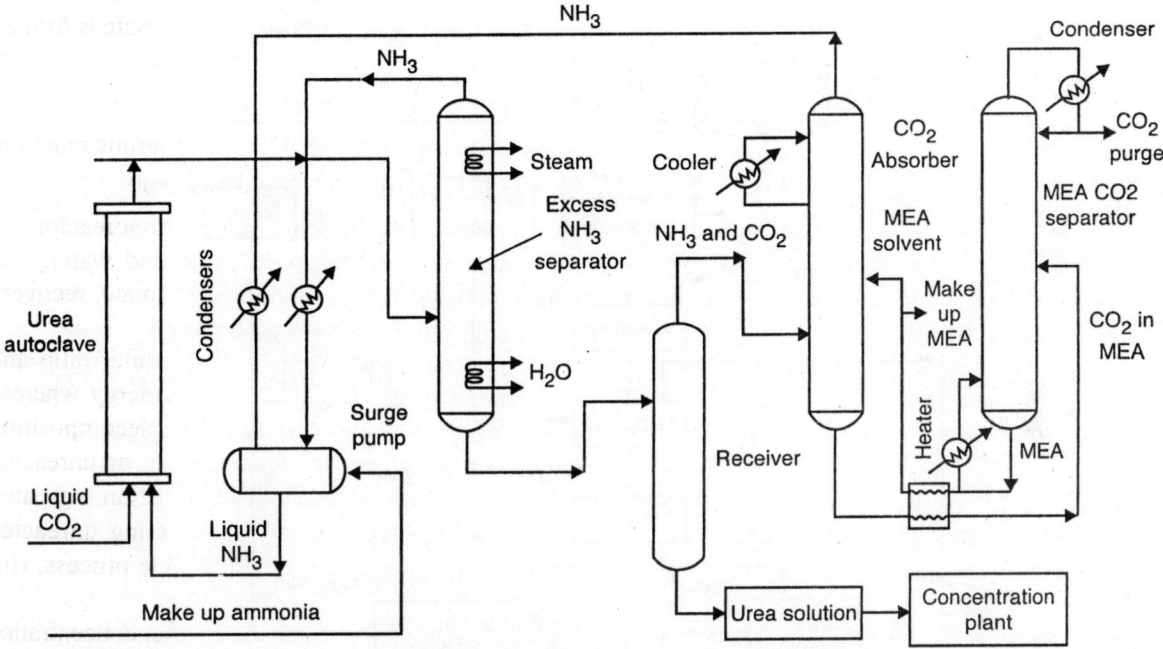

Fig. 9.17. Production of urea solution.

Countercurrent washing of gypsum is employed to recover as much phosphoric acid as possible without excessive dilution of the acid. Dilute acid is recycled to the extraction step. The acid from the filter is concentrated either by direct contact with combustion gases in submerged combustion equipment or by indirect heating by steam in a vacuum evaporator. The flowsheet of the wet process of phosphoric acid manufacture is given in Fig. 9.18.

Thermal furnace phosphoric acid

In this process nodules or other lump phosphate material, silica pebble and coke are mixed and fed in furnace. The electric current passing through carbon or graphite electrodes fuses the rock and silica and carbon in the coke reduces the phosphate. A mixture of elemental phosphorus vapour and carbon monoxide gas issues out continuously from the furnace. The phosphorous is condensed—which is normally in liquid state at the temperature of operation. Elemental phosphorus produced is burnt in air and dissolved in water to form phosphoric acid. The advantage of this process is that it can use low grade phosphate, rock phosphates and produce pure phosphoric acid (Fig. 9.19). Table 9.3 also indicates properties and analysis of phosphoric acid.

Table 9.3. Typical analysis of merchant grade wet process and furnace phosphoric acid in per cent.

	Wet process acid	*Furnace acid*
P_2O_5	54.0	54.0
H_3PO_4	74.6	74.6
CaO	0.1	0.0503
$Fe_2O_3 + Al_2O_3$	0.7	0.0037
SO_3	3.0	0.0030
F	0.5	0.0070
Suspended solids	5.5	—

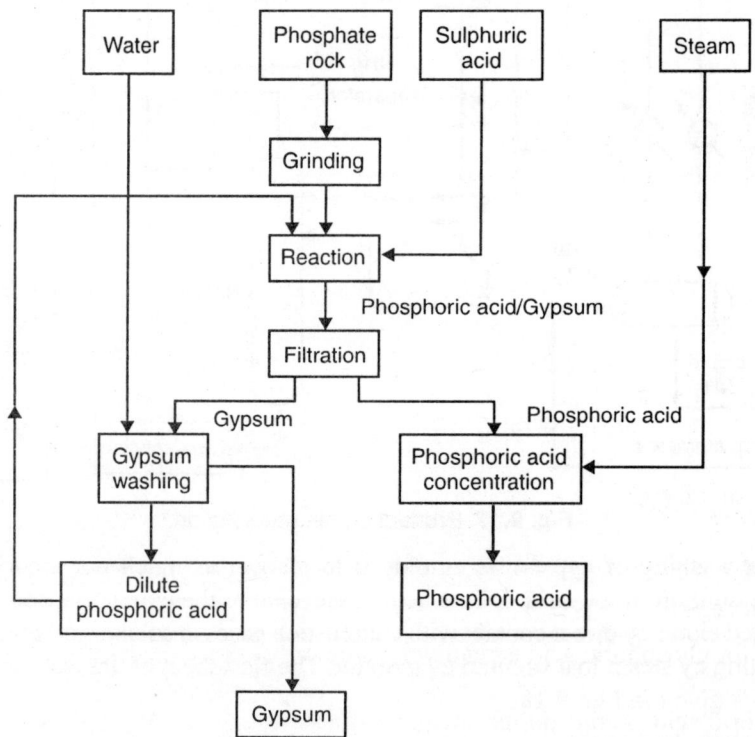

Fig. 9.18. Flowsheet for manufacture of phosphoric acid.

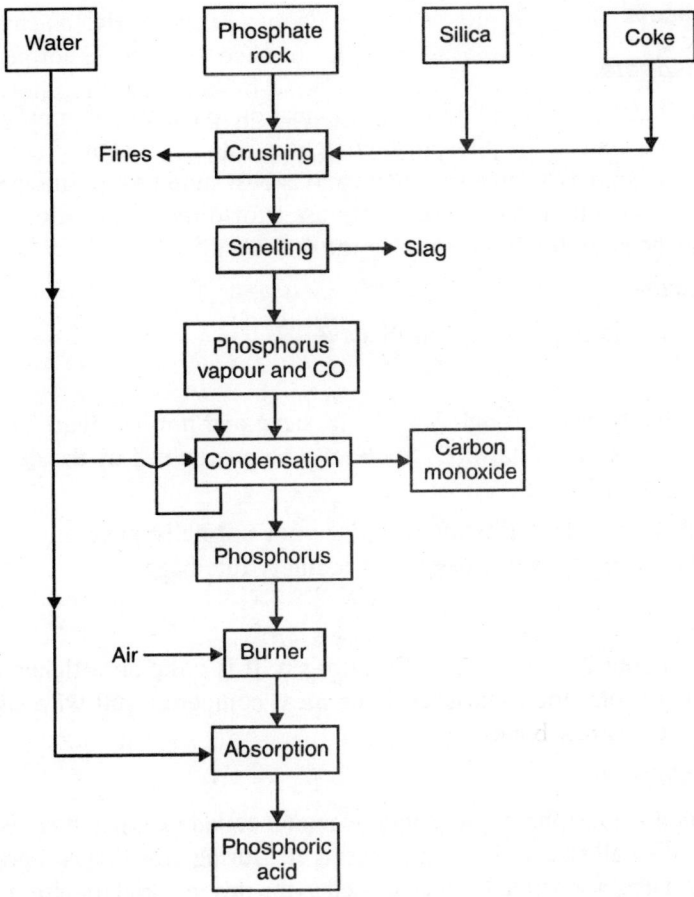

Fig. 9.19. Flowsheet for manufacture of furnace phosphoric acid.

Properties

Chemical formula	H_3PO_4
Molecular wt.	98.0
Colour	Colourless, liquid
Specific gravity at 18.2°C	1.834
Boiling point, °C	213
Solubility in water gm/100 gm at 20°C	2.340

Handling and storage

Storage facility for phosphoric acid is relatively more expensive than that for the solid fertiliser product made from it.

The phosphoric acid is transported in rubber-lined steel tank or in special cars made of stainless steel, glass carboys and oak barrels. The clarified acid is being transported over long distance in special ocean going ships.

Phosphatic Fertilisers

Ground rock phosphate

Finely ground phosphate rock is used for direct application to the land in most countries where fertiliser is used. Only 5–10 per cent of the phosphate content is soluble in neutral ammonium citrate solution. The material is usually sold as a soil amendment and is best suited for pastures and forage crops and for acid soils. The exact quantity of rock phosphate used for direct application to soil is not known. Its effectiveness is enhanced by finally powdering the materials.

Method of manufacture

Rock phosphate is ground to pass 6.3 mm IS sieve.

Particle size

Material shall completely pass through 6.3 mm IS sieve and not less than 20% of material shall pass through 150 microns IS sieve. Total P_2O_5 content to be guaranteed by the dealer.

Handling, storage and packing

Powered rock phosphate is free flowing and does not normally pose any problem in handling and storage. Powdered rock phosphate is bagged in ordinary jute bags.

Bone meal

Bone meals contain about 20% P_2O_5 and 2% nitrogen. It is quite an efficient phosphatic fertiliser on acidic soil. In terms of total food nutrients, bone meal compares well with superphosphate. The raw materials for bone meal is raw bones.

Method of manufacture

Bones contain tricalcium phosphate, some organic matter called ossein and fat. Bone meal is obtained by crushing the bones to small size and then powdering it. During crushing and grinding of dry bones, the fibrous organic materials get separated, which is used as the raw material for glue and gelatine manufacture. The bone meal which has only limited citrate solubility—is used as fertiliser.

Crushed bones are steamed in autoclaves, when most of the fat and organic matter is extracted. Fat is recovered and the extract is used for glue manufacture. The steamed bones are ground and used as fertiliser. Citrate solubility of steamed bones is considerably improved.

Specifications

	Bone meal	
	Raw	Steamed
Moisture per cent by wt. (max)	8.0	7.0
Total phosphorus (as P_2O_5) per cent by wt. (dry basis) min	20.0	22.0
Phosphorus as (P_2O_5) soluble in 2% citric acid solution per cent by wt. (dry) min	8.0	16.0
Acid in soluble per cent by wt. max.	12.0	—
Particle size	The material should pass wholly through 240IS sieve of which not less than 70% shall pass on 80 IS sieve	Not less than 90 per cent by wt. shall pass through IS sieve 120.

Handling, storage and packing

Ground bone meal—whether raw or steamed—poses no problem in handling, storing and packing. It is free-flowing. Both raw and steamed bone meals are marketed in ordinary jute bags or any containers agreed between the supplier and purchaser.

Single superphosphates

Single superphosphates (SSP) is the most important phosphatic fertiliser produced in India. It contains about 16.0 per cent P_2O_5 in water soluble form and has traditionally been very popular in various parts of the country (Fig. 9.20 for SSP process).

Raw materials required to produce 'SSP' are rock phosphate and sulphuric acid.

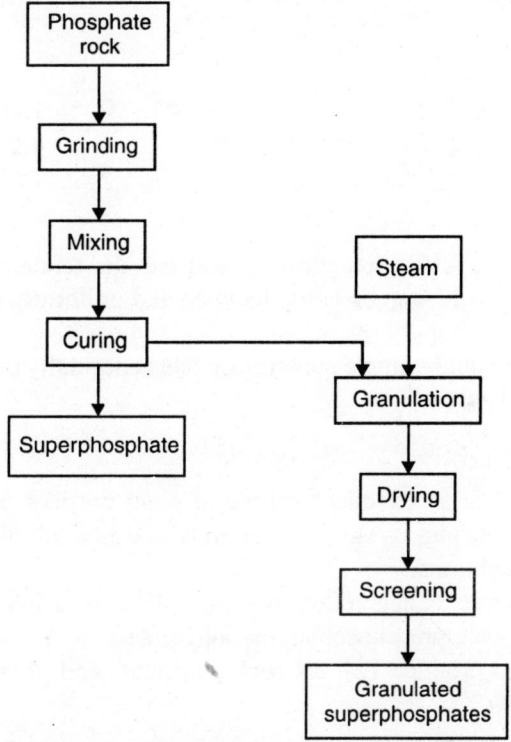

Fig. 9.20. Flowsheet for manufacture of single superphosphate.

Methods of manufacture

Ground rock phosphate (90% passing through 100 mesh) is mixed with sulphuric acid (55–75.0 %) in a specially designed mixer which discharges the product on a wide reaction belt conveyor where the reaction is completed.

The reacted mass is sent to a curing shed where the product is stored for 3–4 weeks for curing and drying. The cured and dried product is excavated, milled, screened and either bagged for marketing or sent for granulation.

Granulation is done in the presence of steam in a granulator and the finished product is then bagged.

$$\underset{\text{Rock phosphate}}{Ca_{10}(PO_5)\,F} + 7H_2SO_4 \longrightarrow \underset{\text{Superphosphate}}{3CaH_4(PO_4)_2} + 7CaSO_4 + 2HF$$

A flowsheet of the process is given in Fig. 9.20.

Properties

P_2O_5 content per cent by wt.	16–18
Free phosphoric acid	3
Critical relative humidity	93.7
Bulk density, kg per cum	961.1
Angle of repose	26°
Calcium sulphate per cent	50

Specification

Moisture per cent by wt. (max)	12.0
Free phosphoric acid (as P_2O_5) per cent by wt. (max)	4.0
Water soluble phosphate (as P_2O_5) per cent by wt. (min)	16.0

Handling, storage and packing

Powder single superphosphate is not free flowing and has the tendency to cake. Granulated single superphosphate, on the other hand, can be easily handled and uniformly distributed in the field without any problem.

Due to the presence of free acids, single superphosphate is normally bagged in India in polyethylene-lined gunny bags or woven bags.

Triple superphosphate

Triple superphosphate (TSP) is an excellent source of plant nutrient phosphorus with P_2O_5 content ranging from 44 to 52%. Its phosphorus content is entirely in water-soluble form and readily available to most plants under all soil conditions.

This product is also known as 'concentrated superphosphate' or 'triple superphosphate'. The product generally consists of 'mono-calcium phosphate monohydrate'.

Raw materials required to produce TSP are rock phosphate and phosphoric acids.

Method of manufacture

The rock is ground to a fine mesh and it is reacted with concentrated phosphoric acid (50–54% P_2O_5). Normally, mixing of rock and acid is carried out continuously in a cone mixer and fed to a continuous disintegrator.

The product is disintegrated after discharge from the mixer and piled for curing. After a specified time of curing the 'run of pile' material is bagged and sold as fertiliser.

$$\underset{\text{Rock phosphate}}{Ca_{10}(PO_4)_6\,F_2} + \underset{\text{Phosphoric acid}}{14H_2PO_4} \longrightarrow \underset{\text{TSP}}{10Ca(H_3PO_4)_2} + \underset{\text{Hydrogen fluoride}}{2HF}$$

The cured material, if required, is granulated, dried, screened and bagged for despatch. A typical flowsheet of the 'TSP' process is shown in Fig. 9.21.

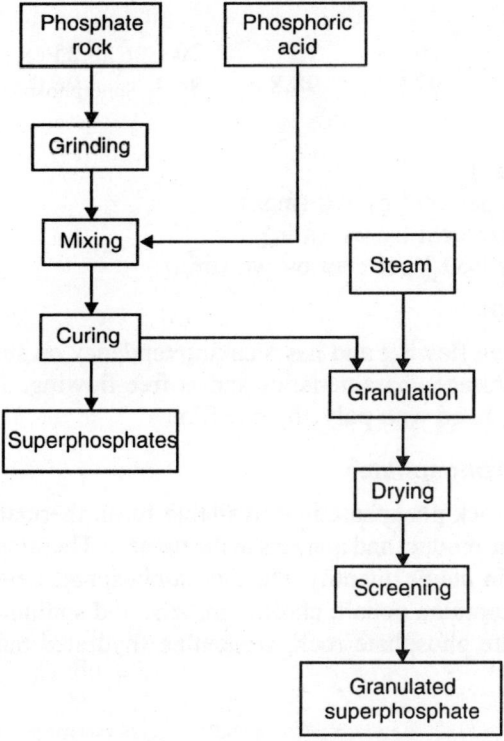

Fig. 9.21. Flowsheet for manufacture of triple superphosphate.

Properties

Typical composition	Per cent
P_2O_5	46.29
N	0.20
SO_3	3.40
F	1.56
CaO	19.99
MgO	0.38
Fe_2O_3	1.36
Al_2O_3	1.93
SiO_2 and acid insolubles matter	4.42
Free moisture	3.30
Total moisture	5.12
Minimum water soluble P_2O_5	40.00
Citrate soluble P_2O_5	6.14
Available P_2O_5	37.41
Bulk density, kg/cum	800–881
Angle of repose	45°

Critical relative humidity

Temp. °C	10	15	20	25	30	40
Critical relative humidity	97.9	98.8	94.1	96.0	93.7	94.5

Specification

Moisture per cent by wt. (max)	12.0
Free phosphoric acid (P_2O_5) per cent by wt. (max)	3.0
Total phosphates (as P_2O_5) per cent by wt. (min)	46.0
Water soluble phosphates (as P_2O_5) per cent by wt. (min)	42.5

Handling, storage and packing

TSP in powder form is not free flowing and has a caking tendency on storage. The granulated product has excellent handling and storage characteristics and is free-flowing. The material is packed in jute bags or multiwall paper bags, lined with polyethylene film.

Fused calcium magnesium phosphate

To convert phosphate in the rock phosphate into available form, thermal methods are adopted. Indian Thermophosphates produce the product and markets in the name of Thermophosphates. Besides containing P_2O_5, CaO, MgO and SiO_2 in major quantity, the thermophosphates also contain manganese, boron, vanadium, copper, nickel, strontium, cobalt, chromium, zinc and sodium in minor quantities.

Raw materials required are phosphate rock, serpentine (hydrated magnesium sulphate) or olivine (magnesium iron silicate).

Method of manufacture

Ground rock phosphate and ground serpentine or olivine are fused in an oil-fired furnace and then the material is quenched in cold water. The quenched material, after draining, is dried ground to 70 per cent by passing through 100 mesh and marketed.

Properties

Products	Total percent
P_2O_5	19.7–23.6
CaO	28.3–35.5
MgO	11.5 – 18.5
SiO_2	19.2 – 23.8
F	16.7– 22.2
Oxides of Al and Fe	12.1 – 22.2

Specifications

Moisture per cent by wt. (max)	5.0
Phosphate as (P_2O_5) soluble in per cent citric acid, by wt. (min)	16.5
Fineness:	
Retained on 150 micron IS sieve per cent by wt.	Nil
Passing 100 micron IS sieve per cent by wt. (min)	80

Handling, storage and packing

The product is in the form of a fine powder and does not cake. It is nonacidic and non hygroscopic. It is free flowing and does not pose any problem in handling and storage. The product is packed in jute bags.

POTASSIC FERTILISERS

Potassium is one of the primary nutrients considered essential for plant growth like nitrogen and phosphorus. Both murate of potash (potassium chloride) and sulphate of potash are used as fertilisers, the former being more common. Potassium occurs in nature as insoluble potash bearing silicates and as highly soluble salts like potassium chloride in underground deposits and in sea water. The earlier source of potassium was ashes from the burning of wood and plant wastes.

The principal potash minerals are sylvinite, a mixture of sylvite (KCl) and halite (NaCl) in different proportions, carnallite (KCl, $MgCl_2.6H_2O$), kainite (KCl, $MgSO_4$, $3H_2O$), langbeinite (K_2SO_4, $2MgSO_4$) and to a limited extent as nitrate (KNO_3). India has no exploitable deposits of potassium. Two known potential sources of potash in the country are the bitterns left over after salt recovery and molasses/distillery slop. However, our present requirement is met through import.

Potassium Chloride

Potassium chloride (KCl) called as muriate of potash (MOP) in the fertiliser trade—contains 60 per cent plant food as K_2O. It is the most important potassic fertiliser and is used either directly or in conjunction with nitrogenous and phosphatic fertiliser in mixtures, or in the granulated 'NPK' complex fertiliser. Main raw materials for the manufacture of potassium chloride are potash minerals or brine.

Methods of manufacture

The main potash mineral is sylvinite : a mixture of potassium chloride and sodium chloride. The potassium chloride is recovered either by flotation or by hot leaching.

In the flotation process, a slurry of the ore is treated with aliphatic amine (a collector) which selectively coats KCl particles. Air is then bubble through the slurry and the air bubbles attach themselves to the coated particles and floating them to the surface, while the uncoated particles sink. The flotation KCl is centrifuged, dried and packed. The flotation process is much cheaper than leaching and hence used more extensively in the industry. A flowsheet of the process is indicated in Fig. 9.22.

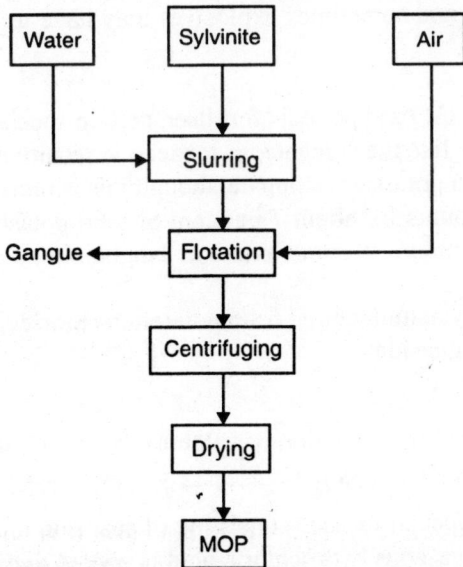

Fig. 9.22. Flowsheet for manufacture of potassium chloride 'MOP' by flotation process.

A certain amount of KCl is recovered from brines from Dead Sea (Israel) and from Great Salt Lake and Seerles Lake in USA. The brine is evaporated in large ponds precipitating NaCl. The brine is transferred to another pond and further evaporation precipitates carnallite, ($KCl, MgCl_2.6H_2O$) together with NaCl. The mixture is collected and the $MgCl_2$ is leached with regulated amount of water to dissolve the $MgCl_2$. Then the KCl is separated from NaCl by hot brine leaching when the KCl is dissolved. On cooling, it gets precipitated, and it is then centrifuged, dried and packed.

Properties

K_2O content, per cent	About 60
Colour	White crystalline
Sp. gr.	1.98
Melting point °C	772
pH	7
Solubility gm/1000 gm	37
Water	
Critical relative humidity at 30° in per cent	84

Specifications

Moisture per cent by wt. (max)	0.5
K_2O content, per cent by wt. (min)	50.0
Sodium (as NaCl), per cent by wt. (on dry basis), max	3.0

Handling, storage and packing

Potassium chloride is imported as bulk cargo and transported to NPK fertiliser and mixing plants in open trucks or in bags. It is stored in bulk in closed storage yard. The crystalline potassium chloride is free-flowing and does not normally pose any problem in handling and storage. However, in presence of impurities caking occurs when humidity is high and the mass tends to become like rock. Retrieval from such storage may pose problems and sometimes explosives may have to be used.

Potassium sulphate

When potassium chloride is the most used potassic fertiliser, certain specialised crops like tobacco need chloride-free potassium fertilisers like the sulphate, as tobacco is sensitive to chlorides. For this reason tobacco is generally fertilised with potassium sulphate, though this is more expensive than the chloride. The sulphate form of potash accounts for about 7 per cent of total potash consumption in India.

Potassium sulphates occur in nature to some extent as langbeinite, a double salt with magnesium, $K_2SO_4.2MgSO_4$.

Potassium sulphate is generally manufactured from potassium chloride, although some production is based on its double salts with magnesium.

Method of manufacture

Potassium sulphate is manufactured by the action of sulphuric acid on potassium chloride:

$$2KCl + H_2SO_4 \longrightarrow K_2SO_4 + 2HCl$$

The reaction is carried out in special furnace consisting of cast iron muffle with rotating ploughs to agitate the reaction mixture. The gaseous hydrochloric acid is cooled and absorbed in water.

Properties

K_2O content, per cent	54
Colour	White crystalline
Sp. gr.	2.662
Melting point, °C	1.067
pH	7
Solubility in water at 30°C gm/100 gm	13
Critical relative humidity at 30°C, per cent	96.3

Specifications

Moisture per cent by wt. (max)	1.5
Potash contents K_2O per cent by wt. (min)	48.0
Total chlorides per cent by wt. (dry basis), max	2.5
Sodium as NaCl, per cent by wt. on dry basis, max	2.0

Handling, storage and packing

It is imported as bulk cargo and transported to NPK fertiliser and mixing plants in open truck or in bags. It is stored in bulk in closed storage yards. The crystalline potassium sulphate is free-flowing and does not normally pose any problem in handling and storage.

Potassium schoenite

Potassium schoenite is the double sulphate of potassium and magnesium (K_2SO_4, $MgSO_4$. $6H_2O$) containing about 27 per cent K_2O. The Central Salt and Marine Chemical Research Institute (CSMCRI), Bhavnagar, has developed floatation technique to recover potassium shcoenite from mixed salt obtained as a deposit from the bitterns of salt works in the Saurashtra area. Based on the CSMCRI process a commercial plant has been put up in the Tuticorin Salt works area in South India. It is estimated that some 75,000 tonnes/yr potassium sulphate can be recovered from the mixed salts obtainable from the salt bitterns in the country. Potassium sulphate can be recovered from the schoenite by treatment with lime or by extraction with organic solvent. The main source of potassium schoenite is salt bitterns.

Methods of manufacture

Mixed salt obtained from bitterns of salt works is agitated with water so as to dissolve magnesium chloride. On adding more of the mother liquor, the coarser particles of sodium chloride settled down. The finely divided suspension of potassium schoenite crystals are decanted.

Properties

K_2O content, per cent	27
Colour	White crystalline

Specifications as given in FCO

Moisture, per cent by wt. (max)	1.5
K_2O (on dry basis), per cent by wt. (min)	23.0
MgO, per cent by wt. (max)	10.0
Total chlorides, per cent by wt. (max)	2.5
NaCl, per cent by wt. (max)	1.5

Handling, storage and packing

Like other potassic fertilisers, this also does not have any special handling and storage problems.

COMPLEX FERTILISERS AND GUIDELINES FOR MIXING FERTILISERS

Ammonium Phosphate Sulphate

This is composed of ammonium sulphate (about 60 per cent) and ammonium phosphate (about 40 per cent) with a nitrogen content of 16 per cent and P_2O_5 content of 20 per cent in the 16–20–0 grade. In the 20–20–0 grade some urea is added to increase the nitrogen to 20 per cent. The 19.5–19.5–0 grade is also produced in India.

Raw materials required to produce ammonium phosphate sulphate are ammonia, phosphoric acid and sulphuric acid.

Methods of manufacture

Two methods are in use—one is by direct neutralisation of a mixture of sulphuric acid and phosphoric acids by ammonia and granulating the resulting slurry in a blunger and the other is to add ammonium sulphate solution from the gypsum ammonia carbonate process to phosphoric acid and then ammoniating the mixture. Urea can be added in the blunger for the 20–20–0 grade. A flowsheet of the process is given in Fig. 9.23.

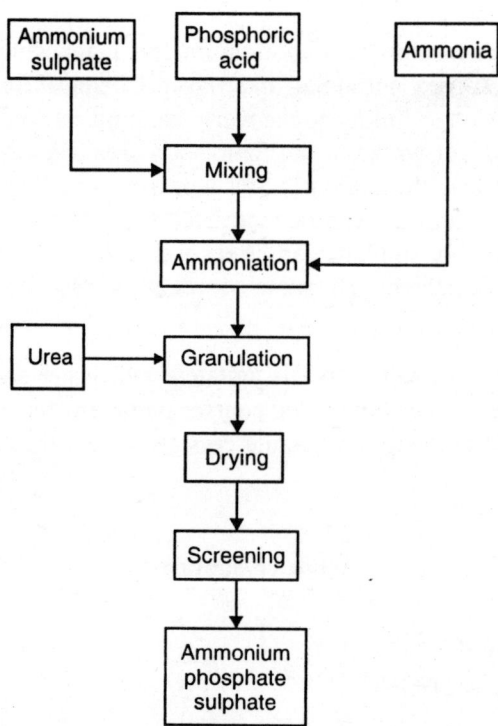

Fig. 9.23. Flowsheet for manufacture of ammonium phosphate sulphate.

Specifications

Typical composition	16–20–0	20–20–0	19.5–19.5–0
Moisture (max)	1	1	1
Ammoniacal nitrogen (min)	16	18	19.5
Urea nitrogen (min)	—	2	—
Total phosphate, min	20	20	19.5
Water soluble, P_2O_5 (min)	19.5	17.0	17.5

Handling, storage and packing

The grades are free-flowing and do not usually pose any handling and storage problems.

Diammonium phosphate

Diammonium phosphate (DAP) 18–46–0, is a fertiliser containing 18 per cent by wt. of ammoniacal nitrogen and 46.0 per cent by wt. of P_2O_5—the latter mostly water-soluble. It is used fairly widely all over India. Imported 'DAP' is used to augment production in the country.

Phosphoric acid of 40–54 per cent P_2O_5 and ammonia are the raw materials.

Method of manufacture

'DAP' is manufactured by reacting two moles of ammonia and one mole of phosphoric acid :

$$2NH_3 \quad + \quad H_2PO_4 \quad \longrightarrow \quad (NH_4)_2HPO_4$$

Ammonia   Phosphoric acid   DAP

The preliminary neutralisation is done in a pre-neutraliser and then the slurry containing a mixture of 'DAP' and 'MAP' is metered into the desired mole ratio of ammonia and phosphoric acid. Any unreacted ammonia gas is then scrubbed with water and weak phosphoric acid are returned to the pre-neutraliser. The granulator discharge is then dried, screened, cooled, conditioned by coating agent, if necessary, and bagged. The process is indicated in the flow diagram in Fig. 9.24.

Properties

Nitrogen per cent	18
P_2O_5 per cent	46
Bulk density kg/cum	993 (62 lbs/cft)
Angle of repose	31–33°C
Sp. gr.	1.619
Critical relative humidity at 30°C in per cent	82.8
pH of 10 per cent solution	7.8

Specification

Moisture, per cent by wt. (max)	1
Total nitrogen, all in ammoniacal form, per cent by wt. (min)	18
Total phosphates (as P_2O_5) per cent by wt. (min)	46
Water-soluble phosphates (as P_2O_5) per cent by wt. (min)	41

Handling, storage and packing

DAP—after cooling normally remains free flowing with little or no bag set. In order to achieve this condition, it is necessary that all process variables be maintained under careful control and the product be dried to a low moisture content.

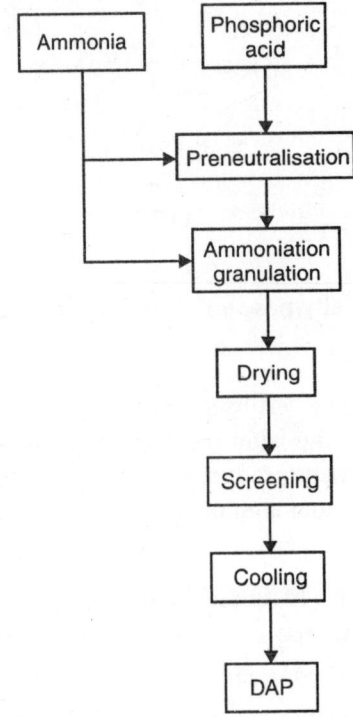

Fig. 9.24. Flowsheet for manufacture of diammonium phosphate.

Nitrophosphates

The term 'nitrophosphate' covers the range of fertilisers containing the nutrients 'N' and 'P' (along with potassium, sometimes) obtained from nitric acid treatment of phosphate rock. Ammonium nitrate is one of the principal components of the product.

The three important raw materials required to produce nitrophosphates containing N, P are nitric acid, phosphate rock and ammonia. Depending on the process adopted, other materials like DAP, phosphoric acid, sulphuric acid and ammonium sulphate will be needed. Potassium is added as necessary ingredient for NPK grades.

Methods of manufacture

The basic principle of nitrophosphate manufacture is the acidulation of ground rock phosphate with nitric acid (53–65 per cent concentration) in a series of reactors. The reaction mass contains calcium nitrate and phosphoric acid. This mixture can be converted into a solid granulated fertiliser by three or four important methods and depending on the method, the water soluble P_2O_5 content and the 'NP' ratio in the final product can be varied significantly:

$$\underset{\text{Rock phosphate}}{Ca_{10}(PO_4)_6F_2} + \underset{\text{Nitric acid}}{20HNO_3} \longrightarrow \underset{\text{Phosphoric acid}}{6H_3PO_4} + \underset{\text{Calcium nitrate}}{10\,Ca(NO_3)_2} + \underset{\text{Hydrogen fluoride}}{2HF}$$

In the carbo nitric process, the mixture of calcium nitrate and phosphoric acid is ammoniated and carbonated in the presence of stabilisers at controlled pH conditions to yield a product containing dicalcium phosphate, ammonium nitrate and calcium carbonate (Fig. 9.25). All the P_2O_5 will be citrate soluble.

$$6H_3PO_4 \;+\; 10Ca(NO_3)_2 \;+\; 20NH_3 \;+\; 4CO_2 \;+\; 4H_2O$$

Phosphoric acid   Calcium nitrate

$$\longrightarrow\; 6CaHPO_4 \;+\; 20NH_4NO_3 \;+\; 4CaCl_2$$

Dicalcium phosphate   Ammonium nitrate   Calcium chloride

1. *Trombay process:* In the second process, as adopted at Trombay, to the acidulated mixture is added some 'DAP' or phosphoric acid and the reaction mass is ammoniated to provide ultimately a slurry of ammonium nitrate, dicalcium phosphate and ammonium phosphate. The water soluble P_2O_5 can be adjusted up to 30 per cent. The slurry is dried in a spherodiser to yield granulated product of 20–20–0 or potassium salt can be added to the slurry before spherodisers to yield products of the type 18–18–9 and 15–15–15 ('suphala'). The process is indicated in the flowsheet as per Fig. 9.26. The following reactions take place:

$$Ca_{10}(PO_4)_6F_2 \;+\; 20HNO_3 \;+\; 7(NH_4)_3HPO_4 \;+\; 10NH_3$$

Rock phosphate   Nitric acid   DAP

$$\longrightarrow\; 20NH_4NO_3 + NH_4H_2PO_4 + 9CaHPO_4 + CaF_2$$

MAP   Dicalcium phosphate

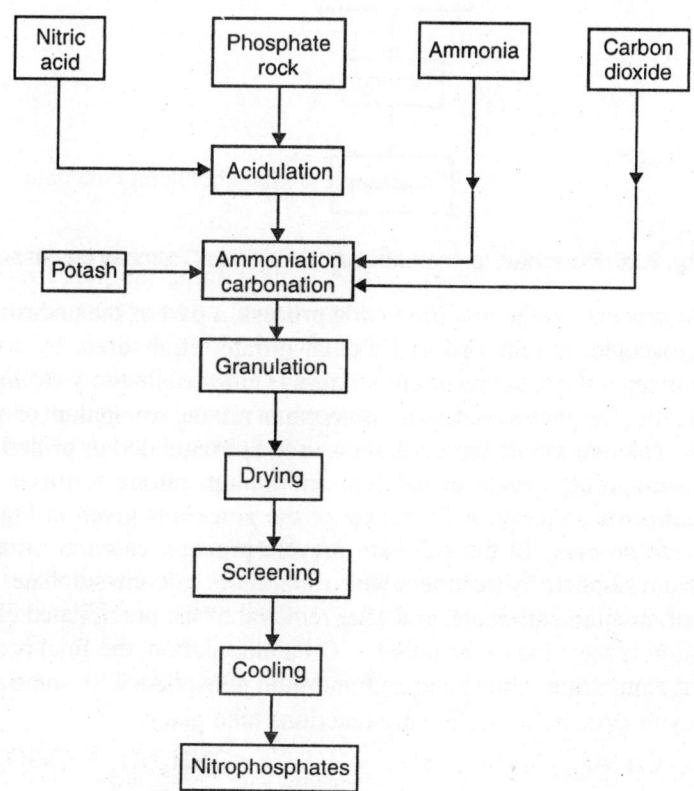

Fig. 9.25. Flowsheet for manufacture of nitrophosphates by carbo-nitric (PEC) process.

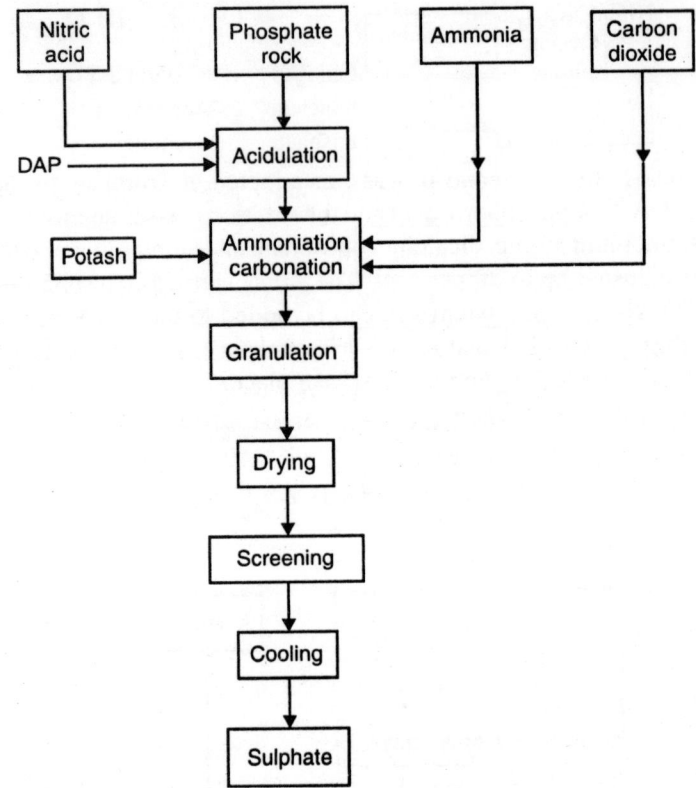

Fig. 9.26. Flowsheet for manufacture of sulphala (Trombay process).

2. *Modified odda process:* In the modified odda process, a part of the undesirable calcium nitrate, which is Hygroscopic, is removed as Calcium nitrate tetrahydrate by cooling the acidulated mass. After this separation, the resulting mixture is ammoniated to yield high water soluble (up to 85 per cent) nitrophosphates containing ammonium nitrate, ammonium phosphate and dicalcium phosphate. The calcium nitrate tetrahydrate is used as granulated or prilled fertiliser, as such, or converted to ammonium nitrate or calcium ammonium nitrate fertiliser after treatment with ammonium carbonate solution. A flowsheet of the process is given in Fig. 9.27.

3. *Sulphate recycle process:* In the sulphate recycle process, calcium nitrate is converted into insoluble calcium sulphate by treatment with ammonium calcium sulphate. It is then filtered out, reacted with ammonium carbonate, and after removal of the precipitated carbonate, ammonium sulphate solution is recycled to the process. On ammoniation, the final product consists mostly of a mixture of ammonium nitrate and ammmonium phosphate with some dicalcium phosphate.

In the sulphate recycle process the following reactions take place.

$$Ca(NO_3)_2 + (NH_4)_2SO_4 \longrightarrow 2NH_4NO_3 + CaSO_4$$

$$CaSO_4 + (NH_4)_2CO_3 \longrightarrow (NH_4)_2SO_4 + CaCO_3$$

Water soluble P_2O_5 content in the nitrophosphates can be up to 90 per cent.

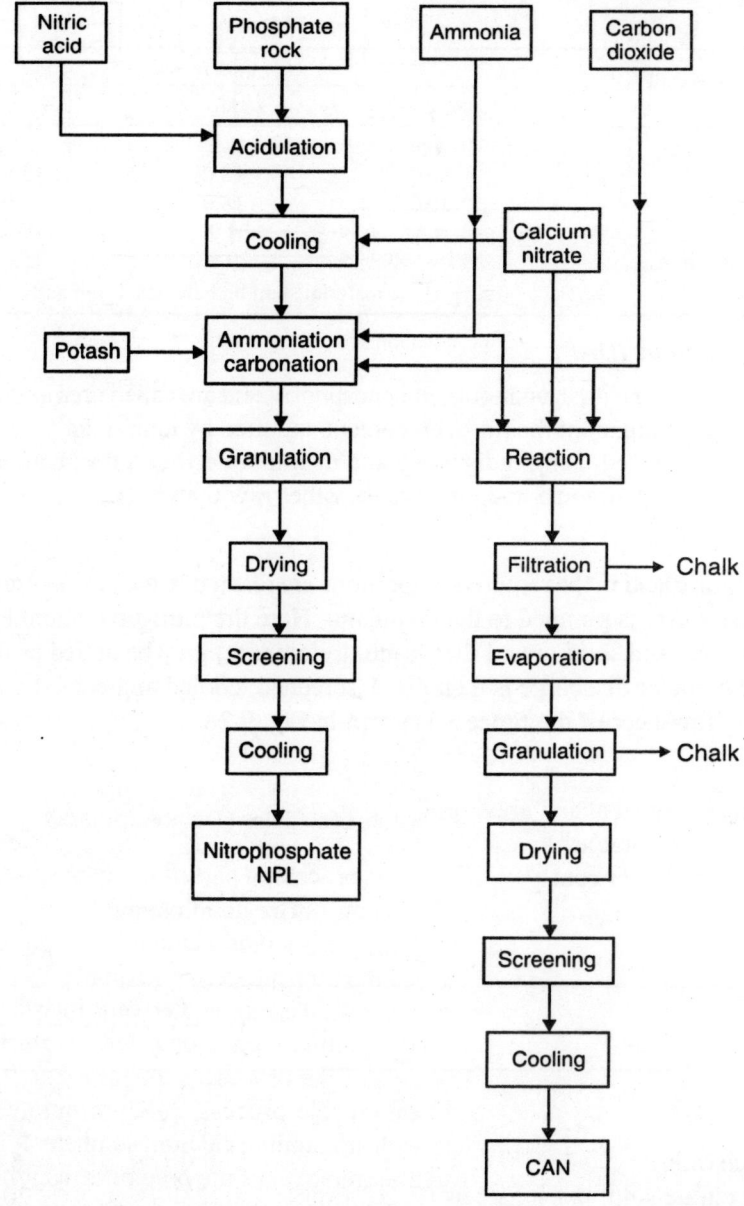

Fig. 9.27. Flowsheet for manufacture of NPK and CAN by modified ODDA process.

Properties

Nitrogen content	Partly ammoniacal and partly nitrate
Phosphate content	Partly water soluble and partly citrate soluble
Bulk density kg/cum	1100
Critical relative humidity at 80°C per cent for	20–20–0 51
	15–15–15 45–50
	18–18–9 50

Specifications

	Per cent by wt.		
	20–20–0	*18–18–9*	*15–15–15*
Moisture (max)	1.5	1.5	1.5
Total nitrogen	20.0	18.0	15.0
Citrate-soluble P_2O_5 (min)	20.0	18.0	15.0
Water-soluble P_2O_5 (min)	5.4	4.9	4.0
Water-soluble potash K_2O (min)	—	9.0	15.0
Particle size	90% of the material shall be between 4 mm and + 1 mm IS sieve		

Urea ammonia phosphates (UAP)

Fertiliser manufactured by using urea, ammonia and phosphoric acid are called urea ammonium phosphates (UAP). Solid urea helps to augment the nitrogen content supplied by ammonia.

Phosphoric acid is produced using indigenous and/or imported rock phosphate and sulphuric acid based on imported sulphur. Ammonia and urea are the other raw materials.

Methods of manufacture

Ammonia and phosphoric acid in the required proportions are reacted in the pre-neutraliser. The resulting ammonium phosphate slurry is pumped to the granulator. Here the nitrogen content is further increased by adding more ammonia and solid urea. Filler (sand or dolomite) may be added as desired, depending upon the grade. The granular discharge is then dried, screened, cooled and coated with a coating agent to prevent caking. A flowsheet of the process is given in Fig. 9.28.

Properties

All the nitrogen is in ammoniacal and urea form	—
Most of the P_2O_5 is water-soluble	—
Critical humidity at 30°C in per cent	57
Colour	Grey and cream

Specifications

	Per cent by wt.	
	28–28–0	*20–20–0*
Moisture	1.0	1.0
Total nitrogen (min)	28.0	28.0
Ammoniacal nitrogen (min)	9.0	6.4
Neutral ammonium citrate soln. phosphate as (P_2O_5) (min)	28.0	20.0
Water-soluble P_2O_5 (min)	25.2	18.0
Particle size for 28–28–0	All materials shall pass through 2.35 mm IS sieve and be retained on 1.18 mm IS sieve. The particle size of the material is such that 90 per cent of the material shall pass through / 3.35 mm IS siever and be retained on 1.18 mm IS sieve.	

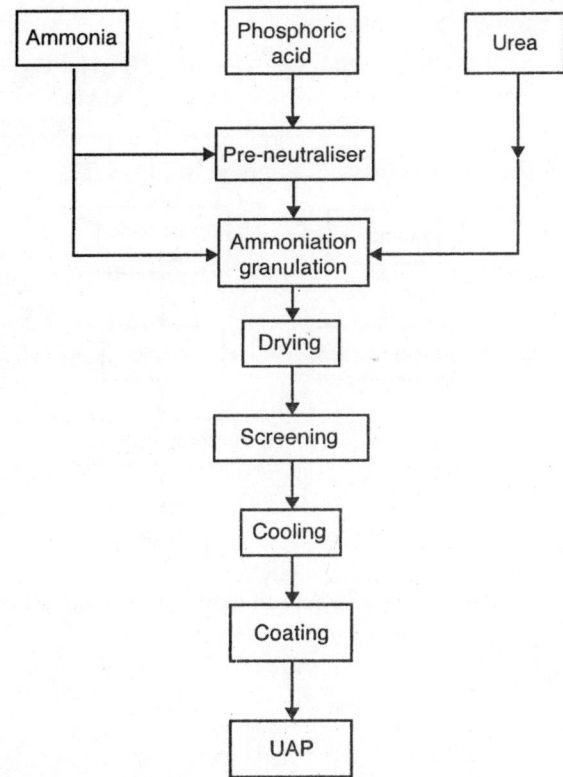

Fig. 9.28. Flowsheet for manufacture of urea ammonium phosphate (UAP).

Handling, storage and packing

Granular urea ammonium phosphates are free-flowing and do not normally pose any problem during handling and storage. Coating agents are used to make them resistant to caking. However, exposure to long periods in regions of high humidity can cause caking. They are bagged in polyethylene-lined gunny bags or coated with bitumen to prevent any possible seepage of moisture from the atmosphere which tends to promote caking.

Monoammonium phosphate (MAP)

Monoammonium phosphate (MAP) is a rich fertiliser intermediate with high P_2O_5 content of about 55 per cent and nitrogen content of 11–12 per cent. It is produced in a powdered form, as it is primarily meant as an intermediate to produce 'NP' and 'NPK' grade mixtures and granulated fertilisers. It is, however, not manufactured in India.

MAP require phosphoric acid and ammonia as raw material.

Methods of manufacture

Vapourised ammonia is injected into a reactor containing wet process phosphoric acid of 45–52 per cent conc. The NH_3; H_3PO_4 ratio is maintained at 1 : 1. The reaction is complete with release of considerable quantity of steam.

The chemical reaction can be represented as

$$H_3PO_4 + NH_3 \longrightarrow NH_4H_2PO_4$$
$$\text{MAP}$$

The concentrated MAP solution flows to a spray drier. The powdered material is collected at the bottom and packed. A flowsheet of the process is given in Fig. 9.29.

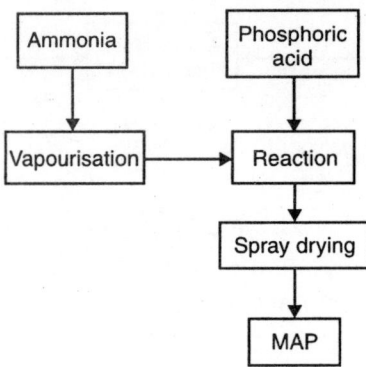

Fig. 9.29. Flowsheet for manufacture of mono ammonium phosphate (MAP).

Properties

Physical state	White crystals
Sp. gr.	1.8
pH of 10 per cent soln.	3.5–4.0
Critical relative humidity at 30°C per cent	91.6
Bulk density, kg. per cum (powder with 0.1–1.5 mm range)	848.98 (53 lbs/ft^3)

Handling, storage and packing

MAP is a free-flowing and non-hygroscopic material. The powder stores and handles well. It can be transported in bulk.

NPK Complex Fertiliser

NPK complex fertilisers are solid fertilisers in the form of uniform granules commonly referred to by a sequence of those numbers, the first of which represents the per cent nitrogen expressed as N, the second the per cent phosphorus expressed as available P$_2$O$_5$ and the third the per cent of potassium as soluble K$_2$O. These fertilisers are very convenient to use, because they contain all the three primary plant nutrients in the desired proportions.

The required raw materials are phosphoric acid which is manufactured from phosphate rock and sulphur. Ammonia, imported potash and urea are required—where necessary—to increase the nitgoren content. Locally available filler (sand, dolomite etc.) and coating agents (clay and soapstones etc.) are also requried for certain grades.

Methods of manufacture

Ammonia and phosphoric acid in the required proportions are metered to the pre-neutraliser and the resultant slurry is pumped to the granulator which can be a blunger or a rotating drum. Here the nitrogen content is increased by adding more ammonia and feeding in urea, whenever necessary. Filler (sand or dolomite) and potash are also added here to make up the required product formulation. The granulator discharge is then dried, screened, cooled and coated with a coating agent (clay or powdered soapstone) to improve the storage properties.

Properties

Colour—uncoated condition	Grey
Colour—coated	White or cream
Bulk density, kg/cum	865
Angle of repose	30–32°
Critical relative humidity at 27°C in per cent	60 to 50
Sp. gravity	1.54

Handling, storage and packing

Granular NPK complex fertilisers are free-flowing and do not normally pose any problem during handling and storage. Coating agents are used to make them resistant to caking. However, exposure to long periods in regions of high humidity can cause caking. 'NPK' complex fertilisers are, therefore, bagged in polyethylene-lined gunny bags coated with bitumen, to prevent any possible seepage of moisture from the atmosphere which tends to promote caking. Table 9.4 shows the specification of different NPK fertilisers.

Table 9.4. Specification of different NPK fertilisers.

	17–17–17	14–28–14	11–22–22	14–35–14	22–22–11	10–26–26	12–32–16	14–36–12
Moisture (max)	1.0	1.0	1.0	1.0	1.5	1.0	1.0	1.0
Total nitrogen (min)	17.01	14.0	11.0	14.0	22.0	10.0	12.0	14.0
Ammoniacal N_2 (min)	5.0	8.0	6.5	14.0	7.0	10.0	12.0	14.0
Nitrogen in the form of urea (max)	12.0	6.0	4.5	—	15	—	—	—
Neutral ammonium citrate soluble phosphate as (P_2O_5) min	17.0	28.0	22.0	35.0	22.0	26.0	32.0	36.0
Water soluble phosphate as (P_2O_5) min	15.3	25.2	19.8	29.0	19.6	22.1	27.2	30.6
Water soluble potash (K_2O) min particle size	17.0	14.0	22.0	14.0	11.0	20.0	16.0	12.0
	Shall be such that 90 per cent of the material will be between 4 mm and 0.85 mm IS sieve.				Shall be such that 90 per cent of the material will pass through 4 mm IS sieve and shall be retained on 1 mm IS sieve.			

Guidelines for mixing fertilisers

The term mixed fertilisers or 'mixtures' is used to denote multinutrient fertilisers containing two or three primary nutrients. Dry mixtures are made by mechanical mixing of fertilisers material to produce the grades of products required. In dry mixing there is no significant chemical reaction. The product may be in pulverised form or in granulation form.

Mixed fertilisers have certain advantages over single nutrient fertilisers, especially where farmers are not well-informed. On the basis of soil test and crops to be produced it is necessary to prescribe definite grades of mixtures to meet local needs.

Of the two forms, it is simpler to make pulverised mixture as it requires very little equipment. The ingredients are weighed and charged in a rotary mixer which can be a batch or a continuous type and the product is straightway bagged. In this form, however, the mixer suffers from certain disadvantages.

It is difficult to apply pulversied mixers evenly to the soil due to segregation of ingredients which may have different bulk densities and particle size. Since the powder does not flow properly, application through a seed drill is not uniform. During the application, there are more losses due to dusting. It is far easier to adulterate the pulversied mixtures. Because of the above limitations, physical mixtures are generally converted into granules by moistening during the granulation period.

The moistening may be done by injecting a small quantity of moisture or steam. A scheme showing the production of granulated fertiliser is shown in Fig. 9.30.

The different fertiliser materials are stored in process bins from where proportionate quantities are fed to a weighing hopper. These are then ground, mixed in a rotary drum mixer and marketed either as a powdered product or sent to a rotary drum granulator where a small quantity of moisture is added and the material granulated. After granulation, the product is dried and screened. The oversize is crushed and returned to the granulator and the undersize is recycled. The product is generally stored and marketed as bagged.

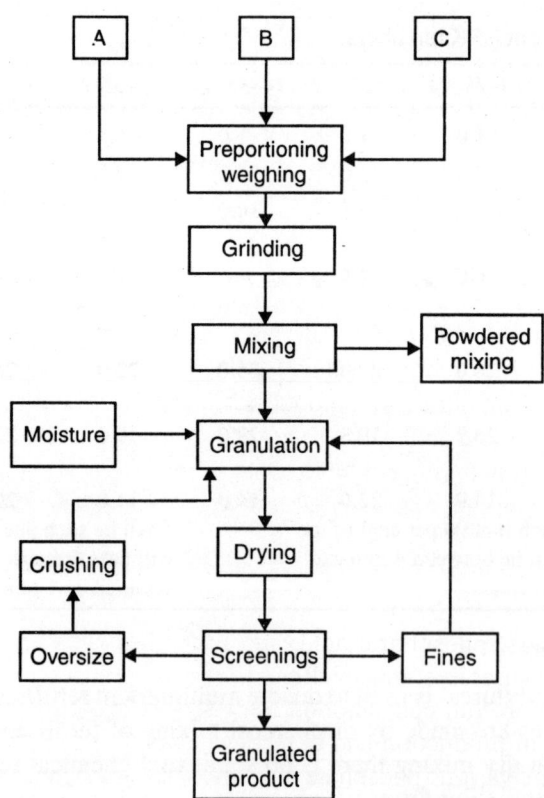

Fig. 9.30. Flowsheet for granulation of mixed fertilisers when A, B and C are three components.

There are a number of such mixing and granulation plants all over India. Principal raw materials are solid fertilisers like single or triple superphosphates, DAP, urea, muriate of potash etc. The grades produced are usually of medium nutrient content.

Fertiliser mixtures, whether in powder or granulated form, serve the same purpose as complex fertilisers in that either of them contains two or more primary plant nutrients. Fertiliser mixtures have a few additional advantages:

1. The mixture contains plant foods in proportion and tailor-made to suit local conditions and ready for application. This saves the farmer the bother of buying, storing, handling and applying individual fertilisers.

2. Micronutrients, which are required in small quantities by plants under well-defined conditions, can be incorporated in mixtures as necessary.

3. Mixtures ensure balanced fertilisation leading to increasing yields of crops of higher profits.

4. Unlike complex fertilisers, which are normally produced in very large scale plants, the fertiliser mixtures can be produced on small or medium scale, as such there is considerable flexibility in producing a variety of NP, NK, PK and NPK grades.

In the preparation of mixtures, the two important points to be borne in mind are the physical and chemical properties of ingredients and their compatibility with each other. Different fertiliser materials are available in different forms, e.g., powder, crystals, granules and prills. Because of their different particle size and bulk densities, mixing of such ingredients may result in segregation. Further, in some cases, mixing some materials may result in increased hygroscopicity while, in others, there may be even loss of nutrient. Hence a judicious selection of the ingredients becomes necessary.

Nutrient content of fertilisers

The primary nutrients for plant growth which fertilisers supply are nitrogen, phosphorus and potassium. Their concentration in a fertiliser is expressed as percentage of N, P_2O_5 and K_2O. In the case of nitrogenous fertilisers, the nitrogen may be in the ammoniacal, nitrate or amide form. Ammoniacal form of nitrogen is contained in fertilisers like ammonium sulphate, ammonium chloride etc. Nitrate nitrogen is contained in fertilisers like ammonium nitrate, ammonium sulphate nitrate, CAN etc. The amide nitrogen is in urea.

The phosphate fertilisers may be in the water-soluble form or available form. When phosphate fertilisers are soluble in water, the product is called water-soluble phosphates. If it is not soluble in water but 2 per cent neutral ammonium citrate, the product is called citrate soluble phosphate. The sum total of water-soluble and citrate-soluble values is termed as available phosphates.

Fertiliser in which the phosphate is not soluble either in water or 2 per cent neutral ammonium citrate solution is termed insoluble. The sum of the available phosphate and the insoluble phosphate is termed as total phosphate.

PHOSPHORUS

The use of artificial fertilisers, phosphoric acid, and phosphate salts and derivatives has increased greatly, chiefly because of aggressive and intelligent consumption promotion on the part of various manufacturers. However, before full consumption of these products could be achieved, more efficient and less expensive methods of production had to be developed.

During recent decades, the various phosphate industries have made rapid strides in cutting the costs both of production and distribution and have thus enabled phosphorus, phosphoric acid, and its salts to be employed in wider fields and newer derivatives to be introduced. Supplementing the development of more efficient phosphorus industries have been the epoch-making pure chemical studies of phosphorus, in its old and in its new compounds. These phosphates are not simple inorganic chemicals, as was assumed several decades ago, and their study has become a unique and complicated branch of chemistry that may some day be compared with the carbon (organic) or silicon branches of today. The properties of phosphorus chemicals are unique because of the important role of phosphorus in many biochemical processes, the ability of polyphosphates to complex or sequester many metal cations, and versatility in forming various types of organic and inorganic polymers.

Table 9.5 is a compilation of phosphate-rock treatment processes. Tricalcium phosphate in raw and/ or steamed and degreased bones and in basic slag is also used after grinding as a direct phosphate fertiliser. A small percentage of the former chemicals. Large tonnages of phosphate rock are converted to phosphorus or phosphoric acid and their derivatives.

Phosphorus, the twelfth most abundant element, is fairly widely distributed in igneous and sedimentary rocks. The only important source is apatite, $Ca_5X(PO_4)_3$ (X = F, OH, or 0.5 carbonate). The co-ordination number of phosphorus (at. wt 30.98) is 3 or 4, but 1, 2, 5, and 6 are also known. There is one stable nuclide. ^{31}P, and four radioactive isotopes.

Table 9.5. Phosphate-rock processing, products, and by-products.

Process	Raw materials and reagents	Main products and derivatives	By-products
Acidulation	Phosphate rock, sulphuric acid, nitric acid, phosphoric acid, hydrochloric acid, ammonia, potassium chloride	Superphosphate, phosphoric acid (wet process), triple super-phosphate, monoammonium phosphate, diammonium phosphate, monopotassium phosphate	Fluorine compounds, vanadium, uranium (limited)
Electricfurnace reduction	Phosphate rock, siliceous flux, coke (for reduction), electrical energy, condensing water	Phosphorus, phosphoric acid, triple superphosphate, various Na, K, NH_4, Ca salts; phosphorus pentoxide and halides	Fluorine compounds, carbon monoxide, slag (for RR ballast aggregate, fillers, etc.), ferropho-phorus, vanadium
Calcium metaphosphate	Phosphate rock, phosphorus, air or oxygen, fuel	Calcium metaphosphate	Fluorine compounds
Calcination or defluorination	Phosphate rock silica, water or steam, fuel	Defluorinated phosphate	Fluorine compounds

Phosphorus and Phosphides

Properties

Phosphorus is a colourless or white waxy solid, d 1.28 g/cm^3, which melts at 44.1°C to a clear colourless liquid (bp 280.5°C). Commercial white phosphorus (99.9%) is slightly yellowish. Red phosphorus forms from molten white phosphorus at ca 580°C. Amorphous or crystalline black phosphorus (d 2.70 g/cm^3) is made from the white form under high pressure (> 980 MPa or 9670 atm) and at high temperature. The black modifications resemble graphite and are good conductors of electricity.

White phosphorus is more reactive than the red modification, which is more reactive than the black. White or liquid phosphorus ignites spontaneously in air and is usually protected from oxidation by a layer of water. Commercial red phosphorus is more stable and reacts only slowly with oxygen and water vapour. Spontaneous combustion, catalysed by traces of iron, sometimes takes place in storage piles. Oxygen at sub-atmospheric pressures or moist air causes white phosphorus to emit a greenish glow.

Halogens, sulphur, and oxidising acids give halides, sulphides, and oxo-acids, respectively, of phosphorus. Phosphine, PH_3, is formed when phosphorus is heated with aqueous solutions of strong alkalies.

Manufacture

Elemental phosphorus is manufactured either in the electric or the blast furnace; both depend on silica as a flux for the calcium present in the phosphate rock. (Fig. 9.31).

$$4Ca_5F(PO_4)_3 + 18SiO_2 + 30C \longrightarrow 18CaO.SiO_2. \frac{1}{9} CaF_2 + 30CO \uparrow + 3P_4 \uparrow$$

$\quad$ Fluoroapatite $\qquad$ Silica $\quad$ Coke $\qquad\qquad$ Slag

Modern furnaces are three-phase units with potentials of 200-300 V. The outer shell is of welded steel sheets and is often cooled by a water spray. The floor is of monolithic carbon which extends a few metres up the side and above the surface of the molten slag. The upper walls are of firebrick or cast refractory cement and the roof is a cast monolithic structure. Operating data are given in Table 9.6; capacities of 50 MW are common.

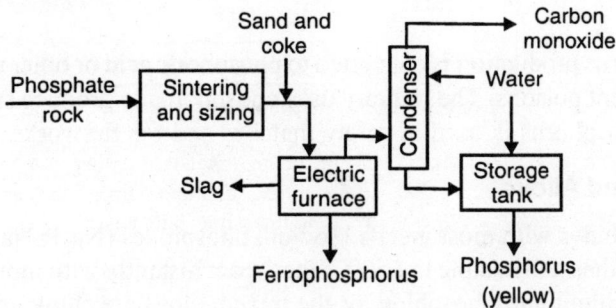

Fig. 9.31. Flow diagram for phosphorus from phosphate rock (electric furnace).

Most furnaces have two tap holes, a lower one for ferrophosphorus, and an upper one for slag. From furnaces with only one tap hole, the molten stream runs into a catch basin from which the lighter slag overflows. The practice of selling granulated slag as a liming agent has been discontinued because of emission, albeit low, of radioactivity. Hot gases are emitted at ca 100 m^3/min; ca 93% is carbon monoxide and the remainder is primarily phosphorus with some silicon tetrafluoride and dust. The phosphorus is condensed in towers equipped with water sprays of 45–55°C temperatures.

The carbon monoxide is usually burned as fuel. Red phosphorus is manufactured by a batch process, although continuous methods have been developed. White phosphorus is heated gradually in a steel or cast-iron vessel equipped with a condenser. The temperature is kept at 400°C for several hours.

After cooling, the product is wet-ground and boiled with sodium carbonate solution to remove remnants of the white modification. The red phosphorus is sieved, washed, and vacuum dried. It can be stabilised by suspension in a 1 wt% solution of sodium aluminate, the slurry being aerated for several hours. Phosphorus is poisonous and can cause bone necrosis and fatty degeneration of the viscera. It must be handled with special safety measures. The vapour is highly poisonous, and the established exposure limit is 0.1 mg/m³ air. Phosphine is much more lethal than elemental phosphorus, but it does not exhibit long-term effects.

Table 9.6. Operating characteristics for a 15-MW phosphorus furnace.

Average potential between electrodes, V	300.00
Power factor	0.97
Raw materials consumed per kg of elemental phosphorus produced	
Power, kWh	14.30
Phosphatic material, kg	10.00
Silica material, kg	1.50
Coke material, kg	1.50
Products formed per kg elemental phosphorus produced	
Slag, kg	4.00
Ferrophosphorus, kg	0.30
Carbon monoxide, kg	2.80
Recovery, based on the element, of the charged phosphorus, %	87.00
Temperature of off-gases, °C	370.00
Temperature of slag at tapping, °C	1500.00

Uses

Nearly all of the phosphorus production is converted to phosphoric acid or other phosphorus compounds. White phosphorus is rodent poisons. The military uses phosphorus to produce smoke clouds or to ignite gasoline bombs. Red phosphorus is used in safety matches and for fireworks.

Metallic Phosphides and Alloys

Phosphorus forms phosphides with most metals. Sodium phosphides (Na_2P, Na_3P, Na_3P_{11}) are reddish-brown to black materials thermally stable to 650°C; they react instantly with moisture to give phosphine. Magnesium phosphide, aluminium phosphide, or the mixed alloys are shiny grey-to-yellow crystalline materials which are stable in dry air but which decompose upon contact with water. They are used with an igniting agent in sea flares. Addition of ferrophosphorus to high strength, low alloy steel eliminates quenching or tempering. Phosphor copper is used for the deoxidation of copper and its alloys, and phosphor tin for deoxidation of bronzes and German silver. Phosphor bronzes are copper and tin alloys containing 1.25–11 wt% tin; 0.03–0.35 wt% phosphorus deoxidises the alloy. Zinc phosphide, Zn_3P_2, is used as a rodenticide. Silver solder contains small amounts of silver phosphide.

Phosphorus Compounds

Phosphorus exhibits oxidation states from −3 to +5. The largest group of phosphorus compounds contains oxygen, and the oxo acids form the basis for the most systematic nomenclature. However, the term phosphorous acid is often used to describe phosphonic acid, $H(H_2PO_3)$, which is commonly written as H_3PO_3 or $(HO)_3P$. The least ambiguous method for interpreting the nomenclature of phosphorus compounds is by referring to the structural formula. Many phosphorus compounds are named as salts with phosphorus being the metallic or electropositive element, i.e., phosphorus trichloride, PCl_3; phosphorus oxybromide (phosphoryl bromide), $POBr_3$; phosphorus triamide, $P(NH_2)_3$; and tetraphosphorus decasulphide (phosphorus(V) sulphide), P_4S_{10}. Some compounds are named as derivatives of the simple phosphorus hydrides, phosphine, PH_3, diphosphine, P_2H_4, and their oxides.

An unshared electron pair, characteristic of many trivalent phosphorus compounds, can be described in terms of the general formula

$$\begin{matrix} R \\ R' \\ R'' \end{matrix} \diagdown \!\!\!\!\! = P$$

where R, R', and R'' can be hydrogen or alkyl, aryl, alkoxy, amino, halo, mercapto, or other groups.

The strength of a complex formed between a phosphorus donor and an electron acceptor depends on the nature of the acceptor. Strong acids provide stronger complexes than weak acids. Of phosphonium halides, only the iodide has a low dissociation pressure, which imparts stability to the compound under ambient conditions. The unshared electron pair on phosphorus reacts with oxidising agents, e.g., hydrogen peroxide, sulphur, or halogens:

$$(RO)_3P: + H_2O_2 \rightarrow H_2O + (RO)_3P{=}O$$
$$R_3P: + S \rightarrow R_3P{=}S$$
$$:PCl_3 + Cl_2 \rightarrow PCl_5$$

Another reaction of the unshared electron pair is quaternisation with alkyl halides :

$$R_3P: + R'X \rightarrow [R_3PR']^+ \, X^-$$

The tetrahedral structure is the most common co-ordination pattern for phosphorus. Frequently, there are distortions in a pure tetrahedral environment that result from steric and electronic effects. Pure tetrahedral co-ordination probably occurs only in species where there are four identical groups and no packing distortions.

Phosphorus sulphides

Phosphorus and sulphur form the binary compounds tetraphosphorus trisulphide (phosphorus sesquisulphide), P_4S_3, mp 171–172.5°C, the most stable; tetraphosphorus pentasulphide, P_4S_5; tetraphosphorus heptasulphide, P_4S_7; and phosphorus(V) sulphide (tetraphosphorus decasulphide), P_4S_{10}. A stable oxysulphide, $P_4O_6S_4$, exists as a colourless, deliquescent crystalline solid, mp 102°C (Table 9.7).

Table 9.7. Physical properties of the phosphorus sulphides.

Property	P_4S_3	P_4S_5	P_4S_7	P_4S_{10}
Mp, °C	171–172.5	170–220	305–310	286–290
Bp, °C	407–408		523	513–515
Sp gr at 17° C	2.03	2.17	2.19	2.09
Colour solid	yellow	sulphur yellow	almost white	yellow
liquid	brownish yellow		light yellow	reddish brown
Solubility at 17°C, g/100 g CS_2	100	ca 10	0.029	0.222
Action with cold water	scarcely attacked		fairly readily decomposed	slowly decomposed
Stability of solid on standing in air	slowly oxidised		decomposes	slowly decomposed

o,o'-Dialkyl or diaryl thiophosphoric acids, e.g., $(C_6H_5O)_2$ P(O)SH, are obtained readily by alcoholysis of phosphorus(V) sulphide. They provide the basis for high pressure lubricants, oil additives, and flotation agents. The insecticides parathion and methylparathion are made from dialkyl thiophosphates.

The phosphorus sulphides are manufactured by direct union of the elements. The pentasulphide, P_4S_{10}, is purified by distillation. Exposed to moisture, it hydrolyses, giving off H_2S; ventilation in storage avoids excessive H_2S build-up and fire and explosion hazards.

Phosphorus halides

Phosphorus forms well-defined halogen compounds of the type PX_3, PX_5, POX_3, and PSX_3. All, except the pentaiodide and the oxy- and sulphoiodides are known. Physical properties of some halides are given in Table 9.8.

The trihalides, obtained by direct halogenation under controlled conditions, are Lewis bases. Examination by electron diffraction has confirmed pyramidal structures for the gaseous molecules.

Phosphorus trichloride is a clear volatile liquid with a pungent, irritating odour. Depending on the mole ratio of $H_2O : PCl_3$, hydrolysis produces phosphonic acid, a mixture of phosphonic and pyrophosphonic acids, or an unstable polymer of indefinite composition. Although it is nearly insoluble in water, PCl_3 hydrolyses rapidly, and a polymer forms at low $H_2O : HCl$ ratios.

Table 9.8. Physical properties of some phosphorus halides.

Compound	Mp, °C	Bp, °C	Specific gravity	Heat of vaporisation kj/mol[a]	Critical temperature °C	Colour[b]	State at STP
PCl_3	−111.8 (−91)	75.5	1.57	30.5	285.5		liquid
PCl_5	167 (under pressure[c])	sublimes	1.6	64.9	372	white to pale yellow	solid
$POCl_3$	−1.2	106.5	1.68	34.7	331.8	liquid	
PBr_3	−40	175.3	2.89	38.9		liquid	
PBr_5	< 100 (dec)	dec				yellow	solid
PI_3	61	dec 120[d]				dark red	solid
PF_3	−151.5	−101.1	3.91	14.6	−2.05		gas
PF_5	93.7	−84.5	5.84				gas
POF_3	−39.4	−39.8	4.65	21.1	73.3		gas

[a] To convert J to cal, divide by 4.184.

[b] Colourless unless otherwise noted.

[c] Sublimes at 101.3 kPa (1 atm) at 159°C; $\log P_{kPa} = 10.159 - 16,100 / 4.57$ K; $(\log P_{mmHg} = 11.034 - 16,100/4.57$ K).

[d] At 2.0 kPa (15 mm Hg).

It also forms upon storage in contact with moist air. Phosphorus trichloride is made by combining the chlorine and phosphorus in the presence of a pre charge of phosphorus trichloride which is refluxed continuously (Fig. 9.32).

Phosphorus oxychloride (phosphoryl chloride), $POCl_3$, is a colourless fuming liquid with a pungent disagreeable odour. It is made by heating PCl_5 with oxalic or boric acid. It is stable above 300°C, yields phosphoric acid upon hydrolysis, and is used extensively for the manufacture of alkyl and aryl orthophosphate esters.

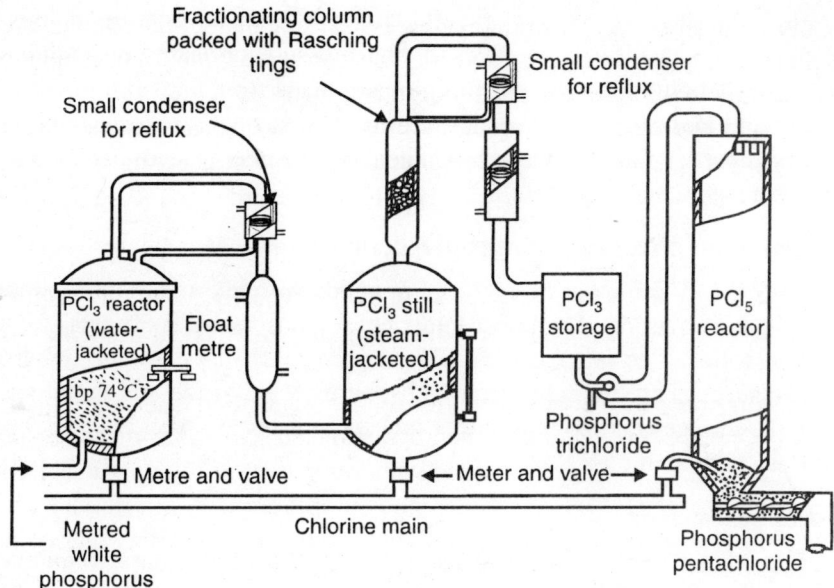

Fig. 9.32. Plant for the continuous production of phosphorus trichloride and phosphorus pentachloride.

Phosphorus pentachloride is a pale greenish solid with a pungent odour. It is made from PCl_3 and chlorine or by burning phosphorus in excess chlorine; reaction with water is violent. It serves as chlorinating agent and catalyst in organic synthesis. Phosphorus halides are shipped in glass containers and must be handled with care. They are all irritating to skin, eyes, and mucous membranes; contaminated clothing must be removed immediately.

Phosphorus oxides

Phosphorus forms five well-defined oxides: phosphorus(III) oxide, P_4O_6; phosphorus(V) oxide (phosphorus pentoxide), P_4O_{10}; phosphorus tetroxide, P_2O_4; phosphorus heptoxide, P_4O_7; and phosphorus nonaoxide, P_4O_9. They are obtained by direct oxidation of phosphorus.

The pentoxide is a very stable, white solid which exists in several crystalline, liquid and amorphous modifications. It is produced commercially by burning phosphorus in a stream of dry air. The usual plant consists of a phosphorus feed system, provisions for drying the air, a burning chamber, and the barn, which is a large room, cooled externally by air or water, where the gas produced is condensed.

Phosphorus pentoxide is treated as a flammable solid because of the explosive violence with which it reacts with water. It is sold in glass bottles which must be cushioned with incombustible packing material. Phosphorus pentoxide is corrosive and irritates eyes, skin, and mucous membranes ($1 mg/m^3$ is suggested as TLV in air). It is an important drying and dehydrating agent and as such is used in the manufacture of methyl methacrylate resins.

Phosphorus acids

Phosphinic (hypophosphorus) acid, $H(H_2PO_2)$ or H_3PO_2, a deliquescent crystalline solid (mp 26.5°C, dec > 133°C), is manufactured by treating white phosphorus with a boiling slurry of lime. Calcium hypophosphite remains in solution, whereas insoluble calcium phosphite precipitates. The reducing capacity of hypophosphorus acid and its salts is utilised in electroless-plating processes, in which nickel salts are chemically reduced to form a smooth adherent surface plate. Phosphonic (phosphorus) acid, $H_2(HPO_3)$ or H_3PO_3, is a white deliquescent crystalline compound, mp 73.6°C, manufactured by the hydrolysis of phosphorus trichloride; the reaction can be violent.

Phosphonic acid and hydrogen phosphonates are used as strong but slow-acting reducing agents, e.g., to precipitate heavy metals from solutions. Aminoalkylphosphonic acids are used in the production of herbicides and water-treatment systems.

Phosphazenes and other phosphorus-nitrogen compounds

The phosphazenes $(NPX_2)_n$, are linear or cyclic compounds with alternating phosphorus and nitrogen atoms; they are toxic and irritating. The lower linear chlorophosphazenes are oils, the cyclic members are crystalline white solids. The latter can be polymerised to chains of highly controlled molecular weight. Chlorophosphazenes are prepared from phosphorus(V) chloride and ammonium chloride. Of the binary phosphorus-nitrogen compounds, only triphosphorus pentanitride has been obtained in a pure state by treating P_2S_5 with ammonia.

Phosphine and its derivatives

Phosphine, PH_3, is produced in a number of ways; however, the inadvertent evolution of phosphine in an otherwise safe reaction is an element of hazard in many procedures involving phosphorus chemicals. Phosphine is conveniently produced by the hydrolysis of an active metal phosphide, e.g., Ca_3P_2 or AlP, or from the acid- or base-catalysed reaction of elemental phosphorus with water. The addition of P—H bonds in phosphines across on olefinic double bond is an economical method to produce alkyl phosphines. Aryl groups are conveniently introduced by Friedel-Crafts reactions of phosphorus halides. Aliphatic phosphines may be gases, volatile liquids, or oils. Aromatic phosphines are frequently crystalline, although many are oils. Because of the bonding ability of phosphines to acceptors, e.g., metals, they are widely used as ligands in catalysts.

A phosphine can block a specific site on a central metal under all conditions, or a site to some species but not to others, or simply make a metal more soluble and thus dispersible in a liquid medium. Phosphine, a central nervous system and liver toxin, is highly toxic and lethal to adults in a 0.5-1-h exposure of 0.05 mg/L. Transport on both passenger and cargo aircraft is forbidden.

Phosphoric Acids and Phosphates

Phosphates may be defined as compounds containing four phosphorus-oxygen (P-O) linkages. Compounds containing discrete, i.e., monomeric PO_4^{3-} ions are known as orthophosphates or simply phosphates, linear condensed P-O-P chains as polyphosphates, cyclic rings as metaphosphates, branched polymeric materials and cage anions as ultraphosphates. Stoichiometrically, phosphates have been represented as combinations of oxides, e.g., H_3PO_4, as $P_2O_5.3H_2O$ and Na_2HPO_4 as $P_2O_5.2Na_2O.H_2O$ (Table 9.9).

The double-neutralisation process is used to purify phosphoric acid during the production of large-volume detergent-builder phosphates. In the furnace process, elemental white (yellow) phosphorus is burned in excess air; the resulting phosphorus pentoxide is hydrated, the heats of combustion and hydration are removed, and the phosphoric acid mist is collected. The processes are called wetted-wall, water-cooled, or air-cooled, depending on the protection of the combustion-chamber wall.

A potential environmental problem is created by the burning of phosphorus, which produces a persistent white cloud of phosphorus pentoxide and phosphoric acid droplets of such high obscuring power that it is used as a screening smoke by the military. Neither precipitators nor Venturi scrubbers reduce the pentoxide content to a level acceptable in states with low plume-opacity regulations; therefore, high efficiency mist eliminators are now standard practice. Physical properties of aqueous solution of phosphoric acid are given in Table 9.10.

Table. 9.9. Traditional classification of sodium phosphates.

Oxide ratio, R ($Na_2O + H_2O_{comp}$): P_2O_5	Designation	General formula	Structure
> 3	Phosphate + metal oxide (includes double salts and solid solutions)		Mixtures
3	Phosphate or orthophosphate, $n \leq 3$	$Na_nH_{3-n}PO_4$	One phosphorus atom
3 > R < 2	Mixture of ortho and pyrophosphates		
2	Pyrophosphate, $n \leq 4$	$Na_nH_{4-n}P_2O_7$	Two phosphorus atoms linear
2 > R > 1	Polyphosphates, $n = 2, 3, 4 \ldots$	$Na_{n+2}P_nO_{3n+1}$	chains
1	Metaphosphates, $n = 3, 4, 5 \ldots$	$Na_n(PO_3)_n$	Cyclic or extremely long chains
1 > R > 0	Ultraphosphates, $0 < x < 1$	$(xNa_2O)P_2O_5$	Cross-linked chains and/or rings
0	Phosphorus pentoxide	$(P_2O_5)_n$	P_4O_{10} or continuous structures

Table 9.10. Physical properties of aqueous solutions of phosphoric acid.

Concentration, wt % H_3PO_4	P_2O_5	Density, 25°C, g/cm³	Bp, °C	Fp, °C	Viscosity, mPas·s (=cP) at 20°C	60°C	100°C
0	0	0.997	100.0	0	1.0	0.48	0.30
5	3.62	1.025	100.1	− 0.8	1.1	0.54	0.33
10	7.24	1.053	100.2	− 2.1	1.2	0.61	0.38
20	14.49	1.113	100.8	− 6.0	1.6	0.78	0.48
30	21.73	1.182	101.8	− 11.8	2.2	1.0	0.62
50	36.22	1.333	108.0	− 44.0	4.3	1.8	1.1
75	54.32	1.573	135.0	− 17.5	15	4.8	2.4
85	61.57	1.685	158.0	21.1	28	8.1	3.8
100	72.43	1.864	261.0	42.35	140	25	9.2
105	76.10	1.925	>300	16.0	600	70	19
115	83.29	2.044	>500			1500	250

Elemental phosphorus produced by the electrothermal process is a distilled product of high purity, which yields phosphoric acid pure enough for most industrial uses. The high heat of combustion of phosphorus 3.05 MJ/mol (730 kcal/mol) can be recovered for process use such as evaporation of water from dilute phosphate solutions. Furnace-grade phosphoric acid is used in metal treatment, refractories, catalysts, and food applications.

Phosphoric Acid

From phosphate rock by electric furnace process

In the electric-arc furnace process, phosphate rock is reduced to elemental phosphorus by the action of coke and heat in the presence of sand. Subsequent oxidation by air to phosphorus pentoxide followed by hydration yields phosphoric acid (Fig. 9.33).

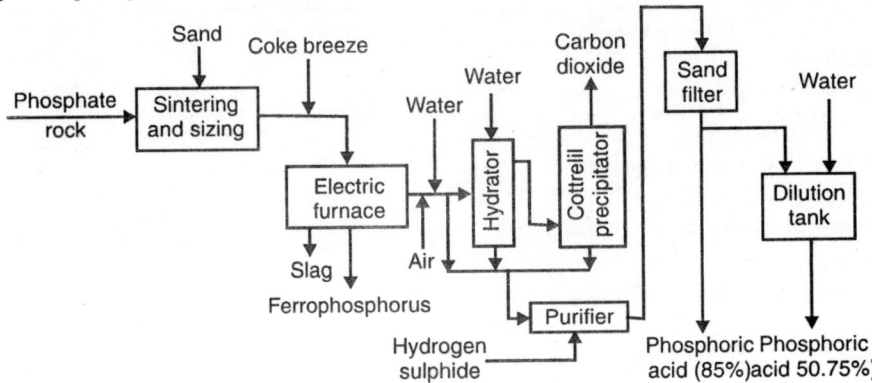

Fig. 9.33. Flow diagram for manufacture of phosphoric acid from phosphate rock by electric furnace.

Phosphate rock is charged into a sintering oven, where it is nodulised to facilitate escape of phosphorus vapours in the electric furnace and to prevent the entrainment of dust or fines (in the vapours). The raw material is sized, and the fines are returned to the sintering oven. Coke (generally in the form of breeze) and sand are added in carefully controlled ratios, determined by rock analysis, to the sintered rock, and the mixture is charged into the shaft of an electric furnace. In the shaft hang three carbon electrodes, which are connected to a three-phase alternating current.

The charge on reaching the level of the arc, is fused at approximately 1315°C, resulting in reduction of the phosphate rock with liberation of elemental phosphorus vapours. Since phosphate rock usually contains fluorides as impurities, or as a constituent of the rock itself, which is often written as fluoroapatite, $Ca_5F(PO_4)_3$, calcium fluoride, fluorosilicates, and silicon tetrafluoride are also formed. The slag (mostly calcium silicate) from the furnace is usually tapped periodically and subsequently crushed for use as aggregate for road construction. Ferrophosphorus (resulting from the iron impurities) runs out with the slag. The amount of this material produced may be increased by adding iron slugs to the furnace charge. The ferrophosphorus is separated from the slag and marketed.

Reactions

$$2Ca_3(PO_4)_2 + 6SiO_2 + 10C \rightarrow P_4 + 10CO + 6CaSiO_3$$
$$P_4 + 10CO + 10O_2 \rightarrow 2P_2O_5 + 10CO_2$$
$$2P_2O_5 + 6H_2O \rightarrow 4H_3PO_4$$
$$\text{Overall: } Ca_3(PO_4)_2 + 3SiO_2 + 5C + 5O_2 + 3H_2O \rightarrow 3CaSiO_3 + 5CO_2 + 2H_3PO_4$$

87 to 92% yield

The gases from the furnace, phosphorus and carbon monoxide, are withdrawn from the furnace by means of a fan. In the one-step system, a current of air is drawn down through the charge by the suction induced by the fan. The two-step method produces phosphorus, which is stored for subsequent processing. The two currents (reaction gases and air) mix in the flue at a temperature sufficient to burn the phosphorus to phosphorus pentoxide (P_2O_5), and the carbon monoxide to the dioxide. The gases pass into a tall, packed tower, where they are sprayed with water (forming a mist of phosphoric acid), and then through a Cottrell electrostatic precipitator made of graphite (to resist the action of hydrofluoric acid) to remove any remaining phosphoric acid.

The crude phosphoric acid (85%) is generally purified with respect to arsenic by the action of hydrogen sulphide. Depending on conditions, the acid may be purified further by adding sulphuric acid to remove calcium salts. Sufficient sulphuric acid is used to precipitate calcium sulphate and also to leave a slight excess to inhibit the corrosive action of the phosphoric acid. This slight excess of sulphuric acid permits the use of lead-lined equipment. Residual hydrofluoric acid may be removed by the addition of finely powdered silica. These purification steps usually take place before arsenic removal. The excess silica, calcium sulphate, arsenic trisulphide, and any suspended material are removed by passage of the acid through a sand filter. The clarified phosphoric acid (85%) may be diluted with water to yield 75 and 50% acid. An overall yield of about 90% is realised on the calcium phosphate content of the rock raw material. The abbreviation BPL means bone phosphate of lime ($Ca_3(PO_4)_2$); 70 BPL means that 100 kg of rock contains the equivalent of 70 kg of calcium phosphate. Engineers are more likely to use notations based on P_2O_5 content; 1% P_2O_5 is equivalent to 2.185% BPL.

Reaction

$$P_4 + 5O_2 \rightarrow 2P_2O_5$$
$$P_2O_5 + 3H_2O \rightarrow 2H_3PO_4$$
$$\text{Overall: } P_4 + 5O_2 + 6H_2O \rightarrow 4H_3PO_4$$
$$94 \text{ to } 97\% \text{ yield}$$

Elemental phosphorus is often converted to phosphoric acid at locations other than the original point of production. The conversion involves oxidation of molten phosphorus to phosphorus pentoxide and subsequent hydration of the oxide to phosphoric acid. (Fig. 9.34)

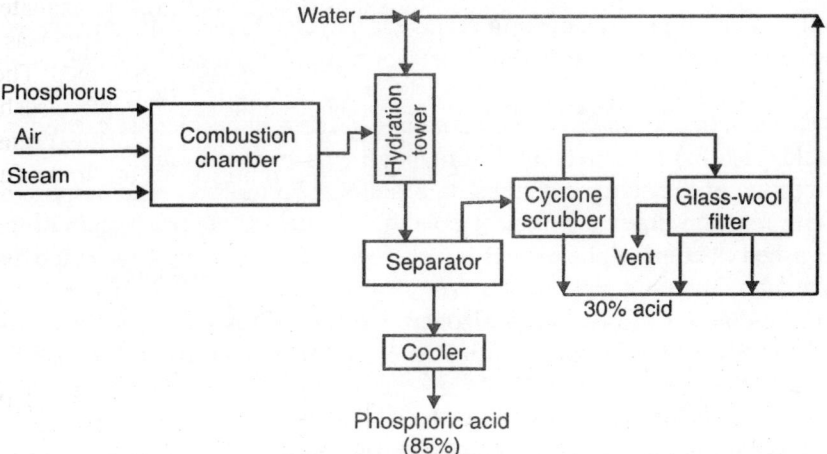

Fig. 9.34. Flow diagram for manufacture of phosphoric acid from phosphorus by oxidation and hydration.

Molten phosphorus is sprayed into a combustion chamber along with air and steam. Flame temperature approaches 1980°C. Chamber design varies from one plant to another, but construction materials are usually limited to acid-proof brick, structural carbon, and stainless steel. Both vertical and horizontal chambers are used. In either case some glassy metaphosphoric acid is formed and drops to the bottom of the chamber.

The effluent gases leaving the chamber (P_2O_5, steam, nitrogen, and some oxygen) are mixed with a spray of dilute phosphoric acid in the hydration tower. If the tower is externally cooled, it too may be built of stainless steel. By proper adjustment of reacting streams, various strengths of phosphoric acid (as high as 116%—equivalent to 84% P_2O_5) may be produced. The usual product is in the range of 75 to 85% phosphoric acid. Some acid mist leaves the hydrator and must be recovered by appropriate means, such as a packed tower or electrostatic precipitator. In the process shown in the flow diagram a combination cyclone scrubber and glass-wool filter is used.

The solution of problems of design and selection of materials of construction is vital to the success of the process. Liquid phosphorus at elevated temperatures tends to form the red, amorphous allotrope which can clog the burner orifices. Build-up of red phosphorus is usually avoided by a design in which rapid atomisation of the phosphorus occurs. Corrosion problems are also severe. Hot P_2O_5 vapours are very reactive, and the phosphoric acid formed from the moisture in the air tends to coat the walls with a viscous, highly reactive layer which attacks most materials. Graphite is used in many installations, unless the units are cooled by sprays, in which case stainless steel is satisfactory.

Wet process

The sulphuric acid or wet process produces phosphoric acid by the action of sulphuric acid on phosphate rock, accompanied by the precipitation of calcium sulphate. The resulting phosphoric acid solution is separated from the precipitated gypsum by filtration, and then concentrated by evaporation (Fig. 9.35).

$$Ca_5F(PO_4)_3 + 5H_2SO_4 + 10H_2O \rightarrow 5CaSO_4 \cdot 2H_2O + HF + 3H_3PO_4$$
$$\text{Fluoroapatite} \qquad\qquad \text{95 \% yield} \qquad \text{Gypsum}$$

Phosphate rock is charged into a special ball mill, where it is ground in a dilute solution of phosphoric acid obtained as a filtrate from an ensuing operation. The slurry is passed into single or multiple tanks, where it is allowed to react with sulphuric acid diluted with sufficient phosphoric acid to yield a 55% sulphuric acid solution. In multiple-tank operations, acid is recycled to maintain uniform temperatures in the several reactors. In both single- and multiple-tank operations vigorous agitation is maintained by propellers or turbines. Heat of reaction is removed by blowing air through the reaction mass or by flash evaporation under vaccum. In both cases water vapour and gaseous impurities are carried to an absorber, where fluosilicic acid (H_2SiF_6) is recovered by spraying the gases with water.

Usually acid digestion of the slurry requires 4 to 8 hours at 75 to 80°C. The purpose of the violent agitation (for uniform reaction time) and close temperature control is the production of uniform easily filtered and easily washed calcium sulphate (gypsum) crystals. If the temperature were too high, anhydrite would form, hydrate later, and plug pipes.

The slurry from the digester passes to a horizontal, rotating, tilting-pan-type vacuum filter, where phosphoric acid (30 to 35% P_2O_5) is removed from the cake. Most plants use tilting-pan filters because liquors from the various washing stages can be kept separate, thus increasing recovery and minimising dilution. Polypropylene cloth makes an excellent filter medium. The washed gypsum cake is slurried with waste-water to a settling pond from which water is ordinarily pumped back to the plant.

The acid filtrate is then evaporated to the desired concentration, usually 54% P_2O_5. Forced-circulation evaporators with outside heat exchangers are most commonly used. Fluorine-containing compounds may be recovered from the flash chamber condensate. Submerged-combustion evaporators are also used, but high-efficiency scrubbers are required to recover P_2O_5 fumes.

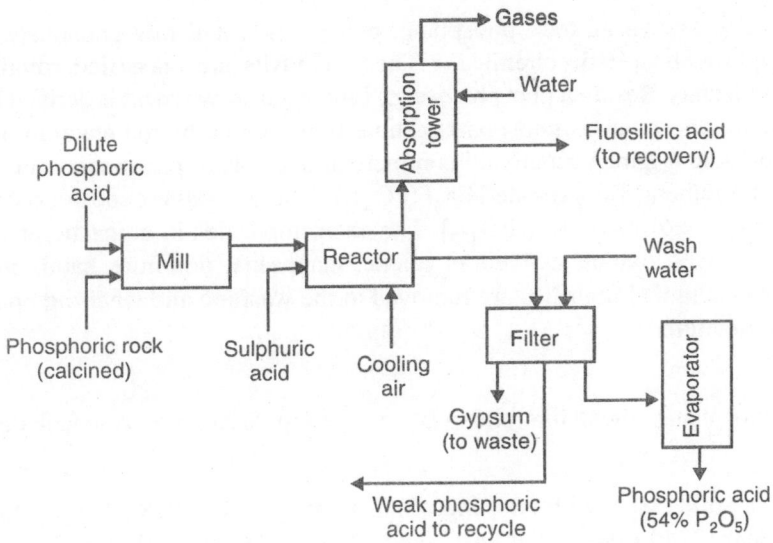

Fig. 9.35. Flow diagram for manufacture of phosphoric acid from phosphate rock and sulphuric acid (wet process).

Throughout the plant, corrosion-resistant materials of construction must be used. The most common ones are structural carbon or nickel alloy for evaporator heat exchangers; rubber or carbon-brick for reactor linings; and polyester-fibre glass in pipes, ducts, and small vessels. The yield of phosphoric acid based on the phosphorus content of raw material is 95%.

Various modifications of the process described, known as the dihydrate process, are in use. In one, conditions are modified to obtain a hemihydrate of calcium sulphate ($CaSO_4$. ½ H_2O) instead of gypsum; this is known as the hemihydrate or HDH process. Advantages claimed for its most recent modification include improved extraction efficiency (99% versus 95 to 97%) of the phosphorus values, higher purity and more concentrated product, 40 to 54% P_2O_5, without the need for evaporation. The calcium sulphate is said to entrain less phosphate and fluoride ions, thus rendering it more suitable as building material. Whatever the modifications used, each strives to produce a rapidly filterable crystal, to minimise P_2O_5 losses, and to produce acid of high concentration.

Phosphoric acid of greater than 100% strength, known as superphosphoric acid, is becoming of greater importance. Acids as high in P_2O_5 content as 83% (115% H_3PO_4) are produced, essentially by a two-step evaporation, preceded by purification to remove contaminants which while not objectionable in weaker acids would interfere with the concentration process. These acids can also be made from phosphorus by using less water to absorb the P_2O_5 and producing the higher concentrations directly; since the water balance is important for conducting away the heat generated, adjustments must be made to compensate for this factor. The superacids contain varying amounts of polyphosphoric acids. Among the advantages claimed for them are lowered corrosivity and reduced shipping costs.

Other phosphorus-containing acids

Phosphoric acid (H_2PO_4) may be converted into pyrophosphoric acid ($H_4P_2O_7$) and metaphosphoric acid (HPO_3) by heating. Pyrophosphoric acid is formed at 250 to 260°C. On heating either the ortho- or pyro- to a red heat, metaphosphoric acid is formed. The latter may also be prepared by treating phosphorus pentoxide with the calculated amount of cold water.

Phosphate rock, also known as rock phosphate, phosphorite, and raw phosphate, is the primary source of practically all phosphatic chemicals. These deposits are classified roughly as residual, replacement, and sedimentary. Residual phosphate (e.g., Tennessee brown rock) is derived from phosphatic limestone. Replacement phosphate is phosphatised limestone formed by the reaction of limestone and phosphoric acid of organic origin. Virtually all commercial deposits of phosphate rock are amorphous, impure varieties of the mineral fluorapatite $[Ca_5F(PO_4)_3]$. The workable deposits contain from 18 to 90% available tricalcium phosphate $[Ca_3(PO_4)_2]$. The chief impurities in domestic phosphate rock are iron, aluminium, and silicon oxides, as well as calcite, magnesite, dolomite, sand, clay, and organic matter. Most of the undesirable impurities are removed in the washing and sintering operations prior to phosphoric acid manufacture.

Uses

Fertilisers, sodium phosphates, dicalcium phosphate and tetrapotassium pyrophosphate.

Properties

Colourless, crystalline solid. Mol. wt 98.00, mp 42.35°C, Sp gr 1.834 (18.2°C), bp loses 1/2 H_2O at 213°C. Soluble in water (2340 g/100 ml at 26°C) and ethanol. Commercial phosphoric acid is a clear, colourless, sparkling, mobile liquid which reaches sirupy consistency at higher concentrations (85%). At 88% concentration, a crystalline hydrate, $H_3PO_4.1/2\ H_2O$, separates at room temperature. Threshold limit value (mg/m^3) 1.

Both the wet process and the electric furnace process are plagued with by-products of low value which require disposal. Calcium sulphate waste from the wet process is not utilised at present, and the slag, carbon dioxide, and ferro-silicon from the electric furnace are only partially utilised. Fluorine is a particularly troublesome by-product of the wet-process. Anti-pollution laws greatly restrict the amounts that can be wasted into the air or into streams. Thus gas streams must be scrubbed and liquid streams neutralised. Some fluorine remains in wet-process acid, but the trend toward higher P_2O_5 concentrations means greater release during evaporation. The calcium sulphate (gypsum) and calcium silicate (slag) by-products are often not marketable, and these sludges also pose disposal problems.

Wet-process acid is mostly (93%) used for fertiliser production; of furnace acid, only about 20% goes to fertilisers, the remainder being used for applications where its higher purity is demanded. Since the principal growth area for phosphoric acid will be in fertiliser production for an increasingly hungry world, expansion in capacity will no doubt be in wet-process installations unless new technology is developed. Phosphate rock itself is too insoluble to be a useful source of phosphorus for agriculture.

Polyphosphoric acids

The only clearly defined crystalline compositions of the $H_2O–P_2O_5$ system are three forms of phosphoric acid and their hemihydrates, pyrophosphoric acid, and crystalline P_4O_{10}. Amorphous condensed phosphoric acids are hygroscopic; they may be viscous, oily, gummy, and may consist of a mixture of glassy and crystalline materials.

Linear-polyphosphoric acids are strongly hygroscopic and undergo viscosity changes and hydrolysis to less complex forms when exposed to moist air. Upon dissolution in excess water, hydrolytic degradation to phosphoric acid occurs.

Pyrophosphoric (diphosphoric) acid, $H_4P_2O_7$, crystallises in two forms (mp = 54.3 and 71.5°C); acid salts are known. Tripolyphosphoric (triphosphoric) acid, $H_5P_3O_{10}$, occurs in varying amounts as a component of condensed phosphoric acids containing more than ca 72% P_2O_5.

The largest use of polyphosphoric (superphosphoric) acids is as an intermediate in the production of liquid fertilisers. Condensed acids of 82–84% P_2O_5 are employed as catalysts in the petroleum industry. Polyphosphoric acid is also used as a dehydrating agent and in the production of phosphoric esters and agricultural chemicals.

Polyphosphates

The condensed phosphates are derived from phosphates by the loss of water. These materials range from simple diphosphates (pyrophosphates) to long-chain polymeric structures with molecular weights in the millions (10^6). Polyphosphates are resistant to chemical attack, but are susceptible to hydrolysis. Short-chain polyphosphates hydrolyse without the concurrent formation of cyclic metaphosphates; in this respect they differ from the long-chain acids. Many polyphosphates form water-soluble complex ions, a phenomenon called sequestration. This property forms the basis for the water-treatment and detergent applications.

Pyrophosphates

The simplest linear condensed phosphates ($M_{n+2} P_n O_{3n + 1}$, where $n = 2, 3, 4...$) are pyrophosphates ($M_4P_2O_7$). A water molecule is eliminated from condensed orthophosphates:

$$2MH_2PO_4 \xrightarrow[\Delta]{\Delta} M_2H_2P_2O_7 + H_2O \quad 2M_2HPO_4 \rightarrow M_4P_2O_7 + H_2O$$

Insoluble pyrophosphates are obtained by treating a soluble salt of the desired cation with a sodium pyrophosphate solution.

Tetra- and disodium pyrophosphates (sodium acid phosphate) are prepared by thermal dehydration of di- and monosodium orthophosphate, respectively. Tetrasodium pyrophosphate is used as a builder in detergent and cleaning formulations, in food applications, and as a deflocculant in drilling muds, dyes and inks. Sodium acid pyrophosphate is used as a leavening and chelating agent (Fig. 9.36).

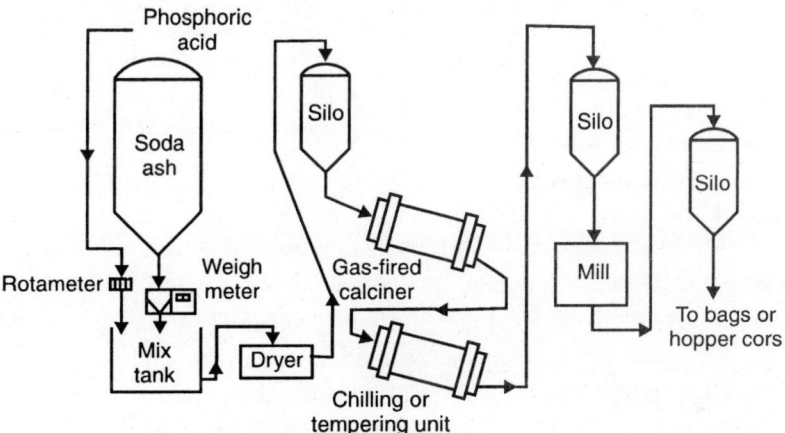

Fig. 9.36. Diagram of a plant for the manufacture of crystalline sodium polyphosphate.

Calcium pyrophosphate exists in three polymorphic modifications; they form progressively upon dehydration of calcium hydrogen phosphate dihydrate. Calcium pyrophosphates are used primarily as abrasives in fluoridated toothpaste.

Sodium tripolyphosphate (STP, pentasodium triphosphate), $Na_5P_3O_{10}$, occurs as the anhydrous forms I (STP-I) (thermodynamically stable) and II (STP-II), the low temperature form; the transition temperature is $417 \pm 8°C$. They are differentiated by X-ray diffraction, IR, Raman spectra, and the temperature-rise test (ASTM D 501, 30). A hexahydrate forms by the addition of either anhydrous form to water or by the hydrolysis of sodium trimetaphosphate, $(NaPO_3)_3$. STP is produced commercially by calcination of a mixture of mono- and disodium phosphates.

The solubility and hydration behaviour of STP are of particular importance in its industrial applications. As a builder for synthetic detergents, STP is the largest-volume phosphate salt for purposes other than fertilisers. Thermal dehydration of monosodium phosphate gives rise to numerous condensed polyphosphates, e.g., Graham's, Madrell's, and Kurrol's salts. In the manufacture of phosphate salts, phosphoric acid is treated with a base, e.g., carbonate, hydroxide, or ammonia, to form a solution or slurry. Most phosphates crystallise readily from solution and are separated by conventional techniques; the solutions or slurries are often evaporated to dryness.

The phosphate glasses are manufactured in refractory-type furnaces, where they are heated to 1000°C, then quenched rapidly to a solid glass; trade names are used for glasses of only slightly different composition. Sodium tripolyphosphate (90-95% pure) is manufactured by drying and subsequent calcination of a solution or slurry with a Na_2O/P_2O_5 mole ratio of 1.67, corresponding to two mole disodium phosphate and one mole monosodium phosphate. Calcining conditions are less critical than drying conditions; combinations of dryer/calcining units are available.

Disodium phosphate, Na_2HPO_4, is marketed as the dihydrate and the anhydrous salt. Tetrasodium pyrophosphate, $Na_4P_2O_7$, is obtained by calcination of disodium phosphate or any of its hydrates. Commercial manufacture is similar to that of STP, except that the final Na_2O/P_2O_5 ratio is adjusted with NaOH; the same equipment can be used.

Chlorinated TSP, the second largest volume sodium phosphate salt, is a complex mixture approximating $(Na_3PO_4.11H_2O)_4$. NaOCl. It is made by the addition of sodium hypochlorite solution to a hot concentrated sodium phosphate solution, followed by cooling, crystallisation, and granulation. Chlorinated TSP is unstable above 40°C and up to 20% of the available chlorine may be lost on heating to elevated temperatures during processing.

In the manufacture of ammonium phosphates, the high partial pressure of ammonia has to be considered, because it rises rapidly with the temperature or the NH_3: P_2O_5 mole ratio. Phosphoric acid reacts quickly with ammonia vapour and is used as a scrubber fluid to prevent ammonia emissions and recover ammonia values. MAP and DAP fertilisers are made in granulation processes from ammonia and wet-process phosphoric acid.

Fire retardancy is second to fertilisers in MAP consumption. Most calcium phosphates are mixtures of several salts; their composition depends on the manufacturing conditions. For fertiliser and animal-feed uses, the primary concern is the CaO and P_2O_5 analysis. Calcium phosphates are used as food additives.

Environmental considerations

Inorganic phosphates present little health hazard to humans and are essential to life processes. However, phosphates can create environmental problems, mainly because they increase the growth of algae in lakes and streams. Problems caused by sewage-borne phosphates are mostly localised to areas that employ lakes as receiving waters for sewage effluents. Average concentrations vary according to industrial input, storm waters, seasonal fluctuations, and population density.

Pharmaceutical and food grades

Pharmaceutical and food grades (purer) of dicalcium phosphate dihydrate for use as a polishing agent in dentifrices, ammonium phosphate for use as yeast nutrient, calcium pyrophosphates, and monocalcium phosphate are manufactured from furnace-grade phosphoric acid. Monocalcium phosphate $[Ca(H_2PO_4)_2.H_2O]$ is manufactured by the crystallising, after evaporation and some cooling, of a hot solution of lime and strong furnace phosphoric acid. The crystals are centrifuged, and the highly acidic mother liquor returned for reuse. This acid salt is also made by spray-drying a slurry of the reaction product of lime and phosphoric acid. This product is used for baking powder.

Sodium Phosphates

Monobasic (NaH_2PO_4), dibasic (Na_2HPO_4), tribasic (Na_3PO_4), tetrabasic ($Na_4P_2O_7$), acid ($Na_2H_2P_2O_7$), meta ($(NaPO_3)_6$), tripoly ($Na_5P_3O_{10}$).

Sodium phosphates are produced by processing the reaction products of phosphoric acid and soda alkalies. Since phosphoric acid (orthophosphoric acid) is a tribasic acid, it forms three sodium salts. Sodium meta- and pyrophosphates are obtained by heating the corresponding sodium orthophosphate. The very important sodium tripolyphosphate is made by calcining a mixture of monobasic and dibasic sodium orthophosphates (Fig. 9.37).

Reaction

$$Na_2CO_3 + H_3PO_4 \longrightarrow Na_2HPO_4 + CO_2 + H_2O$$
$$Na_2HPO_4 + NaOH \longrightarrow Na_3PO_4 + H_2O$$

90—95% yield

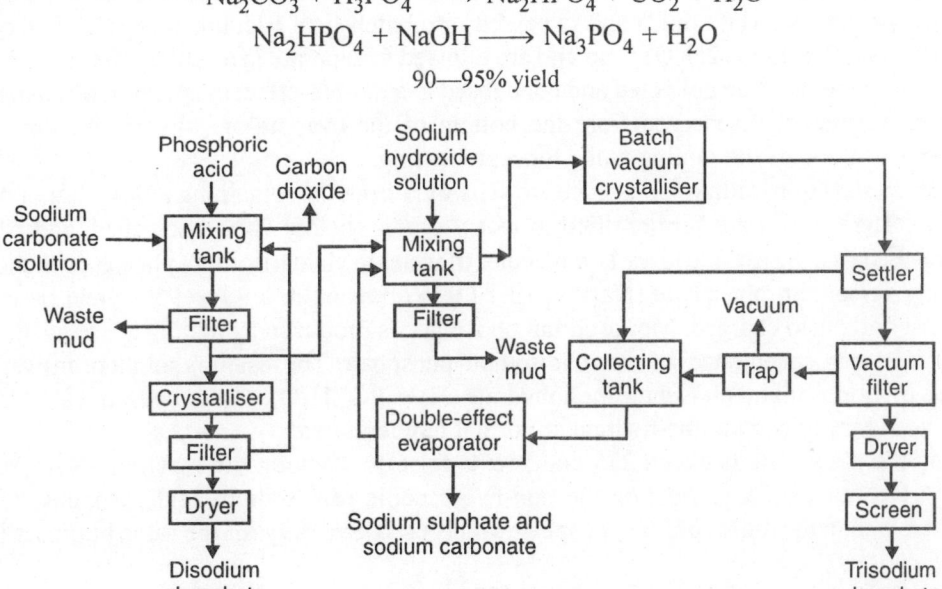

Fig. 9.37. Flow diagram for manufacturing sodium phosphate from phosphoric acid, sodium carbonate, and sodium hydroxide.

Disodium phosphate (Na_2HPO_4) is the ordinary sodium phosphate of commerce and may be considered the base of all the other phosphates. It is the intermediate in the production of trisodium phosphate (Na_3PO_4), sodium pyrophosphate ($Na_4P_2O_7$), and sodium tripolyphosphate ($Na_5P_3O_{10}$). A description of the manufacture of disodium phosphate and trisodium phosphate follows.

Sodium carbonate (soda ash) is introduced into a mixing tank, either as an aqueous solution or as a solution made with hot liquor from the evaporators or mother liquor from the filters. Phosphoric acid (60 to 65% H_3PO_4) is added at the surface of the tank, so that carbon dioxide may be liberated easily. A slight excess of sodium carbonate over the theoretical ratio is added, and the solution is boiled with steam until all the carbon dioxide has been driven off. The resulting disodium phosphate solution is filtered hot (85 to 100°C) and is split into two portions. A small amount of white mud, consisting of silica and iron and aluminium phosphates, remains on the filter and is discharged to waste. If trisodium phosphate is to be made, part of the clear solution of disodium phosphate, containing about 14.5% phosphorus pentoxide and 13% sodium oxide, is pumped to a trisodium phosphate process tank.

The remainder of the solution is cooled in a crystalliser to yield crystals of disodium phosphate, containing 12 molecules of water of crystallisation (60% water). The crystals ($Na_2HPO_4.12H_2O$) are centrifuged and packaged in moisture-proof containers. The mother liquor is returned to the mixing tank. The crystals are efflorescent, losing 5 molecules of water and forming the heptahydrate ($Na_2HPO_4.7H_2O$) on exposure to air. The crystals may be dried to yield the dihydrate ($Na_2HPO_4.2H_2O$, containing 20% water) or the anhydrous salt (Na_2HPO_4). Furnace phosphoric acid is usually used as a starting material.

To the hot sodium phosphate solution in the trisodium phosphate process tank is added 50% sodium hydroxide (caustic soda) solution. The solution is maintained at about 90°C, and control samples are taken and titrated to control the operation. The hot solution is filtered (on a rotating-leaf pressure filter) to remove any insolubles (white mud) and is passed into batch-type vacuum crystallisers. Crystals of trisodium phosphate ($Na_3PO_4.12H_2O$) form and are allowed to separate in a settler. The clear liquor and mother liquor from the filter are collected and condensed in a double-effect evaporator. Sodium sulphate and sodium carbonate are discharged from the bottom of the evaporator, whereas the concentrated liquor is returned to one of the mixing tanks for reprocessing.

The settled crystals of trisodium phosphate are separated from the remaining mother liquor on rotary vacuum filters. The crystals are further dried in rotary dryers (below 70°C), screened, and packaged. By drying above 100°C, the hydrate loses 11 molecules of water to yield trisodium phosphate monohydrate ($Na_3PO_4.H_2O$). Trisodium phosphate ($Na_3PO_4.12H_2O$) is obtained in a 90 to 95% yield based on the weight of phosphoric acid charged. Monosodium phosphate is produced by adding the requisite amount of phosphoric acid to a concentrated solution of disodium phosphate. The resulting solution, an evaporation, yields crystals of monosodium phosphate monohydrate ($NaH_2PO_4.H_2O$). The anhydrous salt (NaH_2PO_4) may be produced by desiccating the hydrate at normal temperatures.

By heating the phosphate between 225 and 250°C for 6 to 12 hour, the reaction $2NaH_2PO_4 \longrightarrow Na_2H_2P_2O_7 + H_2O$ takes place, yielding the non-hygroscopic salt, sodium-acid pyrophosphate. This salt is more stable than the monosodium phosphate and is used generally for the same purposes (baking-powder formulations).

On heating monosodium phosphate to 350 to 400°C, an insoluble metaphosphate is formed. If the monosodium salt is fused to approximately 760°C and quickly cooled [$6NaH_2PO_4 \longrightarrow (NaPO_3)_6 + 6H_2O$], sodium hexametaphosphate results. Tetrasodium pyrophosphate ($Na_4P_2O_7$) is manufactured by dehydrating disodium phosphate in a rotary kiln according to the reaction: $2Na_2HPO_4 \rightarrow Na_4P_2O_7 + H_2O$. Both the anhydrous salt ($Na_4P_2O_7$) and the crystalline decahydrate ($Na_4P_2O_7 . 10H_2O$) are produced, depending on the operating conditions.

Sodium tripolyphosphate (STPP) is made from a mixture of mono- and disodium phosphates. The mixture is made directly in solution in a manner similar to that described for disodium phosphate. The ratio of phosphoric acid to soda ash is so adjusted as to produce a Na_2O/P_2O_5 ratio of 1.67, which corresponds to a mixture of 1 mole of mono to 2 moles of disodium phosphate. After filtration the solution is sprayed into a rotary continuous kiln, where drying, molecular dehydration, conversion to STPP, annealing, and cooling take place in sequence. After grinding, the product is packaged for shipment.

Various other polyphosphates may be produced from mixtures of mono- and disodium phosphates. These materials are generally of rather indefinite molecular composition. Only STPP is crystalline. It is usually sold as the anhydrous salt but also exists as the hexahydrate.

Sodium metaphosphates are also of somewhat indefinite composition. The formula of the compound sodium hexametaphosphate, usually written $(NaPO_3)_6$, is only approximate; generalised formulae of this class range from $(NaPO_3)_x$ to $Na_{x+2}P_xO_{3x+1}$; many are amorphous and insoluble glasses.

Uses

Detergents and water treatment, foods, textiles, dentifrices, feeds, clay processing, drilling muds, etc.

Monosodium phosphate

Sodium phosphate monobasic, sodium acid phosphate, sodium biphosphate, sodium dihydrogen phosphate, monosodium orthophosphate.

Properties

Anhydrous, white, crystalline powder; soluble in water. Formula wt 119.98, mp decomposes at 225 to 250°C to acid pyrophosphate. Monohydrate, large, transparent crystals, acid reaction, very much soluble in water, insoluble in ethanol. Formula wt 138.0, mp loses water at 100°C, Sp gr 2.040.

Disodium phosphate

DSP, sodium phosphate dibasic, secondary sodium phosphate, sodium hydrogen phosphate, phosphate of soda, sodium phosphate, or disodium orthophosphate.

Properties

Anhydrous (Na_2HPO_4), colourless, translucent crystals or white powder; soluble in water. Formula wt 141.965, mp converts to pyrophosphate 240°C. Dihydrate, $Na_2HPO_4 \cdot 2H_2O$—Colourless, translucent crystals. Formula wt. 178.0. Sp. gr. 2.066 (15°C), mp loses H_2O at 92.5°C.

Trisodium phosphate

TSP, sodium phosphate tribasic, normal sodium phosphate, tertiary sodium phosphate, or trisodium orthophosphate.

Properties

Formula wt 380.12; Sp. gr. 1.62 (20°C); mp 73.4°C, with decomposition; bp loses $11H_2O$ at 100°C; pH of 1% solution 12.0. Soluble in water (28.3 g/100 g at 15°C, in all proportions in hot water). Insoluble in carbon disulphide.

Tetrasodium pyrophosphate

TSPP, sodium phosphate tetrabasic, sodium tetraphosphate, sodium pyrophosphate, pyrophosphate of soda, normal pyrosodium phosphate, or alkaline sodium pyrophosphate.

Properties

Crystalline ($Na_4P_2O_7 \cdot 10H_2O$), colourless monoclinic crystals. Formula wt 446.1, mp loses water 94°C, Sp. gr. 1.82. Soluble in water (5.4 g/100 g at 0°C, 93 g/100 g at 100°C). Insoluble in ethanol and ammonia.

Anhydrous ($Na_4P_2O_7$), white crystals. Formula wt 265.9, mp 988°C, Sp. gr. 2.45. pH of 1% solution 10.2. Soluble in water (3.2 g/100 g at 0°C, 40.3 g/100 g at 100°C). Decomposes in ethanol.

Sodium acid pyrophosphate

Disodium dihydrogen pyrophosphate, disodium pyrophosphate, acid sodium pyrophosphate.

Properties

Anhydrous ($Na_2H_2P_2O_7$), white powder. Formula wt 221.95, mp decomposes 220°C, Sp gr 1.862. Soluble in water.

Uses

Electroplating metal cleaning and phosphatising, drilling muds, baking powders etc.

Sodium metaphosphate

Sodium hexametaphosphate, Graham's salt.

Properties

[($NaPO_3$)$_6$], colourless or white flakes. Formula wt 612.10, mp 640°C, Sp. gr. 2.181, bp sublimes > 1,000°C. Soluble in hot and cold water.

Uses

Dental polishing agent, detergent binders, water softening, sequestrants emulsifier, food additive, textile processing, laundering.

Sodium tripolyphosphate

Properties

Anhydrous ($Na_5P_3O_{10}$), white powder or granules. Formula wt 367.88, pH of 1% solution 9.7. Soluble in water (14.5 g/100 g at 25°C, 32.5 g/100 at 100°C).

Uses

Water softening, sequestering, peptising, food additive, texturiser.

POTASSIUM

Potassium is far too reactive metal, but it is widely distributed in minerals, the most important deposits of which occur in places where island seas have undergone natural evaporation. The main commercial source is carnalite, but kainite and chonite and sylvinite are sometimes used. Potassium is present in most clays and silicate rocks, and also in sea water (about 0.05%) and wood ash (about 30% of which is potassium carbonate). Potassium salts must be present in the soil in order to have normal plant growth, hence they are essential components of the fertilisers used in food production throughout the world. Various potassium salts are also basically important as raw materials for many commodities such as soaps, detergents, glass, dyes, gunpowder, and pyrotechnics. The sylvinite is mined and treated to yield high-grade potassium chloride, and langbeinite is processed to make potassium sulphate.

Potassium Chloride

Potassium chloride is manufactured from sylvinite by fractional crystallisation and by flotation and also from searles lake brine.

From sylvinite by fractional crystallisation

Sylvinite (42.7% KCl, 56.6% NaCl), is the raw material for the potassium salts. The particular processes used by the various companies vary in their details but are of two general types. In the oldest plant which probably works the highest-grade ores, the sodium and potassium chlorides are separated by fractional crystallisation; the others use flotation processes (Fig. 9.38).

The feed for the fractional crystallisation process is –4 mesh ore. The 8- to 16-mesh material from the grinding operation is separated on tables to yield a granular 50% K_2O product. For fertiliser use, the potassium content of all potassium salts is expressed as the equivalent percentage of K_2O. For instance, 50% K_2O is equivalent to 80% potassium chloride, and 60% K_2O to 97% potassium chloride. The fines (–4 mesh) are fed to a tank where they are partially dissolved in mother liquor returned from further along in the process. This mother liquor, when heated in the dissolver, remains saturated with respect to sodium chloride but has the capacity to dissolve more potassium chloride. Undissolved sodium chloride is removed from the dissolver, dewatered, washed, and sent to waste. The solution, saturated with both salts at 110°C, is clarified in a thickener. The underflow, containing insoluble mud, is concentrated and washed in counter-current decantation units.

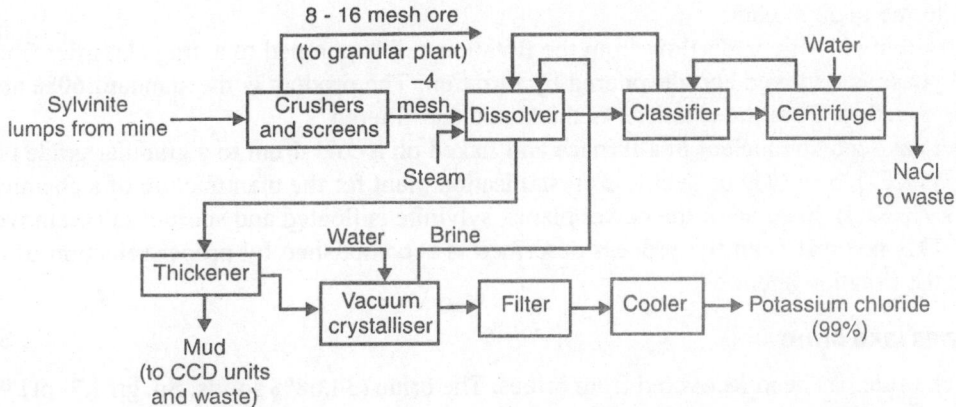

Fig. 9.38. Flow diagram for manufacture of potassium chloride from sylvinite by fractional crystallisation.

The hot clarified solution is then sent to a vacuum crystalliser, where potassium chloride crystallises at 32°C. To prevent sodium chloride from crystallising, water is added to replace that which evaporates. The cooled slurry is sent to a filter from which the mother liquor is recirculated to the dissolving tank. The cake goes to a rotary cooler, and thence to storage.

From sylvinite by flotation

Standard 60% (K_2O) muriate of potash can be produced by separating potassium chloride from sodium chloride by froth flotation of finely ground sylvinite ore. (Fig. 9.39). The ore is obtained by shaft mining and is broken from the mine face by drilling and blasting. About 375 g blasting powder is required per metric tonne of ore. The ore is crushed underground to 13 cm maximum size, hoisted to the surface, and then ground and screened to smaller size.

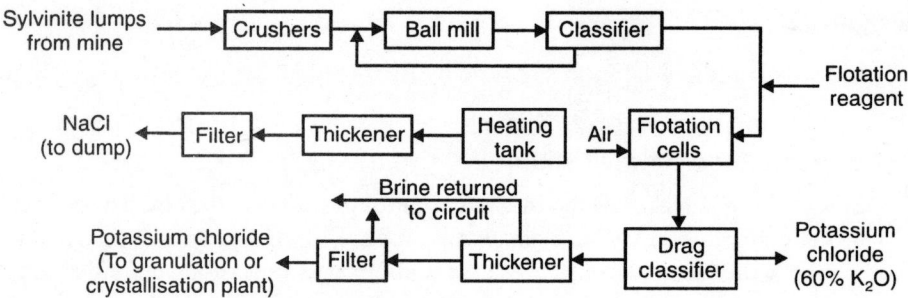

Fig. 9.39. Flow diagram for manufacture of potassium chloride from sylvinite by flotation.

Twelve-centimetre lumps of sylvinite, hoisted from the mine, are crushed to fine granules, pulped with brine, and further ground in a wet ball mill. The milled product is sent to a spiral classifier from which the underflow is returned to the grinding circuit. The flotation reagent mixture is added to the overflow on the way to the flotation cells. A typical flotation reagent mixture is 0.1 kg tallow amine and 0.11 to 0.12 kg of polyalkylglycol/metric tonne of ore. In some cases, cornstarch is also added to the flotation-cell feed to adjust its gravity and viscosity.

In the cell, air is intimately mixed into the feed. The finely divided sodium chloride particles are carried into the froth and overflow from the cells. The froth breaks and is pumped through a heating tank, a thickener, and a filter in series. Tailings of sodium chloride are sent to the dump; the final filtrate is recycled to the main stream.

The potassium chloride underflow from the flotation cell is pumped to a drag classifier from which a wet solid phase is removed and dewatered by filtration. The product is the standard 60% muriate of potash. The overflow from the classifier is thickened and filtered.

The filter cake may be melted in a furnace and flaked on a cold drum to a granular grade 60% K_2O product (97% KCl), or it may be sent to a crystallisation plant for the manufacture of a chemical-grade product (99.9% KCl). In some of the newer plants, sylvinite is floated and sodium salts removed in the underflow. This reversal from the process described is accomplished by proper selection of cell-feed gravity and the flotation agent.

From searles lake brine

Potassium chloride has been recovered from brines. The brine (34.68% solids, Sp. gr. 1.3, pH 9.38) has the following composition.

Constituent	Per cent	Constituent	Per cent
KCl	4.70	Na_3PO_4	0.16
NaCl	16.35	NaF	0.01
Na_2CO_3	4.70	Misc. solids	0.30
Na_2SO_4	6.96	Water	65.32
$Na_2B_4O_7$	1.50		
		Total	100.00

Raw lake brine is mixed with soda products mother liquor and borax mother liquors, and then pumped through a series of condensers and filters to a triple-effect evaporator to effect a crude separation of the soluble salts into a potassium chloride-borax solution, solid crude sodium chloride, burkeite crystals ($Na_2CO_3 \cdot Na_2SO_4$), and solid dilithium sodium phosphate crystals (Li_2NaPO_4).

The hot, concentrated liquor, saturated with potassium chloride and borax, is clarified and then cooled quickly to 38°C in a three-stage vacuum crystalliser. (Fig. 9.40.) The rate of crystallisation of potassium chloride is so much faster than that of borax that an excellent separation is effected. To prevent crystallisation of sodium chloride, enough additional water is added to replace that which evaporates during the vacuum cooling.

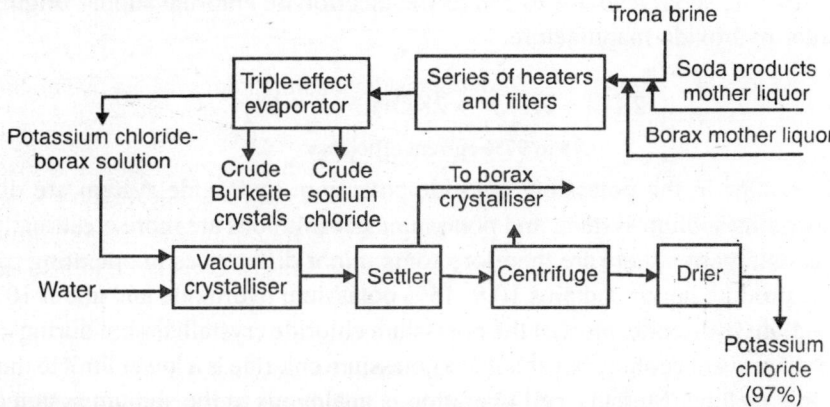

Fig. 9.40. Flow diagram for manufacture of potassium chloride from Searles lake brine.

The suspension of solid potassium chloride crystals in mother liquor goes to a cone settler. The clear overflow liquor is sent to borax crystallisers; the underflow potassium chloride sludge is further concentrated in continuous centrifuges, and then dried in a rotary dryer to yield potassium chloride crystals of 97% purity.

Other processes

Smaller quantities of potassium chloride are recovered from brines other than the Searles lake brine. The natural brine, containing about 1% potassium chloride, is concentrated to 30% in ponds. The crystalline product is then brought to about 95% purity by flotation.

Uses

Agricultural and chemical.

Properties

Colourless cubic crystals, granules, or powder. Low toxicity. Formula wt 74.55, mp 772°C, Sp. gr. 1.988, bp 1411°C. Soluble in water (27.6 g/100 g water at 0°C; 56.7 g/100 g at 100°C). Soluble in ethanol and alkalies.

Muriate of potash (the commercial name for fertiliser-grade potassium chloride) is the principal source of potassium not only for fertiliser but also for the much smaller demand for potassium salts of all kinds, principally potassium hydroxide. It is not, however, the only source; langbeinite, a mixed potassium-magnesium sulphate, is also worked. The greatest demand for potassium salts is for fertiliser; this demand is essentially insatiable as the ever-expanding world population requires more food and fibre. In spite of this basic demand, however, the fertiliser business is notoriously subject to financial cycles depending on weather and general economic conditions.

Various Potassium Salts

Potassium hydroxide

The electrolytic process for production of potassium hydroxide is exactly analogous to that for sodium hydroxide with the substitution of potassium chloride solution for sodium chloride. Obviously, sea water and natural brines cannot be used as starting materials for making-up the potassium chloride solution. Both mercury and diaphragm cells are used, and their relative advantages and disadvantages are the same as for caustic soda. About 4 to 5% of the electrolytic chlorine supply originates as a co-product of potassium hydroxide manufacture.

Reaction

$$2KCl + 2H_2O \rightarrow 2KOH + H_2 + Cl_2$$

95 to 97% current efficiency

Solubility relationships in the potassium chloride-potassium hydroxide system are different from those in the corresponding sodium system; and potassium salt solutions are more electrically conductive than their sodium counterparts. There are therefore some minor differences in operating parameters. In diaphragm cells, the product liquor contains 10 to 15% potassium hydroxide and about 10% potassium chloride. As with sodium hydroxide, most of the potassium chloride crystallises out during concentrating by evaporation and subsequent cooling, but about 1% potassium chloride is a lower limit to the possibilities of purification by this method. Mercury-cell operation is analogous to the sodium system; the product potassium hydroxide is much purer, and is concentrated enough (45%) to be sold directly.

However, the purification afforded by the mercury cathode does not extend to the sodium impurity in the potassium chloride feed; and sodium ends up as equivalent contamination of the caustic potash with caustic soda. A drawback of mercury cells is that they require solid potassium chloride feed; this accentuates the difficulties of purifying the potassium chloride.

Large amounts of caustic potash are used as 45 or 50% solutions; lesser amounts are concentrated to 61% potassium hydroxide, corresponding to the dihydrate $KOH \cdot 2H_2O$, and cast into sticks or pellets. The remainder is concentrated to 90 to 92% and sold as solid flakes or lumps. Reducing the water content further would require melting *in vacuo*. Most technical caustic potash contains up to 1.5% potassium carbonate.

Although, like caustic soda, potassium hydroxide may be safely handled in plain steel tanks, at higher temperatures, approaching 90 to 95°C, it is much more corrosive to steel than is sodium hydroxide, so that when heating tanks to avoid freezing, care must be exercised. Nickel-lined tanks are frequently used for this reason.

Uses

Potassium carbonate, soaps, tetrapotassium pyrophosphate, other potassium compounds, liquid fertilisers, dyestuffs etc.

Properties

Soluble in water (112 g/100 g. at 20°C), ethanol, ether, and glycerine. Insoluble in acetone. Since caustic potash is more costly than caustic soda on the bases of both unit weight and alkali equivalent weight, caustic soda will always be used where only the properties of a strong alkali are required; the uses of caustic potash will be restricted to those where its particular properties are necessary.

Potassium soaps are liquid, while sodium soaps are generally solid; thus potassium hydroxide is required for liquid soaps and detergents. Tetrapotassium pyrophosphate, for the same reason, is used as the builder in liquid detergents. Potassium is required by plants, whereas sodium may actually be harmful if in excess. Potassium salts in general are more soluble than their sodium analogs, and this advantage often outweighs their higher cost.

Potassium carbonate

Properties

It is a white, deliquescent, granular translucent powder, soluble in water, giving an alkaline solution by hydrolysis; soluble in alcohol, sp. gr. 2.428 (19°C); mp. 891°C; bp, decomposes, non-combustible. It can exist as tri, di, and monohydrates, all of which loose their water of crystallisation on heating above 130°C. In its reactions it closely resembles sodium carbonate.

Preparation

1. Alkyl amines or ion exchange resins can be used with potassium chloride and carbon dioxide to yield potassium bicarbonate, which is calcined to carbonate.
2. It can be prepared by using magnesium oxide, potassium chloride, and carbon dioxide, by separating Engles salt ($MgCO_3 \cdot KHCO_3 \cdot 4H_2O$). Decomposition leaves potassium bicarbonate in solution, which can be processed to potassium carbonate.

Uses

It is used to make special glasses (optical and colour TV tubes), potassium silicate, pigments; dehydrating agent; printing inks, laboratory reagent, soft soaps, raw wool washing; general purpose food additive.

Potassium sulphate

Properties

It is colourless or white hard crystals or powder; bitter saline taste. Soluble in water; insoluble in alcohol. Sp gr. 2.66; mp 1072°C.

Preparation

1. By fractional crystallisation from solutions of naturally occurring double salts.
2. By treatment of potassium chloride either with sulphuric acid or with sulphur dioxide, air, and water.
3. From salt lake brines.

Uses

Reagent in analytical chemistry; medicine (cathartic); gypsum cements; fertiliser for chloride sensitive crops, such as tobacco and citrus; alum manufacture, glass manufacture; food additive.

Potassium acid tartrate

Properties

White crystals or powder; soluble in boiling water; pleasant slightly acid taste; insoluble in alcohol. Sp. gr. 1.984 (18°C). Non-toxic.

Preparation

A domestic tartrate industry has been developed by processing pomace, which is discarded from wine fermentors and contains from 1 to 4% potassium bitartrate (from wine lees by extraction with water and crystalisation).

Uses

Baking powder, preparation of other tartrates; medicine; galvanic tinning of metals; food additive.

Potassium nitrate

Properties

It is transparent, colourless, or white crystalline powder; slightly hygroscopic; pungent, saline taste, Sp. gr. 2.1062; mp 337°C; bp decomposes at about 400°C. Soluble in water and glycerol; slightly soluble in alcohol. Low toxicity.

Preparation

It is prepared by adding potassium chloride to a least saturated solution of sodium nitrate:

$$NaNO_3 + KCl \rightleftharpoons KNO_3 + NaCl$$

Because sodium chloride is the least soluble of these substances in hot water, much of it is precipitated from solution and filtered off while hot causing position of equilibrium to move to the right. On cooling the mixture, most of potassium nitrate formed crystallises out, because it is much less soluble in cold water than in hot (the solubility in 100g of water is 13 g at 0°C, and 246 g at 100°C).

Uses

Pyrotechnics; explosives; matches; speciality fertiliser; reagent to modify turning properties of tobacco; glass manufacture; tempering steel; curing foods; oxidiser in solid rocket propellants. It is dangerous, fire and explosion risk when shocked or heated, or in contact with organic materials; strong oxidising agent.

CHAPTER 10

Nitrogen Industry

INTRODUCTION

Humans have increased their supply of food by feeding nitrogen compounds back into the soil. Chemists and chemical engineers have found ways of making nitrogen derivatives out of air in an economical way. The first successful process, the arc process, required much cheap electric energy. However, the present solution to the nitrogen-fixation problem was obtained by reacting nitrogen with low-cost hydrogen to make ammonia under conditions requiring less power and low conversion expense. This made the arc process obsolete. Ammonia has now become one of the heavy chemicals, produced in enormous tonnage around the globe, and at such low prices as to dominate the world supply of nitrogen fertilisers and most nitrogen compounds. This is probably one of the most important chemical engineering achievements in history. The development of a practical synthetic ammonia process was carried out through the chemical engineering efforts of Haber and Bosch and their co-workers, and the growth of this industry has continued. Cyanamide and the ammonia-manufacturing procedures are currently those of technical importance for supplying the world with fixed nitrogen.

The chemical nitrogen industries include not only nitrogen fixed by humans, but also by-product ammonia from coke ovens and such natural nitrogen deposits both of which are subjected to manufacturing processes. Ammonia is used in heat-treating and paper pulping, nitric acid for nitro compounds, high explosives, and propellants. Hydrogen cyanide, made largely from ammonia, is used mainly for acrylates and methacrylates, although a little is used for fumigation. Acrylonitrile uses substantial amounts of nitrogen compounds. Hydroxylamine and hydrazine also are smaller-volume nitrogen consumers. Amines of fatty acids enjoy wide use as surface-active agents and in ore flotation.

CYANAMIDE

Cyanamide, (carbamic acid nitrile; carbamodiimide), $H_2NC\equiv N$, is a weak acid with a very high solubility in water. The major use of cyanamide was at one time agricultural. Its employment as a chemical raw material has become increasingly important, however, the largest chemical use being for the preparation of the dimer dicyandiamide. This dimer is polymerised to produce the trimer melamine, which has applications in plastic resins. Other commercial derivatives are produced by way of a crude calcium cyanide (48 to 50% expressed as NaCN) from a high temperature melt of cyanamide with excess carbon and NaCl. This is used directly for cyanidation of ores and to manufacture ferrocyanides.

335

The use of cyanamide fertilisers is enjoying a minor resurgence, because these chemicals act as pesticides to snails and larvae and also enrich the soil with nitrogen. Cyanamide compounds are used to fire-proof shingles and cotton fabrics.

Reactions and Energy Changes

The essential reactions for the production of calcium cyanamide are:

$$CaCO_3(s) \longrightarrow CaO(s) + CO_2(g) \qquad \Delta H = +43.5 \text{ kcal} \qquad ...(10.1)$$

$$CaO(s) + 3C(amorph) \longrightarrow CaC_2(s) + CO(g) \qquad \Delta H = +103.0 \text{ kcal} \qquad ...(10.2)$$

$$CaC_2(s) + N_2(g) \longrightarrow CaCN_2(s) + C(amorph) \qquad \Delta H = -68.0 \text{ kcal} \qquad ...(10.3)$$

Various catalysts or fluxes are used to increase the rate of reaction or cause it to proceed at lower temperatures. Generally uses of calcium fluoride, and reaction-rate studies have shown that it reduces the temperature of optimum reactivity and increases the velocity of the reaction by 4.5 times at 1000°C. Reaction (10.3) takes place at 900 to 1000°C. Reaction (10.2) is carried out in two 20,000 and two 10,000-kW furnaces at about 2000 to 2200°C. Under normal conditions reaction (10.2) yields up to 90% calcium carbide. Considerable overall energy is needed, principally to secure the high temperature for reaction (10.3) to start when it is self-sustaining and for making the calcium carbide in reaction (10.2). The source of carbon is usually coke. Coal is required to burn the limestone and to dry the raw materials (Fig. 10.1).

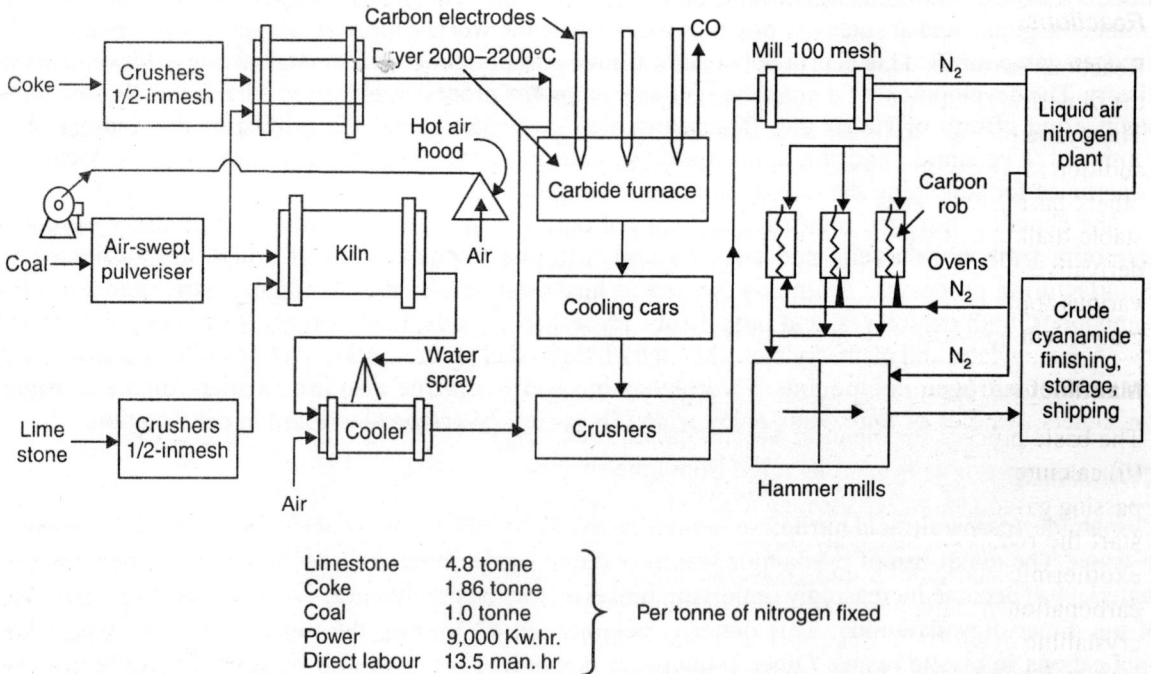

Limestone	4.8 tonne	
Coke	1.86 tonne	
Coal	1.0 tonne	Per tonne of nitrogen fixed
Power	9,000 Kw.hr.	
Direct labour	13.5 man. hr	

Fig. 10.1. Flowchart for the manufacture of calcium carbide and calcium cyanamide.

The following physical operations and chemical conversions are needed to commercialise the reactions on which Fig. 10.1 is based on:

1. Limestone, coal and coke are pulverised, separately.
2. Limestone is calcined to quicklime.
3. Coke is pulverised, dried and mixed with quicklime.
4. Carbide is formed in an electric furnace at nearly 2000 to 2200°C and run out molten.
5. Carbide is cooled, crushed, and finely ground.
6. Air is liquefied by compressing, cooling and expansion.
7. Nitrogen is separated from oxygen by liquid rectification.
8. Calcium carbide is nitrified over the course of 40 hour with 99.9% nitrogen at about 1000°C.
9. Calcium cyanamide is pulverised and treated with a small amount of water to hydrate residual CaO and CaC_2. It may be oiled to reduce dust.

Properties

Cyanamide crystallises from a variety of solvents as somewhat unstable, colourless, orthorhombic, deliquescent crystals. Dimerisation is prevented by traces of acidic stabilisers, e.g., monosodium phosphate. Studies of the infrared and the Raman spectra support the N-cyanoamine structure, NH_2—C≡N. It is completely soluble at 43°C, and has a minimum solubility (eutectic) at –15°C. It is highly soluble in polar organic solvents, and less soluble in non-polar solvents, mp 46°C; bp 140°C (101.3 kPa); d 1.282 g/cm^3; ref. index 1.4418 at 48°C.

Reactions

Reactions are either additions to the nitrile group or substitution at the amino group. Both are involved in the dimerisation to dicyandiamide, which occurs readily at pH 8–10.

The cyanamide anion is strongly nucleophilic and reacts with most alkylating or acylating reagents; addition to a variety of unsaturated systems occurs readily. In some cases, a cyanamide salt is used; in others base catalysis suffices. Alkylation with a variety of common alkyl halides or sulphates gives stable dialkyl cyanamides. The reaction with formaldehyde produces first an unstable hydroxymethyl derivative that resinifies more or less rapidly, passing through a water-soluble stage which permits various applications. Hydrogen chloride gives a dihydrochloride. Chloroform-amidine hydrochloride is a convenient anhydrous form which is easily stored and handled.

Manufacture

The basic process for manufacture comprises four stages: (i) lime is made from high grade limestone; (ii) calcium carbide is manufactured from lime and coal or coke; (iii) calcium cyanamide is produced by passing gaseous nitrogen through a bed of calcium carbide, which is heated to 1000-1100°C in order to start the reaction (the heat source is then removed as the reaction continues because of its strong exothermic character); and (iv) cyanamide is manufactured from calcium cyanamide by continuous carbonation in aqueous medium. For production of commercial 50% solution, and for recovery of crystalline cyanamide, this process is modified to improve purity and concentration, e.g., calcium and iron may be removed by ion-exchange treatment.

Other processes include the reaction of lime with hydrogen cyanide; the reaction of limestone with ammonia; and the reaction of lime and urea to form calcium cyanate which is then converted to calcium cyanurate which gives calcium cyanamide at a higher temperature.

Dicyandiamide

Dicyandiamide (cyanoguanidine) is the dimer of cyanamide and crystallises in colourless monoclinic prisms. It is amphoteric, and generally soluble in polar solvents and insoluble in nonpolar solvents. mp 208°C.

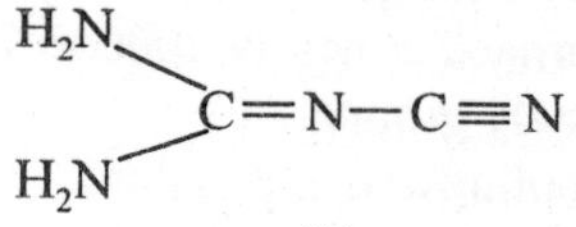

$$H_2N \diagdown \atop H_2N \diagup C=N-C\equiv N$$

(1)

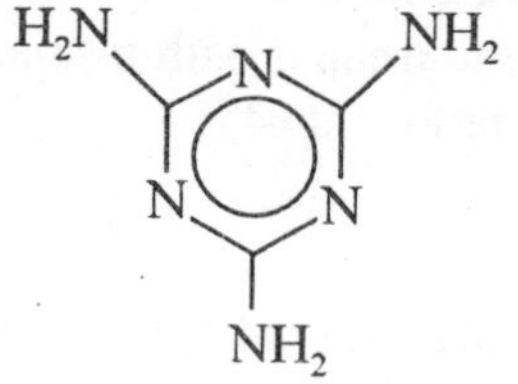

(2)

Reaction

The reactions resemble those of cyanamide. However, cyclisations take place easily and the nitrile group is less reactive. For example, under pressure and in the presence of ammonia, it cyclises to melamine. Guanamines are obtained when dicyandiamide is heated with alkyl or aryl nitriles in the presence of small amounts of alkali. Reaction with ammonium salts gives biguanide salts which react further with the ammonium salt forming guanidine salts; guanidine nitrate is manufactured by this route. Hydrolysis occurs easily at elevated temperatures. It can be treated with formaldehyde to produce resinous compositions of varying proportions.

Dicyandiamide is manufactured by the dimerisation of cyanamide in aqueous solution, with pH adjusted to 8–9 and held at ca 80°C for 2 hour to give complete conversion.

Melamine

Melamine (cyanurotriamide, cyanuramide, 2,4,6-triamino-1,3,5-triazine) is a white crystalline material that is insoluble in most organic solvents. It is appreciably soluble in water, mp 3550°C; d 1.573g/cm^{-1}. Although moderately basic, it is better considered as the triamide of cyanuric acid than as an aromatic amine. Its reactivity is poor in nearly all reactions considered typical for amines. Melamine was formerly manufactured by the conversion of dicyandiamide under heat. The dehydration condensation of urea has displaced the dicyandiamide process.

Health and safety

Manufacture of cyanamide does not present any serious health hazard. Contact or ingestation must be avoided. Dicyandiamide is essentially nontoxic, but may cause dermatitis. Melamine crystal may be handled in ordinary industrial use without special hygienic precautions.

Uses

Cyanamide and calcium cyanamide are used as fertilisers. Calcium cyanamide is used for steel nitridation and desulphurisation; it is the raw material for dicyandiamide. Dicyandiamide is used as a raw material for the manufacture of guanamines, biguanide, and guanidine salts, and various resins. Most of the melamine produced is used in the form of melamine-formaldehyde resins.

SYNTHETIC AMMONIA

Ammonia is one of the truly fundamental raw materials utilised in modern civilisation. The largest consumer of anhydrous ammonia is the fertiliser industry, which utilises it in the production of calcium and sodium nitrate, ammonium sulphate, nitrate and phosphate, ammoniated superphosphates, urea, and aqueous ammonia. Fig. 10.2. highlights the raw materials used for manufacture of ammonia and nitrogen fertilisers. Applied by special machines, anhydrous ammonia has become a very important fertiliser for direct application to soils. In combination with ammonium nitrate, an increasing amount is used each year. Ammonia is the starting point of nearly all military explosives. Scarcely an industry is untouched, since ammonia is required for the making of soda ash, nitric acid, nylon, plastics, lacquers, dyes, rubber, and other products. The product is handled and shipped in two forms, ammonia liquor and anhydrous ammonia. Commercial grades of the liquor usually contain 28% ammonia.

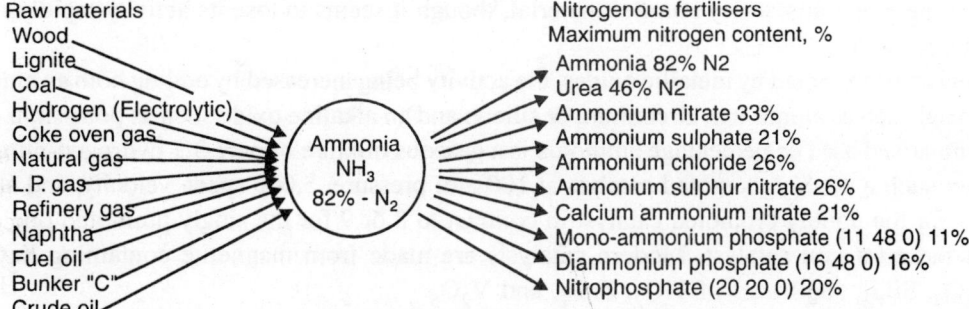

Fig. 10.2. Raw materials used throughout the world to manufacture ammonia and nitrogenous fertilisers produced from ammonia.

Reactions and Equilibriums

For the reaction

$$\frac{1}{2}N_2(g) \; + \; \frac{3}{2}H_2(g) \rightleftharpoons NH_3(g)$$

an equilibrium constant can be expressed as

$$K_p = \frac{P_{NH_3}}{P_{N_2}^{1/2} \times P_{H_2}^{3/2}}$$

Since the volume of ammonia obtained is less than the combined volume of nitrogen and hydrogen, a pressure increase, according to the *principle of Le Chatelier*, gives a higher percentage of ammonia at equilibrium. The conversion percentage increases several fold as the pressure increases from 100 to 1,000 atm, but the percentage of ammonia in equilibrium with the reacting gases decreases continually with a temperature rise up to 1100°C and reaches a minimum at this temperature. When operating below 1100°C the lower the temperature for a given pressure at which an ammonia converter can be run, the larger the possible percentage of ammonia in the ammonia-nitrogen-hydrogen mixture, but the longer the time required to attain equilibrium. Accordingly, temperatures in converters are kept as low as is consistent with a sufficiently rapid reaction rate to be economical with the catalyst used and with the investment required.

Rate and catalysis of the reaction

To be economical, the rate of this reaction must be increased, because hydrogen and nitrogen alone react very slowly. The basis of the commercial synthesis of ammonia rests upon an efficient catalyst to speed up the reaction rate to an economical degree. Such a catalyst has been found in iron, whose reaction rate is promoted by the addition of oxides of aluminum (3%) and potassium (1%). These promoters prevent sintering.

Many reactions are known which occur as a result of one reacting constituent combining with the catalyst to form an intermediate compound capable of reacting with a second reacting constituent to form the product of the catalytic reaction and to regenerate the catalyst. Thus it is postulated that a nitride adduct is possible with the active iron atoms on the surface of the ammonia catalyst. Iron seems to be by far the most satisfactory *catalyst* material, though it seems to lose its activity rapidly if heated to temperatures above 520°C.

The catalyst is promoted by metallic oxides, the activity being increased by adding both an amphoteric oxide of metal such as aluminum, zirconium, or silicon and an alkaline oxide such as potassium oxide. It can be summarised as: The percentage ammonia is a gaseous mixture of pure 3:1 hydrogen-nitrogen gas passed over such a doubly promoted catalyst at 100 atm pressure, 5,000 space velocity, and at 450°C, is 13 to 14 for the doubly promoted catalyst in contrast to 8 or 9 for the singly promoted one and to 3 to 5% for the pure iron catalyst. Modern catalysts are made from magnetite containing K_2O, CaO, MgO, Al_2O_3, SiO_2, and traces of TiO_2, ZrO_2, and V_2O_5.

The promoters make the reduced catalyst more porous. Catalysts more active at a lower temperature would be most valuable. These catalysts are ruined by contact with many substances such as copper, phosphorus, arsenic, and sulphur that affect the electronic structure of the iron; carbon monoxide greatly reduces their activity. Hence much money must be spent in purifying the hydrogen and nitrogen for ammonia synthesis. A detailed mechanism for the catalysis leading to ammonia is as follows, and is itself only speculation, but to a major degree meets most observed facts. The controlling aspect is nitrogen activation.

$$N_2 + 2Fe = 2Fe - N_{ads}$$
$$H_2 + 2Fe = 2Fe - H_{ads}$$
$$N_{ads} + H_{ads} = NH_{ads}$$
$$NH_{ads} + H_{ads} + NH_{2ads}$$
$$NH_{2ads} + H_{ads} = NH_{3ads}$$
$$NH_{3ads} = NH_{3desorb}$$

Space velocity is the number of cubic feet of gases, corrected to standard conditions (0°C and 760 mm), that pass over 1 ft^3 of catalyst space per hour. The space velocity used in commercial operations differs considerably in different processes and in different plants using the same process. Although the percentage of ammonia in the gas stream issuing from a converter goes down as the space velocity goes up, the amount of ammonia per cubic foot of catalyst space per hour increases. Too high a space velocity, however, disturbs the thermal balance of a converter, involves increased cost of ammonia removal because of the smaller percentage present in the exit gases, and makes necessary the recirculation of large volumes of gas.

AMMONIA

Ammonia is a colourless, alkaline gas, lighter than air, and possessing a unique, penetrating odour.

Physical Properties

Bp, $-33.35°C$; fp, $-77.7°C$; critical temperature, $133.0°C$; specific heat (at $0°C$), 2097.2; soluble to the extent of 33.1 wt% in H_2O at $20°C$; specific gravity (anhydrous ammonia) at $-40°C$, 0.690; and density of aqueous ammonia is 0.618 g/cm^3 ($15°C$). The flammable limits of ammonia in air are 16 and 25 vol %; in oxygen the range is 15–79 vol%. Such mixtures can explode, although ammonia–air mixtures are quite difficult to ignite (ignition temperature ca $650°C$). Ammonia is readily absorbed in water to make ammonia liquor. Approximately 2180 kJ (521 kcal) of heat is evolved when 1 kg of ammonia gas dissolves in water.

The alkali and alkaline-earth metals (except beryllium) are readily soluble in ammonia, metallic magnesium only slightly. Iodine, sulphur, and phosphorus dissolve in ammonia; ammonia readily attacks copper in the presence of oxygen. Most fluorides are insoluble in liquid ammonia. Potassium, silver, and uranium are only slightly soluble. Both ammonium and beryllium chloride are very soluble, whereas most other metallic chlorides are slightly soluble or insoluble. Bromides are in general more soluble in ammonia than chlorides, and most of the iodides are soluble. Oxides, hydroxides, sulphates, sulphites, and carbonates are insoluble. Nitrates, e.g., ammonium nitrate, and urea are soluble in anhydrous and aqueous ammonia (this property makes the production of fertiliser nitrogen solutions possible). Many organic compounds, e.g., amines, nitro compounds, and aromatic sulphonic acids, also dissolve in liquid ammonia.

Chemical Properties

Ammonia is comparatively stable at ordinary temperatures but decomposes into hydrogen and nitrogen at elevated temperature. The rate of decomposition is greatly affected by the nature of the surfaces with which the gas comes into contact. Whereas glass is very inactive, porcelain and pumice have a distinctly accelerating effect, and metals, such as iron, nickel, osmium, zinc, and uranium, even more.

At atmospheric pressure, dissociation of ammonia begins at about 450-500°C, whereas in the presence of catalysts, it begins as low as 300°C and is nearly complete at 500-600°C. However, at 1000°C a trace of ammonia remains. Ammonia reacts readily with a large variety of substances.

One of the most important reaction is oxidation giving nitrogen and water. Gaseous ammonia is oxidised by many oxides of the less positive metals (e.g., cupric oxide) when heated to a relatively high temperature. Powerful oxidising agents react similarly at ordinary temperatures, e.g., potassium permanganate :

$$2 NH_3 + 2 KMnO_4 \rightarrow 2 KOH + 2 MnO_2 + 2 H_2O + N_2$$

The action of chlorine on ammonia can also be regarded as an oxidation reaction:

$$8 NH_3 + 3 Cl_2 \rightarrow N_2 + 6 NH_4Cl$$

With a platinum-rhodium catalyst, ammonia can be oxidised to nitric oxide and water which is a means of producing nitric acid. Base metal catalysts are also used to promote ammonia oxidation.

Reducing agents normally do not react. The neutralisation of acids is of commercial importance, e.g., three major fertilisers, ammonium nitrate, ammonium sulphate, and ammonium phosphate, are made from ammonia.

The reaction between ammonia and water is reversible:

$$NH_3 + H_2O \rightleftarrows [NH_4OH] \rightleftarrows NH_4^+ + OH^-$$

With increasing temperature, the solubility of ammonia decreases rapidly. Ammonia is a comparatively weak base and ionises in water much less than sodium hydroxide. Aqueous ammonia also acts as a base in precipitating metallic hydroxides from solutions of their salts and forming complex ions in excess ammonia solution. For example, with copper sulphate solution the cupric hydroxide, which is at first precipitated, redissolves in excess ammonia solution owing to the formation of the complex cuprammonium, tetramminecopper(II) ion, $[Cu(NH_3)_4]^{2+}$. Both potassium and sodium metals dissolve to give the amides, e.g., $NaNH_2$. Other metals, such as magnesium, calcium, strontium, and barium, give the nitrides. Reactions with halogens give unstable trihalides.

Ammonia reacts with chlorine in dilute solution to give chloramine, an important reaction in water purification. Ammonia reacts with phosphorus vapour at red heat to give nitrogen and phosphine. Sulphur also reacts with liquid anhydrous ammonia to produce nitrogen sulphide, N_2S_4. Ammonia and carbon at red heat give ammonium cyanide. Ammonia forms a great variety of addition or coordination compounds, also called ammoniates, in analogy with hydrates. Thus $CaCl_2.6NH_3$ and $CuSO_4.4NH_3$ are comparable to $CaCl_2.6H_2O$ and $CuSO_4.4H_2O$, respectively. Such compounds when regarded as coordination compounds are called amines and are written as complexes, e.g., $[Cu(NH_3)_4]SO_4$. The solubility in the water of such compounds is often quite different from the solubility of the salts themselves, e.g., silver chloride, AgCl, is almost insoluble in water, whereas $[Ag(NH_3)_2]Cl$ is readily soluble. Thus silver chloride dissolves in aqueous ammonia. Similar reactions take place with other insoluble salts such as silver phosphate and cuprous chloride. The reaction of ammonia and carbon dioxide via ammonium carbamate to urea is of considerable industrial importance.

Manufacture

Haber and Bosch in Germany developed the first commercial process for the direct synthesis of ammonia (Fig. 10.3). In the mid 1960s, the ammonia industry grew very rapidly owing to improvements in the energy cycle, including recovery of waste heat to generate high pressure steam, and technical advances in compressors, reformers, and converters. Ammonia is synthesized by the reversible reaction of hydrogen and nitrogen:

$$N_2 + 3H_2 \rightleftarrows 2 NH_3$$

In the design of synthesis facilities, the rate at which the ammonia is formed has to be considered.

The reaction of hydrogen and nitrogen to form ammonia on an industrial scale is always carried out on a catalytic surface based on metallic iron (mostly magnetite) that has been promoted with other oxides. Oxygen and oxygen-containing compounds have a temporary poisoning effect on the catalyst and sulphur, arsenic, phosphorus, chlorine, and heavy hydrocarbons cause permanent poisoning. The following variables affect ammonia synthesis:

1. Synthesis pressure, an increase in pressure increases the equilibrium per cent rate of ammonia and the reaction rate.

2. Synthesis temperature, higher temperatures increases reaction rates but decreases the equilibrium amount of ammonia and increases the thermal degradation of the catalyst.

3. Space velocity (the ratio of the volumetric rate of gas at standard conditions to the volume of catalysts), the percentage of ammonia in the existing gas decreases with increasing space velocity. However, the same volume of catalyst at increased space velocities can produce more ammonia (normal space velocities for commercial operations are 8,000–60,000 h^{-1}).

4. Inlet gas composition, effects vary depending on the concentration of inert materials, ammonia concentration, and hydrogen to nitrogen ratio. Most commercial facilities operate at a hydrogen to nitrogen ratio of 3:1; however, the optimum may be something less than that.

5. Catalyst particle size, catalyst activity increases with smaller particles (most common size 6–10 mm).

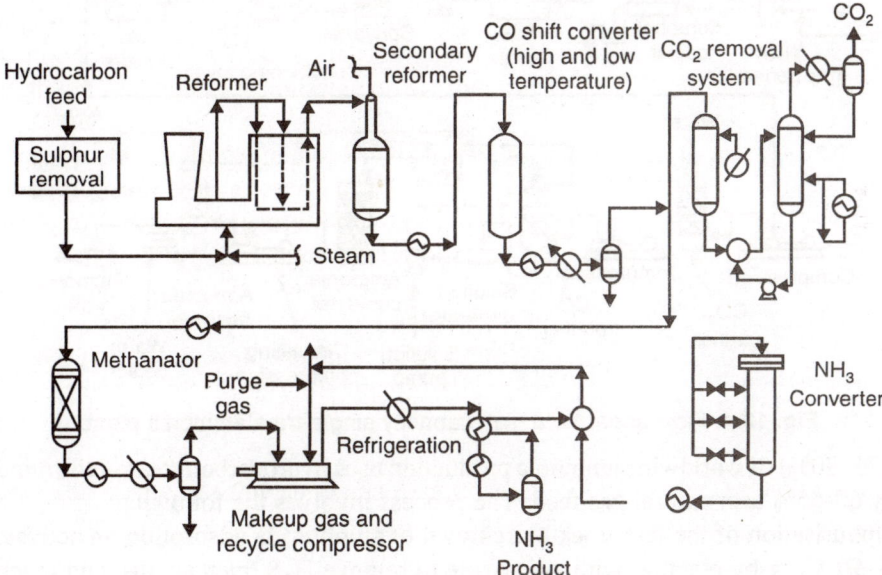

Fig. 10.3. Haber-Bosch process.

The manufacture of synthetic anhydrous ammonia consists of three basic steps: synthesis-gas preparation, purification, and ammonia synthesis. The first two involves the generation of hydrogen, the introduction of nitrogen in the stoichiometric synthesis proportion, and the removal of catalyst poisons (namely, carbon dioxide, carbon monoxide, and water). Ammonia synthesis includes the catalytic fixation of nitrogen at elevated temperatures and pressures and the recovery of the ammonia. Although conditions vary greatly, the chemistry of ammonia synthesis is basic to all commercial processes.

A typical flow sheet for a high capacity single-train ammonia plant is shown in Fig. 10.4.

By and large, the raw-material sources in modern ammonia plants are coal, petroleum fractions, and natural gas. In general, the most economical feedstock has the highest hydrogen to carbon ratio.

The two basic generation techniques for processing these raw materials are partial oxidation and steam reforming. Steam reforming is most often used in commercial plants, but partial oxidation processes are employed where steam-reformable feeds are not available.

The heart of a coal-based partial oxidation process is the gasification step. The two important commercial processes commonly used are the Lurgi, employing a fixed-bed reactor using oxygen and steam at ca 2000-3000 kPa (20–30 atm) and temperatures ranging from 550 to 620°C, and the Koppers-

Totzek, in which gasification takes place at low pressures and much higher temperatures (ca 1480°C). In recent years, much attention has been given to production of synthetic gas via coal-gasification processes. Partial oxidation of hydrocarbon feedstocks for the production of synthetic gas uses the Shell and Texaco processes with operating conditions in the gas generator varying from 1200 to 1370°C and 3200-8370 kPa gauge (465–1215 psig) using heavy oils as hydrocarbon feed.

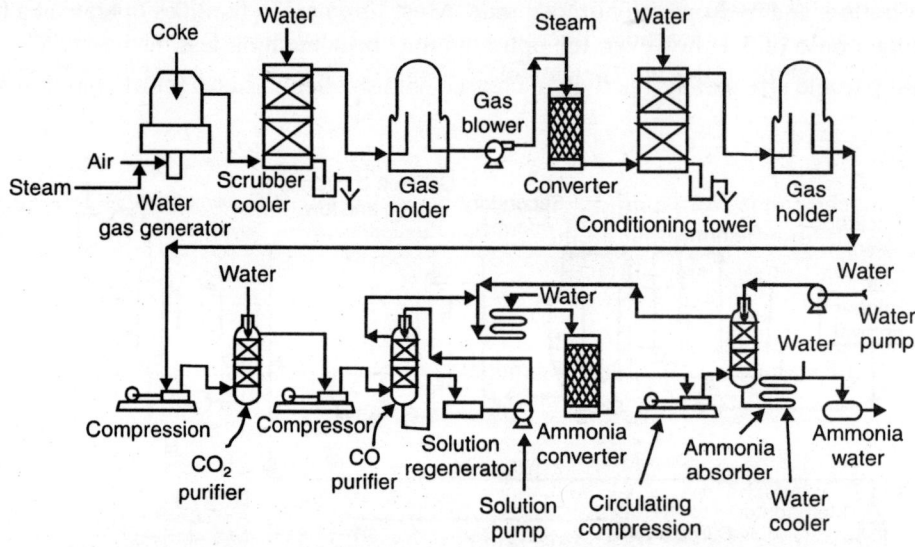

Fig. 10.4. Flow sheet for a high capacity single-train ammonia plant.

The bulk (75–80%) of worldwide ammonia production uses hydrocarbon steam-reforming operations, approximately 60-65% use; natural gas feed. The process involves the following:

1. Desulphurisation of the feedstock by removal of sulphur via adsorption on activated carbon at ca 15–50°C, or by reaction with zinc oxide to remove H_2S, mercaptans, and chlorine.

2. Primary and secondary reforming of hydrocarbon feedstock, carried out in two catalytic stages in the presence of steam: the first producing a partially reformed gas which is then further processed in a secondary reformer to achieve the low methane content desired.

3. Shift conversion, for conversion of carbon dioxide to hydrogen and carbon dioxide.

4. Carbon dioxide removal using reaction systems, e.g., MEA (20% monoethanolamine), promoted MEA (monoethanolamine plus Amine Guard), Vetrocoke (K_2CO_3 plus As_2O_3-glycine), Carsol (K_2CO_3 plus additives), Catacarb (K_2CO_3 plus additives), Benfield (K_2CO_3 plus diethanolamine and additives), and Lurgi (K_2CO_3 plus additives); combination reaction-physical systems, e.g., Sulphinol (sulpholane and 1, 1'-iminobis-2-propanol) and TEA-MEA (triethanolamine and MEA); and physical absorption systems that use various solvents, e.g., Purisol (NMP) (N-methyl-2-pyrrolidinone), Rectisol (methanol), Fluor solvent (propylene carbonate), and Selexol (dimethyl ether of polyethylene glycol).

5. Synthesis gas purification, to remove or convert to inert species residual carbon monoxide and carbon dioxide, either by methanation or cryogenic purification.

6. Ammonia synthesis and recovery, the actual step in which the ammonia is made. Currently, the operating pressure for most modern synthesis loops fall in the range of 13,785 to 34,475 kPa (2000-5000 psi.), plants designed before 1964 tended to favour the region above 31,030 kPa

(4500 psi), whereas new plants favour the lower range. The economics of the more recent large tonnage ammonia plants is more favourable at these pressures.

The question of what is the right synthesis facility varies from project to project with the synthesis pressure the most important consideration, followed by loop configuration, makeup gas-recycle compression, and the converter. The design of modern ammonia plants reflects a high degree of integration between the process and energy systems. Waste process heat is used to provide energy for boiler-feedwater heating and steam generation. Energy recovery in the convection section of the primary reformer includes satisfying similar utility demands of boiler-feedwater heating, and steam generation and superheating.

Health and safety

Effects of ammonia can range from a mild irritation to severe corrosion of sensitive membranes of the eyes, nose, throat, and lungs, depending on the concentration. Because of its great affinity for water, it is particularly irritating to moist skin surfaces.

Uses

The largest market for ammonia is the fertiliser industry. Other important markets include commercial explosives (nitric acid, initially derived from ammonia, is used in manufacture of explosives); and fibres and plastics.

AMMONIUM COMPOUNDS

The ammonium compounds consist of a large number of salts, many of which are of considerable industrial importance, e.g., ammonium sulphate, ammonium chloride, and ammonium nitrate. The ammonium salts are much like salts of the alkali metals in such properties as solubility, but they differ in being considerably less stable. Except for complexes containing metallic radicals, they are completely volatile on heating or ashing. They are more like the salts of potassium than those of sodium in some ways, e.g., in the lesser solubility of the ammonium and potassium perchlorates and chloroplatinates.

Ammonium Chloride

Ammonium chloride, known as sal ammoniac, occurs naturally in crevices in the vicinity of volcanoes as a sublimation product. It has specific heat of 1.55 J/g (0.371 cal/g) in the temperature range 1–55°C. It is soluble to the extent of 29.4 g/100 g H_2O at 0°C. Vapour pressure at 250°C 6.5 kPa; at 320°C 60.9 kPa; and at 338°C 101.1 kPa. The notable high vapour pressure of ammonium chloride at elevated temperatures causes it to sublime readily into a vapour consisting of equal volumes of ammonia and hydrogen chloride.

Like other ammonium salts of strong acids, it has an acid reaction in aqueous solution and the solid tends to lose ammonia and become more acid on exposure and in storage. Its aqueous solutions have a notable tendency to attack ferrous metal and other metal surfaces.

Ammonium chloride manufacture is based on metathesis, or double decomposition reactions of sodium chloride; the ammonium salt is used directly or formed *in situ*. Ammonium chloride is obtained as a by-product of the classic ammonia-soda or Solvay process. Double decomposition using the ammonium sulphate-sodium chloride process is another method used. Preparation by direct neutralisation of hydrochloric acid with ammonia is simple but not as attractive economically.

Ammonium Nitrate

Ammonium nitrate does not occur in nature and was first made in 1659 by Glauber, who called it *nitrum flammans* because of the difference of its yellow flame from that of potassium nitrate. It is the most important of the ammonium compounds from the standpoints of volume of production and major uses.

Ammonium nitrate is a white crystalline solid, d_4^{20} 1.725 g/cm^3. It is extremely soluble in water, and therefore very hygroscopic (Table 10.1). In the solid state, it occurs in five different crystallographic modifications.

Table. 10.1 Solubility of ammonium nitrate.

	Solubility of NH_4NO_3, g	
Temperature °C	*in 100 g water*	*in 100 g soln*
0	118	54.2
20	187	65.2
40	297	74.8
60	410	80.4
80	576	85.2
100	843	89.4

The change from one crystal form to another at various transition points can be detected by time-temperature cooling curves.

The solid salt has a specific heat of 1.70 J/g (0.406 cal/g) between 0 and 31°C. It has a negative heat of solution in water and can be effectively used to prepare freezing mixtures.

Ammonium nitrate decomposes under extreme shock or at elevated temperatures in two widely divergent manners. (Fig.10.5.)

$$NH_4NO_3 \rightarrow N_2O + 2\ H_2O$$

$$2\ NH_4NO_3 \rightarrow 2\ N_2 + 4\ H_2O + O_2$$

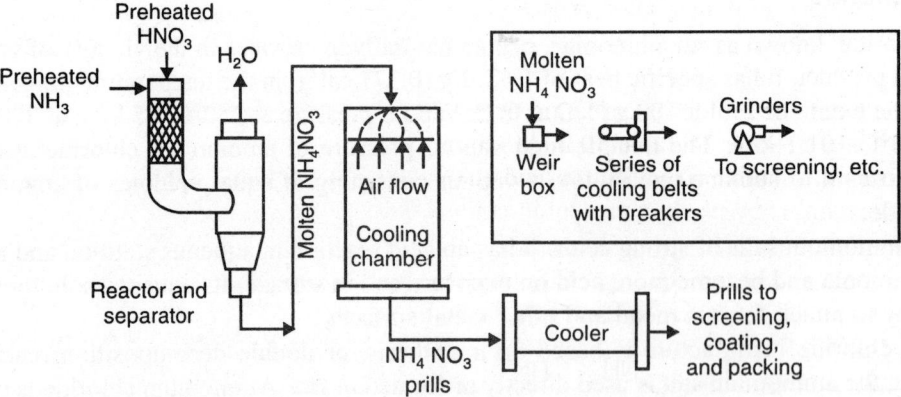

Fig. 10.5. Simplified flowchart for Stengel process for ammonium nitrate manufacture.

The first reaction, which occurs when ammonium nitrate is heated to temperatures from 200 to 260°C, can be carried out safely and can be controlled even when rapid. The second reaction takes place with great rapidity and violence when ammonium nitrate detonates.

Ammonium nitrate acts as an oxidising agent in many reactions at ordinary temperatures and in aqueous solutions it is reduced by various metals.

Ammonium nitrate manufacturing processes depend almost entirely on the neutralisation of nitric acid with ammonia in liquid or gaseous form and involve three essential steps: neutralisation, evaporation of the neutralised solution, and control of the particle size and characteristics of the dry product.

Ammonium Sulphate

Ammonium sulphate is a white crystalline solid, whose crystals are of rhombic structure, d_4^{20} 1.769 g/cm^3. The melting point as determined in a closed system is 513 ± 2°C. Upon heating in an open system, the sulphate begins to decompose at 100°C and yields ammonium bisulphate, NH_4HSO_4, which has a melting point of 146.9°C. The solubility of ammonium sulphate in 100 g water is 70.6 g at 0°C and 103.8 g at 100°C. It is insoluble in alcohol and acetone.

Ammonium sulphate is produced by reaction of by-product ammonia from coke ovens with sulphuric acid. Additional amounts have increasingly been produced by the reaction of synthetic ammonia and sulphuric acid; from the process for production of caprolactam; and via a method using gypsum, $CaSO_4.2H_2O$.

Health and safety

Ammonium nitrate as commonly handled in small amounts is properly considered safe. However, it is a potential high explosive when three conditions are present: priming by a high velocity explosive; confinement at elevated temperatures; and the presence of an oxidizable substance.

Uses of ammonium compounds

Ammonium acetate is used in pharmaceuticals as a diaphoretic and diuretic. Ammonium bicarbonate is used as a leavening agent in the production of certain baked foods. Ammonium chloride is used in the manufacture of dry cells; as a metal cleaner in soldering; as a flux in tinning and galvanizing; and in pharmaceuticals. Ammonium nitrate is used as a fertiliser and as an industrial explosive, as well as in the manufacture of nitrous oxide. Ammonium sulphate is used as a fertiliser, particularly for rice; and as an additive to supply nitrogen in fermentation processes.

UREA

Urea can be considered the amide of carbamic acid, NH_2COOH, or the diamide of carbonic acid, $CO(OH)_2$. Properties are shown in Table 10.2. At atmospheric pressure and at its melting point, urea decomposes to ammonia, biuret (i), cyanuric acid (ii), ammelide (iii), and triuret (iv). Biuret is the principal and least desirable by-product present in commercial urea.

(i)	(ii)	(iii)	(iv)
Biuret	Cyanuric acid	Ammelide	Triuret

Urea acts as a monobasic substance and form salts with acids. Urea and malonic acid give barbituric acid, an important compound in medicinal chemistry.

Table 10.2. Properties of urea.

Property	Value
Melting point, °C	135
Index of refraction, n_D^{20}	1.484, 1.602
Density, d_4^{20}, g/cm^3	1.3230
Crystalline form and habit	Tetragonal, needles or prisms
Free energy of formation, at 25°C, J/mol^a	—197.150
Heat of fusion, J/g^a	251^b
Heat of solution in water, J/g^a	243^b
Heat of crystallisation, 70% aqueous urea solution, J/g^a	460^c
Bulk density, g/cm^3	0.74
Specific heat, J/(kg·K)^a	
at 0°C	1.439
at 100°C	1.887

^{a} To convert, J to cal, divide by 4.184.
^{b} Endothermic.
^{c} Exothermic.

Manufacture

Urea is produced from liquid ammonia and gaseous CO_2 at high pressure and temperature. The formation of ammonium carbamate and the dehydration to urea take place simultaneously, for all practical purposes.

Ammonium carbamate is a white crystalline solid that forms by passing ammonia gas over dry ice. The conversion to urea begins below 100°C. The maximum conversion is attainable at 185°C, ca 53%. The ammonia and CO_2 gases recovered from the effluent mixture in several pressure-staged decomposition sections are absorbed in water and recycled.

$$2NH_3 + CO_2 \leftrightarrows NH_2\overset{\displaystyle O}{\overset{\displaystyle \|}{C}}ONH_4$$
Ammonium
carbamate

$$NH_2\overset{\displaystyle O}{\overset{\displaystyle \|}{C}}ONH_4 \leftrightarrows NH_2\overset{\displaystyle O}{\overset{\displaystyle \|}{C}}NH_2 + H_2O$$
Urea

In the Mitsui-Toatsu total recycle process, the reactor is operated at ca 25 MPA (246 atm) and 195°C at an NH_3 to CO_2 overall mol ratio (fresh feed plus recycle) of ca 4:1. A relatively high conversion of carbamate to urea per pass is reported (67-70%). Many urea plants with capacities of up to 1800 metric tonnes per day are using this process.

In the Montedison urea process, the reactor is operated at ca 20–22 MPa (197-217 atm) and an overall NH_3 to CO_2 mol ratio of ca 3.5:1 (fresh feed plus recycle). A 62-63% conversion of carbamate is reported. In an improvement, based on the new isobaric double-recycle (IDR) technology, the reactor effluent is first stripped with NH_3 gas and then with CO_2 gas. Considerable reduction in steam consumption is reported.

The UTI heat-recycle process offers the following new concepts: An isothermal reactor is provided with an internal open-ended coil for countercurrent heat transfer from the strongly exothermic process of carbamate formation to the endothermic formation of urea; ca 40-50% of the makeup CO_2 feed is injected into the medium-pressure carbamate decomposition section, and more than 70% of the exothermic heat of carbamate formation in the medium-pressure absorption system is exchanged with relatively colder streams within the process. High pressure gas stripping, is based on high pressure CO_2 gas stripping at reactor pressure and relatively high temperature. The unconverted carbamate is decomposed to NH_3 and CO_2 by the stream of gaseous CO_2 passed through the reactor effluent solution at reactor pressure and condensed and recycled. Because of its energy efficiency, the stripping process accounts for almost half of the world's urea production.

In the Stamicarbon CO_2 stripping process, reactor, carbamate decomposer (stripper), and carbamate condenser each operate at ca 14 MPa (140 atm) at an NH_3:CO_2 mol ratio of ca 2.8:1. In the Snamprogetti NH_3 stripping process, a synthesis loop is operated at ca 15 MPa (150 atm) and an NH_3: CO_2 overall mol ratio of 3.8:1. Carbamate conversion to urea per pass is reported to be ca 65–67%.

Waste-water treatment

Under the pressure of government regulations with regard to residual NH_3 and urea in waste-waters, the fertiliser industry made an effort to improve waste-water treatment. It is, however, difficult to reduce the NH_3 and urea content to below 100 ppm. In a fairly efficient method the urea is hydrolysed to ammonium carbamate, and CO_2 and NH_3 are stripped off. The Stamicarbon system uses a hydrolyser for urea and a separate dual-desorption system to strip the ammonia. The residual content of NH_3 and urea is reported to be reduced to about 70 and 80 ppm, respectively.

The UTI hydrolyser stripper (Fig. 10.6) consists of a single stainless-steel tower in which urea is hydrolysed and NH_3 stripped simultaneously by means of steam and CO_2 stripping.

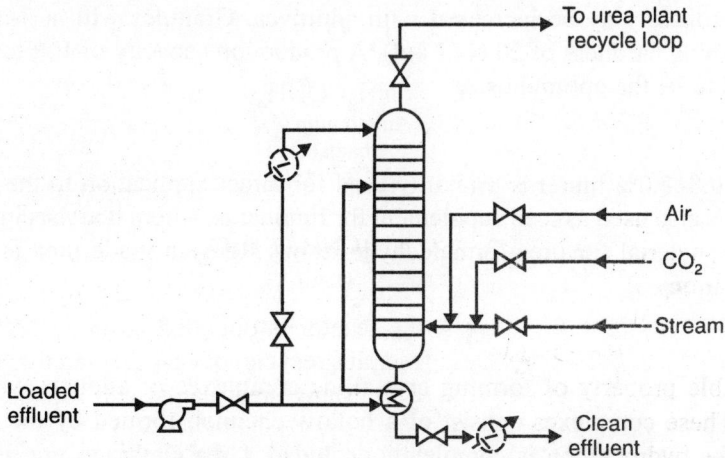

Fig. 10.6. Hydrolyzer-stripper clean-up process.

Finishing Processes

Urea processes provide an aqueous solution containing 70–80% urea. This solution can be used directly for nitrogen-fertiliser suspensions or solutions. The water is evaporated from the steam-heated solutions under reduced pressure with or without the addition of hot air as drying agent.

A combination of crystallisation and remelting removes biuret, which is detrimental in the technical-grade urea used for the manufacture of plastics. The solid urea thus obtained is relatively pure and well-suited for such applications. Until recently, most solid urea was produced by prilling. Molten urea obtained either by evaporation or by crystal melting is sprayed as droplets from the top of a tall cylindrical tower (50–60 m high) and allowed to fall counter-currently through a stream of cold air emanating from a fluid-bed cooler at the base of the tower. Prill towers require large quantities of air, and pollution abatement is very expensive. In addition, granulation offers far more flexibility, and therefore the industries all over the world are shifting to this technique.

In granulation, urea is concentrated to ca 99.5% and sprayed through nozzles onto a combination of falling granules and a cascading bed of recycled fines in a rotating cylindrical granulation drum. The physical characteristics of tne granules are good, with usually < 0.15% H_2O; sizes range between 6.7–4 mm or 3.4–1.7 mm, and the granules are hard and spherical. Biuret content is 1–2%. In pan granulation, a concentrated urea solution is sprayed onto a cascading bed of recycled fines in an inclined rotating pan. The falling-curtain evaporative cooling process is based on TVA sulphur-coated urea technology in combination with some novel approaches to granulation. The process is extremely energy efficient. A full-scale plant of ca 608 tonnes per day is expected to use ca 22.1 kW.h/T of product. Steam consumption, including concentration of scrubbing solution, is expected to be about 75 kg/t.

In the Nederlands Stikstof Maatschappij N.V process, granulation is accomplished in a fluid bed divided into several chambers. Pneumatic atomising nozzles are mounted just above the air-distribution plate and oriented to spray a 95–96% urea solution at 130–135%°C upward into the active bed. Seed particles are introduced into the first chamber. The particles grow by accretion when the atomised droplets of urea solution strike them and the water evaporates as the concentrated urea crystallises; biuret content increases only by 0.03%. In the Mitsui-Toatsu and Toyo engineering process, concentrated urea solution is sprayed onto the surface of particles in a fluidising granulation process; the heat released by solidification of the urea feed is removed by a combination of cooled recycle, fluidising air, and some water evaporating. Hardness   ay be increased with additives. Granules with a 3-mm diameter are reported to have a crushing hardness of 20 N (2 kgf). A production capacity of 100 tonnes perdays per granulator is estimated to be the optimum size.

Uses

Solid urea containing 0.8–2.0% biuret is primarily used for direct application to the soil as nitrogen-release fertiliser. Urea is also used as feed supplement for ruminants, where it assists in the utilisation of protein. Urea is a raw material for urea-formaldehyde resins. Reagent-grade urea is needed in some pharmaceutical preparations.

Clathrates

Urea has the remarkable property of forming crystalline complexes or adducts with straight-chain organic compounds. These complexes consist of a hollow channel, formed by the crystallised urea molecules, in which the hydrocarbon is completely occluded. Urea clathrates are used in petroleum refining for the production of jet fuels and for the dewaxing of lubricating oils.

NITRIC ACID

The essential raw materials for the modern manufacture of nitric acid are anhydrous ammonia, air, water, and platinum-rhodium gauze as a catalyst. Because of its low molecular weight, the ammonia can be shipped economically from large primary nitrogen-fixation plants to various oxidation plants at consuming centers. Furthermore, anhydrous ammonia can be shipped in steel cars, in comparison with aqueous nitric acid, which requires stainless-steel tank cars weighing much more.

Reactions and Energy Changes

The essential reactions for the production of nitric acid by the oxidation of ammonia may be represented as follows:

$$4NH_3(g) + 5O_2(g) \longrightarrow 4NO(g) + 6H_2O(g) \qquad \Delta H_{298°C} = -216.6 \text{ kcal} \qquad ...(10.4)$$

$$2NO(g) + O_2(g) \longrightarrow 2NO_2(g) \qquad \Delta H_{298°C} = -27.1 \text{ kcal} \qquad ...(10.5)$$

$$3NO_2(g) + H_2O(l) \longrightarrow 2HNO_3(aq) + NO(g) \qquad \Delta H_{298°C} = -32.2 \text{ kcal} \qquad ...(10.6)$$

$$NH_3(g) + O_2(g) \longrightarrow \frac{1}{2} N_2O(g) + \frac{3}{2} H_2 O(g) \qquad \Delta H = -65.9 \text{ kcal} \qquad ...(10.7)$$

Several side reactions reduce somewhat the yield of reaction (10.1):

$$4NH_3(g) + 3O_2(g) \longrightarrow 2N_2(g) + 6H_2O(g) \qquad \Delta H_{298°C} = -302.7 \text{ kcal} \qquad ...(10.8)$$

$$4NH_3(g) + 6NO(g) \longrightarrow 5N_2(g) + 6H_2O(g) \qquad \Delta H_{298°C} = -431.9 \text{ kcal} \qquad ...(10.9)$$

$$2NO_2(g) \longrightarrow N_2O_4 \qquad \Delta H = -13.9 \text{ kcal} \qquad ...(10.10)$$

Reaction (10.4) is essentially a very rapid catalytic one, carried out by passing about 10% ammonia by volume, mixed with preheated air, through the multilayered, silk-fine platinum-10% rhodium gauze at a temperature of approximately 920°C or lower here, once ignited, the stream of ammonia continuous to burn. The yield is 94 to 95%. The equilibrium constant for reactions (10.4) and (10.5) is 9.94×10^{14} for NO_2 production at 627°C. This reaction is usually carried out at a pressure of 100 psi. In Europe the process pressure is frequently 40lb and the temperature is about 815°C, with consequent smaller platinum losses and a lower production rate. As can be seen from reaction (10.4), there is only a small increase in volume, so that the *principle of Le Chatelier* does not affect the equilibrium very substantially. However, the increase in pressure, by compressing the reactants, permits greater space velocity to be maintained, resulting in a plant saving until such a pressure is reached that the cost of the increased thickness of the stainless steel required more than counterbalances the saving in volume of equipment per pound produced. Pressure oxidation also furnishes an acid containing 60 to 70% HNO_3 in comparison with the 50 to 55% HNO_3 obtained from atmospheric oxidation.

The speed of ammonia oxidation is extraordinarily high, giving an excellent conversion in the short contact time of 3×10^{-4} s at 750°C with a fine platinum–10% rhodium gauze catalyst. Hence, in the industrial procedure, it has been found economical to mix initially with the ammonia all the air needed for reactions (10.4) and (10.5). The oxidation of the NO to NO_2 is the slowest reaction, but the equilibrium is more favourable at lower temperatures. Hence this reaction is carried out in absorbers of considerable capacity, which are provided with cooling in all the upper trays. Because of a decrease in volume, this reaction is favoured under pressure, according to the principle of Le Chatelier. Although these factors raise the cost of the equipment for carrying out the oxidation of nitric oxide, they increase the conversion. It is necessary in the design of a plant to know how long this reaction will take, in order to calculate the volume necessary for the equipment.

$$3NO_2 + H_2O \rightleftharpoons 2HNO_3 + NO$$

Equation (10.6) is really an absorption phenomenon. This reaction, in the opinion of Taylor, Chilton, and Handforth, is the controlling one in making nitric acid, and its rate was increased by employing an absorption tower under pressure and cooling and with countercurrent graded strengths of acid for the absorption. Warm air is introduced into a short raschig-ring-packed section between the tower and acid trap. This provides for the reoxidation of the NO formed and also desorbs (bleaches) the dissolved nitrous oxides which would colour the acid.

From Ammonia

Air is compressed to approximately 100 psi (0.7 MPa), filtered, and preheated to about 300°C by passing through a heat exchanger. It is mixed with anhydrous ammonia, previously vapourised in a continuous-steam evaporator. The resulting mixture, containing about 10% ammonia by volume, is passed through the reactor, which contains a platinum-rhodium (2 to 10% rhodium) wire-gauze catalyst. The platinum-rhodium alloy is used in the form of a very fine gauze, 80-mesh and 75-μm-diameter wire, packed in layers of 10 to 30 sheets so that the gas travels downward through the gauze sheets. The catalyst temperature is about 750°C, the pressure 100 psi (0.7 MPa), and the contact time about 3×10^{-4} sec. (Fig. 10.7).

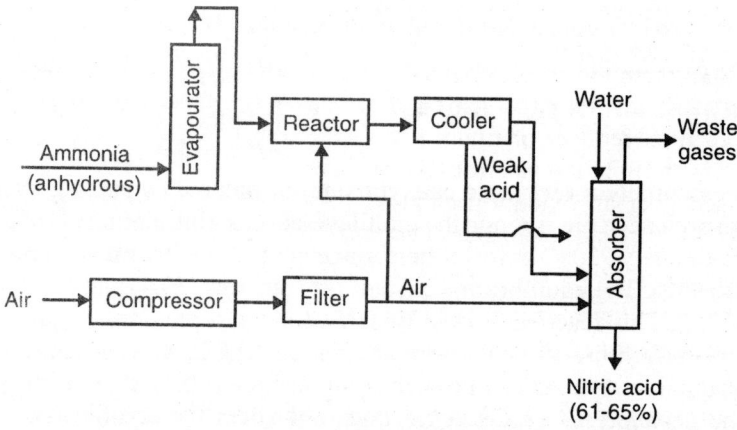

Fig. 10.7. Flow diagram for manufacture of Nitric acid from ammonia.

The hot nitric oxide and excess air mixture (about 10% nitric oxide) from the reactor is partially cooled in a heat exchanger and further cooled in a water cooler. The cooled gas is introduced into a stainless-steel absorption tower with more air being added for the further oxidation of nitric oxide to nitrogen dioxide.

Since the equilibrium of this reaction is more favourable at lower temperatures, most of the reaction occurs in the absorption tower which is continuously cooled externally with water. Small quantities of water are added to hydrate the nitrogen dioxide and partially to scrub the gases. The gas from the top of the tower generally is heated in a heat exchanger and then expanded through a power-recovery compressor for part of the air supply before being released to the atmosphere (Fig. 10.7).

The bottom of the tower yields nitric acid of 61 to 65% strength, which may be concentrated if so desired. This may be done in silicon-iron or stoneware towers by feeding concentrated sulphuric acid, sp gr 1.84, to the top, and 61 to 65% nitric acid below the top. The vapour leaving the tower is 90% nitric acid or more, and the bottoms consist of dilute (70 to 75%) sulphuric acid. The latter may be reconcentrated, or used as is, for example, for making superphosphate fertilisers.

Other processes are available for concentrating tne nitric acid to nearly 100%. In one such process, magnesium nitrate is used as the drying agent. Magnesium carbonate is treated with nitric acid to yield magnesium nitrate; this is fed, with weak nitric acid, to a tower where the magnesium nitrate combines with the water; the magnesium nitrate is dehydrated for recycle in a separate concentrator.

In another technique, nitrogen oxides from the ammonia-oxidation unit are processed directly. Some of these gases are dried and further oxidised to nitrogen tetroxide; a mixture of the tetroxide, weak acid, and air is heated to 70°C at 50 atm (5 MPa) pressure for about 4 h to yield 98 to 99% nitric acid.

These techniques for concentration nitric acid are necessary because simple distillation is not applicable; nitric acid forms an azeotrope with water at about 68% acid.

The process described is termed a high-pressure process, operating at 8 to 10 atm (0.8 to 1 MPa). Low- or atmospheric-pressure and medium-pressure (3 to 5 atm, 0.3 to 0.5 MPa) plants are also used, but to a lesser extent. High-pressure plants have the advantages of lower capital costs, because of the increase in acid strength obtained, a 50-fold increase in the rate of reaction, and a great decrease in tower volume required for oxidation and absorption. These factors reduce the initial plant investment by as much as 50%. Disadvantages of the high-pressure system include greater catalyst loss and lower oxidation efficiencies. Low-pressure designs have the best efficiencies in ammonia conversion and less loss of platinum, and may be preferred where ammonia costs are high. Variations of the described process are chiefly in methods of power recovery. In one process, the gas leaving the burner goes directly to a waste-heat boiler and then to a cooler-condenser where most of the water produced during the oxidation is removed as 40 to 45% nitric acid. A catalyst-recovery filter is usually installed at the cold end of the boiler. Such a filter reduces platinum losses to 170 to 340 mg/metric tonne of 100% nitric acid. In another common variation, tail gas from the absorber, which contains unused oxygen, is mixed with methane and passed over an oxidation catalyst. The resulting hot gases are sent to a power-recovery turbine, which recovers sufficient energy to make the nitric acid plant self-sufficient in power.

Uses

Ammonium nitrate, including military, adipic acid, miscellaneous fertilisers, military other than ammonium nitrate, isocyanates, nitrobenzene, potassium nitrate, and steel pickling.

Properties

Pure 100% nitric acid is a colourless, highly corrosive liquid with powerful oxidising properties. The commercial acid varies from colourless to a yellowish-red colour, owing to the presence of dissolved nitrogen oxides.

Mol. wt 63.01, mp –42°C, sp gr 1.502, bp 86°C (partial decomposition). Threshold limit value, nirtrogen oxides, 5 ppm, 9 mg/m^3.

Soluble in all proportions in water and forms a constant-boiling mixture at 760 mm (100 kPa) pressure of 68% acid and 32% water with a boiling point of 110°C. Solutions containing more than 86% nitric acid are known as fuming nitric acids. The largest application for nitric acid is for conversion to ammonium nitrate; most of this conversion is accomplished on site. Thus nitric acid manufacture is tied very closely to the fertiliser industry, which is notoriously subject to wild swings in demand, although with worldwide food needs the trend must be upward. However, it is not certain that demand for ammonium nitrate will necessarily follow the trend, since alternative nitrogen sources such as urea and anhydrous ammonia may become more economical. However, nitric acid may still be needed for such newer fertilisers as nitric phosphates, ammonium phosphate nitrates, sulphate nitrates, which do not require the separate production and isolation of ammonium nitrate.

Practically all nitric acid production comes from the catalytic oxidation of ammonia. The relative economics of the three types of process—high, medium, and low-pressure—have been discussed.

Although the platinum-rhodium alloy catalyst remains standard, other catalysts have been developed, some using less precious metal and at least one none at all.

SODIUM AND POTASSIUM NITRATES

Sodium Nitrate

Sodium nitrate, $NaNO_3$, occurs in nature associated with sodium and potassium chlorides, potassium nitrate, sodium sulphate, magnesium chloride, and other salts. Chilean nitrate is the only inorganic nitrogen fertiliser produced today, (Table 10.3).

Table. 10.3. Selected properties of sodium nitrate.

Property	Value
Crystal system	Trigonal, rhombohedral
Mp, °C	308
Refractive index, n_D^{20}	
trigonal	1.587
rhombohedral	1.336
Density (solid), g/cm^3	2.257
Solubility in H_2O, molality (± 2%)	
at 0°C	0.62
at 80°C	17.42
Specific conductivity at (300°C), S/cm	0.95
Viscosity^a, mPa·s (=cP) at 590 K	2.85
Heat of fusion, J/g^b	189.5
Heat capacity, J/g^b	
Solid at 0°C	1.035 ± 0.005
Liquid at 350°C	1.80 ± 0.02

^{a} Measurement method: capillary.
^{b} To convert J to cal, divide by 4.184.

Manufacture and Processing

Guggenheim process

The Guggenheim nitrate process demonstrated that caliche ores as low as 7 wt% in nitrate content could be economically exploited. The mined ore is transferred to crushing units where it is crushed in three stages. The crushed ore is leached in 8–10 vats, of which four are used at any given time at an average temperature of 40°C. The leaching cycle lasts 40 hour, the total vat cycle ca 168 hour.

The fines produced during crushing are mixed in the filter plant with liquor containing 150–200 g/l of nitrate, at a mixing ratio of two tonnes of solids per cubic metre ofliquor; the leaching is carried out at 40–50°C. The rich liquor from the leaching vats containing 450 g/l of sodium nitrate is sent to crystallisers where it is chilled to 0–5°C. At the exit of the last crystallisers, the mother liquor and the crystals are pumped out as a slurry to continuous centrifuges.

The sodium nitrate crystals are transferred to the graining plant, which consists of large, oil-fired, reverberatory-type furnaces, where the crystals are heated to 315–325°C. The molten sodium nitrate is pumped to large spray chambers where the spray solidifies as prills. The Guggenheim process recovers 80–85 wt% of the $NaNO_3$ contained in coarse fractions at a total fuel consumption of 0.1 metric tonne per metric tonne of sodium nitrate produced.

In the modified Guggenheim process, the ore is leached countercurrently with water in vats. From the weak sodium nitrate brines, sodium sulphate is extracted by a chilling process that yields Glauber's salt. Rich sodium nitrate solution is passed to the crystallisers and mother liquor to the solar-pond system. Continuous evaporation in the ponds results in concentration increases of potassium nitrate and iodine. Potassium nitrate is removed through crystallisation. Synthetic sodium nitrate is produced by neutralisation of nitric acid with soda ash or caustic soda. The sodium nitrate is melted and prilled.

Health and safety factors

The acceptable daily intake by adults for nitrates (suggested by WHO) is 5 mg/kg in addition to naturally occurring nitrates; large doses are lethal. Accidental ingestion of ca 8–15 g or more causes severe abdominal pain. Poisonings from the ingestion of meat containing sodium nitrate and nitrite have occurred.

Uses

Sodium nitrate is used primarily in agriculture as a fertiliser. The sodium in sodium nitrate corrects soil acidity and liberates phosphates, and is a partial substitute for potassium. The nitrogen from sodium nitrate is available faster than nitrogen from ammonia. The main industrial use is in the manufacture of explosives, e.g., dynamite. Sodium nitrate is also used in the manufacture of glass, fibre glass, enamels, and porcelain as an oxidising and fluxing agent. It is an ingredient in the production of charcoal briquettes. Sodium nitrate has application in the production of certain antibiotics and pharmaceuticals. In curing beef, bacon, and other meats, it is used as a preservative.

CHAPTER 11

Chemical Explosives and Propellants

INTRODUCTION

An explosive is a material which, on proper initiation, becomes rapidly converted into products of greatly increased volume. Ordinarily the original material is solid or liquid in form, whether a single chemical compound or a mixture, and becomes gasified during the explosion reaction. The conversion to gaseous form is accompanied by intense heat, which increases the volume of gases still more. The rapid expansion taking place at the time of explosion gives the accompanying disruptive or propelling effect.

Explosives are destructive in action when they decompose within a completely confined space or in close proximity to other objects. If this decomposition takes place inadvertently, the results may be disastrous. When, however, pre-determined amounts of explosives are used under controlled conditions of time and place, they may become constructive agents for accomplishing definite objectives. By such control of their disruptive power, explosives function in blasting ores of iron and other metals, breaking down coal, mining salt, quarrying limestone for use in road construction, and in performing a multitude of useful tasks.

POTENTIAL ENERGY OF EXPLOSIVES

Explosives are so disruptive in effect that they are commonly reputed to be highly superior reservoirs of energy. The fact is that, pound for pound, commercial blasting explosives contain considerably less potential energy than, for example, liquid fuels. The shattering action of explosives comes from the extremely rapid conversion of solids or liquids to gaseous products; in other words from the oxidation of carbon and hydrogen to carbon dioxide and water vapour. In the case of explosives, the oxygen for this combustion is a chemically combined portion of the material, while with fuels the oxygen of the air is utilised. With most explosives also, there is a considerable content of inert material not convertible into gaseous products. The explanation of the effectiveness of explosives comes in the almost instantaneous release of the pent-up energy. While the combustion of fuels is a relatively slow procedure, the progress of which can be followed by the eye, the decomposition of explosives takes place at tremendous speed. Black powder is considered a slow explosive, but even this has a velocity of combustion, when shot in long cylindrical columns, upward of around 1000 feet per second. The higher velocity dynamites, on the other hand, detonate at rates of between 5000 and 20,000 feet per second.

Data are given in Table 11.1 to show comparatively the temperatures attained theoretically in the combustion of fuels and explosives, the volume of gas produced, both at the maximum temperature and when cooled to 0°, and the energy released per pound of material employed. Because of various assumptions made, the calculated values must be taken as only approximate.

Table 11.1. Energy values of explosives.

Material	Theoretical temp. Combustion °C.	Volume of gases Cu. ft./100 lbs.		Potential energy Ft. lbs./lb.
		Cold	Hot	
Gasoline	3900	2500	83,000	16,000,000
Nitroglycerine	3150	750	14,400	2,050,000
40% Straight dynamite	2620	560	9,100	1,520,000
Black powder	2350	514	5,200	958,000

The volume of hot gases is greater than that of the cold gases both because of thermal expansion and because of the fact that water, which is in liquid state in the cold gases, is in vapour form in the hot gases and has been assumed to follow the gas laws.

PROPERTIES OF EXPLOSIVES

In peace times the explosives industry is chiefly concerned with commercial blasting explosives, and the standardisation and control of their properties are of prime importance. This is particularly true of dynamites. Since these must be applied to all kinds of blasting work, a great many different types must be designed and their properties must be varied widely, to adapt them to do their work effectively.

Strength and Velocity

The two properties which are most important in assuring the proper execution of an explosive in blasting work are strength and velocity. Probably it is a combination of the two properties which determines the shattering effect. This combined effect has sometimes been termed "brisance". Strength is the property on which explosives are graded primarily, and this may be considered to depend on the gas pressure developed from a given weight or volume of material. Velocity denotes the rate at which the effective strength is developed. Obviously, very hard rock can be blasted down only by a high strength explosive and by one which possesses high velocity of detonation. For softer rock and material offering less resistance, lower strengths and particularly lower velocities give more satisfactory results.

Sensitiveness

Sensitiveness to propagation is a property which must be accurately controlled and maintained, as failure to propagate will mean the presence of a certain amount of unexploded material in the borehole, a very hazardous condition. Other properties, such as fumes and water resistance, are likewise important and are matters of routine test by manufacturers of explosives.

Tests

The final judgement as to the efficiency of an explosive must be based on its execution in actual blasting work. Laboratory and proving ground tests have been developed, however, which allow a good preliminary evaluation. Propellant explosives such as smokeless powder, are given very extensive tests, particularly with respect to ballistic properties and stability under storage conditions. It is essential that the explosive gives the desired velocity to the projectile, but that this pressure shall not be developed so rapidly as to burst the gun barrel. Because of the fact that smokeless powder may be stored for years, it is important that there is no deterioration in quality and properties over this period.

Military explosives of the disruptive type used for shell filling and like purposes, TNT for example, are tested mainly for purity and freedom from contamination. Melting point tests and determinations of percentages insoluble in certain solvents, acidity, nitrogen content, amount of ash are the important criteria for such compounds.

Role of Nitrogen in Explosives

The chemistry of explosives is closely bound up with that of nitrogen compounds. The alkali nitrates are the oxidising agents used in black powder compositions, while ammonium nitrate is perhaps the most important ingredient of dynamites, functioning both as a strength producer and as oxygen supplier. Practically all the high explosive compounds important in the commercial and military fields contain nitrogen.

The explanation for the adaptability of the nitrogen compounds as explosives seems to lie in their degree of instability and sensitiveness to shock. A practical explosive compound must be sufficiently stable to permit safe handling and storage, yet must be capable of almost instantaneous decomposition when subjected to impact of definite intensity. In certain nitrogen compounds, the strength of the bond linking the nitrogen atoms to adjacent atoms is such as to give a practical sensitiveness. Since explosive decomposition is essentially a combustion process, the organic nitrates and nitro-compounds fit well into the requirements. Nitrogen itself is usually found in elemental form in the products of explosion, while the attached oxygen is available for the combustion of the carbon and hydrogen present. An additional NO_2 groups are introduced into organic formulae, the sensitiveness of the nitrated compound increases.

MANUFACTURE OF EXPLOSIVES

Black Powder

Black powder is an excellent explosive for the blasting down of coal, as its low velocity gives it a slow, heaving action that does not shatter the coal unduly. For safety reasons, however, black powder cannot be used in mines in which inflammable gases are present or suspected, as its long flame of considerable duration makes it more hazardous than the higher velocity safety dynamites designated as "permissible" explosives.

Composition

The composition of black powder has changed very little over the course of many centuries. An English powder is given the composition of 66.6% saltpetre, 11.1% sulphur, and 22.3% charcoal, while a French powder of 1650 had these ingredients present in the respective amounts of 75.6%, 10.8%, and 13.6%. This latter composition would be satisfactory for a present-day powder. Where potassium nitrate was formerly used exclusively as oxidising agent, it has been replaced in American black blasting powders by sodium nitrate. Because of the higher weight percentage of oxygen in the latter, about 2% less of the sodium salt may be desirably used than of the potassium, for comparable results.

Black powder, therefore, is essentially an intimate mixture of: (i) sodium or potassium nitrate; (ii) sulphur; and (iii) charcoal, in the approximate compositions given above. The replacement of potassium nitrate took place in American powders about the middle of the 19th century. At the present time, all American black blasting powders contain sodium nitrate, which is considerably cheaper as an ingredient than potassium nitrate and possesses greater potential strength. It has the disadvantage of being hygroscopic, hence potassium nitrate is retained in black sporting powders and in fuse powders, where uniformity of burning and of moisture content is essential.

Raw materials for black powder

Sodium nitrate is the ingredient used in much the largest amount in American black powder. Now synthetic sodium nitrate has largely replaced the natural product, and is available free from impurities.

Charcoal is one of the most important ingredients and one that requires most careful control as to uniformity, freedom from undesirable impurities and general excellence. Any market deviation from the standard in the properties of the charcoal will affect the properties of the finished powder. Charcoal may be obtained from various non-resinous woods, for example, maple, and results from a charring and a dry distillation process.

Sulphur should be in a high state of purity, the only specifications generally being that it shall be free from grit and acidity.

Decomposition reactions of black powder

Black powder comprises a combustion mixture of oxidising and reducing agents. Assuming a black powder composition of 73% $NaNO_3$, 13% sulphur, and 14% charcoal, the following reaction would represent the approximate molecular proportions present:

$$2NaNO_3 + S + 3C = Na_2S + N_2 + 3CO_2$$

Actually the reaction is much more complex. Some formation of sodium sulphate and sodium carbonate will take place. Those familiar with black powder blasting know also that some carbon monoxide is always formed. Under conditions of use in mining operations, it is noted that there is substantially no formation of free hydrogen sulphide, this gas being fixed by the alkali element present. Black powder is not completely gasified on explosion, and in the reaction shown the solid residue in the reaction products represents about 1/3 of the total by weight.

The sulphur in the black powder composition is not simply an alternative combustible material. It appears to have a characteristic action in facilitating ignition of the powder and promoting combustion.

Manufacturing operations

The manufacture of black powder comprises a number of steps or mechanical operations which are carried out in separated buildings for safety reasons. These operations briefly are as follows:
1. Preliminary pulverising of ingredients.
2. Incorporation under wheels.
3. Pressing.
4. Cutting of press cake.
5. Granulation, or "corning".
6. Drying and glazing.
7. Packing.

Pulverising

The ingredients are prepared initially in a state of fine sub-division. Ordinarily the sulphur and charcoal are first ground together in a drum containing steel balls, while the nitrate is disintegrated either in a ball mill or under steel crusher rolls. Care is taken at the end of this preliminary grinding that the materials are free from foreign particles of matter, particularly of a metallic nature, and this is accomplished by screening or passing over a magnetic separator.

Wheeling

The finely divided ingredients are given a more intimate and thorough incorporation in the "wheeling" operation. The materials are introduced into a cylindrical bowl, having ordinarily a metal bed plate and wooden side walls. Within this bowl two heavy broad-edged runners or wheels connected by a short horizontal axis rotate continuously. These wheels may be over 7 feet in diameter and around 2 feet in width of rim. At the same time the wheels revolve slowly above their own axis. The wheels are preferably driven from below and have a certain amount of clearance above the bed plate. Metal plows are arranged in front of each wheel to move the mixture from the sides and centres of the bowl under the path of the wheels. Water is added to the charge to ensure the proper final moisture content. The operation is continued for several hours, at which time very intimate incorporation has resulted. The product of this operation is designated "wheel cake".

Pressing

The wheeled and incorporated mixture is subjected to the action of crusher rolls or other devices for breaking up the lumps and is then introduced into hydraulic presses where pressures of over 1200 pounds per square inch may be applied. The pressing operation gives a hard mass, designated "press cake", in which there is no longer any tendency for segregation of the separate ingredients.

Cutting and granulation

The pressed cake is broken up into pieces of about 1 to 2 inches in diameter and passes to the "corning" or granulating operation, in which the fragments of powder are caused to go through consecutive sets of steel crusher rolls, provided with teeth. The powder is gradually reduced to grains of the desired size, and screens beneath the rolls separate the material into the various desired granulations. The corning operation is one of the most hazardous of black powder steps, but in all stages of black powder manufacture care is taken that no workers are present in the buildings while the powder is in course of treatment.

Drying, glazing and packing

If granular black powder is to be produced, the "green grain" powder from the corning operation is run into the revolving drying and glazing drums, which may have a length of 12 feet and a diameter of 4 feet. The powder is in these drums for several hours, during which time the grains have their edges rounded off. Heated air is blown in and the moisture content is reduced to about 0.5%. A small amount of graphite is added during the latter part of the rotation of the drum to glaze the particles and give the grains a polished surface. The finished powder is sorted by means by screens into the desired granulations and is packed in metal kegs, usually holding 25 pounds each.

Pellet powder

Black powder is now prepared in two commercial forms, (i) in the form of free-flowing grains; and (ii) in the form of pellets or cylindrical columns enclosed in a paper wrapper. The preparation of the granular powder has been described in the preceding paragraph.

Pellet powder in its usual form, consists of cylindrical agglomerated pellets 2 inches in length and of diameters up to 2 2/1", each pellet having a longitudinal centre hole. Ordinarily four such pellets are placed end to end and enclosed in a paper wrapper to form a cartridge 8" in length.

In the manufacture of pellet powder, the same materials are taken as for granular powder. The 'green grain' powder resulting from the "corning" step is commonly used and is fed from a hopper into a die or hole in a charging block of the diameter desired for the pellet. A centre pin is introduced previous to pressing of the powder and two plungers compress the powder from above and below respectively, until the desired density has been obtained. About 4.5% of water is present in the powder being compressed. The pressed pellets are transferred from the press house to another building, where they are dried on trays by means of hot air, to a moisture content of about 0.5%. The finished pellets are wrapped in paper, waterproofed by dipping in paraffin, and packed in 25 or 50 pound boxes.

Pellet powder is in a form very convenient for use and very adaptable for accurate measurement of the desired explosive charge. It is safer to handle than granular powder because no pouring of the powder is necessary and spilling is unlikely. In the presence of inflammable gases, however, it is not safer for use in coal mining than the granular form, since it possesses the same long-flame characteristics on combustion.

Potassium nitrate powders

The blasting powders already described have contained sodium nitrate as oxidising ingredient. For certain types of use, fuse powder for example, it is desirable to use potassium nitrate as the oxidising agent.

Black propellant powder for use in firearms, though largely displaced by smokeless powder, is likewise a potassium nitrate composition, since uniformity of pressure development is an absolute requirement here.

Role of Nitration

In black powder the nitrogen of the explosive is supplied as an intimate mechanical mixture. With the development of the studies of chemical reactions it was found that many explosives could advantageously be made by the incorporation of nitrogen in organic molecules. Hence, nitration has assumed a role of paramount importance in this field.

Nitrates and nitro-compounds

Two types of organic high explosive compounds are obtained by the action of nitric acid on organic materials, i.e., by nitration. These are: (i) the organic nitrates or nitric esters; and (ii) the nitro-compounds. Both of these types contain one or more NO_2 groups replacing hydrogen atoms of the unnitrated compounds. In the case of the nitric esters, the NO_2 group is attached to an oxygen atom, while in the nitro-compounds it is linked to a carbon atom. The nitric esters are obtained by the nitric esterification of alcohols, while the nitro-compounds are ordinarily prepared by the nitration of hydrocarbons or their derivatives. Nitroglycerine is an example of the nitric esters,

$$CH_2\!\!-\!\!ONO_2$$
$$|$$
$$CH\!\!-\!\!ONO_2$$
$$|$$
$$CH_2\!\!-\!\!ONO_2$$

As it is apparent from the above formula, the name nitroglycerine is a misnomer, the correct designation being glyceryl trinitrate. Trinitrotoluene (TNT), on the other hand, is a true nitro-compound:

$$
\begin{array}{c}
CH_3 \\
| \\
C \\
O_2N-C \quad\quad C-NO_2 \\
H-C \quad\quad C-H \\
C \\
| \\
NO_2
\end{array}
$$

In addition to the above types of nitrated products, the primary detonating compounds, commonly used in blasting caps as initiating explosives, are likewise nitrogen-containing compounds. Mercury fulminate has the formula $(CNO)_2Hg$, while lead azide is PbN_6.

Nitration processes

Nitration is the term commonly employed to designate the action of nitric acid on organic compounds whereby the nitro group (NO_2) is introduced, usually in place of a hydrogen atom. Thus it includes both the preparation of the true nitro-compounds, where the NO_2 group is attached to a carbon atom, and esterification by means of nitric acid.

The nitration agent in commercial operations is ordinarily one of the following forms:

1. Mixed acid, comprising nitric acid associated with a dehydrating agent such as sulphuric acid.
2. Concentrated nitric acid, for example, around 95%.
3. Nitrogen tetroxide, preferably in vapour form.

Nitration products are important in organic industries generally and particularly as dye intermediates. Many of such nitro-compounds are not explosive compounds, having perhaps one or two nitro groups in their formulae. Generally these compounds have three or more nitro groups that are important as explosives, though ethylene glycol dinitrate is a conspicuous exception. Thus the mononitrophenols and mononitrotoluenes are lacking in explosive properties, but are useful as dye intermediates. The corresponding dinitro-compounds are relatively insensitive explosive compounds, while trinitrophenol (picric acid) and trinitrotoluene are outstanding explosive compounds. The lower nitro-compounds are readily obtained by treatment of the starting material with low strength nitric acid, while the compounds containing three or more nitro groups require a higher nitric concentration in the nitration acid, or higher temperature treatment. The replacement of hydrogen by a nitro group results in the formation of water as a reaction product:

$$R-H + HO-NO_2 \rightarrow R-NO_2 + H_2O$$

Hence the presence of sulphuric acid or other dehydrating agent in the mixed acid helps maintain a high concentration of nitric acid.

EXPLOSIVES MADE BY NITRATION

Manufacture of Nitroglycerine

Nitroglycerine is the ingredient which, from the beginning of the dynamite industry, has given such explosives their characteristic high velocity, as disintiguished from low-velocity black powder, the commercial blasting powder previously in use. Nitroglycerin is obtained by the action of nitric acid on glycerine:

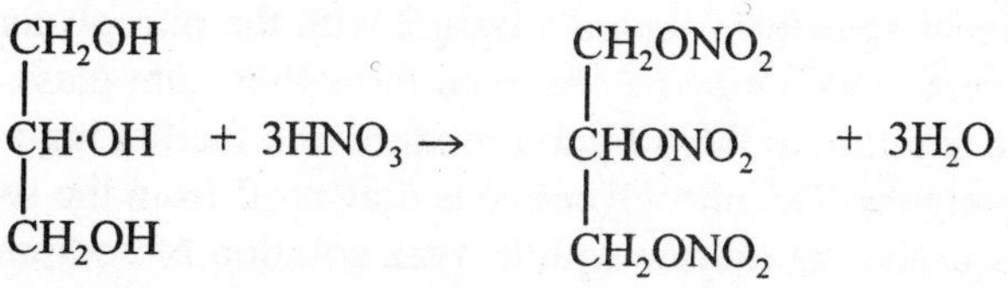

$$\begin{array}{l} CH_2OH \\ | \\ CH{\cdot}OH \\ | \\ CH_2OH \end{array} \;+\; 3HNO_3 \;\rightarrow\; \begin{array}{l} CH_2ONO_2 \\ | \\ CHONO_2 \\ | \\ CH_2ONO_2 \end{array} \;+\; 3H_2O$$

It will thus be seen that nitroglycerine is not a true nitro-compound, but actually is a nitric ester, glyceryl trinitrate.

Nitroglycerine resembles glycerine in appearance, and is a white to yellowish, viscous liquid, having a specific gravity of 1.6.

The raw materials employed in the manufacture of nitroglycerine are glycerine, sulphuric acid, and nitric acid. So-called dynamite glycerine is a high-grade product of around 99% purity, having a specific gravity of about 1.262 and a high viscosity. The mixed acid used in the nitration has a composition of about equal parts of sulphuric acid and nitric acid, with possibly a slight excess of the former, and is substantially free from water. In carrying out the nitration, the proper amount of mixed acid is introduced into the steel nitrating vessel, in which brine coils serve to cool the acid. After agitation has been started, glycerine is added at such a rate that the temperature rise is only moderate. The temperature range is maintained generally between 50° and 37°F. (10 to 3°C). Nitroglycerine itself may freeze at about 55°F. (12.8°C). and its freezing in the nitrator would constitute a hazard, but various conditions present make operation safe at somewhat lower temperatures. Actually it is the practice to make low freezing or difficult freezable nitroglycerine by nitrating, instead of glycerine alone, a solution in glycerine of ethylene glycol, diglycerine or other similar nitratable material. The nitrated product then is a solution in nitroglycerine of ethylene glycol dinitrate, tetranitrodiglycerine or other freezing point depressant. In addition to the formation of glyceriyl trinitrate, therefore, one of the following reactions may be proceeding simultaneously in the nitrator:

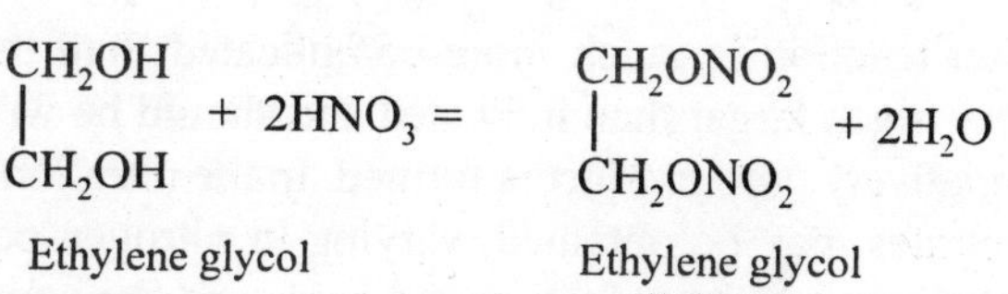

$$\begin{array}{l} CH_2OH \\ | \\ CH_2OH \end{array} \;+\; 2HNO_3 \;=\; \begin{array}{l} CH_2ONO_2 \\ | \\ CH_2ONO_2 \end{array} \;+\; 2H_2O$$

Ethylene glycol   Ethylene glycol dinitrates

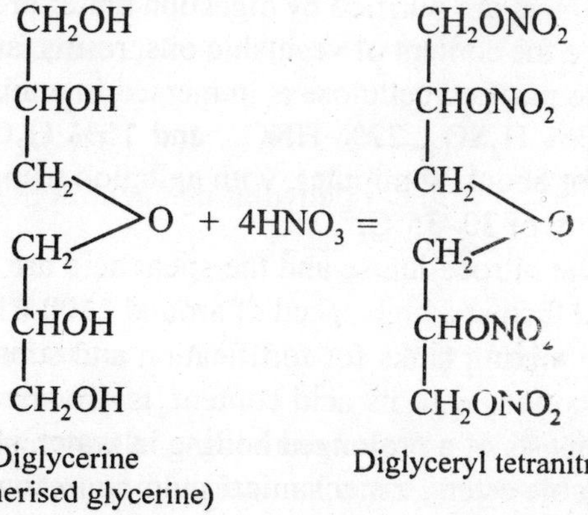

$$\begin{array}{l} CH_2OH \\ | \\ CHOH \\ | \\ CH_2 \\ \;\;\;\;\searrow O \\ CH_2 \\ | \\ CHOH \\ | \\ CH_2OH \end{array} \;+\; 4HNO_3 \;=\; \begin{array}{l} CH_2ONO_2 \\ | \\ CHONO_2 \\ | \\ CH_2 \\ \;\;\;\;\searrow O \\ CH_2 \\ | \\ CHONO_2 \\ | \\ CH_2ONO_2 \end{array}$$

Diglycerine   Diglyceryl tetranitrate
(polymerised glycerine)

The resulting nitroglycerine and spend acid, having a composition of about 76% H_2SO_4, 7% HNO_3, and 17% H_2O, are run from the nitrator into a separating vessel. In the course of about 3/4 of an hour, the nitroglycerine and spent acid form separate layers, with the nitroglycerine on top and with a clear line between the two layers, which may be observed through a sight glass. Usually the separation into layers is hastened by the addition to the nitration mixture of a facilitating material such as a mixture of sodium fluoride and kieselguhr. The nitroglycerine is drawn off from the separator, given a preliminary water wash, and treated in the neutraliser with a weak solution of sodium carbonate during vigorous agitation to remove the last traces of acidity from the nitroglycerine.

It needs to be strongly emphasised that the manufacture of nitroglycerine is a hazardous operation, and one that should never be attempted by inexperienced persons. Such care is taken, however, in commercial production in the control of temperature, the purity of materials and the avoidance of dangerous practices that explosions are uncommon. If there is an unduly rapid rise of temperature in the nitrator, or if, for any reason, a nitration must be interrupted before completion, the entire charge is drowned, that is, dropped into an excess of water directly beneath the nitrator.

The yield is commonly calculated on the basis of glycerine as 100, and the theoretical yield would be 245; in practice, yields of around 230 are commonly obtained under best operating conditions. Charges of around 3000 pounds nitroglycerine and over are frequently obtained in one nitration.

In recent years, continuous processes for the nitration, separation, and washing of nitroglycerine have been adopted in a number of European plants. These have replaced the former batch processes.

Nitrocellulose

The nitration of cellulose follows the same principles as in the case of glycerine, namely the treatment of cellulose with nitric acid and the introduction of nitro groups in place of hydrogen atoms. Nitrocellulose, like nitroglycerine, is misnamed and actually is a nitric ester, cellulose nitrate. Its formation is shown by the following reaction:

$$C_6H_{10}O_5 + 3HNO_3 \rightarrow C_6H_7O_2 (ONO_2)_3 + 3H_2O$$

In reality, the nitration reaction is much more complicated than this, as the size of the cellulose molecule is doubtless very much larger than indicated and should be written $(C_6H_{10}O_5)_n$. In the case of nitroglycerine, a single relatively pure product is formed. In the nitration of cellulose, on the other hand, a mixture of cellulose nitrates may be obtained, varying in nitrogen content and solubility in organic solvents, depending on the composition of the mixed acids and the conditions of nitration.

Prior to nitration, the cotton linters are purified by digestion under pressure with a dilute solution of caustic soda. This serves to reduce the content of vegetable oils, resins, and other removable impurities.

In one method of nitration, the purified cellulose is immersed in a mixed acid, having, for example, the approximate composition 63% H_2SO_4, 22% HNO_3, and 15% H_2O. It is common to nitrate 30 pounds of cotton as one charge for about 30 minutes, with agitation using about 1500 pounds of mixed acid and maintaining a temperature of 30–35°C.

When nitration is complete, the nitrocellulose and the spent acid are run into a centrifugal wringer, which revolves slowly at first and then at a high speed of around 1100 RPM. This causes the separation of the acid, which can be sent to mixing tanks for fortification and subsequent re-utilisation, or to the denitrator. The centrifuged nitrocotton, with its acid content, is drowned in an excess of water.

The purification treatment consists of a prolonged boiling in water, slightly acidified with sulphuric acid in order to break down unstable esters, a mechanical pulping treatment to reduce the length of the fibres, and an alkaline boiling (poacher) treatment to neutralise acid products, followed by several boilings in fresh water and finally a number of cold water washings.

Trinitrotoluene

In the case of nitroglycerine and nitrocellulose, nitration is a one-step process, all the acid and the material for nitration being introduced during one operation. With trinitrotoluene (TNT), either a two step or a three-step process is used, the former separating out the mono-nitro compound first, then going through the *bi* and *tri* stages in one step. The three-step process obtains in separate nitration stages, the *mono*, *di*, and *trinitro* compounds, by successive action of one molecule of nitric acid on one molecule of toluene, or its nitration compounds, with simultaneous splitting off of water:

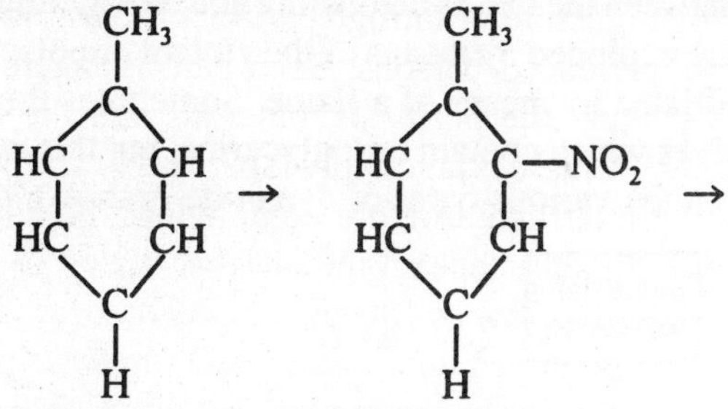

Toluene → Mononitrotoluene →

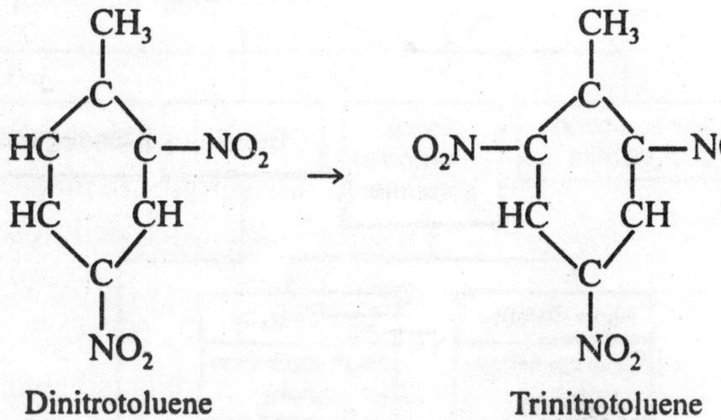

Dinitrotoluene → Trinitrotoluene

The preparation of TNT is not as simple as the outline shown above indicates. Three isomeric forms of the mono-compound exist and all are formed in the mono-nitration, the *ortho*, *meta* and *para* compounds, having melting points of $-10°$, $16°$, and $58°C$, respectively. Six isomers are possible and known for the trinitro-compound. It is the formation of undesirable impurities, both isomeric and oxidation products, that makes the preparation of pure TNT a complicated operation. Commercial TNT consists principally of the 2-4-6-isomer, and the completely purified material melts at $80.8°C$.

In the preparation of TNT in several nitration steps, stronger acids and higher temperatures are used during the contact of mixed acid and nitratable material as the higher stages of nitration are carried out. According to one process, the acid composition for the final nitration step is approximately 79.5% H_2SO_4, 17.8% HNO_3, and 2.7% H_2O.

Other Nitration Procedures

In the three examples of nitration described, mixtures of sulphuric and nitric (mixed acids) have been employed as the nitration medium. While nitric acid is always the active reactant, various other procedures are frequently followed. In the preparation of picric acid from phenol, the latter is commonly first treated with strong sulphuric acid to form phenol sulphonic acid, which is then converted to picric acid by treatment with strong nitric acid. In the case of pentaerythritol tetranitrate, the mixed acid is not suitable for the nitration, and the pentaerythritol should be treated with acid having a content of around 99% HNO_3.

DYNAMITE

Dynamite may be defined as a commercial high explosive. This distinguishes it (i) from the solid organic high explosives which are used for military purposes; and (ii) from low velocity commercial explosives of the black powder type. There is no sharp line as to velocity which separates the high from the low explosives, but the distinction is often made between the two as detonating and deflagrating compositions.

The dynamites or detonating explosives are exploded by means of the violent impulse from a blasting cap, while the deflagrating explosives are initiated by means of a flame. Sometimes the term dynamite has been limited to those commercial explosives which contain nitroglycerine, but this is not necessarily the distinguishing point. The relation between the various types of dynamite is shown in Fig. 11.1.

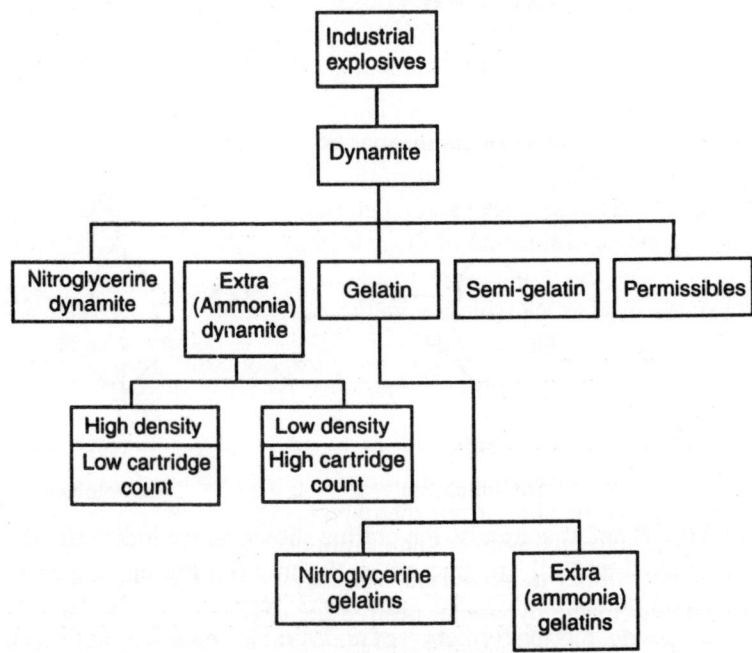

Fig. 11.1. General classes of industrial explosives.

The invention of dynamite dates from 1866, when Alfred Nobel conceived the idea of mixing liquid explosive nitroglycerine with a highly absorbent material. In this way he obtained, in place of a liquid dangerous to transport and to handle, a solid composition relatively insensitive to ordinary shock, but capable of complete detonation by means of a blasting cap.

Nitroglycerine had been discovered by Sobrero prior to Nobel had been of little practical value. The importance of Nobel's discovery to the advancement of civilised nations cannot be too strongly emphasised, and at the same time it should be realised that commercial dynamites are constructive tools. It is this type of explosive which is the most important and the one to which the explosive companies devote the bulk of their equipment and efforts.

Ingredients of Dynamites

Unlike black powder, there is no standard composition for dynamite. Nitroglycerine, probably the most characteristic ingredient, may vary in percentage, for example, from 5 to 90%, and other ingredients may likewise vary widely, depending upon the properties desired in the finished explosive. Not only are there many ranges of composition, but also many types of dynamites, owing to the differing fields of application for such explosives where varied forms of blasting action are needed. A few of the most important ingredients of dynamite compositions will first be considered briefly.

Nitroglycerine

This has long been the most important explosive ingredient of dynamite, though the trend for many years has been continuously toward a reduction of the nitroglycerine content. The preparation and properties of nitroglycerine have been discussed previously in connection with nitration processes.

Ammonium nitrate

While nitroglycerine holds in its place in popular estimation as the outstanding ingredient of dynamites, ammonium nitrate probably should be given first place. It does not possess the explosive properties of nitroglycerine, when used by itself and in small quantities, but it is a powerful explosive compound when once initiated by an external detonation impulse of sufficient magnitude and when used in considerable amounts. It has the advantages of being absolutely safe to handle under ordinary conditions, of being formed readily by the neutralisation of nitric acid with ammonia, two universally available raw materials, and being low in cost. As will be seen in a later discussion of types of dynamites, ammonium nitrate has been increasingly used as an explosive salt to replace a portion of the nitroglycerine formerly used.

Combustible absorbents

Nitroglycerine is retained in dynamite compositions by absorption by combustible ingredients. The absorbent first used by Nobel was kieselguhr, a siliceous material, which was excellently adapted for that purpose but which reduced the strength of the explosive, because it was itself inert in action and could not be converted into gaseous matter. Such combustible materials as wood pulp, starch, and various meals and low density fibres are now used. The choice of absorbent depends on various factors, such as nitroglycerine content of the dynamite, desired density, and sensitiveness, etc.

Sodium nitrate

Sodium nitrate is used as an oxidising agent in dynamite. It is desirable that the explosive be oxygen-balanced so that good fume properties and good blasting execution can be obtained. The combustible absorbents are very deficient in oxygen and the presence of sodium nitrate makes possible a balanced mixture, where all the carbon and hydrogen present will be burnt completely to carbon dioxide and water.

Freezing point depressants

Nitroglycerine has the disadvantage of a relatively high freezing point, about 55°F. (12.8°C). Consequently dynamites containing it will freeze at comparatively mild winter temperatures. When in frozen condition, dynamite is unsatisfactory for use in blasting and must be thawed. Not only is the thawing operation time-consuming, but it is also extremely hazardous unless unusual precautions are taken.

The hazard of frozen dynamite has been removed by introduction into the nitroglycerine of materials which depress the freezing point sufficiently to ensure soft dynamite at the most extreme winter temperatures. At the same time the strength of the explosive is maintained, as the depressant used is itself a highly nitrated explosive compound. The usual procedure is not to dissolve the nitrated compound in nitroglycerine, but to nitrate the solution of the starting material in glycerine.

The introduction of ethylene glycol dinitrate or tetranitrodiglycerine as freezing point depressant has been discussed in connection with the nitration of glycerine. The co-nitration of glycerine with either ethylene glycol or diglycerine (polymerised glycerine) makes it possible to obtain a nitrated product of the desired extremely low-freezing properties. The amount of nitroglycol or tetranitrodiglycerine present in the nitroglycerine can be ascertained by determination of the nitrogen content of the product, as obtained on the nitrometer, together with the specific gravity.

Many other freezing point depressants have been proposed and used, for example, nitrochlorhydrins, nitrotoluenes, etc. No two such depressants are equivalents and, with a change of depressant, formulae must be adjusted so that the explosive properties of the finished explosive remain unaffected.

Ant-acids (anti-acids)

The presence of even a small amount of nitric acid has a slight decomposing effect on nitroglycerine. For this reason, it is the usual practice to include in the formulae of nitroglycerine dynamites a small amount of an acid-accepting material, or ant-acid. Calcium carbonate is commonly used for this purpose, in the amount of 0.5 to 1.0% of the total composition. The regulations of the Interstate Commerce Commission state that, for shipment by freight, explosives containing nitroglycerine or other liquid explosive ingredients, must contain an ant-acid material "in quantity sufficient to have the acid neutralising power of an amount of magnesium carbonate equal to 1 per cent of the nitroglycerine".

Manufacture of Dynamite

The various types of dynamites require different treatments, but the general manufacturing procedure depends upon whether the explosive is: (i) a gelatine dynamite; or (ii) a non-gelatinous dynamite. The operations will be described briefly under these two classifications, and are illustrated on the accompanying flow sheet, Fig. 11.2.

Non-gelatinous Dynamites

The manufacture of this class of pulverulent dynamites from the raw materials consists essentially of two operations; (i) mixing; and (ii) cartridging.

Mixing

The ingredients to be mixed come from two different sources. First, the nitroglycerine from the neutraliser is wheeled to the mixing house in rubber-tyred, hand-operated "buggies", which have copper compartments adapted to hold weighed charges of nitroglycerine, in the amount desired for the dynamite mixing. The second ingredient comprises a mixture of material and is designated as "mixed dope". This mixture comes from what is known as the "dope house", where all the dry non-explosive ingredients, called for by the dynamite formula, have been mixed and screened. This mixture is likewise transported to the mixing houses in the required amount. The mixing vessel ordinarily consists of a circular wooden bowl or trough about 7 feet in diameter and less than 2 feet deep. The bottom of the bowl may be covered with hard rubber. The mixing device consists of two large wooden wheels, preferably rubber-covered, which operate on the edge-runner principle.

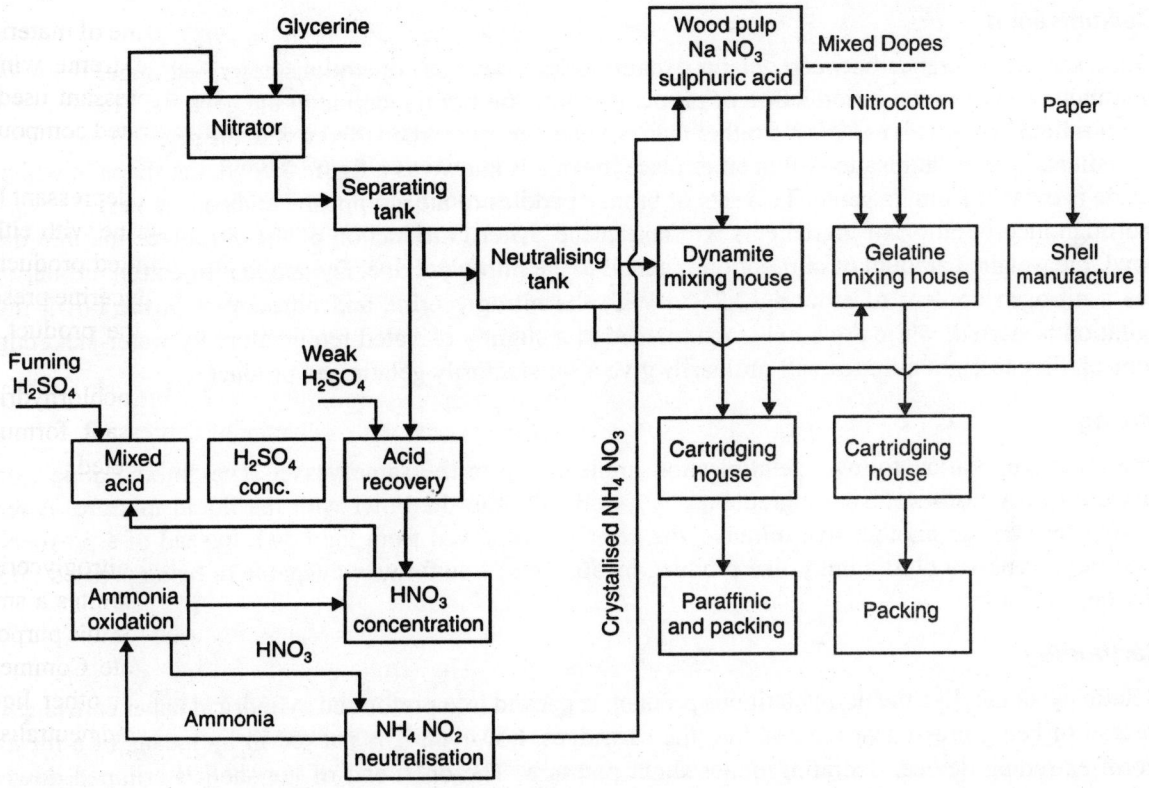

Fig. 11.2. Dynamite and gelatin dynamite manufacture.

When a mixing is to be made, the "dopes" are emptied into the mixing bowl and levelled, and the nitroglycerine is poured from the buggy over the top of the dopes by means of a rubber hose. The mixing wheels are then started, two hard rubber plows following the wheels, one serving to move material from the inside of the bowl under one wheel while the other plow throws the dynamite from the outside edge of the bowl under the other wheel. Between 5 and 10 minutes is ordinarily sufficient mixing time. Mixing of around 1,000 pounds of dynamite are commonly made at one time.

Cartridging

The mixed dynamite is usually of a dry, powdery nature and is transported in barrows to the cartridging or punching house. Here it is loaded into preformed, cylindrical paper shells, open at one end of any desired diameter from 7/8" up, and ordinarily about 8" in length. By means of special machines, the dynamite is packed to the desired density in the paper cartridge, and the open end of the cartridge is crimped down tightly. While a standard size of dynamite cartridge is 8" long by 1 1/4" in diameter, the cartridges may be as large as 7" in diameter by 24" in length. The sticks are packed in wooden boxes for shipment. The cartridged dynamite may be protected from moisture penetration, if necessary, by coating with paraffin.

Gelatin Dynamites

The manufacture of gelatin dynamites includes both operations described under the non-gelatinous dynamites, but with different types of equipment, and one additional operation: gelatinisation.

Gelatinisation

The essential difference between gelatin dynamites includes both operations described under the non-gelatinous comes in the introduction of nitrocotton into the nitroglycerine in the gelatins. This forms a viscous liquid which gives a plastic rather than powdery product when mixed with the dry, non-explosive ingredients. The gelatinisation often takes place in what is known as a figure 8 bowl, the shape of which inside is evident from its name. Two sets of bronze paddles rotate at opposite ends of the vessel so that thorough incorporation of ingredients will take place. After introduction of the nitroglycerine into the bowl, the weighed amount of nitrocotton is added to the nitroglycerine. Dynamite nitrocotton ordinarily has a nitrogen content of around 12.25%. With the nitroglycerine and nitrocotton in the bowl, the agitation is started, while the bowl is maintained at a slightly elevated temperature by water-jacketing control. Five to eight minutes will ordinarily give a satisfactorily gelatinised product.

Mixing

The mixing operation follows gelatinisation immediately in the same vessel. The "mixed dope", or mixture of dry non-explosive ingredients, is introduced into the bowl with the liquid mixture. After mixing for perhaps another five minutes, the gelatin is removed from the bowl. Instead of a powdery mixture, it is now a plastic putty-like product, incapable of free flow, but capable of being pressed into any desired form.

Cartridging

Gelatin dynamite, like the non-gelatinous product, is packed into preformed cylindrical paper cartridges. Instead of being pressed or tamped into the cartridges, however, it is packed in by means of a metal worm extruding device, operating under slight pressure. The open end of the shell is crimped down after the cartridging operation.

Types of Dynamites

There are two conditions which must be satisfied if a dynamite is to prove satisfactory in blasting service. First, it must have the safety properties that allow it to be used without hazard, and, secondly, it must be adapted to the type of blasting in question. Because of the many and varied kinds and hardnesses of material to be blasted and the different uses to which the explosives are put, a great number of types and brands of dynamite have been developed (Fig. 11.1).

Straight dynamites

The straight dynamites are the simplest type of dynamite and may be defined as high explosives containing nitroglycerine as the sole explosive ingredient, a sufficient amount of absorbent material being present to prevent exudation of the nitroglycerine. The straight dynamites are the basis of comparison, as to strength, for all other dynamites. They are made in a series from 15 to 60%, each grade being designated by its nitroglycerine content. For example, 40% straight dynamite contains 40% nitroglycerine. Other types of dynamites are designated as 40%, 60%, etc., when they have the same strength by ballistic mortar determination as the correspondingly named straight dynamite. The standard 40% straight dynamite used in tests by the Bureau of Mines has the following composition:

Nitroglycerine	40%
Sodium nitrate	44%
Wood pulp	15%
Calcium carbonate	1%

The straight dynamites are characterised by high velocity of detonation (over 13,000 feet per second for 40% straight, for example) and a high degree of sensitiveness to propagation. Because of this latter property, they are excellently adapted for use in submarine work and ditch blasting, where the explosion from one stick of dynamite is propagated successively to a large number of other sticks separated from one another by distances of perhaps 18". Because of the relatively high sensitiveness to friction, straight dynamites are not recommended for quarrying and general blasting work, where other types are suitable.

Ammonia dynamites

The ammonia dynamites are dynamites in which a portion of the nitroglycerine of the straight dynamites has been replaced by ammonium nitrate. These dynamites are very widely used, because of their effectiveness and their relatively low price. They are put out in the same strength designations as the straight dynamite but are lower in velocity than the straights for a given grade. The ammonia dynamites are attractive for use because of their insensitiveness to shock and friction and their lack of inflammability in comparison to the straight dynamites. They have one drawback in that they are relatively low in water resistance so that they are not well adapted for use under very wet conditions.

Gelatin dynamites

The gelatin dynamites are those in which the nitroglycerine has been at least partly gelatinised by means of nitrocotton. The highest strength gelatin is "blasting gelatin", which is considered the strongest industrial explosive known. It usually has the approximate composition: 91.5% nitroglycerine, 8% nitrocotton, and 0.5% calcium carbonate. Blasting gelatin is a solid, light-coloured, elastic material, too adherent for mixing with other materials. The common gelatin dynamites are made by the use of a lower ratio of nitrocotton in nitroglycerine, so that a thick, viscous liquid is obtained rather than a solid mass. The other usual ingredients of dynamites are mixed with this viscous liquid or gel. Both straight and ammonia gelatins are made, depending on whether nitroglycerine alone is the explosive ingredient or ammonium nitrate has replaced a part of the nitroglycerine.

The gelatin dynamites are designated by strength marketings also, as 40%, 60%, etc. They are outstandingly excellent with respect to four properties: (i) they possess high bulk density; (ii) they are the most water resistant of all dynamites; (iii) they excel from the point of view of fumes; and (iv) they do not flow, but stick well in holes into which they have been loaded because of their plastic nature. These properties, in addition to their relative insensitiveness to shock, make the gelatin dynamites the preferred explosive for use under wet conditions where a high loading density is desired, as in hard rock blasting, in blasting underground in confined workings where toxic fumes would be dangerous, and in boreholes directed upwardly in hard rock. Because of their relatively high price, the gelatin dynamites are replaced frequently by other explosives of slightly less efficiency.

Semi-gelatins

The so-called semi-gelatins are ammonia dynamites which are intermediate between the ammonia dynamites and the gelatins in the properties of density, plasticity, and water resistance.

Permissible

The 'permissible' dynamites are high explosives adapted to a special use, namely, in gassy coal mines, where black powder and the regular dynamites are not considered safe in the presence of the inflammable gases. A "permissible" may be defined as an explosive similar in all respects to a sample passing certain prescribed tests of the Bureau of Mines, when the explosive is used under the conditions and in the amounts prescribed by the Bureau.

The permissibility tests of the Bureau of Mines include several special determinations of properties which affect the suitability of the explosive for use in coal mines. One such test is the determination of the "unit deflective charge", that is, the weight of the explosive under test which will give a deflection of the ballistic mortar equal to that produced by 1/2 pound of the standard 40% straight dynamite. Tests are also made in their special testing gallery to determine the degree of safety of the explosive when fired in the presence of inflammable gases and of coal dust respectively. A special test is made for fumes also, and permissibles are classified according to the content of toxic gases (carbon monoxide and hydrogen sulphide) in their gaseous detonation products.

The most generally used permissibles today contain (i) a high percentage of ammonium nitrate, in the neighbourhood of 80%; (ii) a relatively low percentage of nitroglycerine as sensitising agent; and (iii) fibrous combustible material for utilising the excess oxygen of the ammonium nitrate and absorbing the nitroglycerine. Black powder is admirably adapted for use in blasting down coal as far as its blasting action is concerned, but it is not safe in gassy mines. Its blasting action characteristically brings down coal in large lumps rather than shattering it into fine fragments, and detonating permissibles have been developed which approach black powder in its heaving action by attainment of very low velocities and very low density by various expedients. A gelatinous type of permissible has been introduced which is more suitable for use under wet conditions of blasting. This type contains a higher nitroglycerine content than the usual permissible, a small percentage of nitrocotton, and a lower content of ammonium nitrate.

COMMERCIAL HIGH EXPLOSIVES CONTAINING NO NITROGLYCERINE

Ammonium Nitrate Explosives

The logical extension of the trend toward the replacement of nitroglycerine dynamites by ammonium nitrate would be to replace the nitroglycerine entirely. The idea is not new in the explosives industry; many formulae have been prepared over a long period of years embodying the combination of ammonium nitrate having an excess of oxygen, with an oxygen deficient but non-explosive sensitising agent. However, ammonium nitrate is a highly deliquescent salt, and mixtures containing it tend to become hard and insensitive, and incapable of propagating the explosion after exposure to moisture.

A new type of blasting agent of this nature was introduced in 1935 designated "Nitramon", which possesses a degree of insensitiveness such that it is incapable of detonating under the influence of a blasting cap. When once initiated by a sufficiently strong booster charge, however, the material is capable of propagating the explosion indefinitely at high and undiminishing velocity when used in large diameters. The disadvantages of moisture effect are overcome by having the material enclosed in a waterproof metal container. A number of such cans of the material, for example, 7" in diameter by 24" long, are loaded into large-diameter vertical quarry holes. The explosion propagates through the metal ends of the cans at velocities around 16,000 feet per second. Because of its insensitiveness to impact, friction, and shock, it is the safest high velocity agent yet introduced.

Nitrostarch Explosives

Nitrostarch is a solid, white explosive material obtained by nitration of cornstarch or cassava starch by means of practically anhydrous mixed acid. The nitrated product is drowned in an excess of water, washed thoroughly, neutralised with dilute sodium carbonate solution, and dried, first by wringing, then in hot-air driers. The commercial nitrostarch is mainly starch trinitrate and ordinarily contains slightly below 13.0% nitrogen, as against a theoretical content of about 13.3%.

Nitrostarch is used in commercial high explosives in America to a limited extent. Since it is a solid explosive material, no freezing can take place. Ammonium nitrate is commonly included in considerable amounts in nitrostarch explosive compositions. Nitrostarch has had some applications in military explosives.

Chlorate Explosives

Explosives containing sodium or potassium chlorate have been made for well over a century and periodically attract renewed attention. The earlier chlorate powders were dry mixtures of chlorate with oxidisable materials such as sugar, charcoal, sulphur, etc. These mixtures were all highly inflammable and sensitive to frictional impact. Sooner or later their manufacture usually led to disastrous results. More recently, chlorate explosives have included a liquid ingredient, or a desensitising material, and such mixtures are less sensitive. In spite of the cheapness of raw materials, however, chlorate explosives have not proved very attractive.

Liquid Oxygen Explosives

Liquid oxygen explosives comprise carbonaceous absorbent materials saturated with liquid oxygen. Theoretically, such explosives appear very attractive, as carbon burns to carbon dioxide completely, if sufficient oxygen for combustion be present, and the explosive is entirely gasified. In making liquid oxygen explosives, the cartridge, consisting usually of lampblack or other carbon black packed in a porous container, is first prepared. This cartridge is dipped into a bath of liquid oxygen and allowed to become saturated.

Potentially, liquid oxygen explosives are of high strength and velocity because of the composition, and this is actually the case under favourable conditions. There is the disadvantage that liquid oxygen evaporates very rapidly, so that the life of an explosive cartridge is relatively short. Also, the strength is not constant because of the steady loss of oxygen. A further disadvantage comes in the high sensitiveness to shock of liquid oxygen explosives, a serious consideration in quarrying operations.

Miscellaneous Explosive Devices

Mention should be made of a type of blasting device which has had some popularity, and which does not depend primarily for its effectiveness on the oxidation of combustible materials to gaseous products. One such device, which has been applied in the mining of coal, is based on the use of a confined supply of liquid carbon dioxide, together with a heating composition, adapted to be ignited at the proper time and to exert the desired blasting effect by converting the carbon dioxide to the gaseous state. The real novelty comes in the use of a specially designed cartridge constructed with a closing disc which will rupture when the pressure exceeds a pre-determined amount. A later development uses the same general type cartridge, but brings about the rupture of the disc and the release of pressure by the use of compressed air from a portable compressor.

A blasting device for use in coal has also been introduced, consisting of an expansible tube of particular design operating under hydraulic pressure.

All these mechanical devices yield a good grade of lump coal, but their ultimate value must still be determined.

INITIATING DEVICES

After nitroglycerine became available as an explosive compound, its high explosive energy could not be utilised satisfactorily until Nobel had devised and introduced into practice a form of blasting cap. It was then realised that, whereas a simple flame was sufficient to initiate the explosion of black powder, the efficient development of the strength of nitroglycerine explosives resulted only when initiation was from the powerful influence of a detonating material.

A group of compounds has been developed which are peculiarly adapted for use in initiating detonating explosives of the dynamite type. Such materials are designated as primary detonating compounds. The secret of their effectiveness comes not in their high temperature of explosion, or the volume of gas produced, but in the fact that they attain their high velocity of explosion with extreme rapidity after initiation by a flame. As a matter of fact, the most effective primary detonating compounds are characterised by small volume of gas formation, and are nitrates or nitrated compounds, as are most of the explosive materials.

Mercury Fulminate

Mercury fulminate, $(CNO)_2Hg$, has long been a very satisfactory primary detonating compound. It is commonly used with a smaller amount of an oxidising agent such as potassium chlorate. It is made by dissolving mercury in a large excess of 68% nitric acid and adding the solution to 95% alcohol, with arrangements for controlling the temperature. The reaction soon causes the solution to boil spontaneously and vigorously and mercury fulminate precipitates in the boiling flask as a fine white powder of high density. One economic disadvantage of mercury fulminate comes in the high price of mercury and the fluctuations in price because of the uncertain supplies. Spain is the principal source of mercury ores.

Lead Azide

Lead azide PbN_6, has become fully as important as mercury fulminate as a primary detonating compound in the last few years. It has the disadvantage of requiring a small ignition top charge because of its relative insensitiveness to flame, in contrast to mercury fulminate. However, it picks up to its maximum high velocity more rapidly than fulminate, hence a smaller weight charge will have the detonating effect of a heavier fulminate charge. An additional advantage lies in the ready availability of the raw materials. It is made by the reaction of sodium azide with a soluble lead salt. Sodium azide in turn is derived from sodamide (from metallic sodium and ammonia) and nitrous oxide.

Other Primary Detonating Compounds

While mercury fulminate and lead azide are the two most widely used compounds of this nature, other materials are also employed; for example, diazodinitrophenol, obtained by reacting concentrated nitric acid with picramic acid or a picramate in the presence of alcohol, and mannitol hexanitrate, a nitration product of the polyhydric alcohol, mannitol.

Blasting Caps

Blasting caps are small cylindrical metal shells, usually about 1/4" inch in diameter by 2" long. These shells are of copper or aluminium and contain a detonating charge of explosive. Most blasting caps are of the compound type and contain a fraction of a gram of a base charge at the closed end of the shell, a charge of a primary detonating explosive upon the base charge, followed by a more readily ignitable top charge if necessary. Mercury fulminate and lead azide are the most commonly used primary detonating charges, while tetranitromethylaniline (tetryl) is an excellent base charge. Blasting caps are used for initiating the explosion of dynamites and are inserted into the explosive at the time of use.

Three types of blasting caps are employed, the first two differing by the method of firing (Fig. 11.3). In the case of ordinary or fuse caps, a length of safety fuse is inserted into the open end and the metal wall is crimped about the fuse. In use, the end of fuse away from the cap is lighted, and the spit of flame from the fuse within the cap fires the ignition charge and brings about detonation. The second type of cap is fired electrically, the two leading wires being connected within the cap by a fine wire of high electrical resistance.

When an electric current passes through the wires, the fine bridge wire becomes heated to incandescence and fires the ignition charge. A third type, designated as a delay cap, is ordinarily fired electrically, but contains a delay charge of a slow burning composition which is inserted in the cap so that the detonation of the cap charge will take place after a definite time interval. By the use of delays, the time interval before a cap will fire, after application of the current, can be varied from 0 to 15 or 20 seconds.

Recently, delays have been improved by the use as delay compositions of charges which give off practically no gas on combustion; for example, mixtures of a combustible metal with an oxidising agent. This permits the employment of ventless caps, so that moisture penetration is prevented, as well as excessive gas pressure variation in the cap previous to firing.

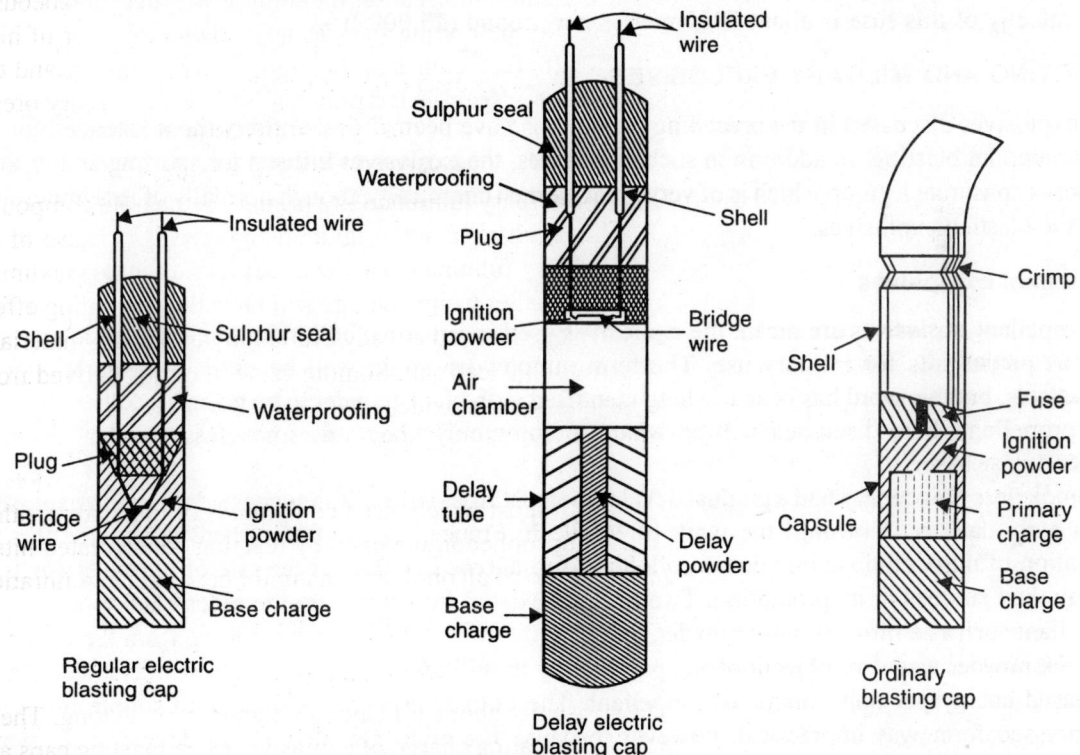

Fig. 11.3. Three types of blasting caps.

Squibs

When deflagrating explosives such as black powder are used, the simple application of flame is sufficient for firing. This is done by using devices similar to blasting caps, except that they contain no detonating explosive, hence explode less violently. Preferably, electric squibs are used and the construction of these is similar to that of electric blasting caps, except for the nature of the explosive charge. Squibs may be in metal or cardboard tubes, and may be either open or closed at the firing ends.

Fuse

Two types of fuse are in general use in blasting. Safety fuse is employed for initiating caps or squibs where electrical firing is not used. Safety fuse consists of a small-diameter core of black powder, enclosed in a covering wrapper of various waterproofed fabrics. It is made to have an approximate burning speed of 30 or 40 seconds per foot. When fuse is used to fire a shot in blasting, a sufficient length is used so that ample time is allowed for the shot firer to reach a point of safety.

Detonating fuse, on the other hand, has a velocity of detonation of over 5000 metres per second (16,500 ft). It is used principally for exploding charges of explosives in deep holes. The usual detonating fuse, called *Cordeau*, consists of a charge of high velocity explosive such as trinitrotoluene contained in a small diameter lead tube. The line of Cordeau is in contact with the charge throughout its length, and thus causes practically instantaneous detonation of the whole charge, regardless of its velocity. A more recent type of detonating fuse comprises pentaerythritol tetranitrate contained in a non-metallic wrapping. The velocity of this fuse is about 7000 metres per second (23,000 ft.).

SPORTING AND MILITARY EXPLOSIVES

The explosives discussed in the preceding paragraphs have been almost entirely those intended for use in commercial blasting. In addition to such explosives, the explosives utilised for sporting and military purposes constitute a group which is of very great interest chemically, though normally of less importance than the blasting explosives.

Propellant Explosives

By propellant explosives are meant the explosives used in firearms, both rifle and shotgun powders, as well as propellants for military use. The term gunpowder might well be used in designating such propellants, but this word has been too long identified with black powder to be given any other meaning. The propellants here discussed will be what are commonly known as smokeless powders having a nitrocellulose base.

Smokeless powder has had a gradual development. Nitrocellulose was suggested for use in explosives at an early date, but not until the work of Vieille in France, about 1886, when he showed that the formation of nitrocellulose into dense colloidal form allowed control of the rate of burning, was there any marked success in its promotion. Two reasons existed for displacing the older black powder by a propellant such as a nitrocellulose powder. First, the dense cloud of smoke resulting from the explosion of black powder was very objectionable, particularly in military use. Secondly, there was a demand for increased energy per unit volume of propellant. The utilisation of nitrocellulose in simple fibrous or compressed form was impractical, however, because the high rate of pressure development would burst the gun. Nitrocellulose has the advantage of raw materials that are of wide distribution and high availability. Cotton is the most suitable source of cellulose for use, and the ordinary form employed is cotton linters, obtained from the cotton seeds as a second cut, after removal of the long-fibred cotton and the first cut of lint. Wood pulp is another possible source of cellulose for nitration, and has been widely used for military purposes by countries shut off from cotton supplies.

Smokeless Powder

Nitrocellulose may vary rather widely in composition, as shown by the nitrogen content. This in turn means a variation in potential, which is an important consideration in smokeless powder manufacture.

Assuming the simplified, but hypothetical formula $C_6H_{10}O_5$ for cellulose, the trinitrate would contain 14.14% nitrogen. The nitrocottons of interest in connection with smokeless powder contain between 12 and 13.4% nitrogen. Pyrocellulose or pyrocotton, an important type for smokeless powder, contains between 12.5 and 12.7% nitrogen and is soluble in a 2 to 1 ether-alcohol solution; gun-cotton, with a nitrogen content of 13.0 to 13.4%, is insoluble in the ether-alcohol mixture; dynamite nitrocotton, with a 12.25% content, is used in gelatin dynamites where it is gelatinised by nitroglycerine; nitrocellulose of nitrogen contents below 12.3% is used for non-explosive outlets such as lacquers, plastics, fabrics, etc.

Manufacture of Single-Base Powders

The term "single-base powder" designates a type of smokeless powder in which nitrocellulose is the sole nitric ester. This is the type of powder used in Army and Navy. The process involved in the manufacture of single-base nitrocellulose powders may be considered under (i) dehydration; (ii) mixing and colloiding; (iii) pressing and graining; and (iv) solvent recovery and drying.

Dehydration

After the wringing of the purified nitrocotton, the material still has a water content of 25 to 30%, in which wet condition it is entirely safe to handle. The drying of this material by warm air currents on trays would be hazardous on a large scale, because of the static susceptibility of dry nitrocotton. In practice, therefore, dehydration takes place by displacing the water with alcohol. Since the following step in the process involves treating the nitrocotton with an ether-alcohol solution, the presence of the alcohol brings in no difficulties. Approximately 1.25 pounds of 95% alcohol are used per pound of dry nitrocotton, the introduction of the alcohol being by means of a hydraulic press.

Mixing and Colloiding

The dehydration operation leaves the nitrocotton in the form of a compressed block wet with alcohol. This is broken up and introduced into a mixing machine, where the material is kneaded by two sets of agitation blades rotating in opposite directions. The nitrocotton here is to be colloided by the ether-alcohol solution containing two parts of ether to one of alcohol, the total weight of solvent being approximately equal to the weight of the dry nitrocotton. Before addition of the ether, diphenylamine is dissolved therein in an amount such that about 1% will be present in the finished powder. The ether is quickly poured into the mixer, where it blends with the alcohol present and the contents are agitated for about an hour. The diphenylamine is added to the other during the mixing operation in order that it may be intimately distributed throughout the blend, so that any nitrogen oxide decomposition products which may be developed during years of storage will react with this stabiliser and thus be removed.

Pressing and graining

The colloidal smokeless powder is now ready for its final treatment. A dense cylindrical mass is formed first by compression of the material in a cylinder by means of a hydraulic press exerting a pressure of around 3000 pounds per square inch. The blocks are then grained by forcing the material through dies provided with pins under a pressure of several tonnes per square inch. The strands of powder are perforated as they pass through the die and then are guided to mechanical cutters. Cannon powders in the United States have seven longitudinal perforations, one in the centre surrounded by six others. Powders for small arms have only one small centre perforation. The object of the perforations in the powder is to obtain a progressive increase in the area of burning surface during combustion. The smokeless powder in its final form is actually a cylinder of variable size, depending on the type of weapon in which it is to be used. Cannon powders in the United States have a length about 2 1/2 times their diameter and are of considerable size, while various powders for small arms are of very small diameter.

Solvent recovery and drying

The solvent recovery process is adapted for partial drying of the "green" powder, with recovery of whatever alcohol and ether are given off. The drying is a continuation of the solvent recovery step except that in the second step, solvent is removed beyond the point where its recovery is justified. Precautions are taken that the removal of solvent is not rapid enough to cause distortion of the powder grains.

Recent Improvements in Nitrocellulose

The last few years have seen the improvement of the nitrocellulose type of propellent powder, particularly in two respects. First, the hygroscopicity of the powder has been reduced by using nitrocellulose of higher nitrogen content and replacing a portion of the nitrocellulose by non-volatile and non-hygroscopic solvents.

Secondly, flashlessness has been obtained in many guns by the use of compositions of suitable potential which are in themselves flashless or by the use of supplemental flash suppressing salts. This means that muzzle flash, due to the secondary ignition of highly combustible gaseous products issuing from the gun, has been prevented from taking place, thus eliminating the usual large luminous muzzle flash which is visible for many miles. This is an important consideration in military operations.

Manufacture of Double-Base Powders

These powders differ because of the fact that, instead of containing nitrocellulose alone as explosive ester, they contain both nitrocellulose and nitroglycerine. A well-known example of this type of powder is the British Cordite. M.D. Cordite contains 30% nitroglycerine, 65% guncotton, and 5% vaseline. The nitrocotton has a nitrogen content of at least 13%, hence is difficultly soluble in an ether-alcohol mixture. Consequently, acetone is used as solvent. The vaseline acts as stabiliser and cooling agent in the powder, as it has a tendency to lower the temperature of explosion.

The Italian Ballistite is another powder of this double-base type, but contains a nitrocellulose of about 12% nitrogen content, soluble not only in an ether-alcohol solvent, but also in nitroglycerine. This type of powder, which contains equal parts of nitrocellulose and nitroglycerine plus 1% of diphenylamine, is made on rolls without the use of acetone or any other volatile solvent.

The nitrocellulose type of smokeless powder propellant has been used by the United States, Russia, England, and France. The nitroglycerine type of propellant is used by Great Britain and Italy. Germany is said to have used to a large extent the nitrocellulose type in their army and the nitroglycerine type in their navy.

DISRUPTIVE EXPLOSIVES FOR MILITARY USE

The explosive compounds up to this point have been mainly nitric esters of aliphatic alcohols and carbohydrates. These compounds are not ordinarily suitable as bursting charges for firing shells and other military devices because of too great a sensitiveness to shock. The aromatic nitro-compounds are generally better adapted for this purpose.

Trinitrotoluene

This explosive compound is prepared by the nitration of toluene, an aromatic hydrocarbon derived from coal tar, or obtained synthetically from petroleum products.

The nitration procedure has been discussed previously. TNT, as it is called, is the most widely used shell-firing explosive, and is well suited for loading into containers because of its low melting point of 80.8°C. Its importance for military use comes from its comparative excellence in the following respects: (i) it is a safe explosive in manufacture, transportation, and storage; (ii) it is non-hygroscopic; (iii) it has no tendency to form unstable compounds with metals, yet; and (iv) it is a violent, disruptive explosive.

Amatol

Amatol is an explosive mixture of TNT with ammonium nitrate, developed during the First World War. A 50–50 mixture of the respective ingredients is sufficiently fluid to allow casting, but an 80–20 ammonium nitrate-TNT mixture was subsequently found suitable for loading into shells by an extrusion process. It was developed because of the imminent shortage of toluene for nitration during the early years of the war. Actually, such mixtures are, in some respects, an improvement on TNT by itself. TNT is deficient in oxygen so that combustion is not complete on explosion and black smoke always results. The detonation products may include carbon, carbon monoxide, hydrogen and methane, all capable of further combustion.

The presence of 80% ammonium nitrate with 20% TNT assures almost an evenly balanced composition with respect to oxygen with resulting increased strength and improved fumes. Compared to TNT, amatol has the disadvantage of being hygroscopic.

Tetryl

Tetryl is the name commonly applied to tetranitromethylaniline, or more properly, trinitrophenyl-methylnitramine. It is made by the nitration of dimethylaniline and has a melting point of around 130°C.

Dimethylaniline   Tetryl
(5 mols)   (5 mols)

$+ 26HNO_3 \rightarrow$ $+ 5CO_2 + 28H_2O + 3N_2$

Tetryl has a very important use in commercial explosives as a base charge for blasting caps. In military explosives, it is too sensitive for use as a main shell filling charge, but is well suited for use as a booster charge, intermediate between the sensitive ignition charge and the main bursting charge.

Picric Acid

Chemically, picric acid is trinitrophenol. It has had considerable use as a shell-filling explosive.

$$
\underset{\text{Phenol}}{\text{C}_6\text{H}_5\text{OH}} \; + \; 3\text{HNO}_3 \;\longrightarrow\; \underset{\text{Picric acid}}{\text{C}_6\text{H}_2(\text{OH})(\text{NO}_2)_3} \; + \; 3\text{H}_2\text{O}
$$

It has a melting point of about 122°C. One objectionable characteristic of picric acid is its tendency, as an acid, to form metallic salts or picrates, which salts are unfortunately dangerously sensitive.

Ammonium Picrate

Ammonium picrate is the salt formed by neutralising picric acid with ammonia. Its relative lack of marked tendency to form sensitive metallic picrates when in contact with metals gives it an advantage over picric acid. It is not altogether free from this tendency, however. Ammonium picrate is claimed to be the least sensitive to shock of all explosives used as shell bursting charges. However, it is inferior to TNT in strength.

Pentaerythritol Tetranitrate

This explosive compound has come to be important since the First World War, and owes its attractiveness partly to the fact that it is obtained from raw materials of almost universal occurrence. This makes the compound of particular interest to those nations which wish to become self-contained. Pentaerythritol tetranitrate, or PETN, melts at around 141°C. It is obtained by the nitration of the irregular tetrahydric alcohol, pentaerythritol, which in turn is prepared by the condensation of formaldehyde with acetaldehyde.

$$
\underset{\text{Pentaerythritol}}{\text{C}(\text{CH}_2\text{OH})_4} \; + \; 4\text{HNO}_3 \;\longrightarrow\; \underset{\substack{\text{Pentaerythritol}\\\text{tetranitrate}}}{\text{C}(\text{CH}_2\text{ONO}_2)_4} \; + \; 4\text{H}_2\text{O}
$$

Trimethylene Trinitramine

This is another post-war high explosive compound, frequently designated hexogen. Like the preceding explosive, it is attractive because of the widely-occurring raw materials. In this case, the explosive is obtained by the nitration of hexamethylenetetramine, a compound obtained from formaldehyde and ammonia. The raw materials in this case give hexogen greater promise than PETN, but the yields obtained are said to be sufficiently low to offset that apparent advantage. Hexogen has a melting point of 200–201°C.

Hexamethylene tetramine

Hexogen

HANDLING AND STORAGE OF EXPLOSIVES

Because of the potential hazard when explosives are present, some generally accepted rules are observed in their handling and storage. The basic principles underlying these rules are that : (i) explosives should be treated with respect; (ii) they should be handled only by those experienced or properly instructed in their use; and (iii) they should be stored under favourable conditions and at positions sufficiently isolated that accidental or malicious explosion will not endanger persons not directly concerned with their use.

In the case of nitroglycerine explosives, the unwrapped explosive should not be handled unnecessarily, since the well-known nitroglycerine headache may result from bodily contact with nitroglycerine or other liquid nitric esters, or from breathing their vapours. This does not occur with ammonium nitrate, nitrocellulose, nitrostarch, or other solid explosive nitrates. The aromatic nitrocompounds possess a certain degree of toxicity and continued contact with them should be avoided, either in handling or inhaling their vapours or dust. For the storage of explosives in magazines there are definite laws and regulations in the various states. A so-called "American Table of Distances" has been prepared which specifies minimum distances at which magazines containing explosives should be spaced from inhabited buildings, railroads, and public highways, the distances increasing with increased amount of explosives stored. Typical components of composite rocket propellants are given in Table 11.2 where as Table 11.3 highlights the characteristics and uses of important explosives.

Table 11.2. Typical components of composite rocket propellants.

Typical components	*Characteristics*
Binders	
Polysulphides	Reactive group, mercaptyl (—SH), is cured by oxidation reactions; low solids loading capacity and relatively low performance; now mostly replaced by other binders
Polyurethanes Polythers, Polyesters	Reactive group, hydroxyl (—OH), is cured with isocyanates; intermediate solids loading capacity and performance
Polybutadienes copolymer, butadiene and acrylic acid	Reactive group, carboxyl (—COOH) or hydroxyl (—OH), is copolymer, butadiene cured with difunctional epoxides or aziridines; intermediate solids loading capacity and better performance than polyurethanes; less than adequate cure stability and mechanical characteristics

(Contd...)

Typical components	Characteristics
Terpolymers of butadiene, acrylic acid acrylonitrile	Superior physical properties and storage stability
Carboxy-terminated polybutadiene	Cured with difunctional epoxides or aziridines; have very good solids loading capacity, high performance and good physical properties.
Hydroxy-terminated polybutadiene	Cured with diisocyanates; have very good solids loading and performance characteristics and good physical properties and storage stability.
Oxidisers	
Ammonium perchlorate	Most commonly used oxidiser; it has a high density, permits a range of burning rates, but produces smoke in cold or humid atmosphere
Ammonium nitrate	Used in special cases only, it is hygroscopic and undergoes phase changes, has a low burning rate and forms smokeless combustion products.
High energy explosives (RDX-HMX)	Have high energy and density; produce smokeless products; have a limited range of a low burning rates.
Fuels	
Aluminium	Almost commonly used; has a high density; produces an increase in specific impulse and smoky and erosive products of combustion.
Metal hydrides	Provide very high impulse, but generally inadequate stability, give smoky products, and have a low density
Ballistic modifiers	
Metal oxides	Iron oxide most commonly used.
Ferrocene derivatives other	Permit a significant increase in burning rate coolants for low other burning rate and various special types of ballistic modifiers
Modifiers for physical characteristics	
Plasticisers	Improves physical properties at low temperatures, and processability; may vapourise or migrate; can increase energy if nitrated.
Bonding agents	Improve adhesion of binder to solids.

Table 11.3. Characteristics and uses of the more important explosives.

Primary explosives

Name	Composition or chemical formula	Density (g/cc)	Detonation velocity[a] (km/sec)	Detonation pressure (kilobars)	Detonation temperature[a] (°K)
Mercury fulminate	$Hg(ONC)_2$	3.6	4.7	220	6900
Lead azide	PbN_6	4.0	5.1	250	5600
Lead styphnate	$C_6H(NO_2)_3$ O_2Pb	2.5	4.8	150	—
Nitromannite (Mannitol hexanitrate)	$C_6H_{58}(ONO_2)_6$	1.73	8.3	300	6000
Dinitrodiazophenol (DDNP)	$C_6H_2N_4O_5$	1.5	6.6	160	—

(Contd...)

Primary explosives

Name	Composition or chemical formula	Density (g/cc)	Detonation velocity[a] (km/sec)	Detonation pressure (kilobars)	Detonation temperature[a] (°K)
Secondary high explosives					
Ammonia gelatin dynamites	30–90% grades same as straight gelatins except for some NG and $NaNO_3$ replacement by NH_4NO_3		1.2–1.5	4–6.5	0.75–1.15
Semigelatin dynamite	15–20% NG 1–2% DNT oil, AN-SN dope		1.2 (depends on diameter)	3.5–5	0.9
Prilled AN-oil	94/6 NH_4NO_3/oil	0.8–0.9	1.5–4	0.81–0.83	
Slurry explosives	TNT	17–40	1.4–2.0	5–8	0.7–1.8
TNT-SE	Oxidiser*	30–65			
	H_2O	12–25			
	Al	0–20			
	Other	0.3–1.5			
Smokeless powder SE	SP	20–40	1.35–1.9	4–7	0.65–1.7
	Oxidiser*	30–60			
	H_2O	3–25			
	Al	0.20			
	Other	0.3–10			

[a] Most important properties of detonators.
* AN, SN, perchlorates, etc.

Sensitivity	Major characteristics	Uses
Very high	Best primary explosive for single-component (fuse) detonators; easily detonated by flame, spark, heat, or friction; easily dead-pressed.	In fuse caps (mixed with $KClO_3$); propellant primer; in fuses for shells; small arms arms cartridges caps.
Very high (higher than NG; less than mercury fulminate)	Powerful detonator but requires strong igniters, e.g., lead styphnate.	Primary explosive in composition (EB) caps; military fuses.
Exceedingly high	Extremely sensitive to sparks, static electricity; explodes rapidly on ignition; good thermal stability.	Igniter in composition caps, military fuses; very satisfactory detonator explosive for fast ignition.
Very high (greater than NG; less than lead azide)	Stronger and more brisant than NG, RDX, PETN.	In composition caps and fuses.
Very high (less than lead azide)	Does not dead-press. About 3/4 as strong as TNT.	In composition caps and fuses.
High	More economical; only slightly less brisant than straight gelatin; exhibits low-order detonation with threshold priming and high pressures.	General small and large diameter blasting in hard rock and under water.
High	Stringy, plastic; easily loaded in 'uppers'; economical; high strength; moderate brisance.	Popular small diameter metal-mining explosive.

(Contd...)

Sensitivity	Major characteristics	Uses
Low (requires booster)	One of the cheapest sources of explosive energy available today; flammable and will explode when ignited under strong confinement; on no water resistance; adaptable to do-it-yourself operations.	Open-pit and underground blasting where dry conditions prevail; most adaptable to soft, easy shooting.
Low (requires booster)	Gel or thick pea-soup consistency; capable of detonation at high pressures, excellent water resistance.	Large diameter, open-pit, small diameter under-ground, oil well, sub-marine, water-filled bore-holes, deep-water bombs.
Low (requires boosters)	Generally similar to TNT slurry.	Large diameters, open-pit blasting.

Name	Composition or chemical formula	(g/cc)	Density (km/sec)	Detonation velocity (kcal/g)	Available energy
Al-SE	Al	1.0–10	1.1–1.5	2–5	0.7–1.3
Slurry blasting agents	Al	0–35	1.1–1.6	2–6	0.7–2.0
	Oxidisers*	50–80			
	H_2O	4–18			
	Other	0.2–10			
Nitrostarch powders	Nitrostarch in place of NG	1.2	4–5	0.8–1.0	
Composition B	40/59/1 TNT/RDX/wax	1.7	7.8	1.1	
Composition B-3	40/60 TNT/RDX	1.73	7.9	1.15	
Haleite or EDNA	$(CH_2NHNO_2)_2$	1.6 (pressed)	7.9	1.2	
Ammonium picrate (Explosive D)	$(ONH_4)C_6H_2(NO_2)_3$	1.56 (pressed)	6.6	0.7	
Nitrostarch	Mixtures of various nitro esters of starch	1.4 (pressed)	6.4	0.95	
Tetryl	$(NO_2)_3C_6H_2CH_3N_2NO_2$	1.45 (pressed)	7.0	0.95	
PETN (pentaerythritol tetranitrate)	$C(CH_2ONO_2)_4$	1.6 (pressed)	7.92	1.31	
Pentolite	50/50 TNT/PETN	1.63 (cast)	7.7	1.1	
Trinitrotoluene (TNT)	$CH_3C_6H_2(NO_2)_3$	1.59 (cast)	6.9	0.9	
		1.45 (pressed)	6.9	—	
		1.03 (Pelletol)	5.1	—	
		0.8 (grained)	4.2	0.8	
Amatols	50/50 AN/TNT	1.55 (Cast)	5–6.5 (depending on diameter)	0.95	
	80/20 AN/TNT	1.0 (loose)	4 (large diameter)	0.93	
		1.45 (pressed)	5.6 (large diameter)		

(Contd...)

Sensitivity	Major characteristics	Uses
From cap sensitive to very low depends on alumnium fineness.	Gelatin; no explosive ingredients.	Small diameter, underground and general blasting.
Low (requires boosters)	Gelatin to thick or thin pea-soup consistency.	Large diameter, underwater, wet- and dry-hole blasting, large bombs.
Moderately high, but less than dynamites	Good "fumes"; fair water resistance; powerful; economical.	Small diameter blasting.
Average	Very high brisance.	Bursting charge and special weapons.
High	Very high brisance.	Experimental standard.
High	High brisance; less sensitive RDX and PETN.	In Ednatols for bursting charges.
Very low	Insensitive to shock and friction; melts with decomposition; shells filled with high-pressure pressing.	Armor-piercing shells.
High	Highly inflammable white powder	Demolition blocks and Trojan blasting explosives.
High	Very sensitive; rapidly reacting; easily pressed with 1–2% graphite; high brisance.	Booster; base charge in caps; in tetrytols for bursting charges.
High	Very powerful and sensitive (more sensitive than RDX, less than NG).	In primacord fuse; base charge in caps.
Moderate	High pressure or brisance; primacord sensitive	Booster and special weapons; commercial booster for prilled AN-fuel oil and slurry explosives.
Low	Easily melted and cast; suitable liquid for slurrying with other explosives; easily pressed into blocks; completely waterproof.	Military; "Nitropel" TNT used in slurry explosives and in filling annulus between charge and bore-hole in water filled holes; in amatols.
Low	Insensitive; hygroscopic, not waterproof; less brisant but stronger than TNT; 50/50 can be cast; 80/20 eitheter pressed or granulated.	Military; oil well shooting; quarrying; dry-hole booster for very low-sensitive either types.

Name	Composition or chemical formula	Density (g/cc)	Detonation velocity (km/sec)	Available energy (kcal/g)
Dinitrotoluene (DNT)	$CH_3C_6H_3(NO_2)_3$	1.28 (liquid) 0.8 (granular solid)	5 2–3.5 (depending on diameter)	0.7
Nitromethane (NM)	CH_3NO_2	1.12	6.2	—
Cyclonite (RDX)	$C_3H_6N_6O_6$	1.2 (loose) 1.6 (pressed)	6.8 8.0	1.32
HMX	$C_4H_8N_8O_8$	1.89	9.1	1.35
HBX	Mixtures of RDX, TNT, aluminum, and wax	1.78	7.5	1.5

(Contd ...)

Name	Composition or chemical formula	Density (g/cc)	Detonation velocity (km/sec)	Available energy (kcal/g)
Plastic explosives (compositions A,C, C-2, C-3, C-4)	Waxed RDX	1.45–1.6	8.0	1.1–1.3
PBX 9404	94/3/HMX/binder/ nitrocellulose	1.84	8.8	1.3
Nitroglycerine (NG)	$C_3H_5(ONO_2)_3$	1.59	7.8	1.41
Ethylene glycol dinitrate (EGDN)	$C_2H_4(ONO_2)_2$	1.48	7.4	1.43
Straight dynamites[b]	20–60% NG, in balanced SN dope 20% grade ≡ 20% NG, etc.	1.3	4–6	0.55–0.85[b]
Ammonia dynamites (and permissibles)	As above except NH_4 NO_3 replaces part of NG and $NaNO_3$	0.8–1.2[b]	1.5–5.5 Depends on AN particle size, NG content.	0.7–0.9[b]
Blasting gelatin	92/8/NG/nitrocotton ("Solidified" NG contains some wood pump to minimise low-order detonation).	1.55 (1.45)	7.5 (7.2)	1.45 (1.4)

[b] Depends on grade.

Sensitivity	Major Characteristics	Uses
Very low	Reddish brown or yellow liquid.	Sensitiser in "Nitramons"; 60/40 NG/DNT in oil well shooting; up to 20% in TNT bursting charges; in FNH (flashless) propellant; 6% in small-arms ammunition (with guncotton).
Moderate	Clear, watery liquid	Special demolition, experimental studies of liquid explosives.
High	High thermal stability in solid state expressively sensitive in pure state; 1.65 times as strong as low density TNT; 1.45 times as strong as cast TNT.	Major ingredients in plastic explosives; one of most brisant explosives in cast explosives; one of most base charge in caps.
High	Better than RDX in all respects.	Same as RDX.
Average	Very powerful.	Underwater explosive.
Moderate	Plastic, easily moulded or pressed.	Specialised military demolition.
Moderate	Plastic bonded.	Specialised military demolition.
Very high (almost a primary explosive)	Oily, toxic liquid; volatile above 50°C; gelatinised by nitrocotton; exhibits low-order detonation with threshold priming.	Shooting oils wells; main explosive in dynamites; used in double-base powders.

(Contd ...)

Sensitivity	Major Characteristics	Uses
Very high	Closely resembles NG; more volatile, toxic, slightly stronger but less brisant (owing to lower density).	Used in solution with NG as freezing point depressant.
High	Cheesy, plastic substance; packed in paper cartridges; may be slit and tamped in borehole for greatest blasting effect; fired by detonator as are all dynamites; heat, friction, shock, and flame sensitive.	Ditching, stumping, other uses where high propagation-by-influence "sensitiveness" is required.
High	Cheaper than comparable grade straight dynamites; must be waterproofed by special additives.	General small and large dynamite blasting; permissible (some grades).
High	Strongest, most brisant dynamite; completely waterproof; exhibits low-order detonation with threshold priming and under high pressures.	Oil well and submarine blasting, tunnel drilling, demolition.

Propellants		
Name	Composition or chemical formula	Sensitivity
Colloidal nitrocellulose (NC) Powders	Pyrocotton: cellulose nitrate with 12.6%N. Guncotton: cellulose nitrate with 13.3% N.	Low
Double-base powders	60–80% Nitrocellulose. 20–40% Nitroglycerine.	Moderate
Cordite	65% N.C., 30% NG, 5% vaseline	Low
FNH (Flashless non-hygroscopic powders)	Either straight N.C. or double-base powders with addition of coolants, etc., to prevent muzzle flash, and decrease water adsorption.	Low
Albanite; DINA powder	Di(2-nitrooxyethyl)nitramine.	Low
Rocket powder (solventless powder)	Nitrocellulose plasticised with about 50% NG, plus stabilisers and potassium salts.	Low
Chemical propellants[c]	Hydrogen perioxide, 80–90% H_2O_2 plus Ca, Na, or K permanganate (solid or aqueous solution). Hydrazine hydrate plus methyl alcohol. Fuming nitric acid-aniline. Mixed acid-monoethylaniline. Liquid oxygen-kerosene.	
Black powder	75% KNO_3 (or $NaNO_3$), 15% charcoal, 10% sulphur.	High

[c] Many new rocket propellants have been described; the best ones are under security classifications.

(Contd...)

Sensitivity	Major Characteristics	Uses
High	Jelly-like substance; powerful, waterproof; exhibits low-order detonation under thershold primering and high pressure.	In hard rock; mudcapping demolition; submarine blasting.

Major characteristics	Important uses
Burning rate controlled by graining, hygroscopic, smokeless flame, with intense flash; gelatinised with alcohol-ether.	Combined with stabilisers and modifiers to make smokeless powders for artillery, small arms, and sporting ammunition.
Pyrocotton and guncotton are usually blended to secure an average of 13.15% N.	Dry guncotton in fibre form is used in primers fired by an electric current.
Very rapid burning rate, controllable by surface area; more powerful and more readily ignitable than straight N.C. powders causes erosion of gun bores; can be detonated and is subject to DDT.[d]	Propellant for mortars and sporting ammunition; not used by U.S. armed forced as cannon powder because of bore erosion
Gelatinised with acetone.	Propellant for large caliber naval guns (English).
Like other smokeless powders, but can be rolled into sheets; flash reduced by DNT, potassium salts, etc.	Propellant for small armor-piercing rockets such as the "Bazooka" (NG base); for naval ammunition (NG base).
Better flashless powder than FNH powders. Very rapid, uniform burning rate; can be made with thick section since no solvent need be removed.	Naval ammunition. For rockets up to 4.5 inches.
Catalytic decomposition into water and O_2 releases about 1000 Btu per pound. Supplies oxygen to burn petroleum fuel.	For driving turbines on submarines: V-2 rocket-fuel pumps; jet motors; launching device for ram-jets.
Rapid combustion; fuel and oxidiser are both liquid. Rapid rate of reaction generates heat and gases.	For torpedo turbine drives. For launching device for ram-jets; jet motors. Rocket motors.
Cheap; excellent "heaving action," persistent smoky flame; very sensitive to friction, spark, and heat; hygroscopic.	Time (delay) fuses for blasting and shell; in igniter and primer assemblies for propellants; pyrotechnics; $NaNO_3$ powder, in commercial black powder and for practice bombs and saluting charges; (it is being discontinued as blasting charge).

CHAPTER 12

Electrothermal and Electrochemical Processes

INTRODUCTION

The electrochemical process has completely revolutionised the production of certain primary products and at such lowered cost as to permit the development of new secondary industries utilising these cheaper raw materials. Chlorine from the electrochemical cell costs about one-third of that produced by the best of the older chemical processes. This has enormously encouraged the utilisation of chlorine in directions that would not have been possible if we were still dependent upon the older non-electrical processes. Metallic sodium may be produced electrolytically at a cost of less than one-tenth that of the old retort method, and this cheap reagent has changed completely the character of development of a multitude of secondary industries.

CLASSIFICATION OF ELECTROCHEMICAL INDUSTRIES

Electro-Thermal Processes

These are high temperature chemical reactions brought about in electric furnaces. Very high temperatures, above 4000°C are attained with the help of electricity. Different devices are adopted for the production of heat, viz., (i) the arc furnace; (ii) the resistance furnace; (iii) the induction furnace; and (iv) the silent electric discharge.

1. *Arc furnace:* In this type of furnace, the arc, placed just above the furnace, is struck between two carbon rods. Sometimes the arc is struck between a carbon rod and the charge. Arc furnaces may be (i) direct; (ii) indirect ; and (iii) multiple.

2. *Resistance furnace:* Here the electrical energy is transformed into heat energy across suitable resistors and a very high temperature may be attained. In direct heating furnace, the charge itself acts as the resistor, while in the indirect heating furnace separate materials, are used for the purpose.

3. *Induction furnace:* This type of furnace works on the principle of electrical induction. Heavy current in the secondary coil produces high temperature. This type is used where conducting materials, like metals, form a part of the secondary load.

4. *Silent electric discharge:* The most common use of the process is the preparation of ozone in Siemens and Halske Ozoniser. Active nitrogen, nitrates, sulphates are prepared by this method.

High temperatures obtained from these different types of electrical furnaces, have made possible the manufacture of various important products. Carborandum, calcium carbide, nitric acid, phosphorous, high speed cobalt steel, ferro-silicon, ferro-vanadium, etc., are some of the examples. Electro-thermal processes are employed for the extraction of phosphorous and metals also. Extraction of iron by carbon reduction (in Stassino-furnace) and extraction of magnesium by heating a mixture of MgO and ferro-silicon by the Pidgeon process are such examples.

Electro-chemical processes

These are processes where chemical changes are initiated directly through the agency of electricity.

These are divided into two classes: (i) processes of the electrolytic furnace using fused electrolytes; and (ii) processes of the electrolytic cell using aqueous electrolytes. In both these processes the chemical reactions are caused by electric currents.

1. *Products of the electrolytic furnace:* This furnace is used in the manufacture of metals like sodium, potassium, calcium, magnesium, aluminium, etc. Lead alloys are also obtained by this process.

2. *Products of the electrolytic cell:* A majority of electro-chemical industries are products of this type of cell as working is easy and mechanical. A list of these industries is given below.

   a. *Electro-winning of metals:* Cu, Au, Ag, Zn, Pb, Sn, Ni, Bi, Sb and Cd are now precipitated on the cathode during electrolysis of their aqueous solutions.

   b. *Electorefining of crude metals:* The crude metal is suspended as the anode and the pure metal as the cathode in a solution of the metal ion. During electrolysis, the impurities are removed by various methods. Thus Cu, Ag, An, Zn and Pb are electrorefined.

INDUSTRIAL APPLICATIONS OF ELECTROCHEMISTRY

Manufacture of Technical Important Compounds

Caustic soda and chlorine by the electrolysis of brine; hydrogen and oxygen by electrolysis of acidulated or alkalinated water are industrially obtained.

Many important compounds are prepared by: (i) anodic oxidation; and (ii) cathodic reduction.

1. During the electrolysis of solutions of different inorganic compounds, reactions, mainly oxidation, take place in the anodic chamber. This is known as anodic oxidation. Per-acids and per-salts of chlorine and sulphur, permanganates, dichromates, chlorates, bromates, iodates, hypochlorites, ferrocyanides, white lead, lead dioxide, etc., are thus produced. Many organic substances like iodoform, bromoform, benzaldehyde, anthraqunione, etc., are also prepared by anodic oxidation.

2. During electrolysis, many reduction reactions take place in the cathode chamber. These are known as cathodic reductions. Mostly organic compounds like aniline, methyl alcohol, etc., are prepared by this process. Inorganic compounds like sodium hydrosulphite, $Na_2S_2O_4$, and other corresponding alkali salts are also produced by the cathodic reduction of alkali sulphates.

Potassium permanganate

Manufacture of potassium permanganate ($KMnO_4$) by electrolytic method involves anodic oxidation of green potassium manganate to pink permanganate. The liquor at the same time becomes alkaline, because of the formation of potassium hydroxide. A cell which has been found suitable has a rotating anode of iron screening through which the liquor is made to flow.

The oxidation reaction is:

$$2 K_2MnO_4 + 2 H_2O = 2 KMnO_4 + 2 KOH + H_2$$

$$MnO_4^{2-} \longrightarrow MnO_4^- + e$$

The complete process of manufacture comprises a fusion of MnO_2 (pyrolusite), with KOH, in a muffle furnace with circulation of air. The furnace product is leached with water, treated to remove excess of KOH, and passed through the cell. The $KMnO_4$ is so insoluble that some of it separate in the cell; on concentration, another part is obtained.

Synthetic graphite

Graphite, an allotrope of carbon, has been known to man from ancient times. It occurs in nature in Siberia, Bohemia, California, Sri Lanka and other places. It is also called Plumbago or black lead. Graphite was used for writing and drawing purposes since the middle ages. It is grey and flaky and forms hexagonal crystals. It is slightly more reactive than diamond. It is oxidised to carbon dioxide in air at 700°C and to CF_4 in fluorine at 500°C. It is also oxidised below 100°C to yellow insoluble graphitic acid, $C_{11}H_4O_5$. Formation of graphitic acid is a test for graphite.

Until 1896 it was thought that graphite was available only in nature. In that year Acheson prepared artificial graphite. The process was discovered during the manufacture of SiC in the electric furnace. The SiC decomposes at a higher temperature into carbon and silicon; silicon volatilises off and carbon in the form or graphite is left behind.

Manufacture

Artificial graphite is manufactured electrically form retort or petroleum coke (Fig. 12.1).

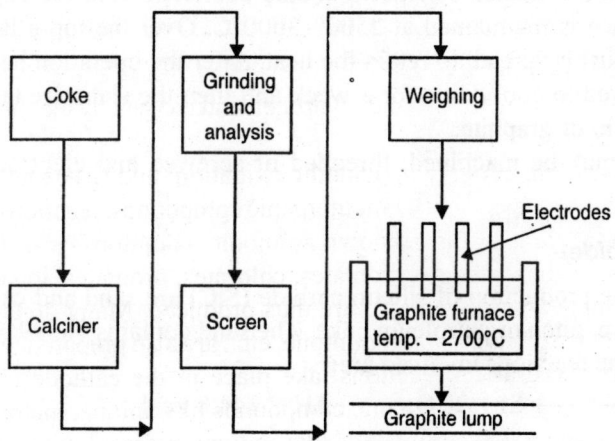

Fig. 12.1. Flowsheet for manufacture of graphite.

The reaction for this allotropic change is essentially

$$C \text{ (amorphous)} \longrightarrow C \text{ (graphite)} ; \qquad \Delta H = -2.5 \text{ cals.}$$

The different steps involved in the reaction are:

1. Coke (petroleum or retort) is selected and shipped to the graphite plant.
2. Carbon material is calcined (at 1250°C) to volatilise impurities.

3. Raw materials are carefully analysed.
4. Raw materials are ground, screened, weighed, mixed with binder and formed into green electrodes and arranged in the furnace.
5. Green electrodes are baked at 900°C to carbonise binder and furnish amorphous electrodes.
6. Amorphous electrodes are converted into graphites in an electric furnace at high temperature (2700°C).
7. Graphite is shaped to industrial demands and the scrap powdered.

In manufacturing graphite, the furnace used consists essentially of core of the coke being converted into graphite, surrounded by a heavy layer of sand, coke and sawdust as insulation. The floor and ends of the furnace are built mainly of concrete, with cooling coils in the ends to reduce the temperatures of the electrodes in contact with air to prevent them from burning. The side walls are torn down after each run. The average charge is from 50,000 to 200,000 lbs of materials. As the coke is converted into graphite, the voltage because of lowered resistance, drops down from 200 to 40 volts. The high temperature employed decomposes, for instance, any siliceous carbide that may be present and volatilises products other than graphite.

Preparation of graphite electrodes from synthetic graphite

Synthetic graphite as prepared is hard. For use as electrodes, it must be made soft so that it may be cut with a knife. The change is made in an electric furnace in which the charge itself forms the resistance to the passage of the electric current.

The furnace is made of concrete—the bed placed on short supports placed close together and ending in two upright pieces, also made of concrete. The furnace is generally 30 ft. long and 12-13 ft. wide. Thick slabs of synthetic graphite are placed on the bed of the furnace—the height of the bed is about 6 ft. The end-pieces protrude 6-8 inches outside to enable connections to the copper terminals. The temperature within the furnace is maintained at 2500°–3000°C. Over the top a layer of a mixture of sand, coke powder and sawdust is spread, to retain the heat. After the operation has ended, the current is shut off, the furnace allowed to cool down for a week and then the slabs are taken out. The power consumption is 2.2 kwh per lb of graphite.

The graphite electrodes may be machined, threaded or screwed and electrodes of all sizes and shapes are in use.

Carborundum (silicon carbide)

The two raw materials for the production of silicon carbide (SiC) are sand and carbon. The carbon is obtained from anthracite, coke, pitch or petroleum coke. The sand contains 98–99.5 per cent silica. The equations usually given for the reactions involved are:

$$SiO_2 + 2C(amorphous) \longrightarrow Si + 2CO; \qquad \Delta H = 144.8 \text{ kcal}$$

$$Si + C \text{ (amorphous)} \longrightarrow SiC; \qquad \Delta H = -30.5 \text{ kcal}$$

The total reaction therefore is

$$SiO_2 + 3C \text{ (amorphous)} \longrightarrow SiC + 2CO; \quad \Delta H = +114.3 \text{ kcal}$$

The flowsheet is given in Fig. 12.2. Sand and carbon are mixed in the approximate molecular ratio of 1 : 3 and charged into the furnace. Sawdust, if added, increases the porosity of the charge to permit the circulation of vapours and the escape of the carbon dioxide produced. The charge is built up on the furnace around a heating core of granular carbon. This core is in the centre of the 30 to 50 feet long furnace and connects the electrodes. The walls of the furnace are loose firebrick supported by iron casting and are taken away at the end of the charge to facilitate product removal. The initial current between the electrodes is 6000 amperes at 230 volts and the final is 20,000 amperes at 75 volts. The temperature at the core is 2200°C. The temperature should not be high as otherwise SiC will volatilise. The time of reaction is about 60 hours 36 hours of heating and 24 hours of cooling. After cooling, the SiC crystals are removed, broken, washed and cleaned by chemical treatment with sulphuric acid and sodium hydroxide. The crystals are then classified and screened.

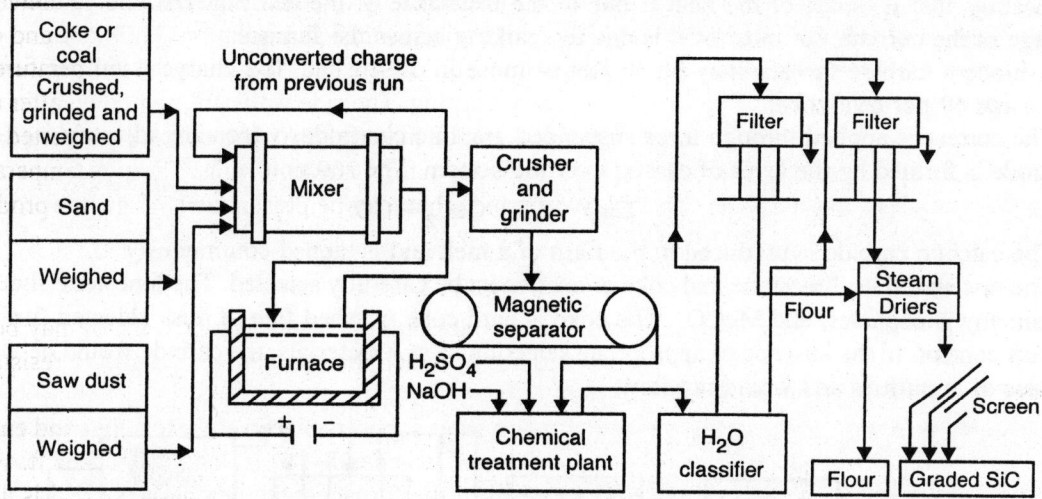

Fig. 12.2. Flowchart for manufacture of Carborundum (Silicon carbide).

Alundum

Artificial emery, a crystalline form of Al_2O_3, is known in commerce under the names of "alundum" and "aloxite". The hardness of emery varies depending on the specimen—it lies between seven and nine of the MOHS scale, on which scale, diamond has a hardness of 10 units.

The hardness of fused and crystallised Al_2O_3 produced in the electrical furnace is uniform, its value is nine, just under that of silicon carbide. More aluminium oxide abrasives are manufactured and consumed than silicon carbide abrasives. The artificial emery contains about 95 per cent alumina, its colour varies from white to yellow brown.

Fused Al_2O_3 is made from bauxite hydrated alumina which contains Fe_2O_3 in amounts varying from 1–20 per cent; it also contains silica and other impurities. The bauxite is first calcined, then mixed with coke and iron borings in amounts depending on the silica present, and subjected to the heat of a combined arc and resistance furnace. The charge may consist of 73 per cent bauxite, 5 per cent fines, 3 per cent coke and 2 per cent iron—the bauxite being 78 per cent pure. The furnace is circular, 7–8 feet diameter at the top and somewhat less at the bottom and 5 feet deep. The furnace is made of steel. Two carbon electrodes of opposite polarity are suspended in the furnace— the current passes in an arc from one electrode to the mass, then passes through the mass and back to the other electrode also through the arc. Increasingly larger amounts fuse with time. During the smelting process, more charge is added gradually until the furnace is filled with the melt. The fused alumina float at the top. The furnace is cooled, turned upside down, the ingot taken out, cooled, crushed to form fine crystals. The current applied is 130 volts, 500 kw and the time for the fusion is 36 hours.

Calcium carbide

Calcium carbide (CaC_2) is a substance which contributed largely to the development of the electric furnace. Calcium carbide is formed when a mixture of lime and coke is heated to a temperature of 2000°C (3632°F) or higher. The carbide furnaces fall into the class of combined resistance and direct arc heating, that is, some of the heat is due to the resistance of the raw material and product to the passage of the current, but most of it is due to sparking across the furnace.

A modern carbide furnace may be 30 feet or more in dimension. The charge is 60 per cent lime (CaO) and 40 per cent coke.

The current is applied through three suspended graphite electrodes which are adjustable —an idle electrode is formed by the layer of carbon over the bottom. The reaction is

$$CaO + 3C = CaC_2 + CO$$

The calcium carbide is produced in the form of a melt and is tapped continuously.

The raw materials, limestone and coke or coal, must be carefully selected. The limestone should not contain any phosphates, and $MgCO_3$. The coke is hard coke, washed free of dust, selected further for low ash content. In the absence of appropriate selection of raw materials, the carbide would be viscous, because of impurities and would not flow.

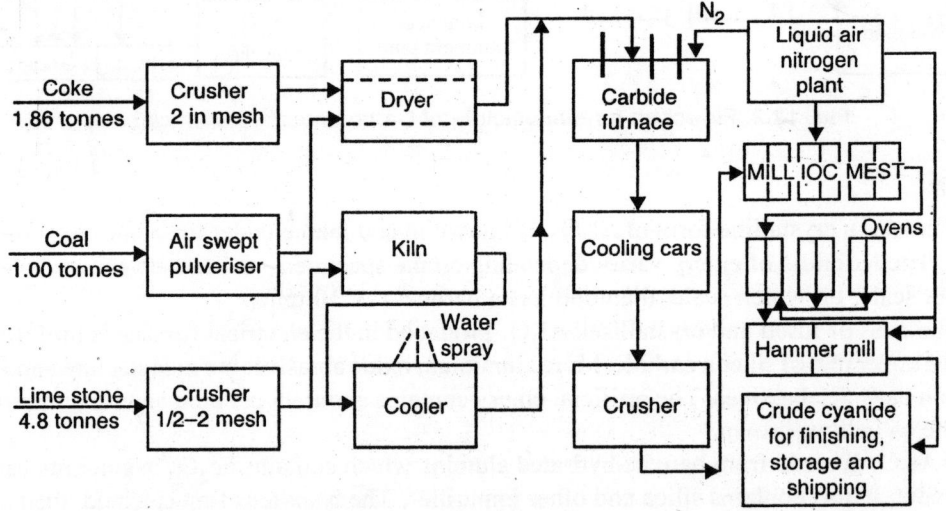

Fig. 12.3. Flowchart for manufacture of calcium carbide and calcium cyanide.

The liquid carbide flows into 'chill cars' of 1000 lbs capacity—it contracts on cooling. If intended for acetylene production, it passes through a jaw crusher and then through very slow rolls to give minimum of dust; the pieces are then screened to size. For making cyanide, the crushed carbide is sent through a continuous pebble mill which gives a 40-mesh powder. These are further powdered to give a 200-mesh powder.

The crude carbide is 83 per cent CaC_2, 14 per cent lime (CaO) and 1 per cent carbon. It contains small amount of silicides, phosphides and sulphides.

A very important point is the number of pounds of carbide produced per kilowatt hour consumed; 4.3 kilowatt hours for 1 kilo of crude carbide. The current efficiency is about 62 per cent.

Products of Electrothermal Processes

The following are the products of electrothermal processes:
1. Artificial graphite.
2. Abrasives: silicon carbide or carborundum, fused alumina or alundum grinding wheels, coron carbide, calcium carbide.
3. Phosphorus.
4. Carbon disulphide is partly an electro-thermal product.

Products of Electrolytic Cells

The following products are obtained from electrolytic cells:
1. Metals like sodium, aluminium, magnesium, etc.,
2. Sodium peroxide, sodamide, sodium cyanide.
3. Potassium permanganate, hydrogen peroxide, ammonium persulphate, sodium chlorate.
4. Potassium dichromate.

Sources of Current

The electrolytic process requires a source of direct current. This may be produced from batteries, though this is most uneconomical except on an extremely small scale. Direct current generators driven by prime movers were frequently a source of such current in early days of the industry but they too have practically disappeared since such units are cumbersome, of relatively low capacity and, therefore, very inefficient. They are still found in a few of the older plants but are gradually being displaced by other equipment described below.

Modern generating equipment is of the multi-phase alternating current type. Distribution lines of any length are always supplied with alternating current. Therefore, in these modern times the electro-chemist is practically restricted, both to large scale generation equipment or to connection with existing transmission lines to accept alternating current even though high processes may force the use of direct current. The electro-chemist, therefore, may expect to receive his electric energy in alternating form and at a more or less high voltage. For conversion of such to direct current he has the choice of several types of apparatus.

A somewhat similar device which combines the motor and generator in a single unit is the rotary converter. This unit however, has the disadvantage that regulation of the direct current is limited to 2 1/2 per cent either side of the designed voltage, which may be an undesirable handicap in many industries.

The most modern of the devices for converting alternating into direct current is the so-called rectifier. It is much less expensive to install than either the rotary converter or the motor generator set, requires reactively little attention and is a sturdy piece of apparatus. It has a very considerable range of regulation in the more elaborate types.

There are several unique sources of direct current available to the electro-chemist. In the electrothermic branch of our industry alternating current is universally used. Since this is the type of current produced by all large power stations and distributed over wide areas, the only conversion to be considered is that to a suitable voltage. For this purpose transformers are used.

Certain types of electrolytic work operate with fused baths. Direct current must be used for the electrolytic decomposition. If the units are large and efficiently designed the heating effect of this electrolysing current may be sufficient to keep the bath molten. We do have examples, however, where the heating effect of the electrolysing current is not sufficient to maintain the bath at the proper temperatures, and it is then customary to superimpose alternating current to assist in maintaining the desired temperature.

There are also electrolytic operations which take place at ordinary temperatures that use superimposed alternating current on the direct current electrolysing circuit. In this case the alternating current and direct current generators are specifically designed to be connected in series.

To avoid influence of electrodes it is practicable to construct furnaces on the transformer principle. A hearth of annular type loops a magnetic circuit, magnetism being induced by a high tension coil fed by alternating current. Current is induced in the molten bath in the hearth and heats the same by resistance. This induction type of furnace has found its principal application in the steel industry.

A modified form of this induction furnace is the high frequency type. High frequency current generated by transformers, condensers and spark gap, or by special generators, is passed through a coil of suitable design inside of which is placed the material to be heated. It is necessary that the crucible holding the material, or the material itself, be a reactively good conductor. The eddy currents induced in this conductor heat the same. Regulation is very easy and very high temperatures are possible in such units but they are of relatively small capacity as compared to the giant furnaces found in the carbide, ferro-alloy and steel industries, a limitation imposed by available high frequency generating equipment.

Carbon bisulphide

Much of the carbon bisulphide produced in the world is made in the electric furnace. Whether produced in the old type retorts or in the electrothermic way the raw materials are the same: sulphur and charcoal. The reaction is a reactively low temperature one and the use of electricity based upon convenience as well as long life of equipment rather than upon any particular inherent properties of the electric current.

In general the electric furnaces are tall cylindrical structures of refractory material divided by an inner cylinder into a central shaft and an annular space. The inside shaft is filled with charcoal and provision is made for feeding sulphur into the annular space. Carbon electrodes project through the walls to contact the charcoal. On passage of the current, charcoal is heated and as this heat penetrates into the annular space the sulphur melts, and as it comes into contact with the heated charcoal carbon bisulphide is produced. By adjustment of voltage on the electrode system the furnace can be made self-regulating. If it tends to become too hot more sulphur will melt and submerge the electrodes cutting off some proportion of the area of the path of the current, thus increasing the resistance and cutting down on the flow of energy. If the furnace tends to become too cold less sulphur will melt and that portion in the furnace will react, forming the volatile bisulphide and thus increases the area of the path of the current. This is a very beautiful example of the convenience with which the electric current can be applied to certain types of reaction. The furnace itself is built of refractories that do not deteriorate under the existing conditions and since the heat is internally generated there is no wear and tear of retort due to external firing. The efficiency of such a furnace is very high.

Carbon bisulphide finds its greatest use in the manufacture of rayon by the viscose process. Alkali cellulose is combined with carbon bisulphide to produce viscose. Carbon bisulphide is also of value as a solvent, as raw material in the manufacture of carbon tetrachloride, in the vulcanisation of rubber, and to some extent in the paper industry.

Phosphorus

At the present time the electrothermic production of phosphorus is one of the most rapid growing of the electrochemical industries. It is true that yellow phosphorus has been produced in the electric furnace for the past fifty years but because of the high cost of the relatively small units used for the greater part of this time, its found it outlet only in the match industry and in a few of the more expensive phosphorus compounds.

Of recent years the development of the phosphorus producing unit has been such as to produce this element on a large scale and at such low cost that it can be oxidsed to phosphoric acid in competition with the older wet methods of production.

Readman process

This was the original of the electric furnace phosphorus processes and the principles upon which it operated are essentially the same as those existing today in the modern high powered furnace.

Tricalcium phosphate in the form of phosphate rock or bone ash is fused with some form of carbon and silica in a closed electric furnace, the reaction being.

$$Ca_3P_2O_3 + 3SiO_2 + 5C \longrightarrow 3CaSiO_3 + P_2 + 5CO$$

The phosphorus vapour and carbon monoxide gases are conducted from the top of the furnace through a condenser which delivers liquid phosphorus. The yield of phosphorus is estimated to be about 2 pounds per electrical horsepower day (18 Kwh).

The Readman furnaces are very small running only a few hundred H.P. Few years ago attempts were made to use much large furnaces of the multi-phase type for the production of ferrophosphorus. In these furnaces scrap iron, phosphate rock, carbon and a flux are smelted together producing a ferrophosphorus running about 20 per cent in phosphorus. The slags made are silicates of lime. It is found that not all of the phosphorus could be reduced into the ferroalloy if the grade was to be above 20 per cent, and that the slags carried material quantities of phosphoric acid as silico-phosphates, unless an excess of carbon is added to the charge.

Under this latter condition the slags could be freed of phosphorus but a considerable proportion is evolved in the elemental state in the form of vapour. By putting a closed top on the furnace and admitting air under the roof these vapours oxidise to phosphoric oxide and this could be recovered from the gases by suitable scrubbing or electrostatic precipitation. The product is a high grade concentrated phosphoric acid since even with electrostatic precipitation it has been found advantageous to introduced steam into the gas current.

More recently developments tended toward feeding a mixture of phosphate rock, clay and silica to a closed type multi-phase furnace. Sufficient coke is added to the charge to reduce the phosphates to elemental phosphorus. This is taken from the closed furnace to condensers and cooled until liquid. In other words this modern development is simply the old Readman process carried out on a greatly enlarged scale.

The liquid phosphorus is run into storage tanks provided with heating coils and is handled through pumps and pipelines much like any other liquid. It is even being shipped in tank cars to various consuming centres. It is a comparatively simple matter to burn the phosphorus in air to phosphoric acid which is recovered through scrubbing as strong acid and of very high purity. Most of the furnaces are three-phase closed type with continuous electrodes. Because of the low content of coke in the charge the voltage is relatively higher than in most electric furnace operations.

Sodium and potassium

Although sodium and potassium are metals, their uses are largely confined to the straight chemical field, so their production is discussed here.

Production from caustic

Sodium is produced on a very large scale. The first successful electrochemical process and one still operated in a number of plants consists in electrolysing a fused bath of caustic soda. The caustic is dehydrated as far as possible by thermal methods and the last traces of water are removed by preliminary electrolytic treatment. When the last of the water has been eliminated, metallic sodium appears at the cathode along with an equivalent amount of hydrogen, and oxygen at the anode. The fusion pots are of iron and both anode and cathode of either iron or nickel. A metallic screen serves as a partial diaphragm for retaining the molten sodium near the cathode so that it can be ladled out of the cell from time to time.

The operation requires very close control for the sodium has a tendency to dissolve in the bath of molten caustic and diffuse to the anode, this tendency increasing sharply with rising temperature. In practice, therefore, if high current efficiency is to be attained it is necessary to operate just as closely as possible to the melting point. Since both hydrogen and sodium appear in equivalent quantities at the cathode the current efficiency computed on the sodium alone can never exceed 50 per cent. At a temperature of only 20° above the melting point of the caustic the actual efficiency will have dropped down as low as 40 per cent, indicating the precise control necessary to obtain the maximum effectiveness of the current.

Production from salt

Because of the low current efficiency of the caustic process, attempts were made over a long series of years to produce sodium by the direct electrolysis of molten salt. The many difficulties have been overcome and probably the large bulk of the world's production of sodium is now made by this process. The cell which is used consists of rectangular box of structural steel lined with refractory material. It is closed at the top, for one of the products of decomposition is chlorine which must be collected. The cathode is made of iron and the anode of graphite and a wire mesh diaphragm surrounds the cathode to retain the sodium which flats on top of the molten bath of the chloride and is removed by a siphon arrangement. Chlorine is collected from the anode compartment, cooled and placed on the market in the usual form. The bath material itself consists essentially of molten salt to which various additions of other salts are made to reduce the melting point as far as possible. However, these counteractions of foreign salts must be limited to relatively small quantities otherwise the sodium will be contaminated. These agents are various fluorides and chlorides of the alkali and alkaline earth metals, such as $CaCl_2$. Such a bath, when molten, has relatively high conductivity and the resistance offered to the passage of the electrolysing current is not always sufficient to maintain the heat losses of the sodium. Provision, therefore, must be made to supplement the energy supplied to this cell by superposing sufficient alternating current to make up for this heat deficiency. Difficulty is experienced with contamination, principally from the refractory lining, and of secondary importance, from the electrodes and the diaphragm. Therefore, it is necessary from time to time to discard these fused baths and replace with fresh material. The life of the bath is relatively short on the average and provision, therefore, must be made for tapping these furnaces and supplying fresh molten bath to them. Chlorine, however, is a credit in this operation. Current efficiencies are relatively high; of the order of 80 to 90 per cent in an operating cell in good condition, for we do not have the same limitation of having to separate hydrogen along with the sodium in this case of direct electrolysis of molten salt.

Production of potassium

For the production of potassium the electrolysis of caustic potash, and similar to that for sodium when using the same cells, is universally used. The quantity of potassium produced, however, is reactively small as compared with the great tonnage of sodium which finds an enormous outlet as sodamide in the cyanide and dye stuff industries, and for the production of the sodium-lead alloy used in producing tetraethyl lead.

Storage Battery

The manufacture of lead storage batteries is now one of the large industries of the country. The enormously increased demand for starting batteries, by the automobile industry, added to the use of batteries for motive power of many types of self-moving vehicles, has called for ever-increasing production.

The overall equation of the reactions taking place in the lead storage battery may be written as follows:

$$PbO_2 + Pb + H_2SO_4 \underset{\xrightarrow{discharge}}{\xleftarrow{charge}} 2PbSO_4 + 2H_2O$$

The left-hand side of this equation represents the charged plates and the right-hand side the discharged plates. Hence, the charged lead cell consists, essentially, of a plate of lead, a second plate of lead peroxide, held apart by a porous 'separator' through which sulphate ions can migrate—all immersed in a sulphuric acid solution. The problem of producing a storage battery of high capacity and efficiency is largely one of incorporating the correct physical and chemical properties to the plates of lead and lead peroxide.

The original lead storage battery was made by suspending two plates of soft sheet lead in a bath of dilute sulphuric acid of specific gravity of about 1.2. One plate was connected to the positive terminal of a source of direct current and the other plate to the negative source. Current was supplied for a short interval of time and then cut off and the two plates short-circuited. This operation was repeated a large number of times until the plates were 'formed'. At first these plates had little capacity for storing up potential chemical energy which was afterwards transformed, on discharging, into electrical energy, but as the operations of charging and discharging were repeated the capacity of the two plates increased materially. Such batteries, however, would have been of no value for the present-day service in automobile starting, for the stored capacity per unit of weight was small and the apparatus itself was incapable of taking the abnormal stating loads of the modern automobile.

The modern battery plate consists of a grill of cast antimonial lead carrying from 5 to 10 per cent of antimony. The grids vary from 2 to 5 mm. in thickness. Into this grid is pressed a paste consisting of lead oxide mixed with sulphuric acid or ammonium sulphate. This paste is packed tightly into the mesh of the grids and left to harden. The plate are then suspended in open tanks, those intended to be anodes connected to a positive source of current and those which are to be cathodes to a negative pole. Under careful regulation of the current these pasted plates are "formed," that is, become active by transformation of the oxide paste of the anodes to lead peroxide and the oxide paste of the cathodes to spongy lead. Great care is needed in the forming of a plate to gain the maximum adherence of the paste without warping the supporting structure.

After formation the plates are washed, dried and stored for assembly. Depending upon the size and rating of the battery a certain number of anode plates are burned to a lead strap; for a 13-plate cell 6 anode plates are grouped together. Similarly the cathode or negative plates are burned to a strap, there being one more negative plate in a battery than positive plate. For assembling the battery, the anode and cathode groups are meshed together with insulating separators between the plates. These separators take various forms, such as thin strips of wood veneer, thin perforated, pieces of hard rubber, or various patented compositions. When put into service, the battery is filled with a high purity dilute sulphuric acid of strength between specific gravity 1.2 and specific gravity 1.3.

The ordinary starting battery contains either 13 or 15 plates per cell with three cells in series. The 13-plate battery has capacity of storing up energy equivalent to 100 ampere hours and the three cells in series will deliver current at 6 volts potential, that is, approximately two volts per cell.

The usual storage battery, when new, will deliver from 80 to 90 per cent of the energy used in charging but as the battery becomes older this efficiency drops off rather sharply. As the battery ages the anodes buckle and shed their paste and the cathodes become clogged with sulphate and lose their porosity. This prevents a diffusion of the sulphuric acid into the plates and hinders charging.

Edison cell

Another type of storage battery, the so-called "iron nickel" or Edison cell, uses a spongy iron anode and a compressed flake nickel cathode, in a solution of 15 per cent caustic soda containing a small amount of lithium hydrate. The chemical reaction involved in this cell is:

$$Fe + 3H_2O + Ni_2O_3 \underset{\xrightarrow{discharge}}{\xleftarrow{charge}} Fe(OH)_2 + 2Ni(OH)_2$$

This battery cannot be used for starting purposes but has given excellent service for motor operation and ignition. It delivers a moderate current but is incapable of furnishing the peak surges for starting a motor. It has the greater advantage of not being injured by complete discharge, which is not true of the lead battery.

Dry cells

The old wet primary batteries have largely disappeared and their place has been taken by the 'dry cell'. This cell is not exactly dry as the name implies, but it at least contains no free liquid. These cells are manufactured by the millions and of all shapes and sizes to meet the wide variety of demands in the radio, flash light, ignition and signal systems. The positive pole consists of a carbon rod. Surrounding this rod is a mixture of manganese dioxide, finely powdered graphite, and ammonium chloride made into paste. The whole is packed in cylindrical container made of sheet zinc which also serves as the negative pole. A cardboard or paper carton surrounds the zinc casing, serving both as a protective coating and an insulating medium. The contents of the cell are sealed in by molter pitch or sealing wax cast into place. Formulation of the active mass varies not alone among the different manufactures, but is also dependent upon the size of the cell and its intended use. Great care must be taken in the purity, size of grain and mixing of the ingredients to insure a long shelf life and quick recuperation in service. A well-constructed dry cell should show 1.5 volts electromotive force and should be capable of shelf storage on open circuit over several months without deterioration.

Electroplating

A branch of electrochemistry that has attained enormous commercial importance, is that of coating the surface of one metal with a thin skin of another either for decorative or preservative purposes. Thus base metal is coated with gold for decorative purposes, iron with nickel for preservative reasons. Practically all of this metal coating is now done suspending the object to be plated as cathode in a suitable bath and electrolytically depositing a coating of the desired metal on its surface. It is not possible here to cover thoroughly all the methods and solutions used in this work, as each establishment seems to cherish its own secret manipulation, a great deal of which secrecy is more imagined than real.

To insure uniform results the base metal to be plated should be sound, free from pores and sand holes, and above all, clean. Cleaning is done by the use of brushes, buffing wheels, acid and alkali baths; the objects to be plated not being touched by greasy hands.

The composition of the baths used are as varied as the establishments, each seemingly possessing its own special formula for each bath. Chemically, the baths are, as far as possible, so constituted that metal is not deposited by simple chemical replacement, as such coatings are usually not strongly adherent. A general idea of the baths used may be gained from the following.

Copper

The solution contains 10–12 per cent copper sulphate and 2–3 per cent sulphuric acid. For plating on iron, this solution, because of chemical replacement, does not give a particularly good adhering coat and recourse is had to first plating a very thin coating of copper on the iron from copper cyanide solution, the object finally being finished in the acid bath, for the reason that regeneration of the cyanide bath by the the of a copper anode is not particularly successful. Such a cyanide bath may be prepared by dissolving a mixture of one part of copper carbonate and three parts of potassium cyanide in twenty-five parts of water.

Silver

A fairly strong solution of the double cyanide of silver and potassium or sodium is most frequently used in plating this metal.

Gold

A similar solution of the double cyanide of gold and potassium is used.

Zinc

Formerly, most of the galvanising or zinc coating of iron was by the so-called hot method; that is, carefully cleaned sheets or objects of iron were dipped in a bath of molten zinc. The surplus adhering to the surface was then wiped off and there resulted the typical bespangled appearance of galvanised sheet. To the uninformed this crystalline appearance denoted a high quality coating. In more recent times it is recognised that an electroplate of zinc of equal or better quality could be produced and more and more of the zinc coating is now produced by the electrochemical method. For complicated batches, as for example screw threads, the electrochemical method permits the formation of a uniform coating preserving all of the characteristics and dimensions and without possibility of warping.

Further, it is now possible to operate the electro-galvanising process mechanically. Small articles are plated in tumbling barrels. Conduit and tubing can be handled on mechanical carriers through the bath. Wire, strip and sheet stock are run continuously through the cleaning, washing, plating and final washing baths.

One of the most interesting examples is the continuous electro-galvanising of wire, the whole operation being mechanical and taking place in a series of tanks through which the wire is fed. An excellent finish can be given by finally running the galvanised wire through a series of dies which give it a hard, polished surface. It is even possible in these highly developed continuous plating operations to use roasted zinc concentrates for maintaining the strength of the bath, and insoluble anodes, so that we have here not only the metallurgical recovery of the zinc from its ore but its direct placement on a finished article.

Nickel

A saturated solution of the double sulphate of nickel and ammonia is largely used for nickel-plating. As nickel anodes do not dissolve in sufficient quantity to recuperate this bath, nickel sulphate is added from time to time to maintain its saturated state.

Chromium

The very valuable properties of chromium; its extreme hardness, its resistance to oxidation and its resistance to scaling, as well as its bright and pleasing appearnace, have created a tremendrous for chromium plated objects. The plating bath consists of a mixture of chromic acid and chromic sulphate. Various forms of insoluble anodes have been successfully operated. Most baths work with lead anodes. For bright plating, it is customary first to nickel-plate and buff before plating with chromium.

Cadmium

Because it gives an excellent protective coating combined with a steel-gray colour and is capable of taking a high polish, cadmium plating has found considerable application on hardware, instrument cases, and aeroplane parts. It is plated from cyanide solutions, made by dissolving cadmium salts in sodium cyanide. The anodes are cadmium.

Brass

A mixture of cyanides of copper and zinc containing varying proportions of the two metals, depending on the colour of the plating desired, will yield a plating of brass. Great skill is needed to use such a bath and much secrecy is maintained regarding the exact composition of a successful brass bath.

Rubber

While properly not a true electroplating operation, nevertheless rubber has been deposited by the use of the electric current. The underlying phenomenon of depositing by use of the electric current is very complex. The latex is suspended in an extremely dilute, slightly alkaline medium which is placed in the cell. As anode, zinc has been found to be the most successful. High voltage is then applied to the cell, creating a strong electrical field between the two electrodes. The suspended particles of latex, which are negatively charged, move toward the anode and there attach themselves. Various solids, such as sulphur, fillers, pigments, and even super-accelerators can be suspended in the solution along with the latex and be deposited with it. By this process rubber can be formed or moulded in its final shape and be readily vulcanised. The deposited rubber has an extremely high tensile strength and elongation, and is in general superior to milled rubber.

After withdrawal from the bath the plated object is thoroughly washed with water, or even alkali if an acid bath has been used. Dull or matte finishes are given to a plated object by the use of a wire or scratch brush. Polishing is done with rough, whiting or other polishing powder on a cloth or leather wheel. In many cases this mechanical manipulation after plating plays a more important part in obtaining a desirably ornamental finish than the actual plating operation itself.

Pigments

The electrolytic manufacture of pigments has attracted a great deal of attention, but up to the present time has attained no great prominence because of the many unsolved technical difficulties. The production of white lead by the electrolytic dissolution of a lead anode and subsequent precipitation of the basic carbonate is in commercial use, in a small way. The electrolyte use consists of a dilute solution of sodium carbonate to which is added a much larger portion of a sodium salt whose acid radical forms a soluble compound with the lead.

Chlorate or nitrate is used for this purpose, it having been found that the use of such a salt in the bath prevents the lead carbonate formed from adhering to the anode and thus insulating it. The caustic formed at the cathode of the cell is immediately neutralised by the introduction of a current of carbon dioxide, thus directly supplying the ingredients used in the process in the form of sheet lead anode and gaseous carbon dioxide.

The greatest difficulty met with in commercial operation is the dropping of small pieces of undissolved anode in to the deposit of white lead at the bottom of the cell, which causes poor colour. Another difficulty is the tendency to form crystalline precipitate, a most undesirably product.

Lead chromate can also be obtained in the same manner by using a bath of potassium dichromate and nitrate, regenerating the solution with chromic acid. The sulphides of cadmium, antimony, etc., have also been made by the use of these metals as anode in a solution of sodium hyposulphite, though the process has never attained much commercial prominence.

Oxidation and Reduction of Organic Compounds

A broad general view of electrolysis would consider an anode dipped into an electrolyte as a reducing agent; each 96,540 coulombs of electricity passing through it causing the production of one gram-equivalent of a substance of unit valence. A cathode is similarly a reducing agent, the passage of the above quantity of electricity causing a reduction of one gram-equivalent of unit valence. We have here an ideal means of reduction, in that no foreign materials, not essential to the reaction, need be added to cause the desired change, and there is hence no contamination of the resulting products to fear. Electrochemical reductions resolve themselves into a choice of suitable electrode materials, current densities and separation of the effects of anode and cathode through diaphragms. In the field of organic chemistry many applications have been made of electricity to the reactions of reduction, where conditions are suitable to their use, but strange to say not many are today on a commercial scale. A few of the prominent commercial processes are the following.

Chloroform

A solution of 20 per cent sodium chloride, to which acetone is added, evolves chloroform on the passage of an electric current. The chloroform vapours are removed from the closed cell and condensed. The direct use of chlorine produced in the caustic soda industry for this purpose is of much greater technical importance than the above direct process.

Iodoform

A solution of potassium iodide to which alcohol or acetone is added, and the whole electrolysed, yield a solution of iodoform, which can be crystallised out by cooling. This process is the basis of an important manufacturing industry at this time, and has largely supplanted the old chemical process.

Sorbitol and mannitol

Probably the most important of the organic processes based upon the chemical action of the electric current, is the reduction of the sugars to the corresponding alcohols. For example, the electrolytic reduction of glucose, that is, the reduction of the aldehyde group in glucose to the corresponding hydroxy group (CHO to CH_2OH), is now carried out on a large scale.

In similar manner mannose is reduced to mannitol. The corresponding sugars are dissolved in a solution carrying free alkali and treated in the cathode compartment of an electrolytic cell. Here the sugars are reduced to the corresponding alcohols which are recovered from the cathode liquor. This is probably the largest commercial example of direct organic synthesis at the present time.

In recent years electrolytic oxidation and reductions of organic compounds have attracted a great deal of attention in the laboratory and thousands of patents covering this field have been applied for. In only a few cases has the electrolytic process been able to compete with the older existing purely chemical treatment, but it is anticipated that there will be a very marked advancement in this art in the next few years. The problems of proper solvents, of diaphragms, of control are slowly being solved in the laboratory and pilot plant and we may expect very greatly increased development in this most interesting field.

Caustic soda and chlorine

By far the most important electrochemical process at the present time is the production of caustic soda, chlorine, hypochlorite and chlorate from common salt.

Oxygen and hydrogen

A very considerable industry has grown up around the electrolytic production of the two gases oxygen and hydrogen. Where essentially pure gases are demanded the electrolytic method of producing them has as yet found no competitor, particularly if both are demanded, since they are produced simultaneously by the electrolytic decomposition of water. The fractionating of liquid air for the production of oxygen in general will produce this gas at a lower cost than the electrolytic process, but its purity rarely exceeds 99.6 per cent, whereas the electrolytic oxygen can be obtained with impurities not to exceed 0.1 per cent. In such electrolytic production of oxygen, hydrogen is also obtained of a purity exceeding 99.9 per cent. Where both gases can be used, costs of production by the electrolytic method compare favourably with any other source in localities where power is relatively cheap.

The electrolytic production of hydrogen on an enormous scale is practised in many countries producing synthetic ammonia, particularly where electric power is cheap and coke relatively costly.

There are two general types of cell construction, the choice depending upon the cost of electric energy. Where this is cheap a low cost type of cell construction of low energy efficiency is usually chosen. When electric energy is relatively more expensive high energy efficiency cells are chosen even though they represent a material increase in capital cost over the low efficiency type.

In general, the low efficiency cell is of the unit type construction, that is, it consists of a single anode suspended in the centre of a closed iron box, the sides of the box acting as cathodes. Between the anode and the cathode is a porous asbestos diaphragm which keeps separate the gas produced at the two electrodes. The anode itself is of iron usually in the form of a grid-like structure so as to permit free access of the electrolyte, and by suitable spacing permits easy evolution of the oxygen. Nickel and nickel-plated anodes are sometimes used, resulting in a slightly increased efficiency of the cell. The electrolyte is a solution of caustic potash in water. Additions of fresh water are required to make up that loss through decomposition and this water is preferably distilled to prevent accumulation of foreign salts in the cell, particularly in the mesh of the diaphragm. Many cell plants operate on raw water for a long time without disturbance, but this must be water of exceptional purity, otherwise the electrodes will be coated and the diaphragm blocked by precipitation.

Low efficiency cells operate with voltage drop of 2.25 to 2.50 volts across the electrodes. Theoretically only 1.69 volts is required for the decomposition of water so that the energy efficiency of this type of cell is only from 68 to 75 per cent. The current efficiencies of any modern cell are usually very high, frequently shown 98 per cent.

The high efficiency type of cell, using a voltage of 1.85 to 2 volts across the electrodes, shows better overall economy where power costs are high. This type of cell uses electrodes much more closely spaced and of a more elaborate design to allow the largest amount of effective surface and to eliminate gas bubbles as rapidly as possible from those portions of the electrolytic which lie directly in the path of the current between the two electrodes. The anodes of this type of cell are usually of nickel or of nickel-plated iron. These cells run at a much lower current density per unit of electrode and cost more to build, so for a given output the investment is somewhat greater than for the low efficiency type.

It will be noted that these hydrogen cells of either class are of low voltage, requiring only from 2 to 2.5 volts over the cell. A current of 1,000 amperes passing through such a cell produces 15.5 cubic feet of hydrogen and half as much oxygen per hour. In order to get the enormous quantities of hydrogen needed by the synthetic ammonia process, it is necessary first to use currents of high amperage.

For purposes of economy, the electrical machines should operate at the highest possible voltage. It requires about 1 sq. in. of copper conductor for each 1,000 amperes carried, so any attempt at paralleling these large machines gets into complications with the current leads. The most advanced practice has standardised at 10,000 amperes at 600 volts per circuit.

CHAPTER 13

Insulating Materials

INTRODUCTION

A good conductor, metallic copper, has a resistivity of 1.7×10^{-6} Ω-cm. A good insulating material has a resistivity on the order of 10^{15} to 10^{18} Ω-cm. The vast difference in resistivity between conductors and insulators makes possible the construction of modern electrical apparatus with closely spaced conductors at greatly different potentials.

Usually, the mechanical, thermal, and cost considerations determine the choice of an insulating material; the electrical losses in many applications are a secondary consideration. They do become important in high-frequency applications and in equipment such as large AC capacitors where it is desirable to minimise heat build-up.

The principal function of an insulating material is to direct current flow through conductors. Cost considerations dictate that the minimum insulation thickness consistent with reliability be used. Other functions performed by insulating materials include the provision of mechanical support, environmental protection (exclusion of moisture, dirt, and chemicals), and heat transfer. The properties important for these functions will be discussed in the following section under the headings: electrical (volume, surface); thermal properties; and mechanical properties. Types of insulating materials will be discussed in a subsequent section under the headings: gases; liquids; and solids.

PROPERTIES

Electrical (Volume)

Ionic conductivity

All insulating materials pass some current, even though very small, when subjected to voltage stress. The conductivity (reciprocal resistivity) is believed to be due to ions carrying either positive or negative charges. The conductivity due to these particles never approaches the magnitude of electronic conductivity in metals since ions are much larger than electrons, and thus more restrained in their motion by surrounding particles. The conduction current generally does not obey Ohm's law as a function of field strength. Consequently, when the "DC conductivity", which is the current density in a unit field, or the "resistivity", which is the reciprocal of the conductivity, is to be measured, one must specify the field strength at which the measurement is made, as well as the time after voltage application.

Insulating materials differ widely in their ionic conductivity, which depends upon the number of ions per unit volume and the mobility of the ions. Common salt, sodium chloride, represents one extreme where all atoms are ionised. But dry crystalline salt at ordinary temperatures is an excellent insulator with very low conductivity, since the atoms are not able to move in the tight crystal structure. It has a slight conductivity as a result of imperfections in the crystal—vacant lattice points, where ions are missing. The smaller sodium ions can diffuse slowly through the crystal by jumping into these vacant spots. If the temperature is raised, ionic conductivity increases for two reasons. First, the number of vacant lattice points increases because there is an increasing probability of their occurrence at higher temperatures. Second, the mobility increases because the ions more frequently acquire enough energy from thermal agitation to jump into adjoining vacant spots. This negative temperature coefficient of resistivity is characteristic of all insulators, in contrast to metallic conductors, which have a positive temperature coefficient of resistivity. In metals, the thermal agitation of the atoms impedes the movement of the small conducting electrons, which move more easily between the atoms when these are colder and quieter. With some exceptions (e.g., sulphur and diamond) solid inorganic insulating materials are ionic crystals or glasses that have amorphous structures. In such structures, the metal atoms are generally ionised, but restrained, so that only a few are able to move. The smaller positive ions—hydrogen, sodium, potassium—move more easily and contribute most to the conductivity when they are present.

In contrast, organic materials primarily contain covalent bonds and relatively fewer ions. The ions present are often impurity materials or products of oxidation or another degradation of the insulation. Therefore, the ionic conductivity of electrical insulation is subject to wide fluctuations, depending on the purity or history of the material. Water will, when absorbed by the insulation, greatly enhance the ionic conductivity by bringing about ionisation of ionisable compounds. The percentage of water in insulation, which is affected by the ambient humidity, grossly influences the ionic conductivity. In organic liquids and plastics, ionic conductivity varies, for the same number of ions, inversely as the viscosity of the material. Thus, mobile liquids are usually more conducting than very viscous liquids and flexible plastics are more conducting than rigid plastics of the same chemical type. Another important factor affecting ionic conductivity is the dielectric constant of the medium containing the ions. A high dielectric-constant material is both a better solvent for impurity ionic compounds and a better medium for dissociating these compounds.

Dipole conduction

Asymmetrical organic molecules or segments in polymer chains have electric dipole moments varying in magnitude with the nature of the atoms in the molecule or chain segment. The dipole is often associated with a particular

$$\overset{\displaystyle O}{\underset{\displaystyle \|}{}}$$

functional group, e.g., —OH, —NH_2, or C—NH_2 (hydroxyl, amine, or amide). Electronegative atoms such as oxygen tend to draw electrons away from less electronegative atoms. This makes one atom of the molecule relatively negative and the other relatively positive. The resulting dipoles conduct for short periods of time by rotating in the direction of the electric field. Conduction stops when they have turned as much toward the direction of the field as their random thermal motion permits. Therefore, dipoles contribute to DC conduction only for a short period after voltage is applied and to AC conduction only at higher frequencies. The time during which dipolar conduction occurs varies inversely as the rate of the dipole turning, or directly as the viscosity. Thus, the frequency at which dipole conduction begins to become appreciable is lower for more viscous materials.

Dielectric absorption and dissipation factor

The current resulting from a limited movement of charge taking place in an insulator when a DC voltage is applied is called the *dielectric absorption*. When the voltage is removed, the ions migrate and the dipoles rotate back to their original random positions. These same effects take place, to the extent that the time of each half cycle permits, when AC voltage is applied. The ratio of the AC conductivity to the product of the angular frequency and capacitance is known as the *dissipation factor*. The amount of heat generated per cubic centimetre of insulation by the AC conductivity of the insulation is a function of the electric field, the frequency, and the dissipation factor:

$$\text{Heat generated } (W/\text{cm}^3) = 5/9 \; \varepsilon' \tan \delta f E^2 \times 10^{-12}$$

where E is the field in V/cm, f is the frequency in cycles per second, ε' is the relative dielectric constant, and $\tan \delta$ is the dissipation factor.

Excessive dielectric heating occurs when the voltage gradient, dissipation factor, or frequency are too high. This may lead to melting or decomposition of the insulation and resulting breakdown. An AC capacitor is an example of equipment that requires a low dissipation factor; voltage gradients are high (450 V/mil) and construction is such that heating occurs in a thick section of insulation of comparatively poor thermal conductivity.

Dielectric constant and loss index

The *dielectric constant* (permittivity) of a material is the ratio of the capacitance of a capacitor with that material between two electrodes to the capacitance of the same capacitor with vacuum or air as the dielectric. Materials that have a large number of dipolar molecules, or parts of molecules, capable of unhindered rotation have a high dielectric constant. The magnitude of the dielectric constant is roughly proportional to the concentration of dipoles and to the square of the magnitude of each. The contribution of migrating ions to the dielectric constant is usually, but not always, smaller than the effect of the dipoles. The product of dielectric constant and the dissipation factor is called the loss index.

Electronic polarisation

All substances exhibit a slight shift of electrons in the atoms with respect to the positive nuclei when an electric field is applied. This occurs rapidly and elastically. This *electronic polarisation* is principally responsible for a dielectric constant of approximately 2 for non-polar substances such as transformer insulating oil and polyethylene.

Ferroelectrics

Some ionic crystals and ceramics exhibit another type of polarisation in addition to electronic polarisation. The ions in these substances shift slightly from their equilibrium positions when an electric field is applied. In substances such as barium titanate and potassium hydrogen phosphate, this elastic shift results in an extremely high dielectric constant because of the polarisation of all the atoms in large domains of the crystal. In analogy to ferromagnetic polarisation this phenomenon has been called ferroelectric polarisation. In most crystals, only a small fraction of the ions are polarised in this way under normal electric fields.

Dielectric strength

Dielectric breakdown in solids, liquids, and gases may result from an electron avalanche in the insulating material. Each avalanche results from the acceleration of a few electrons by the electric field which gives rise to secondary electrons by impact with atoms or molecules. Each of these produces other electrons by the same process. A few free electrons are believed to be present in all materials.

The *intrinsic dielectric strength*, which is the maximum voltage gradient a homogeneous substance will withstand in a uniform electric field, is dependent upon the ability of the material to interfere with the acceleration of electrons. It depends greatly on the density of the material, since more closely packed atoms increase the probability of collisions that slow down the free electrons. It also depends on the electron-trapping ability of the atoms and molecules. Sulphur hexafluoride, for example, at one atmosphere pressure and with a 1-cm gap, has approximately three times the dielectric strength of air under the same conditions, largely because of its greater ability to trap electrons and form relatively stable ions. Some observed intrinsic dielectric strengths are given in Table 13.1.

Table 13.1. Intrinsic dielectric strengths.

Material	V/mil	kV/mm
Air, 1 atm (1-cm gap)	79	3.1
Sulphur hexafluoride, 1 atm (1-cm gap)	200	7.9
Air, 6 atm (1-cm gap)	385	15.2
Sodium chloride crystal	3,800	149.7
Mineral oil	5,000	196.9
Polyethylene	16,500	650.1
Poly(methyl methacrylate)	25,000	985.0
Mica, perpendicular to laminations	24,500	965.3

The dielectric strength values obtained with commercial insulating materials in practical tests are very much lower than those given, respectively, for solids, liquids, and gases in Table 13.1 for several reasons: (i) the existence of defects, holes, and conducting and foreign particles introduced by the method of manufacture; (ii) the presence of a stress concentration at the electrode edges or the points where the breakdown can be initiated because the electric field is much higher than the average; (iii) in AC tests, due to the damaging effect of an electric discharge (corona) during testing; and (iv) because of dielectric heating which raises the temperature and lowers the breakdown strength.

The presence of defects lowers the dielectric strength at points and has the effect, in practical tests, of making the test values smaller as the areas under test are increased. The ratio of electric field strengths in different dielectric materials in series varies inversely as the ratio of their dielectric constants. This effect places a higher electric stress on the low-dielectric-constant constituent of a composite insulation. For example, with oil-impregnated paper, the stress on the oil, which has a dielectric constant of about 2, is approximately three times that on the paper, which has a dielectric constant of about 6. This effect generally lowers the overall dielectric strength of the composite insulation, since, when partial breakdown of one part occurs, the remainder is left under higher stress. When corona (a local breakdown at an edge or point, but not a complete breakdown of the gap) appears at the surface of insulation under electric stress, it erodes the surface by electron bombardment, associated heat, and sometimes secondary effects due to formation of chemical oxidising agents such as ozone and oxides of nitrogen. This effect begins immediately and even fractional seconds of exposure at AC voltages near the breakdown voltage lower the breakdown strength significantly. At lower voltage stresses, with corona still present, breakdown will also occur but the time required for breakdown will be longer. For example, a pair of enameled wires of the type used in small motors, twisted together and tested in a standard manner, might withstand a 1-min AC test of 6000 V. If 1000 V AC is applied continuously, breakdown will occur in something on the order of 100 hr. For this reason, electrical machines are designed on the basis of "voltage endurance" of materials rather than short-time electric strength. Short-time electric strength is a useful guide for quality control. Voltage endurance is determined by measuring the time required for breakdown of insulated samples at several voltage levels. These tests range from less than a millisecond to more than a year in duration. Fig. 13.1 shows a typical voltage-endurance curve, plotted on semi-log paper, in which the dielectric breakdown level is plotted against the log time.

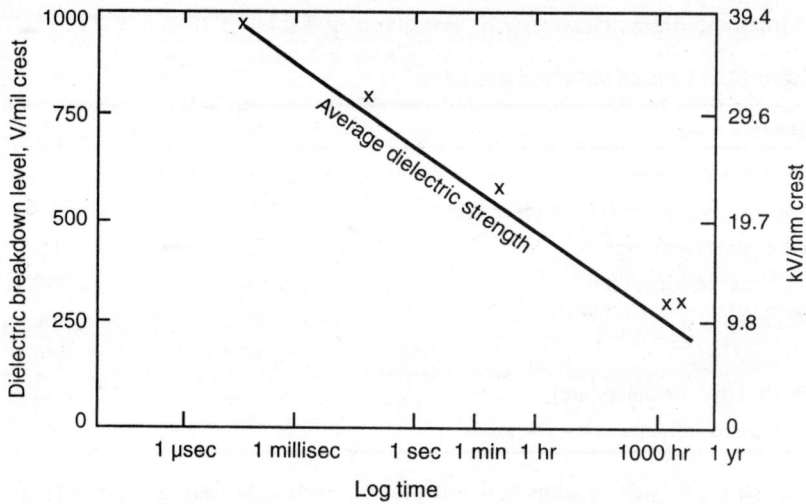

Fig. 13.1. Voltage endurance curve for a modern machine insulation, illustrating reduction in breakdown level at long periods of dielectric stress.

The curve should not be extrapolated to zero voltage because at a stress level where corona is entirely absent, breakdown is not believed to occur in less than fifty years. This conclusion is based, in part, on accelerated voltage-endurance tests in which voltages are applied at higher frequencies, for example, 3000 cps. Under these conditions, the insulation is subjected to as many cycles in one year as it would normally be subjected to in 50 years at 60 cps. This method cannot be used with insulation systems that have a high dissipation factor at the test high frequency, since the time-dependent limination illustrated in Fig. 13.1, most high-voltage electrical equipment such as generators and transformers are operated at voltage gradients on the order of 50 V/mil, which is about 10% of the 1-min dielectric strength and less than 1% of the intrinsic dielectric strength. An exception is provided by power capacitors, which can be designed to stress the insulation uniformly and which are operated at about 450 V/mil stress.

Electrical (Surface)

The exposed surfaces of a dielectric present many opportunities for conduction which do not occur in the bulk of the interior, especially if moisture or dirt is present. Many insulating materials are seriously affected by moisture. When moisture wets the insulating surface, a thin film of water may be formed, rendering the surface conductive. This effect is enhanced if ionisable impurities are also present. Dirt and polar impurities on insulating surfaces generally greatly increase the tendency to attract moisture. A consequence of the resulting low surface resistivity may be a lessening of the effective creepage distance and a flashover failure by breakdown due to the smaller air gap. To minimise the effects of surface moisture or other contamination, the path length of the surface may be increased for example, by putting corrugations in high-voltage porcelain insulators. Because of the earlier-mentioned effect of dielectric constant on the electric-stress distribution of insulators in series, the gas side of a solid-gas interface has a lower dielectric strength than the solid. If breakdown occurs in the gas close to the surface of the solid material, the surface may be partially decomposed and become conducting. The ease with which a surface forms a conducting path (tracks) depends upon the chemical composition of the insulating material, the nature of fillers present, and the current density of the arc acting on the surface. A test using high-voltage low-current arcs is given in ASTM D-495. Inorganic materials such as porcelain and mica have a high arc and tracking resistance. Organic polymers which depolymerise under heat by chain scission, such as methacrylates, have a relatively high arc and track resistance. Thermoset polymers which form carbonaceous residues when heated strongly, such as phenolics, have a poor arc and track resistance. Many polymers such as butyl rubber, epoxy resins, and polyester resins can be made very arc- and track-resistance by incorporation of 50–60% by wt of hydrated alumina filler. It is believed that at the temperature of the discharge any carbon formed by degradation of the polymers reacts with water molecules to form carbon monoxide and hydrogen.

Controlled surface conductivity is sometimes important in dissipating static charges from very good plastic insulators. If the surface resistivity is less than 10^{10} Ω, the resulting charges leak off to ground rapidly enough to cause no problem in most practical cases. If the surface resistivity is higher than this, it can sometimes be lowered by use of a polymeric coating which absorbs enough water from the atmosphere to lower the surface resistivity to a desired level.

Thermal Properties

Some important thermal properties of electrical insulating materials are described below.

Transition-point temperatures

These "internal melting points" in plastic solids are associated with sharp changes in internal degrees of freedom. The most important of these is the heat-distortion temperature, which is related to T_g, the glass-transition temperature. The heat-distortion temperature is that temperature at which a 10-mil deflection occurs in a standard specimen loaded and heated in a standard manner. For some plastics, this point is below room temperature so that care must be taken to avoid high unit-area loads with these materials and thereby avoid cold flow and cut-through.

Thermal expansion

The coefficient of thermal expansion of insulating materials is an important consideration in designs which depend on maintenance of adhesive bonds of insulation to conductors or in which conductors are embedded in a cast-resin structure. The coefficient of thermal expansion of insulating materials varies from 2.13×10^{-4} per °C for thermoplastics like polyethylene to 5.5–13×10^{-6} for inorganic materials like alumina and mica.

Thermal conductivity

With a few exceptions, good electrical insulating materials are also good thermal insulators. In the operation of many types of electrical equipment, it is important to carry away as much as possible of the heat generated in the losses of the magnetic, conductor, and insulation components because machine efficiency and life are lessened if the operating temperature is excessive. Poor contact between the elements of a laminated insulating structure causes the presence of very thin layers of air. Because stationary air is a very good thermal insulator, this markedly reduces the overall thermal conductivity. It is because of this effect that laminated insulation structures based on wrapped tapes, mica splittings, and the like exhibit far better thermal conductivity longitudinally than transversely. Impregnation of this type of insulation with resins or with oil improves the thermal conductivity by displacing trapped air.

Most organic polymers have thermal conductivities on the same order of magnitude. Inorganic materials have much higher thermal conductivities and the thermal conductivities of cast or moulded insulation can be greatly improved by adding large quantities of fillers of this type. Those fillers which have the highest thermal conductivities, such as powdered metals, silicon carbide, and graphite, are also electrically conducting and their use results in composites of low dielectric strength. Beryllium oxide is unusual in that it has both good thermal conductivity and good electrical insulating properties. Its use is limited because of the toxicological hazards associated with it. Substances such as alumina, magnesia, and silica are widely used as fillers in resins to improve thermal conductivity, lower overall coefficient of thermal expansion, and lower cost per unit-weight.

Maximum thermal conductivity in a resin results from the use of a filler which can be added in maximum amount by volume without increasing the viscosity of the composition beyond a usable point. Fillers whose individual particles are in the form of fibres or leaflets give the highest viscosity at low filler-volume contents. The use of filler systems involving a considerable range of particle sizes generally permits higher filler content by volume within a usable viscosity range. An alternative technique for securing a very high filler content in cast-resin units such as small transformers involves filling the space for insulation in the mould with a very coarse filler such as 30-mesh sand and then impregnating the sand with low-viscosity unfilled resin. Very high filler volumes are attainable by this technique and the problem of filler separation is obviated. Still higher filler contents have been obtained by impregnation of a coarse dry filler with a resin containing a suspended finely divided filler. A disadvantage of this method is the long time required for impregnation because of the higher viscosity of the filled resin. When a dry filler of large particle size is to be impregnated, best results can be obtained by introducing the resin at the bottom of the system while evacuating from the top side.

Thermal ageing

The resistance of insulating materials to the effects of ageing at elevated temperatures is a most important property since maintenance of other important properties such as mechanical strength and, ultimately, the life of the equipment is dependent upon it. Thermal ageing is generally considered a chemical property and will be discussed under that heading. However, a physical change which often accompanies thermal ageing of organic polymers, and which should be mentioned here, is shrinkage. Unless careful consideration is given to the possibility of shrinkage in the design of insulation systems it is possible that the thermal conductivity of a system may deteriorate on ageing. For example, taped coil structures, dependent upon intimate contact with magnetic iron for good thermal conductivity and maintenance of the designed operating temperature, might develop small air gaps between the iron and the insulation on ageing, resulting in unacceptably high operating temperatures, accelerated thermal ageing, and runaway conditions.

Mecnanical Properties

The mechanical properties of solid insulations are often more important than their electrical properties. Most organic materials are good insulators if they can be formed into continuous films that exclude moisture and if the film continuity can be maintained through the hazards of manufacturing operations and the thermal cycling and thermal ageing of the equipment concerned.

Insulation in large apparatus must have sufficient mechanical strength to support the weight of the conductors and all the forces these may encounter during operation. In addition, it must have the properties to withstand the abuse which occurs during manufacture. These include flexibility, needed for bending and forming of insulated coils; tear strength and tensile strength for wrapping of taped coils; flexural strength, especially in supporting members such as wedges; and shear strength, needed to provide physical support to the windings and to withstand magnetic and rotational stresses, as well as vibration and shock. Other needed mechanical properties are bond strength between wires, especially on coils which may be subjected to vibration and shock, and abrasion resistance, to prevent insulation damage during assembly. Any serious degradation of the mechanical properties of an insulation system generally leads to electrical failure. The actual mechanism of the failure will vary depending upon the nature of the mechanical degradation and also the environment. An excessively high temperature may lead to plastic flow and cut-through or to loss of flexibility and subsequent crack formation on thermal cycling. The crack may permit the ingress of moisture with a resulting loss of dielectric strength. Or the high temperature might lead to gas evolution and corona erosion breakdown initiated in the gas space.

Effect of structure on mechanical properties of polymers

The general principles relating mechanical properties of polymers to structure have been known for many years. Rubbers, plastics, and fibres, for example, are not intrinsically different materials. Their difference is a matter of degree rather than kind.

If the forces of attraction between the molecular chains are small and the chains do not fit readily into a regular geometric pattern, or lattice, the normal thermal motion of the atoms tends to cause the chains to assume a random, more or less coiled arrangement. These conditions lead to a rubber-like character. In practical rubbers, a few cross-links are added to prevent slippage of the molecular chains and permanent deformation under tension. With such polymers, when the stress is released the normal thermal motion of the atoms causes them to return to a random coiled arrangement. If the forces between the chains are strong and the chains fit easily into a regular geometric pattern, the material is a typical fibre. In intermediate case, where the forces are moderate and the tendency to form a regular lattice is also moderate, the result is a typical plastic.

In general, the tendency to crystallise is determined by two opposing factors: (i) the forces between the molecular chains; and (ii) the geometrical bulkiness of the chains. The forces between chains are determined by the nature of the groups in the polymer. Purely hydrocarbon groups ($-CH_2-$) have the lowest molecular cohesion, while amide groups, containing the electronegative atoms oxygen and nitrogen have the highest. The higher molecular cohesion of the amide groups is attributed to the fact that they have more outer shell electrons available for interaction than do the hydrocarbons and to the effect of hydrogen bonding. Hydrogen, when situated without strain between two electronegative atoms, can act as a weak bond (5% of the strength of a primary carbon-carbon bond) between them. While each individual hydrogen bond is weak, the sum of their strengths in chains, hundreds of units long, exceeds that of primary valence bonds. Thus, polyamides are hydrogen-bonded together as shown.

$$
\begin{array}{ccc}
 & \uparrow & \\
\mathrm{H} & \mathrm{O} & \mathrm{H} \\
| & \parallel & | \\
\mathrm{-C-C-N-C-} \\
| & | & | \\
\mathrm{H} & \mathrm{H} & \mathrm{H}
\end{array}
$$

The effect of the geometric bulkiness of the chains is illustrated by the behaviour of ordinary polyethylene, which contains many side chains due to branching during polymerisation, and high-density polyethylene, which has an almost perfect linear structure capable of close packing into a crystal lattice. The higher-density, more crystalline material has a higher softening point, is more rigid, and has higher strength. The difference is even more pronounced when the amount of branching is increased, as in polyisobutylene, which has two methyl groups (CH_3) attached to alternate carbon atoms. These methyl groups have the effect of spreading the chains apart and making polyisobutylene at comparable molecular weight a viscous tacky liquid rather than a solid. The spatial relationships of substituent groups in polymeric ethylene derivatives such as polystyrene have a marked bearing on the closeness of packing of the chains and on the properties of the polymer. If one considers that the carbon atoms of the polymer chains lie zigzag in a plane with hydrogen and phenyl groups above and below the plane, three stereochemical structures are possible: (i) the phenyl groups are randomly oriented above and below the plane (atactic); (ii) the phenyl groups are all on one side of the plane (isotactic); and (iii) the phenyl groups alternate in a regular manner, above and below the plane (syndiotactic). Ordinary atactic polystyrene softens at about 80°C, but isotactic polystyrene has a crystalline melting point well over 200°C.

Some polymers are used as three different materials—rubber, plastic, and fibre. Polyethylene, for example, is used as a substitute for natural rubber in wire covering, as a plastic in low-loss stand-off insulators and insulating films, and as a fibre in acid-resistant filter cloths, where high fibre strength is not as important as chemical resistance.

The same polyester derived from terephthalic acid and ethylene glycol is used as a fibre and as plastic film. To make the fibre, the filament extruded from the polymer melt is oriented by stretching it 400%. This draws the chains closer together and causes crystallisation in the direction of the fibre axis. This markedly increases the flexibility and the tensile strength of the fibre.

To form the plastic film, the extruded melt is oriented by stretching it in two directions. The orientation is not nearly as complete as is the case with the fibre, and a material with the properties of a plastic results. The lack of complete orientation in such films results in a tendency for recrystallisation to occur in a random manner at elevated temperatures, with resulting embrittlement. Hence, the useful life is often limited by physical change rather than chemical deterioration.

Chemical Properties

Resistance to chemicals

Resistance to chemicals is a requirement in many industrial applications of electrical equipment such as motors. Caustic cleaning solutions in factories, acid fumes from electroplating facilities, the dusts from cement plants and mining operations are examples of some of the hazards encountered. No one material has universal resistance to all chemicals, but some recently developed epoxy formulations are very resistant to a wide variety of chemicals. These may be used as thioxotropic compositions to encapsulate the end windings of motor stators or as solventless varnishes to impregnate motor coils.

Hermetic (i.e., completely sealed) refrigeration systems provide a special case where chemical resistance is required, because of the solvent effect or softening action of the refrigerant on the insulation, which may result in cut-through. There is also the possible chemical reaction with the oil or insulation at higher temperatures. A simple laboratory screening test for hermetic insulation materials comprises sealing the insulation material in a glass tube with oil and refrigerant, strips of copper, steel and aluminium, and ageing at 150°C and examined periodically. A useful laboratory test for studying the softening effect of refrigerants on enameled wire at elevated temperatures involves measuring the resistance to repeated abrasion under load in both the liquid and gaseous refrigerant.

Insulating washers of polytetrafluoroethylene are used in tantalum electrolytic capacitors because of the material's inertness to corrosive dielectric liquids at elevated temperatures. The use of fluorochemical vapours as cooling and insulating media in transformers and the use of sulphur hexafluoride as an arc-interrupting material in circuit breakers are other examples in which chemical resistance in insulation is required.

Outdoor weathering

Although much experience has been accumulated on outdoor weathering of paints and finishes, much less work has been done with structural plastic insulating materials outdoors, particularly under voltage stress. In Europe, some firms are using cast epoxy insulators instead of porcelain ones outdoors with apparent success. Others, using essentially the same compositions, conclude from their tests that the materials are suitable for indoor use, but not outdoor use. Exposure to sunlight and rain results in a gradual loss of gloss on the surface. Undoubtedly, the voltage stress and the type of dirt in industrial atmospheres play an important part in determining the results.

Flammability

Most organic polymers will burn in air, but some, such as polytetrafluoroethylene, will not. Others, such as poly(vinyl chloride), poly(vinylidene chloride), and polychloroprene, can be ignited in air, but are self-extinguishing. If the chlorine content of a polymer is greater than 35% by wt, it is generally considered non-flammable. Fillers, such as antimony trioxide and antimony oxychloride, improve resistance to flammability of resins. Aromatic polyamides and aromatic polyimides ignite with difficulty and are self-extinguishing.

Liquid organic phosphates, such as tricresyl phosphate, are non-flammable, but they are rather poor insulators. Mixtures of chlorinated aromatic hydrocarbons are both fire-resistant and good insulators, and are widely used in capacitors and in transformers which are required to be fire-resistant. Fluorocarbons (i.e. compounds containing carbon and fluorine only) and perfluoroethers (completely fluorinated ethers), used in vapour-cooled and insulated transformers, are also non-flammable.

Hydrolytic stability

Many types of polymers containing oxygen or nitrogen atoms in their chains absorb small amounts of water and then react chemically with it to form smaller molecules, with a marked loss of mechanical properties such as strength and flexibility. This reaction is accelerated by an elevated temperature and by the use of sealed systems which prevent the water from escaping. Polymers formed in condensation reactions which involve the elimination of water are especially susceptible to this reverse reaction.

Thermal and oxidative stability

The relationship between the life of insulation in electrical apparatus and the operating temperature is very complex. Apart from the effect of temperature, the most active agents promoting insulation degradation are usually oxygen and moisture, but air and moisture are not always present and sometimes other factors such as vibration and chemical contamination are present. It has been shown that the reactions leading to insulation deterioration may be treated as a chemical rate phenomenon and a method was proposed of plotting the logarithm of the life in hours, at several temperatures (using an arbitrary criterion for life, such as a limiting value of dielectric strength or mechanical strength), against the reciprocal of the absolute temperature, to estimate the life at any desired temperature. While this method has some limitations, it is widely used, internationally, in evaluating thermal capability of insulation, because it has more validity than other methods which have been proposed and because of its simplicity.

As was mentioned earlier, the usable life of insulation is affected by many factors in addition to temperature effects. Experience has shown that the thermal life characteristics of composite insulation systems cannot be reliably inferred from information concerning component materials alone. For example, it is known that some varnishes have a deleterious effect on the life of enameled wire at elevated temperatures, while others enhance the life. These effects are tied up with such factors as permeability of oxygen through the varnish film, adhesion of the varnish to the enamel, and shrinkage of the varnish film on continued polymerisation brought about by heating. A varnish film on shrinkage can pull enamel from a section of the wire, leaving only air as the dielectric. Laboratory tests on varnished twisted pairs provide a guide to the selection of suitable combinations, but these results need to be verified by evaluation of complete insulation systems under accelerated conditions, in actual apparatus where possible. This type of study is called *functional evaluation* and is made, for example, on fractional horsepower motors, small speciality transformers, and small integral horsepower motors. In the case of large, very expensive machines, such as turbine generators, insulated test bars of the same cross-section as a coil are used to simulate coils in thermal ageing tests. Actual half coils in a typical slot section are used for thermal cycling tests. In all these accelerated ageing tests, insulation deterioration is treated as a chemical reaction and the logarithm of the life is plotted against the reciprocal of absolute temperature.

In all functional evaluation testing, it is recognised that the results obtained are relative, not absolute, and that the life obtained varies with the test apparatus and test cycle, and environments used. By using an insulation system with a known and satisfactory field-performance record as a reference point, usually valid comparisons can be made. In motor tests it is customary to extrapolate to the temperature at which the life would be 5000 hr. In a test cycle which included thermal ageing, thermal cycling, humidification, vibration, and voltage stressing, 5-hp motors, made with a widely used class A insulation system, gave an extrapolated 5000-hr life at approx 105°C. Under the same conditions, a silicone-insulated class H system extrapolated to 5000 hr of life at above 250°C. It is possible to segregate the effects of the factors other than thermal ageing in the motor tests by eliminating them from the test. A study of this sort was made with at least one insulation system, which showed that elimination of moisture, heat shock, and vibrational stresses increased the 5000-hr life temperature by more than 40°C. Functional evaluation provides a valuable tool for selecting insulation systems of improved performance, sometimes with the added benefit of lower cost. No method is known of determining the thermal and oxidative stability of materials quantitatively, without using an experimental approach. Some qualitative observations may be made, however, on the effect of chemical structure on the following properties.

Effect of volatility

The ability of polyesters based on isophthalic and terephthalic acids to be used continuously at temperatures about 25–50°C higher than those based on orthophthalic acid points out the beneficial effect of low volatility. The last-named material is readily converted to its anhydride on heating. Phthalic anhydride sublimes rapidly at 150°C; the other two isomers sublimes at about 300°C.

Effect of cross-linking

Cross-linking increases molecular size and decreases volatility. In wire enamels based on glycol-glycerol-terephthalates, the more highly cross-linked materials show the best thermal stability.

Effect of substituent groups

In general, aromatic groups and groups containing electronegative elements enhance thermal and oxidative stability. For example, phenyl silicones are much more stable when heated in air than methyl silicones. Lubricating fluids based on polyphenyl ethers, $(p\text{-}C_6H_4O)_n$, show excellent stability in air at 250°C, while corresponding fluids based on polyethylene glycols, $(CH_2CHO)_n$, can be used only at temperatures about 100°C lower.

In polyesters, increasing the number of methylene groups in a reactant (thereby diluting the number of electronegative ester groupings) results in a decrease in stability when they are aged at elevated temperatures in air. The very greatly enhanced oxidative stability (250°C) of polytetrafluoroethylene over polyethylene (100°C) illustrates the effect of substituting the very electronegative fluorine atom for hydrogen. In a sense, the molecules containing electronegative groups are already partially oxidised and this makes them less susceptible to oxidation under ordinary conditions.

MATERIALS

Gases

Air is widely used as an insulating material. In many types of low-voltage equipment, solid insulation is used to separate conductors from each other and from the ground, but often these materials are porous and an air gap is the dielectric barrier. The dielectric breakdown strength of small air gaps at atmospheric pressure is shown in Table 13.2.

The breakdown strength per unit of spacing increases with pressure and decreases with increased spacing. A common use of air as a dielectric with high-voltage equipment is in overhead transmission lines and associated apparatus, where large air gaps separate conductors with potential differences as high as 500,000 V. Air is used at pressures greater than 100 psi in some air blast circuit breakers.

Table 13.2. Dielectric breakdown of air (1 atm) for small gaps.

Spacing, mils	Breakdown stress		Voltage breakdown	
	V/mil rms	kV/mm	V, rms	kV/mm
0.3	780	30.8	230	9.0
1.0	330	13.0	330	13.0
3.0	187	7.4	560	22.1
10.0	110	4.3	1100	43.3
30.0	87	3.4	2600	102.4
100.0	76	3.0	7600	299.4

Gases, for most practical purposes, have a dielectric constant of 1.0. The precise values vary somewhat, but not in the first decimal place. The dielectric strengths of gases vary markedly. Helium, for example, has less than one-fifth the dielectric strength of air and a motor designed to operate in, and pump helium for, a gas-cooled nuclear reactor, will require longer insulation creepage distances than if it were to operate in air. Sulphur hexafluoride, on the other hand, has a much higher dielectric strength than air and this is made use of in the design of circuit breakers which utilise its arc extinguishing characteristics, and in transformers which are insulated and cooled by it. The dielectric strengths in a uniform field of some representative gases used in electrical insulation are shown in Fig. 13.2. Most gaseous insulation is used in non-uniform fields. Under these conditions, the breakdown strength of a gas will normally increase with pressure to a maximum "critical" value, beyond which further increase in pressure causes a decrease in the dielectric strength of the gas. The "critical" pressure is also temperature-dependent. The dielectric strength of gases under suddenly applied voltages (surges) of very short duration (microseconds) is higher than that obtained at power frequencies.

In equipment such as large oil-cooled transformers, where oxidation of the oil may impair the life of the insulation system, dry nitrogen is often used as the gas blanket above the boil. These sealed transformers give significantly improved life at the same operating temperatures.

For shipping, large transformers sometimes are filled with carbon dioxide under pressure, after the usual drying and vacuum impregnation with oil, and the draining of the oil in the factory. The solubility of carbon dioxide in mineral oil at room temperature is about ten times that of air or nitrogen. This greater solubility of residual CO_2 makes it easier to avoid trapping gas bubbles in the insulation when the transformer is again vacuum impregnated with oil in the field.

When gases are used to cool apparatus as well as to insulate, usually an economic decision determines the gas selected. Hydrogen has the highest specific heat, the highest thermal conductivity, and the lowest viscosity of all gases. It has lower dielectric strength than air. Large turbine generators are cooled with hydrogen under pressures of 3 atm and higher because this results in lowest windage losses and the insulation function can be managed in other ways. In some high-voltage dry-type power transformers, on the other hand, sulphur hexafluoride is the dielectric gas of choice because of its superior dielectric strength. In thermal conductivity, it is not significantly better than air.

Gases at very low pressures (vacuums of less than 10^{-4} mm of Hg) are good insulators because the mean free path of their molecules is larger than the distance between the conducting electrodes. As a consequence, ionisation by collision is secondary to other sources of charged particles and impurities become important. Vacuum circuit interrupters utilise high vacuum as a dielectric.

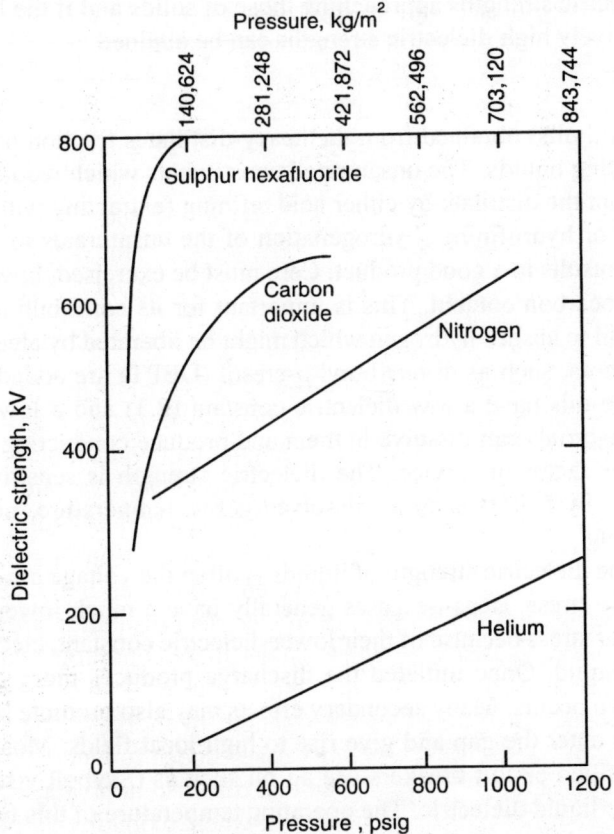

Fig. 13.2. The effect of pressure on the DC voltage strength of representative gases tested in a uniform field. The electrodes are stainless steel plates, and the gap distance is 0.5 in.

Liquids

Insulating liquids are used in equipment such as cables, transformers, circuit breakers, and capacitors. Their advantages over air as a dielectric lie principally in their higher dielectric strengths and better heat dissipation characteristics. For these reasons, liquid-filled equipment, such as transformers, are generally smaller than dry-type units of the same rating and are also less costly, because of the overall reduction in materials used. A very rough rule states that, under practicable conditions, liquids have a dielectric strength an order of magnitude higher than gases, and solids an order of magnitude higher than liquids. It would appear from this that solid insulation systems would be ideal and that there would be little need for liquids, especially when one takes into account other disadvantages, such as flammability of mineral oils. However, advantages of liquids are the low cost of those most widely used (about 2¢/lb for mineral oils and about 15¢/lb for chlorinated aromatic hydrocarbons) and the fact that in liquids defects such as gas bubbles tend to be "self-healing", i.e., the gas bubble will eventually dissolve or migrate to the gas space above the liquid. A similar defect in a solid (unless the solid were a material resistant to corona damage, such as mica) would result in a highly stressed area which would probably eventually fail by corona erosion. Furthermore, cellulosic papers impregnated with mineral oils or chlorinated aromatic hydrocarbons have dielectric strengths approaching those of solids and if the liquid insulation gaps can be kept very small, relatively high dielectric strengths can be attained.

Mineral oils

Hydrocarbon oils (mineral oils) obtained from the heavy-distillates fraction of crude petroleum are the most widely used insulating liquids. The unsaturated constituents which would result in poor oxidation stability are removed from the distillate by either acid refining (extracting with sulphuric acid followed by washing with water) or hydrofining (hydrogenation of the unsaturates to saturated hydrocarbons). Either of these methods results in a good product. Care must be exercised, however, to prevent removal of all the aromatic hydrocarbon content. This is important for its contribution to oxidation resistance and to the ability of the oil to absorb hydrogen which might be liberated by electric discharges in the oil. Small amounts of inhibitors, such as di-*tert*-butyl-*p*-cresol (DBPT) are added to oils to improve their oxidation stability. These oils have a low dielectric constant (2.3) and a low power factor (less than 0.1%). Since very few materials can dissolve in them and produce conducting ions, it is relatively easy to maintain a low power factor in service. The dielectric strength is sensitive to impurities such as moisture, and is affected by factors such as dissolved gases, temperature, pressure, electrode shape, material, area, and spacing.

A limiting factor in the dielectric strength of liquids is often the voltage at which minute gas bubbles are formed under electric stress. Because gases generally have a much lower dielectric strength than liquids and receive greater stress because of their lower dielectric constant, electric discharge is initiated in a gas bubble in the liquid. Once initiated the discharge produces more gas and this progression continues until breakdown occurs. Many secondary effects may also promote breakdown. For example, suspended particles may enter the gap and give rise to high local fields. Most distribution and power transformers and liquid-filled circuit breakers use an oil such as (Saybolt viscosity, 58 sec at 37.8°C; flash point, 135°C) as the liquid dielectric. The operating temperature of this type of equipment usually does not exceed 70–120°C, but some distribution transformers may experience short-time overloads in which the temperature exceeds 200°C for a few minutes.

Chlorinated hydrocarbons

Chlorinated hydrocarbons are used in some transformers, especially in indoor installations, to reduce fire risk, and in capacitors to obtain substantially higher KVAR ratings in a given volume. These liquids are called by the generic name *askarel*. Askarels are defined as non-flammable insulating liquids which evolve only non-flammable gases when decomposed by an electric arc. The askarel used in transformers is a mixture of chlorinated benzenes and chlorinated biphenyls. It has a lower viscosity (42 SUs), a lower dielectric constant (3.6–3.9), and a higher power factor (0.015–0.05) than the mixture of chlorinated biphenyls used in capacitors (200 SUs; 4.2–4.4; 0.015). Because askarels have a higher solvent power than mineral oils, care must be exercised in the selection of plastic materials to be used with them. Askarels are much more oxidation-resistant than mineral oils, which is an advantage in transformer applications.

For capacitor applications, it is important that askarels be free of even trace amounts of hydrolysable chlorine, because this leads to an autocatalytic reaction which results in runaway power factor and thermal breakdown. Tetraphenyltin (0.135%) and epoxy compounds such as phenyl glycidyl ether (0.2%) are sometimes added to askarels as stabilisers. Their reaction with trace amounts of hydrogen chloride render it harmless.

Fluorochemicals

Completely fluorinated hydrocarbons, such as perfluoroheptane, and cyclic ethers, such as perfluoro-3-propyloxane, $C_8F_{16}O$, have been used to cool and insulate small transformers and power transformers of the 15 kV and 34.5 kV classes. These compounds have a high dielectric strength, both in the liquid and in the vapour state and since they boil at approximately 100°C, and have a relatively high heat of vapourisation (21 cal/g), they are very effective for evaporative cooling. Other advantages for this application are non-flammability, a low solvent power (they do not attack other insulation components), a low power factor, and a very good thermal and oxidative stability. Even though they can be used in small amounts, compared with oil, they are still too expensive, with present designs, to find extensive use.

Silicones

Silicone fluids are made up of molecules in which silicone and oxygen atoms alternate in a chain and/or ring structure, and approximately two hydrocarbon or substituted hydrocarbon radicals are attached to each silicon atom. The most common hydrocarbon radicals used are methyl (CH_3–) and phenyl (C_6H_5–). These liquids are available in a wide range of viscosities. Among their outstanding properties are a low viscosity index, a good thermal and oxidative stability, and a very low power factor. Their high cost has limited their use in electrical insulation to highly specialised equipment such as small capacitors.

Solids

The combination of insulating materials used in the insulation of a type of electrical equipment is called an *insulation system*. Materials with different thermal capabilities may be used in an insulation system because not all parts of the equipment (e.g., lead wire, motor windings) are at the same temperature. Some systems are extremely simple and some are complex. For example, in rotating machinery, the insulation system of a battery-operated electric toothbrush motor (1.25 V) is simply the 1.5-mil-thick organic polymer coating (enamel) on the conductor. A washing machine motor, which operates at 115 V, may have 7-mil-thick polyester-film slot cells separating the enameled conductor from the magnetic iron, and in addition will have been treated with one or two coats of a baked-on insulating varnish. As the operating voltage increases, more sophisticated systems are needed. For 24,000-V turbine generators, the stator coils may be insulated with a solid composite of mica flakes and a polymerised solventless resin impregnant.

In a rotating machine or in a transformer, the major insulation between the winding and the magnetic iron core is called the *ground insulation*. The core is usually connected directly to the frame of the machine or the tank of a transformer and is at ground potential. Some types of ground insulation also provide *mechanical support*. In motors, the simplest type of ground insulation is a *slot cell*, which may be paper, film, varnished glass cloth, resin bonded mica flakes or mica paper, or combinations of these. Slot cells are most commonly used in small random-wound motors. Form-wound motors generally employ taped insulation or combinations of taped and wrapper insulation. The wrapper is a sheet-form insulating material applied in one or more layers to the straight portion of the coil to provide a form-fitting tube of high dielectric strength. Flexibility is a prime requirement for this type of material. The shaped end-portions of coils of this type are then finished by taping, either by hand or hand-guided taping machines. Taping is a widely used method of applying insulation to conductors as well as for ground insulation. Automatic taping machines are used to apply insulation such as paper, film, varnished glass, asbestos, etc., to conductors. The insulation which separates conductors from each other is called the *turn insulation* or *conductor insulation*. The voltages involved in turn insulation are generally small compared with the ground insulation. Insulation inserted between layers of conductors is termed *layer insulation*. The principal types of solid insulating materials will be discussed briefly from the point of view of their application in insulation systems.

Enamels

Wire enamels are polymer solutions which are applied to conductors and baked on in a continuous process to give thin (0.5–4.0 mils, or 0.013–0.1 mm), flexible, abrasion-resistant coatings. Hundreds of millions of pounds of conductors are insulated in this way annually with about half a dozen principal types of enamels. These include oleoresinous for low cost, polyurethanes for easy insulation removal in hot solder, epoxy for chemical resistance, nylon for abrasion resistance, polyvinyl formal-phenolic polymers for general-purpose moderate-temperature uses, polyesters and modified polyesters for temperatures to 180°C, and aromatic polyimides for temperatures to 220°C. Selected enamels may also be used to insulate foil and strip conductors of copper and aluminium.

Paper

Cellulose, in the form of papers and pressboard, still enjoys wide use in the insulation of electrical equipment. For conductor insulation, cellulose paper tape, with or without adhesive, is applied with automatic taping machines. A number of papers based on synthetic fibres are available, but most of these, when impregnated with oil or askarel, do not have the high dielectric strength of cellulosic paper, probably because they lack the fibrillar structure of cellulose and its ion-trapping ability. However, mica papers, asbestos papers, and aromatic polyamide papers are sometimes used for conductor insulation for high-temperature motors and dry-type transformers. Papers from olefin fibres (polyethylene and polypropylene), polyester fibres, and glass fibres have been made but these have not yet found much use in electrical insulation. Significant increases in the thermal stability of cellulosic papers in oil can be achieved through the use of stabilising additives.

Film materials

About two dozen kinds of film materials are available for electrical insulation applications but relatively few of these have achieved widespread use. Polyester films are used as slot cells in many fractional horsepower motors, such as washing-machine motors. Here they offer the advantages of low moisture absorption, higher dielectric strength, and perhaps 15°C in thermal capacity over the cellulosic paper they replace.

They are also used in combination with other materials such as mica and varnished glass cloth as slot cells in large industrial motors. Here the film materials imparts toughness and high dielectric strength to the composite while the other materials provide for longtime higher temperature capability. Polyester films are also used in small electronic capacitors.

Recently developed aromatic polyimide films, capable of continuous use at 220°C and intermittent short-time use to 400°C, have most of the properties of an ideal insulation. They retain flexibility at cryogenic temperatures (−269°C), are flame-resistant, have excellent solvent and chemical resistance, are strong mechanically, and have a zero strength temperature in excess of 800°C. Despite their high cost they are finding use as tape-wrapped conductor insulation in apparatus such as large traction motors, dry-type transformers, slot cells in small high-temperature aircraft motors, in high-temperature electronic capacitors, and in printed-circuit and flat-cable constructions.

Fibrous products

Fibres such as glass, cotton, cellulose acetate, polyester, polyamide, acrylic, asbestos, and alumina-silica are used either alone or in combinations as electrical insulation. Glass fibres are available in both staple and continuous filament forms. An alkali-metal-free borosilicate called E glass is used for electrical purposes. Advantages of glass are high strength, thermal stability, fire resistance, and low cost. Combinations of glass fibres and polyester fibres are used where improved abrasion resistance is required. Most fibrous products are available in the forms of lacing cords and tapes, felts, mats, papers, rovings, and woven cloths. Nylon is widely used in the form of lacing tapes and enjoys significant use in phenolic laminates. An aromatic polyamide, is produced in the form of staple and continuous-filament fibres, tapes, papers, and woven cloths. It has outstanding thermal and oxidation resistance, permitting continuous use at temperatures as high as 180°C. It does not melt and is self-extinguishing.

Usually fibrous products are used coated or impregnated with resins. For temperatures up to 155°C, many low-cost varnishes are available for use. These include oil-modified alkyd phenolics, polyesters, and epoxies. For temperatures of 180–220°C, silicone, diphenyl oxide, aromatic polyimide varnishes, or fluorocarbon dispersions are employed. These combinations are used for conductor insulation, lead wire and, in combination with plastic films or mica products, as slot and phase insulation.

Coating powders

If one attempts to insulate a sharp edge or corner, as in the slots of a small motor stator, with a varnish or a paint, it is impossible to insulate the edges to withstand more than a few volts because surface tension causes the liquid to draw away from the sharp edges. This problem has been overcome by the development of solid coating powders, formulated to have a thixotropic character when molten, which are applied to the pre-heated object to be coated. In one method, called the fluidised-bed method, the solid coating powder, which may be thermoplastic or thermosetting, is suspended in air by a stream blown in from the bottom of the container, through a fritted glass filter or other means which distributes the air flow uniformly. The mixture behaves like a liquid in which air may be considered as the solvent or suspending medium. The object to be coated, usually a metallic material of relatively high heat capacity, is pre-heated to a temperature on the order of 135–175°C and is dipped in the fluidised bed and withdrawn. The heat stored in the object is great enough to cause the solid particles to melt, coalesce, and flow to form a smooth continuous coating. If the powder is properly formulated and the treating conditions are properly controlled, the coating does not pull away from the edges and uniform high dielectric strengths are obtained. Another technique is to apply the powders to the pre-heated object to be coated by electrostatic spraying. The technique is applicable to a wide variety of materials including epoxies, polyamides, and cellulose derivatives. Epoxy coatings are used in low-voltage automotive, aircraft, and some fractional-horsepower appliance motors to replace paper and polyester-film slot cells. Advantages of these coatings are lower cost and a higher temperature capability than the previously used materials.

Insulating varnishes

Insulating varnishes are usually blends or reaction products of drying oils and natural or synthetic polymers, dissolved in solvents of low toxicity. Drying oils commonly used are tung, oiticica, linseed, or dehydrated castor. Polymers may be naturally occurring materials such as asphalts, shellac, copal, and rosin, or synthetics such as phenolics, melamine, alkyds, or epoxies. Completely synthetic polymer varnishes are less generally used because they tend to be more expensive.

Oil-modified-phenolic alkyds are widely used because they combine the needed properties of good moisture resistance, thermal capability, and relatively rapid cure with moderate cost. Epoxies, at somewhat higher cost, are used where resistance to chemicals is needed, as in motors for some chemical process industries. Silicone varnishes are used principally for their high-temperature capability in machines which operate at 180–240°C. Silicones cannot be used in totally enclosed rotating DC machines at these temperatures because a few ppm of silicone vapours in the cooling air causes excessively high brush wear. Diphenyl oxide varnishes and polyimide varnishes do not have this adverse effect upon brush wear and also exhibit higher bond strength at these temperatures.

Liquid reactive resins

Liquid polymers, or a polymer dissolved in a liquid with which it can coreact, can be converted to solids by the proper choice of catalysts and conditions. Such materials are often called liquid reactive resins, solventless resins, or solventless varnishes because the solvent, if present, is a coreactant. Examples of solventless resins are liquid epoxies, urethanes, acrylics, unsaturated polyesters dissolved in reactive monomers such as styrene or methyl methacrylate; liquid butadiene polymers; and solventless silicones. By selection of the kinds and amounts of components, materials can be designed for low viscosity, flexibility at high temperatures or low temperatures, low dissipation factor, and high heat-distortion point. Low viscosity and low shrinkage are properties desirable for vacuum impregnation of high-voltage equipment where complete filling of voids is necessary. The Thermalastic insulation system for high-voltage generators comprises mica flakes in tape form, impregnated with a solventless reactive resin. The principal advantages of solventless resins over conventional types are: (i) better fill, which results in better moisture resistance; (ii) better thermal stability; (iii) lower operating temperature because of better heat transfer; and (iv) higher dielectric strength.

Encapsulants

Encapsulants are insulating coatings which conform to the shape of the apparatus being insulated. They are usually liquid reactive resins, formulated with fillers and a small amount of thixotropic agent, such as finely divided silica, to prevent run-off during cure, so that a conformal coating, 5–30 mils thick, results. For high-voltage components, the encapsulating coating acts as the mould and the equipment is then vacuum impregnated with a liquid reactive potting resin.

Casting and potting resins

Casting and potting resins are usually combinations of liquid reactive resins with inorganic fillers, although clear resins are sometimes used. A removable mould is used in casting; in potting, the mould is part of the finished assembly. Fillers help to lower the coefficient of thermal expansion of the composite, making it more nearly like that of the metallic conductors. Fillers also greatly improve heat transfer and lower the cost per unit weight of material. The effect of fillers on cost per unit volume is less because inorganic fillers also increase the density of the composite. Casting and potting resins are sometimes used at temperatures above room temperature to lower the viscosity. Cast-epoxy resins containing hydrated alumina as filler have been used as insulators outdoors and are reported to have good arc and track resistance. Cast-resin bushings, replacing porcelain in some applications, have been used in Europe.

Composite mica insulation

Mica (qv) occurs naturally in a crystalline form which permits fabrication into thin, flexible splittings of high dielectric strength. The two forms of most interest in electrical insulation are muscovite, or white mica, usually obtained from India, and phlogopite, or amber mica, from Madagascar. Muscovite is harder than phlogopite, but it cannot be used at temperatures above 500°C. Above this temperature, it loses water of crystallisation and becomes powdery. Phlogopite will stand 800°C without disintegrating. Inorganic glasslike polymeric bonding materials such as boron phosphate and lithium borophosphates also stand these temperatures. The sheet mica insulation used to support the heating wires in electric toasters is a composite of phlogopite and an inorganic binder such as those just mentioned. In industrial motors designed for 130°C operation, thin laminates of polyester film, polyester fibre mat or varnished glass cloth, and resin-bonded mica are commonly used for slot and phase insulation. For 180–240°C operation, combinations of silicone or aromatic polyimide varnished glass cloth, mica, and aromatic polyimide film may be used. Mica papers are generally made from scrap of naturally occurring mica, although some is made from synthetic fluorophlogopite. Mica papers are available with resin and with fibre reinforcements and also without reinforcement. Rigid laminates prepared with inorganic binders are used as insulating spacers in some vacuum tubes. Mica papers are used in insulating tapes, wrappers, slot wedges, and phase insulation, usually in combination with other materials.

Filament-wound products

Filament-wound products are made by wrapping continuous filament glass impregnated with a solventless resin (usually epoxy) around a mandrel at a selected angle of winding. When enough layers to provide the thickness require have been wound, the part is cured by heat and the mandrel is removed. Filament-wound parts have very high burst (hoop) strength and are used in a variety of electrical equipment such as gas-filled circuit breakers, lightning-arrester and fuse tubes, and bushings and binding bands on motors and generators. Some typical properties for filament-wound structures are given in Table 13.3.

Table 13.3. Filament-wound products.

Specific gravity	2.0
Thermal expansion, per °F × 10⁻⁶	7.0
Hoop strength	
Unidirectional windings, psi	230,000
Helical windings, psi	135,000
Compressive strength, psi	70,000
Flexural strength, psi	100,000
Shear strength (interlaminar), psi	6,000
Modulus of elasticity, tension, psi	6.0×10^6
Dielectric strength, step-by-step, V/mil	400

Laminates, tubes, printed circuits

Laminated products are formed by bonding together two or more layers of material. The layers are generally woven cloths, papers or mats of fibres such as glass, asbestos, cotton, polyester or other synthetic fibre, impregnated with a thermosetting resin (e.g., phenolics, silicones, epoxies, and melamine). Laminates are produced in a number of grades according to the importance of the mechanical, electrical, or chemical requirements. A broad range of properties is made possible by combining various reinforcements with many types of resins, by changing the proportions of resin to base material, by the use of additives, and by arrangement of the plies of the laminate (parallel or perpendicular to each other).

Laminated tubes are prepared by winding impregnated base materials under heat and pressure, layer upon layer, around a heated mandrel corresponding to the inside tube diameter required.

Laminates are used in many insulation structural applications in electrical equipment such as switchboards, switchgear, transformers, generators, and motors. Copper-clad laminates are used in printed or etched circuitry, a technique for producing electronic and electrical circuits (e.g., in radios and computers) accurately at a low cost. The laminate provides insulation and structural support while the copper serves as the conductor. Thin copper-clad materials (0.0035-in. glass-epoxy) are used in integrated or multilayer circuits.

Moulded products

Plastics may be divided into two main types, which describe their processing properties and to a large extent determine temperature limitations of use. Thermoplastics are those which repeatedly can be made to flow under heat and pressure and are processed by extrusion, inaction or compression moulding, thermoforming, blow moulding, rotational casting, etc.

The change with temperature is physical rather than chemical. Most thermoplastics cannot be used for structural insulation applications at temperatures above 60–80°C. A few thermoplastics with very high heat-distortion temperatures are exceptions to this rule. Examples of thermoplastics used in electrical insulation are polyethylene, polypropylene, ethylene-vinyl acetate co-polymers, propylene-modified polyethylene, acetals, acrylics, acrylonitrile-butadiene-styrene (ABS), polystyrene, cellulosics, chlorinated polyether, fluorochemicals, polyamides (nylons), polyimides, phenoxies, vinyls. Thermoplastic materials with fibrous reinforcements have greater impact, tensile, and flexural strengths as well as higher heat distortion temperatures.

Thermosetting plastics undergo chemical change when cured and become infusible and insoluble. They may be cured by heat or at ambient temperatures by the use of catalysts. Thermosetting plastics are usually combined with fillers and/or reinforcing fibres, and are processed by compression and transfer moulding, and by casting. Examples of thermosetting plastics are aminoplasts (urea-formaldehyde, melamine-formaldehyde), phenolics, epoxies, diphenyl oxide, polyurethanes, polyesters, diallyl phthalates, and silicones.

Elastomeric or rubber-like materials are widely used in electrical insulation. The thermoplastic types, such as plasticised vinyls and vinyl co-polymers and fluoroplastics, are used principally in coverings for wire and cable and in flexible sleeving. The thermoset types are used in moulded parts and in encasing complete apparatus such as instrument transformers.

Elastomeric coatings impart abrasion resistance to motor windings used in industrial atmospheres. Examples of thermosetting elastomers are natural rubber, styrene-butadiene, butyl, polybutadiene, ethylene propylene terpolymer, polysulphide, nitrile, polychloroprene, chlorosulphonated polyethylene, polyacrylate, silicone, and fluorochemicals.

Cold-moulded plastics are thermosetting materials, processed in moulds at room temperature under high pressure, and subsequently cured either in ovens or by longer standing at room temperature. The process allows rapid production and low cost. Organic binders used are generally asphaltic or phenolic. Calcium aluminium silicate cements are typical of inorganic binders. Asbestos is the most commonly used filler in both non-refractory (organic) and refractory (inorganic) types. Cold-moulded plastics possess high heat resistance and good arc resistance. Because of these properties and their relatively low cost, they have found many uses. Examples are terminal blocks, fuse blocks, motor starter bases, switch bases, and plugs.

Ceramics

Ceramic insulations have been used for many years because of their low cost, low loss characteristics, high-temperature capability, and durability in outdoor weather. Classical porcelain ceramics are made from clays, flint (SiO_2), and feldspars. Mullite ($3Al_2O_3.2SiO_2$) and alumina (Al_2O_3) type ceramics are used for applications such as spark plugs where high mechanical strength, and high-temperature and thermal-shock capability are needed. Steatites ($3MgO.4SiO_2.H_2O$ with fluxing agents such as feldspars, alkaline earth oxides, alumina or lead oxides) are used for high-frequency applications, because of their low losses. A typical electrical porcelain is composed of 20% ball clay, 25% kaolin, 33% feldspar, and 22% flint. Porcelains are used for both low-voltage and high-voltage applications. Typical high-voltage uses are insulators on transmission lines, circuit breakers, etc. Low-voltage uses include switches, fuses, and hot plates. In addition to steatites, mullites, and aluminas, other ceramics used in electrical insulation are fosterites, cordierites, spinels, magnesia, quartz, zircon, wollastonite, lithium alumino silicates, boron nitride, and silicon nitride.

In microelectric circuits thin film inorganic insulators are applied by vacuum deposition or sputtering. Materials most commonly used are silicon monoxide, magnesium fluoride, zinc sulphide, lanthanum fluoride, cerium fluoride, cerium oxide, silicon nitride, boron nitride and cryolite (Na_3AlF_6).

CHAPTER 14

Cement and Lime

INTRODUCTION

Cement is a substance which in contact with water sets to a hard mass. Portland cement is the most important cement. It is made from (i) calcareous material e.g. lime (70%) from limestone, chalk, marble or ppt. chalk and; (ii) agrillaceous materials of SiO_2 from silica & iron oxide (30%). The sources of agrillaceous materials are clay, shales, slag etc. The main constituents are :

1. Tricalcium silicate $(CaO)_3SiO_2$: C_3S
2. Tricalcium aluminate $(CaO)_3Al_2O_3$: C_3A
3. Dicalcium silicate $(CaO)_2SiO_2$: C_2S
4. Tetracalcium Aluminoferrite $(CaO)_4Al_2O_3Fe_2O_3$: C_4AF

Cement is of various types (i) general purpose; (ii) moderate heat and hardening; (iii) high early strength; (iv) low heat cement; (v) sulphur-resistance cement.

There is also a class of white cement which is iron-free and is made from China clay and limestone $(C_3S, C_3A\ C_2S)$.

Other types of cement in use are:

1. High alumina cement—by heating limestone and bauxite—Monocalcium aluminate.
2. Slag cement. A mixture of granulated blast furnace slag and portland cement clinker.
3. Pozzolona cement—A mixture of cement and pozzolona (a type of cementitious material) react with hydrate lime at ordinary temperature to form cement. Pozzolona occurs naturally from volcanic origin or can be made by calcining clays or shales.

CLASSIFICATION OF THE PRODUCTS

Materials which are used in various mortars may be classified in several ways but for this discussion their properties, manufacture and raw materials will be used.

1. Natural cements are made by calcining impure limestones, containing CaO, SiO_2, Al_2O_3 and Fe_2O_3 at a low temperature which will not cause vitrification. These cements do not slake, and to make hydraulic cement they must be ground.
2. Portland cements are made by calcining to vitrification an intimate mixture of calcareous and agrillaceous minerals, $CaO + Al_2O_3 + SiO_2$. The resulting product does not slake when water is

428

added. When the cement is ground after cooling it forms a cement which will hydrate to form a hard mass in the presence of water.

3. Pozzolan cements are produced by the intimate grinding of a mixture of slake lime with slag, volcanic ash or portland cement clinker and treated slag.
4. Common limes are made by calcining limestone, pure $CaCO_3$, forming CaO. The lime slakes with water but has no hydraulic properties.
5. Hydraulic limes are made by calcining impure limestone, containing silica, alumina and iron, at a low temperature. The resulting product will slake with water and has hydraulic properties.
6. Finishing limes are made by calcining dolomite, high magnesium limestone, at a white heat. The resulting product will slake and has special properties of whiteness and ease of working under the trowel.
7. Plasters are made by heating gypsum ($CaSO_4 2H_2O$) until three-fourths to all of the combined water is lost and then grinding the resulting product.

Typical analyses of the raw materials and methods of treating are shown in Tables 14.1 and 14.2.

Table 14.1. Analysis of raw materials for lime and cement.

Raw Materials	Heat Treatment	Mechanical Treatment	Properties	Classification
Made from relatively pure limestones	Burned at low temperatures 600° – 900° C	Slake on addition of water to burned product.	Not hydraulic	Common limes
Made from dolomite Made from argillaceous or impure limestones	White heat 600° – 900° C			Finishing lime Hydraulic limes
				Natural Roman or Rosendale cement
Made from intimate mixture of argillaceous and calcareous substances in proper proportions	Burned at high temperatures 1400° – 1600° C	Do not slake on addition of water, but must be ground finely for use	Hydraulic	Portland cement
Made from mixtures of slaked lime or portland cement and blast-furnace slag or volcanic ash.	Not burned			Slag or pozzolan cements
Made from gypsum	Burned at from 165° – 200° C		Not hydraulic	Plasters
	Burned at above 200°C			Keene's cement, flooring, plaster etc.

Table 14.2. Analysis of mortar materials.

Materials	SiO_3	Fe_2O_4	Al_2O_3	CaO	MgO	SO_3	CO_2	H_2O
Lime	0.15	0.85	0.85	98.01	0.45	–	0.55	–
Hydrated lime	0.38	0.08	0.06	72.59	0.74	–	2.10	23.11
Hydraulic lime	19.05	0.55	1.60	65.10	0.65	0.30	–	12.45
Plaster of Paris	0.11	0.01	0.03	38.90	0.14	54.81	0.61	4.98
Natural cement	29.92	4.78	11.23	36.50	11.393	–	5.42	–

Selection of Raw Material

Cement contains four essential elements: silicon, aluminium, iron and calcium. Other raw materials whose number is unlimited are used in special cases. In some of the cases no iron is present e.g., white cements, some cements have no aluminium.

In selecting each raw material, the consideration should be: (i) composition; (ii) uniformity; (iii) physical characteristics; (iv) over burden; (v) quantity; (vi) location and transportation; and (vii) units cost.

1. *Composition.* It depends upon the types of cements to be produced which must include the essential elements. It must be possible to proportion these materials so that the essential elements may be present within the desired limits.

2. *Uniformity.* Raw materials are to be uniform as the analysis of the raw material is required every time when it is not uniform.

3. *Physical characteristics.* Some important physical characteristics are:
   (a) Resistance to grinding or grindability (lose chalk or clay).
   (b) Homogenity (has been treated under uniformity).
   (c) Uniform hardness otherwise after grinding the materials segregate and hence the difficulty in clinker formation.
   (d) During blasting in a mine or quarry does it shatter easily or with difficulty.

4. *Over burden.* Cost of quarrying increases.

5. *Quantity.* Continuous supply and expected time of operation.

6. *Location and Transportation.* Location of raw material and of market are important. This helps in deciding the location of the factory, otherwise freight rates, transportation lines, time required for deliveries etc., become a handicap.

7. *Unit cost.* Depends upon the above items.

Raw materials are divided roughly into three categories: (i) primary; (ii) secondary; and (iii) tertiary (Table 14.3).

Primary. Which forms more then 85 per cent of the clinker.

Secondary. Forms less then 15 per cent of the clinker. This is required to modify the percentage of silicon, aluminium or iron.

Tertiary. Are employed for some purpose other than the clinker formation e.g., for reducing water content of slurry, increasing output of grinding equipment or decreasing power, output yields, fluxes for facilitating clinker production.

Table 14.3. Primary, secondary and tertiary raw materials.

Raw material	Primary, secondary or tertiary	Principal element
Limestone	P	Ca
Cement rock	P	Ca, Si, Al, Fe
Marl	P	Ca
Oyster sheel	P	Ca
Calcium carbonate as by-product	P	Ca
Clay	P	Si, Al, Fe
Shale	P	Si, Al, Fe

(Contd ...)

Raw material	Primary, secondary or tertiary	Principal element
Slag	S	Si, Al, Fe
Sand	P	Si
Sandstone	S	Si
Bauxite	S	Al
Diaspore	S	Al
Iron ore, laterite	S	Fe (in iron cement)
Mill scale	S	Fe
Pyrites cinder	S	Fe
Gypsum	T	(Ca) Keen's cement
Slum	T	Al Persian cement
Borax	T	B

Note: P = Primary; S = Secondary, T = Tertiary.

Source of Raw Materials

Raw materials may be quarried, mined or dredged or purchased. It may be a surplus from some industry e.g., screenings from blast furnace, stone or may be a by-product e.g., from alkali (precipitated chalk by Leblanc process) or sulphuric acid industry.

Proportioning of Raw Materials

Perhaps the most conspicuous difference between the procedure employed in the operation of plant producing natural cement and one producing portland cement or other cements where the number of materials are more than one, lies in the great care required in the later to control the composition exactly e.g., a poor limestone or slag may sometimes be enriched by using a purer stone or clay. In making natural cement the native cement rock is calcined just as it comes from the quarry. But for portland cement a small variation in the established ratios of the principal components in the ground rock mixture may be sufficient to change materially the burning characteristics of the mixture or the properties of the cement. For example in the manufacture of cements the temperature available is not sufficiently high to bring all of the materials into liquid state, because the reaction takes place for the production of the final product (between adjacent solids). Therefore, time is required for such reactions to be completed as the duration of time requirement will depend upon several factors such as the chemical composition or nature of the raw materials, the amount of surface provided by the grains and the temperature. Like wise reactions also take place during cooling. Therefore, the rate of cooling will differ with the composition. Before any rock is taken from the quarry the problem that is faced is as to (i) what is the ideal composition of a cement; and (ii) what composition of the raw materials at the disposal can be most advantageously and most successfuly used.

Possible Defects Arising out of Unbalanced Composition

The proper lime content is limited due to the low early strength produced when the lime is too low and sound when it is too high. The old cements were lower in lime than the modern cement, but the strength developed by them was more, though the early strength was lower. In order to increase the strength it was necessary to raise the lime content, or grind finer or both. But higher temperatures are required to burn the high-lime mixtures. There is no advantage in adding the extra lime unless brought into a combination with the other constituents. If specimen crack, curls or expands unduly, it is designated as unsound. When the lime content is raised too high, it becomes impossible to get all of it into combination, regardless of the temperature of burning and the cement is not sound. The high lime mixtures give a cement which is more rapidly hardening but the slightly lower lime mixtures give a cement which continues to become stronger over a longer period of time.

Silica, alumina and ferric oxide are like-wise limited. If the lime content is fixed, and silica becomes too high, which may be accompanied by a decrease in alumina and ferric oxide, the temperature of burning will be raised by the special influence or the high lime is lost. If the lime is too low which means an increase in the alumina and ferric oxide, the cement may become quick-setting and contain larger amount of alumina compounds which appear to be of little value for their cementing qualities. Rapid setting is undesirable because the cement sets in so rapidly that it cannot properly be worked.

The magnesia content is limited, not to exceed 5 per cent because higher magnesia may be dangerous to the soundness of cement specially at the later ages. Pure magnesium oxide, if not heated to too high temperature, possesses distinct hydraulic properties. Light porous magnesia combines with water to form the hydroxide $Mg(OH)_2$, but without setting , while dense magnesia repaired at a full red heat, sets yielding a coherent mass although of low tensile strength, Magnesia does not form solid solution with lime, the two compounds crystallising independently from their fused mixtures The double carbonate, dolomite, if ignited at 300–400°C so as to decompose $MgCO_3$ while leaving $CaCO_3$ unchanged, yields a product which sets with water and becomes vary hard. A mixture of calcium and magnesium oxide prepared by igniting dolomite very strongly does not slake, but if finely powdered; mixed and treated sets rapidly and has been used as substitute for plaster in making cement.

The customary method for expressing the relations is by means of ratios of the several oxides. Numerous formulae have been proposed. Some are based only on empirical results of experience, some on theoretical ideal composition in terms of the probable compounds formed.

The second problem is to determine the combination of raw materials which will most advantageously satisfy the ideal composition range formulae. By knowing the precise nature of raw materials and then by calculation, a correct proportion of the mixture is made. In some cases, the correct proportions cannot be obtained with two types of rock and a third or fourth may be required. Finally, trial runs are made to learn if the physical and chemical characteristics of the materials are correct or not.

Proportioning Formulae

Natural cement

Natural cements are produced by burning, calcining and subsequently grinding clayey or argillaceous limestones, which are natural mixtures of calcium carbonate and clay. These limestones usually carry from 13 to 35 per cent clayey matter ($SiO_2 + Al_2 + Fe_2O_3$), and often a considerable percentage of magnesia, which seems to be interchangeable with lime and to replace the later without disadvantage.

The kilns used for burning natural cement are very similar to those used for burning lime. The best kiln consists of a steel cylinder lined with firebrick and provided with an opening at the bottom through which the burned material may be drawn from the kiln.

The kiln is continuous in operation and the charging is done by introducing at the top alternate layers of limestone and fuel. This latter usually consists of small-size anthracite coal or coke. The temperature of burning is usually between 1000° and 1200°C.

After passing through the kilns the burned material presents the appearance of a soft yellowish-brown mass. It is then ground to a fine powder. As it is quite soft, this is usually done with buhr-stones or in tube mills. However, excellent results can be obtained by grinding the natural cements to an extreme degree of fineness with some of the more modern mills. The grinding of the clinker should be such that 90 per cent will pass through a 100-mesh screen.

Natural cement is not now used for concrete as its strength is far lower and is hardening much slower than portland cement. It is used to some extent for laying brick and stone and as a component of cement used for this purpose. Generally some hydrated lime is added to such a mixture in order to give plasticity, and if some portland cement is also added the mixture will give a stronger mortar. It is also desirable to add some (1 or 2 per cent) waterproofing agent such as stearic acid or calcium stearate.

Portland cement – composition

Investigations on the composition of portland cement indicate that this is a mixture of tri-calcium silicate ($3CaO.SiO_2$), tri-calcium aluminate ($3CaO.Al_2O_3$), and di-calcium silicate ($2CaO.SiO_2$). The most important of these compounds is the tri-calcium silicate and the more of this which is present the better, although the other compounds also have hydraulic value. Most portland cement has its silica about equally divided between the tri-calcium silicate and the di-calcium silicate. All the alumina is present as tri-calcium aluminate and practically all the lime is combined with the silica, alumina and iron. Cement which is sound contains only a trace of free lime. When cement is burned, the CO_2 is first driven off the limestone. The first compounds formed are the $5CaO.3Al_2O_3$ and the $2CaO.SiO_2$. These then gradually unite with the remainder of the CaO to form $3CaO.Al_2O_3$ and $3CaO.SiO_2$, some $2CaO.SiO_2$ remaining in excess. The iron oxide present in ordinary cement acts as a flux and promotes this change.

There is a definite relation between the possible composition of portland cement and the condition under which it is made. By finely grinding the raw materials, burning slowly and having the composition sufficiently high in lime, cements with a high percentage of tri-calcium silicate can be made. These, when finely ground, will be quick hardening and show great strength. The so-called "high-early-strength" cements are obtained by this means. Knowledge gained from the use of the phase rule and phase rule diagrams has done much to produce a better portland cement and one of uniform quality.

The smallness of this field of permissible composition is a challenge to the cement industry, for it calls for very close control of composition and operating conditions to obtain the desired constituent. In recent years the industry has been very successful in meeting this challenge, as evidenced by greatly increased quality and reliability of cement. Iron oxide is also present in cement, so the phase diagram should really be for four components. But in the quantity in which iron oxide occurs in the raw materials it does not greatly affect the size or position of the fields of composition so the three-component diagram is valid for the control of cement manufacture. The chemical composition of portland cement of good quality is usually within the following limits.

Range of composition of Portland cement

Materials	Limits	Average
Silica	20–24%	22.0%
Iron oxide	2–4	2.5
Alumina	5–9	7.5
Lime	60–64.5	62.0
Magnesia	1–5	2.5
Sulphur trioxide	1–2	1.75

Analysis of typical portland cements are reported in Table 14.4.

Table 14.4. Analysis of Portland cements.

Made from	SiO_2	Fe_2O_3	Al_2O_3	CaO	MgO	SO_3	Loss
Cement rock and limestone	21.82	2.51	8.03	62.19	2.71	1.02	1.05
	21.94	2.37	6.87	60.25	2.78	1.38	3.55
Marl and clay	22.71	3.54	6.71	62.18	1.12	1.21	1.58
	21.86	2.45	5.91	63.09	1.16	1.59	2.98
Limestone and clay	21.31	2.81	6.54	63.01	2.71	1.42	2.01
	23.12	2.49	6.18	63.47	0.88	1.34	1.81
	23.56	0.30	5.68	64.12	1.54	1.50	2.92
Blast furnace slag and limestone	22.41	2.51	8.12	62.01	1.68	1.40	1.02
Iron ore and limestone	20.5	11.0	1.5	63.5	1.5	1.0	–

As indicated above, the essential elements in cement are lime, silica and alumina. The magnesia comes from the limestone, some of this being present in all limestone. The amount is limited by the standard specifications to 5 per cent, as it is supposed to be harmful. Iron oxide is present in nearly all clays and shales, and hence is always present in cement. It has a definite advantage, in that it assists in burning and lowers the temperature of fusion. Cement containing no iron is white, but rather hard to burn. The sulphur trioxide for the most part comes from the gypsum which is added to correct the set.

The proportion of a good cement should satisfy the following ratios:

$$\frac{\text{Per cent lime}}{\text{Per cent silica } + \text{ per cent iron oxide } + \text{ per cent alumina}} = 1.9 \text{ to } 2.15$$

$$\frac{\text{Per cent silica}}{\text{Per cent alumina}} = 2.5 \text{ to } 4$$

In the manufacture of portland cement great care is taken to see that the composition satisfies the above ratios. If too much lime is present the cement will be "unsound"–that is, in time concrete made from it will expand and crack. If too little lime is present the concrete will be low in strength and may "set" quickly–that is, harden before the masons have a chance to place it in the forms. Cement in which alumina is high is also apt to be quick setting, and is hard to burn uniformly. High silica cements are usually very slow hardening, and do not attain their full strength for a considerable period.

METHODS OF MANUFACTURE

Raw Materials

The materials from which portland cement is manufactured may be divided into two classes: those which supply the lime and those which supply the silica, iron oxide and alumina. The first are termed calcareous and the second argillaceous. The following groups show the principal materials used in the manufacture of Portland cement.

Calcareous Materials
Limestone
Marl
Chalk
Alkali waste

Argillaceous Materials
Clay
Shale
Slate
Blast furnace slag

The cement rock is an argillacous limestone which contains usually between 65 and 80 per cent calcium carbonate. If it contains more than 75 per cent it is necessary to add clay, shale or slate in order to make a satisfactory mixture for burning. If it contains less than 75 per cent it will be necessary to add limestone for a similar purpose. Limestone is usually mixed with clay or shale; marls and chalks with clay. Blast furnace slag is used with limestone. Alkali waste (or precipitated $CaCO_3$, obtained from the manufacture of caustic soda) is mixed with clay.

Limestones, marls and chalks which are to be used in the manufacture of portland cement should contain less than 3 1/3 per cent magnesia and preferably not more then 3 or 4 per cent silica, iron oxide and alumina combined. Clay, shales and slates should all have a least 2 1/2 and not more then 4 times as much silica as alumina. Exceptions to this are in the case of a high silica limestone, with which a high alumina clay may be used to advantage, since all that is necessary is that the mixture shall satisfy the requirements expressed by the above formulas for the desirable composition.

In rough outline the manufacture of cement consists of mixing the calcareous and argillaceous materials together intimately and heating them to the point of incipient fusion. The intimate mixing of the two materials is accomplished by finely grinding them together. The powder is then subjected to a temperature of from 1400° C–1600° C., when a sintering or semi-fusion takes place and the mixture rolls up into little balls varying in size from that of a walnut down to that of wheat, with an occasional larger piece and some fine sand. After cooling, these lumps or "clinkers" are mixed with a small amount (2–3 per cent) of gypsum and finely pulverised. The resulting powder is portland cement.

Manufacturing Process

The two processes employed for the manufacture of cement are known respectively as the wet process and the dry process. The wet process is the older of the two and is used almost universally. The dry process originated in America and is employed to the greater extent here. The two processes differ only in the treatment of the raw materials and very much the same equipment is used in each. The treatment of the burned clinker is the same in both cases (Fig. 14.1).

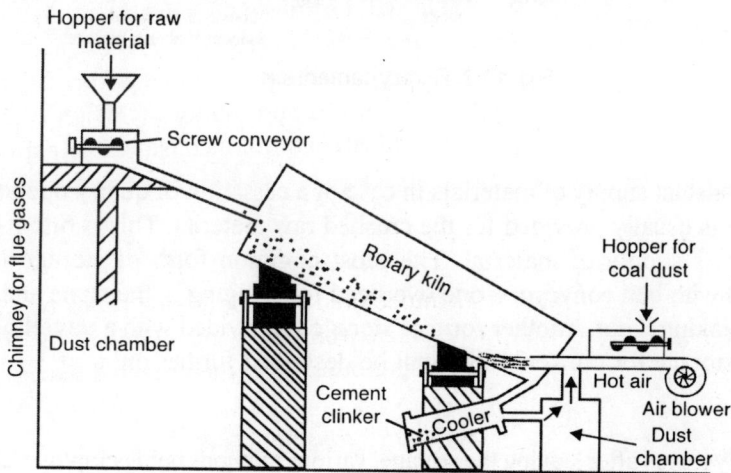

Fig. 14.1. Flow diagram for manufacture of portland cement.

The wet process is always used for marl and clay and the dry process for cement rock and for blast-furnace slag. Both processes are used for limestone and clay or shale. Where applicable, the dry process is the more economical but it is easier to control the composition of the cement by the wet process. This later is also better where the materials cannot easily be dried.

Where the wet process is employed for limestone and shale, the two materials are crushed and stored without drying just as in the dry process. They are then mixed in proper proportions and fed to the grinding machinery, at which point water is added and the materials ground wet. The result is a thin mud or "slurry," as it is called, which is made just fluid enough to flow easily. This slurry, containing from 35 to 40 per cent water, is fed directly into the kilns and burned. In the process, the limestone, cement rock, and shale are usually crushed to about two inches or smaller and then dried. The crushing is nearly always done in large jaw or gyratory crushers which are followed by hammer mills. The dried materials are stored in separate bins or piles and are drawn out of these as desired and mixed in proper proportions by automatic scales. The mixture is then ground and burned. Sometimes the storage and mixing precede the drying. Drying is done by means of rotary driers. These are cylinders of sheet steel from 6 to 8 feet in diameter and from 60 to 100 feet in length. They are unlined, and are usually provided with channel irons bolted to the inside to act as shelves, to carry the rock up and drop it through the hot gases. The driers are heated by a coal fire at the lower end or else by the waste gases from the rotary kilns. They are similar in construction to the rotary kilns described farther on except that they are smaller and are not lined with firebrick (Fig. 14.2).

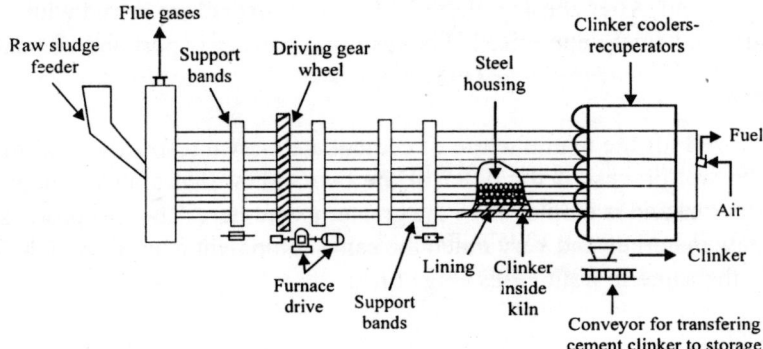

Fig. 14.2. Rotary cement kiln.

Storage

In order to provide a constant supply of materials in case of a cessation of quarry operations, due to bad weather, etc., a storage is usually provided for the crushed raw material. This is often sufficiently large to hold a week or two's supply of material. The most common form of storage is large covered concrete bins provided with belt conveyors–one overhead for bringing in the stone and other in tunnels underneath the bin for taking it out. Another form of storage is provided with a travelling crane and grab bucket, similar to the storage for clinker which will be described further on.

Mixing

The materials are usually mixed after leaving the storage, various methods being employed for proportioning the two different kinds. Hopper scales are used in many of the older mills, while in most of the never ones limestone and shale are proportioned just as they are fed to the grinding mills by means of some type of adjustable feeder such as a "poidometer" or a rotating table feeder. Clay is often first worked up with water and the thin slip formed is proportioned by volume to the limestone just before the latter goes to the ball mill or compeb mill. A revolving wheel with wickets, known as a Ferris wheel, is usually employed for this purpose.

Where the wet process is employed for limestone and shale or clay, no drier is employed and sufficient water to make the ground material flow is added just before the mill.

Grinding

The grinding of the raw materials may be done in one or in two stages. The combinations now most often employed are (i) ball mill and tube mill; (ii) kominuter and tube mill; (iii) hercules mill and tube mill; (iv) griffin mill and tube mill.

In place of the combination of ball and tube mill, a single mill combining the elements of these two mills is now most generally used in the newer cement plants. These combination mills go under different names according to the maker such as "Compeb Mill," "Combination Mill," etc. These mills are somewhat similar to a tube mill except that they are made longer and there are usually three or more compartments, separated by perforated or grid partitions.

The degree of fineness to which the raw material is to be ground depends entirely upon conditions. It is stated as a general rule that it should be sufficiently fine so 90 per cent will pass through the 100-mesh sieve, and in most cases 95 to 98 per cent is required to produce a sound cement. The finer the grinding the more perfect the combination between the silica, the alumina, the iron and the lime during the burning operation. If the raw materials are not finely enough ground, the cement will be unsound—that is, some of the lime will not combine. This yields a cement which disintegrates rapidly.

Flotation

The fine grinding of the cement materials affords the possibility of the use of a unique method for adjusting the composition of the mix. As has been indicated above, the various oxides must be proportioned within fairly narrow limits to give a satisfactory product. This means that, many deposits of natural materials cannot be used because they cannot be proportioned in any way to give the desired chemical composition. Recently the operation of flotation has been used with great success to separate the finely ground mineral constituents from each other. It is now being used to correct mixtures and to make many raw materials usable which were formerly considered unusable. By removing impurities from cheap materials they can be used to replace the more costly ones. Proper control of the flotation operation produces a better mix before burning. This has been one of the principal developments in the cement industry in recent years.

Storage of ground material

In the wet process, the ground materials containing from 33 to 40 per cent water, are stored in tanks or basins which are agitated with either mechanical or compressed air agitators. The materials as ground is usually passed into one set of vats called correction basins from which samples of the slurry are drawn and analysed. When a basin is full, if the composition is not correct, it is adjusted either by stirring in more clay or by mixing the contents of two or more basins. The slurry whose composition has been satisfactorily adjusted is then passed on to a second set of basins known as "kiln feed basins". These are also provided with agitators.

In the dry process, it was quite usual at the older plants to send the material directly from the grinding mills to the kilns. At the newer plants, however, the ground material is sent into large storage tanks where its composition is checked and if found unsatisfactory is adjusted by blending the contents of two or more tanks, etc. When this is done the dry process will give fully as uniform cement as the wet.

Conveying materials

The material is usually carried from one stage of manufacture to another by various types of conveyors. The product of the crushers is conveyed to the granulating mills on belt conveyors, and the product of these latter mills and the tube mills is transported by screw-conveyors. The elevating is done by means of bucket elevators of the link-belt type. Finely ground raw material and cement may also be conveyed by means of the Fuller Kinyon system through a pipe line. Slurry and marl are pumped through pipe lines by means of either plunger or centrifugal pumps, or by a compressed air system.

Burning

In the early days of the American Portland cement industry, the burning was done in intermittent upright kilns, similar to those used for burning lime. These were soon improved, by making them continuous in action in order to economize on fuel. This allowed the charge to receive the waste heat from the clinkering of the cement, and the air for combustion to be preheated, by passing through the fully-burned materials.

The rotary kiln

The rotary kiln, in its usual form consists of a cylinder, from 6 to 12 ft. in diameter and from 60 to 350 ft. long, made of sheet steel and lined with firebrick. The burning of cement is essentially an application of the unit operation of heat transfer, by radiation, and therefore it is necessary to have equipment and conditions which will bring about a large amount of heat transfer from the flame to the solid material.

The steel sheets of the kiln are held together by single-strap butt joints. This long cylinder is supported at a very slight pitch (1/2 to 3/4 in. to the foot) from the horizontal, on two or more tyres made of rolled steel, which in turn revolve one heavy friction rollers. The kiln is driven at a speed of from one-half to one revoluation per minute by a girth-gear situated near its middle, and a train of reducing gears. The power is supplied by either a line shaft or a motor. The upper end of the kiln projects into a brick flue, which is surmounted by a steel stack, also lined with firebrick for its entire height. The flue is provide with a door at the bottom, which serves not only to allow the flue to be cleared of the dust which accumulates in it, but also a damper to control the draft of the kiln.

The material to be burned is usually fed into the kiln through a horizontal water-jacketed screw-conveyor, or else spouted into it through an inclined cast-iron pipe. The raw material feeding device is usually attached to the driving gear of the kiln, so that when the kiln stops the feed also stops.

The lower end of the kiln is closed by a firebrick hood. This is usually mounted on rollers, so it can be moved away form the kiln when the latter has to be relined. The hood is provided with two openings; one for the entrance and support of the fuel-burning apparatus, and the other for observing the operation, temperature, etc., of the kiln, and through which bars may be inserted to break up the rings of material which form, and to patch and repair the lining. The lower part of the hood is left partly open. Through this opening the clinker falls out and most of the air for combustion enters.

Reaction in the kiln

1. 1/4 of the kiln is drying zone and water is driven out.
2. In the next zone, i.e., in the calcination zone organic matters burn away and $CaCO_3$ breaks to CaO and CO_2. Temperature is about 700°C. When reaction is complete and material has travelled 3/4 of the of the kiln decarbonation is complete. After decarbonation the material goes to the hottest zone. The temperature is 1,400–1,450°C. Here 20–30% material is converted to lime and

the material reacts with the silica and sintering takes place in the last zone and clinkers are formed. The clinkers fall out from the outlet of the kiln to clinker cooler.

All these reactions take place in the kiln:
1. Up to 100°C evaporation of water take place.
2. At 500°C the evolution of combined water from clay starts. The reaction is endothermic.
3. Between 900–1,200°C, the main reaction takes place between clay and lime.
4. Between 1,200–1,250°C there is liquid formation and the reaction is endothermic.
5. At 1,200°C and beyond up to 1,450°C further formation of liquid and then sintering takes place. The reaction is endothermic. The clinker is formed.

Heating the kiln

The kiln is heated by a jet of burning fuel, usually powdered coal, but sometime natural gas or fuel oil are used. The necessary temperature of the hottest part of the kiln is about 1400°C, and is rarely less then 1300°C. To maintain this temperature, about 80 lbs. to 160 lbs. of fuel are required per barrel of cement, the actual amount depending on the coal itself, the material to be burned and the dimensions of the kiln. The longer the kiln, the greater the fuel economy. Dry materials require much less coal than slurry. With limestone and shale mixture, in a kiln 100 ft. long by 7 ft. in diameter, the coal consumption will amount to about 90 lbs. of good gas slack per barrel. A kiln 60 ft. long by 6 ft. in diameter will, on the other hand, require about 110 lbs. of coal per barrel. Wet materials require about 30 per cent more fuel.

Grinding of coal

When coal is used for burning, this is pulverised in mills similar to those used for grinding the raw materials. It is, however, first crushed by passing through rolls or roll-jaw crushers, and then dried in rotary driers of special type. The mills most used for coal pulverising are the Fuller mill and the Raymond mill. The coal should be pulverised so that 90 per cent of it will pass a sieve having 100 meshes to the linear inch, and should contain from 30 to 45 per cent volatile matter.

Thermal efficiency

Of the heat supplied to the kiln by the burning of the coal, by far the larger proportion is wasted. About 50 to 75 per cent of it is carried off by the waste gases of the stack, and form 10 to 15 per cent by the hot clinker falling from the lower end of the kiln. The gases enter the stack at from 600°–800°C, and the clinker leaves the kiln at not much under 1200°C. If the kiln could be made to show the same economy as is common in good boiler practice, a barrel of cement could be burned with 25 lbs. of coal.

The gases leave the dry process kiln at about 800°C. In many plants the waste gases are led through waste heat boilers located at the end of the kilns. By so doing about 4 to 5 lbs. of steam are generated per pound of coal burned. The flow of gases through the boilers at a high velocity is one of the requisites for successful employment of the waste gases for steam generation and this is obtained by means of an inducted draft fan. By employing modern turning engines, directly connected to electric generators, enough power may be obtained from the waste gases to operate the entire plant. The gases are sometimes purified and the carbon dioxide reclaimed.

Dust losses

Normally from 3 to 5 per cent of the raw material is carried away in the exit gases of the kiln as dust. Various schemes have been tried with a view to eliminating the dust, such as settling chambers, water sprays and electrical precipitation. The later, Cottrell precipitator, is the only method, however, which is used to any extent. Several installations of this system are now in operation. The dust collected by the latter is found to contain considerable potash for about half the potash in the raw materials is volatilised in the kiln. Some of this potash is water soluble and may be recovered and used for fertilizer.

Forming the clinker

The raw material as it enters the kiln contains about 33 per cent carbon dioxide. For the first 30 ft. of its journey through a 100-ft. kiln, it is merely heated up, and whatever water it contains is driven off. In the next 40 ft. it loses all its carbon dioxide and sticks together, forming small, soft, lemon-yellow balls, which, as they reach the hottest part of the kiln–the last 30 ft.—practically vitrify, become rough and hard.

Properly burned portland cement clinker is greenish-black in colour, of vitreous lustre, and usually, when just cooled, sparkles with small bright glistening specks. It forms in lumps from the size of a walnut to hardly more than dust, with here and there a larger lump. Under-burned clinker is more or less soft, is irregular in shape, and not as black as the well-burned material. It usually show soft-brown centres. Hard brown centres are due to very hard burning.

The cement is brought into the stock house by an overhead conveyor or the Fuller-Kinyon system pipe line and dropped into any desired bin. A screw-conveyor also runs under the floor of the stock house. The bins are provided with gates, and when it is desired to pack from any bin, these gates are opened and the cement is allowed to run into the screw-conveyor. The screw-conveyors then carry it to the packing machines.

Cooling of clinkers

The clinkers through clinker conveyor falls inside the cooler. In the cooler, atmospheric air passes over hot clinker, cooling the clinkers and itself getting heated. The common type of coolers are rotary coolers. The quality of cement largely depends on the rate of cooling.

1. If the rate of cooling is slow no glass is formed but there will be dusty clinkers. Dusting means conversion of dicalcium silicate to powder form and not crystalline form. This is caused by the conversion of B form of SiO_2 which has no binding property and hydrates very slowly.

2. If the rate of cooling is high then the alumina and iron solidifies into glass and there is any formation of crystalline component of alumina and iron. This non-formation of any crystalline component is adverse to the process.

3. If the rate of cooling is such that the melt liquid in the clinker crystallises—there is formation of a large quantity of $3CaOSiO_2$ which gives higher strength to cement.

4. If the rate of cooling is so high that the melted liquid in the clinker turns into glass and if there is magnesia also in the clinker—that will be converted to glass and will not crystallise out as periclase (MgO). This will give beneficial effect to the properties of cement as the crystalline periclase MgO hydrates very slowly—when it is exposed to water and this causes large expansion. If the cement after several years, when magnesia is in the form of glass, it does not expand to a high degree when exposed to water. So, by cooling rapidly the clinker containing magnesia— there will not be excessive expansion of cement after several years.

It is very necessary that the cooling of clinker should be controlled to produce a definite degree of crystallisation of the melted clinker.

Mixing of additives

1. Retarder : Addition of retarder is necessary to prevent quick settling of cement plaster. Such a retarder is gypsum ($CaSO_4.2H_2O$) or plaster of Paris, $CaSO_4. 1/2H_2O$. The quantity required is 2.5–3.0%.
2. Dispersing agent : Small quantity of dispersing agent, e.g., sodium salt or naphthalene or sulphuric acid is added to the cement. These dispersing agents prevent formation of lumps and cakes in the cement. The power cost of cement manufacture is reduced by 30–35%.
3. Additives which cause air entrainment improves durability of concrete, particularly against the alternate freezing and thawing. For this vinsyl resins or Darex may be added. These agents have the property of imparting air into the cement paste.
4. Water porosity or dispersing agents are also added to cement to make then resistant to water porosity.

Grading of clinkers

The clinkers are pulverised to fine grains in tube mill. During grinding, additives are added. The grind powder is packed into bags through automatic packaging machine.

Fineness of cement

The portland cement is so fine that 98–99.0% passes through 200 mesh and 9% through 325 mesh screen.

Constituents of clinkers

(i) tricalcium silicate—$3CaO\ SiO_2$; (ii) dicalcium silicate—$2\ CaO\ SiO_2$; (iii) tricalcium aluminate—$3CaO\ Al_2O_3$; (iv) a solid solution of tetracalcium aluminate—$4CaO\ Al_2O_3\ Fe_2O_3$; (v) MgO; (vi) free CaO.

Theories about settling of cement

There are different theories about setting of cement and these are :

1. *Le Chatelier theory:* The hardening of cement is due to the interlocking of crystal formation during hydration.
2. *Colloidal theory of Michaelis:* The hydration product of silicates form rigid cells and help to the hardening of cement. The theory is that at ordinary temperature the hydrated calcium silicate appears to be non-crystalline, and physical properties of settled cement explains the basis of this theory.

Hydration of cement

By addition of water two types of reaction take place :

Hydrolysis : Water to calcium silicate—it decomposes to silicate of lower base and release CaO which forms $Ca(OH)_2$:

Tricalcium silicate

$$3CaO\ SiO_2 + H_2O \xrightarrow{\text{Hydrolyses}} 2CaO\ SiO_2\ (aq) + Ca(OH)_2 + SiO_2$$

Hydration of dicalcium silicate

$$2CaO\ SiO_2 + H_2O \longrightarrow CaO\ 2SiO_2 + Ca\ (OH)_2$$

Silica and alumina take up water of hydration:

$$2CaO\ SiO_2 \xrightarrow{4H_2O} 2CaO\ SiO_2\ .4H_2O$$

$$3CaO\ Al_2O_3 \xrightarrow{6H_2O} 3CaO\ Al_2O_3\ 6H_2O$$

At least three different hydrated tricalcium aluminates exist:

$3CaO\ Al_2O_3\ 6H_2O$:   —

$3CaO\ Al_2O_3\ 12H_2O$:   hexagonal

$3CaO\ Al_2O_3\ 8H_2O$:   ortho

Hydrate tricalcium aluminate combines with gypsum to form calcium sulphoaluminate :

$$3CaO\ Al_2O_3(aq) + CaSO_4\ 2H_2O \longrightarrow 2CaO\ Al_2O_3\ 3CaSO_4.3H_2O$$

$$3CaO\ Al_2O_3 + Ca(OH)_2 + 12.5\ H_2O \longrightarrow 4CaO\ Al_2O_3.nH_2O$$

Hydration of dicalcium aluminate:

$$2CaO + Al_2O_3 \longrightarrow 2CaO\ Al_2O_3.12H_2O$$

Also $$4CaO\ Al_2O_3\ H_2O \longrightarrow 4CaO\ Al_2O_3.13H_2O$$

Hydration also takes care of pure Al_2O_3 formation

$$4CaO\ Al_2O_3\ Fe_2O_3 \longrightarrow 3CaO\ Al_2O_3 + CaO + Fe_2O_3$$

On hydration there is evolution of heat and there is temperature rise. The evolution of heat occurs in first seven days of hardening. The following compounds are responsible for evaluation of heat in the decreasing order:

(i) tricalcium aluminate. (ii) tricalcium silicate; (iii) tetracalcium alumina ferric; (iv) dicalcium silicate. The heat evolved is used in setting cement. The function of different constituents are :

1. Tricalcium aluminate   :   initial setting
2. Tricalcium alumina silicate   :   for further setting start after seven days.

Cement specification

BSS

1. $\dfrac{CaO}{2.8\ SiO_2 + 1.2\ Al_2O_3 + 0.65\ Fe_2O_3}$ should be greater than 1.02 and less than 0.66.

2. $\dfrac{Al_2O_3}{Fe_2O_3}$ shall no be less than 0.66

3. Wt. of insolubles should not exceed 1%.
4. Magnesia not to exceed 4%.
5. Total sulphur or SO_3 shall not exceed 2.75%.
6. Total loss in ignition—not to exceed 4%.

Other requirement

1. Fineness BS—170 mesh not to exceed 1%.
2. Setting time—Initial 30 mins. Final 12 hrs.
3. Tensile strength—not less than the 300 lb/sq inch after three days, should increase to 375 lb/sq inch after seven days.

ISS specifications

1. $$\frac{CaO}{2.8\ SiO_2 + 1.2\ Al_2O_3 + 0.65\ Fe_2O_3}$$ should not be greater than 1.02 and less then 0.66.

2. $\dfrac{Al_2O_3}{Fe_2O_3}$ not less than 0.66.

3. Insolubles should not exceed 1.5%.
4. Magnesia—not to exceed 5%.
5. Total sulphur or SO_3 not to exceed 2.75%.
6. Total loss on ignition not to exceed 4%.

Other properties

1. Fineness—BS- 170 mesh not to exceed 1%.
2. Setting time—Initial —mins. and Final—10 hrs.
3. Tensile strength not less than 300 lb/sq. inch after 72 hrs. and should be 325 lb/sq. inch after 7 days.
4. Heat of hydration 7 days—65 cals., after 25 days—75 cals.

Analysis of Indian portland cement

(i) SiO_2—23%; (ii) FeO—2.19%; (iii) Al_2O_3—5.93%; (iv) CaO—64.86%; (v) MgO—1%; (vi) SO_3—1.72%; (viii) Loss 0.74%

   Mortars use in building works
   (i) cement/sand/aggregates (size 3 1/2")

$$1 : 1.5 : 3$$

(ii) concrete mixtures type :   1 : 2 : 4

$$1 : 3 : 6$$

TYPES OF PORTLAND CEMENT

By varying the percentage of constituents' changes in the rate of setting, heat evolution and strength characteristics can be varied. The main types of portland cements are :

Type I: General purpose and regular—40–60% C_3S, 10–30% Ca_2S, 7–13% Ca_3 A hardens to full strength in 28 days.

Type II: Modified—Higher C_2S/C_3S to resist sulphate attack.

Type III: High early strength—Attains strength of I in only 3 days, high heat rates—useless in massive structures, higher C_3S and C_3A percentage with finer grinding increases hydration rate.

Type IV: Low heat—Designed for massive structure work, low C_3S and C_3A which are largest contributors to heat of hydration.

Type V: Sulphate resistant—Good for sea water content; C_3A—4%.

Other Types of Cement

High alumina: manufactured by fusing limestone and bauxite, rapid rate of strength development to high values but with high heat liberation, superior resistance to sea and sulphate water.

Pozzuolene: Mixture of volcanic ash, burnt clay, or shale in 2–4 parts with hydrated lime. Mixed with portland cement as cheap extender.

Hydraulic lime: Use for brick mortar composition, low price and strength.

LIME

Lime is nearly pure calcium oxide, CaO; or a mixture of calcium and magnesium oxides, CaO+MgO; sometimes called quicklime. High calcium limes are stronger than those containing considerable percentages of magnesia. They are also better suited for mortar work, as they slake more readily. Magnesium limes, on the other hand, are better for plaster finishing because they work more smoothly under the trowel. Pure lime whether magnesium or not, is snow white. However, a very small percentage of certain impurities such as iron or manganese may give the lime a gray or yellow colour. Through certain methods of burning the ash of the fuel may be introduced into the lime, causing discoloration.

Lime is made by burning limestone in suitable furnaces at a temperature sufficient to drive off all of its carbon dioxide, the reaction being:

$$CaCO_3 + 21,900 \text{ cals.} \rightleftarrows CaO + CO_2$$

Theoretically, 806 gram-calories per gram of calcium oxide and 733 gram calories per gram of magnesium oxide are required to produce this change. At atmospheric pressure the temperature at which calcium carbonate decomposes is stated as 898°C, while magnesium carbonate decomposes at 575°C (Fig. 14.3).

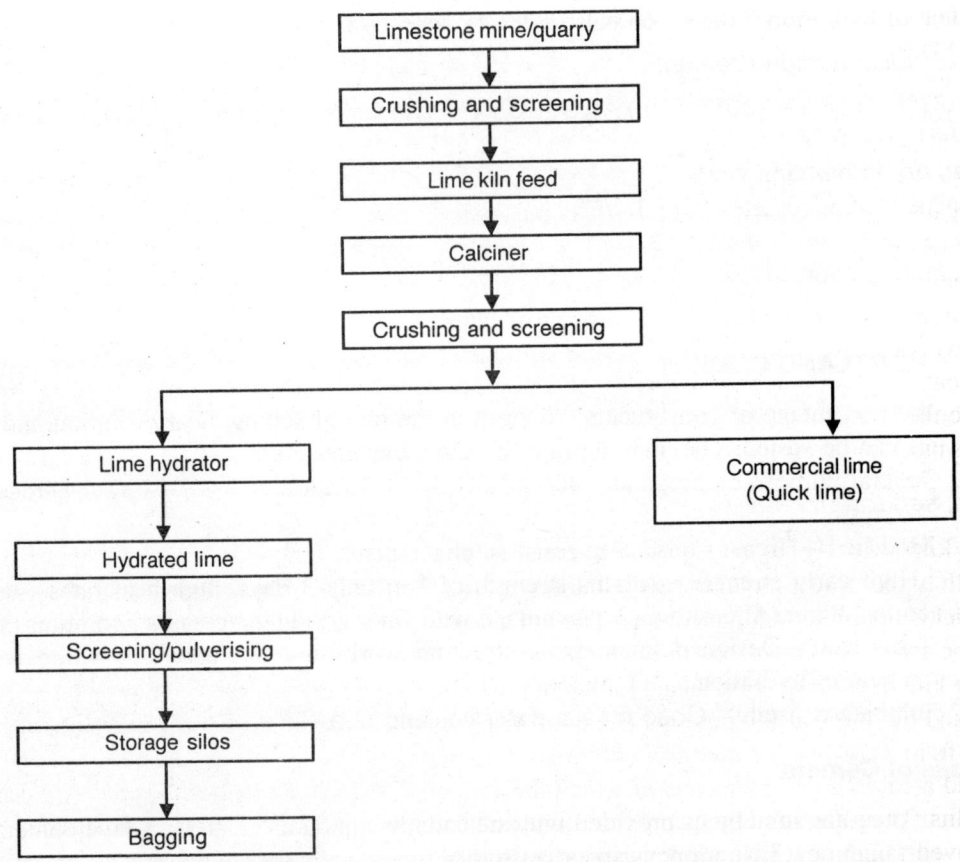

Fig. 14.3. Flow sheet of lime manufacturing operations.

Kilns

Lime kilns are ordinarily operated at from 900° to 1100°C. In the hottest or burning zone. If a temperature much above 1200° C. is employed, the lime will be partially vitrified on the outside of the lumps, due to combination of the CaO with impurities SiO_2 and Al_2O_3, always present in small quantities in even the purest limestone. This causes the lime to be very slow in slaking, which is undesirable, as some of it may escape hydration in the mortar box and later will expand, or "blow" or "pop" in the wall. This manifests itself in small blisters in the finished plaster work.

Intermittent Kilns

The type of kilns ordinarily employed in burning lime may be divided into two classes—Intermittent and continuous. The intermittent kilns are primitive and uneconomical, though they are frequently used by farmers and other small producers of lime. There is a great waste of heat and time in such a kiln, owing to the fact that it must be cooled and reheated each time it is charged. They are seldom, if ever, used in large scale commercial operations.

Continuous kilns

Three different types of continuous kilns are employed: these are :
1. The vertical kiln with mixed feed, in which the limestone and fuel are charged in alternate layers.
2. The vertical kiln with separate feed, in which the limestone and fuel are not brought in contact.
3. The rotary kiln.

Vertical kiln, mixed feed

Vertical kilns with mixed feed are similar to the intermittent ones, except that they are provided with an arrangement whereby the lime may be drawn at regular intervals from below. They are built on the side of a hill, usually of limestone blocks, and are sometimes lined with firebrick. In charging them, first a layer of anthracite coal or coke and then a layer of limestone is fed into the top. Fire is started at the bottom and works its way up. The process of charging and drawing the lime is continuous. These kilns are economical and, for the same size kiln, yield a larger quantity of product than do the vertical kilns with separate feed. On the other hand, the lime is contaminated by ash of the fuel, and the lime burned in these kilns must be carefully sorted in order to discard those lumps to which the fuel ash has adhered.

Vertical kiln, separated feed

The vertical kiln with separate feed usually consists of a steel cylinder lined with firebrick. This is equipped with two or four fireplaces for the burning of the fuel, which are built into the sides of the kiln, so that the fuel is not mixed with the stone. The hot gases of combustion pass from the fire-box into the kiln, while the ash of the fuel drops through the grate bars into an ash-pit below, and does not mix with the lime. The kilns are often constructed with a hopper-shaped cooling chamber, set below the firebox, which is closed by doors at the bottom. The cooling chamber holds about one draw of lime. They are from 6 to 10 ft. in cross-section, and from 40 to 50 ft. in height. They are usually charged by employing an incline and a cable hoist, by means of which the cars of limestone are dawn from the quarry to the top of the kilns. They are sometimes provided with steel stacks in order to induce a better draft.

An improved system of draft employs an exhauster. Where this is done, it is of course necessary to close the top of the kiln and to charge the limestone into the latter through a door or a charging bell somewhat similar to that of a blast furnace. When the kiln gases are drawn off for their carbon dioxide (as in the Solvay process for soda) this bell type of seal is generally employed. A properly installed induced draft will often increase the capacity of a kiln 50 per cent.

·Wood, oil, and coal are employed for burning the wood. Wood is the best fuel, as it burns with a long flame of comparatively low temperature. This is an advantage, as it is essential that the heat should be dispersed at a considerable distance throughout the kiln without having excessive temperatures at the mouth of the fire-box. The steam, which the wood introduces, also seems to be beneficial, and indeed some manufacturers prefer to use green wood because of the greater quantity of steam which it introduces. Wood-burned lime is whiter than that burned with coal.

Where coal is employed as a fuel, it is customarily used wet. A steam jet is also often employed, being inserted below the fire-box. The steam passing through the hot bed of coal is decomposed into hydrogen and carbon monoxide as follows:

$$H_2O + C = H_2 + CO$$

These gases are burned in the kiln itself, and hence carry the heating zone further up the shaft.

With hand-fired kilns the diameter cannot be increased beyond the limit to which the flames from the fire-box can reach effectively or the limestone in the centre of the kiln will not be burned. The limiting diameter for such a kiln is about 6 to 7 ft. With gas, on the other hand, the kiln may be made much larger because the gas may be made to burn in all parts of the kiln. The use of gas also saves the labour of stoking the grates. On the other hand, more skill is required to burn lime with gas than with any other fuel and considerable experimenting is usually required before satisfactory results are obtained. Natural gas is used and producer gas is now being used for kilns of large capacity. Usually the producer gas enters through openings around the side of the kiln and the air through holes in the bottom of the kiln. The air is thus preheated by passing up through the hot lime, which it cools. It then comes in contact with the producer gas in the centre of the kiln, where combustion takes place, spreading back to the gas openings. Gas-fired lime kilns are now built having capacities of from 40 to 60 tonnes per day.

Oil also makes an excellent fuel. When it is used in the kiln described, the burner is placed in the door openings of the fire-boxes and these are bricked in with firebrick leaving openings for the burner and for observing the lime.

The shaft kiln (Fig. 14.4) described above when heated with wood, coal, or oil will produce from 8 to 20 tonnes of lime per day, depending on the kiln and the limestone burned and whether natural or induced draft is employed. When fired with gas the capacity will range from 17 to 60 tonnes per day.

Rotary kilns

Lime is also burned in rotary kilns similar to those described in the section on portland cement. The limestone is first crushed to pieces ranging in size for 2 1/2 ins. down to dust and fed into the kiln, which is heated by producer gas, oil or powdered coal. These kilns are peculiarly adapted to burning highly crystalline stone, which would crumble when subjected to heat and so stop the draft of the vertical kiln, and to supplying lime for chemical and metallurgical purposes.

There was at one time considerable objection to the use of rotary kiln lime, due to small particle size produced. The mason, being accustomed to lime in large lumps, assumed the rotary kiln lime was air-slaked. Now, however, this prejudice has been overcome and much lime is sold in the powdered or granular condition. Powdered lime, made by grinding lump or granular lime which results by decrepitation when certain crystalline limestones are burned, slakes much quicker and is less likely to pit than are most lump limes. Pebble lime is a rotary kiln lime made by burning carefully sized stone usually ranging between 1 to 2 1/2 ins. Better results are obtained as regards the uniformity of the lime if material finer than 1/4 in. is screened form the stone before burning.

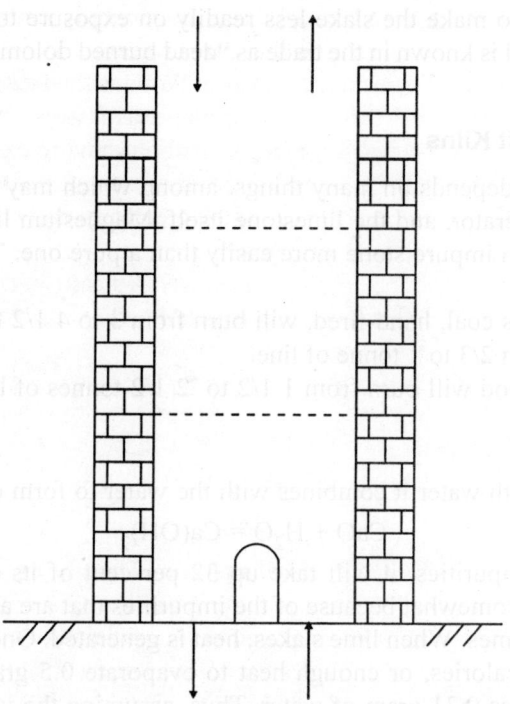

Fig. 14.4. Schematic diagram of a mixed feed shaft kiln.

The rotary kiln is more economical of labour than is the shaft kiln. This applies to both the quarry, where stone for the latter must be carefully sized and all small stones discarded, and to the attention required by the kiln itself. It requires more fuel then the best type of shaft kiln, but the heat in the stack gases may be recovered by installing waste-heat boilers after the kilns, passing the waste gases through the boilers and utilising the steam for power, etc. Rotary kilns are now being quite generally employed for burning building lime where a large output is required. A kiln 8 ft. diameter by 150 ft. long will produced 100 tonnes or more per day.

Rotary kilns are also used to burn lime form various wastes, such as "lime-sludge" from paper mills and sugar purification. Lime is now burned from the calcium carbonate waste from caustic soda manufacture at a number of plants by first passing the waste through some form of dewatering device such as a continuous rotary, drum or disc filter, the waste being obtained from the latter in the form of a wet mud containing about 50 per cent water. The recovered lime is very pure and usually contains some alkali otherwise lost, the percentage varying from 3 to 8 per cent of the lime recovered. This practice not only can be carried out generally for less than the cost of purchasing lump lime but it also disposes of a troublesome waste product.

The rotary kiln is particularly well adapted to burning chemical lime and is to be preferred where a very well-burned lime, free from carbon dioxide, is desired as in the manufacture of carbide and in metallurgy.

A refractory product is now made by grinding together doomite, $CaMg(CO_3)_2$, and a small amount of iron ore, and burning the mixture in a rotary kiln at a somewhat higher temperature than is required for lime. The product so obtained consists of small roughly rounded modulus, hard and dark brown to black the latter compound is to make the slake less readily on exposure to the air and so improve its keeping qualities. This material is known in the trade as "dead burned dolomite" and under various trade names.

Fuel Requirements for Shaft Kilns

The quantity of fuel required depends on many things, among which may be mentioned the kind and quality of fuel, skill of the operator, and the limestone itself. Magnesium limestone burns more easily than high calcium stone and an impure stone more easily than a pure one. The amount of fuel actually required is about as follows:

1 tonne of good bituminous coal, hand-fired, will burn from 3 to 4 1/2 tonnes of lime.

1 barrel of fuel oil will burn 2/3 to 1 tonne of line.

1 cord of seasoned hardwood will burn from 1 1/2 to 2 1/2 tonnes of lime.

Hydrated Lime

When quick-lime is treated with water it combines with the water to form calcium hydroxide :

$$CaO + H_2O = Ca(OH)_2$$

If the lime is free from impurities, it will take up 32 per cent of its own weight of water. This quantity, however, is reduced somewhat because of the impurities that are always, found to a greater or less extent in all commercial limes. When lime slakes, heat is generated. One gram of CaO converted to $Ca(OH)_2$ liberates 270 gram-calories, or enough heat to evaporate 0.5 gram of water at 100°C. The chemical reaction itself requires 0.31 gram of water. Thus, assuming the water and lime both to be at 100°C, 1 gram of CaO could satisfy 0.81 gram of water. In practice appreciably less is employed due to radiation losses, and heat required to bring the materials up to 100°C. Formerly, lime was hydrated, or slaked, by the mason just preparatory to its use. An excess of water was always used, and the calcium hydroxide formed a wet mass called "lime putty". Now, mechanical means of hydration have been introduced whereby the lime is hydrated by the manufacturer with just sufficient water to form the hydrate, leaving none in excess. This hydrated lime is a fine dry powder, practically all of which will pass through a 100–mesh screen. It is packed in paper bags or cloth sacks and will keep indefinitely. It can be stored without danger of causing fire, which is not true of caustic lime. When added to cement, it makes it waterproof to some extent and more easy to trowel.

In manufacturing hydrated lime the lump lime is first ground to small size. It is then mixed with predetermined amount of water, when it falls to a fine powder. The slaked lime is then sieved to separate out the unhydrated lumps or siliceous cores from the latter, or else these cores are ground so fine that they will cause no "popping". The plan adopted in grinding the quicklime is the most successful hydrating plants consists in crushing the lime by means of a swing hammer mill or a Sturtevant open-door crusher. This reduces it to pieces about 1/2 in. and under. Lime which is to be hydrated should not be burned at as high a temperature as is ordinarily used. Fresh lime hydrates much more readily than that which has been allowed to remain for some time in the air. There are a number of processes and machines for mixing the lime with water which have been successfully used in hydrating. The four best known of these are the Kritzer, the Schaffer, the Schulthess, and the Clyde hydrators.

Hydrators

The Clyde hydrator is a batch machine, in which a given quantity of lime, usually one tonne, is placed. The machine itself consists of a revolving pan provided with plows which store up and mix the water and the lime. The water is weighed and added in a predetermined amount. When the operator judges the process to be complete, as is determined by the fluffiness of the powder, the lime is scraped from the pan through an opening in the centre of the same into a hopper under the hydrator.

The *Kritizer hydrator* consists of a number of cylinders one over the other, which are provided with paddles which revolve around a central shaft. The lime is fed into the upper cylinder in a continuous stream. Here a regulated amount of water is spread upon it. The moist lime is worked by the paddles and passes through he upper cylinder to the next lower one, etc., and finally works its ways out at one end of the bottom cylinder. It is now entirely hydrated and dry. The steam from the lower cylinders passes to the upper ones, and helps to hydrate the lime.

In principle of operation the Schaffer hydrators resembles the Kritzer. In its case, however, the working is done by plows in shallow trays which are superimposed one on top of the other.

The *Schulthess hydrator* consists of a stationary horizontal cylinder in the centre of which revolves a shaft with plows. The lime is introduced into a revolving screen at one end of the cylinder and water is sprinkled on it here. As it slakes, it falls through the screen and the plows work it through the cylinder. The advantage of this hydrator is that it will handle lumps and hence no crusher is needed.

Hydrated lime is now packed in paper and cloth bags. Automatic machines have been devised which force the lime through a valve in one corner of the bags, which are pasted shut or tied, except for one corner, before the lime is placed in them.

Finishing Lime

Finishing lime is made in the same manner as ordinary lime in a rotary kiln. Dolomite is used varying in magnesium carbonate content from 10 per cent to 50 per cent. Finishing lime is always supplied to the trade in the slaked form and is called Finishing Hydrated Lime. It is used chiefly as a finishing coat in wall plastering as it gives a very white surface and can be trowelled smooth. Another use is as a lubricant in concrete mixtures. Due to the fact that the lime is finer than portland cement, it acts as a filler and renders the cement watertight. Lime with a high magnesium content has a higher plasticity than other limes.

Hydraulic Lime

Limestones containing amounts of impurities sufficient to give the calcined produced hydraulic properties, but insufficient to take up all the lime present, make, when burned, hydraulic limes. They form an intermediate product between ordinary lime and natural cement. These products range from feebly hydraulic limes to limes which harden quite satisfactorily under water. At one time these limes were manufactured to a large extent in Europe. Until recently they were not manufactured in any quantity in this country, but are now made and used to some extent for laying brick and stone. They are generally sold under the designation—"brick cement". They are made by burning limestone containing from 10 to 17 per cent silica, alumina and iron, and from 40 to 45 per cent lime. Magnesia may replace lime to a considerable extent. Hydraulic lime slakes with water just as does ordinary lime, only much more slowly.

Grappier Cements

These are obtained by grinding the hard cores which are obtained in the manufacture of hydraulic lime, and consist of that portion of the hydraulic lime which does not slake when water is added. La Frage cement is of this class, owing to its light colour and the fact that it does not stain marbles and other building stones as does portland cement and natural cement.

GYPSUM PRODUCTS

Plaster of Paris

Plaster of Paris is sold in the form of white powder. After being mixed with water it sets very quickly. It is used in art work for casting of figures, in dental work, in making moulds, and as a basic material for several plasters. It is made from gypsum by heating the latter to a temperature of between 100° and 204°C., when three-quarters of the water of crystallisation of the gypsum is driven off.

$$2(CaSO_4.2H_2O) = (CaSO_4)_2 \cdot H_2O + 3H_2O.$$

In actual practice the temperatures employed to bring about this reaction are 165° to 199°C. If gypsum is heated above 204°C., it loses all of its water of combination and becomes anhydrous calcium sulphate, the latter begin the basis of hard finish plaster, floor plaster, Keene's cement, etc.

When plaster of paris is mixed with water its sets or hardens very promptly, this change being due to absorption of water, forming gypsum again.

$$(CaSO_4)_2 H_2O + 3H_2O = 2 (CaSO_4)2H_2O$$

A pure plaster of Paris will normally harden or set in from five to fifteen minutes after having been mixed with water. If the gypsum from which the plaster is made contain impurities, the set will be much slower than this. Plaster to be used for building purposes must be slow setting. For ornamental use it must also be white, and since the impurities usually render the plaster slightly coloured, it is the common practice to add retarders to the plaster before placing it upon the market. The materials used as retarders are usually of a colloidal natural, such as glue, sawdust, blood, packing-house tankage, etc. Retarders are usually made by digesting hair with caustic soda. If a very quick-setting plaster is desired, crystallised salts are added, such as common salt, sodium sulphate, sodium carbonate, etc.

Gypsum

Gypsum, the raw material from which plaster is made, is when pure, a hydrous calcium sulphate, $CaSO_42H_2O$. As mined, however, it usually contains a considerable percentage of impurities, the chief of which are clay, calcium carbonate and magnesium carbonate. Impure earthy gypsum.

Gypsum occurs usually in the form of beds, frequently associated with deposits of rock salt, and almost always interstratified with beds of limestone and shale. The beds, of course, vary greatly in extent and thickness, some of them being as thick as 60 ft. though most of them are very much thinner than this. Anhydrite, anyhydraous calcium sulphate, is often associated with gypsum. This mineral is not suitable for the manufacture of plaster of paris.

The usual plan of working the gypsum deposits is by mining, for deposits seldom lies near enough to the surface for quarrying. For the most part the mining methods are crude.

Manufacture of plaster of Paris

The operation of manufacturing plaster of Paris from gypsum consists in first crushing and grinding the gypsum, then calcining the ground product, and finally pulverising the calcined product, after which the retarders are added. In some plants the gypsum is merely crushed and calcined, the calcined rock then being ground very finely. Where the kettle process is employed, the pulverising is done after the calcining.

The gypsum is usually first crushed to such a size that it will pass a 2-in. screen. After the coarse crushers, the gypsum is further reduced by means of a double cone or pot crusher. These crack the gypsum to such a size that most of it will pass a 1/2 in. ring screen. From the pot crusher the gypsum passes to buhr-stones, rock-emery mills or in the newer plants, Raymond mills. These reduce the gypsum so that about 60 to 95 per cement of it will pass a No. 100 mesh sieve, the finer, the better the product. It is then ready to be fed to the kettles.

The gypsum kettle

The gypsum kettle, consists of a steel cylinder set in brick work. The bottom of this kettle is made of cast iron, and is convex in shape. It has a thickness of about 3/4 in. at the edges and 4 ins. at the crown. This kettle bottom is a very important part of the apparatus, and is the part which seems to need repairing most often. This kettle itself is made of boiler plate, 3/8 to 5/8 in. thick. It is from 8 to 10 ft. in diameter and 6 to 8 ft. deep. A kettle of this size will hold from 7 to 12 tonnes of pulverised gypsum and produce from 5 1/2 to 10 tonnes of plaster in one batch. It is provided with from two to four flues, 12 ins. in diameter, placed horizontally about 8 ins. above the crown of the kettles bottom. It is surrounded by brickwork, so the heated gas from the fire may rise around its sides and through the flues. The top is covered with sheet irons, and has a movable door and a vent to carry off the water. Two kettles are usually placed side by side and work in pairs.

It is necessary that the material be continually agitated. For this purpose the kettle is provided with a stirrer, which is actuated by a vertical pinion wheel and line shaft. The stirrer itself consists of a cross arm which is curved to conform to the bottom of the kettle and provided with either paddles or a chain which drages along the kettle bottom. The stirrer is run at about 15 to 20 revolutions per minute, and is so arranged as to throw the materials towards the centre of the kettle. About 15 or 20 HP is required to operate the stirrer. If the gypsum is not stirred the charged will settle down and become hard. The bottom would be also melted out of the kettle.

In starting a kettle, the heat is gradually applied, the crude material is fed in through the charging door and the stirrer put in motion. The material is added gradually until the kettle is full. As the temperature rises, the water begins to be driven off. The mechanically held water is first driven off at a temperature of 100°C., after which the contents remain fairly quite until a temperature of 140°C. is reached, when the material begins to boil and the water of combination begins to be driven off. An expert calciner can tell by the way in which the charge boils when the process is finished. When the material has been calcined sufficiently it is run out into a pit by means of a small grate in the side of the kettle.

In burning pure gypsum the temperature rarely exceeds 170°C. Thermometers may here be used to advantage. In gypsite plants, however, a higher temperature is required, which may sometimes reach as high as 200°C. Here, owing to the complexity of the material, the proper temperature for calcining varies, and thermometers cannot be used. It takes about 100 lbs. of coal to calcine a tonne of plaster.

The gypsum kettles are arranged in pairs, with a pit to each pair of kettles, in order that the pit may be emptied from one charge while the other is being cooked. Sometimes the material is fully ground before calcining. Often, particularly in the older mills equipped with buhr-stones, the grinding is done partly carried form the pit, by means of an elevator, to a bolting reel, where the coarse material, usually amounting to only a few per cent, is screened out, sent back to buhr-stones and reground. From the bolting reel the material is conveyed to a storage bin. There are usually several of these in order to separate the runs of different days.

Rotary calciners

The Cummer system of calcining consists in first partially crushing the material, so that it will pass a 1-in. ring screen. The gypsum crushed to this state is fed mechanically into a rotary drier or calciner.

The material, in passing through the calciner, is heated to a temperature of from 175° to 200°C., the exact temperature depending upon the nature of the rock. From the drier the gypsum is carried into storage bins. They are built of vitrified brick or concrete, and are thoroughly ventilated so that they will not take fire or absorb the moisture given off by the gypsum. The calcining is completed in the bins, the moisture being driven off by the residual heat of the rock itself. After the gypsum has been dehydrated it is drawn out, crushed, ground, pulverised and bolted.

Adding the retarder

The retarder is added to the plaster after it has been finely ground. Usually from 2 to 15 lbs. are required for every tonne of plaster. The Broughton mixer is extensively used for this purpose. An ordinary wall plaster will also contain, in addition to retarder, a certain per cent of finely picked hair or other fibre, in the proportion of about 1 1/2 to 3 lbs. of hair to a tonne of plaster. Wood fibre is sometimes added as a substitute for hair, and such use is growing.

Plaster is usually packed in jute sacks containing 100 lbs., or in paper bags containing 80 lbs. It is customary in the plaster trade, as in the cement trade, to charge for the jute sacks, and give a rebate on the return. Where the plaster is packed in paper bags, a charge is usually made for these.

Hard Plasters

Flooring plaster and hard-finished plaster are also gypsum products made by burning this mineral until all of its water crystallisation is driven off. Flooring plasters are prepared by simply burning gypsum at a high temperature, while the hard-finish plasters are produced by a double burning with the intermediate use of some chemical. Flooring plasters are manufactured by burning pure gypsum, broken into lumps, in a vertical kiln. The kiln is heated by means of a grate, to one side of the kiln, upon which coal is fired. The hot products of combustion pass through, and so heat the gypsum in the kiln. The temperature reached is about 500°C. The time of burning is four hours. These floor platers give a very hard and durable surface, but they must be very carefully used to prevent cracking.

CHEMICAL ADMIXTURE

Admixtures are materials other than water, aggregates, hydraulic cements and fibre reinforcement, used as an ingredient of concrete or mortar and added to the batch immediately before or during mixing. Admixtures vary in composition from surfactants and soluble salts and polymers to insoluble minerals. The properties of concrete in both the fresh and hardened states can be modified by adding admixtures to concrete mixtures. 70 to 80 per cent of all the concrete produced contains one or more admixtures. Therefore we should be familiar with the commonly used admixtures, together with their applications and limitations. Inspite of intrinsic technical and economical advantages of concrete, the corrosion of reinforcement resulting in premature deterioration of concrete, sometimes even collapse, has become a major problem worldover. The pressing problem of deterioration of concrete have necessitated the search for new materials that could make concrete denser, stronger, more compliant, chemically resistant, waterproof and freezer/thaw resistant. The durability aspects of concrete has now become an imperative than a technical novelty. Admixtures are a must. They have led to the development of innovative concretes and mortars for new construction and for repair/rehabilitation of deteriorated infrastructure.

Admixtures are used to modify properties of fresh concrete, mortar and grout so as to:
1. Increase workability without increasing water content or decrease the water content at the same workability/to retain workability for long operations (RMC).

2. Retard or accurate time of initial setting as per requirement.
3. Modify the rate/or capacity for bleeding.
4. Reduce aggregation.
5. Improve pumpability.
6. Reduce the rate of heat evolution.
7. To increase the durability of concrete to specific exposure conditions.
8. Reduce the rate of slump loss.

Admixtures are also used to modify properties of hardened concrete, mortar and grout so as to:
1. Retard to reduce heat evolution during early hardening.
2. Enable to get high strength at an early stage.
3. Decrease permeability of concrete.
4. Control expansion caused by the reaction of alkalis with certain aggregate constituents.
5. Increase bond of concrete to steel reinforcement.
6. Increase bond between existing and new concrete.
7. Improve impact existence and abrasion existence.
8. Inhibit corrosion of embedded metal.
9. Produce coloured concrete or mortar.

In short the advantages of concrete admixture is to get a high performance durable concrete which can be manufactured, transported, placed compacted easily in the plastic state and to get a durable high strength impermeable concrete in the hardened state.

It is very important to know how it functions then only we can effectively monitor desired parameters. These admixtures play a reactive part in the chemical reactions of the hydration system.

All chemical admixtures should be tested adequately for their desired performance in full, before they are used in any large construction admixtures, in line with the requirements, with the same type and grade of cement proposed to be used at site.

Admixtures should be used in prescribed dosage as excess could be detrimental to concrete. It is most important to remember that admixtures are no substitutes for good concreting practice.

Classifications

Admixtures vary widely in chemical composition and may perform more than one function. Therefore, it is difficult to classify them according to their functions. The chemicals used as admixtures can broadly be divided into two types.

Some chemicals begin to act on the cement - water system instantaneously by influencing the surface tension of water and by adsorbing on the surface of cement particles, other break up into their ionic constituents and affect the chemical reactions between cement compounds and water from several minutes to several hours after addition.

The physical effect of the process of these admixtures on rheological behaviour of fresh concrete becomes immediately apparent but it takes several days to several months for the chemical effects to manifest (Fig. 14.3).

Surface - active chemicals or surfactants

They function by interacting at the surface (interface). This class covers air-entraining and water reducing (plasticiser) admixture.

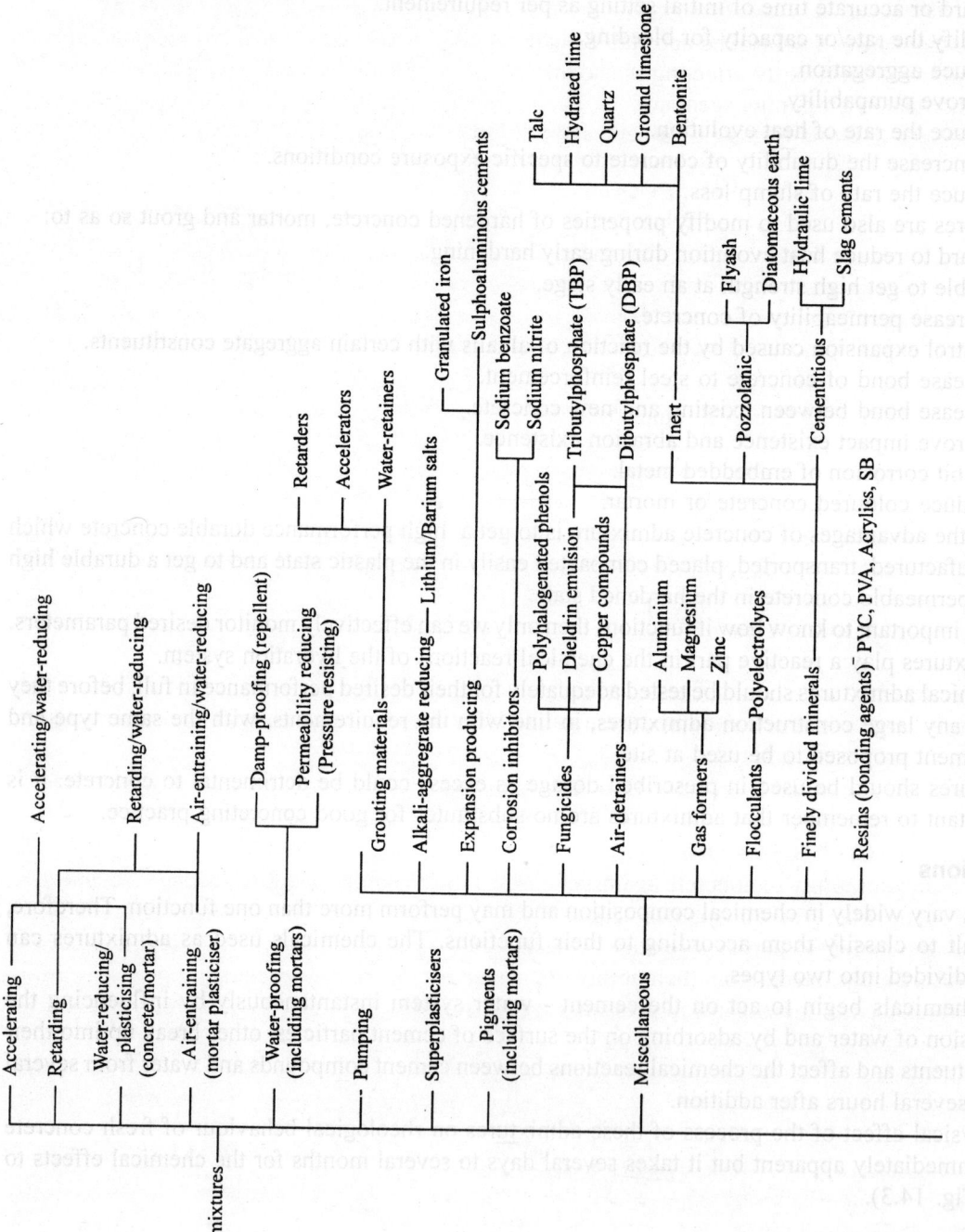

Fig. 14.3. Classification of admixtures.

Air-entraining agent

An admixture that causes a controlled and stable quantity of air to be incorporated during the mixing of a concrete, without significantly affecting the setting of the concrete.

Commonly used air-entraining agent are:

1. Alkali salts of wood resins e.g., abietic and pimeric acid salts.

$$CH_3 \quad COOH$$

Abietic acid

2. Alkyl aryl sulphonates e.g., Sodium dodecyl benzene sulphonate.

$$C_{12}H_{24}$$

$$SO_3Na$$

3. Alkyl sulphates.

For example Sodium dodecyl sulphate, Sodium cetyl sulphate, Sodium oleyl sulphate–

$$CH_3(CH_2)_7CH=CH(CH_2)_8SO_4Na$$

Salts of fatty acids derived from animal and vegetable fats and oils.

For example Sodium oleate

$$CH_3(CH_2)_7CH = CH(CH_2)_7COONa$$

It enhances the durability of concrete against cycles of climatic freezing and thawing and against the affects of de-icing salts. The air bubbles acts as minute compressible ball bearing which assist the movement of the aggregate particles relative to each other. The value of air entraining admixture in preventing frost damage, stems from their ability to provide within the concrete matrix, millions of tiny air bubbles which relieve the expansion pressure (Fig 14.4).

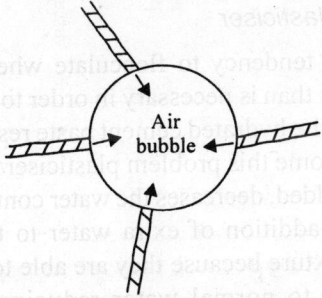

Air
bubble

Fig. 14.4. Air entraining admixtures.

Air-entraining agent involve a physico-chemical process occurring at the surface of the constituent materials in a system. Air entraining agent forms a stable foam with water and concrete, a stable dispersion of bubbles of the specified size and spacing. The production of a foam depends upon reducing the surface energy of the water (Fig. 14.5).

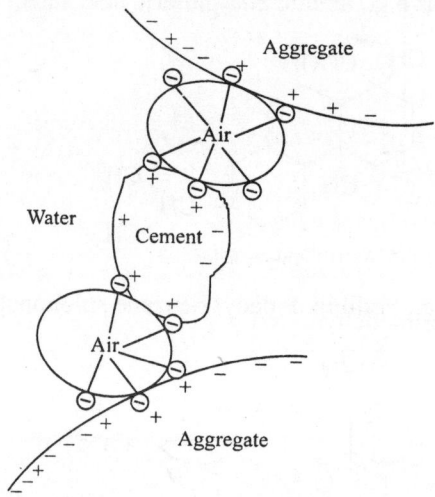

Fig. 14.5. Mechanism by which entrained air voids remain stable within concrete.

Due to hydrogen bonding in water molecule more energy is required to pull them apart. Since it is surface active agent, it functions by interacting at the interfaces between air, water, cement and aggregate in the concrete mixture. At the air-water interface, the polar groups are oriented towards the water phase lowering the surface tension, promoting bubble formation and counteracting the tendency for the dispersed bubbles to coalesce.

At the solid-water interface where directive forces exist in the cement surface, the polar groups become bond to the solid with the non polar groups oriented towards the water making the cement surface hydrophobic so that air can displace water and remain attached to the solid particles as bubbles. The basic structure of air entraining agent is that of a long chain molecule with a hydrophilic head and hydrophobic tail.

This causes the surfactant molecules to orient in such a way that surface tension of the water is reduced and thereby stabilising the bubbles that are formed when the solution is agitated.

Water reducing, plasticiser to superplasticiser

Since cement particles have a strong tendency to flocculate when they are brought in contact with water, it is necessary to add more water than is necessary in order to obtain a certain level of workability, which in turn generate porosity within the hydrated cement paste resulting in weakening of the structures and decrease of its durability. To overcome this problem plasticiser/superplasticiser is added. Plasticiser can be defined as an admixture, when added, decreases the water content without changing the workability or imparts more workability without addition of extra water to the concrete. Superplasticisers also called high range water reducing admixture because they are able to reduce three to four times water in a given concrete mixture compared to normal water reducing admixture. The commonly used superplasticiser are:

1. Sulphonated melamine - formaldehyde condensate

$$\left[H - CH_2NH - \underset{\underset{\underset{SO_3^-Na^+}{CH_2}}{NH}}{\overset{N}{\underset{N}{\bigtriangleup}}} - NHCH_2 - OH \right]_n$$

n is usually in the range of 50–60, giving molecular weight in the region of 20,000. Mainly in the form of sodium salt illustrated, which is very much soluble in water due to the sulphonate groups on the side chains.

2. Sulphonated naphthalene-formaldehyde condensate

$$\left[H - \underset{}{\overset{SO_3^-Na^+}{\bigcirc\bigcirc}} - CH_2 - OH \right]_n$$

n is the range of 5–10, giving a molecular weight of the order of 2,000.

3. Modified lignosulphonates.
   The crude lignosulphonates derived from wood pulps are commonly used as plasticisers. The basic unit of lignosulphonate molecule has a rather complex phenyl-propane skeleton. Substituent groups vary and may include phenolic, carboxylic and methoxy in addition to sulphonate.

4. Acid amides/polysaccharide mixtures and other high molecular weight hydroxylated polymers and copolymers also act as superplasticiser.
   The mechanism by which superplasticisers produce their effect is very much similar to that of normal plasticisers. The former consists of very large molecule (colloidal size) which dissolve in water to give ions with a very high negative charge (anions). It is explained in Fig. 14.6. The sulphonate groups oriented outwards into the water. These anions are attracted to the surface of the cement grains and at the normal levels of admixture usage, are adsorbed in sufficient numbers to form a complete monolayer around them. The combination of electrostatic repulsion and large ionic size (which provides physical separation) brings about a rapid dispersion of the individual cement grains. In doing so, water trapped within the original flocs is released and can then contribute to the mobility of the cement paste and hence to the workability of the concrete.

Set controlling chemicals

Chemicals used to modify the time of setting as desired. They affect the rate of hydration, thus controlling the time of setting.

Accelerating admixtures

These admixture shorten the setting time as they enhance the rate of hydration; reduces the period over which the cement remains workable, leading to an early stiffening of the mix. They cause a more rapid dissolution of the compounds of cement, particularly C_3S (tricalcium silicate) in water and hence facilitates more rapid hydration of these compounds.

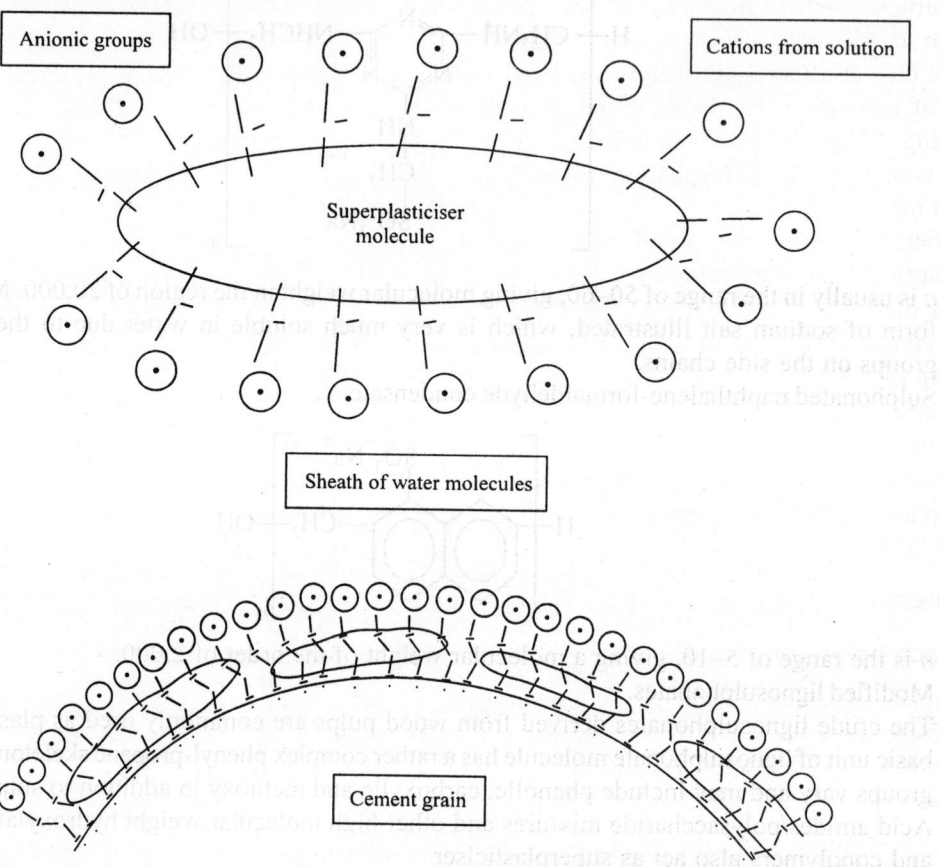

Fig. 14.6. Representation of superplasticiser molecules and mode of adsorption on cement.

Most accelerating admixture do not significantly affect the rheology (flow characteristics) of cement pastes at early ages and hence should not affect the consistency of the mixes. However at later ages, the effect of more rapid hydration of cement and consequent stiffening of the mixes will influence the workability of fresh concrete if there is considerable delay in its transportation and placing. Mixes with accelerators should be placed and compacted in their final positions without undue delay.

Calcium chloride was the most widely used accelerator. It is only suited for plain (unreinforced) concrete (e.g., precast blocks). It is obtained either as a liquid or as flakes, used upto 2 per cent by weight of cement.

Inspite of its effectiveness calcium chloride has now fallen from favour due to its action in promoting the corrosion of any steel which may be embedded in the concrete. Due to this tendency its use was banned for prestressed or post-tensioned, reinforced, embedded concrete. Calcium chloride is replaced by calcium formate. The dosage added is same as that of calcium chloride. Triethanolamine and soluble inorganic salts are also used as accelerating admixture.

Retarding admixtures

Retarders slow down the rate of setting of cement. They slow down the rate of initial hydration of cement particles so that more time is enabled for the placement of concrete. Retarders are more frequently used than accelerators are often combined with other types of admixture such as water reducers. Retarders do not affect significantly the final setting time of cement nor they influence the 28-day strength of concrete. Ready mixed concrete plant deploys retarders with advantage. They assist in transporting of fresh concretes in hot weather over long distances and in placing and consolidating of concrete pours without the formation of cold joints. This admixture maintains concrete in a workable condition for a longer period.

Retarders slow down the rate of initial hydration of cement either by forming a thin coating on the cement particles and thus delay their dissolution and reaction with water or by increasing the intra-molecular distances of the reacting silicates and aluminates from water molecules by forming certain transient compounds in the system. These reactions are not permanent and as, silicate and aluminate hydrate are formed, the influence of the retarder diminishes and the hydration process becomes normal.

The common retarders are:
1. Soluble carbohydrates such as starch, dextrin, casein.
2. Inorganic retarders such as hydroxides of zinc, and lead, alkali bicarbonates, calcium borate, alkaline tetraborate.
3. Sugars.

Miscellaneous admixtures

Grouting admixtures

These admixtures are used to impart special properties to the grout. A wide variety of special purpose admixtures are used to obtain the special properties required. Grout for preplaced-aggregate concrete requires extreme fluidity and non-setting of the heavier particles and usually contain water reducing admixtures alongwith the one required for non-setting of the heavier particles. Nonshrink grout requires a material that will not exhibit a reduction of its volume at placement. Oil-well cementing grouts encounter high temperatures and pressures with considerable pumping distances involved and retarders are useful in delaying setting time. Tile grouts and certain other grouts use materials such as gels, clays, pregelatinised starch and methyl cellulose to prevent the rapid loss of water.

Alkali-aggregate reducing

This admixture is used to reduce the expansion caused by the alkali-aggregate reaction. If the aggregates proposed to be used are alkali reactive, then reactive form of silica in it will react with alkali oxides released from the cement. The reaction results in formation of alkali-silica gel and involves expansive forces, which in turn causes cracking and disintegration of concrete. Soluble salts of lithium, barium, salts of proteinaceous materials, certain air-entraining and some water-reducing set-retarding materials have shown reductions in expansion.

Expansion producing

Admixtures, which react with other constituents of the concrete to cause expansion, are used to minimise the effects of drying shrinkage. The most common admixture for this purpose is a combination of finely divided or granulated iron and chemicals to promote oxidation of the iron. Expansion is greatest when the mixture is exposed alternately to wetting and drying. They are used both in restrained and unrestrained concrete placement. Expensive cements are used on large projects where a predetermined uniform degree of expansion is required.

Corrosion inhibitors

The major contributor to corrosion of reinforcing steel is the presence of chlorides in the concrete. The chlorides may come from circumstances such as exposure of the concrete to saline or brackish waters, saline soil or use of calcium chloride as an admixture constituent. Such corrosion is difficult to control once it has started. Mechanism of corrosion of steel is an electro-chemical process, where concrete act as an electrolyte. In presence of chloride, oxygen and water the transformation of the metallic iron to rust takes place accompanied by an increase in the volume, depending on the state of oxidation. The damage to concrete resulting from corrosion of embedded steel manifests in the form of expansion, cracking and eventually spalling of the cover, which in turn may result in structural damage due to loss of bond between steel and concrete. Chromates, phosphates, hypophosphorities, alkalies, nitrites and fluorides are commonly used corrosion inhibitors. Recently calcium nitrite has been reported as an effective corrosion inhibitor. Calcium lignosulphonate reduces the tendency for corrosion of steel in concrete containing calcium chloride and sodium. Sodium nitrite as corrosion inhibitor for autoclave products.

Since water, oxygen and chloride ions play important roles in the corrosion of embedded steel and cracking of concrete, we have to ensure low permeability by low/water cement ratio, adequate cement content and control of aggregate size and grading.

Fungicidal, germicidal and insecticidal admixtures

The primary purpose of these admixtures is to inhibit and control the growth of the bacteria and fungus on concrete floors and walls or joints. Polyhalogenated phenols, dieldrin emulsion and copper compounds are found to be effective. The effectiveness of these materials, particularly the copper compounds is reported to be of temporary nature. The higher doses, above 3 per cent weight of cement, may have an adverse effect on the strength of cement.

Gas formers

The gas void content of concrete can be increased by the use of admixtures that generate or liberate gas bubbles in the fresh mixture during and immediately following placement and prior to setting of the cement paste. Such materials are added to the concrete mixture to counteract settlement and bleeding, thus causing the concrete to retain more volume, nearly the same volume at which it was cast. They are not used for producing resistance to freezing and thawing. Admixtures that produce these effects are hydrogen peroxide, which generates oxygen; metallic aluminium, which generates hydrogen; and certain forms of activated carbon from which adsorbed air is liberated. Only aluminium powder is extensively used.

Flocculating admixtures

These materials decrease the bleeding capacity, reduce flow, increase cohesiveness and increase early strength. It is believed that these compounds containing highly charged groups in their chains, are adsorbed on cement particles, linking them together. Synthetic polyelectrolytes, such as vinyl acetatemaleic anhydride copolymer have been used.

Bonding admixtures

This admixture is used to enhance bonding properties generally consists of an organic polymer emulsion commonly known as latex. Upon drying or setting, the polymer particles coalesces into a thin film, adhering to the cement particles and to the aggregate, thus improving the bond between the various phases. The polymer also fills microvoids and bridges micro-cracks that develop during the shrinkage associated with curing. This secondary bonding action preserves some of the potential strength normally lost due to micro-cracking. Greater strengths and durability are associated with the lower w/c of latex mixes. The polymer particles act as a water replacement, resulting in more fluidity than in mixes without latex, but having a similar w/c ratio.

Colouring admixtures

These are used to produce the adequate colour without materially affecting the desirably physical properties of the mixture. Carbon black, phthalocyanine green, titanium oxide, and various iron oxides are used as pigments. The addition of any pigment to concrete normally should not exceed 10 per cent by weight (mass) of cement.

Pumping admixtures

The sole function of pumping aids is to improve concrete pumpability. They normally are not used in concrete that is not pumped or in concrete that can be pumped readily. The primary purpose of using this admixtures to enhance pumpability of concrete is to overcome difficulties that cannot be overcome by changes in the concrete mix proportions. As in the use of any ingredient in concrete, the objective is economy.

Many pumping aids are thickeners that increase the cohesiveness of concrete. The Standards Association of Australia identified five categories of thickening admixtures for concrete and mortar as follows:

1. Water-soluble synthetic and natural organic polymers that increase the viscosity of water-cellulose derivatives (methyl, ethyl, hydroxyethyl, other cellulose gums); polyethylene oxides; acrylic polymers; polyacrylamides; carboxyvinyl polymers; natural water-soluble gums; starches; and polyvinyl alcohol.
2. Organic flocculants - carboxyl-containing styrene copolymers, other synthetic polyelectrolytes, and natural water-soluble gums.
3. Emulsions of various organic materials — paraffin, coal tar, asphalt, and acrylic and other polymers.
4. Finely divided inorganic materials that supplement cement in cement paste-fly ash and various raw or calcined pozzolanic materials, hydrated lime, and natural or precipitated calcium carbonates and various rock dusts.

In ACI 212-3R-91, such air-entraining agents or surface-active agents as hydroxylated carboxylic acid derivatives, lignosulphonates and their derivatives, formaldehyde-condensed naphthalene suplphonates, melamine polymers and other set-retarding or water-reducing admixtures, which are widely used in concrete in North America are considered to be normal constituents of concrete, and are not therefore considered specially as pumping aids eventhough they improve pumpability.

Dampproofing admixtures

The term 'dampproofing' implies to prevention of water penetration of dry concrete or stoppage of water transmission through unsaturated concrete. However, admixtures have not been found to produce such effects and the term has come to mean a reduction in rate of penetration of water into dry concrete or in rate of transmission of water through unsaturated concrete.

Dampproofing may reduce the rate of penetration of aggressive chemicals found in water, however, it will not stop them. Dampproofing admixtures also may reduce the penetration of water into concrete, thus delaying the effects of damage caused by freezing and thawing by reducing the amount or rate of mixture entering the concrete.

Admixtures for dampproofing are used to render the concrete hydrophobic and therefore capable of repelling water that is not under hydrostatic pressure. They include soaps, butyl stearate and cretins petroleum products such as mineral oils, asphalts emulsions and certain cutback asphalts. Thus chemical admixtures play a very important role to meet the objective of developing concretes that will last for longer periods under severe conditions. In the long term, the construction chemicals have bright future. Construction chemicals will help to build a high quality structure with saving in cost and time.

COLOURED CONCRETE: A CREATIVE ALTERNATIVE FOR EXTERIORS AND INTERIORS

Concrete masonry today is a combination of colour and structure, a positive interplay of utility and aesthetics. Architects and planners are rather concerned about maintaining rational harmony with the given surroundings and environment. Drab gray cement concrete is no longer a compulsion. Iron oxide pigments may turn the whole visual of cast concrete in pleasing, subtle colours. Aesthetically soothing multicoloured cement concrete is creative and beneficial option available to architects in India as well.

Pigmentation of concrete and mortar or integral colouring of concrete in all conceivable form like ready mix, pre cast concrete mesonary, mortar cement is not common in India as yet, though usage in pavement tiles, superior quality checkered tiles, prefabricated structure, architectural blocks etc. is gradually receiving acceptance in few chosen applications. This low cost colour tool available with architects in the form of pigments has stepped up interest in diverse concrete applications. Beneficial and functional aspects of colour concrete that further compliments are lasting colouration without maintenance, protection form corrosion, stability against UV rays and above all supreme weather resistance. The industrially produced synthetic iron oxide pigments quite favourably combine aesthetics and utility at rational cost that adds life to the concrete. This favourable relationship between quality and cost has added fillip to its popularity worldwide.

Exteriors—Imagination is the Limit

Aesthetically conceivable exposed cement concrete applications in all form of construction activities may be just right for use of synthetic iron oxide pigments. Maintenance free colouration, protection from corrosion and all form of weather borne hazards are rather complimentary advantages. Some of the important exterior concrete applications in both building construction, as well as in industrial and utility constructions are rather innovative and have added aesthetically appealing dimensions that is to be seen to be believed. These include:

1. Pigmentation of underground conduits.
2. Colouration of ingress lanes on highways.
3. Precast structures of fly over, bridges and culverts.
4. All exposed castings of dams and barrages, concrete roads, dams etc.

Exterior applications

Some of the more common exterior applications are:

1. Concrete bricks and coloured mortars.
2. Exterior skirting of buildings.
3. Extruded wall claddings and panels.
4. Architectural blocks.

5. Stucco and slum blocks.
6. Concrete roof tiles.
7. Vertical and horizontal roof castings.
8. Pavement castings and pavement tiles.
9. Pre-fabricated concrete.
10 Ready-mix concrete.

Applications in Interiors

In the interiors, innovative areas of application are basements, car park area, security area dividers, marking of driveway/safety humps, identification/caution makings, etc. In low cost housing, castings of staircase and support railing walls, panels and general covered areas where net cementing may be considered, can also be prepared in coloured concrete. Since coloured cements are still not available in India, exclusive applications of coloured cement in interiors and even for interior decoration may have to wait.

Choice of Colours

The basic three colours are black, red and yellow, while shades of brown and tan can be derived through proper and proportionate mixing of these basic colours. Synthetic red iron oxide pigments used for colouration are available in a range of shades from yellowish terracotta to bluish burgundies, with all possible shades in between.

The black oxide offers a range from light gray to slate black, to core charcoal and yellow oxide provides a wide range of shades within the fold of cream to buff. Other than the above major colours chromium oxide green, cobalt blue, water dispersible carbon black, ultramarine blue, etc. also find applications in cement and other cementitious media like lime and plaster.

Types of Pigments

The pigments for cement or cementitious applications must be alkali stable, UV resistant, weather resistant and light fast. Colour consistency, strong tint and a good yield of colour are also very important for getting considered in this segment of end-use. Synthetic iron oxide pigments more than satisfy all the said pre-requisites.

Maximum level of pigmentation set forth by the American Concrete Institute (ACI) is 10 per cent of the cement content. ASTM identifies the following criteria for acceptance of pigments for use in integrally coloured units:

1. Light fastness: No significant difference between the unexposed portions with the specimen exposed under test.
2. Water wetability: To be easily wetted by water. Must disperse uniformly in the mix and not leach out of the concrete upon weathering.
3. Curing stability: Colour of the pigment must remain unchanged at the time of curing in highly humid atmosphere.
4. Alkali resistance: Significant colour change must not occur when pigments get treated with alkali like sodium hydroxide.

The standard BS 1014:1975 applies for quality of pigments and performance of cement and cement products.

Sourcing Pigments

Iron oxide pigments are available in two types — natural and synthetic. Natural iron oxides are obviously mined, whereas synthetic oxide pigments are produced from basic chemicals.

Natural iron oxide pigment sources are as follows:

1. Red: derived from *hematite*.
2. Yellow to brown (ocher, sienna, amber, etc.): derived from *limonite*.
3. Black: derived from *magnetite*.

Synthetic iron oxide pigments are manufactured by one of the following processes:

1. Thermal decomposition of iron salts or ion compounds.
2. Precipitation of iron salts.
3. Reduction of organic compounds by iron (aniline route).

Natural iron oxides are relatively low in iron oxide content, plus contain some amount of inert materials without any colour value. On the other hand, the synthetic iron oxides are practically all colour in their composition. Other than the low colour content, natural iron oxides have inconsistency in shade and these are not as good in UV resistance or weather resistance compared to synthetic oxides. Cost wise the natural pigments are cheaper, but proper techno-economic evaluation and cost benefit analysis would find the synthetic oxides to be better value for money.

Pigment Selection

The standard parameters commonly perceived for selection of pigments particularly for pigmentation of cement and mortar are shown in Table 14.5.

Table 14.5. Pigment selection chart for concrete and mortar.

Pigment types	*Colour range or tone*
Iron oxide synthetic	
Reds	Terracotta to burgundy
Yellows	Cream to buff
Black	Gray to charcoal
Browns	Varied
Chromium oxides	
Chromium oxide green	Grass green to olive
Hydrated chromium oxide green	Turquoise
Synthetic inorganic complexes	
Nickel titanate yellow	Canary yellow
Cobalt aluminate blue	Sky blue

In this context, a mention of water dispersible carbon black is necessary because unlike black iron oxide it produces intense jet black in concrete with relatively lower loading level. But the extremely fine particle size has the tendency to get washed out of a concrete matrix with the passage of time.

The water dispersible phthalocyanine green in concrete use is good value for money, but is not considered good for long-range permanence. The ultramarine blue in concrete is not dependable in absence of consistent performance record.

In India, usage of iron oxide pigments is confined to anti-corrosive paints and to some extent as flooring oxides. This is difficult to comprehend because the menacing impact of corrosion all over the country, particularly in the coastal belt is responsible for substantial national loss, besides recurring cost of maintenance.

Resistance to weather hazards and stability against UV rays should also have been prime considerations that are being ignored. As a matter of fact, use of synthetic iron oxide pigments in cement concrete applications is still in its infancy that needs to be nurtured for taking advantages of its inherent strength and buoyancy.

UTILISATION OF FLY ASH AS PART REPLACEMENT OF CEMENT IN CONCRETE

Fly ash is finely divided power thrown out as a waste material at the thermal power plants using pulverised coal for generating steam in the boilers. In India there are about 65 thermal power plants producing more than 80 million tonnes of fly ash.

The disposal of fly ash which is occupying the fertile land, in the vicinity of power plants has become an industrial hazards. It constitutes the bulk of the total ash produced and being lighter than the remaining 20 per cent pollutes the air causing serious problems of corrosion of structural surfaces and respiratory troubles to those inhaling it.

Since a huge quantity of Portland cement is being used in concrete constructions and the cost of fly ash is negligible as compared to that of the Portland cement, the utilisation of fly ash in concrete brings about a substantial savings in cement consumption and overall cost, thereby conserving energy and valuable natural resources.

It has been found that the performance of concrete was excellent with fly ash. The significant achievement of the research relates to its excellent mechanical properties and long term durability. Some of these include:

1. The compressive strength of over 100 MPa after 10 years have been obtained with fly ash.
2. Highly resistant to freezing and thawing cycling tests.
3. Fly ash concrete generally exhibits a low permeability.

The experimental study reported here evaluates the compressive strength, flexural strength, split tensile strength and heat of hydration of concrete by incorporating fly ash as part replacement of Portland cement.

Research Significance

The experimental study, examines the performance of concrete by incorporating fly ash as part replacement of Portland cement. The utilisation of fly ash in construction industry is not only economical, but also conserves energy and valuable natural resources.

Methodology

For experimental study it was proposed to use fly ash at 0, 10, 20 and 25 per cent by mass of cement as partial replacement, in concrete mix. The concrete mix 1:2:4 was adopted for control specimens for comparative study.

Materials Used

The constituents of concrete were tested as per Indian Standards (BIS) and reported as follows.

Cement

Birla Super Ordinary Portland cement (53 grade)

Specific gravity	3.15
Consistency	31.2 per cent
Final setting time	600 minutes
Fineness modulus	3066.3 cm^3/gm
Initial setting time	200 minutes
Soundness of cement	1 mm

Fine aggregates

Natural sand (collected from Pune region) passing through 4.75 mm sieve.

Specific gravity	2.70
Water absorption	4.90 per cent
Bulk density	1.75 gm/cm^3
Fineness modulus	3.26
Free surface water	1.91 per cent
Bulking	8.0 per cent

Course aggregates

Crushed basaltstone passing through 20 mm and retained on 4.75 mm I.S. sieve.

10 mm size aggregates

Specific gravity	3.01
Water absorption	1.63 per cent
Bulk density	1.42 gm/cm^3
Fineness modulus	2.10
Free surface water	1.40 per cent

20 mm size aggregates

Specific gravity	3.01
Water absorption	1.63 per cent
Bulk density	1.48 gm/cm^3
Fineness modulus	3.55
Free surface water	1.40 per cent

Fly ash

Source : Thermal Power Plant, Ozar, Nashik
Passing through 75 micron sieve
Fineness modulus : 0.086
Water : Tap water

No superplasticiser was used

The fly ash used in experimental work was tested for chemical composition and results are reported in Table 14.6. The requirement of chemical composition of fly ash as per Indian Standards and the chemical composition of fly ash used in present experimental work are also reported in Table 14.6.

Table 14.6. Chemical composition of fly ash (per cent).

Characteristics	Indian Standards requirement	Fly ash used in present work
Silicon dioxide (SiO_2) plus aluminium oxide (Al_2O_3) plus iron oxide (Fe_2O_3) per cent by mass (min.)	70.00	88.44
Silicon dioxide (SiO_2) per cent by mass (min.)	35.00	57.82
Magnesium oxide (MgO), per cent by mass (max.)	5.00	0.93
Total sulphur trioxide (SO_3) per cent by mass (max.)	2.75	1.89
Available alkalis as sodium oxide (Na_2O), per cent by mass max.	1.50	1.34
Calcium oxide (CaO)	–	1.02
Potassium oxide (K_2O)	–	0.34
TiO_2	–	1.40
Loss on ignition, per cent by mass	12.0	6.21

Testing Programme and Results

A total of four mixtures of concrete were prepared to investigate the properties of concrete with fly ash 0, 10, 20 and 25 per cent as part replacement of cement by mass prepared. The water-to-cement ratio of 0.55 by mass of cement was used for all mixtures. The experimental study was carried out on the specimens prepared from the above concrete mixtures. The adopted mix proportions of different conditions of concrete mixtures are summarised in Table 14.7.

Table 14.7. Proportioning of constituents of concrete.

Mixture No.	Fly ash content (%)	Cement (N)	Fly ash (N)	Fine aggregate (N)	Course aggregate (N) 10 mm	Course aggregate (N) 20 mm	Water litre
N1	0	2452.50	–	4905.00	4905.00	4905.00	137.00
N2	10	2207.25	245.20	4905.00	4905.00	4905.00	123.75
N3	20	1962.00	490.50	4905.00	4905.00	4905.00	110.00
N4	25	1839.38	613.12	4905.00	4905.00	4905.00	103.13

Compressive Strength

The compressive strength of concrete was determined in accordance with Indian standards. For this test in all 36 cube specimens of size $150 \times 150 \times 150$ mm (i.e., 3 specimens for each of 3 days, 7 days and 28 days) strengths for 4 concrete mixtures were casted. The specimens were tested under uniaxial compression after 3 days, 7 days and 28 days of curing. The average compressive strengths of concrete are reported in Table 14.8.

Table 14.8. Effect of fly ash on strength of concrete.

Mixture No.	Fly ash (%)	Compressive strength (N/mm²)			Flexural strength (N/mm²)			Split tensile strength (N/mm²)		
		3 days	7 days	28 days	3 days	7 days	28 days	3 days	7 days	28 days
N1	0	16.29	18.29	21.18	3.9	4.59	4.66	0.89	1.02	1.22
N2	10	14.81	17.29	20.29	3.2	2.59	3.86	0.76	0.85	1.16
N3	20	13.68	14.40	18.22	2.2	2.99	3.39	0.60	0.72	0.99
N4	25	11.11	12.62	16.82	1.86	2.53	2.86	0.56	0.69	0.89

Flexural Strength

The flexural strength of concrete was determined by bending test in accordance with Indian Standards. For this test all 36 beam specimens, that is 9 specimens for each four mixtures of size 100 × 100 × 500 mm, were casted. The beam specimens were tested under universal testing machine with two-point loading after 3 days, 7 days and 18 days age. The average flexural strength are reported in Table 14.8.

Split Tensile Strength

The split tensile test is an indirect method of finding the tensile of concrete. In this test an uniaxial compressive force was applied normal to the two opposite generators of the cylindrical specimen. For this test in all 36 cylindrical specimen i.e., 9 specimens for 4 concrete mixtures each of size 150 mm diagnosis × 300 mm height were casted. The specimens were tested after 3 day, 7 day and 28 days age and the results are reported in Table 14.8.

Heat of Hydration

For this test 9.81 N(1 kg) cement and 300 ml water at 30°C was mixed to prepare the control specimen paste. The test specimens were prepared with varying fly ash by 10 per cent, 20 per cent and 25 per cent as part replacement of cement. The resulting specimen pastes were placed one by one in heat of hydration testing machine for 24 hours. The variations in heat of hydration were reported in Table 14.9.

Table 14.9. Effect of fly ash on heat of hydration.

Mixture No.	Fly ash (%)	Heat of hydration (°C) after time (in hours)												
		0	2	4	6	8	10	12	14	16	18	20	22	24
N1	0	42	42	40	46	51	48.5	42	40	38	37.5	37	36	36
N2	10	36	38	38	38	40	46	50	47	43	40	38	37	37
N3	20	34	38	38	40	44	48	46	42	40	38	37	36	36
N4	25	36	40	38	40	44	46	44	42	40	38	38	37	36

Discussion

1. The compressive strength of concrete was gradually decreased as the percentage fly ash increased. The difference in gain in compressive strength at higher age was less than that at

early age. The 28 days compressive strength of concrete at 10 per cent fly ash was 4.2 per cent less and gradually decreased to 20.60 per cent at 25 per cent fly ash than the strength of control specimen. (Table 14.8 and Fig. 14.7).

2. The flexural strength was gradually decreased as the percentage of fly ash was increased. There was decrease in flexural strength by 17.17 per cent at 10 per cent fly ash and 36.6 per cent at 25 per cent at 28 days strength (Table 14.8 and Fig. 14.8).

3. The split tensile strength was gradually decreased with increase in fly ash per cent. The decrease in strength was 4.9 per cent to 27.05 per cent as the fly ash was increased from 10 per cent to 25 per cent at 28 days strength (Table 14.8 and Fig. 14.9).

4. The heat of hydration was decreased gradually as the per cent age of fly ash was increased. The heat of hydration was maximum (51°C) after 8 hours for control specimen. It was maximum 50°C after 12 hours for concrete with 10 per cent. There was gain in delay in heat of hydration from 8 hours to 12 hours at 10 per cent fly ash replacement (Table 14.9).

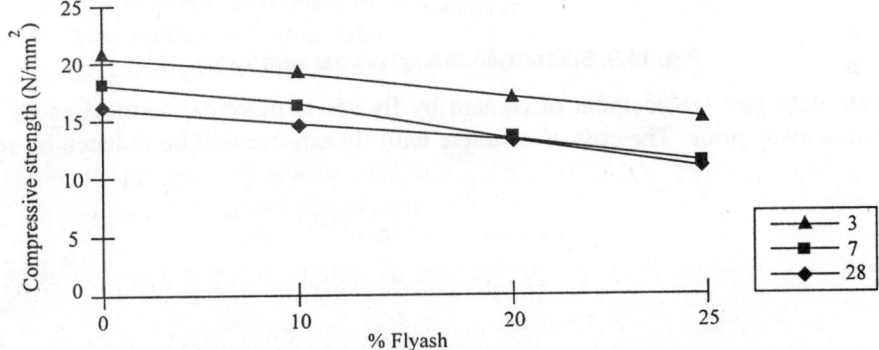

Fig. 14.7. Compressive strength vs per cent fly ash.

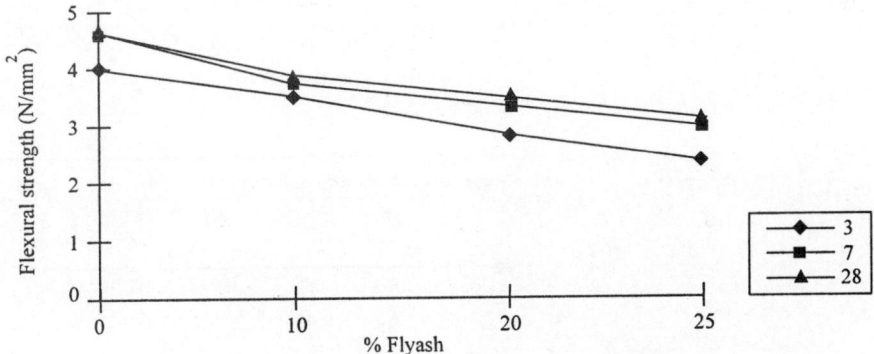

Fig. 14.8. Flexural strength vs per cent fly ash.

Thus from above it is possible to partly replace Portland cement by 10 per cent fly ash in concrete without much loss in compressive strength, flexural strength and split tensile strengths which are with in five per cent limit to fall below the strength of control specimen. The heat of hydration was delayed by four hours at 10 per cent fly ash replacement compared to control specimen. This concrete is more useful for mass concret.

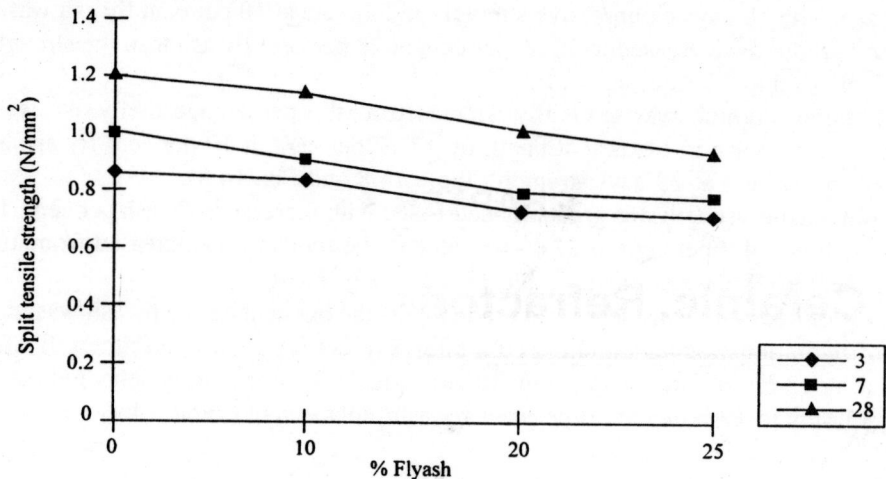

Fig. 14.9. Split tensile strength vs per cent fly ash.

The concrete with part replacement of cement by fly ash is more economical as fly ash is waste available at throw away price. The cost of concrete with fly ash use will be reduced by seven to eight per cent.

CHAPTER 15

Ceramic, Refractories and Potteries

INTRODUCTION

The clay products industry is one of the oldest of the manufacturing industries. Primitive clay vessels were merely sun-baked, and therefore very limited in their uses, but very early man learned that if clay vessels are heated to a red heat in a fire their strength is greatly increased and that the clay no longer softens in water. During the centuries that followed, the utility and desirability of clay wares was appreciated more fully, and clay was used in the manufacture of a larger variety of articles.

Manufacture of ceramic is an ancient art. In general ceramic can be defined as "the art and science of making solid articles by the action of heat on earthen materials, which have inorganic non-metallic materials as their essential component". This definition includes not only materials, such as pottery, porcelain, refractories, structural clay products, abrasives, porcelain enamels, cement and glass but also non-metallic magnetic materials, ferrotrics manufactured single crystals, glass ceramic and variety of products which were not in existence a few years ago.

The major characteristics of ceramics familiar to everyone is that they are brittle and fracture with little or no deformation. This behaviour is in contrast to metal. This behaviour on one hand makes their application very extensive but on the other hand restrictive also. Other properties of ceramic are very low thermal conductivity, low thermal expansion coefficient and thermal capacity. The variation in composition of ceramic products is wide. Therefore, product classification is very difficult. However, for convenience of understanding, ceramic products could be broadly classified as: (i) white wares; (ii) porcelain enamel; (iii) structural clay products; (iv) refractories of various types; (v) abrasives; (vi) special or new ceramics; (vii) pure oxide ceramics; (viii) nuclear ceramics; (ix) electro-optic ceramics; (x) magnetic ceramic; (xi) ceramic nitrides; (xii) metal-ceramic composites; (xiii) ceramic borides; and (xiv) ferro-electric ceramics. The composition of ceramic products varies from product to product, the main constituent being silica and alumina.

MANUFACTURING PROCESS

The mineral raw materials used in the ceramic industry are inorganic non-metallic crystalline solids formed by complex geological processes. Inexpensive silicate and aluminium silicate materials form the backbone of high tonnage products of the ceramic industry. Manufacture of building bricks and tiles does not require extensive beneficiation of the raw material. In contrast, manufacture of fine ceramics require the raw materials of higher quality, normally obtained through various beneficiation processes. For new ceramics such as magnetic ceramics, nuclear fuel materials, electronic ceramics etc., the chemical purification and even chemical preparation of raw materials may be necessary.

The raw materials of widest application are the clay minerals—fine particle hydrous aluminium silicates which develop plasticity when mixed with water. They vary over wide range in chemical, mineralogical, and physical characteristics. Clays perform two important functions in ceramic bodies. First, their characteristic plasticity, ability of clay-water composition to be formed and to maintain their shape and strength during drying and firing is unique. Second, they fuse over a temperature range, depending on composition, in such a way as to become dense and strong without losing their shape at high temperatures.

The most common clay minerals and those of primary interest to ceramists, since they are the major component of high-grade clays, are based on the kaolinite structure, $Al_3(Si_2O_3)$ (OH).

The other raw materials are : (i) talc, soap stone, and pyrophylite; (ii) quarts/quartzite; (iii) feldspar; (iv) mica; (v) kynite, sillimanite and andalusite; (vi) fluorspar; (vii) marble; (viii) magnesite; and (ix) sand (quartz).

In addition to above mentioned raw materials, variety of chemicals in pure form or mineral are used in different colour glazes or to give special appearance to the ceramic products. These chemicals/minerals are MgO, chrome ore/chrome oxide, Mn and its compounds, Pb compounds, Zn compounds, sodium compounds, barium and titanium compounds (carbonates) and borax etc.

Ceramic products are wide ranging. Each of these products need different composition of raw materials different glazing materials and also to be fired to a definite maximum temperature, which differ for each product, and to a definite firing and cooling time temperature schedule. Thus manufacturing processes for production of ceramics are equally wide ranging. However, the basic manufacturing process remains same. In general the manufacturing process can be divided in following steps. The steps are also shown in Fig. 15.1.

The raw materials are procured and mixed in fixed proportions. The ceramic materials are to be moulded in shapes. The moulding can be dry or wet. Most of the time moulding of the materials is done after making slip or paste. To make slip or cake the raw materials are crushed/ground in ball mills and then blunged in blungers to prepare slurry. Either this slip is used for moulding purposes (for bone china potteries) or the slurry is passed through filter press for dewatering to get cake, which is further processed before moulding it into desired shapes. In some cases dry moulding is carried out and the raw materials are ground finely and mixed with some binders for press moulding. Most of the products are initially dried in open or in heated chambers.

The dried products are dip glazed or spray glazed. In some cases the products are first fired in kilns upto a temperature of 900°C. The firing is called biscuit firing and then the material is glazed, dried and re-fired. After glazing and drying the last finishing operations are carried out manually or mechanically and the products are ready for final firing in the kiln. The firing operations are carried out in different types of kilns depending upon the type of the product, fuel available, temperature to be achieved in the kiln etc. Typical product dependent temperature profiles are to be maintained in the kiln.

Type of the kilns used for firing purposes are: (i) down draft kiln; (ii) chamber kiln; (iii) tunnel kiln; (iv) top hat kiln; (v) vertical shaft kiln; (vi) tank furnace; and (vii) shuttle kiln.

Once firing cycle is over the products are allowed to cool slowly within the kiln. The cooling schedule and duration depends upon the type of kiln and product. After cooling and unloading the kiln, the products are inspected for quality check and finally the product is ready for despatch.

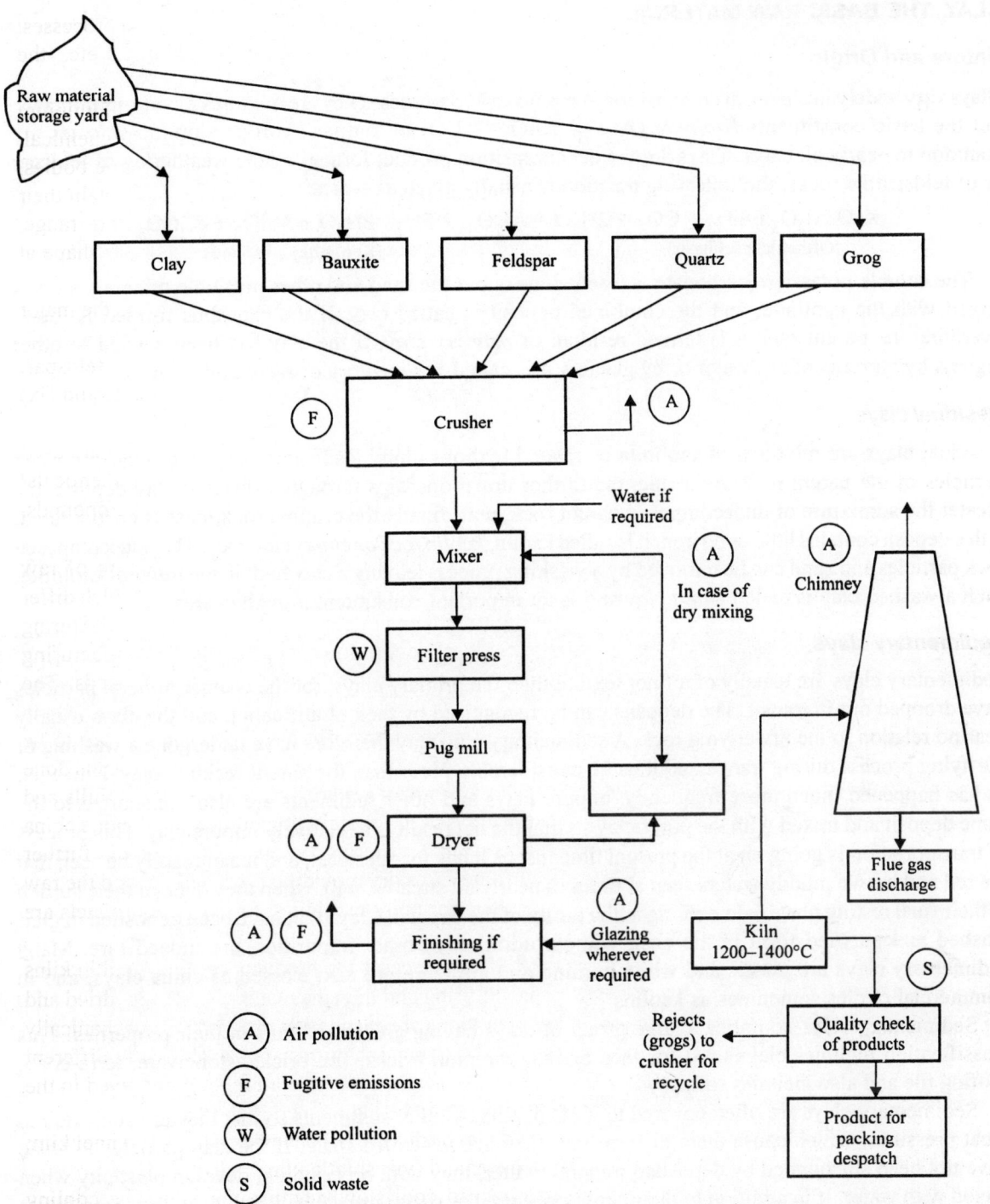

Fig. 15.1. Step involved in manufacturing ceramic products.

CLAY, THE BASIC RAW MATERIAL

Nature and Origin

Clays vary widely in chemical composition and physical properties. They are mixtures of various minerals, but the basic constituents *kaolinite* (Al_2O_3. $2SiO_2$. $2H_2O$) or similar hydrous alumino-silicates are common to nearly all clays. Clays are the decomposition product formed by the weathering of feldspar or of feldspathic rocks; the following reaction is usually given as typical:

$$K_2O.Al_2O_3.6SiO_2 + CO_2 + 2H_2O = Al_2O_3. 2SiO_2 . 2H_2O + 4SiO_2 + K_2CO_3$$

(Orthoclase feldspar)   (Kaolinite) (Sand)

The soluble potassium carbonate is leached away but the sand and other insoluble minerals are left mixed with the kaolinite, and the combined deposit is called clay. If the clay thus formed is found overlying the parent rock it is termed residual or *primary clay*; if the clay has been carried to other regions by currents of air, water or by glaciers it is called *sedimentary* or *secondary clay*.

Residual clays

Residual clays are mixtures of kaolinite or related hydrous aluminosilicates, sand and undecomposed particles of the parent rock. As a rule, the farther down one digs through a residual clay deposit, the greater the admixture of undecomposed parent rock, until finally the original rock mass is encountered. If this deposit contains little or no iron, it is called kaolin, *kaolin rock or china-clay rock*. The undecomposed rock particles and sand can be removed by a washing process leaving a clay high in the mineral kaolinite; such a washed clay is called *china clay* and is an important constituent in whitewares.

Sedimentary clays

Sedimentary clays are usually of a finer texture than the primary clays, for the coarser mineral particles have dropped out in transit. The deposits can be recognised by their stratification and the clays usually bear no relation to the underlying rock. A sedimentary clay may therefore have undergone a washing or purifying process during transit and thus be considerably purer than the parent residual clay. However, as has happened much more frequently, impure clays and other sediments are also transported to the same deposit and mixed with the purer clay so that the net result is a relatively impure clay. This process of transportation is going on at the present time just as it has for ages past, and it can readily be seen that the red and brown muddy waters seen at times in nearly all streams, will, when they deposit their burden at their final resting place, add nothing to the purity of the deposit. Clays that have been deposited in their washed and purified form in the past, without addition of other impurities, are indeed rare. Many sedimentary clays are plastic and white burning and are therefore also classed as china clays, and in commercial circles sometimes as kaolins.

Sedimentary clays comprise a large group of clays having great variations in their properties. This classification includes clays used for face bricks, common bricks, fire bricks, stoneware, terra-cotta, roofing tile and also includes soil clays.

Sedimentary clays are often covered to great depths by other sediments so that they are subjected to great pressures, which cause them to form into firm masses called shale. If the clay particles in shale have not been surrounded by deposited mineral matter, they will, on grinding, develop plasticity when mixed with water. If in addition to the intense pressure the shales are subjected to heat, they may lose a good share of their chemically combined water, thus being transformed into slate and schist.

Chemical Composition

The kind and amounts of minerals in a clay determine to a large extent its suitability for any particular purpose, but the physical states of the minerals also have an important bearing on their usefulness. In the chemical analysis of clays, the various constituents are usually reported as the oxides; such an analysis is called an ultimate analysis and is of considerable aid in forecasting the suitability of a clay for certain purposes. The average ultimate analyses of various clays, are given in Table 15.1.

Table 15.1. Ultimate analyses of various clays.

Types of clays	SiO_2	Al_2O_3	FeO, Fe_2O_3	CaO	MgO	Na_2O, K_2O	TiO_2	Ignition loss
Kaolin	57.84	28.66	0.91	0.18	0.32	0.80	0.0	10.37
Ball-clay	48.61	35.38	0.98	0.39	0.31	2.12	0.50	11.56
Fire-clay	60.75	25.56	1.36	0.29	0.35	1.23	0.96	9.39
Stoneware clay	58.09	20.79	2.48	0.32	0.74	1.67	0.48	7.60
Sewerpipe clay	58.84	21.34	6.12	0.47	1.00	3.42	0.38	0.68
Paving-brick clay	56.00	22.50	6.70	1.20	1.40	3.70	–	7.00
Common-brick clay	58.00	13.60	4.17	6.92	1.35	3.33	0.40	11.36

Many ceramists find more use for a rational analysis, which indicates the amount of different mineral constituents present. The minerals in a high grade kaolin consist chiefly of kaolin, $Al_2O_3.2SiO_2.2H_2O$, orthoclase feldspar, $K_2O.Al_2O_3.6SiO_2$, and quartz, SiO_2. Each of these minerals exerts its influence on the properties of the clay: the kaolinite is plastic and refractory; the feldspar is non-plastic and easily fusible; the quartz is non-plastic and reduces the shrinkage due to kaolin and feldspar. It is therefore desirable that for certain purposes the constituents be held with close limits. A rational analysis helps the clay worker in adjusting the various clays by adding feldspar or sand or both in order to bring about certain desired properties. A rational analysis can best be made with a petrographic microscope supplemented by an ultimate analysis. An approximate rational analysis can be calculated from an ultimate analysis, and methods have also been suggested for determining the rational analysis directly by chemical means. Mistakes can easily be made by judging a clay solely by its chemical analysis. Materials differing so widely in their properties as kaolin, ball clay and coal ash may have practically identical compositions as far as their ultimate analyses will show.

Impurities

In addition to kaolinite, clays contain various other chemical components which are usually classed as impurities, even though they confer very desirable properties to the clay. Free silica as quartz, SiO_2, is present in all clays and is also in the combined form in other minerals in the clay. It increases the refractoriness of clay at low temperatures, and even at high temperatures, if present in large quantities. It diminishes the air- and fire-shrinkage and increases the porosity of the fired product. Alumina, Al_2O_3, like quartz, is present in the combined form in kaolinite, feldspar, micas, and other minerals; it is a refractory component of clay. Iron oxide, FeO, Fe_2O_3, lime, CaO, magnesia, MgO, and the alkalies, K_2O, Na_2O, are classed as fluxes. They lower the fusion point of the clay, which is desirable for certain purposes, but is especially undesirable if the clay is to be made into refractory ware. The iron oxides are the strongest naturally occurring colouring ingredients in clay.

Clays with an iron oxide content up to a little over 1 per cent may burn white; buff-burning clays have an iron oxide content ranging from 0.5 to 5 per cent, and red-burning clays 4 to above 7 per cent. It will be seen from this that the colour produced by iron oxide is not merely dependent on its amount; its physical state and other materials present, influence the colour as well. For instance, lime, CaO, has a bleaching action on iron oxide. If iron is present in the clay in the ferrous state it is changed to the ferric state by the usual oxidising conditions in the kiln. It may be possible, however, that so much carbonaceous material will be present that the oxidation is not completed by the time vitrification sets in, in which case dark colours are obtained instead of a red.

Lime is the cause of considerable trouble in the firing of clays; not only is it an active flux, but it gives a very short vitrification range to the clay, which may mean a large percentage of spoilage in the product of the kiln. If lime is present in the clay as carbonate it will burn to the oxide and may slake later; an expansion accompanies the slaking which is often sufficient to rupture the fired ware. If lime is present as sulphate it may cause efflorescence. Magnesia is not as objectionable as lime, for it gives a longer vitrification range to the clay than the lime does. It is sometimes added to clay to be used for paving bricks, producing a more uniform, tough product. The alkalies are the strongest fluxing ingredients in clay. They are present in the feldspars and micas. It is on account of the alkalies in clay that a relatively low-firing temperature produces the proper degree of vitrification. Feldspar is added to clays for certain wares to lower the fusion point of the clay. Alkalies are undesirable in clays to be used for refractories. Mica is common to many clays, but if not present above 2 per cent it exerts but little influence on the properties of the clay.

Physical Properties

More important to many ceramists than the chemical properties of clays are their physical properties. A very important physical property is that of plasticity. Most clays when wet are plastic; i.e., they will yield under pressure and will hold the new shape when the pressure is released. The degree of plasticity varies widely in the different clays. The kaolins are but slightly plastic (lean) while the ball clays are very plastic (fat) but the difference in plasticity cannot be attributed to difference in chemical composition. If a clay is too plastic, its plasticity can be reduced by adding a less plastic clay or sand, while plasticity of a clay batch can be increased by adding a highly plastic clay. The plastic clays will take up more water than the less plastic ones and on drying will, therefore, be more subject to shrinking and cracking. The practical potter can readily tell by the "feel" whether a clay batch has the right degree of plasticity for his particular piece of work, but modern scientific developments require that the human element be eliminated as far as possible, and more precise methods for determining the degree of plasticity would be desirable. No satisfactory all-around method has yet been devised. Wilson gives a review of various methods that have been used with some degree of success, and a summary of the theories proposed to account for the plasticity of clays. Other important physical properties are texture (fineness) including distribution of sizes of particles, tensile strength, air- and fine-shrinkage, porosity and fusibility.

TYPES OF CLAYS

Certain types of clays have well-recognised properties and have found special uses in the arts. In some cases a clay used for one purpose may also be used for some other, but the following types may be mentioned. A difficulty encountered here is that certain terms through long years of use have changed their meanings considerably or some terms have been used loosely without much regard to their original meaning.

Kaolin

This term has been variously used to designate: (i) a primary clay of low plasticity but with a high degree of purity; (ii) sedimentary white-burning plastic clays of Florida; (iii) an amorphous variety of $Al_2O_3.2SiO_2.2H_2O$ (the crystalline form is called kaolinite); and (iv) china clay. A suitable present-day definition is: (v) a white-firing clay which in its beneficiated condition is made up chiefly of minerals of the kaolinite type.

China Clay

The term china clay can properly be applied to a relatively pure commercial variety of kaolin derived from the original kaolin formation by an artificial or, in some cases, a natural washing process. In this washing process, which may consist of elutriation, levigation or sedimentation, the disintegrated clay mass is agitated with a large amount of water and is then allowed to stand quietly in a pond or large tank. The coarse particles of sand, flint, mica and other parts of the undecomposed parent rock settle out in the washing process, leaving a relatively pure white-burning clay. The clays are used for paper filler and coating, whiteware refractories, and a variety of other purposes. Over three times as much china clay is used in the paper industry than is used in the clay products industries.

Ball Clays

Ball clays are sedimentary clays and have a high degree of sub-division. They derive their name from the shape of the clay masses as they are dug from the deposits, first in somewhat cubical forms which become rounded with handling. On account of the extreme fineness of the particles and because of some organic matter present, these clays are exceptionally plastic. They also have a high drying shrinkage and are relatively strong in the dry unfired state. The ball clays vitrify at lower temperatures than do china clays. Ball clays usually contain an appreciable amount of organic matter which burns out in the kiln leaving a white or cream mass.

The organic material discolours the raw clay which leads to such names as blue ball clay and black ball clay. Ball clays mixed with china clay are used in a large number of whiteware bodies used for making dinnerware, electrical porcelain and wall and floor tile.

Fire Clays

Fire clays will stand a high degree of heat. As a rule clays are expected to withstand temperatures above 1600° C(2912°F) before they are classed as fire clays. In order to withstand such temperatures a clay must have a small percentage of fluxes, such as Fe_2O_3, FeO, CaO, MgO, K_2O, Na_2O and TiO_2. Usually the sum of all these fluxes represents less than 4 per cent of the total, and commonly less than 3 per cent. The SiO_2 content for regular fire clays ranges between 40 and 60 per cent and the alumina between 25 and 45 per cent. High alumina clays may contain Al_2O_3 above 70 per cent. The high alumina clays are highly refractory and are used for refractories intended for severe duty.

Fire clays are commonly divided into the following types: *plastic clays* are soft and easily worked into desired shapes; *flint clays* are hard and rock-like due to great pressures encountered since their deposition and require fine grinding to develop plasticity; *refractory shales* are fire clays that have been converted to shales as described in the following section. Fire clays are used for making refractory wares such as fire bricks, special shapes and muffles, but in some cases are also used for making structural wares such as face brick and hollow building tile.

Shales

These are clay strata that have become hardened by pressure and usually also by heat, but chemical alternation has not taken place. On exposure to weather they soften and become more clay-like and may become fairly plastic after wet grinding. Shales are widely used in the manufacture of structural clay products.

Buff and Red-Firing Clays

These clays, together with shales, form the basis of a large industry manufacturing what is commonly known as heavy clay products or structural ware. The red and dark firing clays contain some form of iron oxide together with other impurities and as a rule will not stand as high a temperature as the buff burning clays, and are widely used for face bricks, common bricks, and hollow tile. The hardness may vary from a shale-like variety to a soft plastic variety. The clays find use for making structural brick and tile, terra cotta, stoneware, sewer pipes and pottery.

Preparation of Clay

When clay is delivered to the plant it must be prepared for the process in which it will be used. The preparation consists of one or more steps which resolve themselves into a number of unit operations.

Disintegration

Weathering improves all clays, but since it ties up capital for a considerable time it is not used to any large extent. The clay in layers 2 to 3 ft thick is exposed to the action of rain, frost, and the sun for a period varying from a few months to several years. In the weathering process some of the soluble salts are leached out, the larger particles disintegrate through slaking, and the plasticity is improved. The process is necessarily expensive and more direct methods are usually used.

Shales and flint clays may require jaw crushers, but dry pans are more commonly used. The pan rotates and as the material passes under the heavy wheels or mullers it is crushed. Centrifugal force throws the material over the perforated bottom to the rim of the pan, the fines falling through to the pit below. The scrapers then move the coarser material back under the mullers for further grinding. The clay and shale to be ground in a dry pan should be dry.

If the clay requires fine grinding, or if it contains small stones, roll crushers may be most satisfactory. Also in use are disintegrators for clay. A disintegrator consists of a large feed roll operated at a low rate of speed and a small disintegrating roll, which is provided with steel cutting bars, driven at a high rate of speed. The combination has the effect of removing successive portions of the clay, and at the same time breaking it up and destroying its original grain or fibre. Disintegrators will handle dry or damp clays. Ring roll crushers and hammer mills are finding wider application for fine grinding of clay and shales.

Prepared clay may be improved by ageing the damp clay in a moist and dark room. Ageing improves the plasticity of the clay because of bacterial growth.

The action is believed to be due to some colloidal formation and due to removal of dissolved or entrapped oxygen in the clay. A poorly plastic clay will improve in its plastic properties in two days, but two weeks to several months is necessary for the maximum effect.

Ageing is not used as much as it was formerly, partly because the process ties up material in production for a considerable time, and de-airing processes, to be described later, in a number of instances obtain similar results in a few minutes.

Tempering

Plastic clays improve in their properties by tempering. This consists of thoroughly mixing the clay and the water content so that the clay becomes a uniform homogeneous mass throughout. Several machines have been developed for this purpose and in some cases grinding and tempering can be performed in a single operation.

Ring pits were formed in common use and are still in use in some yards. The pits are about 25 ft in diameter and from 2 to 3 ft deep. A heavy wheel, usually of iron, is so arranged that it travels around the pit by horse power, in a spiral motion, thus grinding and mixing the clay.

Pug mills, are used in most modern plants. The mill consists usually of a horizontal semi-cylindrical tank, open at the top and containing longitudinally a rotating shaft to which are attached a number of blades set at different pitches. The blades work the clay into a homogeneous mixture and discharge the worked clay at one end. A pug mill takes up little room and power and is continuous in operation.

Wet pans also are used in modern plants for tempering and mixing clays. A wet pan is constructed similar to that of a dry pan except that it has a solid bottom. The wet pan is used for tempering clay or shale after it has been prepared in a dry pan or crusher preparatory to the manufacture of bricks and other heavy clay products.

The term "tempering" is also used in a different sense in connection with brick manufacture, in which case the term implies a partial and uniform reduction in the moisture content of the bricks made by the soft-mud process.

De-airing

If a plastic clay mass is subjected to a vacuum, a good deal of the dissolved and adsorbed air is removed from the clay. For many purposes this de-aired or evacuated clay has markedly improved properties over clay not so treated. De-airing may improve the plasticity, working, drying and firing properties of a clay. De-airing of a plastic clay mass is accomplished in a special de-airing machine, which consists essentially of two pug mills in tandem, with a vacuum chamber between. The first mill extrudes the clay in the form of flat ribbons into the vacuum chamber, where the air leaves the clay. The second mill compacts the evacuated clay into a solid mass and extrudes the clay in the customary bar or column. De-airing is applied to all types of clay products manufactured by the plastic method from heavy clay products, such as bricks, to the finest china bodies. De-airing is also successfully applied to the manufacture of clay wares by the dry-press process and for preparing casting "slip".

Drying

In nearly all forming processes used in the clay industries, the wares after they are shaped, are in a damp condition. The moisture must be removed before the clay can be placed in the kiln for firing, for the rapid drying in the kiln would warp or crack the ware. In some instances the damp articles are merely exposed to the atmosphere until sufficiently dry, in others elaborate drying equipment is used with controlled temperature and humidity.

Shrinkage

Clays, like most porous materials, shrink on drying. In a wet clay, each tiny particle is surrounded by a film of moisture which holds the clay particles apart. As the moisture is removed by evaporation and these films are removed from the particles the latter are drawn closer together and the piece of clay shrinks. When the shrinkage has reached its maximum, there will still be free moisture in the pores throughout the clay, which is removed on further drying. The total moisture in clay is therefore divided into two groups, the shrinkage water, equal in volume to the total volume shrinkage of the clay on drying, and the pore water, equal to the difference between the total moisture content and shrinkage water.

The amount of shrinkage of clay on drying will depend on the amount of moisture present and the nature of the clay. A ball clay, on account of the fineness of the particles, shrinks considerably, while a less plastic clay like a kaolin or a sandy clay, shrinks less. The linear drying shrinkage of clays varies from over 12 per cent down to a negligible amount. A clay with considerable amount of coarse material like sand may shrink very little. Sometimes fired clay is ground to form 'grog' which is incorporated with the clay to reduce its drying shrinkage.

Drying rates

Clay wares cannot dry uniformly throughout, for in order for the moisture to travel to the surface there must be a moisture gradient and the outside will be drier than the inside, and will therefore shrink first. During this time the damp and more plastic parts may yield to the distortion caused by uneven shrinkage, but by the time they dry and shrink the parts that dried earlier have lost their plasticity and have "set" so that permanent deformation and stress are bound to follow. This can be reduced by slow drying so that there never is a great moisture differential throughout the ware. In massive clay articles like some terra-cotta pieces great care must be taken during the drying process. In articles with thin sections only, like tea cups, such care is not necessary.

There are two essential factors in drying; rate of evaporation from the surface, which is governed by the temperature, humidity and velocity of the drying air; and rate of diffusion of the moisture through the material being dried, which depends upon the amount of moisture, the moisture gradient, the temperature of the material and the size and shape of the pores.

The drying rate is accelerated at elevated temperatures because the rate of evaporation from the surface is increased, and the viscosity of the pore moisture is decreased so that diffusion from the interior to the surface takes place more readily. At higher temperatures there is danger that the surface moisture is removed too rapidly, causing an excessive moisture gradient through the ware and therefore uneven shrinkage with the possibility of cracks forming at the surface. The practical upper temperature limit for drying is about 90°C(194°F). In clay wares the limiting factor in the rate of drying is the diffusion rate of the moisture. Therefore, to increase this rate it is desirable to operate near the upper temperature range, but it is necessary to keep the air velocity low and the humidity high to prevent too rapid evaporation from the surface.

Types of dryers

The oldest method of drying which is still in use is to place bricks out in the open to be dried by the sun and wind; sometimes the drying areas are covered with a roof. Drying sheds are also common in which the bricks are placed on boards which are held in tiers by racks. The sheds are more economical in space but are also limited to summer use.

Hot floors are used considerably for drying refractories. The floors may be heated by waste heat from the kilns or by direct heat from coal or oil fires, the hot gasses passing under the floor in flues, or the floors can be heated by steam pipes placed under the floor. On account of closer control of temperature and heat distribution, steam is preferable to other means of heating.

Tunnel dryers

Tunnel dryers are widely used in the brick industry. The bricks loaded on cars are passed through heated tunnels, entering at the cool end and progressing slowly to the heated end as car after car is pushed into the tunnel on a predetermined schedule. For every car of moist bricks entering the tunnel a car of dry bricks leaves it. The tunnels are heated by direct firing in flues beneath the tunnels, by steam coils or by waste heat from kilns that are cooling. In a newer development of the tunnel dryer, the tunnel is divided into sections, each section being under independent temperature and humidity control. In one dryer, for example, there are four sections separated by doors, each section containing four cars of clay wares. Periodically all doors open, including the end doors of the tunnel, and four cars of moist wares are pushed into the dryer at the cool end and four leave at the hot end. Each group of four cars advances to a new section of the dryer under a higher temperature and lower humidity than the preceding section.

Circulating fans or injectors are used to circulate the air around the steam coils and through the bricks. This type of dryer is used for a variety of clay products, including structural wares and sanitary ware. For drying tableware and electrical porcelain it has been common practice to place the damp ware in a room heated with steam coils. A newer method, now widely used, is to use scientifically designed dryers that contain a set of conveyors. The ware to be dried is placed on boards which are placed in the dryers. The conveyor carries these boards up and down over pulleys, gradually moving the boards forward to the other end of the dryer, the boards with their load of wares always being kept horizontal. The specially constructed chamber in which this conveyor operates is heated with steam coils, and humidity control can be installed if desired. The drying time for dinnerware is usually under two hours.

FIRING

Firing is one of the chief steps in producing clay wares, for the firing process gives the clay its strength, hardness, durability and pleasing appearance. In all cases the clay is heated to a red heat or higher. Seldom can a clay be fired sufficiently hard at a temperature below 800°C (1472°F), while some wares like fire bricks may require a temperature of 1500°C (2732°F). Firing temperatures of various ceramic materials are given in Table 15.2.

Table 15.2. Firing temperatures of various ceramic materials.

	Cone	°C	°F
Common brick	012–06	875–1015	1787–1858
Vitrified brick	5–10	1205–1305	2200–2381
Firebrick	5–18	1205–1490	2200–2714
Hollow blocks and fireproofing	03–1	1115–1160	2039–2120
Terra-cotta	010–8	895–1260	1643–2300
Red earthenware	010–05	895–1040	1643–1904
Sewer pipe	1–11	1160–1325	2120–2418
Vitreous floor tile	8–12	1260–1335	2300–2435
Stoneware	6–8	1230–1260	2246–2300
Semi-vitreous tableware, bisque	8–10	1260–1305	2300–2381
Semi-vitreous tableware, glost	4–7	1190–1250	2174–2282
Whiteware, overglaze decorations	019–010	660–895	1220–1643
Vitrified hotel china, bisque	11	1325	2418
Vitrified hotel china, glost	5–8	1205–1260	2200–2300

Stages During Firing

The firing of clay wares takes place in several stages, none of which, however, is definitely divided from a preceding or following stage. The stages or periods are commonly divided as follows :
1. Dehydration
    (a) Mechanical dehydration or "water smoking" 20°–150°C (68°–300°F).
    (b) Chemical dehydration or "chemical water smoking" 150°–600°C (300°–1110°F).
2. Oxidation, 300°–950°C (570°–1740°F).
3. Vitrification, 900°C (1650°F) and above.

Dehydration

During the dehydration period, water not removed during the drying process, and hygroscopic water adsorbed from the atmosphere are removed. In addition to this mechanical water, chemically combined water is also removed, as for instance that of the kaolinite, $Al_2O_3.2SiO_2.2H_2O$. The greatest rate of moisture removal is near 500°C (932°F). Just following the chemical dehydration, the clay ware is at its greatest porosity. The time of water-smoking for bricks varies according to the moisture content, ranging from about twelve hours as a minimum to eight or ten days or even more for dry-pressed bricks which have not been previously dried. Low fires and a good draft to carry off the water vapour are ideal, but it is difficult to obtain a good draft with low fires. It is desirable to connect several kilns to one chimney so that a kiln under full fire aids in the draft of a kiln in the water-smoking stage. During this stage moisture and soot have a tendency to condense in the cooler parts of the kiln.

Oxidation

The oxidation period starts with oxidation of the easily ignited organic matter at about 300°C (572°F) or less, and may continue up to above 900°C (1652°F). During this period the ferrous iron is changed to ferric iron, but before this can be completed the carbon must be oxidised, and prior to this, the sulphur must be oxidised. In other words, the more easily oxidisable sulphur and carbon must be removed before the bright-red of ferric oxide can be obtained, and all this must take place before vitrification begins. During this stage calcium and magnesium carbonates dissociate into their respective oxides and carbon dioxide.

Vitrification

The vitrification period is the period during which some of the minerals melt to a glassy state and on cooling bond the unmelted particles into a firm mass. For some wares, like high tension electrical insulators, vitrification is carried out to completion so that no pore spaces remain. During this stage the clay wares shrink and the amount of shrinkage is often taken as a measure of the progress of the firing. The vitrification temperature varies according to the sand and fluxes in the clay; for common clay the range is between 950° and 1050°C (1742° and 1922°F). The degree of vitrification obtained depends upon the temperature and the time (soaking period), while the degree of vitrification desired depends upon the nature of the wares fired. In common bricks complete vitrification is unnecessary, as a porous structure is not undesirable; paving bricks are more completely vitrified and are classed as vitrified bricks.

Coloration or flashing

By shutting off the air supply from the fires the kiln atmosphere can be made strongly reducing, which if properly controlled may produce a variety of colours in the bricks. A clay that will ordinarily fire red (due to ferric oxide) will flash to brown due to the iron being reduced. Clays that burn buff in ordinary firing will change to a rich yellow or brown during the flashing period.

Cooling and annealing

All clayware should be cooled slowly in the kiln after the temperature is down to a low red heat; then all drafts are shut off completely so that the ware may be annealed properly. If the annealing is not carried out carefully, strains will develop in the bricks or other wares fired in the kiln, which diminishes their soundness and toughness. Cooling cracks may be so small as to escape casual observation, but can be detected by the lack of the ring when two bricks are struck together.

Types of Kilns

Various types of kilns for firing bricks, pottery and clay wares in general are in use. They vary widely in construction and operation and do not admit of simple classification, but the following types of kilns exemplify the chief ones in use today.

Scove kiln

One of the oldest types of kilns for firing bricks is the "scove-kiln or "clamp", which is an intermittent up-draft, open top, and open-fire kiln. The bricks are set in the form of large rectangular blocks about forty courses high, so arranged that fires may be built in openings left at the bottom and that the products of combustion can pass upward through the brick setting. The bricks are surrounded by temporary brick walls daubed on the outside with clay to keep the heat in and to keep drafts out. The top is covered closely with a layer of bricks and any sections showing local heating on account of localised drafts are covered with dirt to check the draft. Such a kiln is easy to construct but is of a temporary nature. This type of kiln is used mainly for firing common bricks.

Up-draft kilns

A common type of kiln especially used in the pottery industry is the periodic up-draft kiln. In the kiln eight furnaces (oil burners in this case) supply the heat. The hot products of combustion pass from the furnaces along under-floor flues. Some of the hot gases in this flue pass up through short stacks or 'bags' and into the kiln, while others pass along the flues to the centre and up through the opening. The hot gases then pass up among the wares in the kiln, and up through openings in the crown and out the stack.

Down draught kiln

Down draught kilns are the most widely used type of intermittent kilns. These kilns may be either circular or rectangular with fire holes in the circular walls or in the two side walls. In these kilns, the combustion gases from the fire hole rise from behind the bag walls into the crown of the kiln and then descend through the setting of bricks/tiles to openings in the kiln floor through which they pass into flues and then to the chimney. The size of the kilns varies considerably, capacities ranging from 10,000 to 1,00,000 bricks, but the more usual capacity is 30,000 to 50,000 bricks. The round kilns are 6 m and above in diameter and 3 to 5 m to the top of the crown. Rectangular kilns are generally 4 to 6 m wide 3 to 4 m high and 8 m or more in length internally. Round kilns usually have 10 to 12 fireholds around th kiln and one wicket entrance. Rectangular kilns have six or more fireholes, depending upon length ar capacity of the kiln, in each of its two longer walls and a wicket entrance in each of the shorter wa'

Fuel consumption depends largely upon the condition of the kiln, the manner of setting the bricks and the control of firing process. The loss of heat varies from one type of kiln to another. Large kilns consume less fuel than small kilns for a given number of bricks. These kilns are comparatively much more efficient than clamps.

Hoffman kiln

Hoffman kiln is generally used for firing roofing tiles. The early kilns of this type too form great circular ring-chamber with massive walls with a chimney at the centre to which 12 underground radial flues converged from the inside wall of the chamber. The chamber was barrel arched—like a railway tunnel—and in the arch were small feedholes through which the coal could be fed into the setting. Twelve wickets were arranged for setting and drawing the bricks, in the outer wall of the chamber alternating with the flues. The annular chamber was thus divided in effect into 12 compartments defined by the wickets and flues leading from them. Once the kiln was lit, it was not allowed to cool out and the sequence of operations is continuous and when the kiln was in full operation two wickets would probably be open and the other ten brick up. Latter to give increased length, the kiln was changed to an elliptical form with main flue in the centre wall connected direct to the chimney at its centre. Also, an extra flue was added so that hot air could be transferred from the cooling zone of the kiln to chambers containing freshly set bricks to dry them (Fig. 15.2).

The Hoffman kiln, now-a-days, has 16 or more chambers without division walls between chambers. The operating principle is exactly the same as that of the original design.

Car tunnel kilns

The car tunnel kiln, has come into prominence in recent years but its introduction antedates that of the Hoffman compartment tunnel kiln. In the car tunnel kiln the wares to be fired are placed on cars that pass slowly through a long tunnel about 130 to 500 feet long which is heated to its maximum temperature near the middle of its length. As the cars enter the comparatively cool charging end of the kiln they slowly approach the hot zone so that water-smoking, dehydration, the vitrification progress in the wares as the cars pass through the zones of corresponding temperatures. After the cars pass the burner zone they enter the cooling zone in which the temperature gradually decreases toward the discharging end so that the wares are cooled slowly. The kiln is full of cars in its entire length, the cars being moved slowly by a pusher or other mechanical device. When the cars leave the tunnel they move at right angles along a short transverse track and then back to the charging end on a track parallel to the tunnel. While the cars are on the return track they are unloaded and re-loaded with fresh wares to be fired. The fuel economy of a car tunnel is about equal to that of a large kiln of the Hoffman type. The hot zone of the kiln is centralised, being in a region of about 35 to 40 feet long; only this zone requires high-grade refractories. The car tunnel kiln is used in a wide variety of plants producing practically all types of fired clay wares. The car is loaded with saggers filled with white ware. The top of the car is about 4 feet wide, and the kiln will hold 40 to 60 cars.

Circular tunnel kiln

The circular tunnel kiln is one of the newest developments in kiln design. In some respects it is similar to a car tunnel kiln, but instead of being straight, the kiln is built in the form of a circle so that the entrance end and the discharge end of the kiln come fairly close together. The ware, instead of being placed on individual cars, is placed on a large horizontal ring which carries a refractory clay slab on its top on which the ware is set. The ring turns slowly carrying the ware through the kiln just as cars carry ware through a straight kiln. As the ware comes out at the discharge end it is removed and new ware is put in its place, the operations being performed as the ring continues to move. This type of kiln is adapted to firing nearly all types of wares from fine china to heavy clay products.

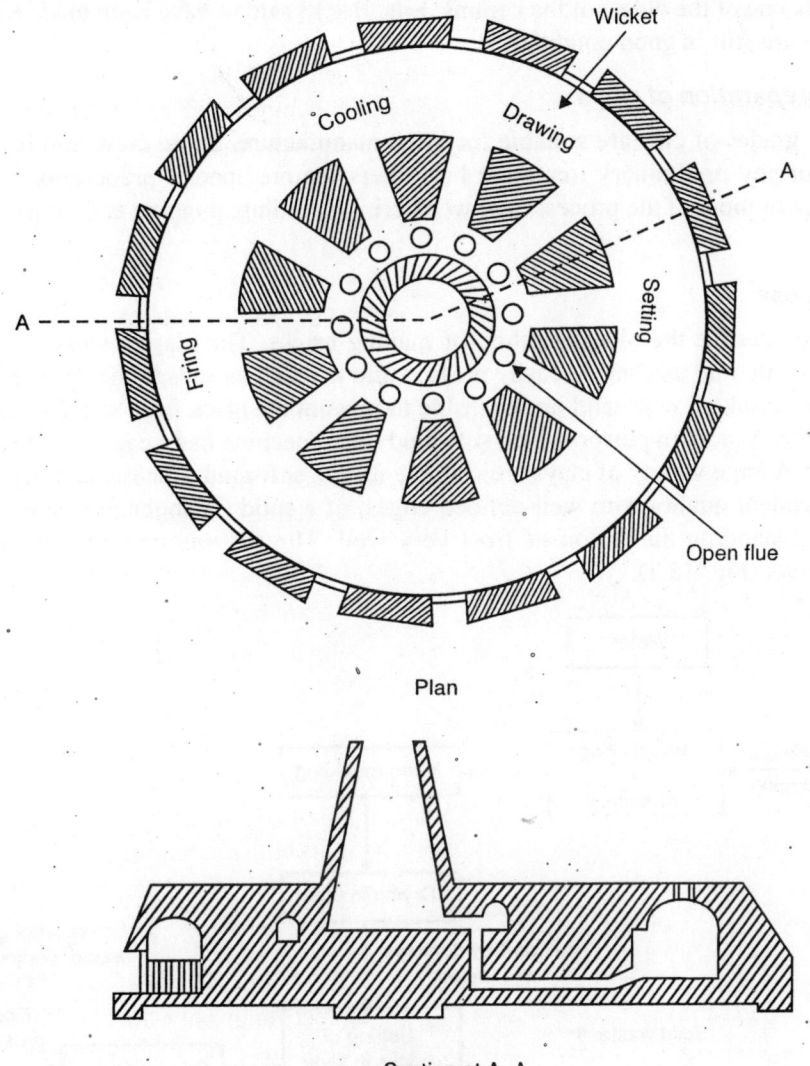

Fig. 15.2. Hoffman kiln–plan view.

MANUFACTURE OF HEAVY CLAY PRODUCTS

The term "heavy clay products" commonly used in the clay industries includes bricks, sewer pipe, drain tile, hollow block, terra cotta, flue linings, conduits, roofing tile and fire clay refractories. These wares are usually made from clay directly as dug except that the clay is ground, and in some cases screened, and tempered. In fire brick manufacture several clays may be mixed together.

The manufacture of all the various kinds of heavy clay products cannot be discussed here but a few types will suffice to give a general view of the processes.

Bricks

Brick-making is one of the oldest of the ceramic arts. Bricks said to have been made by the Babylonians 6000 years ago are still in good condition.

Preliminary preparation of clays

Relatively low grades of clay are suitable for brick manufacture. Some clays can be made into bricks directly without any preliminary treatment but others require special preparation, and therefore are subjected to one or more of the processes of weathering, grinding, pugging and tempering as previously described.

Soft-mud process

The soft-mud process is the oldest method of making bricks. The clay is worked up with sufficient water to make a soft mud that can be easily pressed into wooden moulds, either by hand or by machines. The moulds are sprinkled with sand on the inside to prevent the brick from sticking; this sand shows in the finished brick. A modern power-driven soft-mud brick machine has a capacity of from 1000 to 5000 bricks per hour. A large variety of clays are suitable for the soft-mud process, and the bricks if properly fired are of excellent quality with well-defined edges, of a solid homogeneous structure and have the property of withstanding the action of frost very well. The soft-mud process is especially used in making fire bricks (Fig. 15.3).

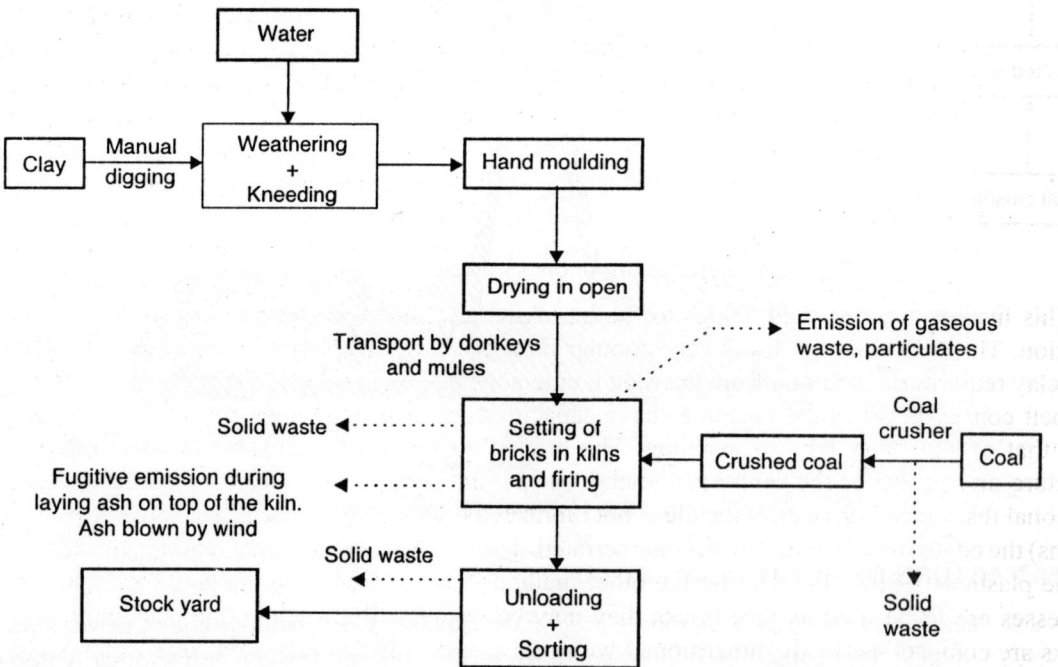

Fig. 15.3. Manufacture of bricks by manual process.

Stiff-mud process

The stiff-mud process is a more modern development of the brick industry. The clay is ground and pugged just as in the soft-mud process, but the amount of water added is much less. The mud is just soft enough to be squeezed through a die under pressure, but stiff enough for the bricks to hold their shape in handling. The mixed and pugged clay mud is fed into an auger machine, in which a tapering auger forces the clay through a die. The clay comes out of the die in the form of a bar whose cross-sectional dimensions are equal to one face of the brick. The clay bar passes along a short belt conveyor onto a cutting table on which a frame with a number of wires automatically cuts the bar into bricks. While the wires are passing through the clay they, with the frame, move in a direction parallel to the motion of the clay and at the same velocity (Fig. 15.4).

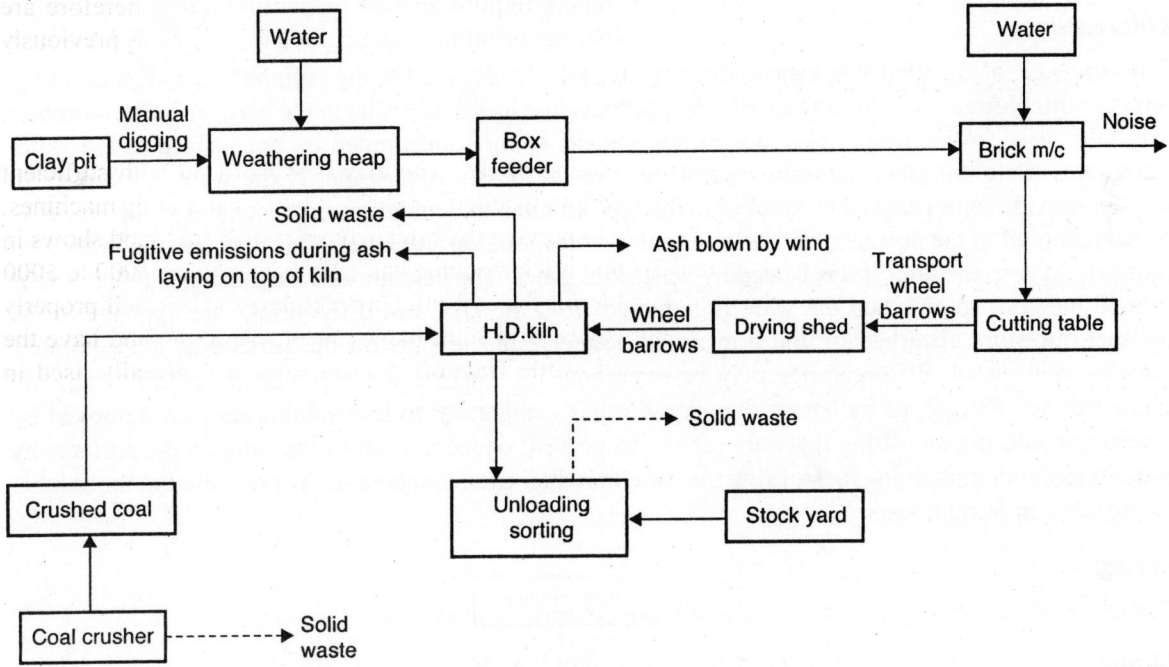

Fig. 15.4. Manufacture of bricks by semi-machined process.

'This insures a cut at right angles to the bar. After each cut the frame moves back to its original position. The stiff-mud process is very popular on account of its automatic features and large output. The clay requires no attention from the time it enters the pug mill until the cut bricks are removed from the belt conveyor. Stiff-mud machines have capacities up to 15,000 bricks per hour, which is greater than that of any other type of machine. The bricks, however have a tendency toward a laminated structure on account of the centre of the clay bar passing through the die faster than the sides where frictional resistance is greater. If the die is not lubricated properly (with steam, oil, water or by electrical means) the edges of the bar of clay become serrated. The stiff-mud process also requires closer regulation of the plasticity of the clay than do the other methods. If the bricks from the soft-mud and stiff-mud processes are to be used as face bricks they may be re-pressed in a repressing machine. Re-pressed bricks are compact, perfectly dimensioned with edges and corners clearly defined. Re-pressing is done in hand presses or in power presses, the latter having a capacity up to 3000 bricks per hour.

Dry-press process

In the dry-press process the dry clay containing usually less than 15 per cent moisture is fed from overhead bins into the press hopper through canvas tubes. As the machine operates, the charger moves over the mould, fills it and withdraws. Top and bottom plungers then move toward each other in the mould, subjecting the clay to very great pressure, producing a dense compact brick with well-defined and smooth faces. The bricks tend to be of a stratified texture if not carefully prepared. The clay must be thoroughly ground and pulverised and should be screened through a 16-mesh or finer screen. The bricks from the dry-press machine, may be set in the kiln directly, but it is considered preferable to give the bricks a preliminary drying. Semi-dry pressed bricks are made in a manner similar to the dry-pressed except that the clay contains more moisture.

Efflorescence

Efflorescence, also called whitewash, drier white, kiln white, wall white, sulphation, and scumming, refers to discoloration on the surface of clay products due to soluble salts in the clay. This discoloration may take place in the driers (drier white) due to salts being concentrated on the surface as the water evaporates. Calcium and magnesium sulphates are the most bothersome. If the ware is exposed to sulphur dioxide fumes caused by sulphur in the fuel, calcium and magnesium carbonates in the clay may be decomposed in the presence of moisture to the corresponding sulphates. This may happen in the kiln (kiln white) or in the drier if it is heated by waste kiln gases. The calcium and magnesium sulphates thus formed may not show when the ware is removed from the kiln, but if the bricks are exposed to the weather, moisture absorbed by the bricks will concentrate the salts on the surface as it evaporates, forming wall white. Efflorescence may be overcome by working the clays before soluble salts are formed by weathering, or by letting the clay weather completely so that soluble salts are removed by leaching; by drying and firing the ware rapidly to prevent concentration of the salts on the surface; by avoiding sulphur containing fuels; firing the ware in reducing atmosphere; or by precipitating the soluble sulphates with barium salts.

Drying

Bricks are dried outdoors, or in sheds, and more commonly in car tunnel dryers.

Firing

Bricks are commonly fired in scove kilns, periodic down-draft kilns, continuous kilns, and car tunnel kilns. The firing range is generally between cones 012 to 06 (875° to 1015°C, or 1787° to 1858°F).

Guidelines for better working in brick kiln industry

Guidelines for better working in brick kiln industry are given as under:
1. Maximum coal feed size should be limited to 10 mm.
2. Coal feeding cycle should not be more than 20 minutes. Coal feeding should be done using mechanical feeder as this would ensure more efficient burning of coal and lower the emissions.
3. Brick kiln should be established preferably at least two km away from residential areas and fruit gardens.
4. Installation of brick kiln should not be allowed in sensitive areas as notified by State Pollution Control Boards under National Air Quality Standard.
5. To avoid clustering of brick kilns in a locality, the distance between two kilns should be more than one km.

6. More and more agro industrial wastes should be used to replace clay so as to conserve good quality solid for agricultural purposes.
7. Use of coal with more than 35 per cent ash content should be avoided.
8. Heavy machinery should be mounted on anti-vibration mounting. Regular servicing and maintenance of plant and machinery, should be ensured. Ear muffs should be used by the workers.
9. The ash layer on the top of kiln should be covered with a layer of fired bricks or special tiles to check disposal of dust and for better thermal insulation which will also reduce discomfort to the workers. Excess ash produced everyday should be removed daily to a close by disposal site.
10. A double wall should be constructed around the kiln and the gap should be used to fill the ash produced from the kiln.
11. The brick works should be within covered space as far as possible.
12. To minimise generation of fugitive emissions, the passage around the bricks should be paved with fired bricks.
13. To improve combustion efficiency and also to reduce source emissions, properly graded coal should be used.

Sewer Pipe Manufacture

Sewer pipes are made mainly from shale tempered with fire clay. Sewer pipes are made in special presses, by the stiff-mud process. The press consists of two cylinders in tandem with axis vertical. The upper is the power cylinder operated by steam; the lower is the clay cylinder from which the prepared clay mixture is extruded downward through a die at the bottom. The process is, therefore, intermittent. A counterbalanced table free to move vertically is placed below the die to receive the pipe as it is expressed. The pipe is cut into proper lengths by a power cutter or by a wire operated manually. Special shapes as elbows and traps are made by hand in plaster moulds.

Some sewer-pipe presses can be fitted with a "goose neck" at the bottom of the clay cylinder. The clay is then expressed horizontally instead of vertically. The "goose neck" may be fitted with dies to make bricks or structural hollow ware.

Sewer pipes are salt glazed; the pipes are set on end in the kiln unprotected from the flames. When the kiln is up to temperature, cone 1 to 11 (1125° to 1285°C, 2057° to 2345°F), salt is put in the fireplaces, a few shovelsful at a time, over a period of several hours. The salt volatilises at the high temperature and combines with the clay, forming a glaze on the surface. All clays are not adapted to salt glazing. Salt glazing is also applied to some stoneware.

Hollow Structural Tile

Included in this classification are fire-proofing terra-cotta lumber, hollow blocks and hollow bricks. The fireproofing materials are those which are employed in floor arches, partitions, and wall furring for girders and columns. Terra-cotta lumber is a soft and porous material produced by mixing sawdust with clay and subsequently burning it out. It is not fired very hard and can be nailed and sawed the same as lumber. Hollow blocks and hollow brick are used for outside walls.

Hollow ware is made by the stiff-mud process by auger machines. Many stiff-mud brick machines have interchangeable dies so that they can be used not only for making bricks, but also for a great variety of hollow ware. Sewer pipe presses can also be fitted with dies for making hollow structural material.

Terra Cotta

Terra cotta denotes hand-made clay products for structural decorative purposes. Semi-fire clay mixed with a low grade of clay or shale and grog (ground firebricks) is commonly used for terra cotta. The ware is formed either by hand modelling or in plaster moulds and requires a great deal of skill. On account of the large bulky shapes frequently made, the drying must be carried out very carefully to prevent warping and cracking. Although most terra cotta is glazed, the body is not given a biscuit fire, but the glaze is sprayed in the form of a slip directly on the dried green ware. Terra-cotta is fired in muffle kilns at temperatures varying from cone 010 to 8 (895° to 1260°C, 1643° to 2300°F).

Tile

Under this heading come roofing tile, floor tile, and wall tile. For roofing tile the clay is prepared similarly as for bricks although more care is used. Simple forms can be expressed from auger machines and cut into proper lengths. Other forms are made by repressing slabs of clay. Floor tiles are made from a body similar to whiteware from fire clays, buff-burning clays or shale. They are fired to a dense body at cone 8 to 12 (1260° to 1335°C, 2300° to 2435°F). Wall tiles are made in dry press machines from a white burning body but fired at a temperature considerably below that of vitrification. The shapes are first given a biscuit fire in saggers and are then glazed and glost fired in a muffle kiln.

FIRE CLAY REFRACTORIES

Fire clay refractories are among the earliest manufactured refractory materials, and at the present time, on the basis of tonnage, represent more than all other types of refractories manufactured.

Methods of Manufacture

Fire clay refractory shapes are manufactured by the various methods used for making building bricks, including the dry press, stiff-mud, soft-mud and hand-made processes and some shapes are made by casting the thin clay slip in plaster of paris moulds. Massive shapes must be dried very carefully to prevent cracking and warping. The dried shapes are fired in various types of kilns such as down-draft, car tunnel and circular kilns.

Characteristics of fire clay refractories

Certain properties are desirable in refractories and these will vary according to the uses to which the refractories are put. For instance, for a boiler setting or a kiln lining, low heat conductivity is usually desired, while for use in muffle, high heat conductivity is desired.

Resistance to high temperatures

As a rule a fire clay refractory should be able to resist temperatures of at least 1600°C (2912°F), without softening and some of the high alumina clays may be put to use in furnaces in which they will encounter temperatures above 1800°C (3272°F). The temperature at which a refractory can be used satisfactorily depends to a large measure upon such factors as load, chemical nature of the charge or slag, oxidising or reducing conditions and the abrasion to which it is subjected. Fire clays that soften between 1500°C (2732°F) and 1600° C (2912°F) are sometimes classed as semi-fired clays or low heat duty clays.

Load-bearing capacity

Refractory linings of furnaces are ordinarily not expected to carry the load of the superstructure of the furnace, such as the boiler in boiler furnaces. This load is carried by common bricks encasing the refractory structure and therefore not subjected to any high temperatures. The refractories usually carry only the load of the other refractory materials resting on them. Fire clay refractories have a tendency to shrink under load at high temperatures, the degree of shrinkage permitted again varying with the nature of the service. In a standard test the refractory is heated under a load of 25 lb. per sq. in. A high heat duty brick should not deform a great deal at 1350°C (2462°F), an intermediate heat duty brick at 1300°C (2372°F) and a moderate heat duty at 1100°C (2012°F). In the earlier specifications the actual deformation permitted for each class was specified, but at present this is not as rigid, except that a high heat duty brick containing 70 per cent or more of silica should not shrink more than 4 per cent or expand more than one per cent at 1350°C (2462°F) under the above conditions.

Spalling

When bricks are subjected to a temperature gradient there will be uneven expansion or contraction with the result that a stress is set up between different parts of the brick or other refractory ware, which may be of sufficient magnitude to cause pieces of the brick to break off. This breaking off, called spalling, may also be caused by compression in the structure due to expansion of the whole from a rise of temperature, or may be due to slag penetration into the brick, thereby causing variations in the coefficient of expansion. Spalling may be reduced by avoiding sudden temperature changes, or by modifying the furnace design so stresses are not set up as the furnace heats. If the spalling is the result of uneven expansion due to slag penetration into the brick, the spalling can be reduced by originally firing the bricks at a temperature greater than their operating temperature. This reduces the pores in the brick and checks slag penetration. However, such a brick is dense and rigid and cannot adjust itself to relieve strains caused by uneven heating. If contact of slag with the brick is avoidable, spalling caused by thermal shock can be reduced by making a relatively porous brick. Such a brick can more readily expand unevenly without spalling.

Chemical properties

Refractory materials may come in contact with slags or other materials in the furnace that will react with the refractory and thereby reduce its useful life. Since it may be impossible to eliminate chemical action between the refractory and the charge, efforts may be successful in reducing the rate of reaction and thereby prolong the life of the refractory. Some of the chief methods for reducing the rate of reaction are the following:

1. Since chemical reactions increase in velocity with rise in temperature a logical attack would be to operate the furnace at lower temperatures. However, most furnace processes require a certain temperature and this method of attack is not practical. It is possible to cool the furnace walls from the outside and thereby reduce the interfacial temperature between the charge and the refractory. This necessitates considerable heat loss but is common practice in glass tanks. After the tank wall has become thin due to the reaction between the glass and the tank blocks, air blasts or water coolers are used to cool the refractory and thus extend its life.

2. If the refractory is of the same chemical nature as the charge or slag, reaction will be slow. For this reason it is generally recommended that acid refractories be used with an acid charge and basic refractories with a basic charge. At moderate temperatures, however, chemical action may be so slow that the refractory and charge may be of opposite chemical nature, as is the case in a lime kiln lined with silica refractories. In these cases other factors such as abrasive resistance become more important than those of chemical nature. Fire clay refractories are acidic in their nature, unless they contain much excess alumina, in which case they may be basic.

3. The rate of reaction depends also upon the surface exposed. If a refractory is porous, slag can penetrate the pores and on account of the large surface for attack reaction may be rapid. A dense brick thus resists corrosive reaction much more readily.

4. If the slag or charge flows rapidly over the brick surface, the reacted material is washed away and fresh surface is exposed for more attack. Thus, a fluid slag that flows down a refractory wall rapidly will wear down the wall much faster than a viscous slag that accumulates and moves slowly.

Heat transmission

Heat may be transmitted through refractories by conduction and by radiation. Convection is of little importance in transmitting heat through furnace walls, but may be of great importance in carrying the heat from the interior of the furnace to the walls. In a dense refractory the heat is carried through the walls by conduction. If the refractory is porous, convection of the gases in the tiny pores is relatively unimportant in aiding heat transmission, but at higher temperatures radiation across the pores becomes important. Since the amount of heat radiation is proportional to the difference between the fourth power of the temperatures, radiation becomes of increasingly more importance with rise in temperature. The more porous a refractory, the greater is its resistance to heat transmission, especially at low and moderate temperatures. At high temperatures radiation across the pores becomes more important and the insulating qualities are reduced. As a rule, it is desirable to have a furnace lined with refractories of low heat conductivity to reduce heat loss, but in some cases this causes the inner surface of the refractory wall to reach such high temperatures that its life will be reduced. Therefore it is desirable in some cases to use walls having good heat conductivity and to cool the walls from the outside. For muffles, in muffle furnaces where the charge and the flame are separated by a refractory wall, a refractory possessing good heat conductivity is desirable.

Use of grog in refractories

As a rule, fire bricks are not made from a single raw clay, for the firing shrinkage would be high and the porosity of the brick would be low. A high firing-shrinkage causes cracking and warping of the refractory and it is difficult to produce the finished product to close dimensions so desirable in refractories. Most refractories are made from a plastic clay and a grog. The grog is a granular, non-plastic material consisting of flint fire clay or of fire clay that has been fired in a kiln and then ground to a granular form. Broken, cracked and warped fire bricks are commonly ground to make grog, but grog is also produced by firing clay especially for this purpose.

Increasing the amount of grog in a refractory up to about 70 per cent continually reduces the firing shrinkage. After this point the addition of grog has little effect on the shrinkage. The size and amount of grog affects the spalling characteristics of the brick. In the range from 2 to 20 mesh the spalling resistance increases with increase in the size of the grog particles. An increase in the amount of grog up to approximately 50 per cent increases the spalling resistance of the bricks.

Insulating Refractories

The use of insulating bricks for backing-up refractories to conserve heat has been common practice in many installations. A number of refractory manufacturers are now making a porous refractory brick that has heat insulating properties itself. Among the advantages of using such insulating refractories may be mentioned saving in heat storage, reduced time in bringing the furnace to temperature (for a given rate of firing a furnace lined with insulating refractories may be brought up to temperature in from 25 to 35 per cent of the time for a furnace with a firebrick lining), and saving in fuel, the saving amounting to 20 per cent and over for a fully insulated furnace.

The heat insulation is obtained by producing a porous brick. Insulating refractories have a bulk density ranging commonly between 45 to 50 lb. per cu. ft., as compared to 120 to 140 lb. per cu. ft. for firebricks. The porosity may be obtained by: (i) adding a combustible material such as coal or sawdust to the refractory mix, which burns out in the kiln; (ii) aerating a powdered body; (iii) bloating the mix in slip form by gas evolved by chemical reaction; or (iv) developing a stabilised foam in a refractory mix in slip form.

Refractory Mortars and Cements

Firebricks are customarily laid up with a mortar in slip form consisting of ground fire clay and water. The mortar is applied with a brush, or the face of the brick may be dipped into the slip. Only enough mortar is used to fill up any irregularities in the surface of the bricks, for the bricks must actually rest on one another.

Cements in plastic form are also used for patching walls and for constructing walls. Such cements may contain a number of ingredients, raw clay and grog being in wide use. For some cements sodium silicate is added. This makes the cement air setting to hold it in place until the first firing. Plastic refractories are used widely for building up refractory walls especially in boiler furnaces, generally referred to as monolithic walls. A refractory cement called "lumnite", consisting essentially of calcium aluminate, has many of the properties of a Portland cement (calcium silicate) and in addition is refractory. It can be mixed with a refractory aggregate such as old firebricks or a variety of other materials ground to proper size to make a refractory concrete that can be placed in the same manner as ordinary concrete, and has similar air setting properties.

CHEMICAL STONEWARE

Chemical stoneware is used widely in industry to resist acid solutions and also mildly alkaline solutions. It is used for pipes, stop-cocks, acid pumps, montejus, chlorinating vessels, condensers, retorts, stills, packing for absorption and scrubbing towers, pickling baskets, and a variety of other purposes. The physical properties of chemical stoneware are summarised in Table 15.3.

Table 15.3. Physical properties of chemical stoneware.

Ultimate compressive strength, lb. per in.^{2}	82,000–115,000
Ultimate tensile strength, lb. per in.^{2}	1,650–4,300
Modulus of elasticity, lb. per in.^{2}	$1.05–5.95 \times 10^6$
Specific heat	0.2
Thermal conductivity, Btu. per ft.^{3}	0.78–2.05
Linear expansion, per °F	$1.9–2.3 \times 10^{-8}$
Ultimate bonding strength, lb. per in.^{2}	5,900–12,500

The chemical composition of stoneware is usually between the following limits: SiO_2 60–70 per cent, Al_2O_3 20–25 per cent, Fe_2O_3 0.8–1.8 per cent, $CaO + MgO$ 0.5–1.2 per cent, $K_2O + Na_2O$ 1.5–3.5 per cent. The best grades of stoneware are considered to contain approximately 70 per cent SiO_2 and 22–23 per cent Al_2O_3. Chemical stoneware is closely related in its composition and properties to porcelain. In fact, for small articles, especially for laboratory use, porcelain is used, and it can be considered at high grade, highly refined chemical stoneware.

Clay for chemical stoneware is carefully selected, and it is usually necessary to blend two or more clays to produce the desired properties. The clay should contain little iron oxide and lime compounds, and $CaCO_3$ especially should be absent. The clay should vitrify at about cone 9 (1250°C or 2282°F) and should produce a dense texture. Grog in the form of ground stoneware is necessary for the manufacture of all but smallest pieces of ware. The grog is ground so that all will pass through a 60-mesh screen, but a good share should be retained on a 200-mesh screen. For some clays it is necessary to add 5 to 10 per cent feldspar as a flux in order to produce a dense body.

The raw materials are crushed and screened and then mixed with excess water in a blunger. On leaving the blunger the slip is passed through a magnetic separator to remove the iron-bearing particles. The slip is then filtered in a press to remove excess water and some of the objectionable soluble salts. The plastic filter cakes, the grog, and any feldspar are thoroughly pugged in a pug mill. The pugged body is then stored in cool moist vaults for two weeks to improve the plasticity.

The plastic material is formed into desirable shapes by one of several operations, such as jiggering, extruding, hand pressing, casting, and dry pressing.

The ware is carefully dried and can be fired in any of the types of kilns used for firing heavy clay products, the down-draft kiln being generally used for this purpose. It is salt-glazed in the kiln, but the inside of the ware may receive a special glaze for better chemical resistance.

Chemical stoneware is not a good heat conductor and cannot withstand severe thermal shock. To obtain sufficient mechanical strength the walls are made relatively thick which further hinders heat transfer. Thus for heating or cooling solutions in stoneware the heat is seldom transferred through the container wall. Cooling or heating coils are placed inside the container. These coils may be made of chemical stoneware, for the wall sections of pipes can be made relatively thin.

CERAMICS—POTTERY

The term ceramics was originally applied to pottery; its scope has now been broadened to include "all products fashioned from silicates or oxides and rendered durable in form and composition by a heat treatment applied at some stage of the process.

The principal types of pottery articles are: table-ware or crockery, other types of white ware, sanitary ware, decorative and artistic ware, and industrial ware. Crockery may be porous earthenware, vitreous porcelain, bone china, and sometimes, stoneware. Hotel china is made from the best porcelain. Sanitary ware is made of vitreous stoneware and porous earthenware. Laboratory articles are made of chemical porcelain. Chemical or acid-proof stoneware is used in process industries.

Earthenware is opaque and hard enough to withstand ordinary usage; it has a porous body and can be glazed. Coarse earthenware articles—jugs, jars, basins, kitchen utensils, etc.,—both glazed and unglazed are common in the rural parts of India. Fine earthenware is stronger but not so strong as stone- or porcelain ware. It has an almost white body and is fairly porous. Glazed earthenware is used for all kinds of table- and toilet ware and for some types of sanitary articles.

Stoneware has a dense vitrified body consisting essentially of a skeleton of unfused material, all the pores and interstices being largely filled by molten glass. It is almost impervious to water, hard and opaque with a concoidal fracture, and resistant to heat and strong acids. On account of the plastic nature of the clay used stoneware can be made in larger sizes than porcelain, and the process of manufacture is simple.

Coarse stoneware has a coloured body. Many types of sanitary articles and domestic ware such as pickle and jam jars, jugs, tea-pots, ink-pots, etc. are of stoneware. Fine stoneware has a light-coloured body, suitable for artistic and decorative work. Chemical stoneware such as tanks, jars, vats, etc., form a special variety of fine stoneware with high mechanical strength and resistance to thermal shock.

Drain or sewer pipes are particular types of stoneware with maximum 5% porosity. The manufacture of drain pipes is a specialised branch of the ceramic industry.

Porcelain has a dense, highly vitreous, white body. Hard porcelain consists of a skeleton of felted mass of microscopic mullite crystals embedded in a felspathic glassy matrix together with some grains of quartz, and is translucent. It is used for table-ware. Chemical porcelain is a special form of hard porcelain used for laboratory work. Electric porcelain used for insulators, switches, and other electrical appliances, possesses high mechanical strength, resistance to thermal shock, and dielectric properties.

China-ware or bone china is a special kind of soft porcelain. It has a translucent vitrified body with a milky white appearance resembling alabaster. The translucency is due to the use of bone ash as a fluxing agent. It is superior to domestic earthenware in hardness, and is extensively used in England for the manufacture of high quality table-ware.

For high tension (H.T.) and low tension (L.T.) electric insulation and electrical fittings, porcelain and stoneware are employed. Glazed fireclay insulators are used for L.T. insulation, and steatite porcelain insulators are required wherever high voltages are involved. Electrical porcelain and stoneware with sillimanite body are employed for H.T. insulators.

Indian Industry

Pottery is one of the most important village industries in India and there is hardly a village which does not have its own potter, or Kumhar, who employs the traditional techniques in making a large variety of vessels required for domestic use. Any locally available plastic clay forms the raw material for his work. The clay is sometimes mixed with non-plastic earth or spent wood ashes or cow dung and kneaded by treading. The mixture is passed through bamboo sieves and aged for about two weeks for 'souring' with kneadings at short intervals. The article is shaped on the potter's wheel. For large articles, a crude pot with the required mouth-size is first made on the wheel; it is removed for drying from the main clay mass by cutting with a thread or wire and partially dried in the sun. The final shape is given by beating on a round polished stone with a slab of wood. The articles are burnt in crude kilns. The locally available fuel is stacked in the kiln, the articles are arranged one upon the other, and the spacings closed with fuel. The upper part of the stack is covered with a layer of clay and the bottom is provided with a number of openings for firing. Once it is ignited it is allowed to burn freely. The kilned goods comprise over-burnt to under-burnt articles. The products are not usually glazed.

Art pottery

Artistic pottery is made throughout the country on a cottage industry scale. The wares are shaped on the potter's wheel by throwing, or by pressing in clay moulds. Important among Indian art pottery are the red-glazed pottery of Dinapore, black and silver pottery of Surajgarh (Bhagalpur), imitation *bidri* of Patna and Surat, painted and gilded pottery of Kotah and Amroha (Rajputana), glazed and unglazed pressed pottery of Madras and glazed pottery of Punjab.

Chunar pottery (vases and toys) are fired at about 800°F and coated with glaze. Lead glaze or leadless glaze made from bottle glass and borax is employed. The coated articles are fired at about 700°F.

Glazed faience pottery is produced in Khurja. Khurja articles comprise flower pots and vases with floral designs, water jugs, powder pots, plates, and tea-pots. The patterns and paintings are of Indo-Persian character; blue and green colours are imparted by cobalt salts and copper oxides in a highly alkaline glaze of soda glass. The designs are painted by hand on dried articles or on a white engobe which is applied before baking-in ceramic colours, coated with a thick coating of transparent blue, green, yellow, red, or dark red glaze, and finally baked in a kiln. The transparent glaze shows the multi-coloured decorations on a white background.

Nizamabad pottery is famous for its lightness and shining black base. The surface is rendered so smooth and polished that the fired ware shines without any glazing. The articles are decorated with designs engraved on the surface and filled with tin or lead amalgam. The neat designs and exquisite workmanship of Lucknow art pottery are comparable to some of the best art potteries of the world.

CHAPTER 16

Glass Industry

INTRODUCTION

Glass is an inorganic substance in a condition which is continuous with, and analogous, to the liquid state of that substance but which has attained so high a degree of viscosity as a result of having been cooled from the fused condition that it is, for all practical purposes, rigid. All known glasses are thus undercooled liquids. This definition excludes all organic substances. Some organic materials such as glucose solutions and certain plastics would otherwise be included in the glasses but since they differ so greatly in composition, methods of manufacture, and properties from the substances which have always been known as glass, they are classified in a different division of chemical technology.

Transparency is frequently considered a characteristic of glass, yet much commercial glassware is either translucent or opaque. It is always true, however, that the lack of transparency is caused by the fact that some material, usually crystalline, is dispaired or suspended in a glassy matrix which is itself transparent. The material may be a colloidal suspension, as in a gold or copper ruby glass; or a suspension of sub-microscopic to microscopic particles, as in some opaque glasses; or microscopic crystals ranging in size from the lower limits of resolving power to fairly coarse crystals, as in some opals and white glasses. In all of these cases, however, the material is essentially glassy.

The dominant characteristic of glass is that it is a rigid material obtained by cooling from the fused condition, with resultant continuously increasing viscosity, but without crystallisation. The definition has been so worded as to exclude the necessity of its being below its proper freezing point, that is, an undercooled liquid. The circumstances that all known glasses are undercooled liquids is not, so far as we known, essential. A material might have its true freezing point below ordinary temperatures and yet be rigid enough for use as glass, and be properly considered a glass. However, no such material is known. The essential feature is that glass has the non-periodic structure of a liquid, and X-ray studies have shown that the atoms in glass form a network which is not periodic and symmetrical as in crystals. In silica crystals the dominant feature is the tendency of each silicon to be tetrahedrally surrounded by four oxygen atoms and each oxygen bonded to two silicons. The orientation of each tetrahedral group to every other is definitely fixed and indenfinitely repeated in space. In silica glass the dominant feature is still the tendency of each silicon to be tetrahedrally surrounded by four oxygen atoms, but the orientation of any one tetrahedral group with respect to neighbouring groups can be practically random. The difference in structure between a crystal and a glass is illustrated in Fig. 16.1.

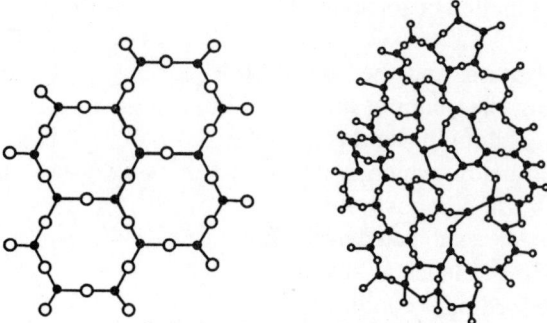

Fig. 16.1. Schematic representation in two-dimensions of the difference between the structure of a crystal (left) and a glass (right).

In silicate glasses the same tendency toward tetrahedral grouping persists. As other oxides are introduced and the number of oxygen atoms to each silicon atom is increased, some oxygen atoms become bonded to only one silicon. Fig. 16.2 shows a schematic two-dimensional representation of the structure of soda-silica glasses in which each silicon is shown surrounded by only three oxygen instead of four as in a spatial representation. Some of the oxygens are bonded to two silicons, others to only one silicon, and as the SiO_2 content is decreased with increase in M_2O and MO the proportion of singly-bonded oxygens increases. The sodium ions occupy holes in the irregular silicon-oxygen networks and in soda-lime-silica glasses the calcium ions occupy similar positions.

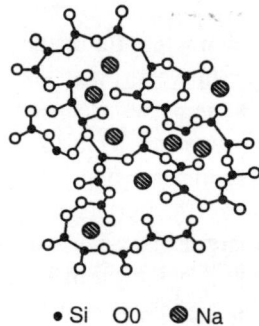

● Si   O O   ◍ Na

Fig. 16.2. Schematic representation in two dimensions of the structure of soda-silica glass.

COMPOSITION OF GLASS

Commercial glassware is for the most part of simple composition. It is a soda-lime-silica glass containing relatively minor quantities of magnesia, alumina, and sometimes boric oxide. This relatively simple composition is essentially that which has been used for glass from the earliest times, and the reason for this narrow range in composition is the basis of glass technology.

The essential feature of glass is that when it is cooled from the molten conditions it does not devitrify or crystallise. Molten glass resembles other molten substances in that on cooling it has a true freezing temperature at which crystals should begin to separate. This is the same as the melting temperature, where the last trace of crystals will dissolve in the melt. However, glass differs from other materials in the readiness with which it undercools, that is, passes through its freezing point without freezing. The temperature at which glass is melted bears no relation to its true melting or freezing point which is called the liquidus temperature.

Devitrification ruins the glass, and the composition must be so adjusted that the glass does not crystallise, especially during the shaping and annealing processes. The two features which are of importance are first, the actual liquidus temperature above which the glass cannot devitrify because it is above its melting point, and below which it will crystallise under certain conditions, and second the viscosity. All commercial glass compositions are melted at temperatures higher than their liquidus temperatures, but most glasses are subjected to shaping operations to temperatures near to and frequently below their true freezing points. Hence at shaping temperatures the viscosity must be great enough to prevent crystallisation under ordinary conditions of working.

Water Glass (Sodium Silicate)

Sodium silicate is manufactured by fusing soda ash with clean sand in furnaces especially suitable for silicate manufacture. These are generally of the tank furnace type meant for continuous production. The heating and the operation are also similar to those of glass furnaces.

$$SiO_2 + Na_2CO_3 = Na_2O.SiO_2 + CO_2$$

The fused mass is taken out from time to time out of the furnace, cooled to solid blocks, and then broken into pieces. Sometime the liquid molten mass is broken to pieces by a stream of cold water. The pieces are then digested with steam and water under pressure in boilers. The solution thus obtained is then evaporated to required consistency and is then known as water glass.

Water glass is an important commercial article and used in fire-proofing, weighting of silk, in preparation of adhesives. It is also used as a filler in cheap type soaps.

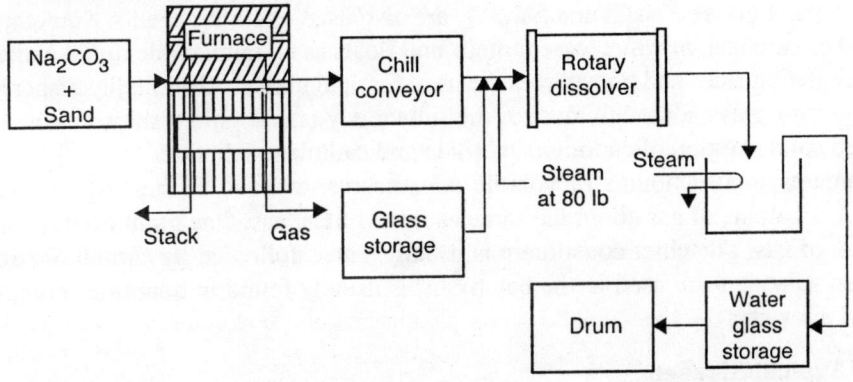

Fig. 16.3. Flowsheet for manufacture of sodium silicate (water glass).

Soda Lime Glasses

Most successful glass compositions are in the field of devitrite, usually in the silica-rich portion of the field, and sometimes a little over the boundary in the tridymite field. The SiO_2 content ranges from about 69 to 72 per cent; CaO, from 12.5 to 13.5; Al_2O_3, from 1 to 4; and Na_2O, from 13 to 15. This general composition range has been used from the beginning of glass manufacture and is used today because it includes the only compositions of matter which are fluid enough at an industrially accessible temperature to be melted on a commercial scale, viscous enough to be worked above the freezing points, and so viscous at their freezing points that they cannot devitrify.

Pure soda-lime glasses are rarely used. Small amounts of other components are always present, either accidentally as impurities or deliberately added for modification of properties. Some or all of the limestone is frequently dolomitic, especially in this country, and the resulting glass is more difficult to devitrify than that containing CaO alone. However, a pure MgO glass is not satisfactory. Barium oxide is often added in small amount to make an easier-melting and more brilliant glass. Sodium oxide is frequently replaced by potassium oxide, especially in optical glasses because of greater brilliance, and glasses containing both sodium and potassium oxides are superior in chemical durability. The use of potassium oxide is limited by its greater cost and the higher viscosity of the potash glasses.

Alumina in small amounts is practically always present, and amounts up to 2 or 3 per cent are common. It improves chemical durability and decreases the tendency to devitrify. The amount which can profitably be added is limited by the great increase in viscosity with resulting high melting temperature and greater attack on refractories. Boric oxide is frequently added to make the glass easier to melt, to lower the coefficient of expansion, to improve chemical durability, and to decrease the tendency to devitrify. Iron oxide is usually kept at a minimum because of its objectionable colour. However, in cases where the colour does not matter, better chemical durability, freedom from devitrification, and ease of melting make it a desirable constituent.

Minor Constituents

Minor constituents of glass are either accidental impurities or substances added to secure some desirably effect. Iron oxide is usually present as an impurity which is undesirable because of its colour. Several oxides, Cr_2O_3, V_2O_5, CuO, U_2O_5, MnO, NiO, CoO, are added as colouring agents. The chemistry of coloured glasses is complex, As_2O_3 is often present as a "fining agent," a substance which aids in the removal of bubbles from the glass, Sb_2O_3 is sometimes present for the same purpose and is used in quantity in some optical glasses. $NaCl$ and Na_2SO_4 are also used as fining agents. Sometimes much of the Na_2O is added as sulphate, in which case it melts and floats as an immiscible liquid and carbon must be added to reduce it. Fluorides and phosphates are used for opal glasses. Colloidally dispersed gold and copper are used in ruby glass, although most of the ruby today is a selenium ruby, in which the colour is due to dispersed solid solution of cadmium sulphide and cadmium selenide.

Glass also contains small amounts of volatile constituents, evolved on heating *in vacuo*, in such amounts as to give a volume of gas about the same as that of the glass. One gram of glass thus contains from 0.5 to 1.5 cc. of gas. The chief constituent is usually water, followed by carbon dioxide, nitrogen and oxygen. When sulphates are used in the batch SO_2 is usually found in quantities comparable to or even in excess of the water.

Composition of Typical Glasses

The compositions of a number of glasses are given in Table 16.1. Numbers 1 to 4 are antique glasses. Prior to the development of glass-working machines glasses of much the same composition were used for most purposes. Representative compositions are No. 5, a window glass made by the old hand cylinder process, and No. 6, a bottle glass. Polished plate glass is high in lime, as shown by No. 7. Modern flat-drawn sheet glass is usually a little lower in CaO and contains some MgO, as shown in No. 8. Modern high-speed machine production of container ware requires a glass not only of controlled viscosity but also having a definite viscosity-temperature curve, and this requirement has caused the development of special glass compositions. They differ from the older glasses chiefly in a reduction in CaO and an increase in the Na_2O and Al_2O_3 contents. A typical container glass is No. 9.

Table 16.1. Compositions of some typical glasses.

No.	SiO_2	B_2O_3	Na_2O	K_2O	MgO	CaO	BaO	ZnO	PbO	Al_2O_3	Fe_2O_3
1.	61.70	—	17.63	1.58	5.14	10.05	—	—	—	2.45	0.72
2.	62.71	—	20.26	20.26	4.52	9.16	—	—	—	1.47	0.96
3.	69.82	—	13.51	2.18	3.09	5.79	—	—	—	1.40	1.80
4.	68.48	—	14.95	2.83	5.28	5.71	—	—	—	0.70	—
5.	72.26	—	14.01	—	—	13.34	—	—	—	1.42	1.42
6.	70.9	—	10.66	—	0.49	14.02	—	—	—	1.35	2.24
7.	72.68	—	13.17	—	—	12.95	—	—	—	0.50	0.07
8.	74.14	—	16.23	—	1.29	9.06	—	—	—	0.84	0.08
9.	74.50	—	15.0	—	4.1	5.5	—	—	—	0.81	0.09
10.	63.0	—	7.6	6.0	0.2	0.3	—	—	21.0	0.6	0.6
11.	73.6	—	17.23	—	3.67	5.37	—	—	—	—	—
12.	75.9	—	7.1	7.9	0.17	8.7	—	—	—	0.14	0.08
13.	64.7	10.9	7.5	0.37	0.21	0.63	—	10.9	—	4.2	0.25
14.	80.5	12.9	3.8	0.4	—	—	—	—	—	2.2	—
15.	74.6	—	9.0	11.0	—	5.0	—	—	—	—	—
16.	70.4	7.4	5.3	14.5	—	2.0	—	—	—	—	—
17.	40.6	—	1.5	7.8	—	—	—	—	43.8	—	—
18.	28.4	—	—	2.5	—	—	—	—	69.0	—	—
19.	34.5	—	—	—	—	—	42.0	7.8	—	5.00	—

The first electric light bulbs were made from a glass containing lead oxide, of which No. 10 is an example. This differs little from a light flint optical glass. It has been supplanted by a glass of the soda-lime type containing significant amounts of both magnesia and alumina, as shown in No. 11.

Laboratory Ware

Laboratory ware requires a high degree of chemical durability and the usual soda-lime types are not satisfactory. One way in which improvement was sought was by increase in silica content and replacement of some soda by potash, as shown in No. 12. The chemical durability was increased by replacing some SiO_2 by B_2O_3, and some Na_2O and CaO by ZnO and Al_2O_3 (No. 13).

These types of ware have been largely replaced by the Pyrex chemical resistant glass (No. 14) which is fundamentally different from the usual soda-lime-silica composition. It is better considered to be a glass in which the melting point of the silica has been lowered by the addition of B_2O_3 and, in smaller amount, of Al_2O_3, with only the smallest possible amount of alkali. It is intrinsically superior to the best glasses of the soda-lime group in its chemical durability and in its resistance to breakage from heat shock or mechanical strain.

Optical Glass

The optical glasses have the greatest diversity of composition because of the lens designer's need for the widest range of optical properties. They are charactersied by almost perfect uniformity of composition and freedom from physical defects, and their quality represents the highest achievement of the glassmaker's art. No. 15 is an ordinary crown, similar in composition to window or plate glass; No. 16 is a borosilicate crown; No. 17 a medium flint; No. 18 an extra dense flint; and No. 19 a dense barium crown, of a type widely used for anastigmat lenses.

The terms "crown" and "flint" are frequently used. Their present usage is inconsistent and they are now chiefly of historical significance. The name "flint glass" comes from the use of flint as a source of silica in a clear glass of good quality, which also contained lead oxide. It now means either a clear, colourless, bottle glass of the type used for dispensing drugs, or a type of optical glass. As applied to optical glass it originally was a glass containing lead oxide, characterised by high index and relatively large dispersion. The term 'crown glass' is of uncertain origin. It probably arose either because glass manufacture was a monopoly of the English crown or because the old method of manufacture of window glass was by the 'crown' process.

It thus denoted a glass of the composition of window glass, and as originally applied to optical glasses was a glass of low refractive index and small dispersion. At present however, the dividing line between optical crowns and flints is wholly arbitrary, and as applied to some of the newer glasses it is devoid of significance. Similarly, the incongruous terms 'crystal glass' refers to a clear, colourless glass, often of slightly higher index than ordinary glass because of a small lead oxide content. Glass No. 10, if used for cut glass, might be termed a "crystal" but its lead content is about twice that more commonly used for crystal. The same glass is often used for tubing but many types of glass are worked in tube form.

PROPERTIES OF GLASS

Chemical Durability

The resistance which glass offers to the corroding action of water, atmospheric agencies (primarily water and carbon dioxide), and aqueous solutions of acids, bases and salts is a property of great practical significance, and is denoted by the term 'chemical durability'. In a large proportion of the uses to which glass is put its power of resisting such attack is the chief reason for its preference over competing materials. An example is the use of glass containers, of which enormous numbers are used for the distribution of commodities ranging from milk to medicine and acids. In this field the superiority of glass leaves it without a competitor.

Even in chemical manufacturing, where the requirements are more exacting, glass is being used to an increasing extent as an engineering material because of the resistance which it offers to surface attack under extreme conditions. In other uses of glass chemical durability is a secondary factor, but the requirement of a chemical durability sufficient for the service contemplated, places a limit on the compositions which may be employed.

Examples of such uses are those in which glass is chosen for its optical properties these uses range form windows to lens systems. Although glass used for such purposes is not subjected to as drastic treatment as in the preceding cases, it is essential that there be no appreciable amount of surface alteration. The necessity of producing glass stable enough to serve the purpose for which it is intended places a practical limit on the compositions which may be employed. The limit set by this requirement differs widely with the use for which the glass is intended.

Many glasses possessing desirably optical or mechanical properties are unsuitable because of their susceptibility to corrosion; others may be suitable for optical purposes in protected lens systems and still be worthless for laboratory uses. The methods of testing glass thus become of fundamental importance.

Testing

The evaluation of the chemical durability of glass is difficult because the process is not simply one of solution but it is one of decomposition as well. There is no true "solubility" of glass in water. All that can be measured is a rate of decomposition, and rates of reaction differ from equilibrium values because they are highly susceptible to slight differences in experimental conditions.

The only true measure of the suitability of a glass in a particular service is its actual use in such service, but it is desirable to have laboratory methods which make it possible to anticipate the results of the service test. Many methods have been proposed, but no one of them is free from serious criticism. All have been developed from the basic thesis that the verdict of actual service may be anticipated by accelerating the corrosive action by the use of more drastic reagents, or by increasing the surface through the use of powders, or by the use of higher temperatures, and by using sensitive methods to detect and measure the resulting action.

The choice of method is determined largely by the use to which the glass is to be put. Methods suitable for differentiating glasses to be used in optical instruments may be far too mild to discriminate adequately between types of chemical glassware. For resistance to attack by acids, the glasses high in SiO_2 and B_2O_3 are best, but they are not necessarily the most resistant to attack by alkaline solutions. Change of temperature at which glasses are tested will frequently alter their order of durability, and may have no relation to their relative excellence in service at ordinary temperature. Our knowledge of the durability of glass, of the factors affecting the durability in the different uses of glass, and of the factors determining the results obtained by the various methods of measuring durability is far too meager to justify positive statements of any but the most general type. The setting up of specifications based on any method of testing should be done only after securing evidence as to correlation of such results with actual service.

PHYSICAL PROPERTIES

The physical properties of glass are determined primarily by chemical composition, but when measured at ordinary temperatures they are affected by the thermal history of the glass. As the glass is cooled from a high temperature some molecular re-arrangement takes place with a general decrease in interatomic distances and possibly an increase in regularity of atomic distribution. At high temperatures and low viscosities this re-arrangement is practically instantaneous.

As the glass is cooled through the upper part of the annealing range it takes from a few minutes to hours for the glass to become stabilised, and below the annealing range impracticably long times are required for the glass to attain the equilibrium conditions. The stabilisation affects all the properties of glass, although it is usually a second-order effect. In a precise consideration of the relationship between composition and property it is necessary to define the thermal history. It is not sufficient to state that the glass was well annealed.

Measurement and specification of properties are necessary in desiging of equipment using glass and for control of process and product. Some properties can be calculated from composition by assuming an additive contributions from each component proportional to the amount of it present, but all such calculations are rough approximations. Most measurements are made at ordinary temperatures, but viscosity and surface tension can be measured only at high temperatures where the glass is in a fluid or molten condition.

Viscosity

The viscosity is of importance in control of glass manufacture. Glass is usually melted at such a temperature that its viscosity is from 10 to 15 poises, and the most favourable temperature for hand-gathering is about 300 poises. The "softening point", an empirically-specified point on the viscosity temperature curve, is the temperature at which a thread of glass 9 inches long and 0.6 millimeter in diameter lengthens under its own weight at the rate of one millimeter per minute when heated in an electric furnace throughout its upper 9.3 centimetres of length. This corresponds to about 4.5×10^7 poises. The annealing temperature determined by Lillie as 4.7×10^{13} poises is such that internal strain will disappear by viscous flow in about 15 minutes, and the strain point (viscosity about 10^{14} poises) is such a temperature that there is practically no viscous yield in a short time. At the strain point the glass anneals in about 16 hours, and below it no permanent strain will be introduced if the glass is cooled as quickly as is possible without breaking.

Surface Tension

Surface tension is the other property usually considered at high temperatures where the glass is molten. The measurements in the literature are not concordant. They usually agree in that the surface tension is not greatly changed by change in composition, and that the temperature coefficient is small and negative, of the order of 0.01 to 0.04 per cent per degree. The best value of the surface tension of ordinary soda-lime-silica glasses is about 300 dynes per centimetre at 1200°C. Glasses high in B_2O_3 or in PbO have lower surface tensions.

Calculation of Physical Properties

Certain of the physical properties in multicomponent glass systems are sufficiently close to being additive that they may be computed by appropriate simple relations.

Density

Density is the property of glass most frequently measured and numerous measurements have been made of glasses of widely differing compositions. Within the narrow compositions range of most soda-lime-silica glasses, calculation of density by means of additive factors should give a fair approximation to the correct value.

Coefficient of expansion

The rate of expansion of glass with temperature is of scientific and technological importance, and the coefficient of expansion is a property of glass which is greatly affected by change in composition. The rate of linear expansion of glass with temperature is almost constant over a temperature interval up to the annealing range of the glass. More accurate measurements, however, require equations of three or more constants to represent them, and when such high accuracy is in question the thermal history of the glass becomes of importance.

Elasticity

When a piece of glass is strained by an applied force it returns to its original size and shape on removal of the force, provided that the temperature is not too high and that the force is neither too great nor applied for too long a time. The property by which the glass regains its original dimensions is called elasticity and is measured by the ratio of the stress to the strain. The constant of proportionality for each type of stress is property of the substance. The stress may be applied in various ways: for example by hydrostatic pressure producing a pure compression in which the elasticity is measured by the bulk modulus K; by torsion, producing a shear, in which case the elasticity is measured by the modulus or rigidity R; or by bending or tension, both of which produce both compression and shear. The modulus of extension in tension is known as Young's modulus, E; the ratio of lateral to longitudinal strain under unidirectional strain is Poisson's ratio, σ. These four quantities are related by the formulae

$$E = 9KR/(3K + R) = 3K (1 - 2\sigma) = 2R (1 + \sigma)$$

Strength

When glass is subjected, for short periods, to stresses within the limits of elasticity, the resulting strain disappears after the removal of the stress, and the piece returns to its original dimensions. With most substances, on addition of load beyond the elastic limit, a condition is reached in which yielding takes place continuously, and if the load is maintained, fracture results. A substance in which there is an initial elastic limit, that is, which is not viscous, is said to be in a condition of plasticity in the region beyond the ealistic limit in which deformation takes place continuously. With glass, there appears to be no region of plasticity and Hooke's law holds up to the breaking point of the glass. There is no accepted relation between the elasticity of a substance and its ultimate strength. A high value of Young's modulus does not necessarily imply a large tensile strength, nor does a high modulus of compressibility imply a large crushing strength. Also, glass may be broken in any one of many different ways, by tension, compression, twisting, or impact; but there is no connection between the numerical values of the breaking strengths as determined by these several methods. Indeed, it is probable that in all the various methods of applying stress to glass, fracture always takes place in tension. A failure in uniform compression is unthinkable, and fracture results from tensile stresses developed by the manner of applying the load.

The strength of a great number of glass samples have been measured, but much uncertainty exists as to the significance of the results. Observed strengths are greatly influenced by the size and shape of the sample. With fine fibres, strengths up to three million pounds per square inch have been observed.

From theoretical considerations there is reason to believe that the strength of glass in tension should be that required to overcome the molecular cohesion, about three million pounds per square inch, but the values found in practice are always far smaller than this hypothetical limit. The reason is usually to be found in the effect of surface flaws or cracks which concentrate the stress at the point of the flaw and thus determine the point of breakage. The strength of glass is determined by the weakness of its surface, and in tempered glass such as is sometimes used in windshields the strength is increased by a heat treatment by which the surface is put under strong compression.

Factors have been proposed for calculating the strength of glass from its composition, but they are of little value. Experience has shown that a tensile strength of about 0.7 kilobar (10,000 lb. per sq. in.) is a fair estimate of the strength to be expected, although larger values are frequently found. In general engineering calculations a factor of safety of 4 or 5 should be allowed when this value of the tensile strength is used.

Thermal endurance

The ability to withstand thermal shock resulting from sudden change in temperature is important in many uses of glass, and methods of measuring thermal endurance find application in the routine testing of glassware. Several formulae have been developed for the calculation of thermal endurance from those properties of glass which affect (i) the development and distribution of strain; and (ii) failure under stress. These properties include thermal expansion, Young's modulus, Poisson's ratio, thermal conductivity and diffusivity, and tensile strength. It is doubtful if the use of any of these formulae in comparing glasses offers any advantage over the comparison of expansion coefficients alone. The coefficient of expansion is more affected by change in composition than are Young's modulus, Poisson's ratio, or the tensile strength. Moreover, since the strength of glass is largely determined by surface conditions, the breakage of glasses of similar expansion properties under thermal shock will be determined largely by the conditions of their surfaces.

Heat capacity

The specific heat of glass is assumed to be an additive function of the constituent oxides, and this is more nearly true for specific heat than for any other property.

MANUFACTURING TECHNIQUES

Basically, manufacture of glass is high temperature conversion of raw materials into homogenous melt capable of fabricating various product.

The mixed batch is conveyed to furnace and melted. The melting temperature is usually around 1450°C. The molten glass is given required final shape in shaping or forming machine. The final product is then annealed to remove the strains generated during the forming operations.

Process and Equipments

Whatever be the type of glass, the manufacturing process can be divided into following phases: (i) batch material handling and preparation; (ii) melting and refining; (iii) shaping or forming; (iv) annealing; and (v) finishing.

Batch Handling and Preparation

Glass industry of today consists of units both modern and old, large and small. There are no universal systems for material handling, storage, mixing and transportation. Each unit has its own set up suited to its batching system. However, some of the larger units have afforded modern material handling methods.

Raw Material Receiving

The raw materials are received in packets or in bulks depending on the value of the materials and the requirements. Smaller plants receive materials such as soda ash and lime in bags while large plants receive in bulk. High value items like selenium, arsenic and cobalt oxide are generally received in small packets.

Unloading

In this operation, manual labour is still predominant. Materials are carried out manually in wheel barrow or wheeled out in small trucks in case of larger units. In case of bulk shipments, manual labour employs shovel and wheel barrow for movement of material to the storage silos or conveyers.

Silos or storage bins in modern units are built up in the elevated position just above the batch house. Various devices for transferring the material to the silos are used depending on the type and location of the bins and amount of material to be handled. A travelling elevator can deliver materials to any bin, while a stationary elevator can deliver only to a particular bin. Bucket elevators, belt conveyers, screw conveyers and pneumatic conveyers are the various materials handling systems which are being used in the industry.

Storage

The method of storing raw materials vary from unit to unit. In smaller until, the raw materials are stored in open floors with partition wall separating the type of ingredients. In large units, raw materials are stored in silos or bins with different ingredients in a different silo or it can be single silo with partition of steel separating the different ingredients. However, minor ingredients are always handled separately in containers, packets and bags.

Cullet Handling

Handling of cullets depends upon the source, i.e., whether generated in house or brought. Cullets are generated internally (waste glass down into a hard mass due to slow annealing). It creates a serious problem for disposal and require size reduction to make fit for recycling. Cullets purchased from outside require processing before its use. The cullets are first screened for removal of unwanted materials like stones, coloured glass, ceramic and ramp particles.

The screened cullet is cleaned in washing machine and then sent to a Hammer mill for size reduction to about 6 mm cubes. Smaller units do it manually.

Mixing

In large glass units, various systems are available for converting large tonnage of raw materials into mixed homogeneous batch. The mixing machine used in glass units are very similar to the conrete mixer. The mixing process can be : (i) manual; (ii) semi-automatic; and (iii) automatic.

Manual System

Generally in the small units operators use shovels for putting the required quantity of materials into a mixer. Here the operator's skill is vital as there are no provision for correcting over weighed material. In the organised units, operator controls the bin gate and allows the required raw materials to fall on the collecting car through a weigh hopper.

Semi-Automatic System

In this system, collecting car is replaced by a conveyer housed in a dust proof enclosure.

Automatic System

The system is similar to semi-automatic excepting that manual operation *in situ* is completely eliminated.

Melting

The melting covers the stage, when mixed batch is introduced into furnace and ends when molten glass, free from solid inclusions air bubbles and homogeneously mixed, is passed into the chamber where it is conditioned before forming.

Furnace

The type of furnace vary widely from unit to unit. Even units manufacturing same type of glass product differ in type and construction of furnace. The smaller units usually have pot furnaces. Pot furnaces have very limited capacity (4 to 5 tonnes of glass) and melting temperature is limited to 1400°C. Fuel consumption is very high and operating efficiencies are low.

Principally four types of coal fired furnaces are employed in the small scale units such as: (i) direct and indirect pot furnace; (ii) bangle making (including slab re-heat belan bhatti) furnace; (iii) pakai bhatti; and (iv) regenerative tank furnaces.

Pot Furnace

In the direct type furnace, coal is fired in the centre of the furnace, the flue gases are reflected back from the roof, pass in and around each pot and are finally exhausted through a common duct.

Tank Furnace

This type of furnaces are mostly regenerative and the production capacity varies from 5–30 T/day of finished product in small scale sector. In the coal fired furnace, coal is burnt in a static producer and then producer gas is used in the furnace. For 20 T/day 'draw of glass normally 4 to 5 static producers are used.

The average glass melting temperature varies from 130°C –1450°C. The molten glass is drawn from the furnace manually and mould into different shapes.

Continuous Regenerative Furnace

Continuous regenerative furnaces are similar to tank furnace. The main difference is in its higher capacity and provision of heat recovery with regeneration or recuperation.

The main type of continuous regenerative glass furnace are: (i) cross port furnace; and (ii) end port furnace.

In both the types, melting and conditioning portions of the furnaces are separated by a refractory wall. A small opening in the wall i.e., throat allows the molten glass to fall effectively removing surface contaminations.

Cross Pot Furnace

Cross port furnaces contain a series of ports on either side of glass wall. Each port may have two or three burners. These burners on the either side of the wall are fired alternatively. The flue gas/air paths are reversed periodically every 15 to 30 minutes enabling air to recover flue gas heat in the regenerator. This type of furnace is most common in sheet and container glass units.

End Port Regenerator Furnace

It is also known as end fired furnace and used in melting of smaller quantities of glass, is most common in shell and tube glass units. Here the burners port are located at the end and the regenerators are situated behind the ports.

Fuels

At one time coal was the most important source of fuel for glass melting. Even coal is converted to producer gas in a producer gas unit situated alongside the furnace.

But oil has almost replaced coal excepting for some unorganised glass units located in and around Firozabad, West Bengal and Bihar. Various grades of oil ranging from L.S.H.S. to extremely viscous residual fuel oils are being used. Natural gas is being used as fuel in Gujarat, Mumbai, Uttar Pradesh and other places.

Refractory for Glass Furnace

Refractory used in glass industry has evolved significantly during the past three decades as a result of continuous evolution of newer and better refractory products.

Depending upon the application areas, refractory can be broadly classified into three categories such as: (i) refractory in contact with molten glass; (ii) refractory for upper structure; and (iii) refractory for regenerator.

Shaping or Forming

Glass may be shaped either by machine or hand moulding. With the development of faster and better machines, the tedious hand moulding process is gradually giving way. The common types of machine shaped glass are window glass, plate glass, float glass, bottles, bulbs and tubes.

MANUFACTURE OF GLASS

Raw Materials

The raw materials of glass manufacture are chiefly sand (SiO_2), soda ash (Na_2CO_3), and lime ($CaCO_3$), with smaller amounts of many other substances.

Sand

Quartz sand is the chief raw material in glass manufacture. It is usually shipped in paper-lined box cars after purification at the plant. The treatment necessary depends on the source of the sand, that is, whether from hard or decomposed quartzite, sandstone, or unconsolidated beds. The harder rocks are mined by blasting, crushed first in jaw crushers, then in gyrating crushers, and finally in a water-fed chaser mill.

The crushed material after passing through revolving screens to remove coarser particles, usually those larger than 20-mesh, is washed to remove organic material, clay, and fines smaller than 100-mesh, then drained, dried, and passed through magnetic separators.

The chief impurities in glass sand are: iron oxide, titania, alumina, calcium carbonate, and zirconia. Of these, iron oxide is the only one usually objectionable. For the best grade optical glass the total iron (expressed as Fe_2O_3) should be less than 0.01 per cent, and for the best grade of tableware, less than 0.035 per cent. As the necessity for freedom from colour decreases, the permissible iron content increases, and for some types of bottles 1 per cent is not objectionable.

Soda ash

The chief source of sodium oxide in the glass industry is soda ash made by the Solvay process. If the sulphate is used in large quantity, carbon in some form is added to reduce it. In smaller quantities it is added to help fining and to reduce scum. Potash usually is added as $K_2CO_3.1.5. H_2O$.

Calcium oxide

Calcium oxide is added as limestone (the carbonate), as burnt lime (CaO), or as unslaked lime [Ca(OH)$_2$]. The choice appears to be based on personal preference. Most limes used are dolomitic, and the presence of magnesia in the dolomite ratio is common, especially in bottle glass. For finer grades, especially optical glass, precipitated calcium carbonate is used.

Miscellaneous Ingredients

Alumina

Alumina is sometimes added as precipitated alumina, made by precipitation from a solution of the sodium aluminate. This is the form of alumina used when a minimum iron content is necessary. In general practice alumina is added as a mineral, usually nephelite, $Na_2O \cdot Al_2O_3 \cdot 2SiO_2$, or feldspar, a mixture of orthoclase, $K_2O \cdot Al_2O_3 \cdot 6SiO_2$, and albite, $Na_2O \cdot Al_2O_3 \cdot 6SiO_2$. The original state of combination of the alumina does not affect its influence on the properties of glass, and the choice of mineral is determined by the ultimate cost, purity, and amount and desirability of other mineral constituents. Alumina in amounts from one-half to two or three per cent is a desirably constituent of glass, especially in its influence in decreasing the tendency toward devitrification, and increasing chemical durability.

Boric oxide

Boric oxide is frequently added to the usual glass types in the amount of one per cent or less, to increase the rate of melting and to improve chemical durability. It is used in large amounts in the borosilicate crown optical glasses, and especially in low-expansion glasses such as pyrex chemical resistant glass. When the soda content is not objectionable it is added as borax $Na_2B_4O_7 \cdot 10H_2O$, or as anhydrous borax, supplied in crystalline form. When the Na_2O content is not wanted, boric acid, H_3BO_3, is used. In either case a high purity product is obtainable.

Nitrate

Sodium nitrate or in some batches, potassium nitrate, is frequently added for its oxidising action. Either product can be obtained in satisfactory purity. Sodium chloride is sometimes present as an impurity in soda ash, and is sometimes added as a fining agent.

Oxides

Lead oxide is usually added as litharge, PbO, as red lead, Pb_3O_4, or as a lead silicate. Red lead is used to secure oxidising conditions in the melt, but the same condition is easily insured by other means. The chief impurity in lead oxide is alkali oxide, usually in amount insignificant for ordinary glass-making. Zinc oxide is used as the sublimed product, easily obtained of adequate purity. Barium oxide when used in large proportion is added as barium carbonate; when used in small proportion barium sulphate is used for the effect of the sulphate as a fining agent. The carbonate is made from the sulphate by reducing to sulpnide, dissolving, and precipitating with sodium carbonate. The remaining sulphide is the chief impurity. Arsenic oxide is added to many glasses as a fining agent, and when purified by sublimation it is satisfactory as purchased.

Cullet

A large proportion of the charge of a furnace consists of broken glass, called cullet. The proportion used varies widely, from 10 to 75 per cent, depending on the amount of scrap glass available and the practice of the operator. It is generally believed that a generous proportion of cullet speeds up melting and produces a more uniform product. When there is not enough scrap being returned, scrap glass is purchased or a furnace may even be run to produce cullet. In this case the glass is powdered by being run with water. Other cullet is crushed, usually with a jaw crusher, but the crushing is only to coarse pieces and often no crushing is necessary.

Equipment and Handling in Glass Manufacture

Equipment

Glass is made either by an intermittent process in which the individual containers, or pots, are heated in suitable furnaces, or by a continuous process in a tank. Pot melting has been used since the beginning of glass manufacture, and is so especially suited for many types of small-scale or discontinuous operation that it probably never will be entirely displaced in spite of the commercial advantages of large-scale production in the continuous tank process.

Glass pots

Glass-making pots are made of clay which is carefully chosen and blended to obtain high plasticity, a minimum of shrinkage on heating, and a maximum of refractoriness and resistance to attack by the glass. They were formerly built up by hand, but are now also made by the slip-casting process. The hand-made pots are more porous and require a long time, from three to six months, to dry, while the dense slip-cast pots are made and dried more quickly. The pots may be made open, or sometimes closed by a hood like top with an opening for gathering. In size they range from small pots used for experimental meltings or for small lots of glass to be used in ornamentation, to large pots for plate glass. These small pots hold from 5 to 25 pounds while the large pots hold one or two tonnes. Most pots are almost cylindrical, with a small taper, and some are oval in cross-section. In the manufacture of plate glass they are removed from the furnace by overhead cranes. The glass is poured onto the casting-table and the pot is retuned to the furnace. For making ordinary blown and pressed ware, the glass is dipped or gathered from the pot and these pots continue in service until they are worn out. In the manufacture of optical glass the pots are used but once. The pot of glass is removed from the melting furnace and cooled in an auxiliary furnace called a pot-arch, after which melt and pot are broken up.

The furnaces used in pot melting are of a wide variety of types, ranging from crude single-pot furnaces without regeneration to regenerative furnaces, either rectangular or circular in cross-section, and holding from sixteen to twenty pots. Plate glass furnaces are usually rectangular, holding as many as twenty pots in two rows. Other furnaces are circular, holding up to sixteen pots, each of which can be set, removed, or worked from, without interfering with the others.

Tanks

Tanks are large furnaces so constructed that the furnace walls themselves are the containers for the glass. The lower part of the tank is built of tank built of tank blocks. These are special refractories of appropriate shapes so that the furnace may be assembled without mortar, and of special composition to offer the greatest resistance to attack by the glass. In recent years the introduction of high alumina and electrocast refractories has so improved the quality of tank blocks as to increase the life of the tank by several fold, which is now often more than a year. Tanks vary widely in size, but a moderately large tank for manufacture of containers may be 16 by 24 feet, and delivers about 90 tonnes of glass per day. Such tanks are separated into two sections by a bridge-wall in which a channel, called the throat, allows the glass to flow from the large melting end to the working end. Tanks for float glass are usually longer in order to give the glass more time to become free from bubbles ("fined" or "planed"), and do not have a bridge-wall. When glass is worked from tanks having no bridge-wall, the "gather" is taken from within annular clay floaters which provide a glass surface free from scum. In all furnaces the top, or crown, is made of silica brick and the structure is reinforced by steel beams and girders. With regenerative firing the checkerwork is below the firing level. The fuel enters the furnace in ports which are arranged to be symmertical with the exit ports, thereby insuring uniformity of heating when the ports are interchanged on the reverse cycle. The firing is either from the side or the rear.

Fuel

The fuel used in glass melting is either oil or gas, with the greater number of gas-fired furnaces using natural gas. The temperature of a tank at the melting end is usually from 1425 to 1550°C, depending on the type of glass and rate of pull, and at the working end about 1200 to 1300°C. The fuel cost is about one-fifth of the cost of glass delivered to the feeder.

Batch handling

The methods used in mixing the batch and feeding it into the furnace range from the crudest hand-mixing and shoveling, which is still practiced in many plants, to largely automatic processes.

The raw materials are delivered in ground or powdered form, and crushing operations are not necessary. Sand, limestone, lime, soda ash, and salt-cake are usually delivered in paper-lined freight cars, from which they are unloaded by shoveling and wheelbarrows, by power shovels and coveyor belts, by suction, or traveling elevator. Materials delivered in barrels or bags, such as borax and boric acid, lead oxide, and nitrates, are usually handled by trucks. Weighing procedures range from hand-barrow operations with platform or flush scales, to systems in which power-operated cars are run successively under the various hopper bins and the requisite weights of ingredient delivered. The batches are delivered to mixers, or the ingredients themselves are collected in portable mixers. The mixed bath is then transported to the feeding floor, in larger plants, by travelling containers. Cullet is usually added in the mixers. Special care must be taken to prevent segregation since uniformity of the batch is highly important for producing good quality glass. In pot operation the batch is fed with scoops which are either hand or machine-operated. Filling of pots is necessarily an intermittent operation. The batch is filled in, melted down, more batch added and the process repeated until the pot is full of glass. Tank operation is more nearly continuous, with filling taking place according to a schedule which depends on rate of pull and time of reversal of the furnaces. Filling takes place from a built-in structure connected with the tank, called a "dog-house," which is usually filled by gravity by a chute from an overhead bin. A recent development is the triangular dog-house, from which the batch is delivered automatically, according to a predetermined time schedule, by a motor-driven pusher in each feeder. The quantity of batch is determined by the size of the batch feeders themselves, the length of the pusher stroke, the number of strokes per minute, and the length of the feeding time.

Annealing

A heat treatment of glass in the viscosity range 10^{13} to 10^{14} poises is important to secure stabilisation. It is in the same range of low fluidity that glass is annealed to prevent the development of permanent strain, or to remove strains introduced by too rapid or non-uniform cooling. When glass is cooled a temperature gradient is introduced. If the gradient is introduced when the glass behaves as an elastic substance, as it does for stresses of short duration at the strain point, the stress resulting from the temperature will remain until the gradient has disappeared and will disappear with the gradient. If the gradient is introduced at the softening point, at which the glass yields quickly under stress, no strain will result from introduction of the temperature gradient, the glass will cool free from strain until the temperature gradient is removed, and the removal of the temperature gradient will produce a permanent stress. The permanent stresses in poorly annealed glasses are developed by the removal of a temperature gradient at low temperatures, and the amount of this stress is the difference between that caused by the removal of the temperature gradient, which produces a compression on the surface, and the temporary strain carried down from the high temperature at which the gradient was introduced. In short, the permanent stress is equal and opposite in sign to the stress lost by flow in the first part of the cooling.

In a typical commercial annealing lehr the change in temperature with distance in a lehr is used for annealing bottles. The front part of the lehr is usually heated by gas. Where the "bloom" on the ware caused by sulphur in city gas is objectionable, either natural or propane gas is used, and electrically-heated lehrs are in development. The delivery end of the lehr is not heated and is usually lower than the front end in order to aid in controlling convection, which usually should be from the cold to the hot end. The ware is carried through the lehr on a woven wire belt.

Glass Products

Hand manufacture

The manufacture of glass throughout the centuries has evolved from a small-scale handicraft to a scientifically-controlled, mechanised, mass-production industry. Many phases of the industry, however, are still dependent on the skill of the glass-blower. Even in the highly-mechanised container industry, special shapes which are used in too small quantity to justify the expense of putting on a machine, are blown by hand.

Free blowing

The process by which glass is shaped entirely by hand-blowing and manipulation of the hot glass without the aid of moulds is known as free-blowing or off-hand blowing. It requires a high degree of skill, and the finest products of the glass-blower's art are free-blown. The molten glass is gathered on a glassmaker's pipe, roughly shaped or "marvered" by rolling on a slab of iron, then given final shape by blowing and hand-shaping with simple tools. Parts not produced by simple blowing, such as stems, handles, and feet, are welded into the main piece.

Blowing in moulds

Much glass is hand-blown in a mould. Paste-moulds of metal, lined with a paste of resinous material, coated with flour or other organic powder which will char and leave a carbonaceous lining, are used for thin ware of circular cross-section. For simple forms the mould is often of one piece, but usually it is of two parts opening on a hinge. The hot gather is formed by marvering, placed in the mould, and the pipe rotated while the piece is blown. The mould is used wet, and the layer of steam formed by the hot glass acts as a cushion which prevents the glass from coming into actual contact with the mould. The product has a good polish and is usually free from mould marks. Unlined moulds of cast iron, usually in two parts, are employed for much large-scale production of shapes that do not require high polish or freedom from mould marks. To prevent wrinkles or ripple marks the moulds are kept as hot as is possible without formation of too much scale. Ware blown in a mould, either paste or cast iron moulds, often requires a subsequent finishing operation. Some ware is gathered by hand on an iron rod, called a pontil, then pressed without blowing.

In the making of hand-drawn tubing a bubble is blown in a gather of suitable size, and the gather is then marvered into cylindrical shape. It is then manipulated by rolling and blowing so that the bubble becomes a cylindrical hole extending almost to the free end of the gather. This free end is then stuck on an iron rod and pulled into a tube by the carrier who walks away carrying the hot mass of glass, while the blower controls the roundness of the tube by blowing and manipulating the gather. Thickness of tube is controlled by the rate of drawing, and local hot-spots at which the tube tends to draw down are controlled by fanning.

Machine-made ware

The mechanical processes used in working glass differ widely according to product. The first branch of the industry to be considered is flat glass, consisting chiefly of window and plate glass, and laminated glass used for automobile glazing.

Window glass

Window glass, formerly made by the cylinder process, is now almost entirely made by drawing in flat sheets directly from the tank. In the Fourcault process the sheet is drawn through a "debiteuse," which is a clay block sunk to a varying depth into the molten glass, and containing a longitudinal slot through which the glass rises under a slight head. The slot is four to six feet long, and the sheet is pulled up by rollers which are as wide as the full width of the sheet. The sheet is drawn at exactly the rate at which the glass rises though the slot under the slight head, thus avoiding the narrowing effect of surface tension, and the thickness is determined by the rate of drawing, which, for a single-strength sheet, 80 inches wide, is 48 inches per minute. The sheet is drawn vertically through an annealing zone and then cut into standard lengths. In another process, the sheet is drawn vertically without the aid of a debiteuse. In the Colburn process, the glass may be drawn directly from the tank over bending rolls, then horizontally on a flattening table, and through an annealing chamber. The product of these continuous sheet processes is greatly superior to the older glass; so much superior that flat-drawn glass is replacing polished plate in some of its uses.

Plate glass

By plate glass is meant a sheet that has been ground and polished. Much plate glass, including all large sheets, is still made in industrial pots, which are lifted from the furnace, the glass being poured onto an iron casting table. A heavy iron roller flattens the glass into a sheet, which is then annealed, inspected, and cut into sheets. These sheets are then ground and polished.

In the Bicheroux process the glass is poured from the pot onto a set of power rollers, and in some plants the glass flows directly from the tank through rollers. Wire glass is made by a process similar to that for plate glass except that a sheet of woven wire is rolled into the glass. "Cathedral" glass is cast with a rough, usually figured, surface and is not ground.

Safety glass

Safety glass is made like a sandwich, with a layer of an organic plastic between two layers of glass. The first laminated glass used cellulose nitrate of cellulose acetate, but this was not entirely satisfactory because of the effect of light in turning it brown and spoiling its plasticity, and because it became brittle at low temperatures. Newer plastics, vinyl butyrol and methyl methacrylate, have overcome these difficulties, and now laminated glass is used in the greater part of the automobile industry, in airplanes, and for bullet-resisting glass.

In the manufacture of laminated glass, the glass and plastic are cut to shape, inspected, washed, an adhesive is applied to the glass if such a step is necessary, and the sandwich assembled. The parts are pressed together at moderate temperature and pressure, and then given a final treatment in an autoclave at about 250 pounds pressure to weld the plastic and glass firmly together. After pressing, the edges are cleaned and given a finishing operation to seal them, except with such plastics as are not affected by moisture. The outermost glasses of the laminated light should be as thin as possible to avoid undue spalling when broken. In some of the production the eliminated glass used is flat-drawn sheet glass, which is not as flat as plate glass and consequently does not given such completely satisfactory vision.

Tubing

The first mechanisation of the manufacture of glass tubing was in the provision of towers for drawing. The large hand gather is marvered as in the hand drawing, stuck on, and then fastened to a vertical elevator which draws the tube at a uniform rate in a closed chamber. This process is much used for thermometer tubing where uniformity of bore is essential. The white backing is marvered into the gather.

True mechanical drawing of tubing takes the glass directly from the furnace. In the Danner process the glass flows onto a hollow revolving mandrel through which air is blown. The glass wraps itself around the mandrel and flows together. The drawing process is started by hand; the tube being drawn off the mandrel and fastened to a pair of asbestos-covered belts which draw it continuously. In another process the tube is drawn upward through a circular clay block floating on the glass, while air is blown upward through the centre of the block.

Sheet Glass

Manufacture of sheet glass is one of the older sector of industry compared to container and lamp glass. Almost all the glass units involved in manufacture of sheet glass are either using conventional Fourcault process or PPG processes. These processes require finishing i.e., grinding and polishing prior to despatch. In the year 1959, newer and better process called Float glass process was invented which eliminated grinding and polishing operations and gained wide spread attention. It is completely displacing the older process. This development has been a milestone in the glass industry and has also afforded uniform, distortion-free glass, cost reduction and flexibility in product range.

Fourcault method

Glass from the furnace is passed into a canal especially shaped floating refractory block called 'debiteuse', made of material which has lower density than glass and therefore it floats, but it is maintained just below the level of the glass surface. The debiteuse contains a slot, which has been accurately cut and the glass is drawn upward through this slot, cooled and then passed vertically through rollers. By the time glass reaches the roller, it is rigid, and fresh hot glass is drawn upward through this slot, cooled and then passed vertically through rollers.

By the time glass reaches the roller, it is rigid, and the whole process can be operated continuously, in that fresh hot glass is drawn up through the debiteuse as fast as the roller pull the rigid glass away. The sheet starts in the first place by allowing a piece of iron bait to contact the molten glass, and this iron is withdrawn upward bringing the glass with it.

Once started the machine pulls glass continuously, and the flat surface becomes rigid by the time, it reaches the first set of rollers. Finally the glass free of roller marks is achieved, if debiteuse is of high quality.

Pittsburgh process

Pittsburgh is a vertical process. The window glass produced from this process is 3 mm to 6 mm thick. Here the floating debiteuse is substituted by a submerged draw bar also called shut off for controlling and directing the molten glass. The shut offs are made of clay and its installation is a tedious operation.

Shells and Tubes (Lamp Glass)

In spite of growth of the industry and undertaking electrification of rural areas on priority basis, electric lamp industry is still at incipient stage. Thus it has scope for future growth.

The lamp glass industry is fairly self-sufficient in respect of raw materials, refractories and control instruments etc. which are all indigenously available. There is no dearth of technical personnel either. However, furnaces used for manufacturing lamp glass-ware are presently either of imported design or fully imported. The fabricating machines are still being imported.

Hollow ware

The various types of glassware, including plates, dishes and cups, tumblers and goblets, lamp chimneys, globes and reflectors, ovenware, and the wide variety of containers, are made by one or two types of machine. One type is the suction-feed Owens machine, the pioneer machine in this field which started the wholesale mechanisation of the industry; the other type is the gobfeeder machine, of which there are several variants.

The Owens machine is a self-contained, synchronised, circular assemblage of a number of units or arms, which takes glass out of an auxiliary circular revolving pot fed from the main tank. In its operation a suction mould is lowered onto the molten glass, a gob of the requisite size is sucked up, and the excess glass cut off. The suction-mould then swings away, leaving a shaped gob called a parison suspended from the neck-mould. The blow-mould then rises to surround the parison, and the bottle is blown. The suction-mould repeats its cycle. The blow-mould delivers a finished bottle and again rises into position to take another parison from the suction-mould. For large ware, each arm has a single mould, but for smaller bottles a double or triple cavity mould which shapes two or three bottles at once, can be used. Most other types of automatic machines makes use of a feeder, which delivers hot glass in gobs of pre-determined shape and weight at regular intervals to a forming mould. Numerous problems arise in accomplishing this apparently simple purpose, and differences in their solution characterise the several types of feeders. Feeders are an integral part of the "blow and blow" and "press and blow" machines on which are made the larger part of all hollow ware. In these latter type machines the gob is formed into a parison by pressing in the feeder mechanism and then blown into a mould. The two units may be separate or integral.

Lamp bulbs

Most lamp bulbs are made on the Corning machine, a ribbon machine which has also been used for bulbs for Christmas tree ornaments and for tumblers. The glass flows in a continuous stream from the forehearth of tank and passes between two rollers which deliver a continuous narrow ribbon of glass. One roller has depressions on its surface, giving a series of uniformly spaced buttons along the ribbon, which fit over the successive openings in an endless metal belt.

The ribbon of molten glass is carried by the belt in a straight line under a row of blow-heads, forming a second endless chain moving in time with, and above, the first. As the belt moves on, the glass bottoms sag through the openings in the belt and the blow heads register above them, after which each lengthening bulb of glass is enclosed from below by a split rotating metal mould which is paste-lined. The bulb is blown as it moves forward in its place in the line by air admitted from the wind-box. After the bulbs has cooled enough to maintain its shape, the moulds open and the bulb is separated from the glass ribbon and directed into an annealing chamber. This machine can produce as high as one million bulb in 24 hours, an excellent illustration of the manner in which mechanism in the industry has greatly reduced the price of an improved product.

Float Glass

In the float process, molten glass is drawn through a channel by a horizontal pair of water cooled rollers, as in plate process. The glass is then passed into high temperature chamber, and floated on the surface of a bath containing molten tin, so that that upper surface in fact is fine polished by the action of molten tin. Thus, surface irregularities are removed and a high polish is produced. As the ribbon floats down the chamber, the temperature is reduced and eventually the sheet becomes rigid enough to be taken by rollers without becoming marked or deformed. It is then transferred to the annealing lehr. In float process, the inside atmosphere should be maintained constant for correct operation.

Bottles, Jars and Other Containers

Today, almost all the container glass are manufactured by automatic machines. A wide range of machines are in use. It requires two stages of forming (or more in special cases), in the first stage, the hot glass is passed into a mould known as a blank mould, where the length and neck are formed. At the same time, a slight hollowing out of the body of the glass is made. At the end of operation, the glass is in a form known as parison. In the second stage, the parison is transferred to a mould which has the shape of the final bottle, and the parison is blown out to form the bottle.

Glass wool

"Fibreglas" is a new product whose manufacture does not resemble any process yet described. Glass fibres have long been known. However, the modern development, by the Owens-Corning Fibreglass Corporation, has been not only one of method of manufacture, but also one of greatly enhancing the usefulness of glass fibres by decreasing their diameter. Much of the older material was little more than thin glass rod. The modern product is smaller in diameter than any other industrial fibre.

Two processes are used in the production of "fibreglas". Continuous filament is produced by mechanical drawing of the molten glass into strands of indefinite length (measured in miles). The average diameter of the individual fibre is 0.00022 inch, but filaments from 102 or more orifices are drawn into a strand without twist. After spraying with a lubricant, one or more strands are twisted together to form yarn, and the yarns are plied by twisting two or more together to provide any desired thread or cord construction. The continuous filaments are smooth and cylindrical, and the yarn is free from fuzziness.

Properties of fibreglas

The fibres owe their properties to their extreme fineness, which is of the order of five ten-thousandths of an inch. The tensile strength of glass fibres increases rapidly as the diameter decreases, and especially as it decreases below two-thousandths of an inch. Strength as high as one and a half million pounds per square inch have been obtained on fibres of about one ten-thousandth of an inch in diameter.

Combined with high strength is extreme flexibility and softness to the touch. It has been calculated that a fibre two ten-thousandths of an inch in diameter has a length of 23 million feet, or 4356 milles per pound, and total area of 1500 square feet. With so large a ratio of surface to mass, the properties of the surface, and especially its chemical durability, become of utmost importance. The composition of the glass must be such as to have unusual resistance to weathering and to attack by chemical reagents. Glasses have now been developed which will give satisfactory service even under extreme conditions. These are usually soda-lime-silica glasses, low in soda, and containing Al_2O_3 or B_2O_3 or both. Some alkali-free glasses are in use.

Staple fibres

In the manufacture of staple fibers, molten glass glowing from a multitude of fine holes is met by jets of high-pressure steam which break each stream into a large number of fibres, each so fine as to be almost invisible, and essentially endless. If it is to be used as a yarn, the fibres are sprayed with a sizing or lubricant, as in other textile operations, then gathered into strands which may contain 60 or more individual filaments. In the manufacture of yarn, several of these strands are combined to form a thread, and two or more of the threads are combined with a reverse twist. The resulting yarn is free from any tendency to kink, and its further treatment is the same as that of any other textile material.

This yarn has many uses. It may be spun into cloth, used for dresses, for non-flammable curtains and hangings, as a filteraing medium in filter presses, or as a separator in storage batteries. Decorative yarns, in which the colouring materials is a part of the glass itself, are available. When used as insulation in electrical motors and generators, it increases their safe operating temperature, increases efficiency, and reduces fire hazards. When the coarser steam-blown fibre is gathered into a mat, it becomes almost useful heat insulator which is finding innumerable applications.

Fibreglass mats

Fibreglass in mat or loose form is used as a heat insulator in houses, Pullman cars, battleships and submarines, refrigerators, and as a filter in air-conditioning. The fibre for this purpose resembles staple fibre but is of coarser texture. In many of these uses it is considered as one type of "mineral wool," a trade name which also includes wool made from rock or slag. These various products all have about the same thermal conductivity, but the presence of unfibrised material, shot and slugs cause the rock and slag wool to have a higher density, about 10 pounds per cubic foot. Glass wool has a natural density of about 1.5 pounds per cubic foot, and when compressed to 10 pounds has a lower conductivity.

Rock wool

The rock or slag wools are made by melting in a cupola, usually with coke as fuel. The molten charge is run into upwardly-directed jets of steam which shred it into wool. The raw product is of variable fibre-thickness, and contains from 10 to 30 per cent of granular material called "slugs," some of which is usually removed before marketing. The raw materials are of widely differing character: argillaceous limestone or calcareous shale, ordinary shales, lead, iron or copper slags. To these may be added limestone, dolomite or quartz rock to give a more easily melted mix. Rock and slag wools differ widely in composition, but a representative one is: SiO_2, 42 per cent; Al_2O_3, 12 per cent; CaO, 34 per cent; MgO, 10 per cent and; Fe_2O_3 1 per cent. When the lime content is too great the wool tends to disintegrate and settle from reaction with moisture, and the presence of sulphides also cause deterioration.

Building blocks

In the production of glass building blocks, the glass must be of high chemical durability, and is usually of the soda-lime type containing alumina or boric oxide. The blocks are hollow, and are pressed in two parts. The molten glass is delivered in units or gobs of the proper size by an automatic feeder to an automatic press. The half-sections from the press are sealed together before cooling, either by dipping the edge in molten aluminium and pressing together, or by transferring the hot sections to an automatic sealing machine.

This is a rotating and indexing machine which carries the hot sections over and above stationary burners, and then moves the hot sections into contact and presses the weld. The subsequent annealing of the rigid, entirely closed, and stressed blocks is difficult, and requires special equipment designed to give the necessary time and temperature-schedules which must be precisely followed.

Chemical engineering construction

During the past few years glass has been increasingly used in chemical engineering construction where corrosion is troublesome or where purity of product is essential. Used originally in the heavy-chemicals industry, particularly against HCl solutions or wet chlorine gases, it has more recently found numerous practical applications in processing fine chemicals, pharmaceuticals, foods and beverages. "Pyrex" glass piping of heat and chemical resistant glass, complete with fittings and accessories, is available in sizes up to 4" I.D. and is designed for maximum working pressures of 100 pounds per square inch. Other glass chemical engineering equipment includes heat-exchangers, bubble-towers, absorption columns, raschig rings, and centrifugal pumps. The average composition of some glasses is given in Table 16.2.

Table 16.2. Average composition of some glasses.

	SiO_2	B_2O_3	Na_2O	K_2O	CaO	MgO	Al_2O_3	Fe_2O_3
Table glass	77.70	—	10.02	—	10.02	—	0.43	0.21
Plate glass	71.80	—	11.10	—	15.70	—	1.26	0.41
Bottle glass	63.51	—	9.50	—	14.32	3.87	2.54	2.36
Jena lab. glass	65.30	15.06	—	—	—	—	3.80	0.41
Sheet glass	72.42	—	13.13	—	11.97	—	2.13	0.35
Corning pyrex lab. glass	80.75	12.00	4.10	0.10	0.30	—	1.50	0.70
Spectacle glass	69.04	0.25	5.95	11.75	12.07	—	—	—

SMALL SCALE SECTOR

Firozabad has the origin of glass making right from the Mughal period and accounts for more than 70 per cent of all glass items produced in the small scale sector of the country. A peculiar feature of glass industry in Firozabad is that in spite of non-availability of raw materials and coal which is the main fuel, majority of the industry's produce is from this district. The industry is pre-dominant by use of traditional skills, marked by craftsmanship transferred over from generation to generation.

In Firozabad, majority of the population is engaged in the glass industry and making a variety of glass and glassware items. The growth of glass industry in Firozabad has been quite substantial.

Mostly soda lime silica glass products are manufactured in Firozabad. However, lead glass and opal glass products are also manufactured for decorative items and laboratory glasswares.

The bulk of the production is of glass and glassware items which include hollow ware and pressed items. The high value added items in this category are headlights for automobiles, wine glasses, glass shells etc.

The produce from Firozabad is supplied directly and indirectly to various segments of the industry. The customer segments include automobiles industry, hotel industry, lighting industry, glass industry, perfumery, medical and educational institutions, research and development institutes, household etc. Manufacturers sell their products to the dealers and their agents.

Manufacturing Technology

The Firozabad glass industry can be classified broadly in two categories depending on the furnace used, which may be either pot furnace or tank furnace. Generally, both type of furnaces are coal-fired. However, the technology which has been quite traditional has not undergone significant changes. The following technology is being used in Firozabad for glass forming operation.

Mouth blowing

The required quantity of molten glass is gathered at one end of an iron tube, placed in a mould and blown by mouth to take the required shape of the mould.

Semi-automatic machine

The required quantity of the molten glass is taken on an iron rod and moved to pneumatically controlled hydraulic press to take the desired shape as per the mould.

Pressing

In this technique, the molten glass is collected on iron rods and the desired shape is achieved by placing into the moulds of the press.

Process of manufacture

The process of manufacture, particularly forming depends upon the product and the composition of glass.

Batching

The major raw materials like silica and soda ash are mixed manually or mechanically in pre-determined quantities depending upon the type of products to be manufactured. Cullets are added while melting the batch of raw materials. In Firozabad, about 40 to 50 per cent of the batch constitutes cullets.

Melting

The batch mixture is charged into the melting furnace and melted at temperature ranging from 1200°C to 1450°C. Two types of furnaces are being employed in the Firozabad glass industry—pot furnace (closed/open type) having a capacity of 4 tonnes/day with melting temperature in the range of 1200°C to 1350°C and tank furnace having capacity upto 30 tonnes per day with melting temperature in the range of 1400°C to 1500°C. These furnaces are utilised to produce a variety of items including tumblers, thermoflask refills, lamp shades, head light covers, laboratory ware and jars etc.

Forming

The molten glass drawn from either pot or tank furnace is formed into required shape of product by the following techniques: (i) blowing and moulding; (ii) drawing; (iii) pressing; and (iv) spiralling.

Blowing and moulding

Mouth blowing in conjunction with moulding is adopted for producing items such as bulb shells, lamp shades, chimneys, beakers, tumblers, thermoflask refills etc.

Drawing

Tubes and rods are made by drawing operation.

Pressing

Products such as auto head light covers, electric light covers, railway signal light covers, ash tray; candle stand and paper weight etc. are manufactured either by hand press or semi-automatic press. Tumblers are also manufactured by semi-automatic press.

Spiralling

This operation is carried out for bangles only.

Annealing

Annealing is carried out in annealing chambers and lehrs.

Finishing

Finishing operation include grinding of edges, cutting, melting (fire polishing), and sand blasing etc. This operation differs from product to product.

POLLUTION IN GLASS INDUSTRY

Glass is a vital component for day to day domestic use as well as of the industry. Glasswares provide a more hygienic way of life style. The greatest advantage of the glass items is that it is absolutely non-polluting as a waste. The whole glass waste materials are recycled and re-used.

Air Pollution from Different Types of Glass Melting Furnaces

For melting of soda lime, borosilicate and other special glass other than lead.

Coal fired pot furnaces

The direct coal fired pot furnaces existing in the small sector at Firozabad, generate air pollution in the form of particulate matter. The particulate matter mainly consists of unburnt coal particles and raw material fines. The higher level of emissions are due to improper firing and carry-over of the raw material. The limits for particulate matter emissions for pot furnaces melting soda lime glass is 1200 mg/nm^3. The emission for sulphur dioxide are generally low as it depends on the sulphur content of the fuel. The average capacity of such furnaces is 4 tonnes per day of melt glass.

Coal fired tank furnaces

These furnaces also fall into the category of the small scale sector. The capacity of such furnaces may range from 5 tonnes glass melt to 30 tonnes glass melt per day. The crude type of static producers are used to produce gas which is burnt in the furnace for glass melting. In these furnaces, particulate matter is the main pollutant. The limits of particulate emission in this furnace is 2 kg/hr. Sulphur dioxide emission should meet the criteria of stack height.

Oil fired tank furnaces

The oil fired tank furnaces are in use for glass melting in small as well as large scale industry for many years. In these furnaces, the emission limits for particulate matter is 2 kg/hr. upto a product draw capacity of 60 MT/day. For the product drawn capacity more than 60 MT/day additional particulate emission of 0.5 Kg/T of product is permitted.

Air pollution from melting lead glass

Lead glass

Furnace of any capacity melting lead glass, should not emit particulate matter more than 50 Mg./Nm^3 and the lead contribution should not be more then 20 Mg/Nm^3.

ROLE OF GLASS IN FURNITURE

Glass used to be perceived as a fragile medium. But with houses shrinking in size by the day, glass provides the ideal solution for creating a spacious ambience. No other material can match the feeling of space that glass furniture imparts to a home or office. Glass today has come of age as a strong and versatile medium for furniture. The basic idea is to render furniture that is artistic as well as practical and sturdy.

Glass tables create a light, ethereal design mood. Glass chairs can be sturdy enough to sit upon and they also liven up a dull corner. Another plus point is that maintenance of glass furniture is not a back-breaking task.

The potential of glass furniture had remained untapped in the country as earlier most glass would be marred by air bubbles and pinholes. Now 100 per cent clear glass can be sourced easily. Designers today enjoy greater leverage because size and thickness too can be varied according to requirement. Available today are glass side tables, centre tables, cabinets, consoles, pedestals and dining tables.

Glass as a medium can be used in a variety of forms and styles. It adds life and transparency to any piece of furniture and makes it look light. It has got a fluid feel to it that can be used very productively.

Coloured glass can be used for making side tables that harmonises with the colour theme of the sofa. Glass furniture breaks the monotony of other heavy pieces in the room. When using glass furniture, make sure that the lighting of the room is good. Glass has always been the mark of beauty. Glass top tables can have polyacrylic figurines as their bases. These acrylic bases may be shaped as flowers, animals or a lady in different poses. Thereby they can serve both a decorative purpose and can also be used for serving things. Plus one can go for vibrant coloured glass tables shaped as a dolphin, panda, horse, cupid, sunflower or guitar.

Glass furniture frees glass from its historical role of ornamental accessory and makes it a protagonist in furnishing an unobtrusive, important presence in people's lives. Curved glass is the result of a challenge to chemical-physical laws, ready to be modelled by designers. Glass wall units are the ultimate in modularity. They enable you to create a living room that transcends the restrictions of space and size.

Glass has the potential to capture the light of the sun, the colour of grass, and the blue of the sky. The transparency of this magical crystal can even take on the hue of your emotion.

Glass as a designer medium can fulfil imagination and style, and can be adapted to artistic shapes suitable for any interior. Glass makes the room much lighter, giving it a minimalist, yet formal look.

People are going in for more glass accessories to add value to interiors," These days while shimmering crystal serves as the table top, the legs are made up of Morano glass in different colours like blue, yellow, green and light brown.

ROLE OF GLASS PLANT AND EQUIPMENT IN CHEMICAL INDUSTRY

Glass is resistant to more chemicals than any other material of constuction. Thousands of kilometres of glass pipelines installed all over the World proved that glass has won and found full application as a construction material in construction of production equipment in all branches of insdustry. Today glass pipelines and glass fittings are in successful service in dairies, food canning, breweries, distilleries and all plants of chemical, pharmaceutical industries etc. With its superior universal resistant to the effects of chemical agents, smooth wall surface and transparency, it has become a traditional construction material in uptodate plants.

Characteristics of Glass

Some of the characteristic properties that make the material so versatile are given below.

Chemical resistance

Inert to practically all known chemicals except hydrofluoric acid, phosphoric acid and hot strong caustic.

Thermo resistance

Coefficient of expansion 32×10.7 per deg. C over 0–300°C.

Specific heat

Average 0.233 between 25–300°C.

Thermal conductivity

0.0027COS over usual operating range.

Working temperature

Can be heated upto 300°C provided it is not subject to rapid thermal changes. The thermal shock should not exceed 120°C per minute. Safe working temperature is 240°C.

Mechanical properties

The breakage strength is as high as 703 m/cm^2. But to allow irregularities and thermal stress 35.2 m/cm^2 forms practical basis for calculating internal working pressure of glass pipelines. All pipelines are safe for use even under full vacuum. The maximum internal working pressure recommended at 25 mm bore is 7 kg/cm^2 and it drops to 0.77 m/cm^2 at 300 mm bore. The dimension etc. are governed by B.S.S. 2958 and it is considered as an international standard. With the availability of glass pipelines and fittings in India, it is now possible to have all glass distillation and reaction units of 200 litres capacity for pilot lab work of high value chemicals. The manufacture of small items such as beakers, test tubes etc. was started in our country nearly fifty years ago. However, the industry did not make much progress for quite a few years. It was only during last ten to fifteen years, that full fledged large scale manufacturers of glass equipment came into existence.

Manufacturing of glass equipment for the chemical industry has made spectacular progress during last few years. This has been partly due to the boom in manufacture of chemicals, drugs and dyes in our country. It is a well known fact that in the manufacture of many chemicals contact with metals has to be avoided. This is the reason why all glass equipment hold a position of importance in the field of chemical manufacture. The transparency of glass equipment is an added advantage over metal equipment. The operator can actually see the progress of reaction. This gives him a better control on the entire process and any likely accident, either due to boiling-over of the reactants or due to building up of pressure can be easily avoided. In the field of manufacture of large equipment meant for use in a manufacturing process, prime importance is given to proper annealing of all glass parts. Glass is fragile and liable to break due to any mechanical shock. This drawback cannot be overcome by any means because of the inherent property of glass which cannot be altered. However, breakage due to heating or cooling can definitely be prevented by proper and careful annealing of all glass parts. The user of glass does not know whether it is properly annealed. That is why the responsibility of supplying properly annealed equipment rests on the manufacturer. Fortunately, instruments to test annealed stage of apparatus etc. are available. The manufacturer can therefore, ensure that this equipment is properly annealed before it is supplied it to the clients.

Pollution Control

The chlorination of organic compounds is an important and widely practiced procedure in the manufacture of solvents, dyes, drugs, disinfectants, fungicides, and a host of allied products. Half of chlorine gas used in chlorination reactions comes out as HCl gas, the disposal of which is a major problem

For example:

$$CH_4 + Cl_2 \longrightarrow CCl_4 + CCl_2$$

$$C_2H_4 + Cl_2 \longrightarrow CH_2 + CH\ Cl\ HCl$$

$$CH_2\ Cl\ CH_2\ Cl \longrightarrow CH_2 + CH\ Cl\ HCl$$

Glass the only answer for HCl gas pollution control

"HCl absorbers" constructed wholly out of borosilicate glass is unattached by HCl and has higher thermal endurance which is vital for heat transfer operations. It has proved to be the most economical answer to the problems of HCl disposal. It is essentially a combination of scrubbing and condensing units. The scrubbers serve the purpose of exposing a large surface of water to HCl gas while condensers are essential to absorb the high heat of solution of HCl gas in water.

The HCl gas is introduced at the bottom of the vertical assembly while water enters the top and flows down. The resulting hydrochloric acid solution collects in the receiver placed below the assembly. By recirculating the acid from the receiver, a strength of 25 to 30% w.w. can be easily obtained.

Chemical engineers in our country have now understood the importance of glass as a material of construction for chemical plant/equipment and have adopted glass in their plant as material of choice. Many multinational companies like—Sandoz., Bayer, Glaxo, Pfizer, Roche, NOCIL etc. and be Public sector organisations like HOCL, HAL, IDPL, Research Institutes like RRL, Space Centre, CCL, IIT, are using glass plants since many years.

Safety Measurements

If proper care is taken, glass though appearing fragile to physical impact, can be used safely. The glass equipment and piping is always supported between 2 or 4 poles of G.I. or M.S. pipes. A nylon net (finishing net) can be wrapped around these poles so that the visibility is not obstructed and at the same time no flying object can touch the glass. The net being very flexible can be wrapped around by hand on any shape of the framework and all service lines such as water vacuum, electrical steam, can come out from the equipment.

The recent development of covering the entire glass equipment by F.R.P. using special high temperature resin has brought the 'danger by physical impact' factor in glass to practically nil. This special equipment is called "Corguard in Foreign Countries and "Armoured glass" in India. The equipment is precovered by the manufacturer by this process. Virtually any shape of glass equipment can be covered with this process. The special resin used in this allows the glass equipment to be heated internally to its normal limiting temperature of 240°C without the external covering getting spoiled.

The 'armoured glass' equipment thus has all the advantages of glass but has safety factor added to it. It also allows the operator to continue the process without stoppage as the external covering does not allow the glass to break in pieces resulting in flowing out of the chemicals in the event of breakage of glass. The damaged glass pipe can be safely changed after the batch is over. It also prevents thermal shock to glass pipeline carrying hot vapours/chemicals from accidental spillage of cold water on it within the plant. Various types of 'Windows' can be formed of the desired size to allow the operator to "See" the process at required points.

Cost

The glass equipment is cheaper than glasslined/teflon lined equipment by approximately 30 to 50% depending upon shape., Through glass is costlier by about 30 to 40% compared to special SS, glass is thought of when SS cannot be used due to its ability to withstand only selected chemicals. When the size of the reaction vessel is large the present tendency is to use reactors of SS or GL and the entire overhead distillation setup is made of glass. The combination makes the entire equipment economical.

CHAPTER 17

Photographic Products Industry

INTRODUCTION

The word photography is derived from two *Greek* words which mean "drawing by light." Vision is the most common method humans have at their disposal for receiving and conveying impressions of the world in which they exist—ample proof of the importance of photography to modern civilisation. There is no field of human activity at the present time, whether it be industry, science, recreation, news reporting, printing, or the mere recording of family histories, which is not, in some phase or other, touched upon by the photographic process. Although it is one of the youngest of the chemical process industries, it has universal appeal.

Photography finds widespread application. The amateur uses it in three major ways, for reflection prints, for home movies, and for small transparencies. Professional usage is more varied: entertainment, education, sales promotion, graphic reproduction in magazines, display advertising, industrial illustration, data recording, and medical and scientific records. Table 17.1 delete lists important photochemicals and their functions.

PHOTOGRAPHY

White crystalline silver chloride turns violet when exposed to sunlight because of a photochemical reaction in which Ag^+ (silver ion in the ionic silver halide crystal) is reduced to Ag (elemental silver). As the photo-chemical reduction continues, elemental silver atoms aggregate and grow into clusters of a colloidal size sufficient to scatter light and produce hue shifts. Photography makes use of this photochemical property of silver halide to form images.

The AgBr, Ag(Br, I), and Ag(Br, Cl, I) microcrystals popularly used in modern photography are nominally sensitive only to electromagnetic radiation with wavelength shorter than 500 nm. Photographically, these crystals are said to be blue-sensitive but green- and red-insensitive. The blue sensitivity refers to the intrinsic sensitivity of the silver halide crystals. To reduce the number of blue photons required to produce a developable latent image, i.e., a catalytic centre, the silver halide crystals are treated with chemical compounds that adsorb to the crystal surfaces and may or may not react with them. Such compounds, called chemical sensitisers, usually contain sulphur or gold, and do not significantly alter the light-absorption properties of the silver halide crystals, but they do alter the efficiency with which the latent image is formed.

Table 17.1. Selected photochemicals used in black-and-white photographs.

Use	Chemical name	Action
Developing agents		
Capable of reducing exposed crystals of silver halide in emulsion to silver without reducing unexposed crystals at the same time	p-Dihydroxybenzene	Slow, powerful
	p-Methylaminophenol	Fast, soft-working
	p-Phenylenediamine	True grain (toxic)
	l-Phenyl-3-pyrazolidone	Forms superadditive mixtures with other developing agents
Activators		
Activate developing agent and control pH	Sodium carbonate	Forms sodium salt
	Potassium carbonate	Forms potassium salt
Preservatives		
Guard against oxidation	Sodium sulphite	Transforms oxidation product into
Prevent staining of emulsion	Sodium carbonate	colourless compounds
	Sodium citrate	
Restrainers		
Prevent chemical fog	Potassium bromide	Lowers degree of ionisation of silver
	6-Nitrobenzimidazole	bromide
	Benzotriazole	
Calcium precipitants		
Prevent sludging	Sodium hexametaphosphate	Forms complex compounds
	Sodium tetraphosphate	Forms complex compounds
Wetting agents		
Facilitate absorption of developing solution	Ethyl alcohol	
Neutraliser	Acetic acid	Stops development by acidification
Fixatives		
Dissolve unexposed silver halides	Ammonium thiosulphate	Converts silver halides into water-
	Sodium thiosulphate hypo	soluble compounds
Hardeners	Alum	
	Glyoxal	Hardens gelatin to insolubilise it and increase mechanical strength
Intensifiers		
Increase density	Polymeric dialdehyde	
Reducers		
Reduce density		Varies with concentration; oxidation process

To achieve photographic sensitivity in the green (500–600 nm) and red (600–700 nm) regions, the silver halide crystals are sensitised with dyes, i.e., dye molecules are adsorbed on the crystal surface.

Today, a broad range of photographic materials is available for amateur and professional uses including x-ray films, graphic-art films, microfilms, and complex multilayer coatings for colour films.

Photographic Crystal

The preparation of light-sensitive microcrystals is referred to as precipitation (Fig. 17.1). These light-sensitive crystals are emulsified silver halide grains. Various techniques are applied during crystal growth to achieve the desired grain morphologies (shapes), size-frequency distributions, solid-state properties, light sensitivity, and catalytic activity. Chemicals are usually added to control crystal-growth rates, ripening characteristics, stability, and light sensitivity. During the crystal-growth process, reactant solutions containing a halide and a silver salt (usually silver nitrate) are mixed in the presence of a peptising agent, preferably gelatin. This produces a suspended solid phase which separates in the form of microscopic crystals of silver halide. The peptising agent adsorbs on the grain surface but does not inhibit continued growth; it prevents coagulation of the microcrystalline grains, maintaining a uniform dispersion.

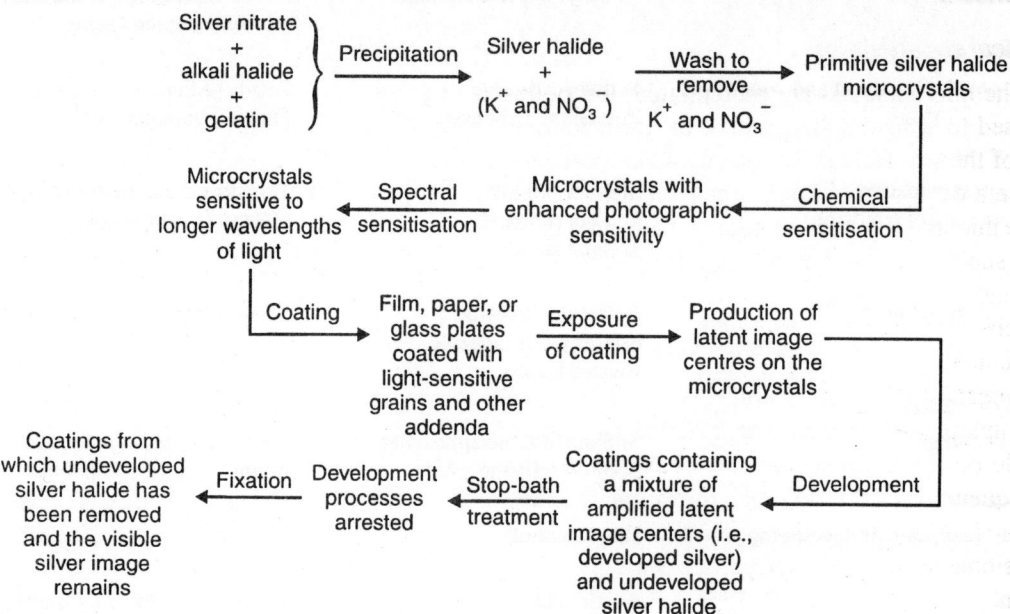

Fig. 17.1. Flow chart of the photographic process.

The ultimate size-frequency distribution of these emulsion grains depends upon the rate of addition of reactant solutions, the temperature, and the presence of growth modifiers or ripeners. The latter are compounds that form water-soluble silver salts or complexes and preferentially dissolve the smallest grains from a given population of crystal sizes and enhance growth in the larger grains. Ripeners, e.g., ammonia, sodium thiosulphate, or sodium thiocyanate, can be added before, during, or after the crystal-growth process.

The reaction to form a silver halide crystal is initiated by nucleation, for which the solution must be supersaturated in silver and halide ions to overcome adverse free-energy effects. As silver halide nuclei form, they provide a substrate or surface for continued growth. When there is a wide range of grain sizes within the reaction vessel, Ostwald ripening can occur. There are three stages in this ripening process; dissolution of small grains, ionic diffusion through the aqueous phase, and finally redeposition of ions on large grains. Ripening is enhanced by temperature increases.

For certain photographic systems, small amounts of inorganic impurities or dopants are added to the emulsion during precipitation to achieve desirable photographic responses. Dopants can have strong effects on solid-state properties of the grains as well as on light sensitivity, contrast, and developability. When crystal growth is completed, the photographic emulsion is a dispersion of silver halide microcrystalline grains in an aqueous gelatin phase. Counter ions, ripeners, and other additives are also present. These by-products must be removed, usually by so-called noodle washing or by flocculation washing. In the former, more gelatin is added after the precipitation and the emulsion is solidified into a jelly after cooling. The more recent technique of flocculation washing requires less time and no additional gelatin and gives a more concentrated emulsion. The gelatin is coagulated by adjusting the pH or by adding salts. The gelatin floccules carry with them the silver halide grains and leave the water-soluble by-products in the supernatant.

Response Enhancement

Chemical sensitisation

After the microcrystals are precipitated, but before they are coated on a support, they are chemically sensitised to enhance their native or intrinsic light sensitivity and spectrally sensitised to increase the range of the wavelength sensitivity. Chemical sensitisation reduces the number of photons required to produce a developable latent-image centre by as much as a factor of ten. Typical chemical sensitisers include thiourea and sodium thiosulphate, gold thiocyanate and potassium tetrachloroaurate, and reducing agents such as hydrogen, *tert*-butylamineborane, stannous ions, or hydrazine. Used in trace amounts (μmol/mol Ag) alone or in combination, chemical sensitisers optimise photographic properties and have beneficial effects on the solid-state phenomena during exposure without affecting amplification during development.

Silver sulphide is generally considered to be the active chemical species resulting from sensitisation with sulphur compounds. The reaction of thiosulphate with silver halide crystals to form adsorbed sulphide on the grain surfaces is activated thermally. If the reaction is allowed to continue too long before quenching or if too much thiosulphate is used, the grains become spontaneously developable (no exposure is required to induce catalytic activity), and image discrimination is lost. The sulphiding reaction is envisioned to be a two-step process involving adsorption followed by a thermally activated chemical reaction:

$$(AgBr)_n + S_2O_3^{2-} \rightleftharpoons (AgBr)_{n-1}[Ag(S_2O_3)]^-_{adsorbed} + (n-1)\ Br^-$$

$$(AgBr)_{n-1}[Ag(S_2O_3)]^-_{adsorbed} + Ag^+ + H_2O \rightleftharpoons (AgBr)_{n-1}Ag_2S + SO_4^{2-} + 2H^+$$

Sulphur sensitisation is an effective and popular method for improving photographic sensitivity. Gold is often used in combination with sulphur, particularly to improve sensitivity to exposures of high light intensity. Gold enhances the catalytic activity of the latent-image centre and reduces the number of photochemically formed silver atoms required for developability, thereby increasing the light sensitivity of the grain.

Spectral sensitisation

The intrinsic absorption, and therefore the intrinsic photographic sensitivity, of silver bromide and silver iodobromide microcrystals falls off rapidly for wavelengths greater than 500 nm. In fact, silver chloride crystals have almost no sensitivity in the visible region of the spectrum. The need to extend sensitivity into the green and red regions is obvious for colour photography. Extending the wavelength sensitivity beyond the intrinsic region is called spectral sensitisation. It is usually done after precipitation but before coating by adsorbing certain dyes to the crystal surfaces. Once the dye molecule is adsorbed, the effects of electromagnetic radiation adsorbed by the dye are transferred to the crystal. The part of a dye molecule that enables the molecule to absorb visible or infrared light is called the chromophore.

Coating Additives

Certain additives are used to facilitate coating operations (e.g., surfactants), reduce spontaneous development in unexposed regions (e.g., tetra-azaindenes, mercaptotetrazoles), and reduce abrasion and permit high temperature processing. Stabilisers include halide ions, benzimidazoles, benzotriazoles, and mercaptotetrazoles. Gelatin cross-linking agents render the coated emulsion layers more resistant to abrasion during handling and improve the thermal stability and hardness of the gelatin.

Emulsion coating

For most applications, the sensitised emulsions must be coated on a base or support of glass, plastic, or paper for convenient handling. Supports are chosen on the basis of dimensional stability, low water permeability, lack of surface irregularities, compactness, cost, and safety. Today, clear plastic film supports made of cellulose esters or poly(ethylene terephthalate) (polyester) are most commonly used. These materials are safe and dimensionally and chemically stable. Paper supports are used for colour or black-and-white print materials. Paper supports may be undercoated with barium sulphate in a gelatin matrix to improve smoothness and enhance whiteness, or water-proofed with impervious resin coatings of polyethylene whitened with titanium oxide.

Emulsion coatings must be uniform in thickness and composition and free of streaks. Multilayered coatings are often composed of more than ten layers, each containing a variety of different chemicals. As light-sensitive materials, they must be coated in the dark. The individual coated, dried layer is generally 1–30 μm thick. The support is transported on rollers past a coating station where the liquid emulsion is delivered to the moving support by pumps or gravity flow from hoppers. In other systems, melted emulsion is held in troughs or trays and the moving support is brought into contact with the liquid emulsion. Intermediate layers are coated between the emulsion and the base to facilitate spreading and improve the adhesion of the gelatin layer to the hydrophobic support. Such interlayers may also contain light-absorbing (antihalation) materials to prevent stray light from reflecting back into the emulsion layer during exposure. Antihalation material, e.g., carbon particles, dyes, or colloidal silver, can degrade the appearance of the final photographic image and must be removed in processing. Overcoating the emulsion layers with a gelatin layer affords protection against image-degrading effects owing to pressure and abrasion.

Exposure

When the image is produced by UV, visible, or irradiation, an optical-lens system is required which focuses the image on the emulsion layers of the sensitive coating. The degree of magnification is a function of the effective focal length of the optical system. In negative-working emulsion coatings, the photoelectrons react with silver ions to form clusters of silver metal on the grain surfaces. These clusters function as catalytic centres for amplification during subsequent development. Negative dye-scale images are produced when the developer molecules that have reacted with the catalytic centre initiate process reactions. In positive working emulsion coatings, the density produced by developed silver or by dye decreases with increasing exposure.

Exposure is a measure of the total incident light energy and is therefore equal to the mathematical product of the light irradiance, I, and the exposure time, t. Positive photographic images can be produced by at least four different exposure-related phenomena: solarisation, the photobleach effect, the Herschel effect, and the Clayden effect. Special development solutions or processing sequences are not required in these four cases.

Development

Developer solutions contain reducing agents, restrainers, and preservatives. The ability to discriminate between exposed and unexposed grains is a property of chemical reducing agents that possess the Kendall structures represented by A—CH=CH and, where n may have zero or integral values and A and B may be hydroxyl, amine, or substituted amino groups. Most, but not all, chemical reducing agents are benzene derivatives with Kendall structures, e.g., hydroquinones, catechols, aminophenols, p-phenylenediamines, and ascorbic acid. Phenidone developing agents and certain thiadiazoles also discriminately reduce silver ions. The activity of development solutions, such as hydroquinone (an important black-and-white developer), depends on hydrogen and halide ion concentrations.

Oxidation

Hydroquinone   Semiquinone   Quinone

Reduction

$$AgBr \xrightarrow{+e} Ag^0 + Br^-$$

The presence of bromide ions in the development solutions restrains the conversion of Ag^+ to Ag^0 by the effect of its concentration on the electrochemical overpotential for the overall redox couple. Preservatives such as sodium sulphide are scavengers for oxidation products. Development can be viewed as an electrochemical redox reaction:

$$nAgBr + H_m D \rightleftharpoons D_{ox}^{(n-m)+} + nAg^0 + nBr^- + mH^+$$

where D and $D^{(n-m)+}$, respectively, correspond to the reduced and oxidised forms of the reducing agent. Development depends upon the diffusion rates of the developer components through the gelatin to the silver halide grain surfaces where the catalytic silver centres are located. Such diffusion processes are rate limiting, and once a reducing agent has diffused to a grain surface, nucleation and growth of an elemental silver phase can be initiated.

In general, the development rate increases with increasing temperature; many modern systems have been designed for high temperature processing which requires specially hardened coatings. Several photographic materials are based on special development techniques such as activator processing, thermal development, monobath processing, and silver-diffusion transfer.

Once satisfactory development has been achieved, the reaction is quenched by the rapid decrease in pH upon transferring the coatings from the alkaline developer solution to an acidic "stop" bath. For both the silver-diffusion-transfer films and the dye-transfer films used in instant photography, development is initiated by spreading a viscous alkaline reagent between an emulsion layer and an image-receiving layer. Recent integral films rely on the timed release of an acid to quench development.

Fixation

In conventional photography, undeveloped silver halide is removed with thiosulphate ions which convert the remaining silver to water-soluble complexes such as argentodithiosulphate and argentotrithiosulphate. Thiosulphate is stable, non-toxic, inexpensive and does not react with gelatin or with the developed silver.

In black-and-white photography, fixation is generally conducted under acidic conditons. Decomposition of thiosulphate is retarded by the addition of bisulphate. The rate of fixation is monitored in terms of clearing time, i.e., the time required for the last visible opacity to disappear. At various post-development stages, the coatings are rinsed to reduce carry-over of chemicals; washing is essential for image permanence.

Environmental Aspects

Before being discharged, photographic effluents require special treatment, such as settling, biochemical degradation, aeration, or chlorination. Silver, usually bound in thiosulphate complexes, can be converted to silver sulphide and removed as a solid sludge.

Silver image

Humidity, temperature, chemistry of the environment, and particularly air oxidation, adversely affect image permanence. The subjective quality of a developed silver image depends upon the colour tone, the brightness reproduction, and the perceived graininess and sharpness. The impression of crispness is achieved when the boundaries and edges of the objects are clear and well defined. The ability of photographic material to record fine details is a function of both development and optical effects.

COLOUR PHOTOGRAPHY

Colour photography is a process by which light of varying wavelength, intensity, and location, collected and focused by a lens system, can initiate chemical reactions leading to a relatively permanent record in two-dimensional space of these variations. All the practical systems of colour photography fulfil at least the following two requirements: the sensitivity of the photographic material must be such that objects producing different colours and different lightnesses form latent images than can be differentiated; the chemical or physical process that makes the latent image visible must maintain these differences and translate them into material images that will modulate some form of general illumination to cause the observer or other sensor to perceive the desired result.

Much of colour photography may be called pictorial in that the end result is a picture that appears natural. In such photography, the sensitive materials must in some way approximate the spectral sensitivity of the eye, and the viewing system must provide a gamut of colours similar to those of the observer's experience.

In technical uses, the image frequently need not be natural to convey information. Energy outside the visible spectrum may be involved: e.g., ultraviolet or infrared. Colours need not be reproduced in a natural or pleasing fashion. The reproduction may be designed for presentation to the human observer or for presentation to some other sensor such as a photoelectric cell.

The fundamental light-sensitive element in most colour-photographic materials, just as in most black-and-white materials, is a silver halide or mixture of silver halides, silver chloride, silver bromide, or silver iodide, dispersed as crystals or grains in a medium such as gelatin. For colour photography a silver halide must be sensitive to all regions of the visible spectrum and sometimes beyond.

Experiments with light and colour and human vision have established that for full colour reproduction the spectrum must be separated into three components, generally in the red, green, and blue light regions. The identity of the separate components must be maintained long enough so that they can be translated into the physical or chemical systems that serve to modulate three components of the viewing illuminant to give the sensation of full colour to the observer. One process called the additive colour process can be illustrated by current colour television systems that use juxtaposed multicolour phosphor elements in the display tubes to generate the gamut of colours.

Most current colour processes depend on silver halide grains which can be selectively sensitised to red, green, or blue and placed in separate layers so that the three records can maintain their identity. The rest of the colour-image-forming process consists of using some reactions associated with the developing silver grain or the silver grain itself to form, destroy, or modify dyes, generally yellow, magenta, and cyan. The latter dyes are termed *subtractive primaries* in that each absorbs or subtracts one of the colours that is the basis of colour vision. Thus yellow absorbs blue but transmits green and red; magenta absorbs green but transmits blue and red; and cyan absorbs red but transmits blue and green. The combination of yellow in superposition with magenta appears red; yellow with cyan appears green; and magenta with cyan appears blue. The three subtractive colours in equivalent amounts reproduce a range of grays through black, depending upon the densities of the three dye images.

Current Basis for Colour Photography

Sensitising dyes

The natural sensitivity of silver halide crystals to ultraviolet and to blue light can be extended to green and red by the adsorption of sensitising dyes to the crystal surfaces. The sensitising dyes absorb green and red light and transfer the energy to the silver halide substrates. In this way the silver halide can form a latent image of light to which it is otherwise insensitive.

The most widely used sensitising dyes are of the cyanine class. These consist of heterocyclic moieties linked by conjugated systems of atoms.

Photographic emulsions

As in black-and-white photography, specially sensitised silver salts are precipitated in gelatin media with which they form emulsions. In order to make use of the ability of such silver salts to differentiate a scene into three-colour records, they must be prepared in separate emulsions, which are most commonly coated as separate layers, in superposition on a supporting film. A variety of other layers are frequently used; a typical colour-photographic material is illustrated in Fig. 17.2.

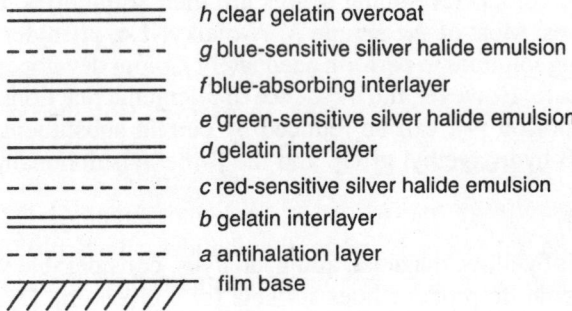

Fig. 17.2. Typical cross-section of a colour-photographic product.

The colour records obtained by exposure of these emulsions may be processed to a colour negative, from which a colour positive can be obtained by printing onto a similar emulsion package, followed by colour processing. Alternatively, the colour positive may be obtained by reversal processing.

Colour development

The chemistry of colour photography involves the conversion of the latent images into dye images. Fischer reported that N,N-dialkyl-p-phenylenediamines are useful developing agents whose oxidation products are capable of forming dyes in proportion to the amount of exposure to a given region of the spectrum. The second component of the dye is the coupler, the choice of which determines the colour obtained with a given developing agent. Fischer's proposal is given schematically as follows:

> Exposed silver salt + developer   → Oxidised developer;
> Oxidised developer + coupler   → Dye.

Colour-Forming Agents

Yellow couplers give dyes with maximum absorption (λ_{max}) between 400 and 500 nm. The compounds suitable for this purpose contain an active methylene group that is generally not part of a ring. By far the most important class of yellow couplers consists of benzoylacetanilides. Ease of preparation and versatility of substituent effects are features of this system.

Magenta couplers are selected to give dyes whose maximum absorption occurs between 500 and 600 nm. Selection of a suitable magenta coupler is facilitated by the wide variety of types available. In general, the magenta dye formers are similar to the yellows in structure except that they give dyes that are shifted in hue as a result of substitution. Such substitution may involve cyclisation, which makes the active methylene group part of a ring structure, or it may involve substituents that extend or intensify the conjugation of the dye structure. Cyan couplers give indoaniline dyes with l_{max} generally between 600 and 700 nm; in almost all cases considerable absorption is noted beyond 700 nm; outside the visual range. Cyan dyes generally also adsorb a considerable amount of green light and usually some blue light.

Colour developers

Important properties of the colour-developing agents are their solubilities, their reactivities, and their effect on the hue of the dyes. Most of the simple N, N-dialkyl-1,4, phenylenediamines are sufficiently soluble in alkaline processing solutions to perform adequately. Colour developers are notoriously allergenic and must be handled with care. However, the incidence of allergenic reactions, resulting in skin irritation similar to that caused by poison ivy, can be reduced by certain substituents. Most prominent among these substituents are the β-hydroxyethyl group and the β-methylsulphonamidoethyl group.

Modification of Dye Hue

Within each of the classes of yellow, magenta, and cyan dyes, considerable variation of hue is possible and, indeed, necessary in order to produce hues suitable for a given use. Dye hues can be modified by the introduction of substituents into either of the dye-forming components, viz, coupler or developer.

Practical Chromogenic Colour Processes

Developer-soluble couplers

The positive silver halide can be developed chromogenically to form a positive dye image without removal of the negative silver image. This principle is used to devise a colour process that was developed in the Kodachrome film. Agfa colour film, like most of the many colour films that have succeeded it, used colour couplers that are incorporated in the emulsion layers. Kodachrome film, on the other hand, depends on soluble couplers in three different colour-developing solutions to obtain the dye images. The complex Kodak process now is used mostly for amateur motion pictures and for 35-mm still transparencies.

Incorporated couplers

A simplification in the colour process is effected when the different dye-forming couplers are incorporated into the appropriate sensitised layers. Most manufacturers use the type of incorporation exemplified by Kodak Ektachrome film. That type involves couplers that are non-diffusible by virtue of long alphatic chains or combinations of short aliphatic chains and aromatic systems without hydrophilic groups. Colour materials containing incorporated couplers can be processed in a reversal mode or in a negative-positive mode.

Reflection prints

By coating the proper amount of emulsion and colour formers on white reflecting supports instead of on transparent film, it is possible to make reflection-print materials.

Non-chromogenic Colour Photography

A number of colour-photographic processes depend upon the black-and-white development of exposed silver salts followed by chemical reactions that result in the destruction of a preformed dye or in the transfer of a dye out of the emulsion to a mordanted receiver. The Cibachrome process is an example.

Quality characteristics of colour materials

Quality characteristics of colour materials that should be considered include colour fidelity—masking, dye stability, colour balance, grain, speed, and sharpness.

Two-Colour Systems

Because of the complexity and cost of arriving at a natural three-colour photographic image, several processes were derived that depend on just two- colour records. Since it is impossible to obtain a full gamut of colours with only two records, a compromise in colour quality is made in two-colour systems for pictorial photography. Two-colour systems were also developed for data-recording uses, where a full range of colours is not needed. Special applications of colour photography include colour television and infrared-recording colour film.

Manufacture of Films, Plates and Papers

In the making of photographic films, plates, and papers, three distinct steps are carried out : (i) preparation of the light-sensitive emulsion; (ii) manufacture of the base, or support, for the emulsion; and (iii) coating of the emulsion on the base.

Emulsions

The so-called photographic emulsion is in reality not a true emulsion, but rather a dispersion of tiny silver halide crystals in gelatin, which serves as a mechanical binder, a protective colloid, and a sensitiser for the halide grains. Many different types of silver halide emulsions are manufactured, the characteristics of each being dependent upon the silver halide used and the details of manufacture. In slow positive emulsions for photographic papers the bromide, chloride, and chlorobromide are chosen. All fast emulsions, usually for negatives, contain AgBr and small amounts of AgI. The iodide is essential for high-speed types, but seldom exceeds 5%. The finished emulsion generally consists of 35 to 40% silver halide and 65 to 60% gelatin. The manufacture of the emulsion may be divided into four principal steps: precipitation, first ripening, washing, and second ripening, or after-ripening.

Precipitation

The gelatin, after soaking in cold water for about 20 min., is dissolved in hot water. The requisite quantities of KBr and KI are dissolved in this solution, which is then ready to receive the $AgNO_3$ solution. In the case of the ammonia process, a 10 to 15% excess of KBr is used, and the amount of gelatin is 20 to 60% of the total required in the finished emulsion. For neutral, or "boiled," emulsions, a 2 to 5% excess of KBr is generally used, and 10 to 20% of the total gelatin. In the ammonia process a $AgNO_3$ solution is mixed with concentrated ammonia until the initial precipitate of Ag_2O redissolves. This solution is added slowly, with thorough mixing to the halide-gelatin solution. The mixing temperature is usually about 40°C. For neutral emulsions no ammonia is used. A 10% $AgNO_3$ solution is added to the halide-gelatin solution at 60 to 80°C. In this mixing operation an excess of KBr is necessary to produce large grain size and to prevent interaction of silver ions with the gelatin. The KBr solution is prepared with a relatively small part of the total gelatin, since too much of the latter would interfere with grain growth during the ripening step. Moreover, the principal part of the gelatin is protected from the harmful influence of heat and ammonia. Careful control of the mixing process is essential, since the concentrations of the solutions, the temperature, and the manner in which the $AgNO_3$ solution is added, all have a pronounced effect on the character of the emulsion.

First ripening

The ripening, or digesting, process is essentially a continuation of grain growth under the influence of heat, the temperature being about the same as that of the mixing step. The excess of KBr in the emulsion increases the solvent capacity for AgBr. The small crystals, being more soluble than the large ones, tend to dissolve and reprecipitate upon the large crystals. Since the sensitivity of the final product is dependent upon the grain size, the ripening step is an important one and warrants careful regulation. If this process is carried too far, the finished plate or film will be subject to fogging. After the first ripening, no more crystal growth takes place.

Precipitation and Washing

At the close of the first-ripening process the warm emulsion is coagulated with a salt solution or an organic sulphonic acid of sulphate. The density of the silver halide-gelatin mixture causes it to settle to the bottom of the vessel, and the aqueous phase is removed by decantation. The emulsion is usually reconstituted by heating the curds in water and reprecipitating. This is sufficient to remove excess halide ions. Careful control of pH, temperature, rate of addition of precipitant, and degree of stirring are all essential to produce a satisfactory material.

After-ripening (Chemical Sensitising)

The emulsion grains are still relatively insensitive to light and are treated to increase their sensitivity. Three general kinds of sensitisers are used: (i) sulphur-containing compounds (thiosulphate or thiourea);

(ii) reducing agents (stannous chloride, sodium sulphite); and (iii) salts of precious metals (gold chloride, gold thiocyanate).

The original gelatin used is inert, so that known quantities of sensitising chemicals can be added and thus produce predictable results. The emulsion grains, after addition of the chemical sensitiser, are digested for some time at 40 to 70°C in the after-ripening or second-ripening tank. The time may vary from a few minutes to several hours.

Before coating the emulsion on the support, it is customary to add chrome alum or formaldehyde as a hardening agent. Phenol or thymol may be introduced to prevent the growth of mould or bacterial attack. The addition of KBr at this stage aids in the prevention of fog. The introduction of amyl alcohol or saponin serves to depress the surface tension of the liquid emulsion, facilitating uniform foam-free spreading on the support. The sensitising dyes for orthochromatic and panchromatic films are here added to the emulsion. The warm fluid emulsion is filtered and is then ready to be coated on the support.

Preparation of the support

The support for the photographic emulsion can be of glass, paper, cellulose acetate, or polyester film. Glass plates are still employed for some scientific and commercial purposes, map making, and some copying and colour printing jobs. Glass has the advantages of freedom from distortion and of not deteriorating with age. The glass plates are automatically cleaned by revolving brushes in a strong soda solution. They are coated with a thin substratum of gelatin containing chrome alum and dried in ovens.

The strong, chloroform-insoluble, white, flaky cellulose acetate for film has an acetyl content of 39.5 to 42.0%. To decrease the tendency of cellulose acetate to stretch in water and shrink after drying, substances such as triphenyl phosphate are generally added. This highly viscous solution is known as "dope." Filtration, aeration, and temperature adjustment are necessary before sending the dope to the film base-coating machines. Each machine has a heated, rotating drum 20 ft in diameter and 4 to 6 ft wide. It is coated with silver. By the time the drum has made one revolution, the solvents have evaporated. The film is stripped off, wound into rolls, and transferred to the cooling rooms. The solvents may be recovered by adsorption on activated carbon. The acetate film is given a thin undercoat, or substratum ("subbing"), of gelatin and chrome alum to increase the adhesiveness of the light-sensitive emulsion.

Paper for photographic purposes, is made from rag or high-grade sulphite pulp for the specific use intended to ensure permanence. This base is coated with gelatin containing blanc fixe (precipitated barium sulphate) in order to present a smooth surface for emulsion coating or to provide the desired surface pattern. Sometimes china clay or satin white (a coprecipitate of calcium sulphate and aluminium hydroxide) is used instead of blanc fixe. Photographic paper supports having polyethylene extruded on both sides are being used for applications where rapid drying and dimensional stability are desired.

Emulsion Coating

The modern method utilises multilayer extrusion coating and up to six layers can be coextruded onto a ramp and then transferred to the film base without intermixing of the layers, which would be a catastrophe in the manufacture of colour film. From an engineering and a scientific standpoint, the multilayer extrusion technique and controls necessary to obtain uniform photographic response are truly revolutionary.

Polyester film

Polyester film support is manufactured by the melt-casting process. The molten polymer is forced through a die to form a continuous sheet. Then this sheet is stretched several fold in both the machine and cross-directions to give it the desired mechanical properties. A final heat treatment is required to stabilise these properties. Both the basic chemical nature and the method of manufacture of polyester support differ markedly from those of cellulose ester film supports which are cast from solvent solutions. The absence of solvent in the making of polyester film support is one of the reasons why polyester-base films show excellent dimensional stability.

SPECIAL APPLICATIONS OF PHOTOGRAPHY

Photomechanical Reproduction for Illustrations

Photography has found one of its most important applications in the reproduction of photographs on the printed page by means of printing inks. These processes may be classified as:

1. *Relief printing,* also referred to as photo-engraving, in which the raised portion of a plate receives the ink for transference to the paper, and so-called *line* plates and *halftone* plates are used.
2. *Intaglio printing*, which includes photogravure, rotogravure, and metal engraving, in which the relief printing procedure is reversed and the hollow regions of the plate or metal cylinder hold the ink.
3. *Planographic printing*, or lithography, which makes use of the inability of a water-wet surface to take ink. The *offset printing process* utilises lithographic plates which are particularly adaptable to illustrative work in colour, and the usual separation negatives are used in their preparation.

Photopolymerisation

Photopolymerisation or polymerisation initiated by light, has the ability (as has silver halide photography) to amplify the effect of light enormously; the multiple effect, however, occurs simultaneously with exposure, rather than in a separate processing step. This method has been used extensively in the printing field.

Printing-out Techniques

Photographic copying of documents reaches back to the earliest days of photography. Although newer methods of photocopying have largely replaced the earlier types in which visible images are produced on exposure of prepared paper, some of these printing-out processes are capable of producing exceptionally fine prints, and others afford a certain control over tone values unequalled by any other process.

A method dependent on the fact that ferric ions are reduced to ferrous ions in the presence of organic matter under the influence of strong light. Paper is coated with ferric ammonium citrate and potassium ferricyanide. When a line drawing is placed over the prepared paper and then exposed to light and treated with water, an insoluble blue image is formed where the light reaches the paper; this image is in prussian blue, ferric ferrocyanide ($Fe_4[Fe(CN)_6]_3$). The cyano-type, or positive blue-print (where lines are blue and background light), uses a more radiation-sensitive ferric mixture and, processed in a potassium ferricyanide solution, yields Turnbull's blue, ferrous ferricyanide ($Fe_3[Fe(CN)_6]_2$).

Photocopying Processes

More and more business offices, governmental units, especially court record rooms, libraries, and archives find copying machines indispensable in disseminating information efficiently, rapidly, and accurately. Large volume users of equipment prefer not only permanent and legible copies, but also inexpensive ones that are almost indistinguishable from the originals, produced by compact, trouble-free machines that operate automatically. Several processes, such as *thin-layer silver halide film*, which has no gelatin coating, *free-radical photography*, which involves the interaction of diphenylamine and carbon tetrabromide, and a high-speed long-distance electrostatic copying system called *videograph*. The current trend is away from coated-paper copiers and toward the use of plain-paper copiers, even though the original cost of coated-paper machines is less.

Photographic miniaturisation systems

Microphotography is the art of making miniature photographic facsimiles of original materials. Its current use in commercial records by insurance companies, banks, and engineers for active use or security purposes, as well as for historical records and newspaper files processed for permanent retention, has brought about a revolution in *microforms*.

The necessity for conserving storage space and for providing physical protection for essential records, and also research in information-retrieval methods, have accelerated development of all facets of microreproduction in which individual copies of any page are available by the touch of a printout button.

Photolithography

Photolithography is used to transfer patterns from a mask containing circuit-design information to thin films on the surface of a silicon wafer. The pattern transfer is accomplished with a photoresist, an ultraviolet light-sensitive organic polymer. A wafer that is coated with photoresist is illuminated through a mask and the mask pattern is transferred to the photoresist by chemical developers. Further pattern transfer is accomplished by appropriate liquid or gaseous etchants.

The photoresist must be resistant to the etches used for patterning thin films of thermally grown silicon dioxide. The pattern in the silicon dioxide can be etched chemically by high purity (electronic-grade) hydrofluoric acid that is buffered with ammonium bifluoride using the exposed and developed photoresist as a mask. Various gaseous plasmas are also used to perform etching. Plasma etching provides better control and minimises undercutting of the photo resist mask. The masking process is usually repeated 5–10 times in the fabrication of an IC.

Photoconversion

In the context of solar-energy conversion, photoconversion refers to the direct production of fuels and chemicals from sunlight and water, carbon dioxide, nitrogen, and simple organic compounds. Typical products are hydrogen, methane, alcohols, ammonia, and organic nitrogen compounds.

The natural photosynthetic process produces large amounts of biomass useful for fuel, but only at modest conversion efficiencies. Photobiological systems require the physical separation of dark reactions from light-driven reactions, as has been demonstrated for both bacteria and green plants, e.g., by immobilisation of the photosynthetic mechanism which produces charge separation. The immobilised system can be coupled to a photoelectrochemical cell.

The production of hydrogen by various phototropic organisms occurs in the presence of light with photosynthetic bacteria, cyanobacteria, or algae. *In vivo* cultures of photosynthetic bacteria offer the best prospect for efficient photobiological hydrogen production. *In vitro* hydrogen generation, employing isolated spinach chloroplasts, ferredoxin, *Chromatium hydrogenase*, and cysteine, was first reported in 1961. Since then, considerable improvements have been made. Photochemical reactions are carried out in either homogeneous or heterogeneous media. Liquid-phase processes are preferred. Synthetic chloroplasts have been produced analogous to the natural chloroplasts in green plant cells. These structures split hydrogen from water and form carbohydrates by combining hydrogen with carbon from carbon dioxide in the atmosphere.

Photoelectrochemistry is based on the properties of photoactive semiconductor electrodes in contact with liquid electrolytes. Cells can be designed to produce either chemicals and fuels or electricity. In the former case, the different oxidation and reduction reactions result in a net change in the electrolyte that creates fuels or chemicals.

Storage of electricity is possible in an electrochemical photovoltaic cell with the help of a third storage electrode or the use of redox electrolytes in a redox battery. Experimental solar-powerered, electrochemical, photovoltaic storage cells have been operated for months, providing an electric current day and night.

Liquid-junction solar cells offer several advantages over conventional solid-state photovoltaic cells. A redox electrolyte has a Fermi level or redox potential and thus can take the place of p-type semiconductor or a metal to form the necessary junction. Liquid-junction cells are much larger and heavier and may require sealing if the electrolyte solution is affected by air.

Photoelectrosynthetic cells that drive reactions thermodynamically uphill are of particular interest since solar energy is thus stored as chemical energy. The photolytic splitting of water, called photoelectrolysis, is of exceptional interest and research is in progress worldwide.

Photogalvanic processes involve absorption of incident light by dye molecules in the cell electrolyte. The photoexcited dye molecules drive redox reactions in the electrolyte, and these redox products then transfer charge to the cell electrodes to produce an electric current.

ELECTROPHOTOGRAPHY

Electrophotography can be defined broadly as a process in which photons are caputred to create an electrical image analogue of the original. This electrical analogue is, in turn, manipulated through a variety of steps that result in a physical image. In electrophotography the primary quantum yield is generally much higher than that of silver halide photography.

The necessary amplification in the case of projection imaging, focusing an illuminated original through a lens onto the photosensitive element, is obtained by some form of physical development. The free energy required to initiate image gain is supplied in the form of an electrostatic potential, typically applied across a photosensitive dielectric layer before exposure, or to a bias electrode (a closely spaced metal plate to which a potential is applied) during development, e.g., by the attraction of relatively large, visible particles in response to only a few surface charges.

Transfer Xerography

The most widely used form of electrophotography today is called transfer xerography. Fig. 17.3 illustrates one cycle of transfer xerography starting at the nine o'clock position.

In transfer xerography, almost all process steps are carried out on the photoreceptor surface. The first step necessary for creating a latent electrostatic image is the application of a uniform charge pattern on the surface of the photoreceptor; at the same time a uniform field from front to back is provided. Charges must be held on the surface. Corona-discharge devices have proven the most reliable means of applying a stable, uniform charge layer to large-area photoreceptors. Typically, these consist of one or more thin wires connected to a positive or negative high voltage supply and backed by a grounded shield that serves to stabilise the discharge.

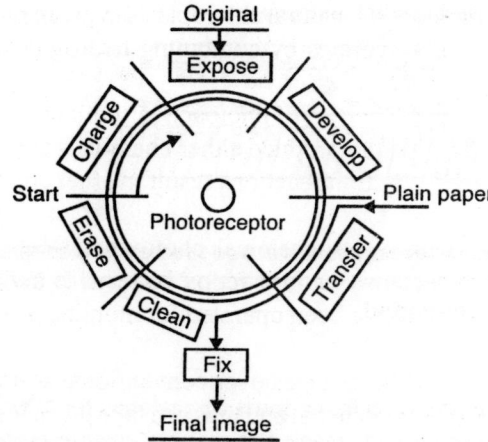

Fig. 17.3. Schematic of one cycle in transfer xerography. The process steps take place on the surface of the photoreceptor, which is coated on the central imaging drum.

As in other photographic systems, the image must be brought to the photosensitive surface with efficiency. Light sources are selected to match the spectral response of the photoreceptor and to provide adequate illumination of the original to be imaged. Automatic copying machines usually require continuous motion of the photoreceptor in order for all other process steps to go on at the same time.

Xerographic development

The electrostatic latent image can be made visible by a variety of techniques. Toner particles can be attracted only where electrostatic field lines extend above the surface of the receptor. In the absence of a development electrode, external field lines appear only along the periphery of charged areas, permitting toner attraction only to the outline of the area. Field lines can be extended outward across the entire charged surface by ringing a development electrode (Fig. 17.4) connected to the back electrode of the photoreceptor, close to its surface. Charged toner particles introduced into this narrow space follow the external field lines and deposit more uniformly to provide solid area development.

For the development process to occur, toner particles must first be charged reproducibly. Conditions in the development zone must be carefully controlled to avoid dusting and contamination elsewhere. The field configuration is crucial to uniform development. Main dry-development systems include powder cloud development, cascade development, magnetic brush development, and touchdown or impression development. Image reversal development can be achieved by dry- or liquid-development techniques.

Transfer of developed image

To be useful, the developed toner image must be transferred to a receiving sheet, usually paper. With the aid of an externally applied field, the toner particles are induced to jump to the paper as it separates from the photoreceptor.

To become permanent, the transferred toner image must be fixed to paper. A thermoplastic resin base in the toner powder permits fixing by fusing onto the paper. In non-contact fusing, a radiant-heat source provides energy to the toner image as the paper passes below it. In contact fusing, heated rollers with good release properties (e.g., silicone or fluorocarbon elastomers), backed by a pressure roller, melt the toner and force it into the paper.

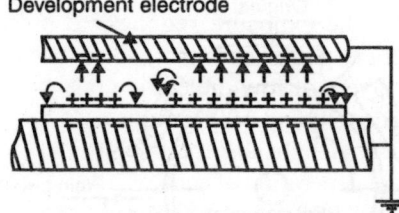

Development electrode

Fig. 17.4. Addition of a closely spaced development electrode forces field lines to extend into the space above the photoreceptor. The field strength becomes everywhere proportional to the local photoreceptor potential, and the entire image area can be developed.

Post-imaging process steps

In cyclic-transfer xerography, the photo receptor must be restored to its original condition after image transfer. Between 10 and 30% of the toned image and residual charge patterns remain on the receptor. Flooding the photoreceptor surface with light and applying neutralising charges aid in removing toner. Other aids are lubricants that are added to the toner or applied directly to the photoreceptor surface. Both help form low friction surfaces, increasing transfer efficiency and cleanability. Fatty acid metal salts and fluorinated organics have been used for this purpose. Toner is removed by rotating brushes made of fur or of natural or synthetic fibres, and is collected by vacuum suction. The photoreceptor can also be cleaned by a conforming wiper blade.

Liquid development

The most widely used liquid-development scheme relies on electrophoretic particle migration. This development system offers very high resolution, approaching that of the latent image itself; 800 lines and spaces (or line pairs) per millimeter have been achieved. Although liquid development has found its most prominent use in direct electrophotography, successful liquid-transfer systems have been devised and marketed recently.

Materials for Transfer Xerography

Photoreceptors

The photoreceptor is an electronic device that forms the electrostatic image charge pattern in response to light. It typically comprises three layers (Fig. 17.5) a conductive base layer which can be omitted if the support of the photoreceptor is metallic; a thin dielectric barrier layer which is required only to prevent charge injection into the photoconductive layer; and the photoconductor layer on top. If a separate base layer is required, as on polymeric film supports, it consists of a sputtered or evaporated metal or conductive metal oxide layer well < 100 nm thick. Deliberately applied barriers may consist of inorganic oxides, sulphides, or thin polymeric films.

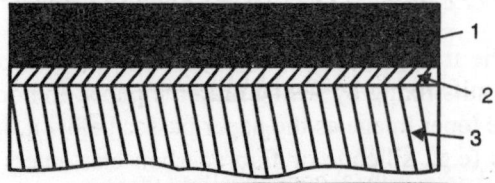

Fig. 17.5. The cross-section through typical xerographic photoreceptor. (1) Photoconductive layer, (2) injection barrier, (3) base electrode.

The optimal thickness of the photoconductor layer is determined by the potential required for development, typically 300–800 V. The exposure required to produce this potential measures the photosensitivity of the device. Practical photoconductors are between 15 and 25 μm thick if they are made up of organic or binder-type layers of low dielectric constant ($K = 3–5$); selenium and its alloy layer ($K = 6–9.5$) are typically between 30 and 100 μm thick.

High photosensitivity requires effective light absorption in the desired spectral region (molar extinction coefficient 10^4–10^5/(mol·cm); efficient generation of carriers, i.e., separation of the photo excited electron/hole pair in the applied field; and subsequent complete transport of the mobile charge through the photoconductive layer. Most photoconductor materials exhibit efficient transport for either electron or holes. Vitreous selenium has long been the basis of the most commonly used photoreceptors. Tellurium may be added to extend red response. Arsenic alloys have been found to enhance both photosensitivity and stability against crystallisation. Dispersion of photoconductive pigments such as cadmium sulphide, zinc oxide, and phthalocyanine in dielectric binders may be used.

Organic photoconductors compete today effectively with selenium. Fig. 17.6 shows a new family of double-layer photoreceptors in which a relatively thin, visible-light-sensitive, carrier generator film is adjacent to a much thicker, transparent, polymeric, active-transport layer. The former supplies photosensitivity; the latter, dielectric strength and voltage contrast. This allows greater freedom in designing photoreceptors that have desirable mechanical properties as well as high photosensitivity.

Developer materials

The toner particles used for developing xerographic images generally consist of 8–15 µm particles of a thermoplastic powder coloured by a dispersion of 5–10% carbon-black particles of less than 1 µm. Cyan, magenta, or yellow colourings may be substituted for use in colour xerography. Pigment concentration and dispersion must be adjusted to impart a conductivity to the toner mass that is appropriate to the development process. For efficient induction development a conductivity greater than 10^{-4} S/cm is found to be desirable; most other development processes require the toner to retain charge applied by contact electrification for extended time periods. The toner thermoplastic is generally selected on the basis of its fusing characteristics; it must melt sharply at the lowest temperature consistent with stability to storage and to vigorous agitation that occurs in xerographic development chambers.

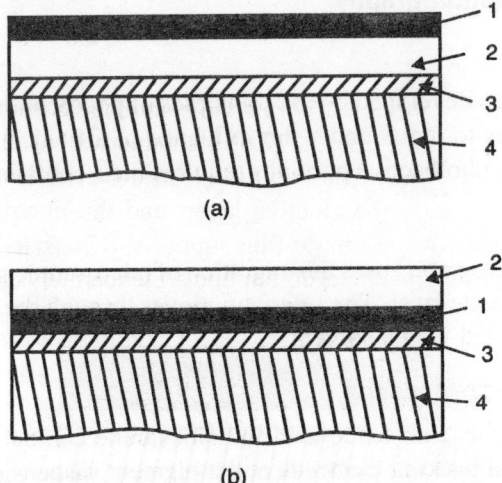

Fig. 17.6. Active matrix photoreceptor with (a) generator layer on top, (b) at interface. (i) generator (sensitising) layer; (ii) active transport layer; (iii) blocking interface; (iv) base electrode.

Direct Electrophotography

Electrofax process

The key innovation in electrofax was to produce images directly on the final sheet in fewer steps and make the photoreceptor an inexpensive consumable item. Electrofax has its widest use for low volume black-and-white copying.

Transparent-film processes

Direct electrophotography on transparent organic photoconductor films has been used to produce transparencies. Since the transfer step is omitted, images resolving 200 line pairs/mm and better can be readily produced, particularly with electrophoretic liquid developers.

Transfer of Electrostatic Images (TESI)

An alternative that obviates the toner transfer and cleaning steps in conventional xerography is the transfer of the electrostatic latent image to a suitable receiving sheet prior to development. Transfer of an electrostatic charge pattern can be accomplished either during photoreceptor exposure or after a stable charge pattern is formed on the photoreceptor by conventional charging and exposure.

Persistent Photoconductivity Imaging

A number of attempts have been made to reverse the usual xerographic sequence, by exposing the *neutral* photoconductor first and then charging it to form a latent image. A process known as Magne-Dynamic is one approach to chargeless electrophotography.

Photoconductography

In photoconductography, a conductivity pattern formed in a photoconductive insulator by light exposure is developed by electrochemical deposition. Photographic gain is achieved by drawing multiple charges through the photoconductor-electrolyte series circuit, or by secondary chemical amplification, or both.

Photoactive-Pigment Electrophotography

Photoconductive toner

A number of processes have been developed on the principles of photoconductivity of a pigment and the ability of such pigment particles to move under the influence of a field. A successful approach to the formation of images by moving photoactive particles requires their containment in a liquid or plastic.

Particle-migration imaging

A totally new form of electrophotography uses the migration of microscopic selenium particles embedded in the top surface of a 10 μm plastic layer. The selenium moves through the layer under the influence of a field when the plastic is softened with a solvent.

Photoelectrophoretic imaging processes

Photoconductive pigments such as zinc oxide and phthalocyanine can be caused to print out directly onto paper interposed between a backing electrode and a pigment suspension covered by a transparent imaging electrode. This process, which is called photoelectrophoresis or PEP, has been applied to an experimental printer capable of converting light input to very clean black print on paper.

Colour photoelectrophoresis

Perhaps the most challenging use of photoactive pigments and their electrophoretic motion is a single-step, full-colour electrophotographic process that uses the spectral response of individual photosensitive particles in a mixture of cyan, magenta, and yellow pigments. Under the influence of an electric field, the particles migrate selectively according to illumination to produce instant full-colour images.

Applications for Electrophotography

By far the most widespread application of electrophotography today is plain-paper copying; fast xerographic duplicators can now produce two copies per second, and sort and collate them into finished booklets. Computer printers can print at high speed by means of a deflected and modulated laser beam. These machines can be operated on-line from a computer or off-line from magnetic tape and disk drives. Such machines can handle very large volumes of individualised printing of high quality (500,000 or more copier per month).

PHOTO-MULTIPLIER TUBES

The photo-multiplier is a highly sensitive detector of radiant energy in the UV, visible, and near-ir regions of the electromagnetic spectrum. The basic sensor is a photocathode located inside a vacuum envelope. Photo electrons are emitted and directed by an electric field to an electrode or dynode within the envelope. A number of secondary electrons are emitted at this dynode for each impinging primary photoelectron. These secondary electrons, in turn, are directed to a second dynode and so on until a final gain of perhaps 10^6 is achieved. The electrons from the last dynode are collected by an anode which provides the output-signal current.

Despite the solid-state revolution, the photomultiplier, because of its high gain and fast time response, continues to be manufactured on a large scale.

Design

Dynodes are shaped and positioned in such a manner that all the stages are properly utilised and no electrons are lost to support structures. The shape of the field should encourage the return of electrons to a centre location on the next dynode. Magnetic fields may be combined with electrostat fields to provide the required electron optics, although most photomultipliers are electrostatically focused. For some applications, the time spread of the electron trajectories must be minimised and strong electric fields are provided at the surfaces of the dynodes to assure high initial acceleraton of the electrodes. Regenerated effects must be avoided.

The original design was a circular array of photocathode and dynodes, as typified by the present 931A tube. In scintillation counting, the photocathode must be of the semi-transparent type located on a relatively large flat glass surface through which the scintillating crystal is coupled. The electron-optical requirements for this case usually call for an increased photocathode-first-dynode spacing. Various dynode configurations are then utilised, including the circular cage, the box-and-grid and the venetian-blind for good electron-collection efficiency, the linear-array cage for short-time response, and the close-spaced mesh-dynode array for use in high magnetic field environments.

A recent innovation in front-end design is the so-called teacup photomultiplier, named after its large first dynode. Secondary electrons from the first dynode are directed to an opening in the side of the teacup and then to the second dynode. Fields between the photocathode region and the first dynode region are separated by a very fine grid structure (Fig. 17.7). Sidewall photoemission results from the light that passes through the semi-transparent photocathode. The increased photoemission and collection efficiency improve the pulse-height resolution in scintillating-counting applications.

Other recent designs include high speed tubes utilising microchannel plates in proximity to the photocathode and anode. Time resolution or pulses initiated by single photoelectrons measured at full width at half maximum (FWHM) is less than 800 ps. Sensitivity to external magnetic fields is much reduced, a fact that is important in some nuclear-physics experiments. Even faster photomultipliers use crossed electrostatic and magnetic fields to direct electrons to repeated stages of secondary emission.

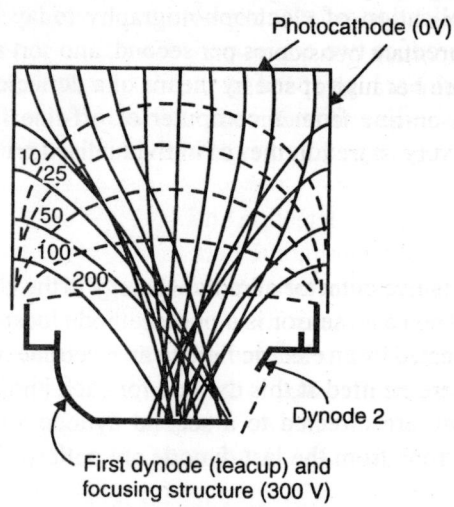

Photocathode (0V)

10
25
50
100
200

Dynode 2

First dynode (teacup) and
focusing structure (300 V)

Fig. 17.7. Teacup photomultiplier showing photoelectorn paths directly from the photocathode and those initiated by light transmitted through the photocathode and striking the sidewall which also has an active photoemissive layer. Equipotential lines are indicated in the region between photocathode and first dynode.

The photoemitters used in these tubes are all semiconductors. Light is absorbed by the valence-band electrons in the semiconductor only if the energy of the photon is at least equal to the band-gap energy E_G. If, as a result of light absorption, electrons are raised from the valence band into the conduction band, photoconductivity is achieved. For photoemission, an electron in the conduction band must have energy greater than the electron affinity E_A, and photoemission can occur only if the photon energy exceeds $E_G + E_A$.

Photoemission from semiconductors can be improved by modification of the energy-band structure. Reduction of the E_A permit the escape of electrons that have been excited into the conduction band at greater depths within the material. Indeed, if the electron affinity is reduced to less than zero, the escape depth may be as much as 100 times greater than for normal material. At less than zero, the vacuum level is lower than the bottom of the conduction bond and, a condition develops described as negative electron affinity (NEA). In secondary emission, the impact of primary electrons rather than of incident photons causes the emission of electrons.

Photocathodes

The most common photocathodes are cesium antimonide Cs_3Sb, multialkali or trialkali ($Na_2KSb:Cs$), and bialkali (K_2CsSb) derivatives. The bialkali cathodes have the greatest thermal stability. Recently, a rubidium-cesium-antimony (probably Rb_2CsSb) has been introduced that has excellent blue sensitivity.

Characteristics

When several secondary-emission stages are coupled, the total gain μ is given by $\mu = \delta^n$ where δ is the secondary emission per stage, and n is the number of stages. Photomultiplier gain is usually present as a function of the applied voltage, and may vary from 10^4 at 500 V to 10^7 at 1200 V. All photomultipliers are sensitive to external magnetic and electrostatic fields. The higher the voltage, the smaller the effect of these fields. The anode current is proportional to the incident radiant flux. The limit of linearity occurs when space charge begins to form, usually between the last two dynodes.

Operating temperatures can be as high as 75°C although, with a Na_2KSb photocathode, 175°C can be tolerated for a short time, an important consideration in oil-well operations. Very low temperatures (below −40°C) may damage the tube because of stresses in the metal-to-glass seals. Operating stability depends on the magnitude of the average anode current, $\leq 1\mu A$ is recommended if stability is of great importance. Photomultipliers should be stored in the dark since light, especially blue or UV, increases dark emission.

At very low light levels, the limitation of detection and measurement is generally the signal-to-noise ratio determined by the fluctuation in the thermionic dark emission from the photocathode. Secondary emission contributes very little to the relative noise output of the tube. The limit to detection can be described by stating the equivalent-noise input or the noise-equivalent power.

Theoretical estimates of photoemission or secondary-emission times for metals or insulators are 0.01–0.1 ps. In NEA semiconductors, the lifetime of internal free electrons having quasi-thermal energy can be on the order of 0.1 ns.

Uses

Photomultipliers are recommended for the detection of very low light or radiation levels. They are also capable of operating over a wide range of light flux, and are used to detect or measure radiant fluxes. Direct applications include oil and gas exploration; blood-sample analysis; the detection and determination of air pollutants; radiation monitoring at nuclear plants; rapid sorting of food and manufactured goods; control of colour-printing machines; gun-fire control; detection of pinhole flaws in steel and paper; laser ranging; photometry; spectrometry; radioimmunoassay; and thermoluminescent dosimetry. Other applications include scintillation counters and the gamma camera for locating tumors or other biological abnormalities. Photomultiplier tubes are used in CAT scanners and the positron camera. Recently, photomultiplier tubes having very large hemispherical photocathodes have been developed for use in proton-decay measurements.

CHAPTER 18

Nuclear Chemistry

INTRODUCTION

Modern civilisation demands energy in many forms. Coal and petroleum are the conventional sources of energy. These are fossil fuels and non-renewable sources of energy. At present we mainly depend upon them for our energy requirement. However, the deposits of fossil fuel may not last more than a few decades to supply the demand of power. Therefore, attempts are being made to explore the non-conventional sources of energy. Though the hydroelectric power is the renewable source, the power generation depends upon the monsoon and cannot meet the ever increasing demand. Solar energy, nuclear energy, wind energy and the tidal energy are the alternate sources of energy. In the years to come the use of solar energy for power generation will be inevitable. But the present technology is inadequate to make use of the solar energy for power generation economically, on large scale. Similarly, the use of wind and tidal energies have also some limitations. Nuclear fuels are potential sources of energy.

Nuclear fuel is a material that undergoes nuclear fission on bombarding with neutrons, releasing enormous amount of energy. ^{233}U, ^{235}U and ^{239}Pu are the nuclear fuels. These isotopes are called *Fissile*. ^{232}Th and ^{238}U do not undergo nuclear fission but on bombarding with neutrons undergo a series of nuclear disintegration reactions and produce fissile materials.

Examples:

$$^{232}_{90}Th + ^{1}_{0}n \longrightarrow ^{233}_{90}Th + \gamma$$

$$^{233}_{91}Th \longrightarrow ^{233}_{91}Pa + ^{0}_{-1}e \quad \text{and} \quad ^{233}_{91}Pa \longrightarrow ^{233}_{92}U + ^{0}e_{-1}$$

$^{233}_{92}U$ undergoes nuclear fission. Hence it is a nuclear fuel.

Similarly, $^{238}_{92}U$ produce $^{239}_{94}Pu$ which undergoes nuclear fission. The material which do not undergo nuclear fission, but can produce fissile materials are called fertile materials. Thus, ^{232}Th and $^{238}_{92}U$ are fertile materials.

1 kilogram of uranium on complete fission can produce energy equivalent to 20×10^6 KWH, while 1 kilogram of coal on complete combustion can produce only 2 KWH of power. The process of nuclear fusion also produces a very large amount of energy. However, at present there are many problems in controlling the nuclear fusion process. The major problem associated with the nuclear fission process is the disposal of the hazardous radio active wastes.

URANIUM AND URANIUM COMPOUNDS

Uranium, at no. 92, At wt 238.03, is a member of the actinide series of transition elements.

Occurrence

Uranium is present in the earth's crust at ca 2 ppm. Acidic rocks with a high silicate content, such as granite, have a uranium content above average, whereas the contents of sedimentary and basic rocks, such as basalts, are below average. However, 90% of the world's known uranium sources are contained in conglomerates and sandstone.

Primary uranium minerals, crystallised from low-melting rocks, are generally black and contain uranium in a valence lower than six. Secondary minerals are yellow, greenish-yellow, bright green, or orange, and contain uranium in the valence +6.

Uraninite and pitchblende differ in physical form; their composition ranges from $UO_{2.0}$ to $UO_{2.67}$. Uraninite is well crystallised; pitchblende is either amorphous or consists of very fine crystals. Gummite, $UO_3.nH_2O$, and becquerelite, $2UO_3.3H_2O$, are typical hydrated oxides. They are fairly common, but commercially unimportant. Carnotite, $K_2O.2UO_3.V_2O_5.3H_2O$, and tyuyamunite, $CaO.2UO_2.V_2O_5.8H_2O$, are the commercially important minerals. Phosphates are represented by autunite, $Ca(UO_2PO_4)_2.8H_2O$, and torbernite, $Cu(UO_2PO_4)_2.12H_2O$. Uranium arsenate minerals are found in European deposits. Numerous organic complexes are found in sedimentary deposits, e.g., thucholite and carburan.

Resources

Reasonable assumed resources (RAR) contain deposits of such size, grade, and configuration that recovery is within the given production cost ranges with the current mining and processing technology. This type of resource is estimated at 1.8×10^6 metric tonnes. Estimated additional resources (EAR) occur as extensions of well-known deposits, little-explored deposits, or undiscovered deposits; they are estimated at 2.6×10^6 tonnes.

Properties

Uranium is a dense, lustrous metal that resembles iron; it is ductile, malleable, and weakly magnetic. In air, it tarnishes rapidly, and in a very short time, even a polished surface becomes coated with a dark-coloured layer of oxide. In the solid state, uranium exists in three allotropic modifications. The thermodynamic properties have been determined with great accuracy (Table 18.1.). Thermal conductivity ranges from 0.251 W/(cm·K) at 309 K to 0.326 W/(cm·K) at 673 K. Spectroscopic properties have been studied in great detail and more than 30,000 lines of the arc-and-sparc-emission spectra have been catalogued.

Table 18.1. Thermodynamic properties of uranium metal.

Function or parameter	Value or equation
Entropy of α-U, J/K[a]	50.21 ± 0.12
Entropy of U(g), J/K[a]	199.6
Enthalpy of fusion, kJ/mol[a]	8.326 ± 0.54
Enthalpy of sublimation, kJ/mol[a]	1062.73

(Contd...)

Function or parameter	Value or equation
Free energy of vapourisation, G, kJ/mol[a]	
solid to gas	$525.3 - 0.137$
liquid to gas	487.6 ± 0.11
Normal (extrapolated) boiling point, K	3818
Vapour pressure of liquid[b]	$\log p \text{ (kPa)} = -\dfrac{(25{,}230 \pm 370)}{T} + (7.72 \pm 0.17)$

[a] To convert J to cal, divide by 4.184.

[b] To convert kPa to mm Hg, multiply by 7.5.

Of the four oxidation states, +3, +4, and +6 are stable enough to be of practical importance; +5 is of minor importance. The alternation between +4 and +6 states is utilised in the extraction of uranium from ores and in purification.

Isotopes

There are fifteen known isotopes of uranium, not counting the isomeric states; ^{234}U, ^{235}U, and ^{238}U exist in nature. The natural isotope composition usually varies. In addition to ^{235}U, an artificial isotope, ^{233}U, was found to be fissionable. Today, it is an industrial product produced in kilogram quantities by the thermal breeding reaction.

$$^{232}\text{Th} + n \xrightarrow{\gamma} {}^{233}\text{Th} \xrightarrow{\beta^-} {}^{233}\text{Pa} \xrightarrow{\beta^-} {}^{233}\text{U}$$

The product is isolated by a radio-chemical process.

Extraction from Ore

Conventional ore-dressing techniques have not been successful in the preconcentration of uranium minerals. Gravity separation is sometimes possible. Electrostatic methods generally give low recoveries in low concentrations. Flotation usually gives satisfactory concentrations.

A high temperature roasting or calcining operation before leaching is frequently desirable, and the characteristics of many ores are thus improved. Treatment with acids or alkalies converts the uranium contained in the ore to water-soluble species. Most mills use acid leaching. Alkaline or carbonate leaching is used for ores with high lime content, which would require excessive consumption of acid.

The crude uranium isolated in the leach liquors requires additional purification. Direct or selective precipitation has not been commercially successful in acid leach systems. However, precipitation may be used to enrich the uranium in side streams. Anionic sulphato or carbonato complexes may be absorbed from leach liquors on anion-exchange resins. The uranium is eluted with a saline or acid solution.

Some uranium ores exhibit extremely poor filtering and settling characteristics after leaching. To avoid large liquid-solid separation equipment; the ion-exchange process has been modified to extract uranium directly from the leach pulp (resin-in-pulp process).

Solvent extraction is widely used in the recovery of uranium from ores. Contrary to ion exchange, solvent extraction can be operated in a continuous countercurrent flow. The preferred extractants are di(2-ethyl-hexyl) phosphate (D2EHPA) and dodecyl phosphate (DDPA). The product of the extraction process is a purified uranium solution. Uranium is usually precipitated directly from solutions resulting from carbonate leaching. From acid solutions, uranium is precipitated by neutralisation with ammonia or magnesia. With the former, a precipitate of composition $(NH_4)_2 (UO_2)_2SO_4(OH)_4 \cdot nH_2O$ is obtained. So-called yellow cake with a higher uranium content is obtained with magnesia.

Conversion and Purification

The crude product from the refinery is purified for nuclear applications. Yellow cake is dissolved in nitric acid and extracted with tributyl phosphate. The uranium is back-extracted with recycled concentrated acid containing $< 1\%$ HNO_3 or with deionised water. Evaporation gives uranyl nitrate hexahydrate, which is pyrolysed to UO_3. Reduction with hydrogen gives the dioxide, which is converted to UF_4 by treatment with HF. Direct fluorination of the tetrafluoride, also called green salt, gives the hexafluoride.

For reactor applications, enriched UF_6 from an isotope-separation plant is converted to other compounds or the metal.

Manufacture of the Metal

Uranium metal is produced by reduction of the tetrafluoride in a bomb at ca 700°C (Ames process).

$$UF_4(s) + \begin{cases} 2Ca \\ 2Mg \end{cases}$$

$$\rightarrow U + \begin{cases} 2CaF_2 & \Delta H° = -560.5 \text{kJ/mol } (-134 \text{ kcal/mol}) \\ 2MgF_2 & \Delta H° = -349.4 \text{kJ/mol } (-83.51 \text{ kcal/mol}) \end{cases}$$

Most nuclear reactors built for the generation of electric power are based on uranium fuel enriched in ^{235}U. Reactors using natural uranium, such as the Hanford production reactors or some power reactors, do not require enriched uranium. Purified natural uranium fabricated in rods (slugs) may be used.

Only the gaseous-diffusion process is used for the separation of ^{235}U and ^{238}U on an industrial scale. Highly purified gaseous uranium hexafluoride is pumped through a series of diffusion aggregates arrayed in cells in a cascade pattern. Enormous amounts of electric power are required. The high capital cost and power consumption of gaseous-diffusion plants have led to the investigation of centrifugal separation of ^{235}U and ^{238}U.

Electromagnetic separation, one of the most powerful methods, so-called calutron separators. The calutron techniques has been used to separate pure samples of ^{234}U, ^{236}U, and stable isotopes of many other elements.

Other methods for isotope separation were developed, but none advanced beyond the pilot-plant stage. Laser excitation appears very promising. Uranium hexafluoride vapour is ionised by means of a tunable rhodamine-B laser in such a manner that ^{235}U atoms are selectively excited to form UF_5 molecules which are collected on a sonic impactor (MLIS-process).

Isolation of specific isotopes

The uranium isotopes ^{232}U, ^{233}U, and ^{234}U are the daughters of the actinide isotopes ^{232}Pa, ^{233}Pa, and ^{238}Pu, respectively, which may be obtained pure. For the production of ^{232}U, proactinium-231 is bombarded with neutrons. The ^{232}U can be separated by an ion-exchange process. Uranium-233 is obtained by bombardment of thorium-232 with slow neutrons. Uranium-234 is the daughter product of plutonium-238 which, in turn, is produced by neutron bombardment of neptunium-237.

Uranium Compounds

Uranium metal, heated to 150–200°C in a hydrogen atmosphere, gives uranium trihydride, UH_3, a black powder. Rare isotopes of hydrogen, i.e., tritium, can be stored in the form of hydrides. Deuterium or tritium are absorbed on uranium turnings heated to 200°C and then may safely be stored as solids. Upon heating to 500°C in vacuum, the isotope is released as a highly pure gas.

Fluorides

Uranium forms seven binary fluorides: UF_3, UF_4, U_4F_{17}, U_2F_9, α-UF_5, β-UF_5, and UF_6. The trifluoride is prepared from stoichiometric amounts of UF_4 and uranium metal. The tetrafluoride is prepared by hydrofluorination of UO_2 with excess gaseous HF at ca 550°C.

The uranium (IV, V) fluorides, U_4F_{17} and U_2F_9, are obtained from the tetrafluoride and the hexafluorides at 200°C and 2.36 kPa (17.7 mm Hg). These two compounds are observed in gas centrifuges as decomposition products. The grayish-white α-pentafluoride is obtained from the tetra- and hexafluorides by reduction with HBr at 80–100°C; at 150–200°C, the yellowish-white β-pentafluoride is obtained.

The hexafluoride is prepared by direct fluorination of the tetrafluoride; its properties are given in Table 18.2. It is produced on a large scale as feed for the gaseous-diffusion, separation-nozzle, and gas-centrifuge processes. A large number of ternary uranium fluorides are composed of a binary fluoride and an alkali or alkaline-earth fluoride.

Table 18.2. Properties of uranium hexafluoride.

Property	Value
Triple point at 151 kPa[a], °C	64.052
Sublimation point, °C	56.4
Density	
solid, g/cm^3	5.09
liquid, g/mL	6.63
Heat of formation, solid at 25°C, kJ/mol[b]	−2158.9
Heat of vapourisation at 64.01°C, kJ/mol[b]	28.899
Heat of fusion at 64.01°C, kJ/mol[b]	19.196
Heat of sublimation at 64.01°C, kJ/mol[b]	48.095

[a] To convert kPa to mm Hg, multiply by 7.5.

[b] To convert J to cal, divide by 4.184.

Chlorides

The trichloride, an olive-green solid, is prepared from the trihydride and HCl at 250–300°C. The tetrachloride, a dark green solid, is best prepared from UO_3 and hexachloropropene.

$$UH_3 + 3HCl \rightarrow UCl_3 + 3H_2$$

$$UO_3 + 3\ CCl_3CCl = CCl_2 \rightarrow UCl_4 + 3CCl_2 = CClCOCl + Cl_2$$

The pentachloride is prepared by treatment of the trioxide with carbon tetrachloride. The hexachloride, a dark green solid, is obtained by disproportionation of the pentachloride in vacuum.

Bromides and iodides

The tribromide, a reddish-brown crystalline material, is prepared from the trihydride and HBr or from the elements. The dark brown tetrabromide is prepared by heating uranium turnings in a stream of nitrogen with bromine vapour. The pentabromide is unstable. The tri- and tetraiodides are made from the elements; the product depends on the temperature and the pressure of the iodine.

Oxides

The uranium-oxygen system is extremely complicated. Alterations of the oxidation state are significant in refining ore concentrates. The dioxide, mp ca 2800°C, is very stable and occurs in nature as pitchblende. The tetrauranium enneaoxide, U_4O_9, forms black crystals and occurs in three modifications. The triuranium octoxide, U_3O_8, is a component of a complicated phase system.

The trioxide, UO_3, exists in six polymorphic modifications that differ in crystallographic properties and colour. It is readily obtained from the thermal decomposition of various uranyl compounds. The monoxide, UO, forms very slowly above 2000°C in mixtures of UO_2 and uranium. The hydrated peroxide, $UO_4 \cdot xH_2O$, is utilised in purification technology. It may be precipitated from uranyl nitrate solution with H_2O_2. Several hydrated oxides have been investigated, e.g., $UO_3.2H_2O, UO_3.H_2O$, and $UO_3.0.5H_2O$.

NUCLEAR FISSION

We have seen that very heavy nuclei have lower binding energies than the nuclei with intermediate masses and hence tend to undergo nuclear fission reactions.

The first nuclear fission reaction was carried out in 1939 by *O. Hahn* and *Strassmann* in Germany by bombarding ^{235}U isotope with slow neutrons. ^{235}U isotope on bombarding with slow neutrons is converted into an unstable intermediate ^{236}U which splits up into two or more fragments and also emitting two or more neutrons with the evolution of enormous amount of energy.

$$^{235}_{92}U + ^{1}_{0}n \longrightarrow [^{236}_{92}U] \longrightarrow ^{140}_{56}Ba + ^{94}_{36}Kr + 2\ ^{1}_{0}n + Energy$$

The process of splitting up of heavier nuclei such as ^{235}U by bombarding with slow neutrons into two or more smaller nuclei of approximately equal masses is known as *nuclear fission*.

A fission process is much more complicated. A given nucleus undergoing fission may split in number of different ways depending upon the energy of the neutrons. For example,

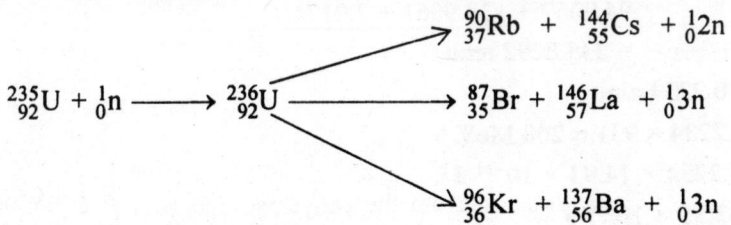

$$^{235}_{92}U + ^{1}_{0}n \longrightarrow ^{236}_{92}U$$

$$\nearrow\ ^{90}_{37}Rb + ^{144}_{55}Cs + ^{1}_{0}2n$$

$$\longrightarrow\ ^{87}_{35}Br + ^{146}_{57}La + ^{1}_{0}3n$$

$$\searrow\ ^{96}_{36}Kr + ^{137}_{56}Ba + ^{1}_{0}3n$$

Important Feature of Nuclear Fission

Chain reaction

During the fission process more number of neutrons are released than the number of neutrons required to initiate the fission. Every neutron so released, in principle, can cause the fission of more atoms and this can continue as long as fissionable material is available. Such a reaction is called as *chain reaction*.

Thus nuclear fission reaction is a self sustained reaction. Since, the fission process is also accompanied by the release of large amount of energy, by properly controlling the process, it can be used as a source of energy (Fig. 18.1).

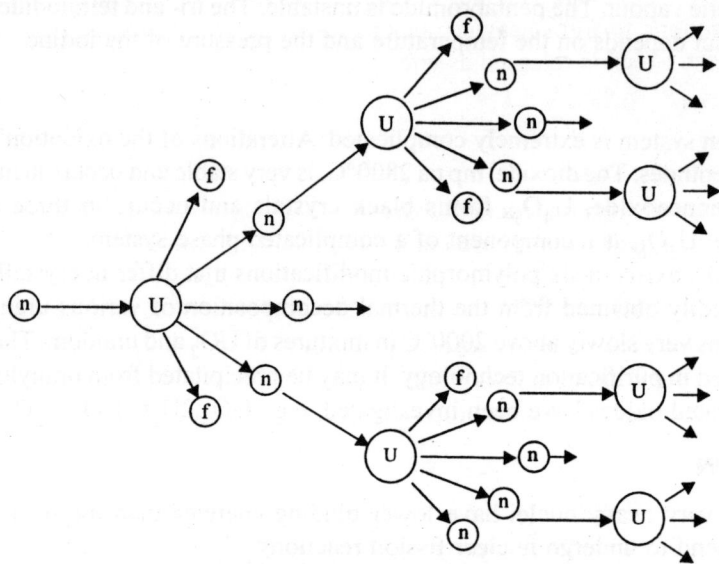

Fig. 18.1. Schematic representation of chain reaction. (n = neutron; u = uranium; f = fission fragments).

Energy released during fission process

The sum of the masses of the fission fragments and neutrons set free during the fission process is found to be less than the sum of the masses of the reactant nuclei. The mass lost is converted into energy and released in the form of heat. This amounts to release of enormous amount of energy during fission process.

$$^{235}_{92}U + ^{1}_{0}n \longrightarrow ^{95}_{42}Mo + ^{139}_{57}La + 2\,^{1}_{0}n + 7\,^{0}_{-1}e + Energy$$

235.0439 + 1.0087   94.9057 + 138.9061 + 2.0174

= 236.0439 amu   = 235.8292 amu

∴ Mass loss (ΔM) = 0.2234 amu

∴ Fission energy = 0.2234 × 931 ≃ 208 MeV

or Fission energy = 0.2234 × 14.93 × 10^{-11} J

            = 33.35 × 10^{-10} J

Fission fragments

Fission fragments are all radio active, which subsequently disintegrate to stable nuclei by a series of β-emissions.

Atomic masses

Atomic masses of fission products ranges from about 70 to 160.

NUCLEAR FUSION

Nuclear fusion is a process of formation of heavier nuclei like helium from those of lighter ones like hydrogen. From the binding energy values it is noticed that there is a sharp increase in binding energy from hydrogen isotope to helium isotope. Therefore, atoms of lighter elements like hydrogen tend to undergo fusion reactions to form heavier nucleus having higher binding energy. The fusion process involves loss of mass which is released as energy.

$$4 \, {}^1_1H \longrightarrow {}^4_2He + 2{}^0_{+1}e + \gamma$$

$\Delta M = 0.02871$ amu

Energy released $= 26.7$ MeV

However, to initiate the fusion reaction, extremely high temperature of the order of 10^8k is required to overcome the coulombic barrier. These reactions are called *thermonuclear reactions*.

Main Features of Fusion Process

1. To initiate the fusion process, extremely high temperature is required.
2. During the fusion process enormous amount of energy is released as heat which makes it possible to maintain the fusion process, i.e., a chain reaction is set.
3. Fusion products are not radio active.

Drawbacks of fusion process

1. It is very difficult to provide the very high temperature to initiate the process.
2. At this temperature the atom is fully ionised, producing a plasma of +ve and –ve ions. It is difficult to confine the plasma state till fusion takes place. One method is to confine a fusion reaction in so called magnetic bottles. The plasma is kept within magnetic field so that it does not come in contact with any material.
3. It is not possible to control the fusion process. Comparison of Nuclear Fission and Fusion process is shown in Table 18.3.

Table 18.3. Comparison of nuclear fission and fusion process.

Fission	*Fusion*
Splitting up of the heavy nuclei like ^{235}U by bombarding with slow neutrons, into two or more light fragments	Formation of one nucleus like He by the combination of two light nuclei like hydrogen.
Fission process is brought about by neutrons	Fusion is brought about by extremely high temperature
More number of neutrons are emitted which causes chain reaction	Fusion is accompanied by the increase in temperature which maintains the fusion process, positrons are emitted.
Enormous amount of energy is released.	Energy released is much higher than the fission process
The fission fragments formed are radio active	The fusion products are not radio active
Fission process can be controlled	Fusion process cannot be controlled
The energy released can be used to produce electricity	At present the process is used in producing nuclear weapons like hydrogen bomb

MASS DEFECT AND BINDING ENERGY

A nucleus of an atom is mainly composed of Z protons, where Z is the atomic number of the element and (A–Z) neutrons where A is the mass number. Protons and neutrons present in a nucleus are collectively known as nucleons. Binding energy is an important characteristic of a nucleus which determine the stability of a nucleus.

Binding energy is the energy which holds the nucleons together in the nucleus. When the nucleons come within the range of nuclear forces a part of their mass is lost resulting in the formation of the nucleus. Thus the actual mass of every nucleus is less than the sum of the masses of the nucleons present in the nucleus. *This difference in mass is known as mass defect* (ΔM). If M is actual atomic mass of the isotope and M' is the actual sum of the masses of constituent protons and neutrons, the mass defect is given by

$$\Delta M = M' - M \qquad \qquad ...(18.1)$$

The mass of an electron being negligibly small compared to the mass of a proton, its mass may be ignored.

If Z is the atomic number and A is the mass number, the number of protons is equal to Z and number of neutrons is equal to (A–Z). Now the equation (18.1) can be written as

$$\Delta M = Zm_p + (A - Z)m_n - M \qquad \qquad ... (18.2)$$

where m_p and m_n are the masses of protons and neutrons respectively.

Binding Energy

When the nucleons are bound in the nucleus, the loss in mass (ΔM) is converted into an equivalent amount of energy, which is called the *binding energy* (binding energy is the energy which holds the nucleons together in nucleus) and given by the Einstein's mass–energy relationship.

Binding energy = $\Delta M.c^2$

where 'c' is the velocity of light, the value being c = 2.998×10^8 ms^{-1}.

Binding energy is expressed in electron volts, or million electron volts (MeV). Generally the atomic masses are expressed in atomic mass units (amu). 1 amu is equal to 1/12th the mass of an atom of C–12 isotope. According to Avogadro's number 1g atom (i.e., 12 g) of C–12 isotope contain 6.02×10^{23} carbon atoms. According to definition, the mass of one atom of C–12 isotope

$$= \frac{12}{6.02 \times 10^{23}} g = \frac{12}{6.02} \times 10^{-26} kg$$

$$\therefore 1 \text{ amu} = \frac{1}{12} \times \frac{12}{6.02} \times 10^{-26} \text{ kg} = 1.66 \times 10^{-26} \text{ kg} = 1.66 \times 10^{-26} \text{ kg}$$

Energy corresponding to 1 amu is therefore

$$E = 1.66 \times 10^{-27} \text{ kg} \times [2.998 \times 10^8 \text{ ms}^{-1}]^2$$

$$= 14.93 \times 10^{-11} \text{ J } [\because J = kg \ m^2 s^{-2}]$$

$$\simeq 931.0 \text{ MeV} \qquad [\because 1.6 \times 10^{-13} J = 1 \text{ MeV}]$$

Thus   1 amu = 14.93×10^{-11} J = 931 MeV   ...(18.3)

$\therefore$ when the mass defect is 1 amu the energy released is 931 MeV

or   B.E = $\Delta M \times 931$ MeV   ...(18.4)

The average binding energy per nucleon is given by the expression

$$B.E / \text{Nucleon} = \frac{B.E}{\text{Total no. of nucleons}} \qquad \qquad ...(18.5)$$

BINDING ENERGY AND NUCLEAR STABILITY

The stabilities of nuclei depend upon their binding energies per nucleon. The nuclei with higher binding energies per nucleon are relatively more stable than those with smaller values of binding energy per nucleon.

The tendency of an atom to undergo fussion or fission can be predicted with the help of the plot of binding energy per nucleon versus the mass number. From the plot three important conclusions can be drawn:

1. The binding energy per nucleon rises sharply from the isotopes of hydrogen to those of next heavier elements like helium or lithium. Therefore, if two atoms of hydrogen are fused to form a heavy nuclei like helium, a large amount of energy is released (Fig. 18.2). Such a process is called nuclear fusion.

2. The binding energy per nucleon reaches maximum at mass number approximately 60 and the elements with intermediate mass numbers 50 to 150 have higher binding energies than those of heavier elements are relatively stable.

3. With the further increase in mass number beyond 60, there is a decrease in the binding energies. Therefore, the heavier elements like uranium tend to split up into nuclei of intermediate masses with the release of large amounts of energy. Such processes are called nuclear fission.

Problems on binding energy:

[Mass of proton = 1.0073 amu

Mass of neutron = 1.0087 amu]

Example 1. The atomic mass of $_2^4 He$ is 4.0017 amu. Calculate the binding energy per nucleon.

Solution: Z of He = 2 and (A – Z) of He = (4 – 2) = 2

$\Delta M = Zm_p + (A - Z)m_n$ - M

$= 2 \times 1.0073 + 2 \times 1.0087 - 4.0017$ amu = 0.0303 amu

BE = $\Delta M \times 931$ MeV

$= 0.0303 \times 931 = 28.20$ MeV

$$\text{Binding energy per nucleon} = \frac{BE}{\text{Total number of nucleons}}$$

$$= \frac{28.20}{4} = 7.05 \text{ MeV.}$$

Example 2. Calculate the binding energy per nucleon in MeV of $_{26}^{56} Fe$. Atomic mass of $_{26}^{56} Fe$ = 55.934 amu. Z = 26, A – Z = 56–26 = 30 and M = 55.934 amu.

Solution: $\therefore \Delta M = Zm_p + (A - Z) m_n - M$

$= (2 \times 1.0073 + 30 \times 1.0087 - 55.934)$ amu = 0.5168 amu

Total BE = $\Delta M \times 931$ MeV

$= 0.5168 \times 931$ MeV

$= 481.14$ MeV.

$$\text{B.E / Nucleon} = \frac{481.14}{56} = 8.59 \text{ MeV} [\because \text{Protons + neutrons = 56}]$$

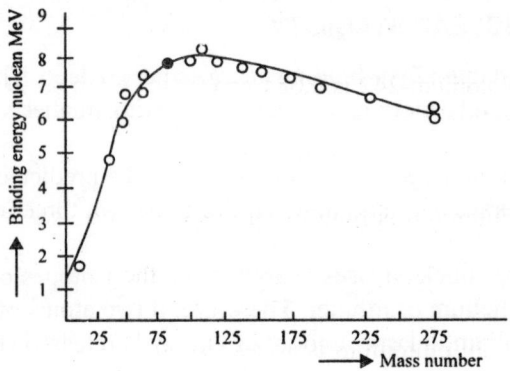

Fig. 18.2. Binding energy curve

PLUTONIUM AND ITS COMPOUNDS

Plutonium

Plutonium, Pu, is element number 94 in the periodic table. It is a member of the actinide series and is metallic. Isotopes of mass number 232 through 246 have been identified and all are radioactive. The most important isotope is plutonium-239; also of importance is plutonium-238.

The large energy release associated with the fission reaction is the most significant property of plutonium. The energy can be applied in electric-power generating reactors, industrial explosives, or military explosives. Large quantities of plutonium are produced in uranium-fueled power reactors. Approximately 200 kg Pu are produced per 1000 MW_e of electric power.

The isotope ^{238}Pu also is of technical importance because of the high heat of its constant radioactive decay. Such radiation is used as fuel in small terrestrial and space nuclear-power sources. Plutonium is unique in that it is the first element to be synthesised.

Sources

From a practical viewpoint, plutonium does not occur in natural ores. All of the fifteen plutonium isotopes have been synthesised and all are radioactive. Technologically, plutonium-239 is the most important isotope. It is characterised by a high fission reaction cross-section and is abundant in irradiated natural uranium.

Commercial electric-power generating reactors generally produce plutonium by irradiating uranium fields to a total neutron exposure of more than 5000 megawatt-days per tonne (MW·d/t). The recoverable plutonium contains a larger fraction of heavier isotopes. A large future source of plutonium will be from fast-neutron breeder reactors.

Physical Properties

Thermal

The expansion coefficient of α-plutonium is exceptionally high for a metal, whereas those of δ- and δ'-plutonium are negative. The net linear increase in heating a polycrystalline rod of plutonium from room temperature to just below the melting point is 5.5 per cent.

Thermodynamic

The thermal conductivity of plutonium-242 is 0.084 and 0.155 W/(m.K) [0.020 and 0.037 cal/(s.cm °C)] for the α- and β-phase, respectively.

Electrical and magnetic

The electrical resistivity of plutonium is high in all modifications as a result of the band structure in metallic plutonium.

Radioactive self-heating

Because of their radioactivity, all plutonium isotopes generate heat. The self-heating of a ^{239}Pu metal sphere has been determined: $(1.923 + 0.019) \times 10^{-3}$ W/g $(1.824 + 0.18$ Btu/(;·g)).

Spectroscopic

Isotope shifts of the 238, 239, and 240 isotopes were obtained in the spark spectrum, reexamined with hollow-cathode spectral data, and supplemented by ^{241}Pu data. Based on the shifts of 20 lines of Pu I, and assuming the 238–240 interval as unity, the shift values obtained are given below :

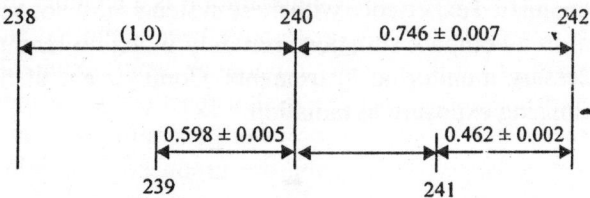

Chemical Properties

Plutonium is the fifth member of the actinide series. Its electronic structure is $1s^2\, 2s^2\, 2p^6\, 3s^2\, 3p^6\, 3d^{10}\, 4s^2\, 4p^6\, 4d^{10}\, 4f^{14}\, 5s^2\, 5p^6\, 5d^{10}\, 5f^{5-6}\, 6s^2\, 6p^6\, 6d^{0-1}\, 7s^2$.

Plutonium is an active metal and, as such, it can be made by the usual methods for active metals such as electrolysis in fused salts, reduction of its halides with an active metal such as Li, Ca, Ba, or reduction of the oxide with Ca in a $CaCl_2$ melt. Plutonium forms compounds with all the non-metallic elements except the rare gases. The halogens and halogen acids form Pu halides, other chalcogens form chalcogenides, and CO forms a carbide. Nitrides are formed with NH_3 and N_2 and hydrides with H_2. The metal dissolves readily in concentrated HCl, H_3PO_4, HI, $HClO_4$, or HSO_3NH_2.

The corrosion of plutonium has been summarised. Plutonium oxidises very slowly in dry air, i.e., $< 10^{-2}$ mm/yr; the rate is accelerated by water vapour.

Extractive metallurgy and conversion chemistry

The production methods used for the separation of Pu from U and fission products are liquid-liquid extraction and ion exchange.

Plutonium Compounds

Plutonium forms compounds with many of the metallic elements and all of the non-metallic elements except the helium-group gases. The refractory compounds are of interest as potential fast-breeder reactor fuels. Commercially important compounds include the oxides, halides, and oxalates.

Health and Safety Factors

The principal hazards of plutonium are those posed by its radioactivity, its nuclear critical potential, and its chemical reactivity in the metallic state.

NUCLEAR REACTORS

Engineered nuclear fission reactors supply an appreciable fraction of the electrical energy used in many countries. Their development and use is in a controversial phase such as experienced in the history of industrial development with the use of coal, steam power for railways and ships, the internal-combustion engine, and even electricity itself.

One remarkable characteristic (established early in the history of nuclear reactors operating at high power density) is that, after a shutdown, it may not be possible to restart the chain reaction for 15–40 hour. This is owing to transient poisoning of the fission-chain reaction by one of the fission products, ^{135}Xe.

Safety Aspects

Much attention has been paid to safety in establishing the technology of nuclear reactors. The training of operators derived from accumulated experience with these systems must not be underestimated. Although a trained mind is essential in a complex situation, certainly in action requires accurate and adequate information from the necessary monitoring instruments. Containment of the fission products is the basic requirement for minimising exposure to radiation.

Heat Removal

The rate of the chain reaction can increase to levels at which the safe removal of the heat from fission and, later, from residual fission products becomes the principal engineering concern. Thus, many coolants or caloporteurs have been used, including gases, especially air, CO_2, He, and dry steam; liquids, especially water, single-phase pressurised water, two-phase boiling water or fog, heavy water, organic liquids, especially terphenyl and hydrogenated terphenyl, which is liquid at room temperature; the alkali metals, sodium, Na-K alloy, potassium, and lithium; and mercury, molten bismuth, and molten salts, especially fluorides.

Fuels

Depending on the purpose of the reactor, many fuels have been used, although UO_2, ThO_2, PuO_2, and mixtures of these as sintered, high density pellets have been used extensively.

Special Materials

Tritium is produced in reactors not only as a direct fission product, but also by the action of fast neutrons on carbon, nitrogen, and other light elements. Helium is troublesome because it promotes the nucleation of bubbles, especially in nickel and nickel alloys. Zirconium hydride is used in TRIGA (Training Reactor, Isotopes General Atomic) as a solid moderator.

CLEAR FUEL RESERVES

Light-water reactors are the chief type of nuclear power reactors in reaction. They are fueled with uranium that has been enriched to ca 3 per cent ^{235}U from the naturally occurring concentration of

0.71 per cent. Some experimental thorium-fueled reactors are in operation, but they are not expected to be in significant use until well after 2000.

Uranium

The earth's crust contains 2–3 ppm uranium, alkalic igneous rocks. Uranium combines readily with oxygen. Its solubility and distribution in rocks and ore deposits depend largely on its valence state: Uranium is highly soluble in the six-valent state and relatively insoluble in the four-valent state. Uraninite, the most common mineral in uranium deposits, contains the four-valent ion.

Resources

WOCA (world outside centrally planned economies) reasonably assumed uranium resources are ca 2×10^6 metric tonnes.

Availability

A principal factor in considering fuel-resources availability is the time and effort needed to discover and develop the resources attainable production levels.

Demand

The greater demand would be predominantly for light-water reaction without fuel recycle; the lower-demand cases would involve substantial employment of liquid-metal fast-breeder reactors.

Low grade uranium resources

Low grade resources include shales, water, mill tailings, and enrichment-plant tails.

Thorium

Resources

The crustal concentration of thorium is 10–20 ppm. The element is dispersed widely in variable amounts in a number of different types. It is commonly associated with rare earths, niobium, titanium, zirconium.

Demand and supply

The demand for thorium in recent years is limited to a small amount used for non-nuclear purposes, e.g., lamp mantles, catalysts, refractories, and in high temperature alloys, significant market for thorium as a nuclear fuel has developed.

WATER CHEMISTRY OF LIGHT-WATER REACTORS

In a (PWR) pressurised-water reactors a closed circuit of high pressure, high temperature water transfers heat from the reactors core to once-through (OTSG) or recirculating U-tube steam generators, where steam is produced. Current secondary-cycle-chemistry control practices parallel those of fossil plants; however, impurity levels are maintained much lower. Initially, secondary water chemistry specified for U-tube generators was based on sodium phosphates. Subsequent to observations of corrosion attack, all volatile treatment (AVT) was adopted at most units. Several new types of corrosion have been observed subsequently. Corrosion of OTSGs has not been as extensive or as serious a problem as with U-tube steam generators. A summary of initial AVT specifications is given in Tables 18.4 and 18.5.

Table 18.4. Initial PWR feedwater specifications for All-Volatile-Treatment (AVT) operation[a].

Property	Babcock & Wilcox	Combustion engineering	Westinghouse
pH	9.3–9.5[b]	8.8–9.2[c]	8.8–9.2[c]
	8.5–9.3[d]	9.2–9.5[e]	up to 9.6[e]
O_2, ppb	< 7	< 10	< 5
Shutdown		100	
N_2H_4, ppb	20–100	10–50	$[O_2] + 5$
Acid conductivity, μS/cm	< 0.5	< 0.5	
Abnormal[f]	> 0.5 to ≤ 1 (24 h)	> 1.5 (4 h)	
	> 1 to ≤ 2 (12 h)		
Shutdown	> 2		
Iron, ppb	< 10	< 10	< 10
Copper, ppb	< 2	< 10	< 5
Total solids, ppb	< 50		
Total silica (as SiO_2), ppb	< 20	< 10	
Ammonia, ppm		< 1	< 0.5

[a] Tabulated values are for normal operation unless noted otherwise.

[b] Carbon-steel feedwater heater tubes or combinations of carbon steel and stainless-steel feedwater and/or reheater tubes.

[c] With copper alloys in feedwater heaters, reheaters, or condenser.

[d] With stainless-steel feedwater heater tubes and stainless-steel or copper-nickel reheater tubes.

[e] With no copper alloys in feedwater heaters, reheaters, or condenser.

[f] Corrective action or shutdown recommended within indicated time.

Table 18.5. Initial PWR recirculating steam generator blowdown specifications for AVT operation.

	Combustion engineering	Westinghouse Freshwater	Seawater or brackish water
pH	8.2–9.2	8.5–9.0	8.5–9.0
abnormal[a]	< 7.5 or 9.5 (4 h)	8.5–9.2 (2 wk)	8.0–9.2 (2 wk)
shutdown	10.5	< 8.5 or > 9.4	< 8.0 or > 9.4
Specific conductivity, μS/cm	< 7		
abnormal[a]	> 15 (4 h)		
Suspended solids, ppm	< 1		
abnormal[a]	> 10		
Free hydroxide, ppm		< 0.05	< 0.05

(Contd ...)

	Combustion engineering	Westinghouse	
		Freshwater	Seawater or brackish water
abnormal[a]		> 0.05 to ≤ 0.34 (24 h)	≤ 0.05 to ≤ 0.34 (24 h)
shutdown	5	> 0.34	> 0.34
Silica, ppm	< 1		
abnormal[a]	> 10 (4 h)		
Cation conductivity, μS/cm		< 2.0	< 2.0
abnormal[a]		> 2 but ≤ 7 (2 wk)	> 2 but ≤ 120 (2 wk)
shutdown		> 7	> 120

[a] Corrective action or shutdown recommended within indicated time.

Boric acid is added to the PWR primary system to compensate for fuel consumption and to control reactor power. A hydrogen overpressure is maintained, and small amounts of base are added to reduce corrosion rates not only to address component-integrity concerns but also to minimise shutdown radiation levels. In a BWR (boiling-water reactor), steam is generated on the surface of the nuclear fuel and is used to drive the turbine generator. Since chemical additions are not made, materials corrosion is dependent primarily on coolant oxygen concentrations, which are governed by radiolytic decomposition of water and the steam-liquid oxygen equilibrium.

BWR chemistry specifications are given in Table 18.6. Because of the importance of maintaining fuel integrity in the direct cycle and of minimising radiation levels on out-of-core surfaces, transport of corrosion products to the reactor by the feedwater is of considerable importance.

Recently, incidences of intergranular stress-corrosion cracking (IGSCC) in stainless-steel piping have reduced the availability of BWR systems significantly. Numerous remedies for the problem are being evaluated and applied.

Table 18.6. BWR water-quality specifications for normal and abnormal conditions.

	General Electric
Reactor water	
Conductivity, μS/cm	≤ 1
abnormal (2 wk/yr allowed)	> 1 but ≤ 10
shutdown	> 10
Chloride, ppm	≤ 0.2
abnormal (2 wk/yr allowed)	> 0.2 but ≤ 0.5
shutdown	> 0.5
pH	5.6 to 8.6[a]
Feedwater	
conductivity, μS/cm	≤ 0.1
pH	6.5 to 7.5
metals, ppb	< 15 total
	< 2 Cu
oxygen, ppb	> 20 but < 200

[a] Inferred from conductivity limits.

ISOTOPE SEPARATION

The difficulty and high cost of isotope separation has limited the use of separated isotopes in nuclear reactors to specific cases where no non-isotopic substitutes are available. The most important example is ^{235}U, the most abundant, naturally occurring, fissionable material. Other isotopes that are separated for nuclear use are ^2H or deuterium, D, which is used as D_2O as a neutron moderator in nuclear reactors and is a probable reactant in thermonuclear reactors; ^3H or tritium, T, which is produced in nuclear reactors and is a reactant in thermonuclear reactors under development; boron, ^{10}B, whose high neutron cross-section is used in control rods and safety devices for reactors; and lithium ^7Li, which is used in reactor cooling water systems because of its low thermal neutron cross-section.

FUEL-ELEMENT FABRICATION

The fuel assembly forms the basic unit of nuclear-reactor core. Fuel assemblies are fabricated by processing fertile and fissile materials into suitable chemical and physical forms, encapsulating the materials in a protective metallic sheath to produce a fuel rod, and assembling the fuel rods into the required configuration. Different reactor types, e.g., water-cooled, gas-cooled, and sodium-cooled breeder reactors, require different fuel-assembly designs.

CHEMICAL REPROCESSING

Fuel elements that are discharged from light-water nuclear-power reactors (LWRs) typically contain uranium, 0.79 wt% ^{235}U and 0.8 wt% ^{239}Pu, which can be recovered by chemical reprocessing. However, the economic incentive for reprocessing fuel for recycle into light-water reactors is marginal. Recycle of breeder-reactor fuel is, of course, essential to the breeder concept. Chemical reprocessing must provide for the high recovery of uranium and plutonium, and their separation from each other and from hazardous radioactive contaminants. Reprocessing implies the efficient separation of fertile materials, e.g., ^{238}U or ^{232}Th; from one another, e.g., ^{239}Pu from ^{235}U and/or ^{233}U; from the highly radioactive fission products; and from materials that are undesirable neutron poisons, including certain fission-product isotopes.

Purex Process

The process used is almost exclusively some form of the Purex aqueous solvent extraction process involving tributyl phosphate. Although other solvents have been employed and other processes, i.e., precipitation, ion exchange, and volatility processes, have been used, the Purex process and a variant for thorium-based materials, i.e., the Thorex process, are almost universally accepted.

The fuel is initially dissolved in nitric acid after an economically optimised period to allow radioactive decay of most of the short-lived isotopes that account for most of the heat and the high intensity gamma radiation associated with spent fuel elements. Preprocessing decay time, which largely eliminates troublesome fission products, e.g., ^{131}I, generally is one to two years for thermal-reactor fuels and three to nine months for breeder fuels.

Dissolution of the solid fuel is accompanied by complete release of the rare-gas fission products, krypton and xenon, into the off-gas system. Since the dissolver overheads also ordinarily contain fractions of some fission products, e.g., iodine, tritium, ruthenium, carbon-14 as CO_2, and tellurium, the off-gas usually is treated to remove or minimise the discharge of these gaseous fractions, as well as entrained particles or aerosols.

The multicycle Purex process depends primarily on countercurrent liquid-liquid extraction techniques that involve the transfers of a large number and variety of solutes between aqueous nitrate dissolver solution and the immiscible, less dense, organic solution. The extraction is repeated several times to achieve the desired degree of purification. Typically, recoveries of > 99.8% of plutonium and uranium and decontamination factors of > 10^6 are achieved.

FAST BREEDER REACTORS

A breeder reactor is a nuclear reactor that produces more nuclear fuel or fissile material than it consumes. Although the emphasis for the past 40 years has been on the liquid-metal fast breeder reactor (LMFBR) and the development of the uranium-plutonium fuel cycle, use of the thorium-uranium fuel cycle for breeder reactors also has been studied, primarily for thermal reactor systems. Consideration has been given to the use of this cycle in fast breeder reactors.

Design

A fast breeder reactor can be designed in core and blanket configurations to produce more fissionable material than it consumes. This is achieved by converting fertile material, e.g., ^{238}U or ^{232}Th, in the blanket into ^{239}Pu and ^{233}U, respectively. Thus, in a typical LMFBR, the central region of the reactor (Fig. 18.3)., i.e., the reactor core, contains the plutonium-bearing assemblies which are comprised of a mixture of plutonium fuel and fertile uranium oxides. The blanket region surrounds the core and holds the uranium-bearing assemblies. Most of the fission occurs in the core, and most of the breeding occurs in the blanket. At intervals, the reactor is shut down to remove selected core and blanket assemblies and to replace them with fresh ones. The spent assemblies then can be reprocessed chemically to reclaim the fuel and the fertile materials for eventual recycling into new and existing reactors.

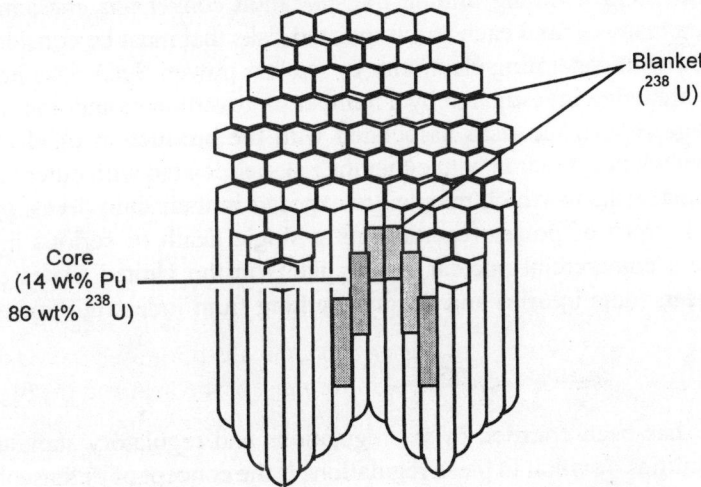

Core
(14 wt% Pu
86 wt% ^{238}U)

Blanket
(^{238}U)

Fig. 18.3. Core and blanket arrangement for a typical LMFBR.

Breeder reactor types

The ability of a reactor to sustain a sufficient conversion or breeding ratio depends on a number of engineering and physical factors that influence the reactor's neutron environment. Average neutron energy is influenced by reactor core geometry, the composition of the fuel, and the presence of other materials, e.g., the coolant and construction materials. Design decisions can reflect the different options available; the various factors often are interrelated, and a number of permutations and combinations are possible. The liquid-metal fast breeder reactor (LMFBR), the gas-cooled fast reactor (GCFR), and the light-water breeder reactor (LWBR), have received the most attention. Primary emphasis has been on the LMFBR, which is the most highly developed of these reactor types, and on the uranium/plutonium fuel cycle; eleven experimental and demonstration LMFBRs are now in operation worldwide.

SAFETY IN NUCLEAR FACILITIES

In many countries, concern for the safety of nuclear power plants is curtailing their growth rate. The Nuclear Regulatory Commission licences and regulates nuclear facilities such that there is no undue restraint to the health and safety of the public. One of many requirements is that each nuclear plant must limit occupational exposures to 5 rem/yr and exposures in unrestricted areas to 0.5 rem/yr. In practice, the public exposure at nuclear plant sites generally is some orders of magnitude less. However, it is generally assumed that, for the purpose of evaluating the risks of radiation, the probability of including cancer is proportional to dose, no matter how small the dose. If the preceding hypothesis is true (and it may not be at very low dose), the benefits of the activity giving the exposure must be weighed against the risks and the costs of the low level exposure. Furthermore, the average public exposure from nuclear power plants is 0.1 mrem/yr, which is very small in comparison to the routine exposure of ca 200 mrem/yr for the average citizen.

Nuclear Fuel Cycle

The nuclear power plant that produces electricity is only one element among many in the nuclear fuel cycle. The other elements include mining, milling, transportation, conversion, enrichment, fuel fabrication, reprocessing, and waste disposal, and each has associated risks that must be considered when assessing the total risk associated with generating electricity by nuclear power. Such risks have been reported in numerous studies and are being investigated by a number of government and nongovernment agencies.

It has been investigated that the risks associated with the production of electric power utilising nuclear energy are small when compared with either the risks associated with other methods of electricity production or the societal risks to which people are exposed in their daily lives. In a recent report on serious accidents in all types of power reactors, not a single death or serious injury resulting from radiation exposure from commercial nuclear power plants in the United States could be identified; although there have been some injuries and deaths resulting from industrial-type accidents in nuclear plants.

Safety philosophy

The safety philosophy has been codified in the regulations and regulatory standards that govern the licensing of nuclear facilities. Implicit in these regulations is the concept of defense in depth, with which three levels of safety are identified. Furthermore, the risk from nuclear plants is shown to be much less than that from most other accepted risks such as fires, plane crashes, dam failures, etc.

The first level of safety is achieved by designing a plant for maximum safety in normal operation and maximum tolerance for system malfunctions. The second level of safety is based on the assumption that incidents occur despite care in design, construction, and operation. Safety systems should be provided to protect operators and the public and to prevent or minimise damage when such incidents occur. The third level of safety is the provision of additional safety systems as appropriate, based on the evaluation of the effects of hypothetical accidents, where some protective systems are assumed to fail simultaneously during the accident that they are intended to control. Other safety considerations in design include quality assurance, surveillance testing, redundancy and coincidence, and indenpendence and diversity.

Safety features

A nuclear power plant abunds with safety features that are installed either to fulfill operational requirements (including safety) of the licensee, or to meet the preceding safety, philosophy as interpreted by the NRC. Some of the more significant safety features are listed in Table 18.7.

Table 18.7. PWR safety features, functions and purposes.

System	Function	Purpose
Reactivity control		
Safety rods	Shut down reactor	Stops fission-heat generation
boron injection	maintain reactor shutdown	prevents fission-heat generation from recurring
Pressure control		
Pressuriser	Maintain system pressure	Maintains cooling in liquid state
relief valves	prevent system over-pressure	prevents equipment leaks and/or rupture
Emergency core cooling		
High-pressure injection accumulators	Maintain coolant at pressure provide cooling during large pipe failure	Maintains cooling at pressure maintains cooling during transient blowdown event
Low pressure injection	Maintain cooling of core after blodwdown	Prevents core from overheating
Emergency heat removal		
Auxiliary feedwater system	Provide heat sink for reactor coolant	Ensures core coolability
Decay-heat removal system	Maintain coolness of core	Prevents core from overheating
Containment		
Isolation system	Enclosure of radiation	Minimises release of radioactivity to the environment
Spray system	Control containment temperature and pressure	Minimises containment leakage
Recirculation system	Cool and cleanup containment atmosphere	Minimises release of radioactivity to the environment
Hydrogen recombination	Recombine hydrogen with oxygen	Minimises explosion potential
Vital auxiliaries		
Emergency power system	Supply power for emergency needs	Permits plant safety systems to function in the event of loss of normal power
Component cooling water	Supply cooling water to equipment needed in an emergency	Permits plant safety systems to function in the event of an emergency
Instrument air	Supply instrument air for emergency needs	Permits plant safety systems to function in the event of loss of normal instrument air

Accidents

The occurrence and consequences of accidents in nuclear power plants is the ultimate measure of the effective application of the previously described safety concepts to reactor technology.

Other Nuclear Fuel Cycle Facilities

Because the nuclear power plant constitutes over 99 per cent of the radiation risk resulting from nuclear fuel cycle, the radiation risk from the balance of the fuel cycle is small and is distributed among many facilities. There are three potential risk mechanisms: inadvertent critically, inhalation exposure, and direct radiation. The last mechanism is accommodated readily by adequate radiation shielding, which is well documented.

SPECIAL ENGINEERING FOR RADIOCHEMICAL PLANTS

Concerns that are unique to the nuclear industry result from the handling of radioactive and potentially fissionable materials. The special engineering considerations are categorised in terms of the design for radiation resistance, the nuclear safety design, the mitigation of occupational radiation exposure, and the minimisation of environmental impact.

Radiation Resistance

Although a spectrum of radiation types, e.g., alpha particles, beta particles, X-rays, gamma rays, and neutrons, is emitted from spent nuclear fuel, the primary equipment-design consideration is the effect of gamma radiation, especially on organic materials.

Specially developed organic materials for use in gamma radiation fields have lifetimes that are improved by a factor of 10–100 over the standard materials intended to function for ca 100–10,000 hour. They generally are used when no design replacement is available. Local shielding can be provided around the organic material as a means of extending its useful life. Inorganic lubricants, e.g., molybdenum disulphide, can also be substituted for organic lubricants.

Safety Design

Fissionable material, i.e., the reusable fuels for nuclear reactors, including ^{233}U from the thorium fuel cycle or ^{235}Pu from the uranium fuel cycle, are capable of spontaneous criticality if accumulated in adequate concentration, volume, and geometry. A safe configuration is an infintely long cylinder, e.g., 11.5 cm dia of solution or 4.4 cm dia of solid uranium or plutonium. Other means of preventing nuclear criticality are by limiting the mass of fissile material present; limiting the concentration of fissile material in solution; eliminating or controlling materials that moderate the neutrons and thus reduce the number required; or using liquid or solid poisons or neutrons absorbers that greatly increase the number of neutrons required to initiate or maintain the chain reaction or nuclear criticality.

Occupational Radiation Exposure

The massive shielding surrounding the process equipment usually is made of concrete and is 1–2 m thick, depending on the intensity of the radiation being shielded and the density of the concrete. Since the shielding thickeness required for alpha, beta, and gamma radiation is a density function, lead and steel are used where the wall thickness must be limited. Visual observation is provided with windows of leaded glass, which are available in densities that are nearly that of steel, with periscopes that penetrate the shielding, or with television cameras. The shielding is provided to protect the plant workers.

Since radioactive gases and fine airborne radioactive particles are generated during reprocessing, the ventilation system for the plant is an important design factor. Conventional ventilation systems maintain a flow of air from the worker-occupied areas to the areas of high contamination to the clean-up systems. In the clean-up systems, particulates are removed and the radioactive gas content is reduced to a safe level before being removed from the plant through the stack.

Environmental Impact

Plant design, as related to environmental impact, refers to the special design requirements that are imposed to ensure that the general population and environment are properly protected. Two key considerations are control and reduction of contaminants from planned or routine releases, and prevention or containment and mitigation of unplanned releases from accident situations.

Remote Operation and Maintenance

The equipment and facilities for radio-chemical use are designed to have high efficiencies, high reliability, and the capabilities, for remote operation and maintenance. The fundamental keys in the analysis of a remote design are a concept of equipment that, for its stated use, produces the required product from a defined input; selection of construction materials that are compatible with the internal environment (initial form, intermediate products, and final product) and external environment (radiation, atmosphere, and temperature) in which the machine operates; the capability to monitor critical functions so that indication of operational status is available constantly; a modular machine design to provide for adjustment and replacement of key machine segments; and space and structural strength to accommodate remote manipulation.

HEAVY WATER FOR NUCLEAR POWER INDUSTRY

Hydrogen has three isotopes, protium or light hydrogen with a mass number of 1, deuterium or heavy hydrogen with a mass number of 2, and tritium with a mass number of 3. The name, 'protium' for the lightest isotope of hydrogen has not become popular. So, we will call this isotope hydrogen in the rest of this discussion though this word actually denotes the element constituting the mixture of all the three isotopes. Hydrogen is quite abundant in nature in the combined form, the most familiar of its compound being water. Deuterium also occurs in nature in very small quantities. Natural hydrogen contains deuterium to the extent of about 1 in 6000. Tritium is a radioactive isotope and exists only in extremely minute amounts in nature, of the order of 1 in 10^{17}, i.e., 1 in a hundred million billion. Just as hydrogen combines with oxygen to form its oxide, H_2O, called water, deuterium combines with oxygen to form its oxide, D_2O which is called heavy water. Heavy water is an isotopic form of water and occurs in natural water to the same extent as does deuterium in natural hydrogen.

Evidence of the presence of more than one isotope of hydrogen came to light as a result of the difference in its atomic weights determined chemically and physically with the mass spectrograph.

Properties of Heavy Water

Isotopes differ in their nuclear and physical propeties, but may be regarded as identical in their chemical properties. The relative mass difference between hydrogen and deuterium is the greatest of all the known isotopes and is about 100 per cent. Consequently, hydrogen and deuterium show a larger difference in their physical properties than do two different isotopes of other element. Light water, H_2O and heavy water, D_2O exhibit similar difference in different isotopes of any other element. Light water, H_2O and heavy water, D_2O exhibit similar differences in their physical propeties. Table 18.8. gives some of the important physical properties of heavy water and light water.

In chemical properties, the differences between heavy water and the light water exist mainly in the kinetic aspects of reaction, i.e., in the rates of reaction. Reactions with heavy water proceed, in general, slower than the corresponding reactions with light water. In biological systems, that is, plants and animals, the substitutions of deuterium for hydrogen can alter the delicately balanced life processes substantially. Hence, it could have a toxic effect on some of the living organisms. Studies have shown that mammals can tolerate replacement of deuterium by hydrogen upto 15 per cent. Severe toxic effects result when the replacement exceeds 30 per cent.

Table 18.8. Properties of deuterium and protium oxides.

Property	Heavy water D_2O	Light water H_2O
Boiling point (°C)	101.4	100.0
Melting point (°C)	3.82	0.00
Specific gravity at 25°C	1.107748	0.997044
Temperature of maximum density (°C)	11.22	4.08
Heat of vapourisation (calories per mole)	9966	9700
Dielectric constant	82.0	80.5
Viscosity at 20°C (centipoise)	1.103	1.002
Refractive index	1.32844	1.33300
Surface tension (dynes/cm)	67.8	72.8

Analysis of Heavy Water

Ordinary water and heavy water have substantially similar chemical properties but differ slightly in physical propeties. A mixture of two or more substances which are chemically different can be easily analysed by means of their varying chemical properties, and the concentration of each component readily and accurately determined. Mixtures of heavy water and light water cannot be so analysed. Hence analysis of heavy water constitutes a challenging task to the analyst. He has to take recourse to very carefully controlled and tedious methods of physical measurements or to the use of expensive, sophisticated instruments such as infrared spectrophotometers or mass spectrometers for accurate analysis. We have seen from Table 18.8 that the density or specific gravity of heavy water is about 10 per cent higher than that of light water. Measurement of density with a pyknometer, under very carefully controlled conditions of temperature, is the absolute method for determination of heavy water in water. One normally employs this technique to standardise a sample and use it as a primary standard for the much easier, relative methods briefly described below.

One of the less familiar physical characteristics in which deuterium oxide differs considerably from hydrogen oxide is in the absorption of infrared light. Infrared radiation forms part of the electromagnetic spectrum of radiation which in the decreasing order of frequency (or increasing order of wavelength) consists of: (i) gamma rays; (ii) X-rays; (iii) ultraviolet rays; (iv) visible light; (v) infrared radiation, and so on. Light with wavelengths less than about 1 μm and greater than about 0.8 μm (1 μm = 10^{-6}m) fall in the infrared region of the spectrum. Heavy water and light water show distinct differences in their infrared absorption spectra. Making use of these differences one can quantitatively measure the concentration of heavy water with a high degree of accuracy. The technique is simple, but employs infrared spectrophotometers which are relatively expensive, sophisticated, optical instruments. A good deal of training is required before an operator can practise this technique to get reliable results. This is a relative method, and analyses are carried out with the help of calibration curves made with sample of known concentrations of heavy water (standards). The other methods for the isotopic analysis include measurements of mass spectra, refractive index, thermal conductivity and viscosity. They vary considerably in the case of operation, accuracy, and sensitivity.

Production of Heavy Water

One way of enriching natural water in heavy water is by electrolysis. Just as the isotopic analysis of water depends on the small differences in the physical properties of the isotopic waters, the methods of separating heavy water (D_2O) from light water also depend on such differences. In addition, the small differences expected in the chemical properties of deuterium and hydrogen lend themselves for use in separating deuterium by means of exchange reactions between deuterium and hydrogen in their compounds. The techniques of enriching heavy water of deuterium fall into two major classes: (i) those based on differences in physical properties, mainly boiling point and vapour pressure; and (ii) those based on exchange reactions between deuterium and hydrogen compounds. The methods that can be employed for commercial production of heavy water includes: (i) fractional distillation of water; (ii) fractional distillation of liquid hydrogen; and (iii) exchange reactions between compounds of hydrogen and deuterium. In all these methods, a parameter of interest is the separation factor which tells us how well the separation of the isotopes occurs during the process. The separation factor is mathematically expressed as follows:

$$S = \frac{(H)_g \ (H)_l}{(D)_g \ (D)_l}, \text{ where}$$

S = separation factor, $(H)_g$ = concentration of hydrogen in gaseous phase, $(D)_g$ = concentration of deuterium in gaseous phase, $(H)_l$ = concentration of hydrogen in liquid phase, and $(D)_l$ = concentration of deutetrium in liquid phase. Table 18.9 shows that the separation factor decreases with temperature.

Table. 18.9. Separation factor for various methods of enriching deuterium.

Method	Separation factor at 30°C
Distillation of water	1.1
Distillation of liquid hydrogen	1.5 at 260°C
Exchange between hydrogen sulphide and water	2.3
Exchange between hydrogen and ammonia	3.5
Exchange between hydrogen and water	3.7
Electrolysis of water	9.5

All the above methods have relative advantages and disadvantages. As they have to be used to concentrate deuterium or heavy water D_2O from its natural concentration of about 0.015 per cent (atom per cent) to nearly 100 per cent, i.e., an enrichment of about 6700 times, all these processes consume tremendous amounts of energy. They are, therefore, expensive and so is the product, heavy water.

As India has chosen nuclear reactors with natural uranium as fuel and moderated and cooled with heavy water as the basic type for its nuclear power programme, and have planned and put up heavy water production plants to cater to the needs of nuclear power industry.

Table 18.9 shows that electrolysis has the highest separation factor. But the consumption of electrical energy is very high if this process is used for enrichment from the natural concentration to 100 per cent heavy water. The chemical exchange between hydrogen and water has the next highest separation factor. This process has the disadvantage that the rate of reaction is too low to enable it to be of practical value; no suitable catalyst to accelerate this reaction has yet been found. Exchange between hydrogen and water is of industrial value as the rate of reaction can be speeded up with potassium amide as a homogeneous catalyst. Exchange between hydrogen sulphide and water proceeds with acceptable speed even without a catalyst and lends itself to ready industrial exploitation. Distillation of liquid hydrogen and water are relatively simpler and proven methods but their, separation factor is low.

In India, at present there are five heavy water production plants in various stages of construction, commissioning and production. They are at Nangal in Punjab, Anushakti near Kota in Rajasthan, Baroda in Gujarat, Talcher in Orissa, and Tuticorin in Tamil Nadu. The Nangal plant uses electrolysis followed by hydrogen distillation. The one near Kota employs chemical exchange reaction between hydrogen sulphide and water. Baroda, Talcher and Tuticorin plants make use of hydrogen-ammonia exchange reactions.

Uses

Among the purely scientific uses of heavy water, the most important is its application as a tracer in the study of reaction mechanisms in the course of physical, chemical and biological processes. Heavy water, D_2O, permits the preparation of many deuterated compounds by direct or exchange reactions with it. Thus, alkali metal oxides such as sodium oxide (Na_2O) and calcium oxide (CaO), when dissolved in heavy water, yield corresponding deuterohydroxides (deuteroxides or deuterated hydroxides), NaOD and $Ca(OD)_2$. The us of heavy water as a 'moderator' in nuclear reactors constitutes its most dramatic technological application. A thermal nuclear reactor works on the fission chain reaction in uranium. Uranium has three natural isotopes: U–238 (99.3%), U–235 (0.7%) and U–234 (negligible). Uranium–238 and U–235 undergo fission reaction when bombarded by neutrons. The nuclei of these isotopes absorb the neutron and become excited compound nuclei. Then they split into the two heavy fragments along with the emission of two or three neutrons. Fig. 18.4. pictorially represents this process. The phenomenon of fission in uranium can be made a convenient source of energy if a chain reaction of fission can be sustained in a controllable fashion. This is precisely what happens in the core of a nuclear reactor.

When a uranium nucleus fissions, it emits two to three fast neutrons. If at least one of these neutrons can be made to cause fission of another uranium nucleus and one neutron from this second fission to cause fission in a third nucleus and so on, a chain reaction with a constant number of fissions per unit time will result. In a lump of natural uranium, neutrons emitted in the fissioning of a uranium nucleus can have three alternative paths: (i) the neutrons can escape from the body of the metal; (ii) they can be captured by other uranium nuclei without undergoing fission (non-fission capture); and (iii) they can fission other uranium nuclei. The escape can be reduced by increasing the size of the lump so that its surface-to-volume ratio decreases. The probability that a neutron emitted in a fission reacts with another uranium nucleus depends on its speed and hence its energy. This probability increases with decreasing neutron energy. The probablility for a nuclear reaction to take place is called the cross-section for that reaction. As the neutron energy decreases, the cross-sections for both fission and non-fission capture increase. However, U–238 cannot be fissioned by neutrons of low energy. So, after having been slowed down below a particular neutron energy, neutrons cannot fission U–238, though they can continue to fission U–235. For a controlled and sustained chain reaction in natural uranium, we have to depend upon its small U–235 content. As the neutron energy decreases, the non-fission capture cross-section for U–238 becomes so high that most of the neutrons from the spontaneous fissions end up in this reaction.

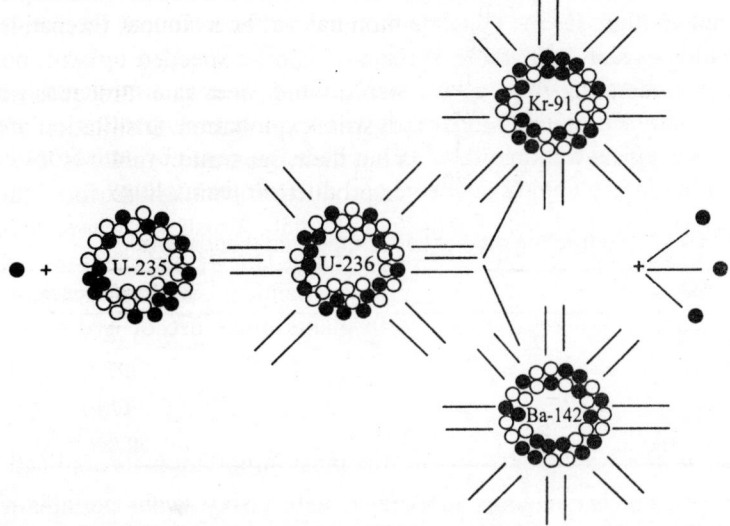

Fig. 18.4. Fissioning of a uranium-235 nucleus by a bombarding neutron.

That is why a lump of natural uranium cannot sustain a fission chain reaction. If this lump is divided into smaller parts and the neutrons from one portion slowed down sufficiently by passing through another substance before they enter other parts, these slow neutrons can produce further fissions in U–235. Such slow neutrons are called thermal neutrons, as they have energies similar to those of the gaseous molecules at ordinary temperatures. The substance used to slow down the neutrons is called a moderator. Fig. 18.5 schematically illustrates these points.

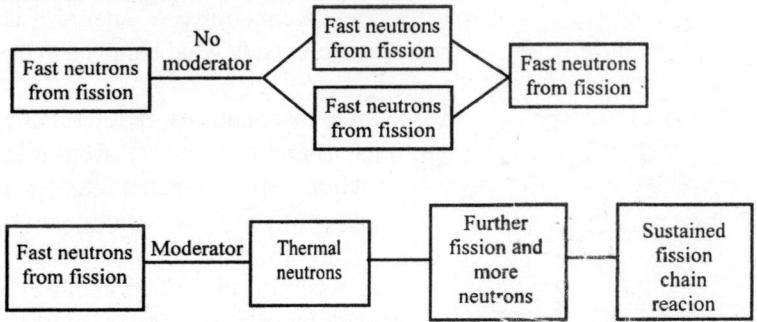

Fig. 18.5. The effect of moderator in bringing about a fission chain reaction in uranium.

A moderator should slow down the neutron sufficiently before it passes to the next part of uranium without interacting with it in other ways. The neutrons lose their energy by colliding with the atomic nuclei in the moderator by elastic collisions in which the kinetic energy is distributed between the colliding nuclei, and the total kinetic energy is conserved. The amount of energy lost by a neutron in colliding with a nucleus is maximum when the target nucleus has the same mass as the neutron. Thus, light water in which the hydrogen nuclei have almost the same mass as neutrons should be the best moderator. But, unfortunately, hydrogen also absorbs neutrons. So, though its moderating power is high, its moderating ratio which is equal to the ratio of the moderating power (a function of the loss of enegy per collision) to the macroscopic neutron capture cross-section (a function of the probability a neutron capture), is not so high. Of the possible moderators for a nuclear reactor, heavy water has the highest moderating ratios as shown in Table 18.10.

Because of its high moderating ratio, heavy water, when used as a moderator, permits a sustained fission chain reaction in natural uranium. Moderators with less moderating ratios require enriched uranium, i.e., uranium in which the concentration of U–235 has been increased by an isotope enrichment process. Such enrichment of uranium is a capital intensive and difficult technology.

Table 18.10. Moderating ratio of some important moderators.

Substance	Moderating ratio
Light water	72
Beryllium	159
Carbon	170
Heavy water	12,000

Another advantage of nuclear reactors moderated with heavy water pertains to the efficiency of utilisation of uranium as a nuclear fuel. One expresses this quantitatively as burn-up in terms of megawatt days per tonne of uranium. The cost of such burn-up is lower with heavy water reactors compared to other types of reactors.

The Rajasthan Atomic Power Station at Anushakti near Kota in Rajasthan has two power reactors with the fission chain reaction sustained by moderation with heavy water in natural uranium oxide fuel.

Just as we can split heavy nuclei to obtain energy so can we make light nuclei to combine to form heavier nuclei and get energy in that process too. Such processes are called fusion reactions. Fusion as a commercial source of power poses many technological problems which are still to be solved. However, this could be a future source of energy to mankind, and deuterium will again play a major role in it.

CHAPTER 19

Metallurgy of Iron and Steel

INTRODUCTION

Although a transition metal of the first series and a member of Group VIII, iron demands a special chapter to itself because of its very great economic and industrial importance. Iron (Fe) has electronic configuration: 2,8,14,2.

Apart from in meteorites, iron is not found as the element, but its compounds are abundant, the commonest being pyrites, FeS_2, which is unsuitable for smelting because of its high sulphur content, haematite, Fe_2O_3, magnetite, Fe_3O_4, limonite or brown iron ore, $Fe_2O_3.H_2O$, and siderite or spathic iron ore, $FeCO_3$. Iron is also present in silicates, clays and soils, and in all living matter, being essential for the production of haemoglobin in blood and chlorophyll in plants.

MANUFACTURE OF IRON

The first stage is the preparation of the iron ores for reduction in a Blast Furnace. Carbonate ores are calcined in kilns in the presence of excess of air, when they are converted into ferric oxide:

$$4FeCO_3 + O_2 = 2Fe_2O_3 + 4CO_2$$

The oxide ores such as magnetite and haematite are crushed and sintered, where necessary, to produce lumps about the size of one's fist. They are also often pre-heated with hot gas from the Blast Furnace in order to drive off moisture and volatile impurities and render the lumps porous.

The coke used must be very hard and must have a low sulphur content. It is made by carbonising specially selected coal in multiple retorts; the volatile by-products such as gas, tar, and ammonia are all recovered and sold. About one kg of coke is required for each kg of iron produced.

The iron ores contain between 20% and 65% of iron, the chief impurities being silica and alumina. Both have such high melting points that they would not melt in the Blast Furnace but would form an ash which would have to be removed mechanically. To overcome this difficulty the ore is mixed with limestone. In the blast furnace this is decomposed into calcium oxide, which combines with the silica and alumina, removing them as a fusible slag:

$$CaO + SiO_2 = CaSiO_3$$

The last raw material to be considered is air, about 6 kg of which is needed for every kg of iron made. The blast of air which gives the furnace its name is produced by huge turbo-blowers, usually driven by steam. Before entering the furnace through the *tuyeres* or *nozzles* this air is normally raised to about 800°C by passing it through Cowper stoves previously heated to redness by burning Blast Furnace gas in them. This pre-heating results in a considerable saving of the expensive coke.

The Blast Furnace itself (Fig. 19.1.) consists of a tapered cylindrical tower about 30 m in height. It is made of steel and lined with refractory bricks. The furnace is fed mechanically in such a way that no gas escapes during charging. The chemical reactions taking place inside are varied and complicated. In the upper part, where the temperature lies between 400°C and 800°C, the oxides of iron are reduced by ascending carbon monoxide as follows:

$$Fe_2O_3 + 3CO = 2Fe + 3CO_2$$
$$Fe_3O_4 + 4CO = 3Fe + 4CO_2$$

The iron so produced is a spongy porous solid; it is often counted with carbon deposited by the decomposition of carbon monoxide thus:

$$2CO \rightleftharpoons C + CO_2$$

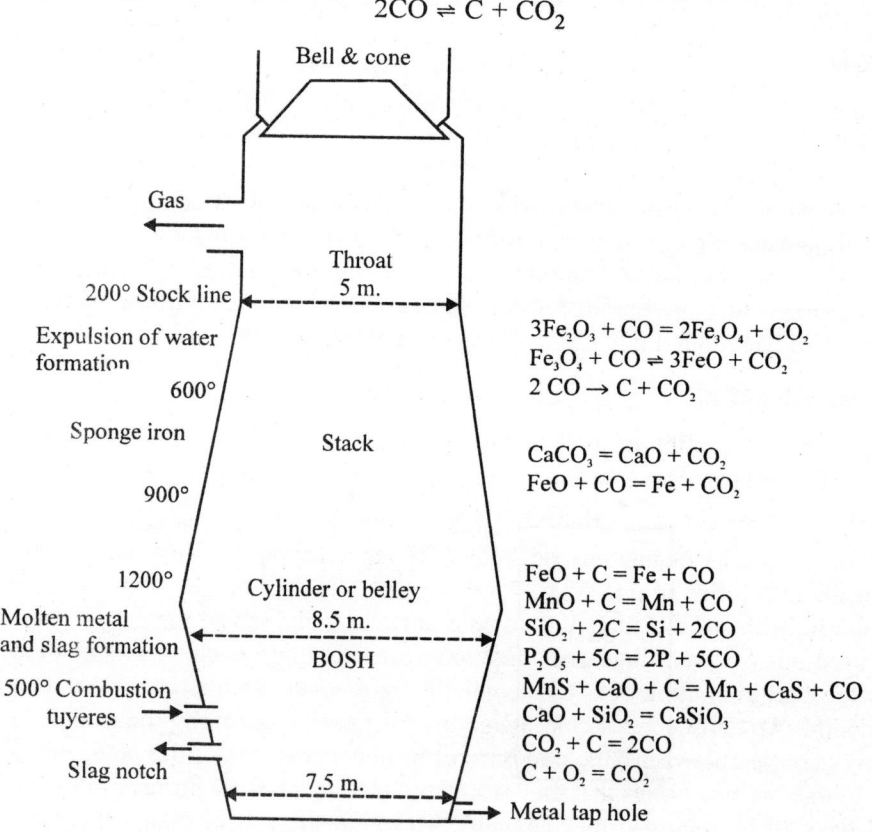

Bell & cone

Gas

200° Stock line — Throat 5 m.

Expulsion of water formation

600°

Sponge iron

Stack

900°

1200°

Cylinder or belley 8.5 m.

Molten metal and slag formation

BOSH

1500° Combustion tuyeres

Slag notch — 7.5 m.

Metal tap hole

$$3Fe_2O_3 + CO = 2Fe_3O_4 + CO_2$$
$$Fe_3O_4 + CO \rightleftharpoons 3FeO + CO_2$$
$$2\,CO \rightarrow C + CO_2$$

$$CaCO_3 = CaO + CO_2$$
$$FeO + CO = Fe + CO_2$$

$$FeO + C = Fe + CO$$
$$MnO + C = Mn + CO$$
$$SiO_2 + 2C = Si + 2CO$$
$$P_2O_5 + 5C = 2P + 5CO$$
$$MnS + CaO + C = Mn + CaS + CO$$
$$CaO + SiO_2 = CaSiO_3$$
$$CO_2 + C = 2CO$$
$$C + O_2 = CO_2$$

Fig. 19.1. Blast furnace.

Lower down the furnace any remaining oxides of iron are reduced directly with carbon according to the equations:

$$Fe_2O_3 + 3C = 2Fe + 3CO$$
$$Fe_3O_4 + 4C = 3Fe + 4CO$$

These reactions and the highly exothermic combustion of coke which takes place in the region where the air blast enters the furnace provide the carbon monoxide necessary for the reactions mentioned above:

$$2C + O_2 = CO_2$$

In the high temperature zone near the tuyeres the iron melts and flows to the base or hearth of the furnace, where it collects and is tapped every 4–6 hours. The molten slag also drips to the bottom and floats on top of the molten iron; it is tapped periodically into large ladles and removed.

Any phosphate in the iron ore will be reduced to phosphorus in the Blast Furnace. It will then dissolve in the iron, and so will any manganese and part of any sulphur present. In the hottest part of the furnace some of the silica will be reduced by carbon and the resulting silicon will also dissolve in the metal. Thus the iron produced by the blast furnace, known as pig iron, is far from pure. It contains all these elements in varying proportion depending upon the ores used and the operating temperature, and, of course, up to 4% of carbon, which is present both as iron carbide and as graphite. Pig iron is hard and brittle; it melts sharply at about 1200°C. The gas from the top of the blast furnace contains about 30% carbon monoxide and about 10% carbon dioxide, the rest being mostly nitrogen. After passage through a dust extractor, it is burned, being used for pre-heating the air blast and the iron ore, for raising the steam for the turbo-blower, and for firing the coke ovens, as shown in Fig. 19.2. A typical blast furnace produces over a million kilogram of iron every twenty-four hours. It is normally run continuously for years on end until the lining needs replacement.

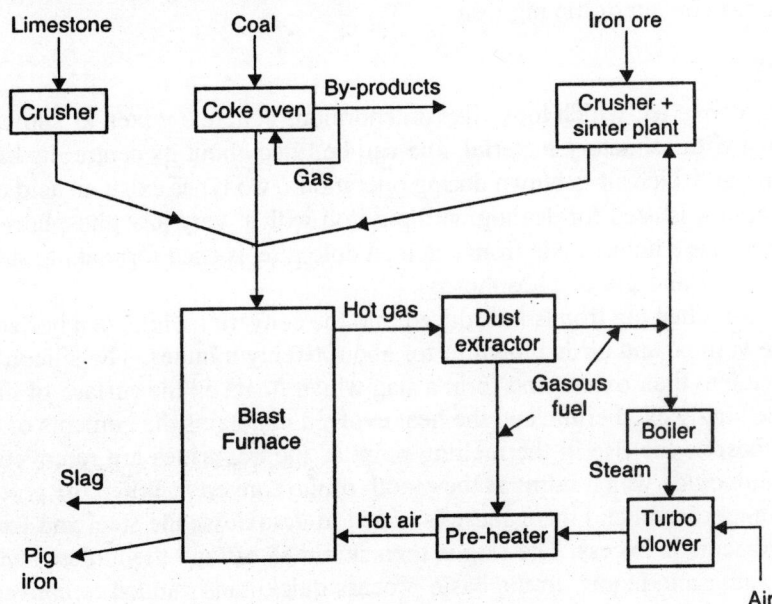

Fig. 19.2. Manufacture of Iron.

TYPES OF IRON

Cast iron is pig iron which has been re-melted, mixed with steel scrap and then cooled in moulds so that it assumes a definite shape. It has the same physical properties as pig iron. It is used for making articles such as drain pipes, lamp posts, and fire grates, where cheapness is more important than strength.

Wrought iron is almost pure iron, made by heating pig iron in a reverberatory furnace lined with haematite and stirring in a current of air until all the impurities have been oxidised. It is soft but very tough and can be easily welded and forged. It is used to make chains and ornamental gates.

Steel is an alloy of iron with carbon and other elements, especially manganese, silicon, and phosphorus. The carbon content usually lies between 0.1% and 1.5%, and the other elements are often present in only small proportions. Alloy steels also exist; they may contain nickel, chromium, vanadium, tungsten, cobalt, molybdenum, etc., sometimes in appreciable amounts. They have special properties such as great hardness or resistance to corrosion which lead to specific uses. Most stainless steels contain 12%–18% of chromium and some nickel; they are used increasingly for cutlery, sink units, car bumpers, machinery, and chemical and nuclear plant. The properties of a steel depend not only on its chemical composition, but also on its heat treatment. Quenching (heating to redness and then cooling suddenly) and tempering (heating to 250°C–300°C and cooling slowly) have a big effect upon its hardness and toughness.

MANUFACTURE OF STEEL

Several processes are in use, but they share the same general principle which is to remove all the impurities from molten pig iron by oxidation and then to add known quantities of carbon and other elements to the molten iron to obtain steel of the desired composition and properties. The various processes chiefly differ in the methods used for removing the impurities. This in turn depends upon the silicon and phosphorus content of the pig iron.

Bessemer Process

This takes place in a converter, which looks like an enormous concrete mixer, as shown in Fig. 19.3. It is made of steel lined with refractory material, and can be tilted about its centre. Its base is perforated with small holes through which air is blown during operation. Two types exist; an acid converter, which is lined with silica bricks, is used for dealing with pig iron with a very low phosphorus content, and a basic converter, which has a lining made from calcined dolomite, is used for making steel from pig iron containing between 1.6% and 2% of phosphorus.

In the acid process molten pig iron is introduced into the converter whilst in a horizontal position. It is then turned to the vertical and air is blown in for about twenty minutes. The silicon and manganese present are converted into their oxides and form a slag which floats on the surface of the molten metal. These oxidations are highly exothermic and the heat evolved maintains the contents of the converter in a molten condition despite the rise in the melting point as the impurities are removed. The carbon is oxidised to carbon monoxide which burns at the mouth of the converter. When the process is complete an iron alloy containing carbon and manganese is added to de-oxidise the steel and leave the required amounts of these elements in excess. The slag is then skimmed off and the molten steel poured into a ladle ready for moulding into ingots. In the basic process quicklime is added to convert the silica and phosphorus pentoxide formed by the air blast into a silicate and phosphate slag, which is poured off before adding the ferromanganese. This basic slag, as it is called, is a useful fertiliser owing to its high phosphorus content.

In recent years the basic Bessemer process has been modified in two important ways. In some plants the air blast has been replaced by a mixture of oxygen and superheated steam. This gives a steel with improved physical properties. Because it is almost free of dissolved nitrogen. The other modification involves replacing the air blast by a jet of oxygen directed forcefully onto the surface of the molten pig iron. The oxygen penetrates into the melt and rapidly oxidises the impurities, giving a relatively cheap and nitrogen-free product. It has the further advantage of being applicable to pig iron of only moderate phosphorus content. These new Kaldo and L.D. converters are fitted with oxygen lances which require a huge supply of oxygen provided by special plants constructed on the site.

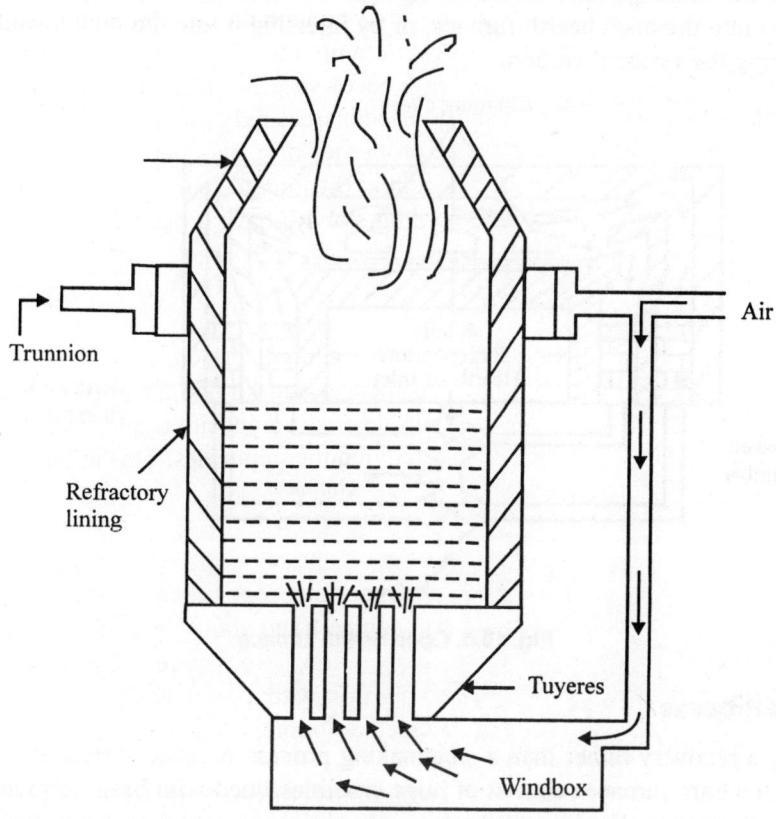

Fig. 19.3. Bessemer process.

Siemens-Martin or Open Hearth Process

A mixture of pig iron, steel scrap, and haematite ore is melted in a shallow hearth furnace as shown in Fig. 19.4. Such furnaces have a capacity of about 2×10^5 kg and are lined with either silica or calcined dolomite depending upon the phosphorus content of the charge.

Producer gas and air are pre-heated to 1000°C–1300°C (by drawing them through firebricks over which the very hot exit gases have recently passed) and burned in the furnace. By reversing the direction of flow at frequent intervals the melt is raised to a temperature of about 1600°C. An excess of air is provided so that the silicon, phosphorus, and manganese in the melt are oxidised and converted into slag (limestone is added to phosphatic charges to facilitate this). The amount of carbon remaining in the iron is lowered to the desired level by adding haematite, which oxidises it to the monoxide. The whole process from charging to tapping takes about ten hours, unless oxygen is used as described below. After tapping, ferromanganese is added to de-oxidise the steel as in the Bessemer process.

The open hearth process accounts for over 80% of the steel made in Great Britain and over half the world production. Its advantages over the Bessemer process are that it is not restricted to pig iron of particular composition, that it uses large amounts of steel scrap, and that its slowness allows better control and more complete separation of the iron from the slag. It has been modified recently by using oxygen to assist in the refining, either for removing most of the easily oxidised silicon from the pig iron before introduction into the open hearth furnace, or by injecting it into the melt towards the end of the process for removing the residual carbon.

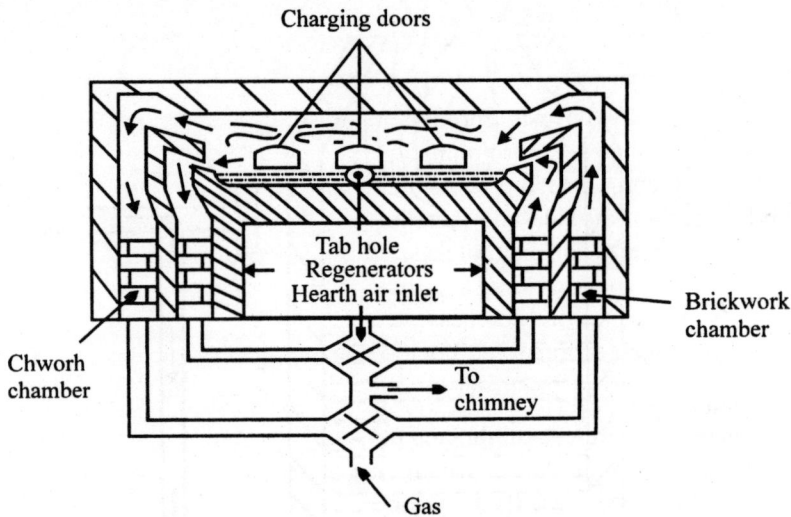

Fig. 19.4. Open hearth furnace.

Electric Furnace Process

This is essentially a recovery rather than a steelmaking process because almost all its raw material is steel scrap. The direct arc furnaces consist of huge crucibles lined with basic refractory materials and fitted with carbon electrodes. The heat of the arc melts the scrap, which is then treated with oxygen to oxidise the impurities. Alternatively, high-frequency induction heating may be used. Electric furnaces are particularly suitable for making high quality alloy steels since the operating temperature can reach 1800°C and the composition of the steel can be accurately controlled.

Spray Process

In this recently developed process, molten pig iron from a blast furnace is poured through high velocity jets of oxygen. The metal is broken up into a fine spray thereby exposing a large surface area to the oxygen and bringing about very rapid oxidation of the impurities. As a result the droplets of steel, dusted with powdered lime to form slag, are hot enough to melt a high proportion of scrap iron in the receiver. The process offers several big advantages and although still in the pioneering stage, it is likely to be widely adopted in future years.

Oxidation is so quick that it can be operated continuously instead of as a batch process; this and the simplicity of the plant leads to low capital cost. The process lends itself to automatic control and to a minimum of refractory wear, thereby reducing labour costs.

Properties of Iron

Pure iron is a white, lustrous, fairly soft metal with a melting point of 1530°C and a density of 7.86 g cm^{-3}. It is malleable and ductile, although the presence of even small amounts of certain impurities (e.g., sulphur) make it brittle. Up to 768°C iron is ferro-magnetic (i.e., it becomes strongly magnetised when placed in a magnetic field), but above this temperature it loses its magnetic properties.

Iron exists in two allotropes, γ-iron, which has the face-centred cubic structure and is stable between 906°C and 1400°C, and α-iron or δ-iron, which is body-centred cubic in structure and is stable below 906°C and over the range 1400°C–1530°C.

Reactions of Iron

On exposure to moist air, iron slowly rusts, forming a surface layer of hydrated oxide, $2Fe_2O_3.3H_2O$, which is unlike those formed by nickel, chromium and aluminium is being porous and not preventing further corrosion. Rusting is a complicated process, being primarily electrolytic in nature. Moisture and oxygen are essential, and carbon dioxide and metallic salts appear to accelerate the process. Various methods of preventing rusting are, for example, painting, galvanising, tinplating, cadmium plating, chromium plating, phosphating, and providing sacrificial protection.

When heated above 150°C in air, iron becomes coated with the oxide, Fe_3O_4. At higher temperatures it burns in oxygen. The reversible reaction of red-hot iron with steam is used for manufacturing hydrogen.

$$3Fe + 4H_2O \rightleftharpoons Fe_3O_4 + 4H_2$$

Iron combines readily with chlorine, sulphur, and phosphorus when heated, but not with nitrogen. It unites with carbon forming a carbide, Fe_3C, which is present in steel and contributes largely to its hardness.

Iron dissolves in dilute hydrochloric and sulphuric acids liberating hydrogen and forming ferrous salts:

$$Fe + H_2SO_4 = FeSO_4 + H_2 \uparrow$$

With dilute nitric acid it forms ferrous nitrate and either ammonium nitrate or the oxides of nitrogen, but concentrated nitric acid has only a momentary action on iron, rendering it passive. Treatment with chromic acid or orthophosphoric acid has a similar effect and use is made of this to prevent corrosion.

If finely divided iron is heated to 200°C with carbon monoxide under 2×10^7 N m^{-2} (200 atm) pressure, it gives a pale yellow liquid, iron pentacarbonyl, which decomposes into very pure iron when heated to 400°C:

$$Fe + 5CO \rightleftharpoons Fe(CO)_5$$

COMPOUNDS OF IRON

Two series exist, ferrous compounds in which the iron is divalent and ferric compounds in which it is trivalent, the two being of similar stability. Thus ferrous compounds are reducing agents, being easily converted into the corresponding ferric form, even by the air, and ferric compounds act as mild oxidising agents, being readily reduced to the ferrous condition.

Iron(II) Oxide (Ferrous oxide), FeO

This is a black powder obtained by heating ferrous oxalate in an inert atmosphere (it inflames in air):

$$(COO)_2Fe.3H_2O = FeO + CO\uparrow + CO_2\uparrow + 3H_2O$$

It is a basic oxide, dissolving in acids to form ferrous salts.

Iron(II) Hydroxide (Ferrous Hydroxide), $Fe(OH)_2$

This is obtained as a white gelatinous precipitate (which oxidises rapidly on standing, turning green then brown), when an alkali is added to a solution of a ferrous salt:

$$FeSO_4 + 2NaOH = Fe(OH)_2\downarrow + Na_2SO_4$$

It is only incompletely precipitated by ammonium hydroxide in the presence of ammonium chloride because its solubility product is usually only just exceeded under these conditions. To avoid this difficulty the solution is boiled with a little concentrated nitric acid at the beginning of Group III of the Qualitative Analysis Scheme, so that any iron present is precipitated as the less soluble ferric hydroxide. Ferrous hydroxide dissolves in acids giving ferrous salts, including a carbonate, bicarbonate, and sulphide.

Iron(II) Sulphate (Ferrous Sulphate), $FeSO_4$

This is obtained as pale green crystals of the heptahydrate, $FeSO_4.7H_2O$, commonly known as green vitriol, by dissolving iron in dilute sulphuric acid and then concentrating and cooling the solution. Use of air-free acid minimises formation of the ferric salt. The crystals slowly effloresce in air and turn yellow owing to partial oxidation to basic ferric sulphate. On heating the following changes occur:

$$FeSO_4.7H_2O \rightleftharpoons FeSO_4.H_2O + 6H_2O$$
$$FeSO_4.H_2O \rightleftharpoons FeSO_4 + H_2O$$
$$2FeSO_4 = Fe_2O_3 + SO_2\uparrow + SO_3\uparrow$$

This decomposition is of historical interest in that the earliest sulphuric acid was made in this way. Ferrous sulphate forms double salts; one of these, ferrous ammonium sulphate, $FeSO_4.(NH_4)_2SO_4.6H_2O$, known as *Mohr's salt*, is used in volumetric analysis for standardising oxidising agents. It is not a complex salt, behaving in solution merely as a mixture of its component sulphates, but it is less readily oxidised by the air than ferrous sulphate.

A solution of ferrous sulphate absorbs nitric oxide readily forming a brown compound, $FeSO_4.NO$, which decomposes when heated releasing nitric oxide again.

Ferrous sulphate is used in the manufacture of ink, in tanning and dyeing, and for killing moss.

Iron(II) Chloride (Ferrous Chloride), $FeCl_2$

The anhydrous salt is obtained by passing dry hydrogen chloride over red-hot iron filings :

$$Fe + 2HCl = FeCl_2 + H_2$$

It is white solid which is deliquescent and very soluble in water. When iron is dissolved in hydrochloric acid in the absence of air and the solution is concentrated, blue-green crystals of the tetrahydrate, $FeCl_3.4H_2O$, are obtained.

Iron(II) Carbonate (Ferrous Carbonate), $FeCO_3$

This occurs naturally as siderite or spathic iron ore. It is obtained as a white precipitate by adding sodium carbonate solution to a solution of a ferrous salt in the absence of air. Although insoluble in water, it dissolves in the presence of carbon dioxide forming the bicarbonate:

$$FeCO_3 + CO_2 + H_2O \rightleftharpoons Fe(HCO_3)_2$$

On exposure to the air this solution gives up carbon dioxide again and the resulting carbonate is hydrolysed and oxidised, depositing red-brown hydrate ferric oxide. It is probable that changes such as these occur during the rusting of iron.

Iron(II) Sulphide (Ferrous Sulphide), FeS

This compound is made commercially by melting together iron and sulphur; when so prepared it is grey-black in colour and usually contains appreciable amounts of uncombined iron. Pure ferrous sulphide is precipitated when a sulphide solution is added to a solution of a ferrous salt:

$$Na_2S + FeSO_4 = FeS\downarrow + Na_2SO_4$$

It is soluble in dilute acids liberating hydrogen sulphide, which is its main use in the laboratory:

$$FeS + H_2SO_4 = FeSO_4 + H_2S\uparrow$$

Ferrous Disulphide, FeS$_2$

This occurs naturally as pyrites and marcasite. It is like brass in appearance, being yellow and lustrous. When heated in air it is converted into ferric oxide and sulphur dioxide, the latter being extensively used for making sulphuric acid:

$$4FeS_2 + 11O_2 = 2Fe_2O_3 + 8SO_2\uparrow$$

Iron(III) Oxide (Ferric Oxide), Fe$_2$O$_3$

This occurs as haematite and as the hydrate, limonite; it is very widely distributed being responsible for the red-brown colour of the soil in many areas. It is obtained by strongly heating ferrous sulphate or ferric hydroxide. It is used as a red pigment and as an abrasive polishing powder (*Jeweller's rouge*).

Iron(III) Hydroxide (Ferric Hydroxide), Fe(OH)$_3$

This is obtained as a red-brown gelatinous precipitate by adding an alkali to a solution of a ferric salt:

$$FeCl_3 + 3NaOH = Fe(OH)_3 + 3NaCl$$

It is a very weak base, dissolving in strong acids to form ferric salts which are extensively hydrolysed in solution.

Tri-iron Tetroxide (Ferroso-ferric Oxide), Fe$_3$O$_4$

This oxide, which occurs naturally as magnetite or lodestone, is highly magnetic. It is made by heating iron in air or stream:

$$3Fe + 4H_2O \rightleftharpoons Fe_3O_4 + 4H_2$$

It is inert, and concentrated nitric acid has no action on it, but it is easily reduced when heated with carbon or carbon monoxide.

Iron(III) Sulphate (Ferric Sulphate), Fe$_2$(SO$_4$)$_3$

This is made by heating ferrous sulphate with concentrated sulphuric acid:

$$2FeSO_4 + 2H_2SO_4 = Fe_2(SO_4)_3 + SO_2\uparrow + 2H_2O$$

When heated strongly it decomposes thus:

$$Fe_2(SO_4)_3 = Fe_2O_3 + 3SO_3\uparrow$$

A solution of ferric sulphate is acidic owing to hydrolysis. It readily forms sparingly soluble double salts called alums of which ammonium iron alum, $(NH_4)_2SO_4.Fe_2(SO_4)_3.24H_2O$, and potash iron alum, $K_2SO_4.Fe_2(SO_4)_3.24H_2O$, are the best known. They are extensively used as mordants in dyeing.

Iron(III) Chloride (Ferric Chloride), FeCl$_3$

The anhydrous salt is made by passing dry chlorine over heated iron. The chloride, which sublimes, at 315°C, is condensed in the cooled receiver. Density measurements show that the vapour consists of Fe$_2$Cl$_6$ molecules at 400°C, but that these dissociate into FeCl$_3$ molecules at higher temperatures. The hexahydrate, FeCl$_3$.6H$_2$O, crystallises when iron is dissolved in hydrochloric acid in the presence of chlorine and the solution is concentrated and cooled. Ferric chloride is highly deliquescent and very much soluble in water, forming a brown solution which is acidic as a result of hydrolysis and which often gives a colloidal solution of ferric hydroxide on standing :

$$FeCl_3 + 3H_2O \rightleftharpoons Fe(OH)_3 + 3HCl$$

Potassium Heracyanoferrate (II) (Potassium Ferrocyanide), K$_4$Fe(CN)$_6$

This is obtained from spent oxide, which contains Prussian blue formed by the absorption of hydrogen cyanide from the crude coal gas. Heating with slaked lime give calcium ferrocyanide, which is leached out and treated with potassium carbonate, giving yellow crystals of the trihydrate, K$_4$Fe(CN)$_6$. 3H$_2$O.

The complex ion [Fe(CN)$_6$]$^{4-}$ is so stable that its solution does not give the reactions of simple ferrous or cyanide ions. For the same reason it is not poisonous. With solutions of copper salts it gives a brown precipitate of cupric ferrocyanide. If potassium ferrocyanide is boiled with dilute nitric acid, crystals of sodium nitroprusside, Na$_2$[Fe(CN)$_5$NO].2H$_2$O, are obtained on neutralising the solution with sodium carbonate.

With solutions of a ferric salt, potassium ferrocyanide solution gives a dark blue precipitate of potassium ferric ferrocyanide, known as Prussian blue:

$$FeCl_3 + K_4Fe(CN)_6 = KFeFe(CN)_6 + 3KCl$$

This reaction is used as a test for ferric ions, and also for manufacturing ink, when the precipitate is dissolved in oxalic acid to give a deep blue solution.

Potassium Hexacyanoferrate (III) (Potassium Ferricyanide), K$_3$Fe(CN)$_6$

This is prepared by the action of chlorine on potassium ferrocyanide solution:

$$2K_4Fe(CN)_6 + Cl_2 = 2K_3Fe(CN)_6 + 2KCl$$

It forms dark red crystals which dissolve freely in water giving a yellow solution. It is an oxidising agent, reverting readily to the ferrocyanide.

If a solution of a ferrous salt is added to potassium ferricyanide solution an ionic exchange first takes place thus:

$$Fe^{2+} + [Fe(CN)_6]^{3-} = Fe^{3+} + [Fe(CN)_6]^{4-}$$

As a result a dark blue precipitate of potassium ferric ferrocyanide, known as *Turnbull's blue,* is obtained. This precipitate, which is used to detect ferrous salts, is not known to be identical with the Prussian blue.

Detection of Iron

Compounds of iron are detected and distinguished as ferrous or ferric by the tests. In remembering the reactions with cyanide complexes it is helpful to realise that the dark blue precipitate is obtained only when the different valency states of iron are mixed, i.e. ferrous with ferricyanide or ferric with ferrocyanide solutions.

The reaction with potassium thiocyanate solution is a particularly sensitive test for ferric ions and use is often made of it to determine when a ferric salt has been completely reduced to the ferrous conditions.

To sum up the world production of iron far exceeds that of all the other metals added together. There are several reasons for this. Its ores are abundant, rich and readily accessible. They are easily reduced to the metal by processes which can be operated continuously on a large scale with consequent low labour costs. The versatility of the product when alloyed with other elements in steel gives it a great range of uses. Its biggest drawback is its susceptibility to corrosion. It is an interesting thought that the principal use of several other metals, notably tin, zinc, cadmium, and chromium, is to prevent this corrosion.

In its chemistry iron is in every respect a typical transition metal. It gives coloured ions in solution, it readily forms stable complexes, it is an active catalyst, it shows variable valency and it is strongly magnetic. Many of its compounds show a resemblance to those of manganese, cobalt, and nickel, although iron differs from them in forming ferrous and ferric salts of about equal stability, whereas in these neighbouring elements the divalent salts are much more stable.

CHAPTER 20

Metallurgy of Non-ferrous Metals

INTRODUCTION

Metals make up a major part of the elements in the periodic table. Metals are divided into ferrous and non-ferrous. The ferrous metals include–iron, manganese, chromium and their alloys. All the rests are non-ferrous metals. Non-ferrous metals are of various types such as copper, lead, tin, nickel, aluminium, silver, platinum, etc. The three metals in Group IB, copper, sliver, and gold, are commonly known as the coinage metals, although other metals are being increasingly used in their place for this propose and they share many of the characteristics of the transition metals.

COPPER

Copper has electronic configuration: 2,8,18,1.

Metallic copper does occur naturally in a few places, but the commonest minerals are chalcopyrite, $CuFeS_2$, which is the main source of the metal, copper glance or chalcocite, Cu_2S, bornite, $Cu_2S.CuS.FeS$, and malachite, $CuCO_3.Cu(OH)_2$.

Manufacture

Most copper ores contain only a few per cent of copper and the first stage is to crush them and concentrate them by flotation methods. The concentrates are roasted in air, which converts at least part of the sulphides into oxides, sulphur dioxide being produced as a by-product:

$$2CuFeS_2 + 4O_2 = Cu_2S + 2FeO + 3SO_2 \uparrow$$

The product is then mixed with limestone and silica and melted in a reverberatory furnace. Most of the iron present reacts with the silica, forming a fusible slag which is run off, leaving a molten mixture of cuprous and ferrous sulphides called matte, which is tapped periodically:

$$FeO + SiO_2 = FeSiO_3$$

Copper is obtained by melting the matte in a converter similar to the Bessemer type (Fig. 20.1)with some flux and subjecting it to a prolonged blast of air. The ferrous sulphide undergoes oxidation first and is removed as a slag, then part of the cuprous sulphide is converted into cuprous oxide, which reacts with unchanged sulphide to give the molten metal:

$$2Cu_2S + 3O_2 = 2Cu_2O + 2SO_2 \uparrow$$

$$2Cu_2O + Cu_2S = 6Cu + SO_2 \uparrow$$

The product is called blister copper because it tends to release bubbles of dissolved gas whilst solidifying. It still contains 2%–3% of impurities, mainly iron and sulphur.

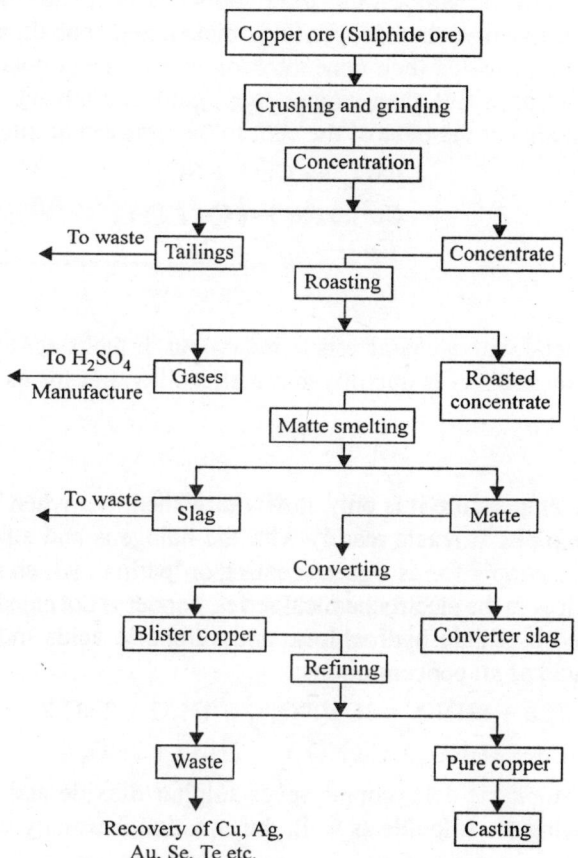

Fig. 20.1. General flow sheet for Copper extraction.

Refining of Copper

For some purposes furnace refining is adequate, but where copper of high purity is required for electrical use or where the crude copper contains appreciable amounts of precious metals, the electrolytic method is used as well.

Thermal method

The blister copper is re-melted in a reverberatory furnace, ca. 2.5×10^5 kg at a time, and exposed to air, which oxidises the sulphur to sulphur dioxide and the other impurities to a slag. This is skimmed off the surface, leaving molten copper saturated with oxygen. The latter is removed by stirring the molten metal for several hours with poles made from green wood, which releases reducing gases in the melt. This gives copper of about 99.5 per cent purity.

Electrolytic method

A cell is constructed using slabs of impure copper as anodes and thin sheets of pure copper as cathodes, the electrolyte being copper sulphate solution acidified with sulphuric acid. On passing the current, copper goes into solution from the anodes and pure copper is deposited upon the cathodes. Those metals which are more electropositive than copper (e.g., nickel, zinc, cobalt, and iron), go into solution as ions and accumulate there, provided their concentration is low, the copper ions being preferentially discharged. Metals less electropositive than copper (e.g., gold and silver), fall from the anode as it dissolves and collect as a sludge at the base of the cell, to be removed at intervals for further refining.

$$CuSO_4 \leftrightharpoons Cu^{2+} + SO_4^{2-}$$

At the cathode: $\qquad\qquad Cu^{2+} + 2e \longrightarrow Cu \downarrow$

At the anode: $\qquad\qquad\quad Cu \longrightarrow Cu^{2+} + 2e$

Properties of copper

It is a fairly soft, lustrous metal with a characteristic red colour. It melts at 1083°C and has a density of 8.94 g cm^{-3}. Copper is renowned for its ductility and malleability and for its high conductivity of heat and electricity.

Reactions of copper

On standing in air at room temperature it is only slowly tarnished, but when heated strongly in air it is converted into its oxides rapidly. It reacts readily with the halogens and sulphur. On exposure to the atmosphere for long periods, copper forms a green coating or 'patina', which is probably a basic copper sulphate or chloride. Being low in the electrochemical series, copper is not capable of displacing hydrogen from acids. It is unattacked by dilute hydrochloric and sulphuric acids in the absence of air, but it dissolves readily in nitric acid of all concentrations:

Dilute acid: $\qquad\qquad 3Cu + 8HNO_3 = 3Cu(NO_3)_2 + 4H_2O + 2NO \uparrow$

Concentrated acid: $\qquad Cu + 4HNO_3 = Cu(NO_3)_2 + 2H_2O + 2NO_2 \uparrow$

With hot concentrated sulphuric acid, copper gives sulphur dioxide and copper sulphate, but side reactions also occur producing the sulphide as well; the equation is usually written as:

$$Cu = 2H_2SO_4 = CuSO_4 + 2H_2O + SO_2 \uparrow$$

Copper also dissolves in the presence of air in ammonium hydroxide and potassium cyanide solutions, with the formation of complex ions.

Uses of copper

The multifarious uses of copper spring directly from its workability, its high thermal and electrical conductivity, its resistance to corrosion, its attractive appearance, and, above all, its readiness to form alloys with other metals. Thus it is used to make electrical conductors of all kinds from the very thin wires used in electronics to the busbars used in power transmission. Sheets of the metal are used for roofing large buildings, and it is widely used for plumbing and for making boilers and condensers in locomotives and ships. When alloyed with other metals copper provides a great range of materials. For instance, at least three-fifths of every piece of brass is copper, the rest being mainly zinc. Bronzes, essential as bearing metals in machinery, are made from copper and tin, sometimes with other metals added or even small amounts of phosphorus (phosphor-bronze). Copper also alloys readily with nickel (one such alloy is used for British 'silver' coinage), and with aluminium, beryllium, and gold (9 carat gold is nearly two-thirds copper).

Compounds of copper

Copper forms two series of compounds, cuprous, in which it is univalent and cupric, in which it is divalent. Cuprous compounds are formed either electrovalently by the loss of the outermost electron, or much more commonly, covalently, by a sharing of that electron and one from another atom. Cupric compounds are formed either by a double covalency involving the outermost electron and one electron from the penultimate shell, or, less commonly, but the loss of two electrons from each atom giving the ion Cu^{2+}. Both cuprous and cupric ions show a strong tendency to form complexes by co-ordination of suitable ions or molecules, the complex ions so formed being much more stable than the simple copper ions from which they are derived.

At high temperatures cuprous salts are more stable than cupric, and if heated strongly cupric oxide, sulphide, fluoride, chloride, and bromide all change into the corresponding cuprous compounds.

The two ions are related to each other in solution in the following way:

$$2Cu^+ \rightleftharpoons Cu^{2+} + Cu$$

For example, soluble cuprous salts such as the sulphate decompose rapidly in solution into the cupric salt and copper, which is deposited, causing the equilibrium to move completely to the right. Insoluble cuprous salts show no such tendency because in their case the concentration of cuprous ions in solution is extremely low.

Cupric salts containing weakly electronegative anions (e.g., I^-, CN^-, CNS^-), are unstable; when they are formed by double decomposition they spontaneously change into the corresponding cuprous salts.

Copper (I) oxide (cuprous oxide), Cu_2O

This is best prepared by the reducing action of glucose on boiling *Fehling's* solution, when it appears as a red precipitate which can be filtered off, washed, and dried. *Fehling's* solution, which is deep blue in colour, is made by adding copper sulphate solution to a solution of sodium potassium tartrate in excess of sodium hydroxide solution. This reaction is used in organic chemistry as a test for reducing agents such as aldehydes and sugars.

Cuprous oxide is insoluble in water, but it dissolves readily in ammonia solution and in acids forming cuprous salts. Some of the latter are unstable in solution and decompose into the corresponding cupric salts and copper:

$$Cu_2O + H_2SO_4 = CuSO_4 + Cu \downarrow + H_2O$$

With hydrochloric acid, however, cuprous chloride dissolves in excess of the acid, forming cuprochlorous acid thus:

$$Cu_2O + 2HCl = 2CuCl \downarrow + H_2O$$

$$CuCl + HCl = HCuCl_2$$

Cuprous oxide is used for making red glass and in the manufacture of metal rectifiers.

Copper (I) chloride (cuprous chloride), CuCl

This is prepared by heating cupric oxide and concentrated hydrochloric acid with metallic copper:

$$CuO + Cu + 2HCl = 2CuCl \downarrow + H_2O$$

When the product is poured into a large excess of cold water, cuprous chloride is thrown down as a white precipitate which can be filtered off, washed with water containing a little sulphur dioxide (to prevent oxidation), and dried in a vacuum desiccator.

Alternatively, sulphur dioxide can be used to reduce a mixture of cupric sulphate and sodium chloride solutions:

$$2CuSO_4 + 2NaCl + SO_2 + 2H_2O = 2CuCl \downarrow + 2NaHSO_4 + H_2SO_4$$

Although it is almost insoluble in water, it dissolves readily in those reagents which form complex ions with cuprous copper, e.g., in concentrated hydrochloric acid forming $HCuCl_2$ and in ammonia solution forming $Cu(NH_3)_2Cl$. These solutions are used to absorb carbon monoxide, with which they form the compound $CuCl.CO.2H_2O$. Ammoniacal solutions of cuprous chloride also give a red precipitate of cuprous acetylide with acetylene.

Cuprous chloride is predominantly a covalent substance. It is a very poor conductor when fused, its vapour density at 1700°C corresponds to the double molecules Cu_2Cl_2, and its crystal structure is similar to zinc blende and diamond, suggesting covalent linkages between its atoms. It is an important reagent in the Sandmeyer reaction. Oxidising agents convert it into cupric chloride:

$$2CuCl + 2HCl + O = 2CuCl_2 + H_2O$$

Copper (I) iodide (cuprous iodide), CuI

This is precipitated as a buff solid when potassium iodide solution is added to copper sulphate solution:

$$2CuSO_4 + 4KI = 2CuI \downarrow + I_2 + 2K_2SO_4$$

This reaction is used for estimating copper salts volumetrically by titrating the liberated iodine with standard sodium thiosulphate solution.

Copper (I) cyanide (cuprous cyanide), CuCN

This is obtained when potassium cyanide solution is added dropwise to copper sulphate solution:

$$2CuSO_4 + 4KCN = 2CuCN \downarrow + C_2N_2 \uparrow + 2K_2SO_4$$

A greenish-yellow precipitate of cupric cyanide first appears, but this decomposes rapidly into a white precipitate of cuprous cyanide and the poisonous gas cyanogen.

If excess of potassium cyanide solution is added, cuprous cyanide re-dissolves forming the complex cuprocyanide or tetracyanocuprate (I) ion:

$$CuCN + 3KCN = K_3Cu(CN)_4$$

This ion is important in copper plating and in qualitative analysis because the high stability of the cuprocyanide ion ensures a very low concentration of simple cuprous ions in equilibrium with it:

$$K_3Cu(CN)_4 \rightleftharpoons 3K^+ + [Cu(CN)_4]^{3-}$$

$$[Cu(CN)_4]^{3-} \rightleftharpoons Cu^+ + 4CN^-$$

Thus if hydrogen sulphide is passed into a solution containing both copper and cadmium salts in the presence of excess of potassium cyanide solution, no precipitate of cuprous sulphide is obtained because the concentration of copper ions is too low to exceed its solubility product, but cadmium sulphide will be precipitated because the corresponding cadmi-cyanide ion is much less stable and provides an appreciable concentration of simple cadmium ions:

$$K_2Cd(CN)_4 \rightleftharpoons 2K^+ + [Cd(CN)_4]^{2-}$$

$$[Cd(CN)_4]^{2-} \rightleftharpoons Cd^{2+} + 4CN^-$$

In this way it is possible to detect cadmium in the presence of copper, which is difficult under normal conditions because the yellow precipitate of cadmium sulphide is masked by the black precipitate of copper sulphide.

Copper (II) oxide (cupric oxide), CuO

This is a hygroscopic black solid made by heating cupric hydroxide, carbonate or nitrate, or by heating the metal to about 800°C in air or oxygen:

$$2Cu(NO_3)_2 = 2CuO + 4NO_2 \uparrow + O_2 \uparrow$$

It decomposes above 1000°C into cuprous oxide and oxygen:

$$4CuO \rightleftharpoons 2Cu_2O + O_2$$

It is a mild oxidising agent. For example, on heating it oxidises hydrogen to water and carbon and carbon monoxide to carbon dioxide:

$$CuO + H_2 = Cu + H_2O$$
$$2CuO + C = 2Cu + CO_2$$

Use is made of these reactions in organic analysis.

Cupric oxide is insoluble in water, but being a basic oxide it dissolves readily in acids giving cupric salts and in ammonia solution forming a complex ion.

Copper (II) hydroxide (cupric hydroxide), Cu(OH)$_2$

This is formed as a gelatinous pale blue precipitate when sodium hydroxide solution is added to a solution of a cupric salt:

$$CuSO_4 + 2NaOH = Cu(OH)_2 \downarrow + Na_2SO_4$$

When suspended in water and boiled, it changes to black hydrated cupric oxide. It dissolves in ammonium hydroxide solution, giving a deep blue solution called *Schweitzer's reagent,* which contains the complex ion $[Cu(NH_3)_4]^{2+}$. This reaction is used as a test for copper in analysis, and commercially as a way of making rayon or artificial silk because this solution dissolves cellulose. The solution of cellulose is squirted through fine jets into a bath of acid when the cellulose is precipitated as long threads of artificial fibre which are washed, dried and made into cloth.

Copper (II) chloride (cupric chloride), CuCl$_2$

The brown anhydrous salt is made by the action of dry chlorine on heated copper. If cupric oxide is dissolved in concentrated hydrochloric acid and the solution is concentrated, green crystals of the deliquescent dihydrate, $CuCl_2.2H_2O$, are obtained. When cupric chloride is dissolved in concentrated hydrochloric acid it gives a brownish yellow solution because of a preponderance of $[CuCl_4]^{2-}$ ions under these conditions. On adding water the colour of the solution changes to green because of the presence of a second complex ion $[Cu(H_2O)_4]^{2+}$, which is blue in colour. On further dilution the solution turns pale blue because the large excess of water causes the hydrate ion to predominate and show its characteristic colour. The changes can be reversed by adding concentrated hydrochloric acid.

In conc. HCl	In dilute HCl	In water
$Cu^{2+} + 4Cl^-$	$Cu^{2+} + 4Cl^- \; Cu^{2+} + 4H_2O$	$Cu^{2+} + 4H_2O$
$\updownarrow$	$\updownarrow \qquad \updownarrow$	$\updownarrow$
$[CuCl_4]^{2-}$	$[CuCl_4]^{2-} \; [Cu(H_2O)_4]^{2+}$	$[Cu(H_2O)_4]^{2+}$
yellow	yellow and blue, i.e., green	blue

Copper (II) sulphate (cupric sulphate), CuSO$_4$

This is the most important copper salt. It is made commercially by spraying hot dilute sulphuric acid on scrap copper in the presence of a current of air, when the metal slowly dissolves:

$$2Cu + 2H_2SO_4 + O_2 = 2CuSO_4 + 2H_2O$$

In the laboratory it is prepared by dissolving copper oxide in dilute sulphuric acid and concentrating the solution, when deep blue crystals of the pentahydrate, known as blue vitriol, are obtained. These behave in the following way when heated:

$$CuSO_4.5H_2O \overset{100°C}{\rightleftharpoons} CuSO_4.3H_2O \overset{250°C}{\rightleftharpoons} CuSO_4.H_2O \overset{750°C}{\rightleftharpoons} CuSO_4 \longrightarrow CuO + SO_3$$

These changes are reversible, the colourless anhydrous salt turning blue on contact with water to form crystals of $[Cu(H_2O)_4]^{2+}SO_4^{2-}.H_2O$.

Copper sulphate solution gives precipitates of insoluble copper salts with many reagents:

$$CuSO_4 + 4KI = 2CuI \downarrow + I_2 + 2K_2SO_4$$
$$2CuSO_4 + 4KCN = 2CuCN \downarrow + C_2N_2 + 2K_2SO_4$$
$$CuSO_4 + NaHCO_3 = CuSO_3 \downarrow + NaHSO_4$$
$$CuSO_4 + 2NaOH = Cu(OH)_2 \downarrow + Na_2SO_4$$

It is used in large amounts as fungicide and timber preservative. Its solution with slaked lime, known as *Bordeaux mixture*, is used as a spray for preventing blight in potatoes and vines. Copper sulphate is also used for electroplating and as the starting point in the preparation of many compounds of copper.

Copper (II) nitrate (cupric nitrate), $Cu(NO_3)_2$

This is made by the action of nitric acid on the metal, or on its oxide, hydroxide, or carbonate. It forms deep blue crystals of the trihydrate, $Cu(NO_3)_2.3H_2O$ which are deliquescent and decompose when heated strongly:

$$2Cu(NO_3)_2 = 2CuO + 4NO_2 \uparrow + O_2 \uparrow$$

Detection of copper

When saturated with hydrogen sulphide solutions of copper salts give a black precipitate which is insoluble in dilute hydrochloric and sulphuric acids:

$$CuSO_4 + H_2S = CuS \downarrow + H_2SO_4$$

Solutions of cupric salts gives a brown gelatinous precipitate of cuprous ferrocyanide with potassium ferrocyanide solution:

$$2CuSO_4 + K_4Fe(CN)_6 = Cu_2Fe(CN)_6 \downarrow + 2K_2SO_4$$

Copper salts can also be recognised by their green flame colouration by their greenish blue borax bead (in the oxidising flame), and by the characteristic deep blue colour formed with excess of ammonium hydroxide solution.

SILVER

Silver, Ag, electronic configuration : 2,8,18,18,1. Some silver occurs native, but it is chiefly found as the sulphide called argentite or silver glance, Ag_2S, either alone or in association with other sulphides (e.g., those of lead, copper, and zinc) particularly in Mexico and South America. Deposits of horn silver AgCl, are also known.

Manufacture

Silver is obtained from its low grade ores by leaching them with a dilute solution of sodium cyanide, when the silver compound dissolve forming the soluble complex argentocyanide ion:

$$Ag_2S + 4NaCN \rightleftharpoons 2NaAg (CN)_2 + Na_2S$$

Air is blown into the mixture to oxidise the sodium sulphide to sulpha and thereby make the reversible reaction proceed completely to the right. The silver is precipitated from the solution by adding some electropositive metal, usually zinc or aluminium.

Considerable quantities of silver are recovered as a by-product during the smelting of argentiferrous lead and copper ores despite the fact that the proportion of silver in these ores rarely exceed 0.1%. Copper is usually refined electrolytically, the silver being recovered as a by-product from the anode mud, but two distinct processes are in use of recovering silver from the ores of lead:

Pattinson process

The molten mixture of lead and silver is cooled slowly. At first pure lead crystallises and is removed, the melt becoming progressively richer in silver until it contains about 2.6%, when a eutectic mixture of this composition solidifies at 303°C. This is then subjected to *cupellation,* i.e., it is melted in a bone ash crucible and exposed to a sustained blast of air. The lead is converted into a scum of litharge which can be skimmed off the surface of the melt, leaving a residue of silver and gold which can be refined electrolytically.

Parkes process

This depends upon the partition law. Advantage is taken of the immiscibility of molten lead and molten zinc and the fact that silver is very much more soluble in the latter. Thus when a little molten zinc is added to the molten argentiferous lead and the mixture is stirred, most of the silver and a little lead passes into the upper zinc layer. When cooled slightly this layer solidifies and is removed; the zinc is distilled off and used again and the silver separated from the remaining lead by *cupellation.*

Silver is refined electrolytically by a method similar to that used for copper. The impure silver is made the anode and silver nitrate solution the electrolyte. Copper and lead impurities pass into solution and accumulate there, only pure silver being deposited on the cathode. Gold which is usually present in small amount, collects beneath the anode.

Properties and reactions of silver

It is a lustrous, white, very ductile metal melting at 960°C. Its density is 10.5 g cm^{-3}. It is the best conductor of heat and electricity and it readily forms alloys with other metals.

Silver is not attacked by air or moisture at room temperature, but when molten it can dissolve considerable amounts of oxygen. As the metal solidifies the gas is expelled again causing the metal to 'spit' violently. The metal tarnishes rapidly when exposed to sulphur compounds, forming a layer of black silver sulphide. Thus silver articles are affected by traces of hydrogen sulphide in the atmosphere in industrial districts, and silver spoons by sulphur compounds present in eggs.

Being low in the electrochemical series, silver is not attacked by hydrochloric acid, nor by dilute sulphuric acid. Oxidising agents such as nitric acid or hot concentrated sulphuric acid dissolve the metal readily, so do solutions of alkali cyanides in the presence of air, but molten alkalies have little effect on it:

$$Ag + 2HNO_3 = AgNO_3 + H_2O + NO_2 \uparrow$$
$$2Ag + 2H_2SO_4 = Ag_2SO_4 + 2H_2O + SO_2 \uparrow$$

Uses of silver

In the past the main uses of silver have been for coinage (usually alloyed with copper), for making jewellery and mirrors, and for electroplating (see below), all uses arising from its high lustre and its resistance to corrosion. In recent years, however, the metal has become increasingly important in other ways, as a catalyst in the manufacture of acetaldehyde and ethylene oxide for example and as a constituent of special low temperature brazing alloys with zinc, copper, and sometimes cadmium. Alkalis and organic acids are often contained in silver-lined vats owing to their resistance to attack and the metal is also used for electrical contacts in switches and in silver-zinc batteries. Much silver is converted into its compounds, which are of considerable commercial importance particularly in photography.

Silver plating

The article to be electroplated, which is usually made of German silver (50% copper, 30% zinc, 20% nickel), is made the cathode of a cell in which the electrolyte is a solution of potassium argentocyanide and the anodes are pure silver. Argentocyanide is used because this complex ion is in equilibrium with an extremely small concentration of simple silver ions, which are discharged at the cathode during the electrolysis and form a firm, coherent, lustrous deposit:

$$KAg(CN)_2 \rightleftharpoons K^+ + [Ag(CN)_2]^- \rightleftharpoons K^+ + Ag^+ + 2CN^-$$

The supply of silver ions in the electrolyte is maintained by solution of silver from the anodes. If silver nitrate is used as the electrolyte, the high concentration of silver ions in its solution results in the rapid displacement of silver by the more electropositive metals present in the article to be plated and leads to the production of a soft, spongy deposit.

Compounds of silver

In all its important compounds silver is univalent. These argentous compounds are formed either by the loss of the single electron from the outermost shell of the silver atom, where they are electrovalent and contain the Ag^+ ion, or by the sharing of a pair of electrons with another atom in a single covalent linkage. They are mostly colourless and insoluble in water. Complex ions are also formed.

Silver oxide, Ag₂O

This is obtained as a brown precipitate when sodium hydroxide solution is added to a solution of silver nitrate.

$$2AgNO_3 + 2NaOH = Ag_2O \downarrow + 2NaNO_3 + H_2O$$

It decomposes rapidly at 200°C, which prevents its preparation from the nitrate by heating:

$$2Ag_2O = 4Ag + O_2 \uparrow$$

It is a basic oxide, dissolving in acids to give silver salts, and turning litmus blue. Because of this and because most silver halides are insoluble, moist silver oxide is much used in organic chemistry for replacing halogen atoms by oxygen or hydroxyl groups. Silver oxide dissolves in ammonium hydroxide solution forming the complex ion $[Ag(NH_3)_2]^+$. This solution is called *Tollens's reagent* or ammoniacal silver oxide. It is prepared by adding ammonium hydroxides solution drop by drop to silver nitrate solution until the precipitate of silver oxide has also entirely re-dissolved. It is an important reagent in organic chemistry for detecting reducing agents such as a lustrous mirror on the inside of the test tube.

Silver chloride, AgCl

This is obtained as a white precipitate by adding hydrochloric acid or any soluble chloride to silver nitrate solution:

$$AgNO_3 + HCl = AgCl \downarrow + HNO_3$$

The precipitate coagulates into large lumps when boiled or shaken vigorously and gradually runs purple on exposure to light. Although insoluble in water and nitric acid, it dissolves readily in potassium cyanide solution forming the argentocyanide ion $[Ag(CN)_2]^-$, in ammonium hydroxide solution forming the complex cation $(Ag(NH_3)_2)^+$, and in sodium thiosulphate solution forming the complex anion $[Ag(S_2O_3)_2]^{3-}$. Silver chloride is reduced to the metal by heating it in hydrogen or by the action of zinc and hydrochloric acid:

$$2AgCl + H_2 = 2Ag \downarrow + 2HCl$$

The silver halides are of great importance in photography owing to their sensitivity to light, which brings about their reduction to metallic silver. Silver iodide is used in artificially inducing rainfall because its crystals when scattered from a high-flying aircraft provide suitable nuclei for the crystallisation of ice and the formation of raindrops.

Silver nitrate, $AgNO_3$

This is made by the action of nitric acid on metallic silver, colourless crystals appearing when the solution is concentrated and cooled. It melts at 208°C and decomposes when heated strongly thus:

$$2AgNO_3 = 2Ag + 2NO_2 \uparrow + O_2 \uparrow$$

Silver nitrate is important as the commonest silver salt which is soluble in water. Its solution is used volumetrically for estimating solutions of chlorides, bromides, iodides, cyanides, and thiocyanates, and in qualitative analysis for detecting those anions which give precipitates with silver ions, e.g., halides, cyanides, sulphides, oxalates, chromates, phosphates, etc. It is also used to prepare insoluble silver salts by double decomposition, and to reveal the presence of arsine in the *Gutzeit test*. Its use to prepare ammoniacal silver oxide solution for detecting reducing agents has already been discussed.

Detection of silver

When treated with hydrochloric acid, solutions of silver salts give a white precipitate which is insoluble in nitric acid but which dissolves readily in solutions of ammonium hydroxide, sodium thiosulphate, or potassium cyanide. With potassium chromate solution, soluble silver salts give a crimson precipitate of silver chromate which is insoluble in acetic acid but which dissolves readily in dilute nitric acid:

$$2AgNO_3 + K_2CrO_4 = Ag_2CrO_4 \downarrow + 2KNO_3$$

GOLD

Gold, Au, electronic configuration: 2,8,18,32,18,1. The metal occurs naturally, usually as very fine particles embedded in quartz.

Extraction

The first stage consists of crushing the gold-bearing minerals and concentrating the gold in them by purely mechanical means, taking advantage of its high density. The concentrate is then leached with a very dilute solution of sodium or potassium cyanide, which dissolves the gold by forming a soluble complex ion, from which it is precipitated by adding zinc. Any residual zinc is removed from the precipitate by treatment with dilute sulphuric acid, and the product is then melted into plates and refined electrolytically. The impure gold is made the anode, chloroauric acid the electrolyte, and a sheet of pure gold the cathode. Only gold is deposited upon the cathode, silver being precipitated as its chloride and platinum remaining in solution in the electrolyte. Some gold is obtained as a by-product during the refining of copper and silver.

Properties and reactions of gold

It is a soft, lustrous, yellow metal of m.p. 1063°C and density 19.3 g cm^{-3}. It is extremely ductile and malleable and is a very good conductor of heat and electricity. Being very low in the electrochemical series, it is not affected by air or water, nor by any single mineral acid. It does dissolve, however, in aqua regia (1 volume concentrated nitric acid +3 volumes concentrated hydrochloric acid), forming a solution of chloroauric acid. It also dissolves in alkali cyanide solutions forming aurocyanides, and in mercury forming a liquid amalgam. Gold combines readily with moist chlorine forming gold chloride solution:

$$2Au + 3Cl_2 = 2AuCl_3$$

Uses of gold

Most of the world's gold is stored in the various national banks, the metal being accepted as an international currency. It is used for jewellery and coinage, but since pure gold is too soft it is usually alloyed with other metals such as copper and silver, the gold content being expressed in carats where twenty-four carats corresponds to pure gold. It is also used for electroplating cheaper metals, for filling teeth and for making compounds of gold, which are used in photography and in making ruby glass.

Compounds of gold

The element forms two series of salts, aurous or gold(I), and auric or gold (III), in which it is univalent and trivalent respectively. When gold is dissolved in aqua regia and the solution is concentrated, yellow crystals of hydrated chloroauric acid, $HAuCl_4$. $4H_2O$, separate on cooling. Heated to 200°C in chlorine, these give red crystals of auric chloride, $AuCl_3$, which decompose when heated to 175°C in air, forming aurous chloride and chlorine. Gold readily forms complex anions with high concentration of CN^- and Cl^- ions, e.g. $[Au(CN)_2]^-$ and $[AuCl_4]^-$. When a solution of chloroauric acid is treated with a strong reducing agent (e.g., formaldehyde, tartaric acid, or stannous chloride), gold is obtained as a colloidal solution which is blue, red or purple in colour, depending upon the particle size.

Thus, copper, silver, and gold are very similar in many respects. All three elements are feebly electropositive metals; they are so resistant to atmospheric corrosion that they occur native, they do not displace hydrogen from acids, and they are displaced from solutions of their salts by more electropositive metals such as zinc, iron, and aluminium. The metals themselves are all crystallise in cubic close-packed structure. They are highly lustrous and capable of taking a high polish, they melt at about 100°C, they are extremely malleable and ductile and are excellent conductors of heat and electricity. They are all attacked by chlorine with the formation of a metallic chloride.

All three elements form colourless univalent compounds. They also exert a higher valency to form coloured compounds in which the penultimate shell is incomplete. Some of their compounds are predominantly covalent in character because their high nuclear charges and the weakness of the screening provided by their penultimate shells causes them to have small atomic volumes and to exert strong attractive forces upon electrons at the peripheries of their atoms. For the same reason they encourage the formation of complex ions by co-ordination of other ions (e.g., Cl^- or CN^-) or groups of atoms (e.g., H_2O or NH_3) which have electron-donating tendencies.

Despite these marked resemblances and the similar electronic configurations which give rise to them (there is a single electron in the outermost shell of each atom), these elements differ in certain ways. The oxides of copper are much more stable than those of silver and gold, for example, and silver oxide is alone in being strongly alkaline. Copper and silver dissolve readily in nitric acid, but gold does not, and silver nitrate decomposes at a much higher temperature than copper nitrate. Again, copper and silver chlorides are much more stable thermally than the chloride of gold.

The most striking difference is in the greater stability of silver in the univalent state compared with gold. For example, silver(I) compounds such as argentous nitrate and argentous sulphate are stable substances showing no tendency to decompose to the metal. On the other hand, compared with copper and gold, silver is much less stable in its higher valency state, the fluoride being the only common argentic or silver(II) compound.

ZINC, CADMIUM, AND MERCURY

Group IIB of the periodic classification comprise the elements zinc, cadmium, and mercury, which are of particular interest because they are the last members of their respective series of transition metals.

ZINC

Zinc, Zn, electronic configuration: 2,8,18,2. The metal does not occur native, but zinc compounds are widely distributed, the commonest ores being zinc blende, Zinc and calamine, $ZnCO_3$. Other minerals are zincite, ZnO, franklinite $ZnO.Fe_2O_3$, and siliceous zinc ore, $Zn_2SiO_4.H_2O$.

Manufacture

A typical flow sheet for zinc extraction is shown in (Fig. 20.2). Three methods are in use : (i) vertical retort process; (ii) blast furnace method; and (iii) electrolytic process.

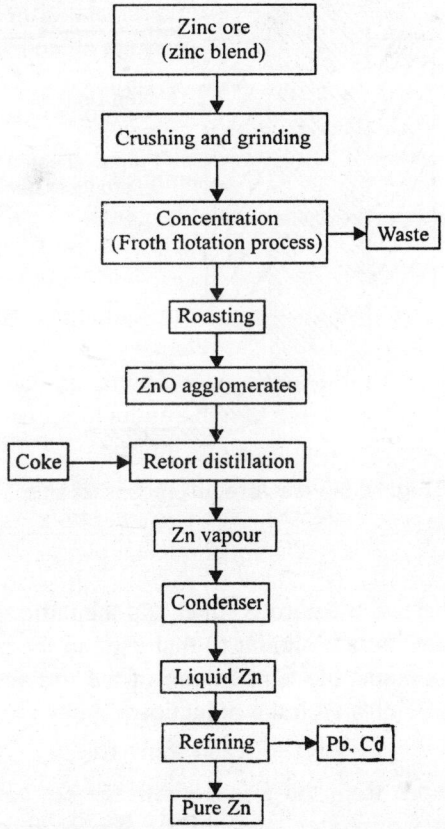

Fig. 20.2. Typical flow sheet for Zinc extraction.

Vertical retort process

The first stage consists of concentrating the ores and converting them into zinc oxide. This is achieved by heating calamine strongly or by roasting zinc blende in air in reverberatory furnace, the sulphur dioxide being a valuable by-product which is used for making sulphuric acid, giving about two kg of sulphuric acid for every kg of zinc:

$$ZnCO_3 = ZnO + CO_2 \uparrow$$
$$2ZnS + 3O_2 = 2ZnO + 2SO_2 \uparrow$$

The second stage consists of reducing the zinc oxide by heating it 1350°C with carbon:

$$ZnO + C = Zn + CO$$

This is done by mixing the oxide with anthracite or coke and making briquettes which are fed into a vertical retort heated by producer gas as in Fig. 20.3. The zinc distils over and condenses to a liquid in the receiver. After solidification the crude zinc, called spelter, is purified by melting, the product still containing about 1% of lead. Zinc of at least 99.8% purity can be obtained from this by re-distillation.

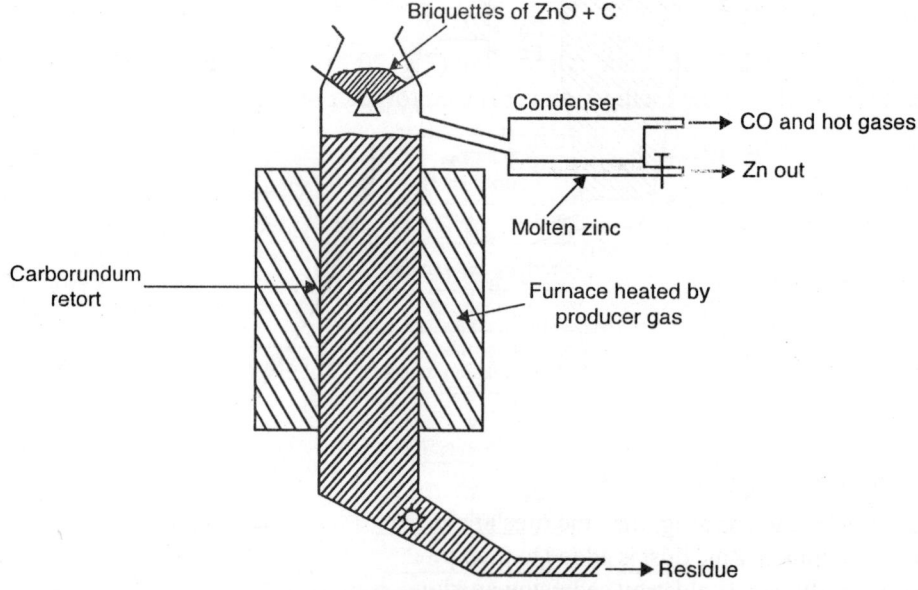

Fig. 20.3. Vertical retort process for zinc.

Blast furnace method

The concentrated ores are first roasted to a porous sinter and then after pre-heating and mixing with hot coke are fed into the top of a blast furnace similar to that used in the manufacture of iron, but smaller (Fig. 20.4). The burning coke maintains the temperature of the furnace above 1000°C and provides a supply of carbon monoxide which helps with the reduction:

$$ZnO + CO \rightleftharpoons Zn + CO_2$$

The zinc vapourises and escapes from the furnace with the hot gases. It is cooled very rapidly to 600°C by passing it through sprays of molten lead; this condenses and dissolves the zinc and prevents it from being re-oxidised by the back reaction. The liquid from the condenser is run off and cooled to about 450°C at which temperature the zinc is much less soluble in the molten lead and mostly separates into an upper layer which is decanted, leaving the lead to be used over again in the condenser.

This process has been widely adopted because it is so economical; it is continuous in operation and uses internal heating and it works well with low grade ores including those containing lead and iron. In fact, any lead present is reduced during the process and collects at the base of the furnace where it can be tapped off periodically with the slag, so the tendency in the last few years has been to use this process to produce both metals from ores containing zinc and lead.

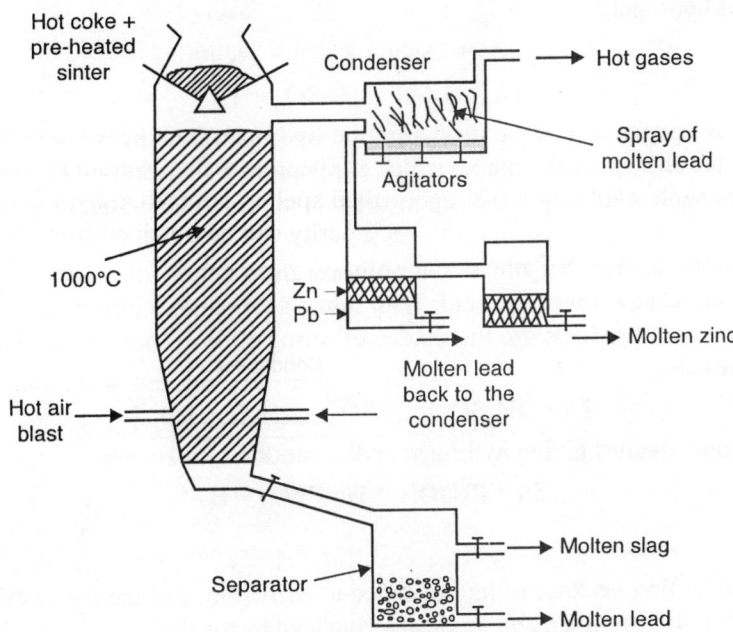

Fig. 20.4. Blast furnace process.

Electrolytic process

After concentration and roasting, the zinc ores are leached with dilute sulphuric acid, giving an impure solution of zinc sulphate. Zinc dust is added to precipitate the less electropositive metals by displacement, and the resulting solution is electrolysed using an aluminium cathode and a high current density. Under these conditions only zinc ions are discharged at the cathode and the pure metal is deposited. Despite their lower electropositivity, hydrogen ions are not discharged because of the overpotential of the cathode.

The electrolytic method offers several advantages over its rivals. It supplies zinc of high purity which does not need refining, it can be applied to low grade zinc ores, and it permits the recovery of any precious metals which are present as impurities. It has the disadvantage of requiring an abundant supply of cheap electric power.

Properties of Zinc

It is a bluish-white, lustrous metal, m.p. 419°C, b.p. 907°C, ρ 7.13 g cm^{-3}. It is brittle at room temperature and above 200°C, but at 100–150°C it is malleable and ductile. Granulated zinc is obtained by pouring the molten metal into cold water; it is used in the laboratory because of its large surface area for a given weight.

Reactions of zinc

In moist air it gradually assumes a dull grey appearance becoming coated with the oxide and basic carbonate, which inhibit further corrosion. It burns to the oxide when heated strongly in air, and it reacts with steam at red heat liberating hydrogen. Zinc combines exothermically with the halogens, phosphorus, and sulphur when heated, but it does not combine directly with nitrogen. At red heat it decomposes ammonia forming the nitride, Zn_3N_2.

Provided that the zinc is not extremely pure, it dissolves in dilute hydrochloric and sulphuric acids with the evolution of hydrogen:

$$Zn + 2HCl = ZnCl_2 + H_2 \uparrow$$
$$Zn + H_2SO_4 = ZnSO_4 + H_2 \uparrow$$

Impurities such as copper or arsenic discharge the hydrogen ions in the acid by setting up small galvanic couples on the surface of the metal, which explains why the addition of a few drops of copper sulphate solution has such a striking effect upon the rate at which acid attacks a sample of very pure zinc.

Zinc reacts with nitric acid giving zinc nitrate solution and a variety of products depending upon the temperature and the concentration of the acid. With dilute acid, ammonium nitrate is formed, but with concentrated acid the main products are the oxides of nitrogen. With hot concentrated sulphuric acid, zinc gives sulphur dioxide:

$$Zn + 2H_2SO_4 = ZnSO_4 + 2H_2O + SO_2 \uparrow$$

It dissolves in strong alkalies giving hydrogen and a solution of a zincate:

$$Zn + 2NaOH = Na_2ZnO_2 + H_2 \uparrow$$

Uses of zinc

The metal is used for roofing because of its resistance to corrosion, and the die castings. It is also used for making the casing of dry batteries, and for desilvering lead by the Parkes process and for precipitating gold and silver after cyanide leaching. Zinc forms a large number of alloys, particularly brasses, where the zinc content is often as high as 40%. The metal is used to make zinc compounds, many of which are of commercial importance. Zinc dust is a useful reducing agent. Plates of zinc are frequently used in the lithographic method of printing.

Large amounts of zinc are used for galvanising iron and steel to prevent rusting. In the hot-dip process the iron is cleaned by pickling it in acid, washed, coated with flux (usually ammonium chloride), and then dipped into molten zinc. Alternatively the article may be sprayed with powdered zinc or coated electrolytically. Even if the layer of zinc is removed at one place by scratching, rusting will not readily occur because the zinc, being more electropositive than the iron sacrificially protects the latter. A galvanic cell is set up around the scratch and the zinc corrodes preferentially, going into solution as zinc ions and leaving the iron intact. A coating of zinc hydroxide and basic zinc carbonate soon forms, inhibiting further corrosion. This protective action is in direct contrast to tinning, where a scratch on the tinplate results in accelerated rusting of the iron because the latter is more electropositive than the tin. Galvanised vessels cannot be used for storing food, however, because of the danger of zinc poisoning.

Zinc Oxide, ZnO

This is made commercially by burning zinc in air; it is prepared in the laboratory by heating the carbonate, hydroxide, or nitrate. It is a white solid which turns yellow when hot. It is insoluble in water, but being amphoteric it dissolves readily in acids and alkalies forming salts. Zinc oxide is used as a pigment in paints and as a filler for rubber. The very pure product is used in cosmetic powders and creams and in medicinal ointments.

Zinc Hydroxide, Zn(OH)$_2$

This is obtained as a gelatinous white precipitate when a solution of a zinc salt is treated with an alkali:

$$ZnSO_4 + 2NaOH = Zn(OH)_2 \downarrow + Na_2SO_4$$

It is amphoteric, dissolving in acids to give the corresponding zinc salts and in excess of alkali to give a solution of a zincate:

$$Zn(OH)_2 + H_2SO_4 = ZnSO_4 + 2H_2O$$
$$Zn(OH)_2 + 2NaOH = Na_2ZnO_2 + 2H_2O$$

It also dissolves in solutions of ammonia or ammonium salts, forming the complex ion $[Zn(NH_3)_2]^{2+}$.

Zinc Carbonate, ZnCO$_3$

This occurs naturally as calamine. It is obtained as a white precipitate when sodium bicarbonate solution is added to a solution of a zinc salt. Addition of sodium carbonate solution gives a precipitate of basic zinc carbonate of composition $2ZnCO_3.3Zn(OH)_2$. Zinc carbonate is present in calamine lotion, used for treating inflammation of the skin.

Zinc Sulphate, ZnSO$_4$

This is obtained as colourless crystals of the heptahydrate, $ZnSO_4.7H_2O$, known as white vitriol, by concentrating a solution of zinc or its oxide or carbonate in dilute sulphuric acid and cooling below 39°C. It is isomorphous with the corresponding sulphates or magnesium, iron, manganese, nickel, cobalt, and cadmium. When heated it changes thus:

$$ZnSO_4.7H_2O \xrightarrow{100°C} ZnSO_4.6H_2O \xrightarrow{250°C} ZnSO_4.H_2O \xrightarrow{750°C} ZnSO_4 \longrightarrow ZnO + SO_3$$

Zinc sulphate is chiefly used to manufacture lithopone for paints.

Zinc Chloride, ZnCl$_2$

The monohydrate is made by dissolving zinc or its oxide, hydroxide, or carbonate in excess of hydrochloric acid and evaporating the solution. The anhydrous salt is made by the action of dry chlorine or dry hydrogen chloride on heated zinc. It is a very deliquescent, white solid which dissolves in water exothermically giving a solution which is acidic by hydrolysis. It is also soluble in alcohol, ether, and acetone. The low conductivity of molten zinc chloride and its solubility in organic solvents suggests that it consists of two forms in equilibrium, one ionic and the other covalent.

Anhydrous zinc chloride is used in organic chemistry as a dehydrating agent. Mixed with moist zinc oxide it gives a hard mass of zinc oxychloride, Zn(OH)Cl, suitable for filling teeth. In solution zinc chloride is used for preserving timber and as a flux in soldering.

Zinc Sulphide, ZnS

This compound is dimorphous, occurring naturally as zinc blende and as wurtzite. It is precipitated when solutions of zinc salts are treated with ammonia sulphide solution. Although white when freshly precipitated, it tends to turn yellow on exposure to light.

Zinc sulphide is a constituent of lithopone. This important pigment, which is extensively used in white paints, is made by mixing solutions of zinc sulphate and barium sulphide, filtering off the precipitates, and heating them to about 700°C in the absence of air:

$$ZnSO_4 + BaS = BaSO_4 \downarrow + ZnS \downarrow$$

Lithopone is preferred to basic lead paints because it is cheaper, less poisonous, and does not darken when exposed to hydrogen sulphide but it is inferior in covering power and durability, particularly for outdoor use. In recent years, however, titanium dioxide has largely replaced both, because of its superior qualities as a pigment.

Detection of Zinc

Solutions of zinc compounds give a white precipitate when treated with ammonium sulphide solution. This precipitate is soluble in dilute hydrochloric acid but insoluble in acetic acid. Zinc hydroxide, precipitated by adding sodium hydroxide solution a drop at a time to a solution of a zinc salt, dissolves readily in excess of alkali, which distinguishes it from the hydroxide of manganese.

When heated on a charcoal block zinc compounds give an incrustation which is white when cold and yellow when hot. When moistened with cobalt nitrate solution and re-heated, the residue turns green.

CADMIUM

Cadmium, Cd. Electronic configuration : 2,8,18,18,2. Zinc ores invariably contain amounts of cadmium compounds as impurities.

Manufacture

The metal is obtained as a by-product during the extraction of zinc. Being more volatile than zinc, the cadmium distils over in the early stages of the distillation process and condenses with the zinc dust in the receiver or is recovered as flue dust form the furnace gases. It is purified by fractional distillation or by electrolysis.

Properties and reactions of cadmium

It is a lustrous white metal, m.p. 321°C, b.p. 767°C. Although malleable and ductile, it is too soft for constructional use. When exposed to moist air it turns dull grey, forming a coating of oxide which prevents further corrosion, but it burns when heated strongly in air.

Uses of cadmium

The chief use is for plating iron and steel to prevent rusting, even a thin film providing excellent protection; it cannot be used for food containers, however, because cadmium compounds are poisonous. The metal absorbs neutrons very readily and rods of it or its alloys are used in slowing down the activity of atomic reactors.

$$^{113}_{48}Cd + ^1_0n = ^{114}_{48}Cd + \gamma$$

Cadmium is increasingly used for making nickel-cadmium batteries which have a long life and are very compact.

Compounds of Cadmium

These closely resemble the compounds of zinc; except for the oxide and sulphide they are colourless and mostly soluble in water. When solutions of cadmium salt are treated with zinc dust, cadmium is precipitated because the metal is less electropositive than zinc.

The oxide, CdO, and hydroxide, $Cd(OH)_2$, dissolve in acids giving cadmium salts, but not in dilute alkalies. Like the corresponding compounds of zinc, they dissolve in ammonia solution forming the complexion $[Cd(NH_3)_4]^{2+}$. The sulphide, CdS, is a bright yellow solid used as a pigment. The sulphate, $3CdSO_4.8H_2O$, is used in Weston standard cells. The cyanide, $Cd(CN)_2$, is precipitated from solutions of cadmium salt by adding potassium cyanide solution. Like zinc cyanide it dissolves in excess of the reagent, giving a complex ion $[Cd(CN)_4]^{2-}$, which is used for electroplating.

Detection of cadmium

When saturated with hydrogen sulphide, solutions of cadmium salts give a yellow precipitate which is insoluble in dilute hydrochloric acid and in ammonium sulphide solution but which dissolves in warm dilute nitric acid and in hot dilute sulphuric acid. Cadmium is detected in the presence of copper by precipitating its sulphide from cyanide solution by hydrogen sulphide.

MERCURY

Mercury, Hg. Electronic configuration: 2,8,18,32,18,2. Although small amounts of the element have been found native, the only important source is the mineral cinnabar, HgS.

Manufacture

The ore, which usually contains only a few per cent of mercury, is roasted in retorts with air or iron :

$$HgS + O_2 = Hg \uparrow + SO_2 \uparrow$$
$$HgS + Fe = Hg \uparrow + FeS$$

The mercury distils over and is condensed in water-cooled earthenware tubes. It is purified by filtration through chamois leather, by washing with very dilute nitric acid, and by distillation under reduced pressure.

Properties of mercury

At room temperature it is a lustrous, silvery-white liquid. It freezes at $-39°C$ and boils at $357°C$. Its density is 13.59 g cm^{-3}. Its vapour is monoatomic and very poisonous.

Reactions of mercury

It is hardly affected by the atmosphere at room temperature, but when heated in air it is gradually converted into its oxide, HgO. It combines readily with sulphur and the halogens and 'tails' when exposed to ozone. Mercury is not attacked by hydrochloric acid, nor by dilute sulphuric acid, but it dissolves in nitric acid giving nitrates and oxides of nitrogen. Hot concentrated sulphuric acid attacks it, evolving sulphur dioxide, and it dissolves readily in aqua regia forming the chloride, $HgCl_2$.

Mercury forms alloys known as amalgams with many metals, especially with silver, gold, and the alkali metals. For example, small pieces of sodium dissolve in mercury forming a liquid amalgam which reacts steadily with water giving nascent hydrogen and is therefore a useful reducing agent.

Uses of mercury

The metal is used in mercury vapour lamps and rectifiers; and in scientific equipment such as thermometers, barometers, and high vacuum pumps. It is also used in the manufacture of mercury compounds, many of which are of considerable importance, and to make amalgams, particularly for filling teeth, and in the manufacture of sodium hydroxide.

Compounds of Mercury

Two series of compounds exist. In mercurous compounds the metal appears to be univalent, whereas in fact it is divalent; X-ray analysis and other physical evidence shows that each molecule contains two mercury atoms linked covalently to each other, so that the structure of the mercurous ion is ^+Hg —Hg^+ or Hg_2^{2+} and mercurous chloride, for example, is Cl^- ^+Hg—Hg^+ ^-Cl, written Hg_2Cl_2 not $HgCl$. In mercuric compounds the metal is also divalent; the mercuric ion has the structure Hg^{2+}, but the formula of mercuric chloride, for example, is written Cl–Hg–Cl or $HgCl_2$, since like most mercuric compounds it is largely covalent.

Mercurous chloride (calomel), Hg_2Cl_2

This is obtained by adding dilute hydrochloric acid or a soluble chloride to mercurous nitrate solution:

$$Hg_2(NO_3)_2 + 2HCl = Hg_2Cl_2 \downarrow + 2HNO_3$$

Alternatively it can be made by heating a mixture of mercuric chloride and mercury, when the mercurous chloride sublimes:

$$HgCl_2 + Hg = Hg_2Cl_2$$

It is a white solid, insoluble in water and dilute acids. It is easily reduced to mercury when heated with carbon or sodium carbonate or when treated with an excess of stannous chloride solution :

$$Hg_2Cl_2 + SnCl_2 = 2Hg \downarrow + SnCl_4$$

With sodium hydroxide solution, mercurous chloride gives a black precipitate consisting of mercuric oxide and mercury, and with ammonium hydroxide solution it gives a black mixture of metallic mercury and mercuric amido chloride, NH_2HgCl.

Mercuric oxide, HgO

This is slowly formed as a red powder by heating mercury in air to about 350°C; stronger heating causes it to dissociate again into its elements (descriptions of the classical experiments of Priestley and Lavoisiter in 1774 will be found in most elementary textbooks of chemistry):

$$2Hg + O_2 \rightleftharpoons 2HgO$$

If sodium hydroxide solution is added to mercuric chloride solution, the oxide is obtained as a yellow precipitate:

$$HgCl_2 + 2NaOH = HgO \downarrow + H_2O + 2NaCl$$

As X-ray analysis shows, the red and yellow forms differ only in their particle size. Mercuric oxide is insoluble in water, but being weakly basic it dissolves in acids giving mercuric salts.

Mercuric chloride, $HgCl_2$

This salt, which is commonly known as 'corrosive sublimate', is made either by dissolving mercuric oxide in hydrochloric acid or by heating a mixture of mercuric sulphate and sodium chloride in a retort, when it sublimes.

It is also formed by passing dry chlorine over heated mercury. It can be purified by re-crystallisation, since it is much more soluble in hot water than in cold.

$$2NaCl + HgSO_4 = HgCl_2 \uparrow + Na_2SO_4$$

It is a very poisonous, white solid, soluble in alcohol and ether. Its aqueous solution is acidic by hydrolysis. It is a weak electrolyte, only slightly dissociated into ions in solution, as shown by its slow conductivity. For this reason it does not give hydrogen chloride when boiled with concentrated sulphuric acid and it is only incompletely precipitated as silver chloride by adding silver nitrate solution.

Mercuric chloride is noticeably more soluble in hydrochloric acid and solutions of chlorides than in water because it readily forms complex ions such as $[HgCl_4]^{2-}$. It is easily reduced to mercurous chloride, or even to metallic mercury, by reducing agents such as stannous chloride or formaldehyde.

$$2HgCl_2 + SnCl_2 == Hg_2Cl_2 \downarrow + SnCl_4$$

When boiled with ammonium hydroxide solution, mercuric chloride solution gives a white precipitate of mercuric amido chloride, NH_2HgCl, which is known as 'infusible white precipitate' because it decomposes below its melting point. However, if an excess of ammonium chloride is also present, a white precipitate of mercuric diammino chloride, $Hg(NH_3)_2Cl_2$, is obtained which is known as 'fusible white precipitate' because it melts without decomposition.

Mercuric chloride is used as an antiseptic and disinfectant, and for preserving timber and hides, and for making fungicides.

Mercuric iodide, HgI$_2$

This is obtained as a yellow then scarlet precipitate when potassium iodide solution is added dropwise to mercuric chloride solution:

$$HgCl_2 + 2KI = HgI_2 \downarrow + 2KCl$$

The precipitate dissolves in excess of potassium iodide solution forming potassium mercuric-iodide, K_2HgI_4.

Mercuric iodide is dimorphic, being stable in the yellow form above 126°C and in the scarlet form below. Although only sparingly soluble in water, it is appreciably soluble in organic solvents. So slight is its dissociation into ions in solution that it gives no precipitate of silver iodide with silver nitrate solution. It is used in ointments for treating skin infections.

Mercuric sulphide, HgS

This salt exists in two forms. One occurs naturally as cinnabar and can be made by heating mercury with potassium pentasulphide solution:

$$Hg + K_2S_5 = HgS \downarrow + K_2S_4$$

It is scarlet in colour and is used as an artist's pigment (vermilion). The other form, which is black, is prepared directly form its elements or by saturating a solution of a mercuric salt with hydrogen sulphide:

$$HgCl_2 + H_2S = HgS \downarrow + 2HCl$$

Mercuric sulphate, HgSO$_4$

This is obtained as colurless crystals of the monhydrate by dissolving mercury or mercuric oxide in hot concentrated sulphuric acid and evaporating the solution:

$$Hg + 2H_2SO_4 = HgSO_4 + SO_2 \uparrow + 2H_2O$$

It is hydrolysed in solution and readily forms a basic salt. It is used industrially as a catalyst for converting acetylene into acetaldehyde.

Nitrates of mercury

Mercurous nitrate is made by treating an excess of metallic mercury with dilute nitric acid. It crystallises as the colourless dihydrate, $Hg_2(NO_3)_2.2H_2O$, which effloresces on standing in air. It is very much soluble in water giving a solution which is acidic by hydrolysis and from which a basic salt is precipitated on dilution. Mercuric nitrate is made by dissolving mercury in hot concentrated nitric acid. It crystallises from the evaporated solution as colourless deliquescent crystals of monohydrate. Its solution is acidic by hydrolysis.

Detection of mercury

When saturated with hydrogen sulphide, solutions of mercury salts give a black precipitate which is insoluble in dilute nitric acid or in ammonium sulphide solution. With dilute hydrochloric acid solutions of mercurous salts gives a white precipitate which turns black when treated with ammonium hydroxide solution.

Thus, all three elements are fairly volatile, divalent metals. They occur naturally as their sulphides, which are insoluble in water. They form many similar compounds, their oxides are easily reduced to the metals, and they readily form basic salts, e.g., carbonates. Their electronic configurations are very similar, each atom possessing eighteen electrons in its penultimate shell and two in its outermost shell.

Two other characteristics common to all three elements are their tendencies towards covalency and complex ion formation. In these elements the nucleus exerts a strong attraction upon the electrons at the periphery because of the high nuclear charge and the small size of the atom. The penultimate shells, although they contain ten additional electrons, possess much less screening power than the penultimate shells in the alkaline earth metals, and consequently many of the compounds of the Group IIB metals are predominantly covalent and are hydrolysed to give acidic solutions. For the same reason they readily accept co-ordinated groups such as ammonia, forming complex ions.

Unlike the majority of the transition metals, they do not show variable valency or paramagnetism, and solutions of their ions are mostly colourless. This is easily understood when it is remembered that these are the very properties which depend upon 'incomplete' penultimate electron shells, whereas in the atoms of these elements the penultimate shell contains the highest possible number of electrons.

Despite all these similarities, the group is characterised by several notable differences between its members. They differ considerably in electropositivity, zinc being strongly electropositive, cadmium less so, and mercury being so weakly electropositive as to be almost noble in character. This is apparent from the widely differing heats of formation of the oxides, from the inability of mercury to liberate hydrogen from acids, and from the thermal instability of mercuric oxide. Moreover, zinc oxide is amphoteric, whereas the other oxides are basic.

Mercury is of especial interest because of its unique features. It is the only metal which is liquid at room temperature, which accounts for many of its uses. Mercurous compounds contain the very unusual covalent linkage between the atoms of the metal, giving the illusion of univalency. No hydroxide of mercury is known, and when mercury compounds react with ammonia they tend to form substitution products rather than complex ions like those of zinc and cadmium.

NICKEL

Nickel occurs in nature mostly in combination with arsenic, antimony and sulphur. The important sources of nickel are *Pentlandite*, which is a nickel sulphide ore, *Garnierite* is chiefly a hydrated double silicate of magnesium and nickel, *Niccolite* is also known as *Kupfernickel* (NiAs) or nickel arsenide.

Pyrrhotite is a ferrous sulphide with nickel as an impurity. The *Pentlandite* or *Sudbury* ore is, however, the principal source of nickel for extraction purposes, although considerable quantities of nickel ore are obtained as a by-product in the electrolytic refining of copper.

Extraction

The extraction of nickel from its ore involves the following steps:

Concentration of the ore

The sudbury ore deposits have about 14–15 per cent of pentlandite mixed with other metallic ores. The ore is concentrated by froth floation method to separate the pentlandite. (Fig 20.5).

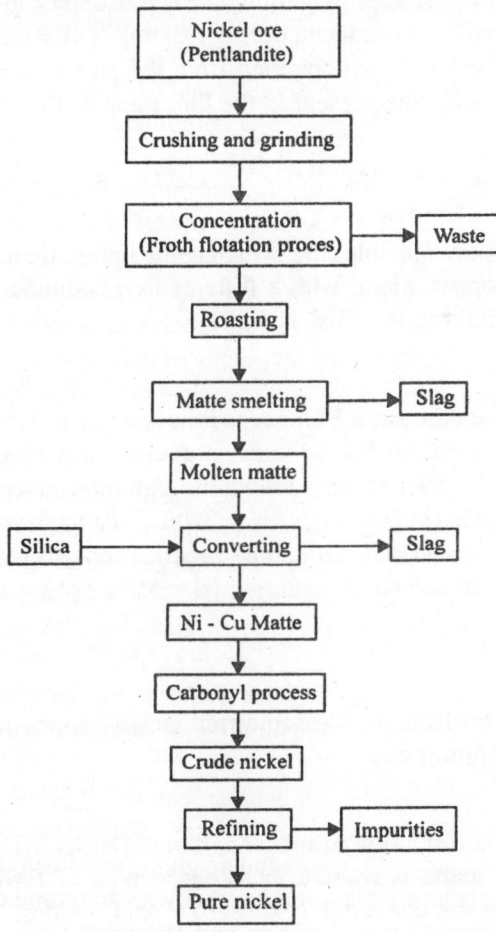

Fig. 20.5. General flow sheet for Nickel extraction.

Roasting

The concentrated ore is then roasted in tall furnaces, called heaps (which are about 30 ft high and 20 to 22 ft wide) by means of wood fires. The roasting is continued for several weeks during which time most of the iron sulphide present in the ore is oxidised to ferrous oxide and sulphur dioxide, but the sulphides of copper and nickel (present in the ore) remain practically unaffected.

$$2FeS_2 + 5O_2 \longrightarrow 2FeO + 4SO_2$$

By roasting, the amount of sulphur is diminished and about 38% of the sulphur in the concentrate is removed as SO_2.

Smelting

The roasted ore is now smelted in a long reverberatory furnace or in a small blast furnace together with coke and fluxing materials such as limestone and quartz. The fluxing materials render the slag fusible and remove proper proportion of iron and basic oxides, considering that SiO_2 is present in the charge. Most of the iron sulphide, which escaped oxidation during roasting is now converted into ferrous oxide. Any ferric oxide that may have been produced during roasting is also reduced by coke to ferrous oxide. The ferrous oxide thus obtained and also obtained from the previous operation (roasting) reacts with silica to form ferrous silicate and lime present in the limestone as fluxing material reacts with silica to form calcium silicate.

$$FeO + SiO_2 \longrightarrow \underset{\text{Slag}}{FeSiO_3} \qquad\qquad CaO + SiO_2 \longrightarrow \underset{\text{Slag}}{CaSiO_3}$$

$FeSiO_3$ and $CaSiO_3$ both form a fusible slag, which being lighter floats on the molten mass consisting of sulphides of nickel and copper, along with a little of iron sulphide. This heavier mass containing sulphides of nickel, copper and iron is called as *matte*.

Bessemerisation

The molten or fused matter from the Blast Furnace is now run into a Bessemer converter having a basic lining and fitted with tuyeres to admit hot air under pressure. Now requisite amount of silicious flux, such as silica is added and a hot blast of air is blown through the converter under pressure. As a result, sulphur is oxidised to iron oxide (FeO) which reacts with silica to form ferrous silicate ($FeSiO_3$) slag and is easily removed. The Bessemerised matte, thus obtained consists mainly of sulphides of nickel and copper. The composition of Bessemerised matte is : Ni = 55%–56%, Cu = 25–30%, S = 14–17% and Fe = 0.2–0.4%.

Refining

The process of extracting nickel from the Bessemerised nickel-copper matte is known as refining. The most important methods of refining are:

Mond's process

This process depends upon the formation of nickel carbonyl $[Ni(CO)_4]$ at 60°C and its decomposition at 180°C. The Bessemerised matte is roasted in a special type of furnace, as a result of which the sulphides of nickel and copper are converted (oxidised) to their oxides. The mixture of these oxides is then digested with dilute sulphuric acid at 80°C. As a result of this treatment, copper oxide dissolves giving copper sulphate, while nickel oxide remains unaffected.

$$CuO + H_2SO_4 \longrightarrow CuSO_4 + H_2O$$

The solution is evaporated to get crystals of $CuSO_4$ which is a by-product. The nickel oxide that remains unchanged is first dried and then introduced into a reducing tower, which is about 7 metre high and provided with horizontal shelves and mechanical rabbles to facilitate the movement of the ore. The upper half of the reducing tower is maintained at a temperature of 300–330°C by means of producer gas and the lower half is cooled. A current of water gas ($CO + H_2$) is led up the tower so that the oxides are reduced to metallic state.

$$NiO + H_2 \longrightarrow Ni + H_2O \qquad\qquad NiO + CO \longrightarrow Ni + CO_2$$

The finely divided metal obtained from the reducing tower is now fed into another tower, called the volatilising tower, which is similar to the reducing tower and is maintained at about 60°C by exothermic reaction. A current of carbon monoxide is passed up through the tower when volatile nickel carbonyl is formed due to the reaction of metallic nickel and carbon monoxide.

$$Ni + 4CO \longrightarrow Ni(CO)_4 \uparrow$$

The nickel carbonyl vapour from the volatilising tower is now charged into another tower, called decomposer. It is a closed one tower heated externally by producer gas to maintain the temperature at about 180°C. The decompose is filled with nickel shots or granulated nickel maintained at 180°C. The carbonyl vapour when enters the decomposer tower through a perforated centre tube comes in contact with hot nickel shots and thus decomposes to give pure nickel which gets deposited on nickel shots which grow in size. The granules or shots are removed periodically from the bottom of the decomposer and screened to separate the oversized shots. These are removed and the smaller sized shots are re-introduced into the decomposer until they grow into large sized shots. Nickel obtained by this process is about 98.8% pure.

$$Ni(CO)_4 \rightarrow Ni + 4CO \uparrow$$

The resultant carbon monoxide gas passes out through various water cooled outlets situated between different sections of the decomposer. This gas is again used in the volatilising tower.

It should be noted that residue of the volatilising tower is subjected to repeated operations in the reducing tower and volatilising tower. The carbonyl vapour from all these operations are treated in the same decomposer to get good yield of nickel. The final residue contains copper, small amount of nickel and iron, along with traces of precious metals. This residue is calcined and treated with CO under pressure. The exit gas is cooled, as a result of which most of the carbonyl condenses into liquid. The residual gas is then decomposed in the decomposer. The liquid carbonyl is vapourised and treated in decomposer. The residue containing precious metals is treated to recover them.

Electrolytic method

In the electrolytic method, the Bessemer matte is first roasted and treated with dilute sulphuric acid, at 80°C to remove copper as explained in the Mond's process. The residue consisting of 65% nickel sulphides and 30% copper is cast into rods and made anodes in the electrolytic cells. The electrolysis is carried out in a series of such cells, each of which is provided with an iron plate to act as a cathode and crude metal anode is placed in a bag of special canvas. In this manner each cell is divided into a cathodic and anodic compartment. The electrolyte is nickel ammonium sulphate solution which enters the cell near the cathode and leaves it from the anode compartment. The solution leaving the anode compartment contains nickel sulphate and copper sulphate. The solution is introduced into a tank, where it is treated with waste anode material. The nickel in the anode material displaces copper from the solution of copper sulphate.

$$CuSO_4 + Ni \longrightarrow NiSO_4 + Cu$$

The solution free from copper thus contains nickel sulphate only. This solution enters the electolytic cell near the cathodes. As a result of electrolysis, pure nickel is deposited on iron plates acting as cathodes. A corresponding amount of nickel and copper, however, pass into solution at the anode. The anode mud may be treated subsequently to recover gold, silver, platinum and palladium.

Orford method

The method is based on the fact that copper sulphide is soluble in fused sodium sulphide, while nickel sulphide is not soluble in fused sodium sulphide. Thus the Bessemerised matte containing mainly the nickel sulphide (NiS) and copper sulphide (CuS) is heated or smelted with a mixture of coke and crude sodium sulphide (salt cake), as a result of which sodium sulphide is formed *in situ*.

$$Na_2SO_4 + 4C \longrightarrow Na_2S + 4CO$$
Fused sodium sulphide

The molten mass gets separated into two layers, the upper layer, called Orford top, having nearly all the copper and the lower layer, called Orford bottom having nearly all the nickel. In other words, copper sulphide dissolves in fused sodium sulphide and the solution being lighter, floats above, while the heavier nickel sulphide which does not dissolve in fused sodium sulphide remains below. The lower heavier nickel sulphide layer is tapped out and washed with water to dissolve out sodium sulphide and then roasted in air to oxidise nickel sulphide into nickel oxide which is finally reduced to the pure metal by heating the oxide with coke.

$$NiO + C \longrightarrow Ni + CO \uparrow$$

German method

In this method Bessemerised nickel copper sulphide matte is heated with carbon monoxide at 525°K under 200 atmospheric pressure for 3–4 days. As a result, about 85% of the nickel present in the matte gets converted into the nickel carbonyl which is cooled to liquefy it. It is then purified by fractional distillation. The purified nickel carbonyl is heated in an empty tower at about 180°C when nickel is obtained as a powder.

$$Ni(CO)_4 \longrightarrow Ni + 4CO \uparrow$$

The cupriferrous residue from the carbonylating autoclave contains the precious metals, such as Ag, Au, Pt and Pd, originally present in the ore.

Properties

Nickel is a silvery white or greyish white metal, highly ductile, malleable and tenacious. Its density, melting point and boiling point are 8.902, 1452°C and 2730°C respectively. It is ferromagnetic metal though much less than iron. Nickel is magnetic below its curie point (353°C). Nickel is resistant to oxidation, similar to cobalt, under ordinary atmospheric conditions. It shows much more resistance than iron to the action of air and water. It does not rust. However, moist air converts it slowly into nickel oxide (NiO), which forms a film on the surface and protects the metal from any further action. Concentrated nitric acid renders nickel passive, but dilute nitric acid is one of the best solvents for nickel. Both molten nickel and finely divided nickel absorb hydrogen, the latter is used as a catalyst in hydrogenation of oils. For this purpose Raney nickel is used as a catalyst. Nickel also displaces hydrogen from acids gradually giving rise to nickel salts. Nickel dissolves in aqua regia and not attacked by alkalies even when fused. Hence it is used for making crucibles, spatulas and tongs required in the laboratory.

ALUMINIUM

The principal ore of aluminium is bauxite. It is composed of hydrated oxides of alumina. The minerals present in bauxite are: Gibbsite ($Al_2O_3 \cdot 3H_2O$), and Bohmite ($Al_2O_3 \cdot H_2O$).

Bauxite may contain trihydrate alumina, monohydrate alumina or the mixture of two. There may be other associated impurities such as oxides of iron, silicon and titanium composition of bauxite having trihydrate alumina. In combined form alumina occurs in feldspars, micas and clay minerals such as kaolin. These minerals contain spinal $MgAl_2O_4$ or $MgO.Al_2O_3$.

Manufacture

Bayer's process

In this process bauxite is digested with hot caustic soda solution. Alumina goes into solution as sodium aluminate. Sodium aluminate solution is separated from insoluble matter. From this solution alumina trihydrate $Al(OH)_3$ is precipitated and is calcined to produce pure alumina. Alumina produced is reduced to give aluminium (Fig. 20.6). Bayer's process of producing aluminium consists of the following steps.

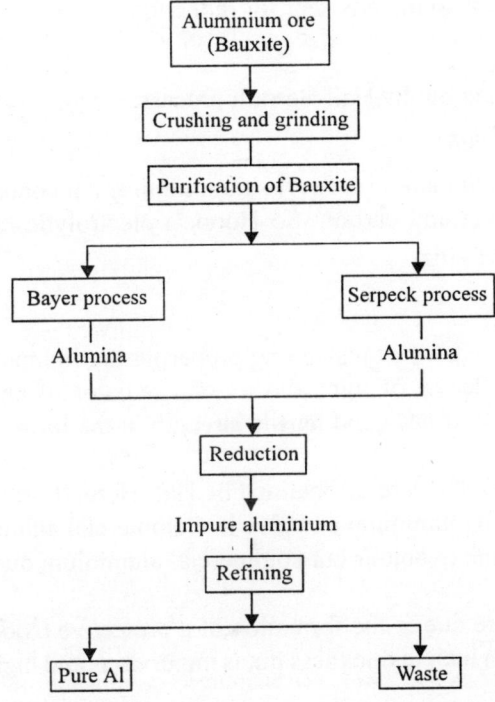

Fig. 20.6. General flow sheet for Aluminium extraction.

Concentration

Bauxite is concentrated.

Crushing and grinding

Bauxite is crushed and washed with water to remove clay, etc., and is then dried. Dried bauxite is finely ground.

Digestion

It is charged into digesters containing spent caustic liquor from previous digesting. Aluminium is dissolved in caustic soda solution in the form of sodium aluminate. The solution containing sodium aluminate is separated from the insoluble residue by settling and filtration. The residue contains a very high percentage of iron oxide.

Precipitation of aluminium hydroxide

The solution containing sodium aluminate is sent to precipitation tanks. It is seeded there with alumina trihydrate obtained from the previous charge. The solution is cooled slowly. By hydrolysis of sodium aluminate, aluminium trihydroxide is formed in the presence of crystalline seed and precipitation of aluminium trihydroxide takes place. When the precipitation has taken place to a certain required degrees, aluminium trihydroxide is separated by means of filtration.

Calcination of alumina trihydrate

Alumina trihydrate $Al(OH)_3$, obtained in the granular form is calcined to remove free and combined water. Calcination is carried out at high temperature of about 1000°C to convert alumina into non-hygroscopic alumina. Calcined alumina is sent for reduction.

Reduction of alumina

Reduction of alumina is carried out by Hall Heroult process.

Electrolytic refining of aluminium

Metal produced by Hall Heroult method is 99.9% aluminium and it contains small quantities of iron and silicon from the bath, alumina and carbon. So Hoope's electrolytic refining process is adopted for refining the metal to 99.99% purity.

Properties of Aluminium

Aluminium is a silver white metal. Its outstanding properties are lightness, good electrical and thermal conductivity. It is a good reflector of light and a good radiator of energy. It is non-magnetic. It is resistant to atmospheric attack. It has good tensile strength in the form of alloys. Due to its ductility it can be easily worked.

Aluminium is generally 99.9% pure as obtained by Hall Heroult process and impurities of iron and silicon present form alloy with aluminium and give it. Commercial aluminium is 99.0% to 99.3% pure. Pure aluminium is silvery white in colour but commercial aluminium due to impurities has got a bluish tinge.

It is resistant to atmosphere due to the formation of a protective oxide film. This oxide film is very thin, less than a millionth of an inch in thickness but is impervious and highly protective. On heating, this film increases in thickness.

Heat of combination of aluminium with oxygen is very high. Finely divided powdered aluminium burns in air.

With oxygen or water, the reaction which takes place is as follows:

$$4Al + 3O_2 \longrightarrow 2Al_2O_3$$
$$2Al + 3H_2O \longrightarrow Al_2O_3 + 3H_2$$

Intimate mixture of aluminium and iron reacts vigorously with production of heat, after igniting the mixture.

$$2Al + Fe_2O_3 \longrightarrow Al_2O_3 + 2Fe$$

Fresh aluminium surface will react with water and oxygen to form H_2O_2.

$$2Al + 6H_2O + 3O_2 \longrightarrow Al_2O_3 . 3H_2O + 3H_2O_2$$

Aluminium combines with free halogens in presence of organic solvents and in the absence of oxygen-containing compounds.

$$2Al + 3Cl_2 \longrightarrow 2AlCl_3$$
$$2Al + 3I_2 \longrightarrow 2AlI_3$$
$$2Al + 6HCl \longrightarrow 2AlCl_3 + 3H_2$$

Aluminium reacts with sulphur, phosphorus and nitrogen at elevated temperature.

$$2Al + N_2 \longrightarrow 2AlN$$
$$Al + P \longrightarrow AlP$$
$$2Al + 3S \longrightarrow Al_2S_3$$
$$2Al + 3FeS \longrightarrow Al_2S_3 + 3Fe$$

At elevated temperature carbon and carbon compounds react with aluminium forming carbides and free carbon.

$$6Al + 3CO \longrightarrow Al_4C_3 + Al_2O_3$$
$$4Al + 3C \longrightarrow Al_4C_3$$
$$4Al + 3CO_2 \longrightarrow 2Al_2O_3 + 3C$$

In hot and boiling water, oxide film contains monohydrate. Water reacts very quickly with aluminium amalgam as protective film is not formal on the amalgam.

$$2Al + 4H_2O \longrightarrow Al_2O_3 . H_2O + 3H_2$$

Concentrated hydrochloric acid attacks quickly, dilutes slowly and by heating the reaction is accelerated. Cold concentrated sulphuric acid has no action, hot reacts as follows:

$$2Al + 6H_2SO_4 \longrightarrow Al_2(SO_4)_3 + 3SO_2 + 6H_2O$$

Nitric acid oxidises the surface and forms a protective coating of oxide over it. If, however, no protective oxide film is formed, for example as in amalgam, then the reaction which will take place is as follows:

$$Al + 4HNO_3(conc.) \longrightarrow Al(NO_3)_3 NO + 2H_2O$$

$$8Al + 30HNO_3 (dil.) \longrightarrow 8Al(NO_3)_3 + 3HN_4NO_3 + 9H_2O$$

Sodium and potassium hydroxide react to form aluminates:

$$2Al + 2NaOH + 2H_2O \longrightarrow 2NaAlO_2 + 3H_2$$

Anodic coating

By anode treatment of aluminium, oxide film of sufficient thickness and abrasion resistant can be formed in certain electrolytes which are acidic in character. Anodic coatings are amorphous in nature and they have got firm adherence to the metal surface. Electrolyte used generally is 15% sulphuric acid.

Thickness of oxide film is between 0.0025 to 0.025 mm depending upon the quantity of current employed. Chromic acid and oxalic acid are also used as electrolyte.

As the anode coating is minutely porous, it should be made non-absorptive to prevent staining, etc. So sealing of pores is done by dipping the anodised surface in hot water. Such sealed surfaces are not stained by coffee or other coloured liquids.

Anodised surfaces may be impregnated with corrosion inhibitor by treating with chromate solutions. Chromate gets absorbed in coating and protects the coating from any corrosive attack. Coatings which are sealed with chromate solutions form an excellent paint base and are used very much for protecting aircraft.

Anodic coatings can be coloured by impregnation with organic dyes and with mineral pigments. Coatings coloured by organic dyes are permanent indoors but not outdoors in sunlight. They fade in sunlight. Coatings which are impregnated with mineral pigments do not fade even in sunlight.

Oxide coatings can also be formed on aluminium by chemical treatment. Such coatings are neither so thick, nor as hard nor as abrasion resistant as anodic coatings. A hot solution of sodium carbonate and potassium or sodium dichromate will produce a greyish green oxide coating.

Uses

Aluminium is resistant to many mineral and organic acids, salt solutions, organic compounds, sulphur and many other substances. Aluminium is available in different fabricated forms and it can be assembled and finished by different processes. Due to all these reasons it is used for fabricating equipment for chemical and food processing industries. For these very reasons it is used for making cooking utensils. Cookers and steam jacketed kettles, etc. are produced from this. Aluminium is used in the metallurgy of iron and steel as it is a powerful deoxidiser and reduces the dissolved and combined oxygen content of molten steel. Metallic aluminium is also used to reduce oxides of metals such as iron, chromium, vanadium, and molybdenum. Aluminium is a fine alloying metal in ferrous metallurgy, in steel for nitriding and in iron alloys where certain electrical magnetic and oxidation resistant properties are desired.

Due to its light weight and high tensile strength it is used for the construction of airplanes, buses, trucks, trains and ships. Resistance of aluminium to the weather makes possible the use of aluminium in architecture for constructing such parts as roofing, sheathing, windows, spandrel sils, etc. Stair rails and furniture, etc., are also made out of aluminium.

It is used in the manufacture of cable. In cables steel wire core is surrounded by aluminium conductors that carry the current. The strength of the core and the light weight of the cable permit long spans.

TIN

Tin is said to occur in the native state in Siberia. From the commercial point of view the important ores of tin are *cassiterite* or *tinstone* (SnO_2) and tin pyrites, $SnS_2.Cu_2S$, FeS (35% Sn). Tin existing along with pyrites ores of copper, iron and tin is however rarely extracted out of them. The extraction of metal is mainly carried out from tinstone, SnO_2 or *cassiterite* which occurs as placers deposits, i.e., alluvial deposits and as lode deposits, i.e., rocks containing grains of *cassiterite* in their veins.

Metallic tin is used in the manufacture of tin alloys and in solders, bronzes, bell metal, gun metal and phosphorus bronze etc. The most important use of tin is in the manufacture of tin plates.

Extraction

The extraction of metal tin from *cassiterite* or *tinstone* (SnO_2) which contains about 78% tin, is conducted through the following steps. It should be noted that tinstone is found mixed with various impurities, such as silicious matter (gangue), wolframite pyrites, arsenical pyrites, arsenic sulphide, antimony sulphide, chalcopyrites etc. It is essential to remove these impurities before the smelting of pure ore, otherwise much tin Fig. (20.7) would be lost as tin sulphate, tin iron alloy and tin silicate. The impurities are removed as follows:

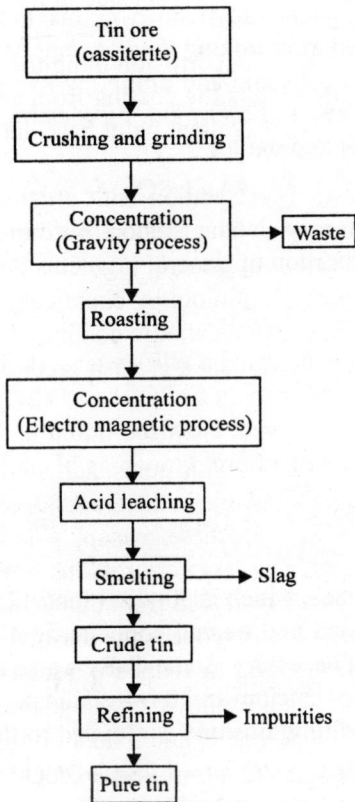

Fig. 20.7. General flow sheet for Tin extraction.

Concentration

Tinstones are crushed and powdered in ball mills and worthless gangue material is separated by washing in a stream of water. The heavier ore particles settle to the bottom while the lighter silicious material is washed away. This process is called gravity separation process in which powdered ore is washed in a current of water in Wilfley table. The placer deposits are concentrated by hydraulic mining followed by gravity separation by allowing the slurry to flow down inclined rifled troughs. The placer deposits have also been dredged and then concentrated. In case of lode deposits the ore is crushed and separated by gravity separation methods, such as Wilfley table, jig etc.

If the tinstone contains wolfarm as an impurity, it cannot be removed by gravity separation, because of the same density of tinstone and wolfarm. In this case, electromagnetic separating is resorted. It consists of a travelling endless belt moving over the electromagnetic rollers. The powdered ore is dropped over the travelling belt through the hopper. The moving belt carries the ore under two large electro magnets. The wolfarm being magnetic, falls in a separate heap near to the magnets, while SnO_2

remains unaffected. The ore, free from magnetic matter moves towards the roller at the end and falls into a different heap placed a little away from the magnet.

Roasting

The concentrated ore is roasted in a Dwight Lloyd sintering machine which is an inclined revolving furnace. The concentrated ore is roasted after mixing the concentrate with Na_2CO_3 or in a free supply of air when sulphur is removed as SO_2, arsenic and antimony as volatile As_2O_3 and Sb_2O_3. Cu_2S is converted into CuO and $CuSO_4$ and FeS_2 as Fe_2O_3. During roasting copper and iron pyrites are thus oxidised to the corresponding sulphates and oxides.

$$S + O_2 \longrightarrow SO_2 \qquad\qquad 4As + 3O_2 \longrightarrow As_4O_6 \text{ or } 2As_2O_3$$

Concentrates containing 25–30% tin may also be roasted in a current of hot air (without addition of Na_2CO_3) in order to double the concentration of tin.

Washing

The roasted ore is first cooled and then washed on the Wilfley table (a rifled vibrating inclined table) with a stream of water when soluble impurities such as $CuSO_4$ and $FeSO_4$ dissolve and lighter oxides like CuO, Fe_2O_3 are washed away and thus removed. Now the liquid is allowed to stand. Tinstone, being heavy, settles down at the bottom. The washed ore, known as black tin contains 60–70% of tin as tin oxide, SnO_2.

Smelting

Smelting is done in a reverberatory furnace, which is 30–40 ft long 12 ft wide and lined inside with fire bricks. The charge consists of the roasted and treated concentrate 4 parts, crushed anthracite or low ash coal one part, limestone and sand, if necessary. Some CaF_2 is also added. The aim of the addition of fluxes is to form a basic slag consisting of calcium and ferrous silicate, because the chief impurity in the concentrate is Fe_2O_3. As a result of smelting, tinstone is reduced to the metal.

$$SnO_2 + 2C \longrightarrow Sn + 2CO$$

Limestone or fluorspar is added as a flux only if silica is present as an impurity. Excess of flux should be avoided because it may react with tin to form calcium stonnate. The impurities are slagged off by the flux. The temperature of the furnace is maintained at about 1200–1300°C. The molten metal collected at the bottom of the furnace is drawn off and cast into ingots. Tin so obtained is called block tin.

Reactions

The chief reaction that takes place is the reduction of SnO_2 with carbon to form metal tin, CO and CO_2. Ferric oxide gets reduced to FeO and forms ferrous silicate slag. CaO formed from the decomposition of $CaCO_3$ (limestone) forms $CaSiO_3$ slag. Some FeO is reduced to Fe and forms Fe-Sn alloy with tin. Some SnO_2 is also reduced to SnO and the latter forms stannous silicate and enters the slag. It is therefore clear that a large amount of tin is wasted which passes to the slag during smelting.

The slag is therefore retreated to recover the tin. The slag is mixed with anthracite coal, limestone and CaF_2. The lime reacts with stannous silicate to form $CaSiO_3$ and SnO which is reduced to tin by carbon. This tin contains large proportion of tin iron alloy because of the formation of Fe from FeO as a result of reduction.

$$SnO_2 + 2C \longrightarrow Sn + 2CO \qquad\qquad SnO_2 + C \longrightarrow Sn + CO_2$$

$$Fe_2O_3 + CO \longrightarrow 2FeO + CO_2 \qquad\qquad FeO + CO \longrightarrow Fe + CO_2$$
$$FeO + C \longrightarrow Fe + CO \qquad\qquad\qquad CaCO_3 \longrightarrow CaO + CO_2$$
$$CaO + SiO_2 \longrightarrow CaSiO_3 \qquad\qquad\quad FeO + SiO_2 \longrightarrow FeSiO_3$$
$$SnO_2 + C \longrightarrow SnO + CO \qquad\qquad\quad SnO + SnO_2 \longrightarrow SnSiO_3$$

Smelting of the slag reactions:

$$SnSiO_3 + CaO \longrightarrow CaSiO_3 + SnO \qquad\quad SnO + C \longrightarrow Sn + CO$$
$$FeSiO_3 + CaO \longrightarrow CaSiO_3 + FeO \qquad\quad FeO + CO \longrightarrow Fe + CO_2$$

Refining

The block tin obtained from the reverberatory furnace is about 96.5% pure and contains iron, sulphur, arsenic and tungsten as chief impurities. The metal is refined to get pure metal in two stages.

Liquation

In this process bars of block tin are slowly melted by heating on a sloping hearth of a reverberatory furnace. The temperature is adjusted. As the m.p. of tin is low, it melts and flows down while impurities like iron, tungsten, along with some tin are left behind unmelted. First stage of liquation takes place just above the melting point of tin. After separation of tin, the residual impurities are heated to slightly higher temperature when some more tin liquefies and is collected separately. Arsenic and sulphur burn and volatilise off as oxides. Tin obtained by first liquation state is called pig tin which still contains some impurities. The process of liquation is again repeated in another similar reverberatory furnace when pure tin melts more readily at 232°C and flows down the hearth and is collected in a cast iron container.

Poling

The molten metal obtained by liquation process is next purified by poling process in which it is heated in a big pot called kettle (6–10 tonnes) much above the melting point of tin. At this point the molten metal is stirred with a green pole or logs of wood. Due to destructive distillation of the green wood torrents of gases and steam are produced. Any tin oxide left unreacted is now reduced by the hydrocarbons of the wood. Because of bubbling and boiling caused by the gases, tin globules are shot up in the air and as the particles of tin come in contact with oxygen of the air, the oxidisable constituents in the metal are oxidised as these impurities are brought to the surface by the evolving hydrocarbons. The oxides as impurities collect on the surface of the metal forming the dross which is skimmed off and pure metal is collected. It is 99.96% pure.

Electrolytic refining

Tin may be further refined by electrolysing a solution of stannous sulphate containing excess of sulphuric acid and cresol-phenol sulphuric acid (or hydrofluorosilicic acid). The anodes are made of block tin (95% crude tin plates) and cathodes are made of pure tin sheets. Operating temperature and current density are 35°C and 10 amp/sq ft respectively. When a current of electricity is passed, tin is dissolved from the block tin and an equivalent amount from the electrolyte is deposited on the cathode. Tin obtained is 99.98% pure.

Properties

Tin is a lustrous white metal which does not tarnish on exposure to air. Its atomic weight is 118.70. Its melting and boiling points are 232°C and 2260°C respectively. It is quite heavy metal and its sp. gravity is about 7.3. It exists in three allotropic forms—grey, white and rhombic. Tin is weakly electronegative and relatively less active because its reduction electrode potential is low. It does not tarnish in dry or moist air at ordinary temperatures. However, when heated strongly, it burns with a bright flame forming SnO_2.

It is hardly attacked by water at ordinary temperatures, because its normal potential does not much differ from that of hydrogen. However, steam reacts with molten tin, liberating hydrogen.

It is not attacked by organic acids. Thus it is used in tinning of cooking utensils. It is slowly attacked by dilute HCl and H_2SO_4.

Uses

1. Tin is used in the preparation of a large number of alloys such as solder, pewter, babbit metal, bell metal, type metal, bronze, gun metal, rose metal etc.
2. Tin is also used for the preparation of collapsable tubes for toothpaste and ointments.
3. Tin foils are used for wrapping cigarettes and other articles.
4. In the tinning of household utensils.
5. In tin plating sheets of iron and steel.

LEAD

Lead does not occur in nature as the native metal. The most important ore of lead from the point of view of its extraction is *galena*, which is the sulphide of lead (PbS). Other ores of secondary importance are sulphate of lead—*anglesite* ($PbSO_4$), *cerussite* ($PbCO_3$), and *matlokite* ($PbCl_2.PbO$).

Galena is a lead-grey coloured mineral. It is heavy and its sp. gravity is 7.5. Its hardness is 2.5–2.75. It contains 6–8% of lead and rest being sulphides of zinc and antimony and small amounts of silver (0.1%).

Extraction

Benification or concentration

The ore is first crushed and then ground in ball mill, with water (Fig. 20.8). The water carrying the suspended ground ore is then introduced into a classifier. The overflow (200 mesh) is concentrated by the froath floation process which consists in suspending the finely ground ore in water with a calculated amout of pine oil in a big vessel and the mixture is vigorously stirred with a blast of compressed air. The lead sulphide particles accumulate in the froath, while the gangue particles which are wetted by water sink to the bottom and are rejected. The froath that rises up carries the ore and is skimmed off. The coarse ore is re-ground and re-classified and then again subjected to froath floatation process. In place of pine oil, potassium ethyl xanthate and cresylic acid are also added for the same purpose. The *galena*, preferentially carried in the froath is collected by thickening and tailings are conditioned with $CuSO_4$ and mixed with more potassium ethyl xanthate and cresylic acid in another vessel and again treated with compressed air. As a result, ZnS is carried in the froath and ZnS concentrate collected by thickening. The composition of *galena* concentrate is Pb 74%, Zn 6.5%, Fe 1.5%, Sb 0.02%, S 16%, and Ag 25.3% oz per tonne. Lead is then extracted from the concentrate by two methods.

Self reduction process

This process is used for richer ores, i.e., for those ores in which lead content is high and no silver in the ore is present. In this process, the concentrate is roasted at moderate temperature in excess of air in a specially designed reverberatory furnace which is provided with various doors. The temperature of the furnace is controlled by regulating the supply of air through these doors. During roasting, *galena* (PbO) is partially oxidised to PbO and $PbSO_4$ by the oxygen of the air inside the furnace. A portion of PbS also remains unaltered. After about 4 hours of roasting, the air supply is cut off and rate of fuel burning is increased to raise the temperature. More of *galena* is also added before raising the temperature. At this temperature smelting operation takes place in which both PbO and $PbSO_4$ react with PbS, resulting in the formation of molten lead by self reduction process.

Roasting : $2PbS + 3O_2 \longrightarrow 2PbO + 2SO_2$ $PbS + 2O_2 \longrightarrow PbSO_4$

Smelting : $PbS + 2PbO \longrightarrow 3Pb + SO_2$ $PbSO_4 + PbS \longrightarrow 2Pb + 2SO_2$

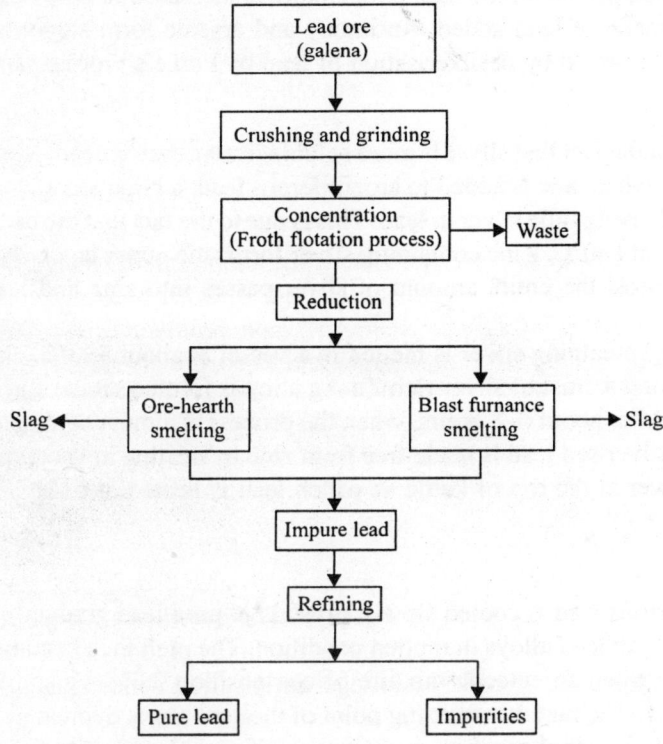

Fig. 20.8. General flow sheet for Lead extraction.

Carbon reduction process

In this process, roasting and smelting processes are performed separately and this process is useful when the ore contains less percentage of lead. Here roasting is done by blast roasting (not by hearth roasting) and smelting is done in blast furnace.

Refining

Lead obtained by above methods is hard and brittle due to the presence of impurities as such copper, silver, tin and iron. Removal of these impurities from lead is called softening of lead because pure lead is soft. A number of methods have been used for the purpose.

Liquation

The impure metal is heated on the sloping hearth of a furnace, when lead melts at a much lower temperature than many of the impurities and flows down the slope. The impurities remain sticking to the hearth at the temperature which is slightly greater than the m.p. of lead.

Oxidation or fire refining

The crude metal is taken in a kettle and heated to 380–400°C and puddled. The impure metal may also be melted on the hearth of a shallow flat bottomed reverberatory furnace and air is passed over the molten mass. The impurities of Cu, Sn and Fe are oxidised and form a scum or dross on the surface, from where it is skimmed off. Next, the bath temperature is increased to 600–650°C, compressed air is passed and 2 lbs per tonne of lead added. Antimony and arsenic form slag which is removed. After softening, the silver is removed by desilverisation of lead by Parke's process, and Pattinson's process.

Parke's process

This method is based on the fact that silver is more soluble in zinc than in lead. Molten lead is immiscible with molten zinc. Thus when zinc is added to argentiferous lead, a large amount of silver passes readily into solution with zinc leaving little silver in lead. This is due to the fact that the distribution ratio between zinc and lead is 300 : 1 at 880°C. Zinc containing silver forms the upper layer. By repeating the process three or four times, almost the entire amount of silver passes into zinc and lead becomes free from silver.

In this process lead containing silver is melted in a vessel at about 550°C. Some zinc is added and bath is stirred. On cooling a crust of silver rich Zn-Ag alloy is formed on the surface which is skimmed off from time to time. After about 6–8 hours, when the process is almost complete, lead is squeezed off from the crust. The desilverised lead is made free from zinc by heating in vacuum. The zinc is collected on a bell shaped receiver at the top of kettle in which lead is heated at 615°C. Lead thus obtained is 99.995% pure.

Pattinson's process

When molten argentiferous lead is cooled slowly, crystals of pure lead gradually separate out in stages leaving behind a silver rich lead alloys in molten condition. The melt thus becomes richer in silver until it contains 2.6% silver when an eutectic mixture of composition 2.6% Ag and 97.4% Pb solidifies at 303°C. This is because of the fact that freezing point of molten lead is depressed due to the presence of dissolved silver in lead. In actual practice, molten argentiferous lead is added to argentiferous lead of some composition in one of the iron pots in a row.

The separated lead crystals are transferred to a pot placed on the right hand side and the silver rich liquid alloys to the left of the iron pot in which cooling is done. The liquid alloy is further cooled to get more crystals. The liquid portion is transferred to left and crystals to right. In this manner, liquid richer and richer in silver travels to the left until it contains 2.6% Ag. At last pure lead accumulates in the last pot on the right.

Electrolytic refining

The desilverised lead may be further refined by the electrolysis of a solution of lead fluorosilicate ($PbSiF_6$) and hydrofluorosilicic acid (H_2SiF_6) with a little gelatine. Crude lead is made the anode and pure lead sheets are made the cathodes. The temperature of the bath is kept at about 35°C. Usually 20 anodes and 21 cathodes are used and electrically connected by multiple system. The electrolytic cell is a concrete lined with asphalt rectangular tank. Current density is 14–16 amp./sq ft. The process is known as Bett's process.

On passing the electric current, pure lead is deposited on the cathode, while the equivalent amount of lead passes into the solution from the anode. The less electropositive elements, such as antimony and copper, settle down as anode mud, while more electropositive metals, such as iron, tin etc., remain in the solution. Ag, Au, Bi, Sb etc., form an insoluble deposit below the anode as anode slime. Lead remains in the solution which is deposited on the cathode. Ag and Au are recovered by removing the base metals by oxidising and fusing on the hearth of a reverberatory furnace.

Properties

Lead is a soft grey metal which is bright and lustrous when freshly cut. It may be rolled and extruded through dies. On exposure to air, it becomes dull due to the deposition of a thin layer of oxide on it. It is malleable, ductile and marks black on paper. Its specific gravity is 11.34. Its m.p. and b.p. are 327. 4°C and 1760°C respectively.

Lead is not very reactive, because its reduction potential is low (–0.126 volts). Dry air has no action on lead. However, moist air tarnishes its surface because of the formation of thin layer of basic lead carbonate which protects it from further attack. When heated in air or oxygen, it is oxidised to litharge (PbO), but at higher temperature it is oxidised to red lead (Pb_3O_4).

It is not attacked by pure air free water, but steam is decomposed by it at high temperatures. Water containing dissolved air may, however, dissolve lead, forming water soluble lead hydroxide. Hence soft water should not be conveyed through lead pipes. This solvent effect of water is called plumbo-solvency. It is slightly affected by dilute HCl or dilute H_2SO_4 (below 78% strength). It is dissolved by alkalies with the formation of the corresponding plumbite.

Uses

The major uses of lead are:
1. In the manufacture of storage batteries.
2. In the manufacture of lead pigments such as red lead (Pb_3O_4) and litharge (PbO), chrome red, white lead, chrome yellow etc.
3. In the manufacture of lead chambers for the manufacture of H_2SO_4.
4. In the preparation of lead sulphide photo cells.
5. In making ammunitions such as bullets, shots etc.
6. It is also made into sheets and tubes and lead pipes are used in buildings.
7. In the manufacture of pipes and protective sheaths for telegraph and telephone wires which are to be buried under the earth.
8. In making a number of alloys, such as brass bronze, casting metal, type metal, soft solder, pewter etc.

CHAPTER 21

Corrosion and Its Prevention

INTRODUCTION

Most metals and alloys are attacked by oxygen, moisture and acids. Some are attacked by alkalies. The process of attack is called corrosion. Corrosion may result in a uniform attack which is generally not very serious; it may attack preferentially, particularly at grain boundaries, resulting in severe weakening of the metal without much visible deterioration; it may attack locally at areas where conditions are varied resulting in perforation; finally it may produce a passive oxide layer as on Cr and Al which gives protection from further corrosion. The rusting of iron is a familiar example of corrosion, which is catalysed by moisture. The phenomenon of metal corrosion is manifested by a heterogeneous chemical or electrochemical surface reaction, which results in a change of the metal to its oxidised state. The most important foundations of corrosion science are, therefore, based on two allied disciplines, namely those of metallurgy and physical chemistry. In the matter of corrosion prevention, knowledge of the properties of various protective coatings, organic and inorganic, also plays an important role. The mechanical and design aspects of corrosion prevention are also of no lesser importance. In addition, because of its tremendous economic significance the corrosion and its control at the present time, exceeds the limits of physical chemistry and physical metallurgy, and has become a particularly fascinating independent branch of science. Besides, it is one of the major problems of many industries, and the consequences of corrosion losses have been well recognised. Hence, corrosion control considerations are no longer subordinate to aspects like good operation and mechanical maintenance of plant and equipment. Corrosion is deterioration or decay occurring when a material reacts with its surroundings or the fluid being transported or contained.

CORROSION AND ITS VARIOUS FORMS

Corrosion of metals occurs in various forms. Most are electrochemical in nature. Their characteristics are well known, and generally the various forms are easily recognised. Extensive information is available regarding measures that have been and can be taken for minimising or preventing the various forms of corrosion in many of the specific environments encountered in industries. It is important to recognise the various forms of corrosion in order that the available information can be applied effectively to overcome corrosion problems. The following description and discussion should assist in identifying the various forms of corrosion and their general causes.

General Corrosion

Uniform thinning and loss of metal not accompanied by localised action such as pitting, cracking, or erosion may result from direct chemical reaction or combination such as that which occurs where metals are corroded in gases or by electrochemical action like that which occurs with metals and alloys corroding in liquid. Practically all metallic materials used in industry have sufficient heterogeneity or local point-to-point differences to produce minute anodic and cathodic areas in their exposed surfaces. Typical are inclusions, microstructural composition differences, localised stresses, and mechanical imperfections such as surface roughness. As the result of electrochemical reaction between these areas, corrosion is accelerated at anodic areas and retarded or completely prevented at cathodic areas, at least temporarily. The metal heterogeneity and resulting electrochemical corrosion cause a constant shifting of the anodic and cathodic areas in the exposed surfaces so that the overall corrosion appears to be uniform. The rate at which this type of corrosion occurs depends upon specific conditions involved and varies widely for the various materials.

General corrosion probably causes the greatest amount of damage to metals. It has been the subject of such extensive study and research that today it is fairly well understood and readily recognised and equipment life can generally be estimated accurately. In most cases it is possible to select and use resistant metals, coating, or linings to minimise or eliminate this form of corrosion. Where problems of general corrosion are encountered, consideration should also be given to the possibilities of decreasing corrosivity of the environment by reducing or changing temperature, pressure, velocity, and/or composition. In some instances it will be practical and economical to use inhibitors to solve the immediate and, perhaps, the long-range problem.

Localised Corrosion

Pitting and crevice corrosion are also electochemical in nature but result in loss of metal at localised anodic areas. The anodic areas occur in the case of pitting as the result of localised breakdown of a film on the surface or by mechanical or chemical action. In the case of crevice corrosion, the anodic areas generally occur because of differences in electrolyte composition in the crevices or shielded areas compared with the surrounding electrolyte. This form of corrosion is highly localised and general corrosion is frequently slight, even in instances where perforation of the metal occurs as a result of this cell-type action.

Pitting is the most serious of these localised cell types of corrosion because it can and frequently does occur very rapidly and may not be detected until failure occurs. The anodic areas do not change as in general corrosion and corrosion progresses at one localised spot. Pitting corrosion may occur in any metal, but the most common and spectacular occurrences are in aluminum and stainless alloys in aqueous environments containing metal chlorides. The amount of chlorides required is small; even a few parts per million may cause this type of corrosion. Where the chemical agents or conditions causing the localised breakdown and pitting cannot be eliminated, use of inhibitors should be considered as a possible means of correcting the problem. Treatment with dichromates is frequently effective where chlorides are the cause of the trouble.

Crevice corrosion and related types such as concentration cell, deposit, and contact corrosion generally occur where some of the electrolyte is confined or restricted in a small area such as under gaskets, bracket supports, or any solid in contact with the metal so that a crevice or pocket is formed. The electrolyte within the pocket changes in composition with respect to metal-iron concentration, oxygen, etc., with the result that a difference in potential occurs and the metal is preferentially corroded in the anodic area. Inhibitors generally are not effective under these conditions, and change in mechanical conditions to eliminate the crevice or change in materials is the most likely means of eliminating this type of corrosion. Difficulty at gasketed joints can be avoided or minimised by using non-porous gaskets.

Stress-Corrosion Cracking

Cracking often results from the combined effects of residual or applied stress and chemical action without noticeable loss of metal through uniform corrosion. This form is insidious since it generally occurs rapidly once action has started and is frequently not detected until failure occurs. It is usually preceded by a fine pitting, with the cracks starting in a pit. Time required for cracking may vary from a few minutes to a few years after initial exposure. It usually takes the form of transgranular penetration or cracking, although some alloys exhibit both transgranular and integranular stress–corrosion cracking. This has been observed in almost all metals or their alloys but each requires certain environmental conditions to produce this form of corrosion cracking. As in the case of pitting, there may be no readily apparent evidence of activity for a period of months or years before failure occurs. Typical examples include the season cracking of cold-formed brass in environments containing ammonia, the cracking of the austenitic strainless alloys in the presence of chlorides, the cracking of Monel in hydrofluosilicic acid, and the caustic-embrittlement cracking of steel in caustic solutions. This form of corrosion may in some instances be prevented by the elimination of high stress. Fabrication stresses including welding are the most frequent sources of trouble, and stress relieving or annealing after fabrication should always be considered where these metals are going to be exposed to these environments. Temperatures and concentration are important factors in all cases.

The presence of chlorides generally does not cause cracking of the austenitic stainless alloys where metal temperatures are below about 50°C., regardless of concentration. It has been found by extensive experience, however, that where metal temperatures are high enough to cause concentration of chlorides on the metal surface, cracking may occur even where chloride concentration in the surrounding media is only a few parts per million. Typical of such conditions and failures is the cracking of stainless heat-exchanger tubes in the crevices at rolled joints and under scale formed in the vapour space below the top tube sheet in vertical heat exchangers. The cracking of austenitic strainless-steel equipment under insulation where moisture causes leaching of chlorides and concentration on the hot surfaces is another example of failure due to stress-corrosion cracking.

Experience with handling caustic in steel indicates that, if the temperature is held to a maximum of about 120°F., as-welded steel equipment can be used without developing, stress-corrosion cracking. If the temperature is higher, and particularly if concentration is above about 30 per cent by weight, cracking at and adjacent to non-stress-relieved welds frequently occurs and the time to failure decreases with increase in temperature.

Corrosion fatigue is a type of stress corrosion that occurs under dynamic or alternating stress conditions in a corrosive environment. Because of the combined effects of cyclic stress and corrosion, cracking-type failures occur with stresses well below the normal fatigue limit.

Hydrogen Blistering and Cracking

Micro-cracking and blistering with loss of ductility are often caused by the entrance and diffusion of atomic hydrogen. The most frequent occurrence of this form of attack is in steel equipment handling solution containing hydrogen sulphide. Under these conditions corrosion of the steel generates atomic hydrogen which penetrates the steel and at submicroscopic discontinuities or voids changes to molecular hydrogen with development of pressures high enough to cause cracking or blistering. Steel plates, piping and forgings containing laminations frequently blister and high-tensile bolting in sulphide service frequently fails by cracking. The latter is a type of stress corrosion, and reduction of the applied tensile stress will minimise such failures. The occurrence of the cracking phenomena increases with increase in hardness and reductions in ductility and is attributed to the inability of the steel to deform or move to accommodate the high and descriptive forces that occur at the discontinuities within the steel. Where steel is used for handling solutions containing hydrogen sulphide, amine-type inhibitors have been found effective in preventing corrosion which could cause cracking and blistering.

Similar phenomena are seen in the cracking failure of hardened-tool-steel items that have been electroplated. In this instance the difficulty can be effectively avoided by removing hydrogen by low-temperature heat-treatment, immediately following the electroplating operation.

Hydrogen attack of steel may also occur under dry conditions at elevated temperatures and pressures, but this is not generally considered to be due to corrosion. It is believed to be caused by the reaction of hydrogen with iron carbides in the steel and the resulting formation of methane. Resistance to this type of attack in steel increases with chromium because of the increased stability of chromium carbides.

Intergranular Corrosion

Localised electrochemical attack occurs and progresses preferentially along the grain boundaries of an alloy, usually because the grain boundary regions contain material that is anodic to the central region of the grains. This type of attack may penetrate completely through the metal section with essentially complete loss of a strength, although the apparent general attack may be slight. Many alloys are susceptible to this form of corrosion under specific conditions, but materials that are most frequently involved are the austenitic stainless steels, high-nickel alloys, and aluminum alloys. This form of corrosion was a common occurrence in austenitic stainless-steel equipment used in acid services 15 to 20 years ago before the effects of carbide precipitation were understood. It is less of a problem today because of the reduced carbon content of the alloys and an understanding of its cause and means of prevention. It is now generally accepted in the case of the austenitic stainless steels that some of the chromium combines with the carbon to form chromium carbide which is precipitated at the grain boundaries when the alloy is heated or cooled slowly through the range of 800° to 1500° F. The rate and extent of the formation of chromium carbide are a function of time, temperature, and carbon content. It occurs during welding in the base metal adjacent to the deposited metal. As a result of the localised impoverishment of chromium at the grain boundaries, preferential corrosion may occur at the grain boundaries in some acidic environments. This form of corrosion can be prevented in stainless steel by annealing (heating at 1950°C to 2050°F. followed by rapid cooling) after welding operations, by the use of columbium or titanium-stabilised grades of stainless steel or by the use of the low-carbon (0.03 per cent maximum) grades of stainless steel. Critical amounts of chromium carbides do not precipitate at the grain boundaries in the stabilised or low-carbon grades during welding; hence annealing after welding is not required as a precaution against this form of corrosion.

Galvanic Corrosion

The more noble of two metals in contact in an electrolyte causes electrochemical attack of the less noble metal. Perhaps the best example is the use of zinc to protect steel equipment. The zinc, being anodic or less noble than the steel, corrodes in most water and atmospheric exposure conditions while the steel being cathodic and more noble than the zinc is protected. This protection of the steel is obtained as a result of sacrificial corrosion of the zinc. The extent or severity of galvanic corrosion depends not only on the difference in potential of the two metals but also upon the relative surface areas involved. If the area of the cathodic metal is large compared with that of the anodic metal, the extent of galvanic or accelerated corrosion of the anodic metal will be much greater than it would be if the relative areas were reversed. Bronse valves and fittings are commonly used in steel water lines without much difficulty due to localised corrosion of the steel at the fitting. But when a steel fitting is used in a copper line, relatively rapid corrosion of the steel fitting generally occurs. Although galvanic corrosion is frequently troublesome, the principle of galvanic corrosion is used to advantage to protect a large amount of equipment. Typical of this in addition to galvanising is the use of magnesium and zinc anodes to protect burred steel pipe lines.

Selective Corrosion

Removal of a constituent of an alloy by corrosion without apparent loss in volume but with serious loss of strength is called 'selective corrosion'. Dezincification is typical of this form where the zinc constituent present in brass alloys is selectively removed, leaving only sponge copper in the original shape and volume of the uncorroded metal. This corrosion may be general or it may occur in localised areas. The latter is called 'plug-type' dezincification. This effect is generally limited to brass compositions containing more than 15 per cent of zinc. The addition of small amounts of arsenic,' phosphorus, or antimony to the alloy will generally inhibit, if not prevent, this form of corrosion.

Another example is the graphitic corrosion of cast iron where galvanic corrosion occurs between the graphite and iron constituents in the cast iron, with the resultant formation of a spongy mass of corrosion products. The original shape and volume are generally retained but the strength of the metal is essentially lost as the corrosion proceeds through the cross section. This is rather common in buried cast-iron lines, particularly where moist and slightly acidic conditions prevail.

Erosion corrosion

Accelerated corrosion may result from erosion which removes normally protective films. This is most likely to occur in liquid systems that contain solids. Items such as pump impellers, agitator blades, and pipe-line fittings are particularly subject to this form of corrosion. It can be minimised by reducing velocities, by changes in environment by addition of an inhibitor to reduce corrosiveness of the solution by the use of more corrosion-resistant materials, and in some instances, by the use of harder material.

Impingement is a type of corrosion similar to erosion corrosion except solids are seldom involved and specific effects are more localised. It occurs where the normally protective film is not maintained in the flowing fluid or gas. Perhaps the most common occurrence of this form of corrosion is in non-ferrous alloy condenser tubes near the inlet ends. A reduction in velocity or turbulence will prevent or minimise this difficulty. It can be overcome also by use of alloy materials having a greater resistance to this type of attack. The use of ferrules of plastic or alloy materials inserted in the inlet ends of condenser tubes has been effective where the attack was confined to an area near the inlet end.

Cavitation may also be considered as a type of corrosion by erosion even though it may occur under essentially non-corrosive conditions. It occurs in metals handling moving liquids and results specifically from the formation and collapse of cavities or vapour bubbles in contact with the metal surface and the associated pounding and wiping action at the point of cavity or bubble collapse. Its appearance varies from that of surface roughness in strong ductile materials to one of pitting in low-strength non-ductile materials like cast iron. It is a common occurrence in impellers of all types. Where the temperature pressure conditions cannot be changed to reduce bubble formation, change in alloys is generally required as a corrective means.

Fretting Corrosion

Corrosion between two surfaces accelerated by or resulting from the mechanical removal of a protective film of corrosion products is called 'frettizng' or 'chafing' corrosion. This occurs at the interface of metals when they are clamped or fitted closely together and subjected to small vibratory motions. Its most common occurrence is on machine parts with small relative motions and high unit loads. Use of rust-inhibiting oil or grease is generally beneficial under atmospheric exposure conditions.

High Temperature Corrosion

Chemical reaction of metals at elevated temperatures with one or more constituents of a gaseous environment often results in corrosion. The most common type is oxidation where metal oxides form by chemical reactions and losses occur in the form of scaling. Rate of metal loss under oxidising conditions increases with increase in temperature and generally is greater under cyclic fluctuations in temperature because of loosening of scale and loss of its protective value. Chromium is considered to be the most beneficial alloying addition for increasing oxidation resistance. The minimum content is 20 per cent for good high-temperature oxidation resistance. Since nickel is also quite beneficial in combination with chromium, most of the alloys designed for high-temperature use contain 20 per cent or more of this element. Another type of high-temperature corrosion is sulphidation, where at elevated temperatures sulphur bearing atmospheres cause an intergranular attack and penetration along the grain boundaries. This is limited for the most part to nickel and alloys of high nickel content. The rate of attack is some what greater under reducing conditions than under oxidising conditions. Nickel is subject to sulphidation attack in sulphur-bearing and reducing atmospheres at temperatures above 600°F. The addition of chromium greatly improves resistance of nickel alloys to sulphur-bearing gases.

Biological Corrosion

Deterioration of steel or iron may occur directly or indirectly as a result of the metabolic activity of micro-organisms. The anaerobic sulphate reducing bacteria are most frequently involved in this form of corrosion. They contribute to corrosion by affecting changes in surface film resistance, creation of corrosive environments, or creation of a surface barrier so as to cause concentration cell or deposit type corrosion. After removal of the corrosion products, the appearance is generally that of pitting, either isolated or overlapping in stringer effects. The most common occurrence is in clay or boggy soils. Coatings are generally used as a preventive means where biological corrosion is suspected.

Factors influencing corrosion

A number of factors influence the occurrence and rate of corrosion, the more important being temperature, velocity, pH, oxidising and reducing conditions, and moisture. As a general rule the rate of corrosion increases with increase in temperature, the extent depending upon specific conditions. The increase in rate of corrosion with increase in temperature is, for instance, not so great for steel in alkaline environments as it is for steel in acid environments. In a few instances, however, a decrease in temperature will increase corrosion due to a change in environmental conditions. An example of this is in flue coolers where steel is exposed to acidic gases which upon cooling reach the dew point, with the resulting condensation causing more corrosion than occurred with the gases at the higher temperature.

Velocity normally increases the rate of corrosion, the amount again depending upon specific conditions. This may be due to removal of protective films or scale or simply to the continued supply of corroding media. Impingement corrosion previously mentioned is caused by velocity effects. Pitting, on the other hand, is most severe under quiescent conditions. It may not develop in parts of equipment where flow conditions exist even though severe pitting occurs in areas of low or negligible velocity.

The pH value of a solution is not a controlling factor with respect to corrosion of metals, although the relative rate of corrosion is a function of the solubility of corrosion films or products and this in turn may be a function of pH value. As a general rule, decrease in pH value will result in increase in corrosion of metals subject to corrosion in dilute acid solutions and an increase in pH value above the neutral range will result in increase in corrosion of metals subject to attack in alkaline solutions.

Oxidising conditions are generally favourable for the stainless steels, which owe their corrosion resistance largely to the existence of a passive oxide film on the surface. The corrosion of many of the non-ferrous metals, however, particularly copper and nickel and their alloys, is greater under oxidsing-acid conditions than under neutral or reducing-acid conditions. It is frequently possible to decrease corrosion of stainless steels in acid solutions by aeration or addition of an oxidising agent and similarly it is possible to decrease corrosion of copper and nickel in acid solutions by inert-gas purging or blanketing to maintain neutral or reducing conditions. Corrosion of ferrous-base materials is generally greater under oxidising than under reducing conditions. Moisture may be a critical factor, particularly where its presence results in change from anhydrous conditions and acidic components are formed by hydrolysis. It is also important in gaseous systems since its presence can cause condensation and a severe corrosion condition. In general, the occurrence or presence of moisture in an environment will cause an increase in corrosion compared with a moisture-free condition.

COMBATING CORROSION

The following items should be considered as means of combating corrosion.

Select Proper Material

Selection of materials for process equipment should preferably be based on experience with materials under similar conditions. Where such experience or information is not available it is desirable to obtain advice and assistance of persons experienced in corrosion engineering work who are familiar with the chemical resistance characteristics and limitations of construction materials. Corrosion tests may be necessary to develop information to permit selection of adequate materials. Where such tests are made, procedures and conditions should be as realistic as practical to simulate operating and exposure conditions with respect to temperature, aeration, and velocity as well as chemical composition. Hot or cold wall effects should be considered where heat-exchange equipment is involved. The possibility of the occurrence of one of the forms of corrosion other than general corrosion should be considered. Tests should then be designed and made to determine suitability of the materials of interest where they are known or suspected to be susceptible to a particular type of corrosion under similar environmental conditions.

In selecting materials of construction for equipment which would be exposed to corrosive environments, careful consideration should be given to the practical aspects and limitations of design, friction, installation or maintenance. For instance, if field welding or cold working must be done on equipment for installation or maintenance reasons, and subsequent heat-treatment is not practical, then it is necessary to select a material of construction not susceptible to attack such as intergranular corrosion or stress-corrosion cracking in the non-heat-treated condition.

Permissible corrosion rates are an important factor and will vary with equipment. Appreciable corrosion can be permitted for tanks and lines if anticipated and allowed for in design thickness, but essentially no corrosion can be permitted in fine-mesh wire screens, orifices, and other items where small changes in dimensions are critical. In many instances use of non-metallic materials will prove to be attractive from an economic and performance standpoint and they should be considered where their strength, temperature, and design limitations are satisfactory.

Proper Design

Design considerations, with respect to minimising corrosion difficulties, should include the desirability for free and complete drainage, elimination of crevices, and ease of cleaning and inspection. The installation of baffles, stiffeners, and drain nozzles and the location of valves and pumps should be made so that free drainage will occur and washing can be accomplished without hold-up. Means of access for inspection and maintenance should be provided wherever practical. Butt joints should be used wherever possible. If lap joints employing fillet welds are used, the welds should be continuous on the process side. The use of dissimilar metals in contact with each other should generally be avoided, particularly if they are widely separated in their nominal positions in the electromotive series. If they are to be used together, consideration should be given to insulating them from each other or making anodic material area as large as possible. Equipment should be supported in such a way that it will not rest in pools of liquid or on damp insulating material. Porous insulation should be weatherproofed or otherwise protected from moisture and spills to avoid contact of the wet material with the equipment. Specifications should be sufficiently complete to ensure that the desired composition or type of materials will be used and the right condition of heat-treatment and surface finish will be provided. Inspection during fabrication and prior to acceptance is desirable.

Alter Environment

Simple changes in environment may make an appreciable difference in corrosion of metals and should be considered as a means of combating corrosion. Oxygen is an important factor and its removal or addition may cause marked changes in corrosion. The treatment of boiler feed water, for instance, to remove oxygen greatly reduces the corrosiveness of the water on steel. Inert-gas purging and blanketing of many solutions, particularly acidic media, generally minimise corrosion of copper and nickel-based alloys by minimising air or oxygen content. Corrosiveness of acid media to stainless alloys, on the other hand, may be reduced by aeration because of the formation of passive oxide films.

Reduction in temperature will almost always be beneficial with respect to reducing corrosion. Reduction in velocity and turbulence will generally result in reduced corrosion, an exception being where solids may collect on surfaces and cause pitting. Where pH values can be modified it will generally be beneficial to hold acid level to a minimum. Where acid additions are made in batch processes, it may be beneficial to add them last so as to obtain maximum dilution and minimum acid concentration and exposure time. Alkaline pH values are less critical than acid values with respect to controlling corrosion. Elimination of moisture can and frequently does minimise, if not prevent, corrosion of metals, and this possibility of environmental alteration should always be considered.

Inhibitors

The use of various substances or inhibitors as additives to corrosive environments to decrease corrosion of metal in the environment is an important means of combating corrosion. This is generally most attractive in closed or recirculating systems where annual cost of inhibitor is low. However, it has also proved to be economically attractive for many once-through systems, such as those encountered in petroleum-processing operations. Inhibitors are effective as the result of their controlling influence on the cathode or anode-area reactions.

Typical examples of inhibitors used for minimising corrosion of iron and steel in aqueous solutions are the chromites, phosphates, and silicates. These minimise corrosion by increasing anodic polarisation and are called anodic inhibitors. Organic sulphide and amine materials are frequently effective in minimising corrosion of iron and steel in acid solution. In this instance they control cathodic polarisation and are called cathodic inhibitors

The use of inhibitors is not limited to controlling corrosion of iron and steel. They frequently are effective with stainless steel and other alloy materials. The addition of copper sulphate to dilute sulphuric acid will, for example, essentially stop the corrosion of stainless steels in hot dilute solutions of this acid whereas the uninhibited acid causes rapid corrosion.

The effectiveness of a given inhibitor generally increases with increase in concentration, but those considered practical and economically attractive are used in quantities of less than 0.1 per cent by weight.

In some instances the amount of inhibitor present is critical in that a deficiency may result in localised or pitting attack with the overall results being more destructive than where none of the inhibitor is present. Consideration for use of inhibitors should therefore include review of experience in similar systems or investigation of requirements and limitations in new systems.

Cathodic Protection

Two methods of providing cathodic protection for minimising corrosion of metals are in use today. These are the sacrificial-anode method and the impressed-e.m.f. method. Both depend upon making the metal to be protected the cathode in the electrolyte involved. Examples of the sacrificial-anode method include the use of zinc, magnesium or aluminium as anodes in electrical contact with the metal to be protected. These may be anodes buried in the ground for protection of underground pipe lines or as attachments to surfaces of equipment such as condenser water boxes or on ship's hulls. The current required is generated in this method by corrosion of the sacrificial-anode material. In the case of the impressed e.m.f., the direct current is provided by external sources and is a passed through the system by use of essentially non-sacrificial anodes such as carbon, non-corrodible alloys or platinum buried in the ground or suspended in the electrolyte in the case of aqueous system. The requirements with respect to current distribution and anode placement vary with resistivity of soils or electrolyte involved, and the assistance of experienced personnel should be obtained when considering possible application of cathodic protection. The practical problem increases as the resistivity of the electrolyte decreases. It is generally impractical, for instance, to use cathodic protection where dilute acids are involved as the corroding solution or electrolyte. It is also impractical where the dimensions or configuration of equipment to be protected are such that adequate placement of anodes cannot be provided; e.g., heat-exchanger tube bundles.

Coatings and Linings

The use of non-metallic coatings and lining materials in combination with steel or other materials has and will continue to be an important type of construction for combating corrosion.

Organic coatings of many kinds are used as linings in equipment such as tanks, piping, pumping lines, and shipping containers and they are often an economical means of controlling corrosion, particularly where freedom from metal contamination is the principal objective. One principle that is now generally accepted is that thin brush or spray applied non-reinforced paint like coatings of less than 10 mils thickness should not be used in services where full protection is required in order to prevent rapid attack of the substrate metal. This is because most thin coatings as commercially applied contain defects or holidays and these lead to early failures due to corrosion of the substrate metal, even though the coating material is resistant. Spark testing of coating-type lining is always desirable for immersion-service applications in order to detect holiday type defects in the coating.

The most dependable linings for corrosive services are those which are bonded directly to the substrate and are built up on multiple-layer or laminated effects to thicknesses greater then 100 mils. These include the glass fibre reinforced resin systems and the elastomeric and plasticised plastic systems. Good surface preparation and thorough inspections of completed lining, including spark testing, should be considered as minimum requirements for any lining applications.

Ceramic or carbon-brick linings are frequently used as a facing lining over plastic or membrane linings where the service temperatures exceed those which can be handled by the unprotected materials. This type of construction permits processing of materials that are too corrosive for handling in metal constructions.

Metal and glass-lined steel has been available for some time as a method for avoiding the high cost of a solid alloy. Nickel, for instance, can be deposited by electroplating or by electroless processes. Such deposits, however, tend to be porous. Steel clad with an alloy is another approach to the problem and a number of large process plants have numerous pieces of equipment made from stainless-clad steel. As for glassed steel, it has all the benefits of glass plus high strength.

SCOPE OF TUNGSTATE AND MOLYBDATE AS EFFECTIVE CO-INHIBITORS FOR COOLING WATER SYSTEMS

The use of corrosion inhibitors is one of the foremost methods of combating corrosion in cooling water systems. The corrosion engineer should identify the problems precisely and explore the use of a right inhibitor combination. He should consider the economics and compatibility of the inhibitors with the processes and should apply the inhibitor under the conditions that produce maximum efficiency under feasible condition.

Toxicity of chromates has restricted its use in recent years. Based on the similarity in chemical structure and periodicity between chromate and other group VI ions, researchers have focused their attention on molybdates and tungstates. The first published information on corrosion inhibition by tungstate and molybdate appeared as a patent describing its use in organic antifreeze solution.

The first reported experiment on corrosion inhibition by these inhibitors was carried out by *Robertson*. Many studies are conducted on tungstate and molybdate as they are most appropriate for cooling water systems. These two inhibitors were also reported to be effective in various other applications like in engine coolants, paint pigments, conversion coatings, boiler water, aqueous slurries, batteries, etc.

This section outline the current research trends on corrosion inhibitor and also examines the validity of the claims and the gap between academic claims and their industrial applications. The emergence of tungstate, molybdate and HEDP (1-Hydroxy ethane-1, 1-diphosphonic acid) based inhibitor combinations as potential co-inhibitors for widespread applications are also highlighted here.

Mechanism of Inhibition

The mechanism of inhibition by tungstate and molybdate, the inhibitors that are highlighted are found to be associated with the formation of γFe_2O_3. In near neutral solutions, the corrosion process of metals results in the formation of sparingly soluble surface products such as oxides, hydroxides or salts. The inhibitors increases the protection qualities of the oxide surface or other surface layers from the aggressive anions. The fundamental step involves the displacement of pre-adsorbed water molecules by the inhibitor, followed by chemical or electrochemical reaction at the surface.

According to *Robertson*, the mechanism of inhibition does not involve oxidation of the Fe^{2+} to Fe^{3+}, but rather involves adsorption of MoO_4^{2-}/WO_4^{2-} on the metal surface to form a metal-inhibitor complex passive film. *Pryor* and *Cohen* suggested a uniform mechanism for chromate, molybdate, and tungstate. They predicted the initial formation of γFe_2O_3 film due to the oxidation of already formed Fe_3O_4 with water becomes thicker by further oxidation of Fe^{2+} to Fe^{3+}, irrespective of the nature of the inhibitor added. *Myrolyubou* predicted that passivation begins immediately upon immersion of the iron sample in the inhibitor solution.

Both molybdate and tungstate require sufficient dissolved oxygen for their effective inhibition. *Abd Kadher* in his recent studies showed that inhibition by tungstate requires its simultaneous presence with oxygen in the solution and formation of an orderly arrangement of these two species on the metal surface. The adsorption of oxygen is very strong when the inhibitor co-exists with it, eventhough both of them are held in the lattice by physical forces. Adsorption of these ions increased with increase in original inhibitor solution. A threshold concentration of the inhibitor should be maintained in the electrolyte to ensure complete passivation and to avoid any localised attack. The threshold concentration may vary depending on the nature of the system. Also the presence of excess inhibitor does not bring any effect either in the strength or thickness of the passive film. However, a little higher concentration above the threshold concentration is needed for effective protection. It is not feasible to use tungstate or molybdate alone as a corrosion inhibitor, due to its low oxidising ability and high cost. The successful formulations combine these inhibitors with one or more other inhibitors. Thus it is suggested that there is a potential scope of tungstate and molybdate to emerge as efficient co-inhibitors for effective protection of steel, aluminium and copper in neutral and alkaline media.

Criteria of Selection of the Right Inhibitor

The objective of cooling water treatment is prevention of corrosion, scaling and fouling of coolers and condensers to increase the heat transfer efficiency and material and energy conservation. Since corrosion and other related problems cannot be eliminated, but perhaps minimised, by adsorption of appropriate design and selection of suitable materials, the developments in cooling water additives (corrosion inhibitors, biocides, antiscaling agents and dispersants) become significant. After choosing the right treatment procedures, the next step is to ensure proper application of the chemicals and timely response to abnormal events. Fig. 21.1 sketches the process of selection of inhibitors formulation.

Process parameters

The corrosivity of water is significantly influenced by concentration of the dissolved species including gases (CO_2, O_2, NH_3 etc.), pH, temperature, suspended matter and bacteria. An inhibitor should be effective under a wide range of conditions of temperature, pH, heat flux, flow conditions and water quality. It should not produce deposits on the metal surface. It is better if the inhibitor can prevent formation of scales and has the ability to combat biological activity.

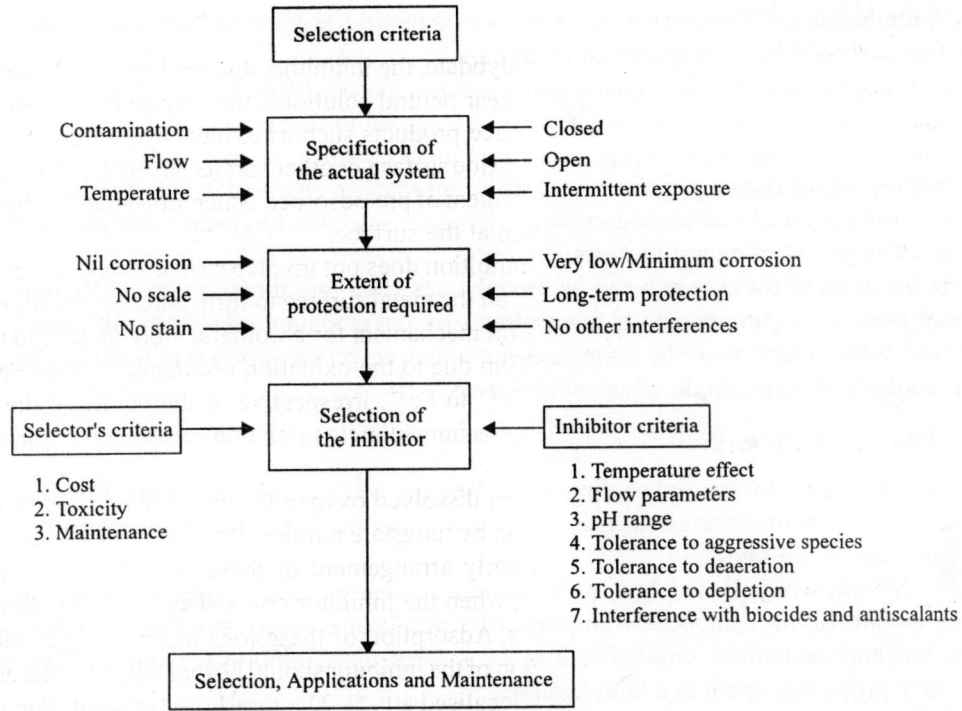

Fig. 21.1. The process of selection of the right inhibitor combination for a specific system.

Cost

The inhibitor should be attractive in terms of the cost factor. The inhibitor, actually, brings benefits in terms of reduced cost of operation and maintenance. Using cheaper material with effective inhibitors is better than using costly corrosion resistant materials without inhibitors.

Toxicity and disposal

There has been increasing concern about the toxicity, biodegradability and bio-accumulation of inhibitors discharged into the environment, considering the importance of marine life and preservation of eco-system. An inhibitors should be safe in terms of its non-toxicity for disposal. Standardised environmental testing protocols are being developed.

Tolerance and maintenance

An inhibitor should be effective for longer durations. It should be capable for self-healing of any film damages, besides effectively tolerating aggressive ions. An inhibitor should maintain the protective film even under adverse circumstances that are generated unintentionally like depletion, deaeration and minor contamination of aggressive species, etc. of the inhibitor solution.

Current Research Trends

With the availability of sophisticated modern computerised electrochemical and surface analytical instrumentations, corrosion studies were directed more towards probing mechanism of corrosion inhibition processes. A combination of electrochemical studies with modern surface analysis techniques is essential for elucidating the mechanism of corrosion inhibition.

The methods like electrochemical quartz microbalance, electrochemical noise analysis, SEM, ESCA and various spectroscopic analyses are applied for studying inhibition processes. Development of semi-empirical criteria such as hydrophobicity, use of Hamment parameters, quantum chemical calculations and the Hansch equation also emphasise the scientific approaches involved in the selection of inhibitors.

With the development of electronic computers, inhibitor selection has been made accessible to users through the development of expert systems.

However the areas of the research can be broadly classified into three categories, for convenience. A substantial portion of current reports concentrates on development of already developed inhibitor based systems. While a section of the current research focuses on substitution of toxic inhibitors, many reports are available on exploration of newer inhibitors.

Renewal of the old inhibitors

The cheapest corrosion inhibitor for cooling tower systems was calcium carbonate. Probably the most effective corrosion inhibitor for cooling water systems was the chromate. Phosphates represent the second major class of inhibitors.

Another inhibitor, which is widely used, is sodium silicate. Organic inhibitors are not used to any appreciable extent. Some compounds employed before 1960's were chromoglucosates, glycerine derivatives, polyamines, tannins, emulsified oils, etc. Among these inhibitors, sodium silicate possesses many attractive properties and it is a better choice.

From very earlier times, multi-component inhibitor combinations have been used for better results. Silicate-phosphate combinations are one such example. 2-Mercaptobenzothiazole was also used as a very effective corrosion inhibitor for copper from earlier time itself.

Substitute of toxic inhibitors

The most common inhibitors, such as chromates and nitrites in aqueous systems, are now considered unsafe from an environmental point of view and their use is banned by government agencies. Hence, studies were focused on development of alternates including molybdates and tungstates. They are less toxic than chromate and eco-friendly in nature.

They prevent corrosion of copper and aluminium also. They can tolerate chloride and sulphate in cooling waters and can work without polymeric dispersant. Most reports show that these anions are very effective, especially as co-inhibitors with other organic/inorganic inhibitors even at very low concentrations.

Exploration of other inhibitors

Among the organic inhibitors that have been recently reported, phosphonates, polycarboxylates, amide derivatives and azole derivatives seem to be important.

Organophosphonates have better hydrolytic stability and they are stable over a wide range of pH and operating temperatures. They are good deposit control agents and chelating agents. Widely used phosphonates include HEDP and ATMP.

There are several polycarboxylic acids of low molecular weight, which are used as corrosion inhibitors in many inhibitors formulations. A polyacrylic acid/polyacrylamide mixture system effectively inhibits corrosion of steel and their effectiveness is attributed to the formation of a polymer-polymer complex.

Few reports are available on amide and thioamide derivatives. Other organic compounds used widely are triazoles, tetrazolium salts, calcium gluconate and sodium lingosulphate. Reports are also available on natural products such as curcumin, cordia latifolia, etc. as inhibitors. Many other surface-active substances and inorganic and organic complexes are also reported as inhibitors.

Exploration of new organic compounds, polymers and complexes, as corrosion inhibitors, is needed. But exploration of entirely new type of inhibitor systems cannot find immediate feasible commercial application. Also lack of comprehensive evaluation by academicians is a challenging problem in such cases.

Claims and Validity

Most academic reports are confined to conventional evaluation techniques and not extended towards actual or simulated systems, including parameters like extent of protection, long-term effect, tolerance to contaminations, effect of unintentional depletion and deaeration, etc.

Based on research experience with many inhibitor systems, it is believed that there is a tremendous scope of exploration of very effective inhibitors like HEDP as co-inhibitors with MoO_4^{2-}/WO_4^{2-}. In this similar approach, newer inhibitors can also be synthesised, explored and evaluated for formulation of effective MoO_4^{2-}/WO_4^{2-} based inhibitor combinations.

Numerous natural products have been explored and evaluated as corrosion inhibitors. However the evaluation of such inhibitors under wide conditions and their co-inhibition performance are found lacking because natural products will be effective as co-inhibitors with MoO_4^{2-}/WO_4^{2-}, though they are not good enough when used alone.

Gap between Academicians and Industrialists

There is much gap between academic claims and their application in industries. In certain cases, it is seemed to be a gap of 30–40 years from the time of exploration and the time of actual application. Same is the case with tungstate and molybdate.

While the extent of inhibition is the prime objective of many investigators, industrialists view things in a different way and are more concerned about the conditions under which the results are supposed to be valid.

Hence studies that were carried out under simulated and wider experimental conditions receive much attention than those were conducted in the other way.

Need of the Hour

Highly efficient and safe inhibitor formulations having high corrosion inhibition is the need of the hour. Elaborative studies with molybdate/tungstate, especially as co-inhibitors with other organic/inorganic inhibitors needs attention. Academicians should comprehensively evaluate the inhibitor formulations, so that they can have industrial validity.

Not only claims, but also solid proof is required. Potential and sharp development of effective combinations of MoO_4^{2-}/WO_4^{2-} with other inhibitors will reach the industry soon. Though many organic compounds have been reported as inhibitors, it is more useful to pay attention to inorganic inhibitors. 1-Hydroxy ethane-1, 1-diphosphonic acid (HEDP) is supposed to be in the front line as co-inhibitors with MoO_4^{2-}/WO_4^{2-}. Based on the review it is predicted that HEDP included molybdate/tungstate inhibitor combination are going to the recognised as the most suitable and efficient combination for wide industrial application.

Thus this section highlighted the criteria of selection of the right inhibitor for the specific system. The current research trend on corrosion inhibitors is outlined. The gap between academic claims and their industrial applications has been discussed, highlighting the validity of academic claims. Comprehensive evaluation by the academicians rather than narrow investigation on the extent of inhibition efficiency is the need of the time.

It has been substantiated that the scope of any inhibitor to find immediate industrial use not only depends on its individual efficiency, but also on many other factors. Reviewing the current research trends and claims, the section also highlighted the potential scope of HEDP and molybdate/tungstate based inhibitor combinations that emerge as efficient and most preferable inhibitor system. These are going to be the first choice of the industry in near future.

CORROSION IN INDUSTRIES

Corrosion in Boiler Plant

An increasing awareness of the need for the more efficient use of fuel has in the past two decades led to the development of steam plants of greatly increased size, operating at increased steam pressure. In water-tube-boilers steam is generated on the surface of the tubes by the hot gases and flames produced by the combustion of the fuel. The steam travels through the tubes and is delivered to the boiler drums.

High-pressure boilers which drive turbines operate on a recirculatory system, and only make use of so-called 'make-up' quantities of water, which are very carefully treated by means of modern ion-exchange methods. Low-pressure boilers, particularly those that provide either hot water or process steam, use less well treated water. The major boiler troubles, due to the use of unsuitable water, may be classified as: (i) corrosion; (ii) scale formation; and (iii) caustic embrittlement.

Corrosion

Corrosion is one of the most serious problems created by the use of untreated water. Boiler tubes, economisers, superheaters and condensers are the most affected parts. The corrosion problem extends even to parts which are not directly in contact with boiler water because gases like O_2 and CO_2 are released during heating of water which have corrosive effects. Hence in studying the corrosion phenomena in a boiler installation, it is necessary to distinguish between two entirely separate aspects:

1. *Corrosion on the water-side of the boiler:* Water used for steam raising often contains dissolved solid and gaseous impurities. These can cause scaling and corrosion in the boiler plant. Corrosion can be attributed to the following:
   (a) Dissolved oxygen
   (b) Mineral acids
   (c) Dissolved carbon dioxide
   (d) Galvanic cell formation
2. *Dissolved oxygen:* Oxygen dissolved in water is mainly responsible for corrosion in boilers. Accordingly, the higher the pressure, the greater will be the dissolved oxygen content, and the higher the temperature the lesser the oxygen content. The amount of oxygen dissolved is also influenced by the other dissolved matters present, solubility of oxygen becoming less in water containing dissolved matters. In a boiler, oxygen is introduced through the raw make-up water supplied through it and also through the infiltration of air into the condensate system. As the water is heated in the boiler the dissolved oxygen is liberated and iron is corroded.

Corrosion inside a boiler can only be restricted by keeping oxygen concentration very low; this stops the cathodic reaction from taking place. The dissolved oxygen can be removed by:

1. *Mechanical deaeration methods* using (i) distillation; (ii) steam scrubbing; (iii) desorption; and (iv) flash-type deaeration techniques. These techniques permit oxygen concentration to be reduced to about 0.01 ppm.
2. *Chemical treatment* often follows the above methods for the final reduction of the oxygen concentration to virtually zero. This is done by adding certain reducing agents such as hydrazine, sodium sulphite (oxygen scavengers).
3. *Ion exchange techniques* are also able to reduce the oxygen concentration in boiler water to very low values.
4. *Mineral acids:* Most of the natural waters are alkaline. Waters in the mining areas are often acidic. Sometimes, in industrial areas the water may become acidic because of the discharge of acidic industrial wastes into the surface water. Some of the inorganic salts may hydrolyse to produce acidity causing corrosion of the boiler tubes. This attack is normally accompanied by hydrogen embrittlement caused by hydrogen formed in the acid attack, penetrating the steel so that failure occurs very quickly.
5. *Carbon dioxide:* Water contains some dissolved carbon dioxide. This gas will also be produced by the decomposition of some bicarbonates if present in the boiler water. Carbon dioxide coming in contact with water produces carbonic acid, H_2CO_3, which causes local corrosion called pitting.

$$Fe + CO_2 + H_2O \rightarrow FeCO_3 + H_2$$

$$4FeCO_3 + O_2 + 10H_2O \rightarrow 4Fe(OH)_3 + 4H_2O + 4CO_2$$
$$\text{(dissolved)}$$

$$4Fe(OH)_3 \rightarrow 2Fe_2O_3 + 6H_2O$$

Carbon dioxide in water can be removed by the addition of lime or by heating the water.

6. *Galvanic cell formation:* Corrosion (pitting) can also be due to simple galvanic cells which iron forms with some of the boiler fittings made of other materials or with impurities. This can be avoided by suspending zinc plates when zinc is anodic to iron and gets dissolved first and iron is saved.

Corrosion on furnace side of a boiler plant: It can be prevented by two general methods:

1. *Purification of the fuel:* Methods of removing sulphur from heavy fuel oil are as follows:
   (a) In hydro-desulphurisation process sulphur present in the oil is converted into hydrogen sulphide by reaction with hydrogen using cobalt molybdenite as a catalyst.

$$H_2 + S \rightarrow H_2S$$

   (b) The Cat-Ox system, developed in the USA utilises a miniature sulphuric acid contact plant for removal of sulphur.
2. *Addition of chemicals:* Dolomite has been used with success, as it reacts with free sulphuric acid to give high fusion point sulphates with light friable deposits instead of heavy bonded and corrosive scales. Pure magnesia or a higher proportion of magnesium carbonate (% > 45 per cent), though more expensive, can give better results.

Scale formation

From the standpoint of the chemical constituents present in water, a boiler may be considered to be a huge concentration of impurities. Water containing impurities is fed into the boiler and pure water, in the form of steam, is removed, leaving behind the deposits of impurities inside the boiler tubes. Some of the deposits stick to the metal surface and are known as scales. If they are in the form of soft muddy deposits, which can be flushed out easily, or in the form of suspensions, they are known as sludges.

Scale formation may be prevented by the following methods:

1. *External treatment:* This treatment is given outside the boiler before the feed water enters it. Attempt is made to remove or reduce the amounts of those substances, mainly calcium and magnesium salts and silica, which form deposits.

   For the removal of Ca, Mg and other ions the following processes viz. (i) lime-soda process (hot and cold); (ii) zeolite process; (iii) demineralisation process; and (iv) sequestration are used.

   Silica can be removed by special ion exchange techniques.

2. *Internal treatment:* In spite of previous treatment some salt often remains in the boiler. Attempt is made, by adding chemicals in the boiler, to convert the remaining salts into more soluble salts or such salts that the deposit will be in the form of non-sticky sludge which can be easily removed.

   (a) Inorganic and organic materials such as sodium silicate, soda ash, kerosene, glycerine, tannin, etc. are added to boilers. These substances form a coating over the scale forming particles which prevents their coalescence; it results in the formation of scales which are not very adherent or are suspended in water and can be easily removed by blowing off.

   (b) Calcium is removed from water of the boiler as calcium carbonate or calcium phosphate which settles down as a sludge. This prevents the formation of calcium sulphate scales.

$$CaSO_4 + Na_2CO_3 \rightarrow CaCO_3 \downarrow + Na_2SO_4$$
$$3CaSO_4 + 2Na_3PO_4 \rightarrow Ca_3(PO_4)_2 \downarrow + 2Na_2SO_4$$

   The best results are obtained when phosphate treatment is done at pH 10.5.

   (c) Sodium hexametaphosphate (calgon) designated by the formula $Na_2(Na_4P_6O_{18})$ may also be added to the boiler as a sequestering agent. This compound forms a complex with calcium having the formula $Na_2(Ca_2P_6O_{18})$ due to which calcium is not present in the form of free calcium ion and cannot be precipitated as calcium sulphate which forms adherent scales.

Caustic embrittlement

Embrittlement is the name that has been given to boiler failures due to development of certain types of crack resulting from excessive stress and chemical attack. These cracks are intergranular which distinguishes them from cracks resulting from strain without chemical attack which are predominantly transgranular (corrosion fatigue). In steam boiler operation, the chemicals that are believed to be responsible are caustic soda and silica. Caustic soda as such as not found in natural water but often originates from sodium carbonate, generally added for water treatment purpose

$$Na_2CO_3 + H_2O \longrightarrow 2NaOH + CO_2$$

a reaction which takes place at elevated temperature. Any minute cracks or grain boundaries in a stressed section of a boiler may be affected.

Steam boilers, which were made by riveting construction, rather than being welded as they usually are today, are particularly liable to damage by caustic alkali solutions (caustic cracking). This cracking is not due to corrosion and the cracks have the appearance of a brittle fracture due to which it has been given the name 'caustic embrittlement'. Following are the conditions under which caustic embrittlement takes place: (i) the metal of the boiler is highly stressed due to internal stresses or due to inaccurate fitting of riveted parts; and (ii) the stressed metal, when it comes in contact with a solution capable of causing the embrittlement.

Principal methods of preventing caustic embrittlement

The principal methods of preventing caustic embrittlement are as follows:

1. *Addition of sodium sulphate:* This chemical is added to the boiler water in a sufficient quantity to ensure that the weight ratio, $Na_2SO_4/NaOH$, always exceeds 2.5. Sometime other sulphuric acid or magnesium sulphate is added to boiler water already containing free NaOH when sodium sulphate is produced *in situ*.

$$2NaOH + H_2SO_4 \longrightarrow Na_2SO_4 + 2H_2O$$
$$2NaOH + MgSO_4 \longrightarrow Na_2SO_4 + Mg(OH)_2$$

   Sodium sulphite is often added to water to reduce the oxygen concentration and this also produces sodium sulphate,

$$2Na_2SO_3 + O_2 \longrightarrow 2Na_2SO_4$$

   The mechanism by which sodium sulphate prevents caustic cracking is that, instead of sodium hydroxide alone penetrating the crack, it is now a mixture of sodium sulphate and sodium hydroxide which enters. Long before the sodium hydroxide solution reaches its critical concentration of above 10 per cent the sodium sulphate crystallises thus barring the entry of any further quantities of solution into the crack or intergranular space. Thus it becomes impossible for the solution to reach its critical concentration of 10 per cent NaOH. The disadvantage of sodium sulphate treatment is that calcium sulphate may be formed which is one of the main scaling agents of boiler plant.

2. *Addition of sodium phosphates:* Polyphosphate $(NaPO_3)_6$ is added to boiler water and can react as follows:

$$(NaPO_3)_6 + 6H_2O \longrightarrow 6NaH_2PO_4$$

   The sodium dihydrogen phosphate formed then neutralises the calcium and sodium hydroxides introduced in the water softening process. Free calcium ions in solution form sparingly soluble calcium phosphate which is precipitated without scale formation. Polyphosphate treatment is carried out in such a way that there is always an excess of about 10 per cent of NaH_2PO_4 in solution in excess of any hydroxide present. Too much of NaH_2PO_4 is, however, to be avoided otherwise the pH of the boiler water may fall below the recommended level (8.5 to 9.0).

3. *Addition of sodium nitrate:* For boilers operating at pressures between 7 and 40 bar, sodium nitrate is added to the boiling water so that the weight ratio, $NaNO_3/NaOH$, is maintained above unity. Sodium nitrate appears to inhibit caustic cracking extremely effectively, as it makes portions liable to be affected passive.

4. *Addition of organic agents:* For low pressure boilers (< 20 bars) caustic cracking can be avoided by the addition of certain organic reagents such as tannin, lignin, quebracho, etc. The function of these is same as that of sodium sulphate mentioned in above.

5. *Use of crack-resisting steels:* Certain steels, such as those which have had a reasonable quantity of aluminium added during manufacture to eliminate traces of intergranular iron oxide formation, appear to be resistant to caustic cracking.

Corrosion in Chemical Industries

Selection of suitable materials of construction which can resist corrosion under severe corrosive conditions is of considerable importance for economic operation of chemical plants. Any material used has to satisfy certain basic requirements of physical, mechanical and corrosion resistance properties under the operating conditions. We shall mainly confine here, without reference to any specific chemical industry, to the (i) reactions dependent on environmental conditions; (ii) different types of corrosive media and their behaviour; and (iii) engineering materials and their corrosion resistance.

Environmental factors

It is important that the equipment should have proper corrosion resistivity at the operating environmental conditions frequently met in the process industries. One of the fundamental properties of the environment is its ability to form or destroy a protective film on the metal. Change in the hydrogen ion concentration of the solution is one of the most critical factors affecting the rate of corrosion.

As an example of the influence of pH on the corrosion velocity, we can cite the rapid increase in the dissolution rate of Fe, Zn, Mg and a number of other metals by a transition from neutral media to solutions of non-oxidising acids or by an increase in the concentration of the acid (e.g. HCl). Indirect influence of pH on corrosion consists of change in the solubility of the corrosion products and the possibility of protective film formation with change in pH. In this respect the metals can be divided into three groups:

1. Metals (noble) which are usually quite stable in acid as well as in alkaline solution.
2. Metals, e.g. Zn, Al, Pb and Sn—the oxides (amphoteric) of which are soluble in acids as well as in alkalies—are not resistant to acid or alkaline solutions.
3. Metals whose oxides are easily soluble in acids but are insoluble in alkalies, such as Ni, Cu, Co, Cr, Mn, Cd, Mg and Fe.

Oxidising acids

Oxidising acids (HNO_3, conc. H_2SO_4) passivate some metals, viz. Fe, Co, Ni, Al, W, Ti, etc. under certain conditions, and corrosion practically ceases. Besides oxidising acids, a number of other reagents, viz. $HClO_3$, $K_2Cr_2O_7$, $KMnO_4$, etc. as well as oxygen can serve as passivating agents. Among activating salts are salts of halogen acids, giving F^-, Cl^-, Br^-, and I^- ions on dissociation. These ions (anodic accelerators) often prevent the formation of a passive film on metals.

Corrosives

For this purpose a chemical may be said to either: (i) dissolve a material uniformly, the rate depending on pH (uniform corrosion) or (ii) non-uniformly leading to pitting corrosion. Various corrosives may, in general, be classified as:

1. *Mineral acids:* With respect to processes of corrosion, mineral acids may be grouped into two types, viz. oxidising acids and non-oxidising acids. Non-oxidising acids may be defined as those in which the cathodic process occurs only by hydrogen depolarisation and oxidising acids as those in which the predominant cathodic process is oxidation depolarisation, i.e., reduction of the acid anions. In practice, however, there are possible intermediate class of gradual transition from one type of action of the acid on the metal to another type dependent on concentration, corrosion potential, temperature, passivity.

2. *Acids:*
   (a) Sulphuric acid is used directly or indirectly in nearly all industries and is a vital commodity in our national economy. Corrosion problems occur in plants, viz., HCl, fertilisers, dyes, drugs, pigments, explosives, rayons, textiles, petroleum refining, rubber etc., where it is utilised under variety of conditions. Sulphuric acid is a strong acid and is normally non-oxidising. Its maximum corrosiveness occurs at concentrations of 60 to 70 per cent. Concentrated sulphuric acid appears to be less corrosive, particularly to iron and steel, than other concentrations at room temperature. Cast iron and steel can safely handle concentrated sulphuric acid for this reason.
   (b) Hydrochloric acid is more active than sulphuric acid and its handling is a problem of great difficulty in industrial plant operations. Factors which help to explain its higher corrosivity are the greater solubility of chlorides, less tendency to form insoluble basic salts, and higher mobility of the chloride ion. Most concentrations of the acid are severely corrosive to most of the common metals and alloys. Increase in temperature, concentration and velocity all tend to accelerate the attack. Many unexpected failures in service often take place due to the presence of minor impurities like oxygen (aeration), oxidising agents like HNO_3, ferric or cupric chlorides, etc.

   In consideration of the above, the limits for so-called acceptable corrosion rates are usually raised when considering materials of construction for handling hydrochloric acid. Materials that show very low rates of corrosion are often not economically feasible. Good judgement is required to obtain a balance between service life and cost of equipment. Where contamination is a problem, e.g. in steam tubes for heating chemically pure acid, expensive materials such as tantalum are the only ones that can be utilised. Molybdenum is an important constituent of the alloys generally used as corrosion resistance materials to hydrochloric acid. Durichlor is a high-silicon iron containing Mo and is excellent corrosion resistant to hydrochloric acid. This alloy is used in industry for all concentrations of hydrochloric and muriatic acids at moderate temperatures. Aeration does not affect corrosion resistance. Durichlor pumps give satisfactory service in 30 per cent HCl and also sludges containing 10 per cent acid at ambient temperatures. Nickel-base alloys with large Mo contents, viz. Chlorimet-2 and Hastelloy–B show good corrosion resistance to all concentrations of hydrochloric acid up to boiling temperatures. These alloys, however, are attacked if aeration or oxidising ions are present. Chlorimet-3 and Hastelloy–C are good in dilute acids at moderate temperatures and exhibit better resistance to oxidising environments because of their high chromium contents. Copper, the various bronzes, cupronickels, Monel, Inconel, Ni-Resist, Hastelloy–D and stainless high alloys are susceptible to influences other than the acid itself and must be used with caution with a definite knowledge of the specific conditions. Ordinary carbon steels, cast irons, aluminium and its alloys, lead and its alloys are never used for hydrochloric acid service.

   Non-metallics have found widespread use in hydrochloric applications because of their good resistance and immunity to attack by oxidising ions. Subject to their relative temperature limitations most of the plastics and rubbers are suitable for all concentrations of hydrochloric acid. Rubber-lined steel can be used for many years for vessels and piping for hydrochloric acid service. Wood also finds application for dilute acid.
   (c) Hydrofluoric acid is similar to hydrochloric acid in many respects. Aeration and the presence of other oxidising agents increases the corrosivity of the acid. Temperature has similar effect.

Magnesium resists attack by hydrofluoric acid and some shipping containers for this acid are made of Mg. Steel is suitable for handling concentrations from 60 to 100 per cent and aqueous mixtures below 60 per cent evolve hydrogen. Wrought Monel can resist all concentrations of HF at all temperatures up to boiling point.

(d) Phosphoric acid is more like sulphuric acid than hydrochloric acid in terms of corrosivity. Aeration and other oxidising agents increase the corrosiveness. The impure acid frequently contains fluoride salts and oxidising compounds which also tend to increase the rate of attack. Thus, the high-silicon irons, ceramics, and tantalum are appreciably affected by the presence of hydrofluoric acid.

Two of the most widely used alloys as materials of construction up to 85 per cent concentration of H_3PO_4 and at boiling point are type 316 stainless steel and Durimet-20. Lead and its alloys are also used at temperatures up to 475°K at concentration up to 80 per cent for pure and 85 per cent for impure acid. High-silicon irons, glass, and stoneware show good resistance to pure acids. High nickel-molybdenum alloys exhibit good resistance to pure acid in absence of oxidising impurities and aeration.

(e) Nitric acid is a strong oxidising agent and influences the corrosion rate by oxidising the polarising film of hydrogen. Only metals which form protective oxide films, such as Al, Ti, the stainless steels, etc. are resistant.

The choice of metals and alloys for nitric acid services is quite limited. High silicon (14.5 per cent) alloys (duriron) find useful application for handling nitric acids. Chromium (17 per cent) iron alloys are also known for their considerable resistance to the oxidising acids, 18–8 S.S. is the most widely used of all the materials for handing nitric acids. Aluminium has been used for transporting nitric acid as its corrosion rate is considerably less at low temperatures in very low and very high concentrations of nitric acid including fuming acid. Aluminium equipment is used in the manufacture and handling of strong nitric acids including the fuming varieties. Teflon shows outstanding corrosion resistance to nitric acid and glass-filled Teflon finds wide application for rotating rings in mechanical seals for nitric acid pumps.

Impurities like sulphuric acid enhance the corrosion rate. Ferrous metals and alloys may be used for handling concentrated nitric acid when mixed with sulphuric acid.

3. *Alkalies:*

Alkalies such as caustic soda and caustic potash are not particularly corrosive and can be handled in pure iron and steel. Corrosion problems, however, arise during high temperature evaporation of concentrated alkali solutions. Pb, Zn, Al and Sn are readily corroded in sodium or potassium hydroxide solutions and form plumbates, zincates, aluminates and stannates. Nickel, Monel metal and nichrome are relatively unattacked with caustic solutions and have extensive applications in caustic solutions.

Corrosion by alkalies is often characterised by pitting and localised attack. Another undesirable phenomenon often connected with alkali corrosion is a form of stress corrosion cracking at certain temperature and concentrations.

A considerable number of gases, in addition to those already discussed, are encountered frequently in the chemical and process industries. These include O_2, H_2, N_2, CO_2 and CO. All of these gases are corrosive in one form or another at elevated temperatures. O_2 causes oxidation of the metal while H_2 penetrates the metal at elevated temperatures and can cause embrittlement. N_2 forms nitrides on the surface of the metal, and CO_2 and CO may carburise the metal.

4. *Corrosive vapours:* In addition to above, in chemical plants which produce acids, the vapours of hydrochloric, nitric, sulphurous and other acids influence severely the corrosion of metals. Steel and iron pipes can be employed for handling dry chlorine and hydrochloric acid gas. Cast iron pipes are used for handling hot SO_2 gases and lead may be employed when the gases become sufficiently cooled. For the construction of apparatus for the synthesis of ammonia it has been reported that low-carbon steel containing 2 to 2.5 per cent Cr can be used with sufficient corrosion resistance.

Thus, it is a problem of great difficulty to select a required material suitable for the construction of equipment for chemical plants which can withstand corrosion, as the corrosion rate is generally dependent on various factors like the temperature, concentration of corrodant, aeration and velocity, etc.

Materials of construction

A brief outline for construction of materials for chemical plants is given in Table 21.1.

Table 21.1. Materials of construction in chemical industries.

Material of construction	Applicable	Not applicable
Ferrous metals	Nitric acid	
Chromium steels (17 to 27% Cr)	Sulphur gases	
Chromium-nickel steels (18% Cr, 8%, Ni)	Atmospheric exposure. High temperature stresses	
High silicon-iron alloys (14 to 16% Si)	Mostly acids	Halogen acids Fuming sulphuric acid, alkalies
Steel	Concentrated sulphuric acid	
Non-ferrous metals		
Copper	Acetic acid	Oxidising acids, non- oxidising acid with air
Copper-nickel alloys (Monel metal)	Non-oxidising acid (except HCl) Hydrofluoric acid	Oxidising acids
Nickel and nickel alloys	Alkalies	
Aluminium	Nitric acid, C > 85% formaldehyde, synthetic resin. Distilled and deionised water storage tanks, piping and condenser.	Alkalies
Titanium	Hot, concentrated oxidisers.	
Lead	Dilute sulphuric acid.	HCl, HNO_3, caustic alkalies.
Alloys of Co, Cr, W or Mo	High resistance to many chemicals, to high temperatures to abrasion.	(high temperature)
Ta	Extreme resistance to chemicals	Expensive
Non-metallic materials		
Porcelain enamel	Distillation of citric, maleic, lactic acids	
Cement lined steel pipe	Salt water, sulphur bearing oils.	
Chemical stoneware	Acids, alkalies	
Glass	Acids	HF

(Contd...)

Material of construction	Applicable	Not applicable
Fused silica	Acids (hot and cold). Condensing, cooling and concentrating apparatus.	HF, H_3PO_1 above 700°K Alkalies fragile expensive.
Wood	Weak acids	Strong oxidising agents. Alkalies, crystallising salts.
Natural rubber	Lining material for steel tank for chemical equipment.	
Synthetic rubber	Many chemicals, acids, HF	

Corrosion in Petroleum Industry

The petroleum industry contains a wide variety of corrosive environments. For example, oil fields are situated in tropical areas where high humidity, salt bearing winds and air borne sand take the toll of structures and equipment. Costly pipelines convey the crude oil—often itself actively corrosive towards iron and steel—to long distances, either to refineries or to coastal installations where ocean-going tankers may be loaded via a submarine pipeline. In the refineries, the vast quantities of cooling water required for their operation, often necessitate the use of sea water, so that intake lines, condensers and coolers all require special protection against corrosive attack. Finally, the refined products must be distributed giving rise to special corrosion problems in ocean going tankers and underground pipelines. It is convenient to group all these environments in relation to corrosion problems, mainly into three broad areas, viz. (i) production; (ii) transportation and storage; (iii) refinery units. The corrosion experienced in these areas may be divided into two classes; viz.: (i) that due to the fluid being produced and hence usually internal; and (ii) that due to environment in which the equipment is placed and hence usually external.

Production

The internal corrosion experienced in typical oil and gas wells is normally associated with hydrogen sulphide, carbon dioxide and organic acids present in the oil, brine or gas. Internal corrosion is normally referred to as being sour (from 'Sour Oil Wells') or sweet (from 'Sweet Oil Wells') according to the higher or lower sulphur content (mainly H_2S) of the oil. The gas may be present in the well or may result from the activity of sulphate reducing bacteria. Sulphide corrosion results in large deep pits and heavy iron sulphide scale. This attack is not restricted to the well equipment only, but continues on into the pipelines and tankers also. Corrosion in the absence of hydrogen sulphide is most frequently associated with carbon dioxide as the chief corrosive agent, with organic acids contributing to the attack. Three methods are used to mitigate this corrosion, viz. coated tubing, inhibitors, and special alloy steels. Coated tubing has found the most favour, and air-dried and baked epoxy resins are now being used in increasing amounts for almost all coating installations.

The external corrosion of well casings is now recognised as a major problem, due to the huge repair costs involved. The most common causes of casing corrosion are due to (i) sulphate reducing bacteria; and (ii) local concentration cells. A variety of corrosion prevention methods are used to mitigate casing corrosion.

1. Adding inhibitors so that these are uniformly dispersed over the entire casing.
2. Cathodic protection, with sacrificial anodes or impressed currents, of the casing.
3. Using protective barriers like cementing the casing.
4. Insulating the well from its flow line.

Transportation and storage

Petroleum products are transported by tankers, pipelines, railway tank cars, and tank trucks. The most severe internal corrosion problem occurs in tankage. If the crude is sour, early perforation of the fixed roof sheets is likely. The use of floating or aluminium roofs, and coatings are the most common preventive methods used. The coatings in most common use are coal tar based. The tank may also be subjected to external corrosion attack. This can be prevented with coatings or by using cathodic protection. Whether the attack is external or internal or both, providing the tank with a concrete bottom prevents further corrosion.

Ordinary carbon steel which is used for the construction of tanker is exposed to an aggressive natural environment of salt water and marine atmosphere. Most serious exposure of steel to sea water occurs during the return (ballast) voyage when the tanks are void of cargo. Although corrosion during cargo voyage is probably relatively small, the nature of different types of cargo has a profound effect on the overall corrosion. Gasoline-carrying tankers present a more severe internal corrosion problem than oil tanks because the gasoline keeps the metal too clean. Oil leaves a film which serves an effective barrier against general corrosion. For ballast tanks a galvanic system (Mg anode) is always used.

Internal corrosion of storage tanks is due chiefly to saline water which settle sand remains on the bottom. Coatings based on vinyl or epoxy resins and cathodic protection are mainly used. For domestic fuel oil tanks alkaline sodium chromate (or sodium nitrate) is an effective inhibitor.

Paint is the traditional material for protecting hulls and super structures of steel ships. In more recent years chemical resistance coatings have been developed which neither affect cargoes nor are affected by cargoes.

Rust formation on the internal walls of the pipelines caused by water precipitated from products may reduce the line throughout and may give rise to contamination of the product. Formation of the rust may be prevented by lining the pipe or rust formation may be inhibited by the injection of inhibitors (a few parts per million) such as amines and nitrites into the product stream. Corrosion of the external walls of the pipelines varies enormously according to the nature of the soil or water, the temperature, access of oxygen and other factors.

To combat soil corrosion, both coatings and cathodic protection are used. The three basic types of coating and wrapping now finding use in oil industry are: (i) hot-applied coal-tar enamel and asphalt coatings; (ii) plastic tapes; and (iii) coal-tar/epoxy coating. Two principal tapes at present offered for buried pipe-line protection are those of polyvinyl chloride, and polyethylene. In the past few years, the application of cathodic protection to products transmission lines have become normal practice. It is now generally recognised that the combinations of good coating and cathodic protection is the best method of ensuring a long leak-free life to buried pipes.

Refinery units

Corrosion problems in the refinery tubular heat exchange equipment can be solved by mainly relying on the selection of materials. Crude oil always contains impurities which frequently lead to severe corrosion problems in processing. Condensed products from the distillation processes are frequently contaminated with such substances as sulphuric acid, naphthenic acid, hydrogen sulphide and hydrogen chloride. Considerable corrosion is, therefore, liable to occur on the product side of the condenser and cooler tubes. Corrosion generally takes the form of uneven general wastage and insoluble corrosion products such as copper sulphide are frequently formed.

Pitting may proceed rapidly beneath these non-protective deposits. The most widely used tube materials are brasses with a high zinc content. When sea water or brackish water is used for cooling, there is a possibility of corrosion on the water-side in the heat exchange equipment. Carbon steel, stainless steel or monel tubes cannot be used, owing to their susceptibility to pitting in sea water. Brass, arsenical Admiralty metal, red brass, and cupronickels are used satisfactorily. On the project-side of heat exchange tube, corrosion can be minimised by injection of alkali or ammonia to keep the pH at a controlled figure in conjunction with the addition of one or other of film forming amine inhibitors. On the cooling water-side, benefits may be obtained by the application of proper systems of cathodic protection in condenser and cooler channels and heaters. Zn or Mg can be used as sacrificial anodes.

Sea water cooling with circulating pumps in refineries provides a major problem in corrosion control. Protective measures employed involve the use of devices to overcome galvanic couples which encourage local attack and pitting. When cooling water is through steel pipes, a combination of thick coatings based on coal tar pitch plus cathodic protection is effective. Cooling water circulating pumps are made of bronze and cast iron, which are fitted with cathodic protection from an external source of impressed current.

For tubing in stills and gas-cracking tubes austenitic stainless steels are used. In some cases, a single tower is lined with two or three different materials to take care of the changing corrosiveness from the top to the bottom of the tower. Corrosion by sour crudes increases with temperature and with increasing sulphur content. Chromium is the most beneficial alloying element in steel for resistance to sulphur compounds. Thus, the Cr content of steel is increased with increasing sulphur and temperature starting as low as 1 per cent Cr. Experience indicates that 2.5 per cent Cr, 1 per cent Mo steel is generally adequate for less than 0.2 per cent H_2S in the gas stream while higher amounts of H_2S require 5 per cent or higher amount of Cr in the steel. Four to 6 per cent Cr, 0.5 per cent Mo steels are widely used in refineries.

Non-metallic materials are resistant to chemical corrosion and are free from contamination of the product. These render them very attractive to refinery and petro-chemical industries. Natural rubber has been used as a structural material and as a lining for vessels to prevent contamination and corrosion. Flexible pipes are widely used as temporary connections in the storage and transport of materials. Graphite has a very good thermal conductivity, and this property combined with its resistance to steam and highly corrosive fluids opens up the field of heat-exchange applications, in addition to its use in vessels, valves, pumps and piping.

Petro-chemical corrosives

These may be classified into two general categories, viz. (i) those present in crude oil; and (ii) those associated with these processes.

Table 21.2. gives an outline of the general nature of these corrosives. In conclusion, it may be said that petroleum industry has much in common with the chemical industry since most of the corrosion problems in refineries are due to inorganics (water, H_2S, CO_2, H_2SO_4, NaCl) rather than organics.

Corrosion in Fertiliser Industries

Like other chemical and process industries, the fertiliser industry also requires special materials of construction to suit the corrosive nature of the media handled and the various process parameters like temperature, pressure, pH, velocity as well as prevention of product contamination. Various types of stainless steels find wide application on account of their excellent corrosion resistance, strength and toughness, good fabrication properties and fine surface finish. The various chemical processes adopted for production of different types of fertilisers, process parameters involved, types of corrosion and corrosion problems encountered in these process industries are discussed briefly.

Table 21.2. Petro-chemical corrosives and their actions.

Name	Nature
Water (in crude oils)	Acts as an electrolyte. Hydrolyses salts (e.g. chlorides producing acids)
Carbon dioxide (in gas wells)	Chief corrosive agent with organic acids.
Salt water (in oil wells and in crude oil)	Salts of $CaCl_2$, $MgCl_2$ and NaCl. Forms acid due to hydrolysis which is removed by ammonia addition. Removed by desalting methods
Sulphide compounds (viz. H_2S, RSH etc.) (in crude and gases)	Reaction leading to severe attack (Sulphide corrosion) in refining process. Removed by caustic addition
Nitrogen (in some crude) (from air in burning operations)	Ammonia. Damage heat exchanger made of copper-bearing alloy Cyanide interferes with diffusion of hydrogen into steel. Controls pH of water
Oxygen (air) (in refinery operation)	Responsible to oxidation
Sulphuric acid (in refinery operation)	Attack stainless steel in presence of sludges containing carbonaceous materials. Copper-base alloys can be used.
Hydrochloric acid (in refinery operation)	Formed through hydrolysis. Present in distillation columns and in condensed petroleum fractions.
Caustic and lime (for H_2S removal and for neutralisation of HCl)	Causes deposits, clogging, and stress corrosion
Naphthenic acid (in oil)	Corrosive above 800°K. Stainless, steel can be used

Nitrogeneous fertilisers

Nearly 70 per cent of India's nutrient consumption is in the form of nitrogen. Major fertilisers in this category are ammonium sulphate, nitrate, urea etc.

1. *Ammonium sulphate:* The major part of ammonium sulphate is produced as a by-product from the high temperature carbonisation of coal. The process involves handling of weak sulphuric acid of 1 to 20 per cent concentration, ammonium sulphate crystal slurry, coke oven gas laden with sulphuric acid mist, ammonia vapours etc. The corrosion problems encountered are due to handling of highly corrosive low concentration sulphuric acid, ammonia vapour and liquor, presence of H_2S and HCN in coke oven gas and abrasion due to ammonium sulphate crystal slurry. The materials of construction used are lead and acid proof brick-lined mild steel, S.S. conforming to AISI-316, cast iron alloy-20 etc., for various equipment, viz. saturator, evaporator, ammonia column, tanks, centrifuges, driers, etc.

2. *Urea:* Urea is now the most common nitrogeneous fertiliser. This is primarily due to its low production cost. Modern urea plants are both energy efficient and make maximum use of the raw materials—ammonia and carbon dioxide. In addition, advanced designs have resulted in lower maintenance requirements and longer service life. A significant factor in the design has been development of economical corrosion resistant material.

The principal steps in the manufacture of urea are: (i) the synthesis of ammonia from synthesis gas mixture; and (ii) processing ammonia into finished fertiliser.

In a nitrogeneous fertiliser factory there would be two main plants, viz. ammonia plant and finished fertiliser plant, e.g. urea plant where urea is the finished product. Another major plant that must exist along with the above two plants is the steam generation plant. Here all discussion are centred on a modern Naptha Based Nitrogenous Fertiliser Plant with urea as the finished product since most of the information on corrosion problems in fertiliser industry have been regarding ammonia and urea plant.

Synthesis of ammonia

Major steps involved in a typical naphtha based ammonia plant are: (i) desulphurisation of naphtha; (ii) naphtha steam reforming; (iii) purification of synthesis gas; and (iv) ammonia synthesis. Besides these, there are ammonia cracking section and water cooled heat exchanger. The ammonia process is considered to be moderately corrosive. The corrosive environments in different sections are given in Table 21.3.

Table 21.3. Corrosive environments in various sections.

Sections	Corrosives	Materials of construction
Hydrodesulphuriser	S, Organic sulphur compounds like mercaptan, mono and poly sulphides, thiophenes.	Carbon steel, 13% Cr and 18-8 Cr-Ni austenitic steel for sulphur compounds.
Catalyst steam reforming (Primary and secondary)	High temperature CO_2, H_2 and O_2	HK 40-25% Cr, 20% Ni, 0.35–0.45% C and rest Fe (more popular); IN-519-24 Cr, 24Ni, 1 Si, 1 Mn, 1.5 Nb, 0.25–0.35 C, rest Fe IN-657-49 Cr, 49 Ni, 1.5 Nb, 0.5 Si, 0.1 C Paralloy CR 32 N-20 Cr, 32 Ni, 1.3 Nb, 1 Mn, 0.8 Se, 0.1 C, 0.03 S and 0.03 P, rest Fe
Carbon monoxide conversion (high temperature and low temperature)	Hot CO-CO_2-H_2 gas mixture	Carbon and low alloy steels
Decarbonation	Hot aqueous CO_2 solution	Austenitic stainless steels, primarily types 304 and 316 in both wrought and cast form
Methanation	Nil	Plain carbon and low alloy steels
Compression	Erosion-corrosion	Type 410, Type 430
Ammonia synthesis	Hot compressed H_2 and condensed ammonia	Type 304 for catalyst cartridge and integral heat exchanger. Low alloy high strength stainless steels for piping and equipment around the converter.

Corrosion problems

Numerous types of corrosion problems have been identified in the ammonia process. These are stress corrosion, pitting corrosion, crevice corrosion, galvanic corrosion, intergranular corrosion, end grain corrosion, high temperature corrosion like carburisation, oxidation, decarburisation, sulphidation and nitriding.

The overall problem in the reformer, where catalytic reforming of the feed occurs at elevated temperatures and pressures is, in particular, the materials used for the tubes inside this reformer and the support and ancillary pipe work that may be exposed to high temperatures and pressures. The pipe work system expands at this severe operating conditions aiding crepe occurrence and loss of metal elasticity. The most commonly used tube alloy in the modern ammonia plants is similar in composition to type

310—nominally 25 per cent Cr and 20 per cent Ni, balance iron—but with a carbon content in the range of 0.35 to 0.45 per cent for improved high temperature crepe and stress rupture strength (HK–40).

Corrosion failure

The following corrosion failures in reformation section may be mentioned as typical examples :

1. Leak caused by intergranular microcracks in the header assemblies in the steam/naphtha reformation section due to formation of ferrite and carbide, the material having been DIN/ 1.4863. Presence of ferrite and carbide caused microcrack which propagated and caused leak due to high heating rate during start up, viz. 550°K per hour as against the standard practice of 320°K per hour and high frequency of shut down and start up.
2. Failures with deformation and bulging of reformer tubes made of centrifugally cast HK-40 alloy and operating at 12 Kg/cm^2 pressure and 1100°K temperature. The failure was due to carburisation, sigma phase formation which caused microcracks, and overheating led to propagation of microcracks, deformation and bulging.
3. Failure in the form of longitudinal crack on the outer side of the outlet pig tail of reformer tube (21 Cr, 31 Ni, 0.07C, 1.5 Mn, 0.25 Si, 0.35 S) due to carburisation followed by oxidation, sulphidation and overheating.
4. Chloride/caustic stress corrosion failures of inlet piping materials (HK-40) of reformer section.

Potash corrosion in H.T. shift converter in a problem is the carbon monoxide conversion unit.

Change of direction of the flow of fluids often causes deposition in dead ends. At times, potash gets leached out from the catalyst and gets deposited in dead legs. There have been instances of failure of the snuffing steam line by potash corrosion attack and corrective action of potash corrosion is to radiograph the line during shut down, to detect potash deposits and/or crack caused by the deposit and to take pre-emptive repair action.

Failures have been reported to occur, by erosion in the lean liquor draw off line from the reboilers in the decarbonation section. Incorporating stainless steel sheath around the zone of attack considerably reduces the rate of attack.

Preheater coils made up of AISI-321 type stainless steel have been found to fail due to nitriding caused by overheating of the tubes much above the maximum allowable temperature (550°–650°K) in the Ammonia Cracking Section during the start up of the fertiliser plant.

Corrosion in ammonia synthesis section offers the greatest problem. The ammonia synthesis converters are pressure vessels containing a removable cartridge filled with catalyst. The unreacted gas at high pressure and temperature can be destructive to carbon and low alloy steels and other metals due to hydrogen embrittlement, hydrogen attack and nitriding.

Condensed ammonia is also a corrodent and in the anhydrous state can cause stress corrosion cracking of stressed carbon and low alloy steels. The remedial measures for these types of corrosion are proper selection of materials of construction, proper design and strict adherence to set control process parameters.

Type 304 is specified for catalyst cartridge and integral heat exchanger. Austenitic stainless steels are resistant to nitriding, hydrogen embrittlement and hydrogen attack. Increased Ni and Ni-base alloys offer greater resistance to nitriding. Corrosion problems in high pressure steam generators are the same as those with high pressure boiler systems.

Corrosion in water-cooled heat exchanger systems also pose many problems. Achieving desired ammonia production depends to a very large extent on the efficient heat transfer/heat dissipation from hot process fluids through metallic surface of heat exchangers into cold cooling water (recirculating (open) cooling water system). Corrosion encountered in ammonia plant cooling system are: (i) electrochemical corrosion; (ii) fouling and fouling-induced corrosion. Although these phenomena are very much similar to other cooling systems, the problems in ammonia cooling systems have got some unique features. These are:

1. The coolers and condensers are mostly made from carbon steel, for economic reasons, which are most susceptible to aqueous corrosion.
2. There are huge number of coolers and condensers in the system and due to their widely scattered locations, uniform water distribution and uniform maintenance of water velocity through all the equipment in the circuit are rare phenomena; as a result coolers deprived of proper water flow repeatedly suffer from corrosion failures.
3. Some of the coolers have water in the shell-side which create inherent stagnant zones, e.g. near baffle corner, shell wall, longitudinal baffle wall etc.
4. Ingress of ammonia, oil/naphtha from the system cause fouling due to precipitation or oil sludge, supply abundant nutrients to micro-organisms leading to microbiological problems, cause wide pH fluctuations, e.g. pH rises when ammonia enters and falls when nitrification occurs; nitrification also produces corrosive nitrate ions; oil also makes corrosion inhibitors ineffective by film formation.

Electrochemical corrosion in the cooling system is brought about by dissolved oxygen and the product of corrosion is FeO or Fe_2O_3. Since cooling water is maintained near neutral or alkaline, corrosion products (Fe_2O_3) accumulate on the metal surface unless water velocity is very high or efficient surface active agents-cum-antifoulant are employed; and it is found that electrochemical corrosion ultimately leads to deposition in practice unless corrosion rate is too low. Corrosive mineral ions, temperature, and other common factors increase the rate of electrochemical corrosion.

In cooling system, failure due to uniform or general corrosion is rare. Problems occur due to fouling-cum-deposit and underdeposit corrosion which makes the control of cooling system failures extremely difficult. Under the deposit catastropic rate of corrosion and unimaginable rapid failure of equipments due to the combined effect of the following phenomena are found to occur.

1. Hydrolysis of Fe_2O_3 produced by corrosion or iron bacteria, to give H^+ ions, the concentration of which develop very high, giving rise to acid corrosion sometimes again giving ferric oxide or salt as the corrosion product; and the cycle of hydrolysis and acid corrosion continues.
2. The area beneath the deposit shelters anaerobic bacteria which produces weak acids like H_2CO_3 and H_2S which corrodes iron surface rapidly by different mechanisms and in multiple ways.
3. Overheating of the metal surface below the deposit associated with differential oxygen concentration make the area below the deposit strongly anodic with respect to surrounding surface thus leading to underdeposit electrochemical corrosion. Also overheating weakens the metal structure by high temperature intergranular corrosion.
4. Deposit also creates stress on the metal surface leading to stress corrosion.
5. Deposits inhibit corrosion inhibitors from reacting on the metal surface to impart protection and electrochemical corrosion between inhibited and non-inhibited zones occurs.

Fouling met in ammonia cooling system are of the following types:

1. Crystallisation fouling caused by precipitation of scale forming salts like $CaCO_3$, $Ca_3(PO_4)_2$ etc.

2. Particulate fouling caused by deposits of water-borne suspended particulate matters like suspended matters, dust etc.
3. Microbiological fouling caused by slime forming bacteria, fungi and algae.
4. Oil/Hydrocarbon fouling caused by oil, naphtha etc.
5. Corrosion fouling caused by in-site deposits of corrosion products.
6. Foreign material fouling caused by ingress of foreign materials, e.g. wood piece, pebbles, aquatic insects, etc. into the system.

Keeping metal surfaces in contact with water clean and free from deposits is the ultimate solution to get rid of the problem. This can be achieved by:

1. Proper microbiological control and monitoring.
2. Proper maintenance of water velocity through equipments and monitoring.
3. Avoiding stagnant zones in 'water in shell soda coolers' by reverting to latest design modification.
4. Incorporating suitable antiscalant (e.g. organic phosphonates) and antifoulant (e.g. polyacrylates) along with chemical inhibitory treatment.
5. Applying on steam desludging and chemical cleaning whenever deposits are felt to form.
6. Arresting ingress of foreign debris into the system, e.g. by incorporating modification of strainer system in the cooling tower.

Only when the heat transfer surfaces are kept clean and free from any fouling and deposit, the corrosion in the cooling system may follow predictable electrochemical mechanisms and the conventional corrosion inhibition vocabulary become applicable.

In conclusion, it may be remarked that proper selection of materials of construction, proper design of equipments and strict process control are the primary requirements and must be given due consideration before starting fabrication of plant equipments. In most of the sections, corrosion is due to process-side environments and once materials of construction and design are fixed nothing could be done excepting controlling recommended process parameters which then also become inefficient to resist corrosion of faulty equipments. In other areas such as boilers and heat exchangers (water cooled) where corrosion occurs due to externally controllable aqueous environments, faulty selection of materials and faulty design would give rise to permanent source of corrosion failures and it can not be arrested even with the most efficient treatment programme of the corrosive aqueous environment.

Production of urea

The rapid growth of the urea industry has been associated with a strong demand for a cheap fertiliser with a high nitrogen content. There are two basic urea process types on the market today. One is the 'total recycle process', while the other is the 'stripping process', offered by Snam Progetti (ammonia stripping) and Stamicarbon (carbon dioxide stripping). In all these processes the basic chemical reactions are the same; ammonia and carbon dioxide are reacted at high temperature and pressure to form ammonium carbamate.

$$2NH_3 + CO_2 \rightarrow NH_2COONH_4 + heat$$

The ammonium carbamate is dehydrated upon subsequent heating to form urea and water.

$$NH_2COONH_4 + heat \rightarrow NH_2CONH_2 + H_2O$$

In practice it is not possible to obtain complete conversion to urea. The processes essentially differ in the methods by which urea is separated from the ammonium carbamate and the by-products are recycled. An important contributing factor in the growth of urea industry is the introduction of total recycle process, where separation of urea from carbamate and dissolved gases is performed by stepwise reduction of pressure in combination of heating. The residual gases are condensed and feed back to the reactor at high pressure.

Generally speaking urea is not corrosive to stainless steels (the usual material of construction for urea reactor), but ammonium carbamate solutions are. For this reason, most of the corrosion problems in urea plants are encountered where carbamate solutions at high temperature and concentrations are handled, e.g. reactors, strippers, condensers and decomposers. Probably the most severe condition in an entire urea plant can occur in the reactor.

As to the selection of materials of construction, besides corrosion resistance, consideration is to be given to factors like mechanical properties, workability, weldability and economies. Selecting the proper materials of construction, is therefore, a highly specialised job. Stainless steels used extensively for this duty are the austenitic grade AISI 316L, 317L, Sandvik grade 2RE69 and UHB 725 LN etc. Roughly 2 per cent Mo content (as in 316L, 317L and UHB 725 LN) imparts rapid passivation in the carbamate solution. Stainless steel owes its corrosion resistance to the presence of a protective oxide layer on the metal. Corrosion of stainless steel occurs during contact with liquid phase. Under equilibrium conditions the oxygen content of the liquid is determined by the partial pressure of the oxygen in the corresponding gas phases and the Huey Coefficient for the system O_2—process liquid. It is stated that in the Stami carbon process 0.1 to 3.0 per cent volume of oxygen reduces corrosion to a negligible level.

Stainless steels with high Mo content are prone to intermetallic phase formation in the heat affected zone. Formation of ferrite strings in welds are to be avoided by the use of proper welding technique as it may give rise to selective attack in carbamate containing media. In order to minimise the chances of carbide precipitation and reduce changes of intercrystalline corrosion due to ammonium carbamate, the steels used must be either stabilised or should have carbon content of less than 0.03 per cent. In the Toyo recycle process, the urea reactor is a titanium lined multilayer carbon steel vessel. Injection of air in titanium clad reaction vessels helps in protection through passivation. It is, however, susceptible to certain problems, for example erosion-corrosion. All equipment in the finishing sections such as centrifuges, driers, cyclones, melters, prill heads, coolers etc. are made up of type 304 S.S. that are resistant to corrosion by urea up to its solution. Most of the corrosion problems are usually carefully monitored by regular NDT test methods, more so during turn around and thus costly replacements are avoided.

Phosphatic fertilisers

Phosphoric acid is a intermediate for fertilisers like ammonium phosphate, triple super-phosphate, nitro-phosphates, high analysis NPK fertilisers, and liquid fertilisers. The wet process, mostly followed today for the production of phosphoric acid either by dihydrate or hemihydrate or anhydrite process, consists of reactors, filters, heat exchangers, evaporators and pumps. Phosphate rocks are normally not corrosive, though mildly abrasive, but the impurities, fluorine, chloride, and silica (reactive) are the constituents which influence corrosive characteristics of phosphate rocks directly or indirectly. A low silica content allows free hydrofluoric acid to remain in the reaction slurry and thereby making it more corrosive. Thus the unpredictable nature of phosphate rock and its impurities, make the selection of materials of construction much more difficult in phosphoric acid and phosphatic fertiliser services.

In the reaction vessel, large percentage of solid is kept in suspension by circling enough liquid from filtration section. This ensures efficient reaction of sulphuric acid with phosphate rock. The equipments involved are reaction tanks, pumps, piping, valves, fittings etc., and require careful selection for their constructional materials. Reasonably suited to this purpose are stainless steels conforming to AISI 316L, 317 and 317L. In case if the chloride content of phosphate rock exceeds 0.5 per cent, use of AISL317L is preferred to AIST 316 and 317. But these materials are not suitable for use in concentration section.

By and large, reaction tanks are constructed in RCC or mild steel with natural or synthetic rubber lining, though concrete tanks require less maintenance cost. Trouble free service is obtained by using rubber lined pumps for handling slurry medium. The filter cloth is polypropylene and polyesters. For most of the pipe lines and instruments carbon steel with rubber lining is commonly used.

Other special alloys which have been found suitable for most of the plants in particular application are Aloyco-20, Carpenter 20, Cooper FA-20, Durimet-20, HV-7, HV-9, HV-90 and other equivalent alloys developed by different producers. Though it is difficult to generalise the suitability of a particular material for all the purposes, materials like HV-9, Narloy, and CAA series have been found satisfactory under severe corrosive conditions existing in phosphoric acid and phosphatic fertiliser plants.

Potash fertilisers

The most commonly used potash fertilisers are potassium chloride, sulphate nitrate, hydroxide and carbonate. The entire demand of potash fertilisers of this country is met from import till this date since suitable raw materials for the production of these fertilisers are not available indigenously.

Corrosion of Other Metals

The steel-zinc data are valid only for two years exposure. The corrosion of zinc is approximately linear with time, but the corrosion rate of steel is not. If other steel data are compared to this, it should be calculated from the corrosion for two years.

If a corrosion problem exists with other metals such as aluminium or magnesium, it should not be assumed that the steel-zinc data estimate the severity of the environments for alloys of these metals. However, if outdoor testing is done with a specific metal or alloy, it will be useful to test zinc and steel at the same time, or to test the metal in question at a site that has been calibrated in order to take advantage of a possible comparison with the existing data.

Solving Corrosion Problems

Outdoor corrosion can be stopped by isolating the metal surface from the moisture. Or, it can at least be inhibited to an acceptable degree by restricting the rate at which it reaches the surface. This is done with impermeable corrosion-resistant metals (nickel/chromium), slightly permeable coatings (paints) or sacrificial coatings (zinc or cadmium). Specifications should be developed by the pollution engineer for acceptable coating thicknesses and quality testing assurance to protect abatement systems.

Unfortunately, most system designs cannot wait for securing long-time data. When such is true, then existing data as close to expected service as possible should be examined. In the meantime it is advisable to start an outdoor testing programme.

MODERN PROTECTIVE COATING TECHNOLOGY

Paints and coatings are usually of all concern both in specification and actual job site practice. They are usually the first area from which the dwindling overall capital budget is cut. As a result, they are often the most obvious area of failure in a waste or water treatment or air pollution control facility.

A small pump, which might control the operation of a facility, can fail and not be noticed until too late. Moreover, when paint begins to chip and peel and rust appears, poor housekeeping is readily evident to everyone. Also, unless frequent recoating is done, equipment, tanks, and other steel must be replaced prematurely. Paint and coating technology has advanced rapidly in recent years. Proper selection of paints and coatings, carefully worded specifications, skilled application, and close field inspection greatly enhance plant performance. At the same time, they reduce both initial capital expenditures and ultimate maintenance costs.

Selection of Paints and Coatings

There are many conditions which must be considered when making a specific end use recommendation for a protective coating system. It is important to remember that the conventional system is not necessarily optimal either in price or performance.

The first area of consideration should be the type of service involved. Will it be immersion, splash, spillage, fumes, weathering, or an environment of salt, air, or industrial chemical fallout graded as to severity? The substrate (steel, aluminium galvanising, concrete, masonry or wood) over which the coating will be applied must be considered. Then, a decision must be made on the type of surface preparation. It must be decided whether the work will be done in the fabricating shop or at the job site.

The physical and mechanical abuse to which the coating will be subjected must be known. For floor coatings the severity of foot traffic, steel-wheeled traffic and rubber-wheeled carts affects the coating choice. The temperature to which the coating will be exposed as well as thermal shock and extremes of both heat and cold are important.

Another area of consideration is the wear factor. This is becoming especially important in pipe and tank linings. There will be abrasion resistance requirements for a coating inside a pipe, slurry tanks, precipitator scrubber, or clarifier mechanism. This relates to the size and nature of the particles, chemical and physical makeup, and the flow rates. In general, it relates to what will be the wear phenomenon which would reduce the life of the coating.

Finally, after all of these performance characteristics have been defined, the cost of the protection enters into the equation. This includes both the initial cost in terms of capital and the ultimate yearly maintenance costs. Maintenance costs are becoming more important.

The selection of protective coating for steel surfaces, especially in new construction, is governed by the degree of surface preparation allowable or tolerable. It is desirable in new construction to specify the form of surface preparation, making certain directions are followed.

The life of a coating system is proportionate to the quality of the surface preparation over which it is applied. It is unwise to apply paints or coatings over mill scale-bearing steel. The steel under the mill scale will be in contact with water and oxygen. Technically with conventional paint systems, mill scale absorbs water which has penetrated the coating. The mill scale is 'popped' from the steel when iron oxide forms. The protective coating also breaks loose. Corrosion undercutting then proceeds at a fast rate. The Steel Structures Painting Council estimates that a coating applied over a blast-cleaned surface will last approximately three and one-half times longer than the same coating applied over power tool-cleaned steel. Therefore, in order to minimise future maintenance costs, it is always recommended that blast cleaning be used wherever possible.

Note that hand blast-cleaning figures do not include costs of scaffolding and removal of sand necessary if this were done in the field. From this, it is apparent that automatic blast cleaning or modified acid pickling is the most economical method of preparing steel.

When shop blasting is specified, priming of the steel is essential. The primer must be capable of protecting the steel during fabrication, and through the final erection and topcoating phases of the project. It must withstand physical abuse during shipment and erection of the steel. Today, primers are readily available which permit the primed steel to be welded or bolted with high-tension bolts. They have excellent corrosion resistance and require negligible touchup in the field. These materials can save up to 50 per cent in painting costs when properly specified and applied.

Selection of Shop Primers for Steel

Steel corrodes through an electrolytic cell reaction. In order for this to happen, there must be an anode, a cathode, and an electrolyte which become a small electrical cell. Anodes and a cathodes may themselves be found on the surface of the steel. Variations in amounts of iron, carbon, or other materials in an alloy, placement of iron oxides as mill scale, or surface contaminants cause this. When water as an electrolyte is present, an electric cell is completed and corrosion begins.

One common way of protecting steel is to remove the water by forming a barrier over it. Protective coatings are a barrier. Another way is to protect it cathodically which is similar to galvanising. Zinc, the most commonly used metal for this type of protection, will sacrifice itself to protect the steel underneath. Inorganic zinc primers, which are used widely, rely on sacrificial protection.

Protective coatings generally are of two basic types, organic and inorganic. Organic coatings include alkyd (often called red lead, zinc chromate, or iron oxide), epoxy polyamide, vinyl, chlorinated rubber and organic zincrich. Inorganic zinc primers, which protect cathodically, fall into the second category. Inorganic zinc primers are classed as alkali silicate inorganic zincs, or ethyl silicate-based inorganic zincs. Of these, the alkali silicates, which are often called water-based, are less desirable. The ethyl silicate-based inorganic zinc primers have a reduced tolerance for surface preparation. Most alkali silicate-based inorganic zincs require a minimum of a near white metal blast cleaning for adequate adhesion over steel. Ethyl silicate-based inorganic zinc can be applied over mill scale, brush-off blasted steel, or commercially blasted steel. As mentioned earlier, shop applied primer must protect the steel during shipment and construction phases prior to topcoating. A properly chosen primer will require minimum touchup prior to topcoating and have a long protective life.

Selection of Field Topcoats

The major physical and chemical properties of inorganic zinc primers have been covered up to this point. It should be stated that inorganic zinc primers may be used in virtually all areas of a water or waste treatment plant or an air pollution control facility. In order to complete the coating systems, a decision must be made on topcoats to be used over the inorganic zinc primer. These general recommendations serve as a guide for topcoat selection.

Non-immersion service

There is a variety of topcoats which may be used over inorganic zinc primers. However, for enclosed steel surfaces, topcoats are not required since aesthetics and direct exposure to chemicals are not considerations.

General weathering

Where resistance to corrosive fumes is not required or in a mild fume environment, an acrylic latex is an excellent topcoat over inorganic zinc. This type has excellent weathering and chalking resistance, and retains both colour and gloss. It is water-based and available in a wide variety of architectural colours.

Acid or caustic fumes

In areas of acid and caustic fumes, a high-build vinyl or chlorinated rubber topcoat is recommended over the inorganic zinc primer. These have been successfully used for a number of years in waste-water treatment facilities. However, they have normally been used over their own generic type primers. This has proven to be costly due to premature failure of the primer and high cost of replacement.

Solvent spillage

In solvent spillage areas, it is recommended that the inorganic zinc primer be topcoated with an epoxy polyamide topcoat. These topcoat are resistant to mild acids and alkalies to heavy solvent conditions. However, these catalysed materials are more inconvenient to use than single component vinyl and chlorinated rubber topcoats commonly used in municipal water and waste treatment facilities. The epoxy polyamide high build materials have greater acceptance in industrial waste treatment facilities.

Severely corrosive environments

These exposures exist in advanced waste-water treatment facilities of the tertiary type or in industrial waste treatment facilities. For recommendations, it is wise to consult with the protective coating suppliers. Generally, modified phenolic and modified epoxy phenolic coatings, applied over inorganic zinc primers, perform well when exposed to acids, alkalies and solvents. However, conventional painters have difficulty applying them, and they are more costly than vinyls, chlorinated rubbers, or epoxy polyamide materials.

High temperature steel coatings

Steel, when operating at temperatures above 250°F normally does not require coating since water will not condense at this temperature. Unfortunately, continuous operations above this temperature cannot be guaranteed.

Most high-temperature stacks, incinerators, and furnaces cycle up and down in temperature. Weekend shutdowns of incinerators are common, causing a cool-down of the entire plant. During thermal cycling and periodic plant turnarounds, the steel cools down, moisture condenses on the steel, and rust forms. For this reason, the steel must be protected.

Surface preparation consists of blast cleaning the steel to a near white metal finish. Next, the steel should be shop primed with 3 mills of an ethyl silicate-based, partially hydrolised inorganic zinc primer. A silicone topcoat is applied in the field. Pure silicones are used for temperature ranges between 1000° and 1200°F.

Silicones may be modified, but the degree and type will determine the temperature range, performance, and cost. Some silicones are modifies for use between 750° and 1000°F, others for use between 500 and 750°F, and still others for use between 250° and 500°F. Since costs increase in proportion to the quantity of silicone resin in the formulation, the system used should be recommended for the temperature range specified.

Immersion service

In immersion service, a white metal blast-cleaned surface conforming to National Association of Corrosion Engineers Standard NACE #1, or pickling to a white finish in accordance with SSPC SP 8-63 with manufacturer's modifications should be specified. This is followed with a 3-mil partially hydrolysed, ethyl silicate-based inorganic zinc primer. Topcoats may then be selected for a given service.

Potable water service

The lining of potable water tanks is commonly done with four or five coats of low solids, vinyl materials. While these materials hold up very well in service, they have poor abrasion resistance. During winter months, ice chunks fall down from the tops of tanks and damage the bottoms. They are very costly because of the labour of applying multiple coats.

A better recommendation is high film build, high abrasion resistant, hard durable coatings for the interior of potable water tanks and process equipment. These are catalysed epoxy materials, either epoxy polyamide or epoxy amine. Normally, they are applied in two coats at dry film thicknesses between 4 and 6 mills per coat. They are applied over inorganic zinc primers to take advantage of shop surface preparation and shop priming. Epoxy polyamides are preferred to epoxy amines as topcoat over inorganic zinc primers.

Waste-water and process water

Here, the topcoat recommendation is a coal tar epoxy material. As a rule, coal tar epoxies are applied in two coats, 7 to 9 mills each, over the inorganic zinc primer.

Other environments and corrosives

It is recommended that a quality coating manufacturer be contracted for recommendations on specific services other than those outlined.

Topcoats for scrubbers

Flyash from both fossil-fired electric generating stations and incinerator waste service systems are acidic and abrasive which must be considered when selecting systems. Most organic tank lining materials used for scrubber installations are not recommended above 250°F. If higher temperatures will be encountered, materials such as acid-proof brick or mortar linings should be used. Such conditions are generally encountered prior to the quenching of scrubber gases. For gases that have been quenched, it is generally possible to line scrubbers with organic coatings.

If organic zinc primers are to be used in scrubber installations, the primed surface should be brush blasted prior to the application of lining material. This will leave a very thin coat of inorganic zinc over the steel, or none at all. Preferably, the scrubber lining material should be applied over practically bright white metal steel. A thin film (approximately 1/2 mill of inorganic zinc) would be allowable as a primer in order to hold the blasted surface. The surface should be cleaned prior to application of the protective lining system.

Materials which have found their greatest application in scrubber lining and demister lining work have been bisphenol–A fumerate, flakeglass polyester lining materials. These are normally applied either by spray or trowel methods. The trowel-applied material must be rolled to align the glass flakes after they are trowel applied. Spray-applied materials do not require this extra step, and are normally less expensive to install. Application generally consists of two coats, 20 mills per coat. Some manufacturers have insisted upon a third coat to yield a dry film thickness of 60 mills for additional safety in scrubber linings and demister lining application areas. The same cautions and design parameters as taken in tank linings should be observed in this application.

Scrubber ductwork

The ductwork of a scrubber system and the stacks coming off it present a different problem from the normal scrubber. The duct stacks are generally made of very light gauge steel. This steel is subject to rapid thermal shocks and severe flexing. Because of this, it is necessary to use a chemically resistant, abrasion resistant, elastomeric material rather than a rigid flakeglass polyester coating. The selection of this elastomeric materials is very critical, since there are many on the market which have experienced failures. Fortunately, there are some which are specially formulated for this service and have proven successful.

The most successful used elastomeric material for steel ductwork has been an abrasion-resistant, chemically-resistant, polyurethane lining. This lining is applied in two coats (20 mils each) over an epoxy polyamide tie coat and inorganic zinc primer. This system has now been accepted and is performing satisfactorily on scrubber installations. Earlier installations of rigid materials, such as flakeglass polyester, had failed in both stacks and ductwork lining areas.

Galvanised steel surfaces

The procedure has always been to passivate the galvanised steel by pretreatment with a material such as two-package wash primer, phosphoric acid, or a vinegar wipe. Then, it is primed with an inhibitive primer, followed by a tie coat and a finish coat. This four-step operation is costly and, in fact, has only marginal performance. The galvanising itself protects the steel very well because of the cathodic protection of the zinc. It is painted, essentially, to improve the appearance. There are available on the market today self-priming vinyl copolymers with high film build characteristics. They are applied at a dry film thickness of 3 mills, and only one coat is necessary. Surface preparation consists of solvent wiping the galvanised surface to remove grease and contaminants. These one-coat systems perform far better than the traditional multiple-coat systems. Over well-aged galvanising, acrylic latex coatings show excellent performance characteristics. Here, the state of the coating art has progressed from an expensive four-step to an easy one-coat system. The vinyl system may also be used over brass, bronze, and aluminium with the same reduction in cost as experienced with galvanised surfaces.

CORROSION-RESISTANT LININGS FOR STACKS AND CHIMNEYS

Power plants and industries discharging gases from sulphur-bearing fuels have always had a need for corrosion-resistant linings in chimneys and stacks. With lower operating temperatures changes in fuels, and growing use of wet scrubbers to met environmental regulations, corrosion resistance is even more critical. In the past, independent liners constructed with acid-proof brick or steel have proven very successful. With the decrease in flue gas temperature and an increase in moisture content, condensation is higher inside the chimney and corrosion is accelerated. Unprotected steel liners are no longer considered suitable. Independent brick linings will still meet these new requirements. A new generation of corrosion-resistant mortars has been developed for use in power plants, refineries, steel mills, incinerators, and chemical plants.

Mortars based on soluble silicate comprise some of the original corrosion-resistant cements used. The first silicate mortars were simply mixtures of fillers, such as silica, quartz, gannister, clays, or barytes, and sodium silicate solution. Mortars of this type harden by loss of water and require exposure to air or heat to set. Construction with such a mortar is extremely low.

Although thin joints are used, the fluid mixture squeezes out if more than three or four courses of brick are laid at one time. In most cases, not over 6 ft. of brickwork per day can be installed. Very careful drying is also necessary. A 30-day period is usually recommended before putting the structure in service. Air-drying mortars are no longer used for brick linings due to these drawbacks and the development of improved mortars.

Chemical Setting Mortars

Chemical setting sodium silicate mortars developed utilise a setting agent which reacts with the soluble sodium silicate to cause the mixture to harden. The setting agent may be either an acid or a compound which will decompose and liberate acid to accelerate the cure. Typical setting agents are: ethyl acetate, zinc oxide, sodium fluorosilicate, glyceral diacetate, hexamethylene tetramine, formamide, and other amides and amines.

Chemical setting mortars are supplied as two-component systems. They consist of liquid sodium silicate solution and filler powder incorporating selected aggregates or as one-part systems in powder form to be mixed with water. Chemical setting mortars take initial sets in 15 to 45 min., and final sets in 24 to 96 hr. or longer, depending on the temperature. Continuous bricklaying is possible because of chemical reaction, and does not require exposure to air or heat. Large quantities of chemical setting sodium silicate mortars have been successfully used in industry for the past 40 years. Mortars of this type are still employed today for many types of acid service.

Potassium Silicate Mortars

New chemical setting mortars have been developed using potassium silicates instead of sodium silicates. Several fundamental properties of potassium silicates combine to make them preferable to sodium silicates. Potassium silicate mortars have better workability due to their smoothness and lack on tack. They do not stick to the trowel nor run or flow from the joints of the brickwork. They possess greater resistance to strong acid solutions as well as to sulphation, and have greater refractoriness. Moreover, they do not effloresce or bloom and have less tendency to form hydrated crystals in the hardened mortar.

Potassium silicate mortars are supplied in two parts—the silicate solution and the filler powder. Mortars are available which utilise organic or inorganic setting agents or a combination of the two. The properties of the mortar are determined by the setting agent used. Such properties as absorption, porosity, strength, and water resistance are affected by the choice of setting agent. For example, organic setting agents will burn out at low temperatures, increasing porosity and absorption. The organic setting agents are water-soluble and can be leached out if the mortar is exposed to steam or moisture. Due to crystal structure formation, the mortars take a longer time to gain strength, remaining in a plastic state for 96 hour or more.

Modified Silicate Mortars

As previously mentioned, the one-part powder from chemical setting silicate mortars have been commercially available for some time. They have not been used extensively because of their higher cost. These one-part silicate mortars are similar to the two-component mortars except they have somewhat lower physical properties. Research and field testing have recently produced new modified silicate mortars with characteristics not previously available in either one- or two-part systems. These new modified silicate mortars utilise different classes of setting agents. The properties of the mortar are dependent upon the setting agent selected.

Monolithic Linings

Many chimneys and stacks have linings of calcium aluminate cements or refractory concrete applied by Guniting, cast-in-place, or troweling. Calcium aluminate cement linings are unsatisfactory if the pH of acid condensate is below 3.5. The alumina gel common to these cements dissolves rapidly in such an environment. Refractory concrete is not acid-resistant, and its additional cost is not warranted when operating temperatures are below 500°F.

Monolithic corrosion-resistant lining materials have been used successfully to restore and repair brick linings in concrete chimneys and steel stacks. These monolithic linings may be applied to both new and existing steel liners by guniting. The original acid-proof concrete lining is an inorganic silicate composition which resists all acids except hydrofluoric, water, oil, most solvents, and temperatures to 1750°F. It is recommended for use over a pH range of 0.0 to 7.0. Supplied in two parts, the powder and liquid are mixed together when used. The liquid is substituted for the water normally used in Guniting.

Modified silicate-base cements designed for use as monolithic linings, to be applied by the cement gun process, are now also being supplied. These are single-component systems which produce a high-strength acid-proof lining when mixed with water. They have extremely good adhesion to concrete, brick and steel, and require only a minimum of surface preparation. These linings are not affected by acids, except hydrofluoric and acid fluoride salts from the lowest pH to 9.0.

CHAPTER 22

Electroplating and Electroless Plating

INTRODUCTION

Electroplating is the electrodeposition of an adherent metallic coating on an electrode in order to form a surface with properties or dimensions different from those of the basis metal.

Electroplating is a surface treatment. The material (work) being treated is made the cathode in an electroplating solution, or bath. Such baths are almost always aqueous solutions, so that only those metals that can be reduced from aqueous solutions of their salts can be electrodeposited. The only major exception at present is aluminium, which is plated on a semicommercial scale from organic electrolytes. Some of the refractory metals (e.g., niobium, tantalum) can also be deposited from fused salts as coherent plates, but little commercial use has resulted from this development.

The thickness of deposit applied by electroplating varies with the application: from as little as 0.025 μm for decorative gold deposits through 25–50 μm for standard nickel-chromium plate on exterior automotive hardware, to 1 mm or more for electroforms.

The properties conferred by electroplating include improved corrosion resistance, appearance, frictional characteristics, wear resistance and hardness, solderability, specific electrical properties, and many others. Electroforming (including electrotyping) is used to manufacture articles that cannot be made as economically in any other way.

SUBSTRATE IN ELECTROPLATING

There is much more to electroplating than the final step of laying down a coating of the plating metal, and much more has to be considered than the properties of the plated metal. Thus it consists not of the deposit alone, but of deposit plus substrate. It is the properties of this combination that often determine the right metal to plate and the right solution from which to plate it.

Preparation of the Substrate

Before a useful electroplate can be deposited on a surface, the surface must be in condition to receive. Useful means, among other things, adherent. The preplating treatments necessary to prepare the surface to accept an adherent deposit are generally aspects of cleaning. The ideal surface would be one consisting entirely of atoms of the metal to be plated upon, and having no foreign material at all. This is virtually impossible to attain, even in the laboratory. A practical definition, then, of a satisfactorily clean surface would be a surface containing no foreign material that interferes with the formation of an adherent deposit. In general, this connotes the removal of gross dirt and soil, heavy oxide or tarnish films, and in some cases surface skins of damaged metal produced by prior mechanical operations.

Cleaning

Choice of the proper cleaning or preparative cycle depends primarily upon the nature of the substrate to be prepared and the nature and amount of the soils to be removed. Ferrous and nonferrous metals generally require different types of cleaners. The more active metal such as aluminium require special techniques to prepare them for plating. And obviously, the more contaminated the surface the more cleaning it will require. A typical cleaning cycle includes the following steps: pickling to remove gross scale; any mechanical preparation such as polishing or buffing; cleaning to remove oils, greases, shop dirt, and polishing and buffing compounds; acid dipping to remove oxide films; and rinsing.

Rinsing

Adequate rinsing between all steps in cleaning and plating is of the utmost importance. Rinse waters should be clean and should not be allowed to become contaminated by drag-in of preceding solutions. Countercurrent rinsing is often employed as a means of conserving water while ensuring that the last rinse is relatively pure. Hot water is more efficient than cold for removing contaminants; on the other hand it entails the risk that the work may dry before entering the next operation. Quality of the available water supply is often of importance; softened, deionised, or distilled water may be required for final rinses in many instances.

Adequate and efficient rinsing assumes additional importance in relation to waste disposal. To the extent that cannot be fed back to the plating cycle, the last rinse constitutes effluent form the plant. Almost universally it must be treated to reduce harmful contaminants forbidden by EPA regulations, and the less of these it contains the less costly waste treatment will be.

Acid dipping

Acid dipping, which generally follows cleaning, serves two purposes: it removes slight tarnish or oxide films formed in the cleaning step, and it neutralises the alkaline film that even good rinsing cannot completely remove from the surface. It is thus particularly important when the plating solution is acid. The acid dip is usually a 10–30 vol% solution of hydrochloric or sulphuric acid, the latter solution being more common.

Drag-out and drag-in

Every solution in a plating cycle is contaminated, to a greater or lesser extent, by the solution that precedes it in the cycle. How serious this situation is depends on many factors, some controllable and some not. The shape of the parts and how they are positioned in the tank is of great importance; drain times vary, and some solutions are more free-rinsing than other. Contamination caused by this drag-in may be serious or inconsequential. The same is true of drag-out which also must be considered in connection with the problem of waste disposal.

Special preparation cycles

For plating on the common substrates such as ferrous metals and copper and its alloys, the preparation cycles briefly outlined above usually suffice. Other substrates require more specialised treatment. Among those requiring special preparation are aluminium and magnesium; zinc-base die castings; refractory metals like titanium, zirconium, tantalum, niobium, molybdenum, and tungsten; and non-metallics like synthetic plastics, leather, wood, and plaster.

Electroplating Process

The operations of electroplating, including the cleaning, rinsing, plating, and postplating treatment, can be carried on manually or with almost any desired degree of automation. Parts may be hung in the plating tank on wires or on racks; when many small parts are to be plated, they may be contained in wire baskets or, more commonly, in barrels that rotate in the plating tank. Movement from one operation to another may be by hand or by machine.

The necessary d-c power is derived from motor-generators or, increasingly, by rectifiers. Solid-state rectifiers may be of three types: selenium germanium, or most important, silicone. Power is conveyed to the plating tanks by bus bars; the anodes are hung into the tank from the positive bus bar, usually along the two sides and the work to be plated from the negative or cathode bar, usually down the centre. Tank voltage is read from a voltmeter, and current from an ammeter: these two instruments should be available for each plating tank. A third instrument, an ampere-hour meter, is often helpful as a means of regulating the thickness of deposit and for general control of the operations.

Some agitation of plating solutions or of the work is usually helpful. The oldest and simplest method consists of an operator merely swishing the work around at intervals, but automatic cathode-rod agitation is preferable.

Temperature control is almost always desirable in plating operations because the characteristics of plating solutions, of the deposit, or of both usually depend to a large extent on the temperature of operation. Heating or cooling coils in the tank itself may be used, or the solution may be circulated through a heat exchanger. Electric immersion heaters may be used. Occasionally, plating tanks are heated by open gas flames beneath the tank.

In addition to the basic equipment–power source, plating, cleaning, and rinsing tanks, and bus bar–most plating installations require one or more of the following: filters, for either continuous or intermittent purification of solutions; drying facilities, which may range from a simple jet of compressed air to large ovens; racks or design appropriate to the part being plated, and a racking station where the work can be conveniently hung on these racks and unracked after plating; one of more stripping tanks for stripping faulty deposits so that parts can be reworked and for stripping the plating racks themselves; reclaim tanks if the drag-out is valuable and worth reclaiming; portable pumps for transferring solutions; and at least one empty tank so that a plating tank can be emptied and worked on.

In addition to the equipment required for electroplating and allied processes, the modern plating shop must include apparatus for waste treatment, metal recovery, or both.

Continuous plating

Electrolytic processes are well-suited to plating continuous coils of strip or wire. The substrate can be uncoiled, pickled, cleaned, plated, given postplating treatments, and recoiled in one continuous operation.

Materials of construction

Plating tanks and auxiliary equipment are constructed of materials resistant to the particular solution involved. This usually means plain steel for alkaline solutions and rubber-or synthetic-lined steel for acid solutions.

Safety

Hazards in plating operations arise from the nature of the materials routinely handled, many of them highly toxic. Thus, certain minimum precautions are absolutely necessary. Most of the normal hazards of plating room can be adequately handled by a combination of proper ventilation and appropriate protective clothing.

Waste disposal and metal recovery

There are two basic approaches to waste treatment: destruction and recovery. Original emphasis was on the former; the objectionable metals were precipitated as sludges and disposed of in landfills, or similar means. As landfills become both less available and more restrictive in their acceptance of such sludges, emphasis has shifted to recovery of the metal for reuse in the plant or for sale to a refiner.

The disadvantages of precipitation methods have turned attention more and more to recovery of the valuable metals from effluents before their discharge. Among the methods proposed for recovery systems are reverse osmosis (RO); evaporative recovery; ion exchange, and various combinations of two or more techniques. Most of these methods are technically feasible and the principal snag is their relatively high cost, large capital expense for equipment, energy for evaporation, membrane life for RO, etc.

Plating Solutions

Plating solutions are usually aqueous. Every plating solution contains ingredients to perform at least the first, and usually several of the following functions: provide a source of ions of the metal(s) to be deposited; from complexes with ions of the depositing metal; provide conductivity; stabilise the solution against hydrolysis or other forms of decompositions buffer the pH of the solution; regulate the physical form of the deposit; aid in anode corrosion and modify other properties peculiar to the solution involved.

Current density range

Current density is the average current in amperes divided by the area through which that current passes; the area is usually nominal area, since the true area for any but extremely smooth electrodes is seldom known. Units used in this regard are A/m^2 (92.9 mA/ft^2). Current densities at the anode and cathode are both important; they may differ considerably although not by so large a factor as to make necessary very great differences between the anode area and the cathode area. Most solutions exhibit a range of current densities within which deposits are satisfactory and outside of which they are not.

Throwing power

Except in the special case of concentric anode and cathode, the current density over the electrodes varies from point to point. In general, the area on the cathode nearest to the anode receive a higher current density then those more remote.

Thus, more metal is plated on the projections than in the recesses. Many plating solutions, however, have the ability to moderate this condition to some degree. Throwing power may be defined as the improvement in metal distribution over primary current distribution on a cathode. (Primary current distribution is the distribution of the current that depends only on the geometry of the cell).

Acidity

Plating baths are either acid, neutral, or alkaline, and for most of them close control of the pH is essential to successful operation.

Anodes

Anodes in a plating bath perform two functions: they act as the positive electrode, introducing current into the solution; and, in most cases, they replenish the metal deposited at the cathode, thus maintaining the balance of the bath.

Temperature

Control of temperature is important in almost all plating processes; each solution is characterised by a range of temperatures within which it gives best results. Temperature affects almost all the variables of the solution: conductivity, current efficiencies, nature of the deposit, and stability.

Current efficiency

Faraday's laws predict the amount of metal that is deposited at the cathode or dissolved at the anode, but these amounts are not always obtained; the deficiency is owing to evolution of hydrogen at the cathode or oxygen at the anode, or to other side reactions. In practical plating operations, current efficiency is not usually of direct concern so long as it is known, but it is often important to equalise the efficiencies at cathode and anode so that the bath remains in balance.

Purity

Plating processes differ in sensitivity to the presence of impurities in the bath. Specific means are available for purifying most plating baths by chemical treatment; filtration through activated carbon is a generally useful method for removing organic contaminants.

Bright plating

Most deposits from simple plating solutions are mat unless the substrate is bright and the plate very thin. If the use is purely functional, this is no drawback, but for decorative purposes a bright appearance is usually desired. This brightness was formerly obtained by mechanical treatment, buffing after plating, but today most metals can be bright as they come from the bath, and require little or no further treatment, at least as regards appearance. This bright appearance is produced by the addition to the plating bath of small amounts or one of more addition agents or brighteners, usually organic compounds.

Maintenance of plating baths

All plating baths require more or less chemical control. They must be analysed, often or seldom, depending on individual circumstances, and adjustments must be made. For some baths a determination of specific gravity suffices for routine control; other baths require frequent and complete analysis.

Individual Plating Baths

Plating baths include cadmium, which affords good corrosion protection to steel and other substrates in marine atmospheres; chromium, the final finish on the great majority of plated consumer items; cobalt, used for its magnetic properties; copper, used most often as an undercoat for other metals; gold, which has become extremely important in the electronics and computer industries because of the low contact resistance, corrosion resistance, and solderability of the metal; indium, used in bearings; iron, which has minor electroforming and other engineering uses; lead, used in battery parts and chemical construction; nickel, used in a host of engineering uses; platinum-groups metals, used for decorative applications as well as for engineering purposes; silver, used for its electrical and mechanical characteristics; tin; zinc, plated on all types of hardware; and various kinds of alloy plating.

Nonelectrolytic Plating Processes

There are many other ways than electroplating to deposit a coating of metal on a substrate. Hot dipping, vacuum evaporation, chemical vapour deposition, and various aqueous processes not requiring current are some of the best developed. Nonelectrolytic aqueous deposition includes immersion plating and chemical, autocatalytic, or what has come to be known as electroless plating.

Postplating Treatments

Postplating treatments include chromate conversion coatings for zinc and cadmium phosphate treatments for iron, and various passivating and brightening dips.

Applications

Applications for plating are classified according to the principal function of the plate; thus the plating may be applied mainly for appearance, protection, special surface properties, or engineering or mechanical properties.

ELECTROLESS PLATING

Electroless plating is the controlled autocatalytic deposition of a continuous film by the interaction in solution of a metal salt and a chemical reducing agent. Electroless deposition can give films of metals, alloys, compounds and composites on both conductive and non-conductive surfaces. The theory and practice of electroless plating parallels that of electrolytic plating. The main difference is that the electrons used for reduction are supplied by a chemical reducing agent present in solution. This means that electroless solutions are not thermodynamically stable because the reducing agent and the metal salt are always present and ready to react.

Electroless solutions contain a metal salt, a reducing agent, a pH adjuster or buffer, a complexing agent, and one or more additives to control stability, film properties, deposition rates, etc. Only a few metals —nickel, copper, gold and silver—are used on any significant scale. The ideal electroless solution would deposit metal only on an immersed article, never as a film on the sides of the tank or as a fine powder. Many electroless copper and electroless nickel baths now closely approach this ideal.

A great advantage of electroless solutions is their ability to give conductive metallic films on properly prepared nonconductors, and their ability to uniformly coat any platable object.

Equipment

Plating tanks are preferably plastic or lined with a plastic or rubber coating. The tank linings must be stripped of metal deposits with acid at periodic intervals. Tank and rack design are less important than tank loadings and rack coatings, especially when plating nonconductors. The critical point is not to overload the plating bath. Electroless baths rapidly become overactive and decompose when too much surface area is being plated at once. Most vendors recommend a maximum loading of ≤ 365 cm^2/L (1.5 ft^2/gal).

Safety and Waste Disposal

Most solutions are not particularly hazardous although they must be highly acidic or basic. All reducing agents (solids or liquids) should be stored separately from oxidising substances such as chromic acid, in case of spills. Waste treatment should always conform to current rules. The vendors' recommendations should be followed for specific cases.

Plating on Metals

The first large-scale process was the Kanigen electroless-nickel process which used a hypophosphite reducing agent. The Ni–P alloy (1–15 wt per cent P) has good corrosion resistance, lubricity and especially hardness. It can be heat-treated to a hardness equivalent to electrolytic hard chromium and its lubricity is comparable. Thus, it is not surprising that many applications for electroless nickel are in replacement of hard chromium.

The advantages over hard chromium include safety of use, ease of waste treatment, plating rates of as much as 40 μm/h, low porosity films and the ability to uniformly coat any geometric shape without burning or use of special anodes. Electroless nickel has superior corrosion and erosion resistance as compared to electrolytic nickel. It is used extensively on moulds, pistons, pump parts, oil-field equipment, dies, compressors, tanks and piping. Electroless-nickel systems based on dimethylamineborane or sodium borohydride reducing agents have advantages in the plating of titanium for the aircraft industry. They are also used to minimise the use of or completely replace gold in connector applications in the electronics industry. Specialised nonmagnetic electroless nickels are becoming the coatings of choice for aluminum memory disks in the computer industry.

Plating on Nonconductors

All nonconductors can be electrolessly plated but only a few can be plated to give good adhesion and appearance.

Plating of plastics

This market began on a very small scale using electroless copper. It was soon discovered that ABS engineering thermoplastic was the easiest plastic to plate. Plated ABS has about 90 per cent of the market; most of the remainder is modified polycarbonate and modified poly(phenylene oxide). Automotive items make up over 60 per cent of the market on a plated-area basis; the remainder is hardware, plumbing and decorative items. The largest new growth market for plating on plastics is the field of EMI/RFI (Electromagnetic Interference/Radio Frequency Interference) shielding. Pure electroless nickel, pure electroless copper and combination coatings of copper plus nickel are being used. A typical process cycle for plating on plastics is shown in Table 22.1.

Table 22.1. Plating on plastics[a].

Step	Solution	Temperature °C	Time, min
Etchant	CrO_3–H_2SO_4	60	5–10
Neutraliser	Mild Cr^{6+} reducer	25	1/2–2
Catalyst	$PdCl_2$–$SnCl_2$–HCl	25	1–5
Accelerator	Mild acid or alkali	50	1–2
Electroless nickel bath	Nickel salts, complexer, buffer, hypophosphite; pH 8–10	25	5–10
Electroless copper bath	Copper salts, complexer, buffer, formaldehyde; pH 12–13	25–40	5–10

[a] As in all plating processes, thorough rinsing is necessary after each step. This is typical ABS plating cycle; for other plastics plating differs mainly in the pretreatment or etchant and neutraliser steps. The electroless metal deposit is typically 0.15–0.5 μm thick; electrolytic plating follows.

Printed circuits

This is by far the most diverse field of electroless plating. Numerous variations exist in solution process, and techniques, both in laboratory and commercial form and are used to create a great variety of products. All have the basic purpose of producing a layer of highly conductive copper in specified areas. Modern electroless copper films have a ductility and conductivity identical to that of electrolytic copper.

Printed-circuit boards are composed of epoxies, phenolics and other heat-stable dielectrics; flexible films of polyester and polyimide are also plated. The ratio of rigid to flexible surface areas plated is about 10:1. They are used in communications, instruments, control, consumer markets, military, aerospace and business applications. Multilayer printed circuit boards are rapidly increasing their market share, especially for computer applications.

Other Applications

Mirror production is the primary applications of electroless silver. The other important commercial glass-plating application is for production of architectural reflective glasses. An electroless-nickel matrix is used commercially to securely bond diamonds to cutting tools.

ALKALINE NON-CYANIDE ZINC AND ELECTROLESS NICKEL PLATING

Electroplating and electroless plating techniques can be widely employed for superior corrosion protection. Developments in plating technology have led to more versatile, economical and environment-friendly processes. Two types of plating processes, alkaline non-cyanide zinc and electroless nickel, which have several industrial applications, are discussed here.

Plating of metals on steel and other substrates has been widely employed not only for decorative purposes but also for a number of functional reasons such as improving corrosion, abrasion and galling resistance, imparting hardness and solderability, etc.

Technologists in the field of metal finishing have been active in developing economical and less polluting plating processes.

Surfaces can be plated with cadmium, chromium, copper, gold, nickel, palladium, silver, tin, zinc and alloys like brass, bronze and solder. Of the variety of plating material available, non-cyanide alkaline zinc and electroless nickel are of particular interest to engineers from functional and environmental considerations.

Alkaline Non-cyanide Zinc Electroplating

Electroplating of zinc has many positive features compared to other methods (such as hot dip galvanising and metal spraying) of coating zinc on steel. Fine dimensional limits can be maintained and there is no filling up of slots and screw holes, nor are threads and other fine work details obliterated. Thicknesses can be more accurately controlled, effecting cost savings. These finishes can be very attractive as well.

Zinc is an electronegative metal and provides sacrificial protection when coated on steel. Even if the zinc finish is scratched, galvanic action continues to prevent ferrous metal from rusting. Zinc is very economical to plate and world wide supplies of the metal are high. It is also non-toxic and the plating processes are relatively simpler. The trend is zinc plating has moved away from the traditional cyanide based processes to acid chloride and non-cyanide alkaline processes, which, besides being more environment-friendly, offer other advantages as well.

Cyanide based processes, while simpler to operate, are increasingly finding disfavour due to environmental reasons. Acid chloride processes require more elaborate equipment and control, and have the disadvantage of rather uneven thickness distribution in the electro-deposits. Non-cyanide alkaline processes had a major disadvantage of yellowing and blistering, but advances in plating technology have helped reduce these problems considerably.

Process overview

The pre-treatment prior to plating of non-cyanidealkaline zinc is similar to that required by most electroplating processes and involves mainly cleaning and rust/scale removal. Cleaning is best carried out by immersion in properly formulated soak cleaners followed by electrocleaning. This would help remove soils, machine and other oils, grease, polishing and buffing residues, etc. Rust removal can be effected by sulphuric acid solutions in which the attack on the base metal is retarded by inhibitors. If scales due to heat treatment or other causes are present, electrolytic descaling solutions can be employed.

Typical composition and operating parameters for alkaline non-cyanide zinc plating are summarised in Table 22.2. Articles such as washers, bungs, nuts, bolts, screws, etc., can be conveniently plated in revolving barrels. The minimum thickness to be plated would depend on service life expectancy and cost considerations. For better corrosion resistance, thicknesses of 8–12 microns are usually specified.

Table 22.2. Alkaline non-cyanide zinc plating: Typical bath composition and operating parameters.

Zinc metal	8 – 10 g/l
Sodium hydroxide	80 – 100 g/l
Proprietory additives	As per supplier's specifications
Temperature	20 – 27°C
Current density	1–3 A/dm^2(vat)
	0.5 – 1.5 A/dm^2 (barrel)
Anodes	99.9% pure zinc

Chromating ('passivation') post treatments enhance the corrosion resistance of zinc plating. Various types of chromating processes are available, affording different degrees of corrosion protection. Neutral salt spray (ASTM B117) results for these chromating treatments are given in Table 22.3. Corrosion resistance of zinc plated articles can be further enhanced by means of water solution dip lacquers after chromating.

Table 22.3. Neutral salt spray test (ASTM B117) results for alkaline non-cyanide zinc.

Thickness of electroplate: 10 microns.

	Number of hours elapsed before first appearance of rust	
Type of chromate	*White rust*	*Red rust*
Clear (blue)	24	96
Black	72	180
Yellow	108	240
Olive Drab	200	300

Properties

An important property of alkaline non-cyanide zinc plating is the excellent thickness distribution provided. Thus, recessed areas are also covered properly and overall thickness requirements may be brought down. The grain structure of the electrodeposit also helps achieve maximum corrosion protection compared to deposits from acid chloride and even cyanide based processes. The deposits are bright and fairly ductile, have a hardness in the range of 80–150 HV_{100} and possess good formability.

Drawbacks

Relatively difficult metal control in the baths and lower cathode efficiencies (50–80 per cent) are major drawbacks of the alkaline non-cyanide zinc plating processes. Cathode efficiency is the current actually used for electrodeposition against the total current passed through the system. Some of the current is used up in electrolysis with the consequent evolution of hydrogen gas at the cathode. The cathode efficiency decreases at higher current densities and thus, the recommended upper limits for current densities are 1 A/dm^2 for barrel plating and 2.5 A/dm^2 for rack electroplating. Working temperature of the baths needs to be maintained below 25°C for best results.

Electroless Nickel Plating

Electroless nickel deposits are amorphous alloys of nickel and phosphorus, produced by chemical reduction of nickel ions over autocatalytic or catalysed surfaces. While electroless nickel processes are more expensive and require stricter control compared to many other plating processes including electrodeposited nickel, they have several distinct advantages. An important feature of electroless nickel deposits is their remarkable uniformity, enabling tolerances as close as 1.2 microns to be obtained in 'plated' conditions.

The superior corrosion resistance of electroless nickel makes these coatings suitable for application in many chemical industries including fertilisers, petrochemicals, oil and gas exploration, pulp and paper, textiles and food processing. Electroless nickel coatings are now replacing chromium coatings, which are traditionally plated in hazardous hexavelent chromium baths.

Unlike electroplating, which can be carried out only on conducting surfaces, electroless plating can be carried out on a wide variety of substrates. These include carbon steels, iron, aluminium alloys, magnesium, copper, brass, bronze, plastics and ceramics. The percentage of phosphorus which is codeposited and consequently the properties, vary with bath composition and operating parameters. A few of these are listed in Table 22.4.

Table 22.4. Operating parameters and deposit characteristics for different types of electroless nickels.

	Plating on plastic	Low phosphorus	Medium phosphorus	High phosphorus
Nickel metal, g/l	3.3	6.0	5.8	5.8
Plating rate, micron/hr	2–3	6–10	15–17	8–10
Operating temp., °C	25–35	75–80	85–90	85–90
Operating pH	9.0–9.5	6.1–6.4	4.6–5.2	4.5–5.0
Phosphorus content wt%	–	1–3	7–9	10–12
Hardness as 'plated', HV_{100}	–	700	550	500
Hardness after heat treatment, HV_{100}	–	1,000	950	950
ASTM BN117 salt spray, hours (25 micron thickness)	–	96	168	500+
Magnetic properties	–	Magnetic	Slightly magnetic	Non-magnetic

Process overview

Electroless nickel can be plated directly onto clean steel surfaces. Metals like aluminium first need to be given a zincate coating. Plastics like ABS and ceramics need to be given an etching treatment followed by treatment in palladium-colloid 'activator' solutions to make their surfaces catalytic for the nickel reduction.

Electroless nickel baths formulated for providing thickness above 5 microns generally are operated at temperatures above 85°C and need to be agitated by oil-free compressed air. Process tanks and all equipment coming in contact with the electroless nickel solutions need to be constructed of suitable non-metallic materials such as polypropylene.

Post plating heat treatment in an inert atmosphere may be performed to increase hardness if required, at a temperature of 400°C for one hour. Baking of the deposits to enhance adhesion may be done at temperatures of 250–300°C for steel substrates and 150–200°C for aluminium substrates.

Properties

1. Essentially amorphous as 'plated' structure.
2. Deposits can be plated to tolerances of upto 1.2 micron.
3. Bond strengths of 2,800–4,200 kg/cm^2 (steel) and 1,050–2,450 kg/cm^2 (aluminium) (Ring Shear method).
4. Hardness, Vickers (HV$_{100}$)

              : 500–700 (as plated)

              : 900–1,000 (after heat treatment)
5. Corrosion resistance (number of hours required to fail in neutral salt spray test (ASTM B 117): more than 500.

Applications

1. Chemical process industries: Heat exchangers, filter units, pumps, impellers, housings, mixing blades, tubing, valve components.
2. Other process industries (e.g., pulp and paper, textile): Cylinders, rolls, bobbins, pumps, frames.
3. Oil and gas exploration: Packers, tubulars, valve components, drills, collars, plugs, valve components, fire tubes.
4. Food processing: Gang knives, slicing blades, conveyor chains, bowls, pumps, mixing blades, packaging equipment.

Drawbacks

Build up of reaction products leads to deterioration of deposit properties after prolonged use. The baths have essentially limited lives and are expensive to operate.

Thus plating processes can provide solutions to many protection problems. Alkaline non-cyanide zinc plating provides a cost effective and aesthetic corrosion resistant finish, while electroless nickel, though relatively expensive, has several remarkable properties. These plating processes are versatile and available to meet specific end uses, and are economic and eco-friendly.

CLEANER PRODUCTION OPTIONS TO ELECTROLESS PLATING PROCESS

In the printed circuit board manufacture, plated through hole technology is common for making drilled holes conductive. This is an important process step and conventional electroless process is employed.

During electroless plating process, a thin seed layer (0.2–0.3 micron) of conductive material, copper is deposited on insulating area to facilitate subsequent deposition of higher thickness of copper by electroplating. In this process copper is reduced from its ionic state in solution by means of a reducing agent. The reducing agent reduces the metal ions in solution to neutral atoms for deposition. The process provides a continuous build up for copper coating on the substrate. The electroless plating line typically contributes a significant percentage of a PCB manufacturers overall waste volume. The use of electroless copper chemistry generates two forms of liquid waste i.e., pollution within rinse waters and solution bale out which is treated as an item of spent chemistry. There are alternate methods of making through holes conductive which are:

1. Direct metallisation under which three different processes have been developed using carbon or graphite.
2. Palladium methods are of two types viz. organic stabilised method (also known as Neopact process) and the tin stabilised method.
3. Conductive polymer method which deposits a conductive polymer layer on the substrate surface of the hole.
4. Non formaldehyde based electroless copper where the hazardous reducing agent is replaced with alternative chemistry.
5. Conductive ink technology which is effective in double-sided surface mount applications.

ELECTROLYTIC MACHINING METHODS

One of the promising novel machining methods for extremely hard alloys is the atom-by-atom removal of a metal by anodic corrosion which has been called electrochemical machining (ECM). Because the metal is removed atom-by-atom, ECM affords the opportunity to machine a given workpiece without work-hardening, burring or smearing the metal, without regard to the hardness of the metal being cut and virtually without tool wear.

In a chemical reaction, electron transfer between the reacting species occurs by the oxidation (loss of electrons) and the reduction (gain of electrons) taking place at the same site and the energy liberated appears as heat and possible light. For an electrochemical reaction to occur, the oxidation must take place at a site remote from the reduction. This situation is accomplished by interposing an electrolyte between the conductor (anode) at which oxidation occurs and that (cathode) at which reduction occurs.

When iron dissolves, ferrous ions enter the electrolyte by donating electrons to the anode. Eventually, the anode becomes so negative that further dissolution of iron is prevented by the electrostatic attraction between negatively charged anode and the positive ions in solution. If a sink for the electrons on the anode is provided by a battery that is connected in an external circuit between the anode and cathode, continuous dissolution of the iron anode can be obtained. Current is carried in the circuit by electrons (electronic conduction), and internally by ions (electrolytic conduction) through the electrolyte. The dissolution of metal at an anode driven by an external source of current, such as a battery or rectifier, is termed anodic corrosion.

Principles of the ECM Process

In the ECM operation, the metal workpiece to be shaped is the anode and the tool that produces the shaping is the cathode. The electrodes are connected to a low voltage source of direct current (dc). The anode and cathode are held in close proximity (0.5 mm) by a properly designed fixture and a solution of strong electrolyte is pumped at thigh rates (2–6 m/s) between the two electrodes.

If there are no side reactions, the passage of each Faraday of electrical charge (96,500 C) results in the dissolution of an equivalent weight of metal. Under these conditions, very high current densities may be passed (50–500 A/cm^2) so that high metal removal rates [2 cm^3/(min.kA) for 4340 steel] may be obtained with good tolerances, with integrity of the geometry and with excellent surface finishes. Voltages applied between the anode and cathode are normally 5–25 V.

Types of ECM operations are shown in Fig. 22.1. Innovations in ECM operations include static fixture-finishing and sizing, embossing and broaching.

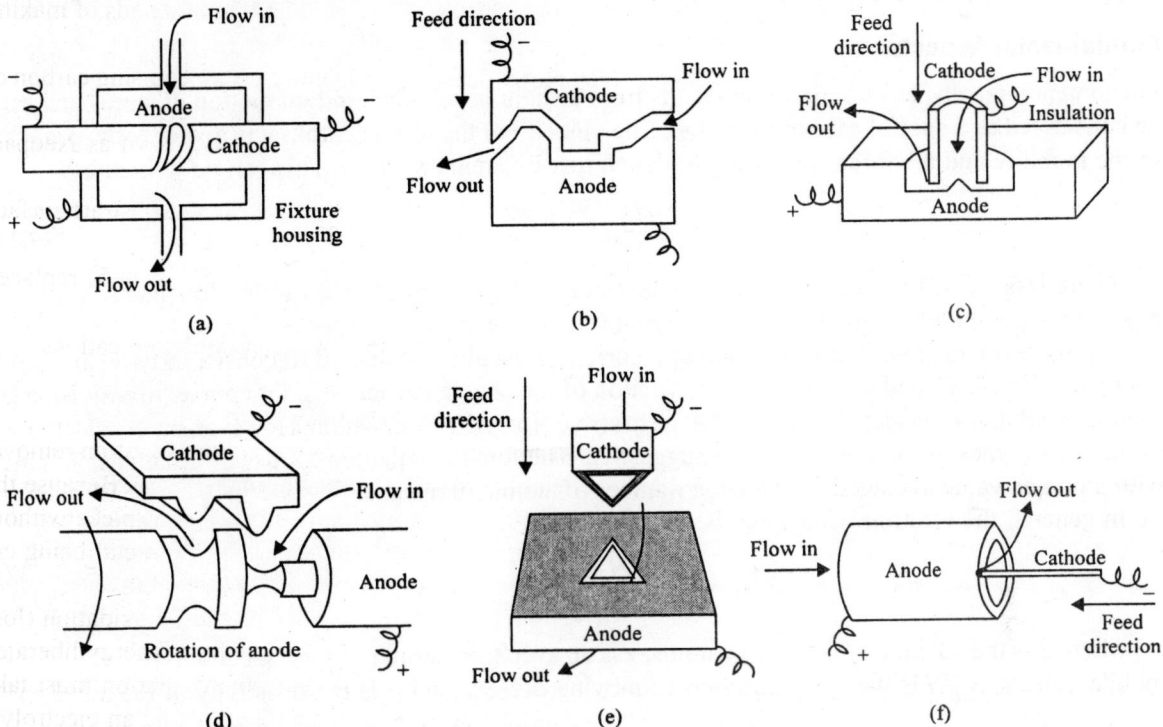

Fig. 22.1. Various ECM operations (a) External shaping; (b) Cavity-sinking; (c) plunge-cutting; (d) turning; (e) trepanning; and (f) internal grooving.

ECM is commonly employed in the transportation and aerospace industries. Turbine blades, vanes, disks and transmission parts are made by ECM. Electrochemical machining is also used in the manufacture of precision guidance parts for missiles.

ELECTROMIGRATION

Electromigration or electrotransport refers to the net motion of atoms owing to the passage of an electrical current through a metallic conductor in either the solid or liquid state. Simple metals may be considered to be constituted of atoms stripped of their valence electrons, existing as ions surrounded by a sea of electrons. In the presence of an electric field, the ion cores are subjected to a force directly resulting from the field, as in an electrolytic solution. However, in electromigration there is, in addition to this direct field force, a force resulting from the friction between the numerous and rapidly moving electrons and the ion cores. This force, known as the electron-wind force, is in the opposite direction to the field force in electron conductors and for many metals is considerably greater than the field force. In metals with complex electronic structures that are not electron conductors but are hole conductors, the concept of electron-wind force is extended to refer to the friction resulting from the motion of the charge carriers regardless of their sign. Whereas in electrolytic solutions the current results entirely from the motion of ions, in usual electromigration phenomena the fraction of the current contributed by ion motion is quite negligible in comparison to that fraction of the current caused by the motion of the charge carriers, either electrons or holes.

Fundamental Aspects

Phenomenologically, electromigration results from a slight bias in the random motion of atoms and can be considered as a special case of diffusion. The velocity of the moving atom v is given by the product of the mobility and the force, according to the Nernst-Einsteing relation. This can be written as:

$$v_i = \frac{D_i}{kT}.F_i \qquad\qquad ... (22.1)$$

where D is the diffusion coefficient, F is the force on the moving atoms, k is the Boltzmann constant, T is the temperature, and the subscript i refers to the atoms of the ith species.

The force on an atom in the presence of a current is usually considered to consist of two parts, one owing to the field F_e and one owing to the friction of the charge carrier F_{wd}. Of course, this division is somewhat arbitrary and ideally one would hope to arrive at a single formulation. However, the consideration of those two types of force is intuitively suggestive. The force in equation 22.1 is of a statistically kind with a unique value averaged over a large number of atomic jumps.

In general, the electron-wind force is expressed as:

$$Z_{wd}^* = -Z.\gamma.\frac{N.\rho d}{N_d.\rho} \qquad\qquad ... (22.2)$$

where Z is the valency of the matrix atoms, γ is an averaging term, ρ_d is the specific resistivity of the mobile defects, N_d/N is the concentrations of moving defects, and ρ is the resistivity of the metal.

Common experimental techniques

For most solid materials, the diffusion coefficients and the effective charges are sufficiently small so that experimental measurements of electromigration transport can be made only under circumstances that maximise the effect of both terms. In practice this means that electromigration experiments on bulk samples, as distinguished from thin films, are conducted at the highest current densities that are compatible with the requirements that the samples should not melt as a result of Joule heating. Generally, the current densities used are of the order of 10 kA/cm². Although one finds a variety of experimental techniques, it may be said that most of these conform to one of the three different types: the drift technique, the marker-motion technique, and the steady-state technique.

Purification by Electromigration

The extensive and successful use of electrolysis methods for the purification of metals such as Pb and Cu stimulated attempts to use electromigration in purification of other metals. Many early reports on electromigration used the term solid-state electrolysis although the two processes are basically different.

Purification by electromigration differs from classical chemical separation processes in that redistribution of an impurity occurs within one phase rather than by partition between two different phases. Electromigration purification has so far been a relatively slow process, restricted to small portions of a metal and requiring large amounts of current. It has been used mostly with very reactive metals that have such a high affinity for carbon, nitrogen and oxygen that common chemical methods have failed. Most of the metals that have been purified were used in research investigations. For this purpose, the small amount of metal processed per run and the high processing cost are not overwhelming considerations.

Formal basis of purification

The steady-state concentration profile

The usual form of a specimen to be purified by electromigration is a rod through which a direct electric current is passed longitudinally. This simple geometric shape is both the easiest to model with mathematical relationships and also seems to be so far unsurpassed in achieving the highest purity. The migration of a solute in this specimen, if the solute does not escape, must result in an increase in concentration in one end and a depletion at the other. For a rod of uniform cross-section at constant temperature and with an initial uniform concentration, the concentration profile will change with time. Ultimately, a steady-state condition will be reached such that the flux owing to back diffusion will be equal to the electromigration flux.

The potential of electomigration as the basis of a purification method has been evaluated for a number of metal-impurity systems. For available electromigration and diffusion data, the average purity of the purest half of a rod was calculated by use of realistic values of the length of the specimen and the electric field. In many systems, the ratio of the electric mobility to the diffusion coefficient is large enough that very substantial purification can be achieved if the steady state can be reached.

Most of the study of purification by electromigration has been on interstitial solutes in solid metals. For many reactive metals, it has been established as a simple and effective purification method for small quantities of materials. In solid metals fast-diffusion substitutional impurities can be removed but the time required for solutes with a normal coefficient to diffusion would be excessive. Liquid metals could be purified quite effectively if mixing in the liquid could be suppressed.

Thin Films

The most prevalent manifestations of electomigration have been the deleterious effects discovered in microcircuit technology. It was found that thin-film conductors of aluminium used in planar silicon circuits can become discontinuous after prolonged passage of a direct current. This discovery set the stage for a period of intense activity in the investigation of electomigration phenomena that are almost entirely specific for thin films. The problems considered below are related exclusively to the metallic conductors that lead the electric current to, or away, from transistors; they do not concern the behaviour of the semiconducting devices themselves.

Failure formation and failure mechanisms

Cracks causing electrical discontinuity are certainly the most dramatic failures caused by electromigration, yet hillocks or whiskers may result in short circuits between conducting lines lying side by side, or lines superimposed with a separating insulation layer (e.g., sputtered silicone oxide). Both mass depletions (cracks and holes), and mass accumulations (hillocks and whiskers) result from some specific discontinuity or divergence in the atomic flux along the length of a conductor.

Lifetime tests, failure models

Technological necessities require that at a specified time the rate of failure in electronic devices should remain lower than some maximum, which can be chosen to be as small as desired. The selection of acceptable rate and time values varies according to the type of application considered. Practically, on account of time limitations, this means that one must conduct accelerated failure tests at high temperatures and at high current densities as well as with a limited set of samples.

The results thus obtained are extrapolated to normal use conditions (low current densities, low temperatures, long times and very large device populations). Such extrapolations require an understanding of the laws relating failure times to current density and temperature and of the statistical distribution of these failure times.

METAL TREATMENTS

Metal treatments are operations performed on consolidated metals and alloys. Most of these are mechanical and/or thermal. Mechanical treatments involve shape changes by forming or machining. Forming entails plastic deformation which changes the microstructure and therefore, properties. In thermal treatments, heat is applied to alter structures and properties. Metal treatments such as joining and coating of metals are not discussed here.

Mechanical Treatments

Forming processes and techniques available for a particular alloy depend upon workability, which is the ability to be plastically deformed.

Hot working

Hot working involves plastic deformation at temperatures sufficiently high that strain hardening does not result. The temperature range for successful hot working depends on composition and other factors such as grain size, previous cold working, reduction, and strain rate. The lack of strain hardening is due to sufficient thermal energy for recrystallisation, which refers to the formation of new grains.

Cold working

Cold working involves plastic deformation well below the recrystallisation temperature. Required stresses for cold working are greater than for hot working and the amount of strain without heat treatment is limited. Advantages are close dimensional control, good surface finish, and increased low temperature strength because of strain hardening. Grain refinement can be achieved by annealing, which entails heating after cold working to temperatures where recrystallisation occurs.

Primary forming process

These operations are usually hot-working operations directed toward converting case ingots into wrought blooms, billet, bars, or slabs. In primary working operations, the large grains typical of cast structures are refined, porosity is reduced, segregation is reduce, inclusions are more favourably distributed and a shape desirable for subsequent operations is produced.

The principal operations used for ingot breakdown are forging, extruding and rolling. Extrusion differs from forging and rolling in that more deformation occurs in one pass. Forging and rolling include many passes and some reheating.

Secondary forming processes

The objective of these processes —either cold or hot-working—is to form a shape. Such processes include rolling, open and close-die forging, upset forging, extruding, roll forging, ring rolling, deep drawing, spinning, bending, stretching, stamping, drawing and high velocity forming.

Thermal Treatments

Annealing

In annealing, a cold-worked material is heated to soften it and improve its ductility. The three stages of annealing are recovery, recrystallisation, and grain growth (Fig. 22.2). Recovery occurs at relatively low temperature and may result in some softening caused mainly by the arrangement of dislocations into a more favourable distribution. Recrystallisation is the formation of new grains with a relatively low dislocation density and little internal strain which replaces strained grains with high dislocation densities. At increasing temperature, the newly formed grains exhibit grain growth. Prolonged exposure at a given temperature also tends to promote grain growth.

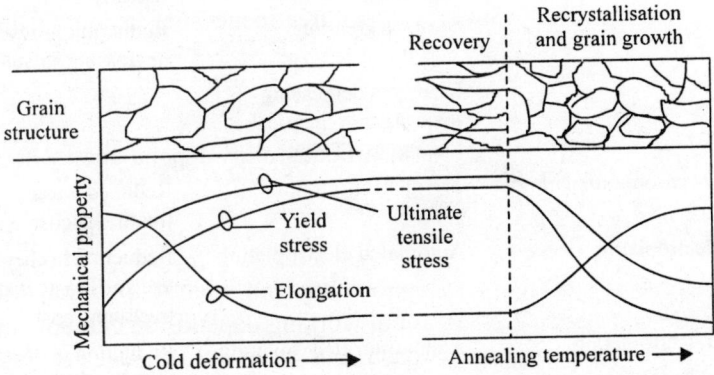

Fig. 22.2. Variation of tensile properties and grain structure with cold working and annealing.

Precipitation hardening

Also called age hardening, this process involves fine particles precipitated from a supersaturated solid solution. These particles impede the movement of dislocations, thereby making the alloy stronger and less ductile. In order for an alloy to exhibit precipitation hardening, it must exhibit partial solid solubility and decreasing solid solubility with decreasing temperatures. An examples of the many alloy systems satisfying these requirements is the aluminum–copper system. At about 500–600°C, an alloy with 4.5 wt per cent Cu consists only of alpha, a solid solution of Cu in Al. Below 500°C, the phase $CuAl_2$ exists in addition to alpha.

The objective of precipitation hardening is to distribute the second phase ($CuAl_2$) as fine particles which are effective in blocking dislocation motion. Precipitation hardening consists of dissolving 'solutioning' (which entails heating above the solvus temperature in order to form a homogenous solid solution), quenching [(rapidly to room temperature to retain in solid solution a maximum amount of the alloying elements (Cu)], and ageing [(heating the alloy below the solvus to permit precipitation of fine particles of a second phase ($CuAl_2$)].

The solvus represents the boundary on a phase diagram between the solid-solution region and a region consisting of a second phase in addition to the solid solution. Cleaner technologies for metal finishing industry are given in Table 22.5.

Table 22.5. Cleaner technologies for metal finishing industry.

Unit process	Conventional technology	Low pollution technology	Advantages	Stage of development
Descaling	Descaling of wire rod coils by: Dippling in a hot acid bath Rinsing Lime treatment	Mechanical descaling of wire, rod, coils	Greater safety and reliability	Pilot plant
Degreasing	Trichloroethane degreasing	ISC-108, an alkali cleaning compound, degreasing	Reduction in pollution	Pilot plant
		Biodegradable 'Bioact degreasing	Environment friendly	Laboratory scale
		Steam degreasing	Reduction in waste degreasing solvent	Pilot scale
		Detergent degreasing toxic organic chemical loading to ETP		
Plating	Hexavalent chromium plating	Trivalent chromium plating	Reduced waste treatment cost	Pilot plant
	Manual electroplating	Automated electroplating	Reduction in chemical use and waste water treatment cost	Pilot plant
	Cadmium electroplating	Aluminium electroplating	Reduction in operating and disposal cost	Pilot plant
	Chromium electroplating with large anode to cathode distance	Chromium electroplating with smaller anode to cathode distance	Reduction in chromic acid requirement	Pilot plant
	Electroplating in an open electroplating bath	Polypropylene balls floated on top of electroplating bath	Less amount of energy and chemicals requirement	Pilot plant
	Cynamide dip and chromic acid bright dip passivation	Sulphuric acid and stabilised hydrogen peroxide dip passivation	Reduction in waste disposal volume	Pilot plant

CHAPTER 23

Mechanism of Catalysts and Biocatalysts

INTRODUCTION

A catalyst is a substance that increases the rate of a chemical reaction without itself being used up; and can be recovered unchanged chemically at the end of a chemical reaction. A catalyst therefore provides an alternative path, usually of lower activation energy, for the reaction to proceed at an accelerated rate. Consider, for example, the formation of oxygen by heating potassium chlorate. The reaction is very slow and takes place as follows:

$$2KClO_3 \xrightarrow{\text{heating}} 2KCl + 3O_2$$

A small addition of manganese dioxide to potassium chlorate accelerates the rate of the reaction and can be recovered unchanged at the end of the reaction. Thus, manganese dioxide acts as catalyst. Various other examples of catalysts are:

1. In the manufacture of ammonia by Haber's process, iron acts as a catalyst.

$$N_2 + 3H_2 \xrightarrow{\text{Fe}} 2NH_3$$

2. In contact process for the manufacture of sulphuric acid, platinum acts as a catalyst.

$$2SO_2 + O_2 \xrightarrow{\text{Pt}} 2SO_3$$

3. Combination of hydrogen and chlorine takes place in the presence of water vapours.

$$H_2 + Cl_2 \xrightarrow{H_2O} 2HCl$$

In these examples, the catalysts accelerate the rate of a chemical reaction and are, therefore, sometimes termed as *positive catalyst*.

On the other hand, there are substances which when added to a chemical reaction retard its reaction rate and are thus called *negative catalysts* or *inhibitors*. Some common examples of negative catalysts are:

1. Auto oxidation of benzaldehyde is strongly inhibited by traces of some sulphur compounds.
2. Decomposition of H_2O_2 is retarded by the presence of a small quantity of sulphuric acid.
3. Oxidation of chloroform is retarded by traces of alcohols.

CHARACTERISTICS OF CATALYSTS

1. A catalyst remains chemically unaffected at the end of a chemical reaction. The catalyst does not undergo any chemical change, although there may be change in its physical state such as particle size or change in the colour of the catalyst etc.

2. Small quantity of a catalyst is usually required to bring about a reaction. A very small amount of a catalyst is sufficient for reactants to combine together. This is because the catalyst is not used up in the reaction. Thus 10^{-4} g of molybdic acid is sufficient for the oxidation of HI by H_2O_2. However, sometimes in many homogeneous catalytic processes, the rate of a catalytic reaction is proportional to the concentration of the catalyst. For example, in the inversion of cane sugar hydrochloric acid acts as a catalyst.

3. Presence of a catalyst does not affect the position of equilibrium in a reversible reaction. This is true when a small amount of the catalyst is used. The catalyst helps in attaining the equilibrium more quickly by increasing the rates of both the forward and the reverse reactions to the same extent. If, however, the catalyst is present in large amount, the same is not true. Some instances are known where the equilibrium constant changes. For example, hydrolysis of ethyl acetate in the presence of varying amounts of HCl, which acts as a catalyst, changes the value of equilibrium constant.

4. A catalyst does not initiate a reaction but only increases or decreases its speed. Generally, a catalyst speeds up the reaction which is already occurring slowly in its absence. However, this is not true in all reactions. Many reactions are known to occur only in the presence of a catalyst.

5. The action of a catalyst is specific. A catalyst can catalyse only a specific reaction and cannot be used for every reaction. For example, manganese dioxide can catalyse the decomposition of potassium chlorate but not potassium nitrate or other substances. Change of a catalyst also changes the nature of the reaction, e.g.,

$$
HCOOH
\begin{cases}
\xrightarrow{\text{Cu or ZnO}} H_2 + CO_2 \\
\xrightarrow{\text{Al}_2\text{O}_3 \text{ or TiO}_2} H_2O + CO
\end{cases}
$$

$$
CO + H_2
\begin{cases}
\xrightarrow{\text{Ni}} CH_4 + H_2O \\
\xrightarrow{\text{ZnO}} CH_3OH
\end{cases}
$$

6. A catalyst has an optimum temperature at which the action of the catalyst is maximum.

7. A catalyst is poisoned by the presence of traces of certain substances which destroy the catalytic activity and are called *catalytic poisons*. Some of the catalytic poisons are arsenious oxide, carbon monoxide, hydrogen cyanide, etc.

8. The activity of a catalyst is enhanced by the presence of a substance called *promoters*. For example, in the synthesis of ammonia by Haber's process, molybdenum is used as a promoter to the catalyst iron.

Types of Catalysis

There are generally two types of catalysis: (i) homogeneous catalysis; and (ii) heterogeneous catalysis.

Homogeneous catalysis

In homogeneous catalysis, the catalyst is present in the same phase as the reacting substances. Many homogeneous catalysed reactions have been studied in the gas and liquid phases.

Some common examples of such catalysis in gas phase are:

1. In the lead chamber process for the manufacture of sulphuric acid, nitric oxide gas catalyses the oxidation of sulphur dioxide.

$$2SO_2 + O_2 \xrightarrow{NO} 2SO_3$$

2. Decomposition of acetaldehyde is catalysed by iodine vapours.

$$CH_2CHO \xrightarrow[\text{Vapours}]{I_2} CH_4 + CO$$

3. Nitric oxide acts as a catalyst in the combination of carbon monoxide and oxygen.

$$2CO + O_2 \xrightarrow{NO} 2CO_2$$

Examples of homogeneous catalysts in liquid phase

Important examples of homogeneous catalysis in liquid phase are acid-base catalysis. The most common acid catalyst in water is the hydronium ion and the most common base catalyst is the hydroxyl ion. If an acid catalyses a reaction, the reaction is said to be the subject of acid catalysis.

Inversion of cane sugar and hydrolysis of esters are some examples of acid catalysed reactions. However, it has been found that different acids have different catalytic activity; hydrochloric acid has a greater activity than acetic acid.

So it is evident that the actual catalysts are H^+ (or H_3O^+) ions. The rates of reaction are found to be proportional to the concentrations of H_3O^+ ions and the concentration of the reacting molecules or ions.

$$C_{12}H_{22}O_{11} + H_2O \xrightarrow{H_3O^+} C_6H_{12}O_6 + C_6H_{12}O_6$$

$$CH_3COOC_2H_5 + H_2O \xrightarrow{H_3O^+} CH_3COOH + C_2H_5OH$$

Such reactions which are catalysed by certain acids (or H_3O^+ ions only) are said to be *specific acid catalysis*. Similarly, there are reactions which are catalysed by OH^- ions only and hence are said to be *specific hydroxyl ion catalysis*.

Conversion of acetone into diacetonyl alcohol or the decomposition of nitroso-triacetoneamine are examples of hydroxyl ion catalysis.

$$CH_3COCH_3 + CH_3COCH_3 \xrightarrow{OH^-} CH_5COCH_2C(CH_3)_2OH$$

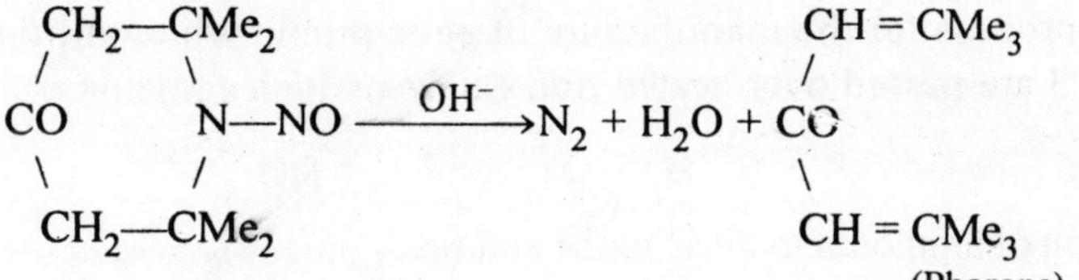

$$
\begin{array}{cc}
\begin{array}{c}
CH_2\!-\!CMe_2 \\
/ \qquad \backslash \\
CO \qquad N\!-\!NO \\
\backslash \qquad / \\
CH_2\!-\!CMe_2
\end{array}
\xrightarrow{OH^-} N_2 + H_2O + CO \!
&
\begin{array}{c}
CH = CMe_3 \\
/ \\
\\
\backslash \\
CH = CMe_3
\end{array}
\end{array}
$$

(Phorone)

There are many reactions in which both H_3O^+ ions and OH^- ions simultaneously act as catalysts, probably along with water. The mechanisms of hydrolysis of ester can be expressed as follows:

1. With H^+ ions as catalyst

$$CH_3-\underset{\underset{R}{|}}{\overset{\overset{O}{\|}}{C}}-O + H^+ \longrightarrow CH_3-\underset{\underset{R}{|}}{\overset{\overset{O}{\|}}{C}}-\overset{+}{O}-H \xrightarrow{H_2O} CH_3-\underset{\underset{R}{|}}{\overset{\overset{O^-}{|}}{\underset{H_2O}{\overset{|}{C}}}}-\overset{+}{O}-H$$

$$\downarrow$$

$$CH_3-\overset{\overset{O}{\|}}{C}-OH + ROH + H^+$$

2. With hydroxyl ions as catalyst:

$$CH_3-\underset{\underset{R}{|}}{\overset{\overset{O}{\|}}{C}}-O + OH^- \rightarrow CH_3-\underset{\underset{H-O}{|} \underset{R}{|}}{\overset{\overset{O^-}{|}}{C}}-O \rightarrow CH_3-\underset{\underset{H-O}{|}}{\overset{\overset{O}{\|}}{C}} + RO^-$$

$$\downarrow H_2O$$

$$CH_3COO^- + ROH$$

Heterogeneous catalysis

In heterogeneous catalysed reactions, the catalyst is present in a different phase from the reactants. In a number of cases, the catalyst is the solid phase and the reactants are gaseous in most cases or liquids in others. The catalysts which are commonly used are metals like platinum, nickel, copper and iron and certain metal oxides such as ferric oxide, zinc oxide, molybdenum oxide, etc. Some important examples of heterogeneous catalysis are:

1. In Contact process for the manufacture of sulphuric acid, sulphur dioxide is directly oxidised into sulphur trioxide by atmospheric oxygen in the presence of platinum as catalyst.

$$2SO_2 + O_2 \xrightarrow{\text{Pt}} 2SO_3$$

2. In Haber's process for the manufacture of ammonia in which nitrogen and hydrogen in the ratio of 1 : 3 are passed over heated iron catalyst which contains a promoter (molybdenum).

$$N_2 + 3H_2 \xrightarrow{\text{Fe}} 2\,NH_3$$

3. The oxidation of ammonia to nitric oxide and finally to nitric acid in the presence of a mixture of ferric and bismuth oxide.

$$4NH_3 + 5O_2 \xrightarrow[\text{Bi}_2\text{O}_3]{\text{Fe}_2\text{O}_3} 4NO + 6H_2O$$

4. Hydrogenation of unsaturated hydrocarbons in presence of nickel as a catalyst.

$$—R—CH = CH — R' + H_2 \xrightarrow{Ni} R—CH_2—CH_2—R'$$

5. Oxidation of HCl by oxygen in presence $CuCl_2$ as catalyst.

$$4HCl + O_2 \xrightarrow{CuCl_3} 2H_2O + 2Cl_2$$

Enzyme Catalysis

Enzymes are complex protein molecules with three-dimensional structures. These are responsible for catalysing the chemical reactions in living organisms. The diameter of the enzyme molecules fall in the range of 10–100 nm. Enzymes are often present in colloidal state and are extremely specific in their catalytic functions. Various enzymes catalysed reactions are known. Some important examples are:

1. Urease, an enzyme that catalyses the hydrolysis of urea but has no effect on the hydrolysis of substituted urea, e.g., methyl urea.

$$NH_2CONH_2 + H_2O \xrightarrow{Urease} 2NH_3 + CO_2$$

2. Peptide, glycyl-L-glutamyl-L-tyrosine, is hydrolysed by an enzyme known as pepsin.
3. Hydrolysis of starch into maltose by diastase.

$$\underset{\text{Starch}}{2(C_6H_{10}O_5)_n} + nH_2O \xrightarrow{Diastase} \underset{\text{Maltose}}{nC_{12}H_{22}O_{11}}$$

4. Conversion of glucose into ethanol by zymase present in yeast.

$$C_6H_{12}O_6 + H_2O \xrightarrow{Zymase} 2C_2H_5OH + 2CO_2$$

5. Conversion of maltose into glucose by maltase.

$$C_{12}H_{22}O_{11} + H_2O \xrightarrow{maltase} 2C_6H_{12}O_6$$

6. Oxidation of alcohol to acetic acid by micoderma aceti.

$$C_2H_5OH + O_2 \xrightarrow[\text{aceti}]{\text{micoderma}} CH_3COOH + H_2O$$

Almost all enzymes fall into one of the two classes, the *hydrolytic enzymes* and the *oxidation-reduction enzymes*. The hydrolytic enzymes appear to be complex acid-base catalysis which accelerate the ionic reactions mainly due to the transfer of hydrogen ions. The oxidation-reduction enzymes catalyse electron transfer perhaps through the formation of an intermediate radical.

Mechanism of enzyme reactions

The mechanism of an enzyme reaction was proposed by *Michaelis* and *Menten* and can be represented in the following manner : Let E represent the enzyme and S the substrate it acts on, then the overall reaction is

$$E + S \underset{k_{-1}}{\overset{k_1}{\rightleftharpoons}} [ES] \xrightarrow{k_1} E \cdot P$$

It is to be noted that in the formation of the product P, the enzyme does not undergo any change. The rate of formation of the product depends on the concentration of the enzyme. In the above scheme *ES* denotes the intermediate between the enzyme and the substrate which decomposes into the product with a first order rate constant k_2. The rate of formation of the product is given by

$$\frac{d[P]}{dt} = k_2[ES] \qquad \qquad ...(23.1)$$

In order to solve the equation (23.1) it is necessary to know the concentration of *ES*. This can be calculated through the steady-state principle.

$$\frac{d[ES]}{dt} = k_1[E][S] - k_{-1}[ES] - k_2[ES] = 0$$

or

$$[ES] = \frac{k_1[E][S]}{k_{-1} + k_2}$$

$$= \frac{[E][S]}{\dfrac{k_2 + k_{-1}}{k_1}}$$

$$= \frac{[E][S]}{K_m} \qquad \qquad ...(23.2)$$

where K_m is often referred to as Michaelis constant.

In this equation, the quantities $[E]$ and $[S]$ are the concentrations of free enzyme and free substrate. If $[E]_0$ and $[S]_0$ are the initial concentrations of the enzyme and the substrate respectively, then we can write

$$[E]_0 = [E] + [ES]$$

or

$$[E] = [E]_0 - [ES]$$

and

$$[S]_0 = [S] + [ES]$$

Since only a little enzyme is added, hence $[ES]$ is very small in comparison to $[S]$, therefore,

$$[S]_0 \cong [S]$$

Substituting the value of $[E]$ in equation (23.2), we get

$$[ES] = \frac{\{[E]_0 - [ES]\}[S]}{K_m}$$

$$[S + K_m] [ES] = [E]_0[S]$$

$$[ES] = \frac{[E]_0[S]}{[S] + K_m}$$

Consequently, the rate of formation of products is

$$\frac{d[P]}{dt} = k_2[ES]$$

$$= \frac{k_2[E]_0[S]}{K_m + [S]} \qquad \qquad ...(23.3)$$

According to equation (23.3) if $[S] \ll K_m$, then the rate of enzymolysis varies linearly with the enzyme and substrate concentrations, i.e., the reaction will be first order in E and S. However, if $[S] \gg K_m$

$$\frac{d[P]}{dt} = \frac{k_2[E]_0[S]}{[S]}$$

$$= k_2[E]_0$$

the rate of the reaction will be independent of substrate concentration and will be first order in E.

A plot of $\dfrac{d[P]}{dt}$ versus $[S]$ for constant enzyme concentration yields a curve (Fig. 23.1) from which it is possible to calculate the value of k_2 and K_m. Further, when the rate is half the maximum value

$$\frac{d[P]}{dt} = \frac{k_2[E]_0}{2} = \frac{k_2[E]_0[S]}{K_m + [S]}$$

or

$$K_m + [S] = 2[S]$$

$$[S] = K_m$$

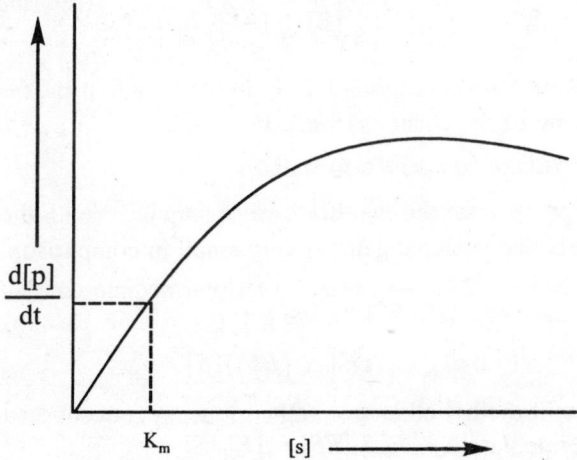

Fig. 23.1. Plot of Michaelis-Menten equation

Rates of enzyme catalysed reactions are slowed down by compounds which are structurally related to the substrate. These compounds combine with the active sites of the enzyme and thus cause inhibition. In cases where the substrate and the inhibitor compete for the active sites, the rate may be increased by taking larger concentrations of substrate. An enzyme reaction has an optimum pH value of which the catalytic activity of the enzyme is maximum. The rate of the reaction decreases as the pH is raised or lowered from the optimum value. At extreme pH's they are irreversibly denatured.

Rate versus temperature graph for an enzyme catalysed reaction shows that rate is maximum at a certain temperature. Above this temperature, the enzyme is denatured and hence the rate decreases.

Theory of Catalysis

As stated earlier, the essential requirement for a reaction to occur is that the reacting molecules must acquire sufficient energy. In case a catalyst is added to the reaction, the energy required to activate the molecules is less than in the absence of a catalyst. Due to lower activation energy, more molecules will take part in the reaction and hence the rate of the catalysed reaction would increase. The action of a catalyst can be explained by two different mechanisms, viz. (i) intermediate compound formation theory; and (ii) adsorption theory.

Intermediate compound formation theory

In this theory, essentially two steps are involved :
1. Combination of the catalyst with one or more of the reactants forming intermediate compound.
2. Decomposition of the intermediate compound or its combination with other reactants yielding the product and the catalyst back. Consider a reaction between the reactants A and B giving the product, viz.,

$$A + B \rightarrow AB$$

This reaction is very slow and is catalysed by the presence of a catalyst X. The reaction will therefore proceed as

$$A + X \rightarrow AX \qquad \text{(Intermediate compound)}$$
$$AX + B \rightarrow AB + X$$

The formation of an intermediate compound AX is an easy reaction and needs low energy of activation thereby accelerating the rate of the chemical reaction.

Some examples of intermediate compound formation

1. In lead chamber process for the manufacture of sulphuric acid, the catalyst NO first forms an intermediate compound with oxygen

$$2\,NO + 1/2\,O_2 \rightarrow 2NO_2 \qquad \text{(Intermediate compound)}$$

and then

$$NO_2 + SO_2 \rightarrow SO_3 + NO$$

2. In the preparation of diethyl ether from ethanol using concentrated H_2SO_4, $C_2H_5HSO_4$ is first formed as an intermediate.

$$C_2H_5OH + H_2SO_4 \rightarrow C_2H_5HSO_4 + H_2O$$
$$C_2H_5HSO_4 + C_2H_5OH \rightarrow C_2H_5O\text{---}C_2H_5 + H_2SO_4$$

3. The formation of water by combination of hydrogen and oxygen in presence of copper as a catalyst is as follows:

$$2Cu + 1/2\,O_2 \rightarrow Cu_2O$$
$$Cu_2O + H_2 \rightarrow H_2O + 2Cu$$

Limitations of intermediate compound formation theory

This theory does not explain the cases of heterogeneous catalysis in general and more specifically, the deactivation by a catalytic poison and the activation by a promoter.

Adsorption theory

A large number of gaseous reactions take place in the presence of solid catalysts. The surface of the catalyst has certain active centres due to the unsaturation of valencies. Appreciable quantities of the reactant molecules are adsorbed or retained by solid surfaces at these active centres and the reactions occur at the surface of the solid. For this reason, this type of catalysis is sometimes referred to as the contact catalysis. The adsorbed molecules form some sort of an activated complex on the surface, which then decomposes forming the products. The products are ultimately desorbed from the surface. A catalytic reaction involves the following steps :

1. Diffusion of the reactants from the bulk on to the surface.
2. Adsorption of the reactants on the surface of the catalyst.
3. Activation of the adsorbed reactants leading to a reaction in the adsorbed phase.
4. Desorption of the products from the surface of the catalyst.
5. Diffusion of the products away from the surface of the catalyst.

Any one of the steps may be slowest and consequently the rate determining but generally step (3) is the rate controlling step.

Due to adsorption, the concentration of the reactants tend to increase on the surface of the catalyst and according to the law of mass action, the rate of the reaction will increase. Furthermore, adsorption being an exothermic process, the heat evolved during adsorption is utilised in the activation of the surface reaction. Adsorption may also lead to proper orientation of the reacting molecules, partial loosening of the bonds in the adsorbed state and thus requiring only small energy to form the activated complex.

Adsorption theory can explain the enhanced catalytic action of a catalyst in the finely divided state. It is due to the larger surface area available for adsorption and also the formation of more active centres.

$$
\begin{array}{ccccc}
| \quad | & & | & & | \\
-Ni-Ni- & & -Ni- & & -Ni- \\
| \quad | & \rightarrow & | & + & | \\
-Ni-Ni- & & -Ni- & & -Ni- \\
| \quad | & & | & & |
\end{array}
$$

Action of promoters which enhances the catalytic activity can also be explained in terms of this theory. A promoter generally increases the number of active centres by the adsorption on the surface of a catalyst. Similarly, poisoning of a catalyst results due to the adsorption of the catalytic poisons on the surface and thereby reducing the number of active centres on the surface of a catalyst.

BIOCATALYSTS

Conventional immobilisation techniques, using insoluble biocatalysts are limited in their efficiency due to the mass transfer restraints in reactions involving insoluble or high molecular weight substrates and/or products. In view of this, recently, a new immobilisation technique employing reversibly soluble biocatalysts has been explored. The solubility of a biocatalyst can be changed by conjugating it with a polymer/hydrogel exhibiting reversible solubility in response to minor changes in its environment. These polymers/hydrogels have been termed as 'Smart materials' and have been utilised for a wide variety of applications such as biocatalysis, bioseparation (affinity precipitation), biosensors, site-directed drug delivery, etc. Small changes in environmental factors, such as pH, temperature, ionic concentration, solvent composition, light, electric field, pressure etc. can stimulate changes at the molecular levels of such smart materials. This section summarises the recent work in this field by classifying it on the basis of the environmental *stimuli* used for the phase transition. The review encompasses the state of the art in reversible biocatalysis.

The term immobilisation has been defined as the confinement or localisation of enzyme molecules during a continuous enzyme catalysed process. Immobilisation of enzymes offers several advantages including possibility of continuous operation, improvement of enzyme stability, application of immobilised enzymes as biosensors etc. However the major rationable behind immobilising an enzyme has mostly been simple recovery. Immobilisation allows repeated use of the catalyst by easy separation of the enzyme from the product. The localisation of enzyme molecules can be brought about by diverse means such as covalent attachment to water insoluble polymers, adsorption to water insoluble inorganic or organic (inert or active) supports, entrapment within gel-matrices or semipermeable membrane dependent devices such as microcapsules and liposomes, modification into an insoluble aggregate by chemical cross-linking or as reverse micelles. Majority of the work has been concentrated on developing effective means of rendering the enzyme insoluble.

However, these conventional means of immobilising enzymes have led to several limitations such as:

1. Substrate/product diffusion limitations arising due to water insoluble enzyme support resulting into large reductions in the enzyme catalysed reaction rates.
2. Heterogeneous biocatalytic reactions generate additional mass transfer resistance since insoluble catalyst increases the overall resistance throughout the system.

In addition, such systems have several engineering problems such as plugging, irregular channeling or temperature gradients in packed bed reactor, risk of mechanical instability in continuous stirred tank reactor for deformable particles such as gels of polysaccharides (dextran or agarose).

In short, the advantages of immobilising enzyme for recovery and reuse may be overshadowed by the limitations due to diffusion in such as system. The quest to solve these problems led to development of a strategy for manipulating of enzyme solubility using external stimulus. The technique keeps the enzyme in solution during the course of the reaction and out of it after the reaction for easy recovery and reuse. It is achieved by conjugating the enzyme covalently to entities or carriers, capable of undergoing reversible phase transition in response to an external stimulus. Such enzyme systems, with regulated phase transition, were termed as Soluble–Insoluble (S-IS) enzyme systems.

REVERSIBLY SOLUBLE CARRIERS

The reversibly soluble entities or the carriers have generally been high molecular weight polymers and hydrogels. The applications of the inherent properties of the smart polymers for enhancing enzymatic reactions have been a subject of interest. The reversible soluble polymers undergo fast, reversible changes in the microstructure, from a hydrophilic to hydrophobic state (soluble to insoluble). These changes are triggered by small alterations in the environment and are reversible.

Various modes of external *stimuli* have been attempted to initiate the phase transition in the smart materials. The major factor governing the phase-transition in a polymer has to be its chemical nature and physical susceptibility to undergo change. An appropriate balance of the hydrophobic and hydrophilic nature of the polymer is required for the phase transition to occur. According to the thermodynamic principles the hydrophilic-hydrophobic balance should be such that the entropy of the system is considerably reduced to effect phase transition.

The external stimulatory forces may bring about this reduction in entropy. Phase transition in a system can be brought about using different principles such as:

1. Reduction in the net charge on the polymer thereby effecting aggregation, either by change in pH or addition of a counter ion. e.g., pharmaceutical polymers like polysaccharides show such a behaviour in response to pH changes.
2. Alteration in the solubility of uncharged polymers by change in temperature or ionic concentration. e.g., poly (N-isopropylacrylamide) polymers exhibiting temperature dependent phase properties.
3. Formation of reversible, non-covalent cross-linking of one or more polymeric systems into an insoluble system. e.g., bi- or multi-functional agents like calcium or borate.
4. Reversible isomerisation or change in the geometric conformation of the original polymer network into an insoluble entity. e.g., photo-responsive smart materials such as poly (spiropyran containing methacrylates).

It should be noted that the magnitude of these *stimuli* must be critically controlled in sensitive biological systems. A few recent examples of such reversible soluble biocatalyst exhibiting reversible solubility in response to pH, temperature and photo-irradiation have been discussed here.

Enzyme Immobilisation on pH-Responsive Polymers-Hydrogels

The conformation of macromolecules plays a vital role in its integrity and functionality. A number of factors are responsible to maintain a macromolecule in a particular conformation. These can be van der Waals forces, inter and intramolecular hydrophobic interactions or ionic interactions, depending on the chemistry of that macromolecules. When the molecule bears a net charge or is ionic in nature, a significant effect on the conformational status may be brought about by the environmental pH change. A range of charged polymers exhibit pH-dependent solubilisation and have been used as enteric coating polymers for pharmaceutical dosage forms (tablets, capsules etc.). These polymers undergo rapid and reversible phase transition with small changes in the environmental pH. Various enteric coating polymers like, Hydroxy Propyl Methyl Cellulose-Acetate-Succinate (HPMC-AS), Carboxy Methyl Ethyl Cellulose (CMEC), Hydroxy Propyl Methyl Cellulose-Phthalate (HPMCP), Cellulose Acetate Phthalate (CAP), Methacrylic acid – Methylmethacrylate co-polymers (Eudragits), etc., have been studied for their ability to be used as reversibly soluble carriers for enzyme immobilisation.

Conjugating the enzyme with a pH-responsive polymer has successfully eliminated the diffusional limitations of the conventional insoluble, immobilised biocatalyst systems. For example, the limitations due to diffusion in the hydrolysis of insoluble chitin were considerably reduced by immobilising chitinase on HPMC-AS (Table 23.1). In another study, β-glucosidase was conjugated with HPMC-AS to hydrolyse phloridzin to glucose and the insoluble product phloretin (Table 23.2).

Table 23.1. Comparative specific activity of chitinase immobilised on HPMC-AS and on insoluble ion-exchange resin (CM-Toyopearl).

	Specific activity (U mg^{-1} protein)		
	Glycol chitin	Colloidal chitin	Native chitin
Free enzyme	2.45 (100%)[a]	3.49 (100 %)[a]	0.78 (100 %)
CH – AS[b]	2.35 (95%)	2.97 (85 %)	0.62 (79%)
CH – IER[c]	1.34 (55%)	1.75 (43%)	0.00 (0%)

[a] Relative activity when activity of free enzyme is 100 %.
[b] Chitinase immobilised to HPMC-AS.
[c] Chitinase immobilised to ion-exchange resin (CM-Toyopearl).

Table 23.2. Comparative specific activity of β-glucosidase immobilised on HPMC-AS and on insoluble AF-Carboxyl Toyopearl.

β-glucosidase preparation	Specific activity (U mg^{-1} protein)		
	Cellobiose	p-nitrophenyl β-D-glucopyranoside	Phloridzin
Native	3.16 (100%)	0.834 (100%)	(100%)[a]
G-AS[b]	1.67 (53%)	0.375 (45%)	(79%)
G-AFT[c]	1.01 (32%)	0.321 (26%)	0.00 (0%)

The hydrolysis was carried out at 50°C and pH 5.0 (0.1 M acetate buffer)
[a] Relative activity when activity of native β-glucosidase was expressed as 100%.
[b] β-glucosidase immobilised on HPMC-As.
[c] β-glucosidase immobilised on AF-Carboxyl Toyopearl.

HPMC-AS has been found to be the most suitable and most popularly used pH-responsive polymer for enzyme conjugation. The phase transition pH-range of HPMC-AS is 4.0–5.2 i.e., it is completely soluble over pH –5.2 and insoluble at/below pH –4.0. This pH-range is found to be suitable for the action of many enzymes. Also depending on the type of HPMC-AS (AS-L, AS-M or AS-H) the range may be varied up to pH –7.0. The loss in the enzyme activity has also been found to be minimal when conjugated with HPMC-AS.

Enzyme Immobilisation On Thermo-responsive Materials

The volume of gels obtained in a specific environment is a result of the balance between the repulsion and attraction of the cross-linked polymer chains in their network. These interactions arise from a combination of four intermolecular, forces, ionic, hydrophobic, van der walls and hydrogen bonding. Hence, the external factors affecting these intermolecular forces, such as temperature, pH, ionic strength or solvent composition, individually or in combination can cause changes in the gel volume. The changes in the hydrophobic-hydrophilic balance induced by increased temperature or ionic strength is attributable to reduced hydrogen bonding efficiency with the rise in temperature. Some polymer have a critical temperature at which hydrogen-bonding efficiency becomes insufficient for macromolecular solubility and thus phase separation occurs. The macromolecules transit from their random coil (soluble) conformation to globule (insoluble) conformation and separate from the solution. This temperature at which a random coil – globule conformation transition occurs is called the lower critical solution temperature (LCST).

The conjugation of temperature sensitive polymers to enzymes has one distinct advantage that the conjugated enzymes after one batch reaction can be thermally separated and recycled to another batch reaction. A α-chymotrypsin conjugated with Poly (N-isopropylacrylamide) (Poly-NIPAAm) – Acrylamido-2-deoxy-D-glucose (AADG) co-polymer could be effectively recycled by cyclic thermal operation between 25°C and 40°C (Fig. 23.2). Thermo-responsive hydrogels have been known to provide a unique opportunity to control diffusion of a solute inside a gel matrix by small changes in temperature (temperature regulated hydraulic pump) thereby reducing the compactness of the gels and increasing the diffusional mass-transfer. When synthetic polymers are conjugated to enzyme molecules, the thermal stability of the enzyme tends to increase. A multi-point attachment of protein molecules to polymer matrices has also been suggested to play a major role in the stabilisation of immobilised or conjugated enzymes. It was shown that on conjugating trypsin to a thermo-sensitive polymer containing a carbohydrate moiety, the autolysis observed in non-conjugated trypsin was significantly reduced.

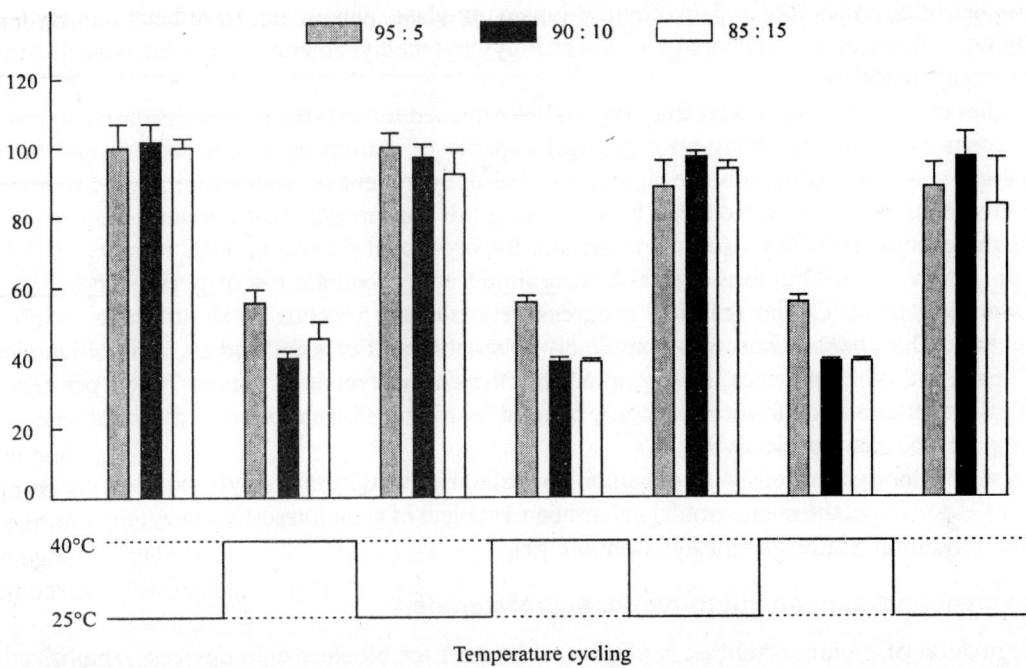

Fig. 23.2. Fractional amount of water-soluble conjugated enzymes remaining in the supernatant after each centrifugation during thermal cycling operation between 25°C and 40°C. NIPAAm/AADG ratios 95:5, 90:10 and 85:15.

Poly-NIPAAm has been, the most commonly used thermo-responsive polymer for enzyme conjugation. The LCST for Poly-NIPAAm is around the ambient temperature (32–34°C). Slightly, above this temperature it exhibits a precipitation behaviour and redissolves at lower temperature behaviour. The LCST phenomenon in Poly-NIPAAm is attributable to the hydrophobic interactions of the isopropyl groups. Poly-NIPAAm has also been co-polymerised with various hydrophilic/hydrophobic functional monomers to control the range of phase transition temperature (Table 23.3). Poly-NIPAAm can be chemical conjugated to enzymes by multiple-point attachment by reacting, functionally activated co-monomers such as N-acroxylsuccinimide in its backbone with primary amine groups of the enzyme molecules.

Table 23.3. Co-polymerisation of NIPAAm with different functional monomers.

Functional monomer	Co-polymer	Enzyme conjugated	LCST
Acrylamido-2-deoxy-D-glucose (AADG)	Poly-NIPAAm – AADG	α-Chymotrypsin	40°C
Methacrylic acid (MAA)	Poly-NIPAAm – MAA	Amylase	42°C
Glucosyoxylethyl methacrylate (GEMA)	Poly-NIPAAm – GEMA	Trypsin	50–70°C
N-acryloxysuccinimide (NASI)	Poly-NIPAAm – NASI	Trypsin, Chymotrypsin	42°C
Glycidyl methacrylate (GMA)	Poly-NIPAAm – GEMA	Alkaline phosphatase	34°C
Glycidyl methacrylate (GMA)	Poly-NIPAAm – GEMA	Amylase	44°C
Methymethacrylate (MMA)	Poly (MMA-NIPAM-MAA)	Papain	40°C
Methacrylic acid (MAA) Styrene (St)	Poly (St-NIPAM-MAA)	–	–

A number of enzymes like trypsin, chymotriypsin, amylase, papain etc. have been conjugated with Poly-NIPAAm. Some other polymers/gels such as Polyvinyl methyl ether have also been used as thermo-sensitive smart-materials.

Examples of a polymer/gel, other than Poly-NIPAAm used for enzyme immobilisation can be quoted from a recent study. In this study an organogel capable of exhibiting a reversible solgel transition phenomenon was used to immobilise lipase, alcohol dehydrogenase, mandelonitrile lyase and horse radish peroxidase. It was observed that by increasing the concentration of the surfactant from 0.2 to 0.3 M in the gelatin-AOT–H_2O-iso-octane system, the organogel formed at 40°C or more ($\eta = 450cp$) melted on cooling to 10°C to form a free-flowing liquid with moderate viscosity ($\eta \sim 6cp$).

On warming to 40°C, the solution progressively became viscous to re-attain its original gel conformation. The phenomenon was completely reversible and was utilised to immobilise *Candida rugosa* lipase and other enzymes. It was found that the enzyme retained almost 70–80 per cent of its original activity after immobilisation and could be used for biotransformations in organic solvents without any changes in the enantioselectivity.

Reverse micellar system consisting of solubilised gelatin in Bis- (2-ethylhexyl) sulphosuccinate sodium salt (AOT)-H_2O-iso-octane micro-emulsion has been a subject of great interest for enzyme immobilisation due to the formation of transparent hydrophobic gel.

Enzyme Immobilisation on Photo-responsive Materials

Photo-regulation of protein activities is of general interest for bioelectronic devices. Amplification of weak light signals, image recording by biocatalystic transformations, reversible biosensor systems and targeted therapeutic materials may be quoted as a few potential applications of reversible photo-regulated proteins. Different methodologies to photo-regulate the activities of enzymes have been developed. These include applications of photo-chromic substrates, chemical modifications by photo-isomerisable components and immobilisation of proteins in photo-regulated polymers.

One of the earliest reports of immobilisation of enzymes in photo-chromic polymer matrices includes that of *Willner* in which α-chymotrypsin was immobilised in a photo-chromic azobenzene co-polymer. It was found that by the immobilisation of α-chymotrypsin in a cross-linked photo-chromic co-polymer of acrylamide and 4-(methacryloyl-amino)-azobenzene lead to complete 'on-off' photo-regulation of the enzyme at a certain composition of the co-polymer matrix. It was also revealed that the photo-stimulated activity of the polymer-encapsulated enzyme originated from the light-controlled permeability of the substrate across the polymer matrices.

Two photo-chromic compounds such as ρ-Phenylazobenzoyl-L-arginine methyl ester hydrochloride (PABE) and its hydroxamide (PABH) have been examined as substrates for trypsin. Various photo-chromic materials such as fulgide dyes, Azo-groups, Spiropyran-groups etc. have been attempted for the synthesis of photo-regulated biocatalysts. Photo-regulated phase transition has also been reported in Poly-NIPAAm gels modified with light sensitive chromophore such as trisodium salt of copper chlorophyllin. It has been shown that by attaching photo-responsive groups (Spiropyran, Azobenzene, etc.) to various organic polymers, their solubility can be changed. In a recent study, this idea was extended to prepare a reversibly soluble subtilisin, regulated by photo-irradiation. A spiropyran-incorporated methacrylate-methacrylic acid co-polymer was covalently coupled with substilisin, which was found to be soluble and active in toluene. On ultraviolet irradiation, the hybrid substilisin precipitated and was easily and quantitatively recovered. The recovered enzyme retained its original activity even after three cycles of precipitation and solubilisation.

Engineering Aspects

Proteins chemists are always interested in alternating the chemical, physical and biological properties of proteins and have been successful in most of their endeavours. However, the promised industrial applications of enzyme chemistry have been repeatedly postponed because it was difficult to transfer the results obtained in test tubes to industrial scale or even laboratory scale production of these modified enzymes or proteins. The main reason is the lack of quantitative description of mathematical modeling of the reaction and also the reversible precipitation/solubilisation phenomena required for extrapolation.

Although the chemistry of the enzyme conjugation to develop S-IS biocatalysts has been adequately investigated, the actual process conditions for the production of such enzymes have been understudied. We would like to stress the requirement of information on the various engineering aspects involved with the production and utilisation of S-IS biocatalysts. A considerable amount of information is available for designing a suitable reactor for producing such enzymes, as the conventional reactors in a chemical plant may be utilised to produce these conjugates by adhering to the prescribed set of process conditions (controlled pH, temperature, mixing fluid, shear etc.). However the more important problem that has remained unaddressed is the lack of quantitative information required, to design a suitable reactor for the application of such reversible biocatalyst. The reactor should be capable of meeting the basic process requirement of incorporation of a stimulus for the phase transition at the end of the reaction.

This requirement can lead to major modifications in designing the reactor. One has to evaluate the questions from not only chemical but also from biological and engineering perspectives.

One has to know beforehand the quantitative description of the kinetics of phase transition as well as the kinetics of such conjugated enzyme-catalysed reactions. The operational factors affecting them also have to be considered.

In a pH-regulated system, the rate of phase transition depends on the concentration of the acid or alkali added and its rate of distribution in the bulk of reaction media. Although a high dose of an acid or base can achieve higher rate of phase transition, the bio-component may undergo inactivation due to localisation effect. This explains the need of efficient mixing without subjecting the conjugated bio-molecule to high levels of mechanical and/or fluid shear. For example, a hydrofoil impeller, with the addition of the acid or base at the plane of the impeller would be ideal to enable rapid distribution of the added liquid throughout the bulk of the reaction mixture. Unfortunately, as described before, the rate of phase transition has not been reported as a function of the gradient in the stimulus. This vital information needs to be generated before one could start designing a reactor achieving such efficient and optimised phase transitions. Also, for a system exhibiting foaming tendencies (especially the enteric coating polymers like HPMC-acetate-succinate) and for reaction systems involving reactants/products susceptible to mechanical shear, it is essential to optimise the impeller speed. This, again, calls for the optimisation of the amount of acid or base to be added to avoid inactivation of the bio-component before it is thoroughly mixed. In certain enzyme-catalysed reactions, involving the production of acid or base, it is necessary to have an additional pH-controlling system to maintain the constant pH of the system for maintaining the required state of the biocatalyst i.e., in solution or in the form of a precipitate.

Similar considerations are required in the case of temperature regulated phase transitions. Generally, the enzyme systems are known to have optimal activities and stabilities in narrow range of temperatures and are susceptible to denaturation at higher temperatures. Hence, it is necessary to select a polymer with LCST higher than the optimal temperature for the enzyme but low enough to avoid enzyme denaturation. From the engineering perspective, it is necessary to optimise the temperature gradient required for the incorporation of heat into the system. In addition, when the solution comes in contact with the heat exchanging fluid at higher temperatures in the proximity of the hot surfaces, it leads to enzymes denaturation on this hot surface. Hence a suitable heat exchanger enabling rapid heat dissipation through out the bulk and working with minimum temperature gradients of the solution should be selected. A shell and tube or immersed-coil type of heat exchanger would be unsuitable for the present application as compared to a jacketed vessel. The intermittent fluctuations in the phase of the system may lead to non-uniform dissipation of heat in a shell and tube or immersed coil type of heat exchanger. A plate type of heat exchanger or an extended surface heat exchanger with a smallest of temperature approach is ideal, though, quantitative information is not readily available to design or even size such equipment for these applications.

Experimentation on the photo-regulated S-IS enzyme systems seems to be at very preliminary levels. A few reports suggests the chemistry of phase transition in response to photo-irradiation, however, there is no report suggesting a particular reactor design for the utility of such enzyme systems on larger scale. For the systems using polymers undergoing phase transition due to irradiation with visible light, reactors with boro-silicate glass windows for light induction should be suitable. However, for systems involving polymers exhibiting photo-responsive solubility under UV-irradiation, boro-silicate glass would be unsuitable as it absorbs most of the UV radiations. In such cases, a quartz window is required, which can considerably influence the economy of the process. One has to determine the intensity of irradiation that is required for the radiations to penetrate throughout the bulk of the reaction mixture and also the effect of this photo-irradiation intensity on the biological activity of these bio-molecules. Also the location or installation of the light windows in the reactor walls is important. Various reactor designs reported for photocatalytic destruction of organic pollutants appear eminently suitable. In brief, extensive laboratory scale studies will be required to generate quantitative information regarding the reactor design aspects, which can be used for the scale-up of these processes.

On the Industrial Horizon

The ultimate aim for any research undertaking is its industrial or commercial acceptance. A project cannot be regarded as successful unless and until it is industrially viable. Any R & D process passes through a series of phases during the transition of a scientific idea into a technological application. One of the most important phases in this transition is the evaluation of the process in terms of its potential improvement and modernisation. The possibility of utilising a reversible S-IS enzyme system for the enzymatic processes involving insoluble or high molecular weight substrates/products, may be considered a significant step towards modernisation of such biocatalystic reaction systems.

A heterogeneous reaction system essentially encounters two mass transfer resistance, viz. (i) resistance to the solubilisation of the insoluble component in the continuous phase of the reaction system; and (ii) resistance to the diffusion of that component within the bulk of the continuous phase.

In a system consisting of a high molecular weight soluble reactant (homogeneous system), although the former resistance is absent, the latter resistance may be high. This resistance may be enough to constitute the same magnitude of overall resistance to mass transfer as that of system consisting of an insoluble reactant. Thus, an overall pseudo-heterogeneous behaviour may be observed in the mass transfer efficiencies for such systems. It is for this reason that, theoretically a homogeneous system is preferred over a heterogeneous or a pseudo-heterogeneous system. However, on practical lines – the heterogeneous system is convenient and economical for easier downstream processing. In the case of the biocatalystic heterogeneous reaction, involving either an insoluble substrate or product, by using an insoluble catalyst, one more resistance gets added to the mass transfer process. Also the down stream processing becomes complicated. Therefore, even with the potential advantage of enzyme reuse, the utility of insoluble immobilised enzymes in such reaction system is low. The industrial potential of the smart biocatalysts is clearly evident from the fact that, in a reversibly S-IS enzyme system this resistance is obviated without sacrificing the advantage of enzyme reuse. Also, the down stream processing is much more convenient as compared to the insoluble immobilised enzymes. One simply has to perform step-wise separation of the components in the order of— unreacted reactants followed by the enzyme and finally the products. For example, *Hoshino* was able to perform simultaneous saccharification and fermentation of reclaimed paper and delignified rice straw for the production of ethanol. He did it by using a combination of two reversibly S-IS enzymes (Xylanase and cellulose) and a pentose fermenting yeast. The potential of reclaimed paper and delignified rice straw had remained unexplored before these studies, due to their insoluble nature, which did not permit the use of insoluble immobilised biocatalysts. It was for the first time that cheaper raw materials such as reclaimed paper and delignified rice straw were illustrated as potential raw materials for ethanol production and this was possible due to the use of reversibly S-IS enzymes.

Chitooligosaccharides, derived from the hydrolysis products of chitin are useful for agrochemical and medicinal purposes. Their production has attracted considerable attention from the related industry. However, the recovery of the enzyme chitinase used for the hydrolysis of chitin is a major concern in the process economics. Immobilisation of the enzymes, a solution in such cases is unfavourable due to the insoluble nature of chitin. *Wang* and *Chio* developed a reversible S-IS chitinase by conjugating a crude enzyme preparation to HPMC Acetate Succinate. Approximately 70 per cent of the original (free) enzyme activity was retained, which was higher than the activity of the insolubly immobilised enzyme. They were able to improve the process economics by implementing a recoverable enzyme with higher activity and also by using a crude enzyme preparation, thereby obviating the enzyme purification procedures. To date there aren't any reports of utilising such modified enzymes by the industry, the main reason for which, as described earlier, may be the lack of mathematical, engineering and kinetic information for the efficient scale-up of a S-IS enzymatic process. However, considering the merits of the S-IS enzyme systems for various reaction systems such as the hydrolysis of insoluble chitin, hydrolysis of Phloridzin to Phloretin, hydrolysis of high molecular weight proteins etc., the industrial utilisation of the S-IS enzymes seems to be inevitable.

Thus the most popularly investigated techniques for the development of reversibly soluble biocatalysts has been discussed. However, the use of certain other methodologies such as enzymes regulated by electric field, solvents, and ions or by specific molecules, cannot be neglected on the mere basis of rarity of literature pertaining to it. Results to date indicate that smart polymers can potentially turn existing biocatalysis protocols into more elegant, efficient and economical methods. The need however, is to generate quantitative information regarding the kinetics of the phase transition and the development of suitable reactors that can meet the prescribed conditions for the S-IS biocatalysis system in terms of time of change and the required gradient in stimulus. It is necessary to develop, to start with empirical correlations for such biocatalyst systems, which relate the rate of change of stimulus and its direct effect on the rate of the phase transition. New phase diagrams need to be constructed which then could be used to develop the necessary design information. One can envisage higher reaction rates due to higher mass transfer efficiencies generated by the transportation of the biocatalyst between two phases similar to the phase transfer catalyst, on manipulation of simple environmental conditions like pH or temperature. The rate at which one observes the work in the field progressing and the speed at which new knowledge is being put to practical use, the day is not far when we can have a whole range of Smart-Biocatalyst with incredible properties, being regularly used industrially.

CHAPTER 24

Pollution Prevention and Waste Minimisation

INTRODUCTION

The population explosion has brought an imbalance in the means of subsistence and nature. This has led to serious economic, socio-political and environmental problems. The environmental degradation consequent to advancement of technology and industrialisation has compelled men to search out the natural resources for its survival on the earth. Changing landscape due to ecological degradation has created havoc on the earth. Moreover, the changing vegetation patterns over the last five decades have been mainly on the declining side. The tree canopy all around has been severely broken by human activities. Barren land is in plenty which has led to serious soil erosion. Because of population pressure, there is a gradual increase in cultivable land, choe terraces and choe valley low lands. All this led to encroachment on deeper forest areas of the hills with high pressure of grazing animals leading to excessive soil erosion. Thus, the ecology has rapidly grown in recent years due to following reasons :

1. Population explosion resulting in excessive burden on natural resources, both renewable and non-renewable.
2. Development of technology which frequently ignored biological laws.
3. Extravagant pattern of consumption which permitted very little or no recycling of waste products.

All the aforesaid factors are responsible for scarcity of natural resources, decrease in productivity of biological systems which sustain man and his domesticated animals, deteriorate climate and cause pollution of air, water and soil. Due to rapid environmental degradation, the ecology has assumed greater importance and the people have begun to believe that political and socio-economic decisions should be based on ecology. Thus, every living organism is facing environmental hazards. There is no doubt that human activities are partly responsible for environmental hazards we are facing now. A few of such activities are misuse of natural resources, rapid urbanisation, deforestation and depletion of forest cover, population explosion, shifting of surface water and ground water in massive amounts, release of heat energy from various industries, emission of particles and trace gases into the atmosphere, release of carbon dioxide into the atmosphere by combustion of fossil fuels, effect of transportation systems on land surfaces, and effect of their emission upon the lower and upper atmosphere, depletion of protective ozone umbrella layer, formation of ozone holes, release of chlorofluoro carbons in to the atmosphere, global warming etc. There are broadly, two types of pollution from industries which need attention as far as their treatment and disposal is concerned: (i) water pollution; and (ii) air pollution.

Industrial waste-waters vary widely in their composition and treatment methods, which have to take into consideration the specific characteristics of wastes. Many treatment practices have followed the approach of mixing the liquid sewage waste with industrial waste and treating the mixture by biological oxidation, broadly as followed for sewage waste.

Low cost methods of treatment, such as lagoons (aerobic and anaerobic), oxidation ditches and aerated lagoons have also been tried with varying degree of success. Stabilisation ponds, activated or modified-activated sludge processes, aerated lagoons, trickling filters and anaerobic biological digestion units have been used for treating these wastes.

But the fact remains that biological treatment, even though cheap in capital and operating costs, may not be best suited for handling industrial waste-water problems. Physico-chemical methods are necessary to remove the chemical ingredients present in liquid effluents.

The re-use of water in processes where the water quality standards are not stringent, is worth considering. A considerable quantity of water is presently being re-used in process industries in India, but a lot more needs to be done in this area.

A survey of the air-quality in major cities in India such as Mumbai, Kolkatta, Delhi, Kanpur and Nagpur, for which detailed ambient-quality monitoring has been done over extended periods, indicates that the level of sulphur dioxide, particulate matter and oxides of nitrogen are much higher than those specified by the Ministry of Environment and Forests/Central Pollution Control Board. The pollution level in ambient air can best be reduced by reducing concentrations in effluent gases from process industries.

POLLUTION FROM MAJOR INDUSTRIAL OPERATIONS

Even though there are many different sources which contribute to air-pollution, industries contribute a major share. There are a number of industries which are responsible for air pollution. Table 24.1 shows the specific pollutants found in the emissions of some typical industries.

Table 24.1. Emission of pollutants in some specific industries.

Source	Pollutants
Fertiliser industry and aluminium manufacturing plant	Hydrogen fluoride, ammonia, flourides, fertiliser dust, and sulphuric acid mist
Lead casting and melting, pigments, etc. Heavy chemical industry like acid plants, synthetic fibres, etc.	Tin, lead etc., fumes and oxides, solvents and thinners, acid fumes
Tanneries and leather industry, cement industry, carbon black manufacture	Mercaptans and sulphides, cement and lime dust, polynuclear hydrocarbons, carbon soot and hydrogen sulphide
Paints and pigments	Nitrobenzene and aniline, thinners, solvents and base material
Coal tar industry	Polynuclear hydrocarbons and aerosols of tar
Paper and paper products	Hydrogen sulphide and mercaptans
Refinery and photochemical industry	Hydrogen sulphide, hydrocarbons, odours of mercaptans
Metallurgical industry	Metallic fumes, dust
Electrolytic manufacture	Chlorine
Coal burning (Power Plants)	Soot

AIR POLLUTION

Air is a natural resource and it is a fundamental element of human life as it makes breathing possible. It is the basis for all forms of terrestrial and two-third of all biological species, also one of the important sources for economic development like agricultural and industrial production, energy generation, heating, cooling and so on.

Air is a mixture of gases that form the atmosphere. The clean air is composed of the gases varying in concentrations (Table 24.2).

Table 24.2. Natural components of the air.

Gas	Volume (%)
Oxygen	20.92
Nitrogen	78.10
Argon	0.9325
Carbon dioxide	0.03
Hydrogen	0.01
Neon	0.0018
Helium	0.0005
Crypton	0.0001

Air becomes damaging to nature and human health when there is an excess of polluting elements in it. Air pollution has become an important factor of environmental degradation all over the world. The increasing agglomeration, in particular, industrialisation, manufacturing units, urbanisation, motorisation and burning of fuel material in households produce a large amount of air polluting substances which have a harmful effect, especially on human health, animal and plant life, and even on the buildings and works of art.

Air pollution may be defined as that quantity of pollutants which is sufficient to cause injury to human beings and other living creatures and damage to objects. It has been observed that under extreme conditions of high concentration of air pollutants, the human deaths have also increased. Diseases like bronchial asthma, lung cancer, irritation reaction, heart and brain damages, etc., are probably due to adverse effects of high concentration of air pollutants.

Air pollution can be categorised into two groups; first kind of pollutants are released into the atmosphere from a specific source, and the second type of pollutants result from chemical changes that take place in the atmosphere. When the amount of such pollutants in the air exceeds a certain level, then the pollution of air is created. Pollutants may be in the form of dust, odours or vapours. The quantities of pollutants which are dangerous to nature have been determined both by national and international organisations.

Major Air Pollutants

Air pollutants exist in various quantities in the atmosphere and they are found in solid, liquid or gaseous forms. According to their mode of formation, pollutants which enter into the air directly are known as *primary pollutants*, and which are created in the air from other pollutants under the influence of electromagnetic radiations from the sun are called *secondary pollutants*. Common types of pollutants can be grouped into two major divisions, i.e., particulate matter and gaseous substances. Dust, flyash, smoke, soot, aerosol droplets, mist, fog and fumes can be categorised under particulate matter, whereas gases and vapours come under gaseous substances. Sub-classification of particulate matter and gaseous substances are shown in Table 24.3.

Table 24.3. Examples of airborne contaminants.

Dust	Fumes	Droplets of mist	Gases	Vapours
Cement	Iron oxide	Sulphuric acid	Sulphur dioxide	Gasoline
Coal	Zinc oxide	Chromic acid	Nitrogen oxide	
Trichlorethylene				
Ores	Lead oxide	Oil	Carbon monoxide	
Perchlorethylene				
Grains	–	Grease (from cooking and smoking of meats)	Hydrogen sulphide	Toluene
Rock sawdust	–	Paint	Chlorine	Styrene

Among these pollutants, carbon monoxide (CO), nitrogen oxide (NO), sulphur dioxide, (SO_2), ozone (O_3), particulate matter and lead oxide cause damage to the environment.

Carbon monoxide

Carbon monoxide (CO) is a colourless, odourless, highly toxic gas, generated when carbon-containing material burns with insufficient oxygen. It is discharged from gasoline engines and burning of coal. It occurs naturally in the atmosphere in trace amounts. Another source of CO is the burning of firewood and wood-waste. CO is also produced in large quantities in thermal power plants as a result of burning of coal. Approximately 250 million tonnes of carbon monoxide is poisoning the global atmosphere every year.

CO can cause adverse breathing effect because it interferes with absorption of oxygen by the red blood cells. It combines with haemoglobin in human blood to form *carboxyhaemoglobin*, which impairs oxygen transport. The normal functioning of the nervous system can be affected by the presence of 2–5 per cent of *carboxyhaemoglobin* in the blood which may occur after breathing air with only 30 ppm of CO. It may cause giddiness and headache in less than one hour at a concentration of 100 ppm and such situations are common during traffic jams. At higher concentrations, it is lethal to man and many other living organisms.

On a global basis, 90 per cent of the total CO originates from the oxidation of methane produced by decaying organic matter. However, man-made CO emission, of which up to 90 per cent arises from the transport sector, mainly automobiles, is responsible for relatively high local concentration in most of the industrialised nations. In Delhi alone, about 250 tonnes of CO is released in the atmosphere per day which reduces the respiratory capacity of the people.

Carbon dioxide

Human activities are responsible for increasing the amount of carbon dioxide (CO_2) in the atmosphere. It is not a pollutant itself and it comes from burning of fossil fuels and the oxidation of carbon stored in trees and soil humus, and released, when forests are fired for shifting cultivation. CO_2 is a natural component of the earth's atmosphere. It plays a vital role in global carbon circulation through the environment and is a main feed stock for the photosynthesis process in green vegetation.

The increase of atmospheric CO_2 has various negative effects, including direct effects on plant life and biosphere, whereas indirect effects are through climatic changes such as increase in global temperature. Increase in global temperature results into further global changes, viz., change in rainfall pattern; aeration of desert like condition; productive land melting of ice in the polar region of the world which will result in rising of sea level; and change in fish production, forest and water supply.

CO_2 concentration in the earth's atmosphere was estimated to be 270 ± 10 ppm in the middle of nineteenth century. At present, the concentration of CO_2 in the atmosphere is approximately 350 ppm, and the annual rate of increase of CO_2 is about 1.5 ppm. In the early to the middle of the next century, CO_2 concentration in the global atmosphere may approximately double than the present concentration. It has been estimated that the concentration of carbon dioxide is increasing at the rate of 0.7 mg/L per year in the atmosphere.

Nitrogen oxides

All the oxides of nitrogen, for example, nitric oxide (NO) and nitrogen dioxide (NO_2) are represented by NO_x. NO_x emission comes about in equal parts from the transport sector and from energy and industrial sectors. The major sources of oxides of nitrogen are motor vehicle exhausts, soft coal burning and acid manufacturing. On a global level, about 60 per cent of NO_x is produced from natural resources, including biomass burning, fixation by lighting, inflow from the stratosphere, chemical conversion from ammonia in the troposphere and loss of gaseous nitric oxide from the soils.

Nitrogen oxides have various direct and indirect effects on human life and environment. They inhibit cilia action so that soot and dust penetrate lungs to cause bronchitis and other respiratory diseases. NO_2 readily combines with water to form HNO_3 which forms a part of acid rain. It plays a great role in the formation of smog in acid and humid conditions. About 10 million tonnes of nitrogen containing gases like NO and NO_2 are entering into the atmosphere every year.

Sulphur dioxide

Various oxides of sulphur have harmful effects on the environment, among which sulphur dioxide (SO_2) is the most damaging. SO_2 is a colourless gas with a suffocating and strong pungent odour. Industrial processes and burning of sulphur-containing coals or heavy oil, particularly in thermal power plants, are responsible for release of SO_2 in the air. In the atmosphere, sulphur dioxide combines with oxygen and water to form sulphuric acid (H_2SO_4), the major component of acidic rain. On a global scale, about half of the ambient SO_2 originates from the oxidation of hydrogen sulphide (H_2S) given off by decaying organic matter.

Long-term or chronic exposures to SO_2 have been linked to the increased incidence of respiratory diseases such as bronchitis, particularly in young children. It causes irritation in throat and eyes, chest constriction, headache, vomiting and death. Monocotyledonous plants are more sensitive to SO_2 and it causes damage to cereal crops, conifer forests and apple orchards. Every year, 90 million tonnes of SO_2 is poisoning the air of our earth.

Particulate matter

Particulate matter includes those air pollutants which may be in the form of solid particles or liquid droplets including fumes, smoke, fog, dust, pollen, bacteria, viruses and aerosols. This category includes about 5 per cent of the weight of all pollutants present in the atmosphere. Natural dust forms about half of the total mass of particulate matter in the air. Dust has a relatively small impact because it tends to be coarse. Fine particulate matter (< 2.5 µg dia) is more harmful to human health than the coarse particulate matter (> 2.5 µg dia). Particulates remain suspended in the air and are transported away over a long distance when the strong wind blows. Small particulates with a size of less than 2 µg have an important impact on health although they contribute only about 1 per cent of the particulate load in the atmosphere.

The sources of occurrence of particulate matter are the sprays, mist, dust from spraying and grinding of building materials, land levelling and clearing, forest fires, harvesting and threshing of field crops, agricultural operations like spraying, volcanic explosion and various construction works. It is estimated that about 8 billion solid particles penetrate into the atmosphere every day.

Diesel run engines generate more particulate matter than the gasoline ones. It is estimated that a diesel engine releases, 62.5 per cent particulate of the total having a size of less than 5µ and 37.5 per cent in the 5–20µ range. Thermal power plants and factories generate aerosols and soots. Aerosols are the fine solid or droplets of liquid particles which are so small that can be inhaled and these penetrate deep into lungs. Their long run effects on the human health are that they create disorders. These particles carry toxic metals and other toxic substances. Fine particulate matter also contributes significantly to visibility reduction.

Ozone

Ozone (O_3) gas is present in significant concentrations in both stratosphere and troposphere. It has a pleasant odour but causes irritation when present in higher amounts. It is used as a disinfectant for air and water, and in industry for bleaching waxes and oil, as well as for organic synthesis. Ozone is produced in the atmosphere through the interaction of nitric oxide, sunlight and hydrocarbons. It is considered to be the major component of *photochemical smog*:

$$\text{Nitrogen oxides + hydrocarbons} \xrightarrow[\text{nitrate}]{\text{Sunlight}} \text{Peroxyacetyl + Ozone (Photochemical smog)}$$
$$\text{(PAN)}$$

Motor vehicles exhaust very high amount of ozone in the atmosphere. It plays a vital role in the photo-chemical formation of air pollutants. Its presence in the stratosphere serves to reduce the amount of short-wave ultraviolet (uv) radiation at wavelength below 310 nm reaching the earth. This natural screening is necessary for human and other living organisms. The uv radiation at 290 nm, for example, is 10,000 to 1,00,000 times more effective in damaging DNA (deoxyribonucleic acid), than is uv radiation at longer wave-length of 320 nm. The main concern about more short-wave uv radiation reaching the grounds is the possible increase in the occurrence of human skin cancer. It might also damage the animal and plant life and possibly rduce the rate and efficiency of photosynthesis, and change in the ozone layer might also affect the global climate.

Lead oxide

Lead is a metal and exists in a variety of chemical compounds with different characteristics. The atmospheric lead contamination comes largely from the automobiles. In general, one litre of fuel contains 2 to 4 g of lead in it. On an average, 70–80 per cent of the lead in the fuel is emitted by the motor vehicles as particulate. Human body gets lead mainly by ingestion and inhalation. About half of the lead taken up by the human body is being absorbed; small amount of lead ingested in food enters the blood. Some part of it is slowly excreted and the rest enters the bones. World Health Organisation (WHO) has prescribed the maximum intake of lead of 3.5 µg per person per week. About 5–10 per cent of the total lead is absorbed by the stomach-intestine tract, while 30–50 per cent is absorbed by the respiratory tract.

In the atmosphere, lead is the most prevalent of a group of *heavy metals* that include mercury, cadmium and manganese. Lead is a contaminant of major concern because of the harm it does to many human tissues and organs, especially the nervous system, the kidney and the cardiovascular system, and particularly in children. Adverse effects have been found in some population groups with blood lead levels as low as 10 µg/dl (microgram per decilitre).

Lead poisoning can lead to anaemia, brain damage and loss of kidney function. Lead can enter the environment from many sources; the mining and smelting of lead bearing ores and metals, lead plumbing and soldering, use of paints and ceramic glass, and the careless disposal of lead-zinc batteries in the open space.

Sources of Air Pollution

The causes of air pollution are urbanisation, industrialisation, motorisation, energy production from thermal plants, burning of domestic fuel, burning of agricultural wastes and development in urban settlements. Parallel to rapid population growth, there has been increased population density in cities causing more pollution and sites for urbanisation and industrialisation have been incorrectly chosen. These factors have exaggerated the pollution problem. Furthermore, ignorance concerning environmental problems and a failure to give adequate attention are gradually adding to the dimension of the problem. Although, a variety of factors play a role, the causes of air pollution may generally be divided into urbanisation, motorisation, industrialisation, agricultural operations and nuclear production (Table 24.4).

Table 24.4. Classification of air pollution sources and emissions.

Source type	Category	Examples	Pollutants
Dust producing	Crushing, grinding, screening	Road mix plants	Mineral and organic particulates
	Demolition	Urban renewal	
	Milling	Grain elevators	
Combustion	Fuel burning	Home heating units and power plants	Oxides of sulphur and nitrogen, carbon monoxide, smoke, flyash, organic vapours, metal oxide, particles, and odours
	Motor vehicles	Autos, buses and trucks	
	Refuse burning	Community and apartment house incinerators, open burning dumps	
Manufacturing processes	Metallurgical plants	Smelters, steel mills, aluminium refineries	Metal fumes (lead, arsenic and zinc), fluorides and oxides of sulphur
	Chemical plants	Petroleum refineries, pulp mills, super phosphate fertiliser plants, cement mills	Hydrogen sulphide, oxides of sulphur, fluorides, organic vapours, particulates, odours
	Waste recovery	Metal scrap yards, auto body burning, rendering plants	Smoke, soot, organic vapours, odours
Agricultural activities	Crop spraying and dusting	Pest and weed control	Organic phosphates, chlorinated hydrocarbons, arsenic, lead
	Field burning	Stubble and slash burning	Smoke, flyash and soot
	Frost damage control	Smudge pots	

(Contd ...)

Source type	Category	Examples	Pollutants
Solvents	Spray painting inks	Automobiles assembly, furniture and appliances finishing, photograving and printing	Hydrocarbons and other organic vapours
	Solvent cleaning	Dry cleaning, degreasing	
Nuclear energy	Ore preparation	Crushing, grinding and screening	Uranium and beryllium dust, fluoride
	Fuel fabrication	Gaseous diffusion	
	Nuclear fission	Nuclear reactors	Argon-41
	Spent fuel processing	Chemical separation	Iodine-131
	Nuclear device testing	Atmospheric explosions	Radioactive fall out (Strontium-90, Casium-137, Carbon-14)

Urbanisation

The rapid urbanisation which has been taking place in India is one of the most important causes of the country's air pollution. The pollution in cities is known to arise to a large extent from fuel burning techniques and poor fuel quality. Among the developments affecting air pollution in cities, many other factors in addition to population, density play an important role. These include incorrect urban settlements not planned according to topographic and meteorological conditions, incorrect plot division, low quality fuel and incorrect burning techniques, dearth of green areas, increase in the number of motorised vehicles and inadequate or improper disposal of wastes.

Motorisation

Traffic also contributes a large part to the air pollution. It is estimated that 50–60 per cent of the total air pollution is caused by running of motor vehicles in Indian cities. This is because of inferior quality of fuel, outdated engines and overloading of motor vehicles. Road traffic in particular is much more damaging to the environment than railway traffic.

Industrialisation

The air pollution caused by industry is principally a result of incorrect selection of site and emission of waste gases into the air without taking adequate technical precautions. Areas largely affected by air pollution at present are in and around the industrial setups like paper, sugar processing, textile, steel, pesticides, leather, and various other chemical industries power plants in India.

Effects of Air Pollution

Air pollution has both direct and indirect impact on human body, animal life, plant kingdom, construction materials, climate and entire ecosystem. These effects have both long-term and short-term implications, and influence on the economy and welfare of the human-beings. Out of these, the effects of air pollution on human health are of great concern.

Effects on human health

Air is a fundamental element of human life as it makes breathing possible. Pollutants enter into the human lungs through the following mechanisms:

1. In the respiratory system, which reacts by the initiation of constructive reaction of a bronchi reflex.
2. In the blood vessels of the bronchus and its branches, which try to reduce the absorption of harmful substances through the bronchial mucosa.
3. In the blood vessels of the lungs, where they react by decreasing absorption from the alveolar (pulmonary) capillaries.
4. In the heart and large blood vessels taking part in the transport of toxic substances.
5. By penetration into organs, tissues or cells and by affecting metabolic processes

Pollution in the air is thus of utmost importance from the view point of human health. Pollutants in the air, such as carbon particles, ozone, carbon monoxide, sulphur dioxide, unsaturated hydrocarbons, aldehydes, carcinogens, etc. disrupt the normal mechanism of the human respiratory tract, causing bronchial infections and stenosis, chronic bronchitis and emphysema (Table 24.5). Polluted air can also cause discomfort, shortness of breath, and is a major cause of cancer.

Table 24.5. Acute and chronic physiological effects of major air pollutants on human beings.

Pollutant	Effects	
	Acute	*Chronic*
Sulphur dioxide	Gives rise to irritation reactions, which cause capillaries to dilate and exude fluid; this leads to tissue fluid accumulation and swelling (oedema), bronchial spasms, and shortness in breath. General physiological reaction to SO_2 is similar to allergic asthma, i.e., with impaired pulmonary function via increased airway resistance, impaired lung clearance, and increased susceptibility to infection.	Contributes to and aggravates lung diseases like chronic bronchitis, pulmonary fibrosis via irritation leading to decreased pulmonary function and increasing stress on the heart.
Particulate matter	Depending on nature and size, particulate matter can cause irritation, altered immune defence, or systemic toxicity.	Depending on the nature and size of the particles, particulate matter can cause decreased pulmonary function and stress on the heart.
Nitrogen dioxide	Incompletely understood, although cell membrane disruption appears to be the principal reason for respiratory tract oedema.	Cell membrane damage and acid-induced irritation leads to or contributes to diminished pulmonary function and right heart stress.
Carbon monoxide	Asphyxiation, heart and brain damage, impaired perception.	Increased red blood cells (polycythaemia) in blood, leading to increased resistance to blood to flow; weakness, fatigue, headaches.
Photochemical oxidants (e.g., ozone)	Decreased pulmonary function and right heart stress	Emphysema, fibrosis, right heart failure, aging of lung and respiratory tissue.
Hydrocarbons	The primary harm of hydrocarbons is in their participation in ozone production. Cancer is one kind of direct primary effect of some organic compounds.	Spread it through both the columns.

Effects on plants

The effects of air pollution are encountered to some extent in other organisms as well on human life. Plants too are faced with the harmful effects of polluted air. Plants have a very close interrelation with the environment through absorbing CO_2 from the atmosphere and releasing oxygen in the air. All forms of life on this earth are directly or indirectly dependent upon plants as a source of food. Plants also act significantly improving our environment ecologically, aesthetically, physically and chemically. In brief, man depends upon plants for his survival, and the level to which he enjoys his life on this earth is greatly affected by the kind and quality of vegetation within the geographical area of his habitation.

Effects on materials

Polluted air damages materials mainly by corrosion of metals, stones, marbles, etc. from acidic compounds in polluted atmosphere. The most important acid forming pollutants are sulphur dioxide and nitrogen oxide. In the presence of moisture, they change into sulphuric acid and nitric acid. Deposition of these acids on the metal parts of buildings, roof covers, down spouts, and other metal equipment results in a considerable loss from atmospheric corrosion in industrial sectors. Also, the property damage due to ozone is due to rubber cracking. Major areas of economic importance are the side walls of tyres and various forms of electrical insulation.

Hydrogen sulphide changes the white lead painted surface into dark grey. Property damage is caused by soiling of surface by pollutants. Cost of cleaning the surface increases the economic burden. Washing, dry cleaning of clothes and other fabrics and repair of buildings add to the cost. Other effects of toxic gases on building materials are presented in Table 24.6.

Table 24.6. Effects of major air pollutants on materials.

Chemical	Primary materials attacked	Typical damage
Carbon dioxide	Building stones, limestones	Deterioration
Sulphur oxides	Metals	
	Ferrous	Corrosion
	Copper	Corrosion to copper sulphate (green)
	Aluminium	Corrosion to aluminium sulphate (white)
	Building materials (limestone, marble, slate, mortar)	Leaching, weakening
	Leather	Embrittlement, disintegration
	Paper	Embrittlement
	Textiles (natural and synthetic fabrics)	Reduced tensile strength, deterioration
Hydrogen sulphide	Metals	
	Silver	Tarnish
	Copper	Tarnish
	Paint	Leaded paint blackened due to formation of lead sulphide
Ozone	Rubber and elastomers	Cracking, weakening
	Textiles (natural and synthetic fabrics)	Weakening

(Contd...)

Chemical	Primary materials attacked	Typical damage
Nitrogen oxides	Dyes	Fading
Hydrogen fluoride	Glass	Etches, opaques
Solid particulates	Building materials	Soiling
(soot, tars, etc.)	Painted surfaces	Soiling
	Textiles	Soiling

Alarming deterioration of the Taj Mahal, which was made of costly marble and stones, through the pollutants released from Mathura Refinery is a typical example in India. The Taj Mahal was damaged with black coating that contained various pollutants. The surface of the Taj Mahal has lost its lustre and has become rough and slightly coloured due to the effect of pollution over the years. Sulphur dioxide is the major pollutant, while the oxides of nitrogen can also cause damage to the historical monument. The effects of sulphur dioxide at higher levels at the domes is much less than that at lower levels at the base.

Effects on climate

When atmospheric conditions are changed by air pollution, they affect the climate. Generally, the average temperature in cities is higher than that in rural areas. Meteorological measurements also show that wind speed decreases in large cities with increased air pollution. This decrease in speed is important because the wind affects heat and moisture. Air pollution is also a cause of increased rainfall in big cities. The abundance of hygroscopic substances, which results from energy heating up the air, increases rainfall by making it easier for the clouds to form. At the same time, the layer of pollution which is formed over a city causes a significant reduction in the amount of ultraviolet light, resulting in decreased exposure to sunlight.

These undesirable developments indicate how air pollution upsets the natural balance of the climate. Other effects of air pollution at international level are the depletion of ozone layer, warming of earth surface, rising of sea level, and acid rain. These are some of the serious issues facing the entire world now-a-days.

Control Measures

The objectives of air pollution control are:
1. To organise the air quality management in the long term.
2. To protect the human beings as well as the plant and animal life, and materials from harmful effects.
3. Reducing air pollution in highly polluted areas.
4. Retaining the existing air quality in areas currently less polluted.
5. Reducing the long-range transport of air pollutants.

Most of our cities exceed the primary standards by a considerable amount throughout the year. While the main reason for our inability to achieve clean air in our country is the failure of strategies other than technological measures. There are various fundamental technological and non-technological approaches to the control of air pollution, which are to be used in one or the other ways. Some of the following remedies can be helpful to minimise/reduce the pollution of air :
1. *Enforcement of air (prevention and control) pollution act, 1981:* Air quality standards (Table 24.7) as recommended by the Central Pollution Control Board must be strictly implemented.

2. *Use of purified fuel:* The basic technology for the removal of sulphur from coal is already known, but it has economic and legal implications. It is also possible to reduce the level of lead oxide in the air by taking lead out of petrol.

Table 24.7. Air quality standards adopted by the central pollution control board.

Category	Permissible concentration ($\mu g/m^3$)			
	SPM	SO_2	CO	NO_x
Industrial and mixed area	500	120	5000	120
Residential/rural area	200	80	2000	80
Sensitive area	100	30	1000	30

3. *Modernisation of out-dated industries and machinery:* We should improve the industrial processes and vehicle engines, and make the overall expenditure of energy more efficient by modifying old machinery.
4. *Installation of air treatment plants:* Various alternative techniques for the purpose are also available in our country. Large number of treatment plants both for air and water purification, must be installed either for individual industry or as common treatment plants. These treatment plants must be run throughout the clock.
5. *Alternative energy sources:* The different sources of energy are wind, water and sun. Each form of energy generation has its unique problems and lends itself in slightly different ways to the control of air pollution. This must be taken into account in the consideration of various solutions to our air pollution problems and energy crisis.
6. *Installation of devices:* The following devices are useful for air pollution control :
   (a) *Filters:* Filters separate out the particulate matter from the stack gases in electric power plants. The smoke passes through a series of cloth bags which trap the particulate matter.
   (b) *Cyclones:* Cyclones also remove particulate matter. The polluted air is passed through a metallic cylinder at high speed. The particulates strike the walls of the cylinder and fall at the bottom. This technique removes 50–90 per cent of large particulate matter along with a few medium and small particulates.
   (c) *Electrostatic precipitators:* Electrostatic precipitators also remove the particulate matter and these are 99 per cent efficient. The particulates become electrically charged when passed through the precipitators. The particulates attach themselves to the wall of the device which has the opposite charge. By frequently turning off the current, the particulates are allowed to fall to the bottom.
   (d) *Scrubbers:* These are used to remove both the particulate matter as well as sulphur dioxide gas. The polluted air is passed through a fine mist of water which traps about 99 per cent particulate matter and 80–95 per cent sulphur oxides.
   (e) *Catalytic converters:* Catalytic converter is a device which runs exhaust gases of automobiles through a bed of alumina pellets coated with platinum or palladium catalyst. This device when attached to the exhaust system of a vehicle converts carbon monoxide and hydrocarbons into carbon dioxide and water, and nitrogen oxides into nitrogen gas. The catalyst is rendered ineffective by lead and, therefore, automobiles with catalytic converters must use lead-free petrol.

7. *Subsidised lead-free petrol:* Lead-free petrol must be available at the lowest rate so that the vehicle users can be motivated to use a non-polluting toxic-free fuel. People can be educated about the benefits of lead-free petrol.

8. *Plantation of trees:* Trees, especially broad leaved plants such as various ornamental trees, forests and fruit trees keep large amount of gases and dust on their leaves, twigs and stems, and these tree parts absorb various pollutants.

9. *Change in life style:* The magnitude of air pollution can be reduced by a drastic change in our life styles and proper planning like using energy more efficiently, relying more on non-combustive sources of energy such as solar and wind energy, restriction of areas in which any industries shall not be installed or installed subject to certain safeguards, planting of pollution tolerant and dust filtering plant species as green belts around industrial and urban areas, maintenance of roads, ideal traffic planning and removal of unnecessary check posts and barriers.

Photochemical Smog

Photochemical smog is initiated by the photochemical dissociation of NO_2 and the consequent secondary reactions involving unsaturated hydrocarbons, other organic compounds and free radicals, leading to the formation of organic peroxides and ozone. This phenomenon takes place during sunny days with low winds and low level inversion. The photochemical smog and the consequent formation of aerosols reduce the visibility, cause irritation to eyes and damage plants and rubber goods.

The oxidation of SO_2 can also take place by interaction with the free radical $HO\cdot$ present in photochemical smog

$$SO_2 + HO\cdot \longrightarrow HOSO_2\cdot$$
$$HOSO_2\cdot + O_2 \longrightarrow HOSO_2 O_2\cdot$$
$$HOSO_2 O_2\cdot \xrightarrow[\text{(sulphate)}]{} HOSO_2 O\cdot + NO_2$$

Chemical oxidation of SO_2 may also take place in water droplets, present in aerosols. This reaction is accelerated in the presence of NH_3 and catalysts e.g., oxides of Mn, Fe, Cu, Ni.

Solid particles, such as soot, bring about catalytic oxidation of SO_2 by providing a heterogeneous phase for contact. Soot is formed during combustion of solid and liquid fuels in domestic and industrial operations and automobile emissions.

Sulphur dioxide is a pollutant responsible for smog formation, acid rains, and corrosion of metals and alloys.

Oxidation of organic compounds

Organic compounds such as hydrocarbons, aldehydes and ketones absorb solar radiation and undergo various photochemical and chemical reactions involving free radicals. Some of these reactions are catalysed by particulate matter such as soot and metal oxides. Some of the intermediates and final products formed contribute to photochemical smog formation.

Greenhouse Effect

The earth is heated by sunlight and some of the heat that is absorbed by the earth is radiated back into space. However, some of the gases in the lower atmosphere, acting like glass in a greenhouse, allow the solar radiations (in the range 300 to 2500 nm, i.e., near UV, visible and near I.R. region, while filtering the dangerous UV radiations, i.e. < 300 nm) but do not allow the earth to re-radiate the heat into space. In other words, these gases in the atmosphere are transparent to the sun-light coming in, but they strongly absorb the infra-red radiation, which the earth sends back as heat. A part of the heat so trapped in these atmospheric gases is re-emitted to the earth's surface. The net result is the heating of the earth's surface by this phenomenon, called the 'Greenhouse effect'. The gases that are responsible for this Greenhouse effect are CO_2, water vapour, CH_4 and man-made chlorofluorocarbons (CFC's). Water vapour strongly absorbs I.R. radiations in the range 4000 to 8000 nm and CO_2 in the range 12,000 to 16,300 nm. The radiations in the range 8000 to 12,000 nm escape unabsorbed and this is known as the region of atmospheric window.

Carbon dioxide is released by volcanoes, oceans, decaying plants as well as human activities, such as deforestation and combustion of fossil fuels. Automobile exhausts account for 30% of CO_2 emissions in developed countries.

Methane is released from coal mines, decomposition of organic matter in swamps, rice paddy cultivation, guts of termites in forest debris and stomachs of ruminants.

Chlorofluorocarbons (CFC's) are used as coolants in refrigerators, propellants in aerosol sprays, plastic foam materials like 'Thermocoles' or 'Styrofoam' and in automobile air-conditioners.

In fact, the 'Greenhouse gases' (particularly CO_2 and water vapour) are responsible for keeping our planet warm and thus sustaining life on the earth. If the Greenhouse gases are very less or totally absent in the earth's atmosphere, then the average temperature on the earth would have been at sub-zero levels. But, however, if the concentration of Greenhouse gases is larger, they may trap too much of heat, which may threaten the very existence of life on the earth. For instance, the CO_2 present in the atmosphere of the planet *Venus*, is about 60,000 times more than that in the earth's atmosphere. Hence the average temperature of *Venus* is about 425°C, thus making the existence of life impossible there.

Oceans and bio-mass are the major sinks for the atmospheric CO_2. Oceans convert CO_2 into soluble bicarbonates. The photosynthetic activity in the green plants increases with increase in CO_2 level in the atmosphere. Forests are the places where lot of photo-synthetic activity occurs. They also act as vast reservoirs of fixed but readily oxidisable carbon in the form of vegetation, wood and humus. Hence, forests maintain a balance in the atmospheric CO_2 level. Therefore, deforestation definitely upsets this balance and increases the atmospheric CO_2 level.

It is estimated that the atmospheric CO_2 content has increased by 25% during the last two centuries. This is mostly attributed to the industrial revolution over these two centuries. This is one of the reasons for the slight increase in the global temperature (about 0.5°C). Since the concentrations of the Greenhouse gases have been continuously increasing because of deforestation, industrialisation, increased burning of fossil fuels, mining, exhausts from increasing number of automobiles and other anthropogenic activities, there is an increasing concern about the possible 'global warming'. Some scientists fear that if proper precautions are not taken, the concentration of the green-house gases in the atmosphere may double within the next 50–100 years. If this happens, the average global temperature may increase by 4 to 5°C. This will increase the evaporation of surface waters, which may influence climatic changes depending upon the pattern of cloud formation. For instance, low-level dense clouds may exert cooling effect whereas high level thin cloud formation may exert heating effect due to increased greenhouse effect.

The projections from computer modelling regarding the climatic changes that could be triggered off due to "global warming" reveal alarming scenarios. Even 1.5°C rise in surface temperature can adversely affect the food production in the world. Thus, the wheat growing zones in the northern latitude may be shifted from the USSR and Canada to the polar regions i.e., from fertile soils to poor soils near the north pole. The biological productivity of the ocean would also decrease due to warming of the earth's surface layer, which in turn, may reduce the transport of the nutrients from deeper layers to the surface by vertical circulation. The computer modelling also indicates the following effects due to 'global warming': melting of the polar ice caps; dry areas becoming drier; humid areas like the Amazon, suffering more intense tropical storms; drastic drop in food production, particularly in lands within 35 degrees north and south of the Equator; increased breeding of pests and diseases due to more humid conditions; shorter, wetter and warmer winters; and longer, hotter and drier summers, particularly in mid-continental areas.

Global warming may also trigger increased thermal expansion of oceans and melting of glaciers, which results in lifting up of the sea-level by 20 cm to 1.5 metres by the latter part of the 21st century. Thus, cities like Mumbai, Miami, London, Venice, Bangkok and Leningrad may become extremely vulnerable. Defences against the rising sea-levels and expanding oceans are very difficult and expensive, which many nations of the world cannot afford.

Further, a global temperature rise, even about 1.5°C, is likely to cause floods, hurricanes, tornadoes, apart from raising of the sea-level due to melting of the polar ice caps and inundating coastal cities like Chennai, Sydney, New York and Boston.

There are differences of opinion among experts regarding the dynamics and effects of 'global warming' due to the complexity of natural phenomena that might be operating simultaneously. More accurate future climatic projections will be possible with better super-computer models, based on greater understanding of the complex natural climatic forces involved. But until that time, the possible devastating effect due to 'global warming' by the 'greenhouse effect' cannot be underestimated.

Some of the steps suggested to minimise the 'Greenhouse effect' include reduction in the use of fossil fuels, encouraging the use of alternative sources of energy (e.g., solar, geothermal, wind, bio-gas, etc.). Conservation of forests, extensive afforestation, encouraging community forestry, reduction in the use of automobiles, research in the development of more efficient automobile engines, ban on CFC's and nuclear explosions, development of environmentally compatible technologies with the help of intensive inter-disciplinary research, effective check on the growth of population and imparting of non-formal and formal environmental education.

Formation and Depletion of Ozone in the Stratosphere

Ozone is an important chemical species present in the stratosphere. At an altitude of about 30 km, its concentration is about 10 ppm. The ozone layer present in the stratosphere acts as a protective shield for the life on earth. It strongly absorbs ultra-violet radiations from the sun in the region 220–330 nm and thereby protects the life on earth from severe radiation damage, such as DNA mutation and skin cancer. Thus only a small fraction of UV radiation reaches the lower atmosphere and the earth's surface.

Ozone is formed in the stratosphere by photochemical reaction:

$$O_2 + h\nu \; (242 \; nm) \longrightarrow O + O$$

$$O + O_2 + M \; (\text{third body, such as } N_2 \text{ or } O_2) \longrightarrow O_3 + M$$

The third body absorbs the excess energy liberated by the above reaction and thereby the ozone molecule is stabilised. Thus, ozone is constantly formed in the stratosphere. However, it is also destroyed by chlorine, released due to volcanic activity and also by reaction with: (i) nitric oxide; (ii) atomic oxygen; and (iii) reactive hydroxyl radical, which are also present in the atmosphere. In atmosphere, nitrogen oxide (NO) comes from chemical and photochemical reactions in the atmosphere, supersonic jets, nuclear explosions, etc; Cl_2 comes from CFC's and volcanoes; and OH comes from biomass burning and from natural water systems by the following reactions:

(i) $\qquad O_3 + NO \longrightarrow NO_2 + O_2$

(ii) $\qquad O_3 + O \longrightarrow O_2 + O_2$

(iii) $\qquad O_3 + HO\cdot \longrightarrow HO_2 + HOO\cdot$

$\qquad HOO\cdot + O \longrightarrow HO\cdot + O_2$

Ozone, in the stratosphere, is also found to be destroyed by man-made chlorofluorocarbons (CFC's), which are used as coolants in refrigerators, air-conditioners, propellants in aerosol sprays and in plastic foams, such as 'Thermocole' or 'Styrofoam'. The CFC molecules, escaping into the atmosphere, decompose to release chlorine in the ozone layer (by photo-dissociation) and each atom of chlorine, thus liberated is capable of attacking several ozone molecules.

$$Cl + O_3 \longrightarrow ClO + O_2$$

This reaction is followed by:

$$ClO + O \longrightarrow Cl + O_2$$

which regenerates Cl atoms, so that a long chain process is involved, which conserves Cl atoms. The environmental hazards of CFC's were recognised as early as 1970. In fact, temporary thinning in the stratospheric ozone layer, leading to the formation of "Ozone hole" was actually detected over the Antarctica during September to November in 1985. Reported increase in cases of skin cancer in South Australia are also attributed to UV radiations reaching the earth, due to depletion of ozone layer, over that part of the world temporarily.

The detection of the "Ozone hole" over Antarctica in 1985 attracted the attention of scientific community in the world. The U.S. immediately banned the use of CFC's in spray cans. Further, in the year 1987, twenty four nations of the world signed the Montreal Protocol, which aims at 35% reduction in the global production of the CFC's by the year 1999. Simultaneously, efforts to produce chlorine-free substitutes have also started. In fact, synthesis of a product called HFC-134a has already been reported as an effective substitute for CFC. The use of hydrofluorocarbons (HFC's), hydrochlorofluorocarbons (HCFC's), and methyl cyclohexane (MCH) as substitutes for CFC's is envisaged for several applications.

Almost all the sulphur present in liquid and gaseous fuels and about 80% of sulphur present in the solid fuels appears as SO_2 in the flue gases. Depending on the sulphur content of the fuel burnt and the conditions of combustion (e.g., % of excess air used), the concentration of SO_x in flue gases varies from 0.05 to 0.4%. However, in metallurgical operations such as smelting of sulphide ores, the SO_2 concentration in stack gases may be 5 to 10%. SO_2 is oxidised to SO_3 in atmospheric air by photolytic and catalytic processes involving ozone, NO_x and hydrocarbons, giving rise to the formation of photochemical smog. Oxidation of SO_2 can take place in presence of catalysts such as NO_x, metal oxides, soot and dust. Under normal humid conditions of the atmosphere, SO_3 reacts with water vapour to produce droplets of H_2SO_4 aerosol which gives rise to the so-called "acid rain" discussed later. The sulphuric acid and sulphate aerosols present in the urban air are smaller than $2\ \mu$ and hence can easily reach the pulmonary region of lungs, causing serious respiratory problems, particularly in older people.

$$SO_2 + O_3 \longrightarrow SO_3 + O_2$$

$$SO_3 + H_2O \longrightarrow H_2SO_4 \longrightarrow (H_2SO_4)_n$$

$$SO_2 + 1/2\ O_2 + H_2O \xrightarrow[\text{metal oxide, soot etc.}]{\text{Catalyst such as}} H_2SO_4 \xrightarrow[\text{aerosol}]{\text{aerosol}} (H_2SO_4)_n$$

Control of SO_x emissions from the anthropogenic activities is contemplated on the following lines:

1. Removing SO_x from flue gases before letting them out into the atmosphere: Chemical scrubbers such as (i) Lime stone or (ii) Citric acid are suggested to absorb SO_2 from the flue gases.

   (a) $2 CaCO_3 + 2SO_2 + O_2 \longrightarrow 2 CaSO_4 + CO_2$

   (b) $SO_2 + H_2O \longrightarrow HSO_3^- + H^+$

   $HSO_3^- + H_2 cit^- \longrightarrow (HSO_3 . H_2 cit)^{-2}$

2. Removing sulphur from the fuels used for combustion: Pyritic sulphur in coal can be removed by grinding and washing in coal washeries. However, organically bound sulphur cannot be easily removed from coals. Research is in progress to synthesise special type of micro-organisms using bio-technology, which are capable of converting organically bound sulphur into soluble form.

3. Utilising low-sulphur fuels.

4. Generation of power by alternative energy sources and discouraging fossil-fuel based thermal power-plants.

Acid Rain

Rain has always been valued by mankind, because good crops and abundant water supplies are possible only by timely and plentiful rainfall. Summer rains refresh people. Spring rains recharge the aquifers and cleanse the groundwaters. Autumn rains and winter snow help cleansing the air. Rain in general, brings with it a sense of hope, vitality and a promise for the future.

Over the last few decades, simple rainfall has taken on a threatening complexity in some parts of the world. In these locales, the rain must pass through an atmosphere polluted with oxides of sulphur (SO_x) and oxides of nitrogen (NO_x). The falling rain and snow react with these oxide pollutants to produce often a mixture of sulphuric acid, nitric acid and water. This is known as *acid precipitation* or *acid rain*.

Rain tends to be naturally acidic with a pH of 5.6 to 5.7 due to the reaction of atmospheric CO_2 with water to produce carbonic acid. This small amount of acidity is, however, sufficient to dissolve minerals in the earth's crust and make them available to plant and animal life; yet not acidic enough to inflict any damage. Other atmospheric substances from volcanic eruptions, forest fires and other similar natural phenomena also contribute to the acidity in rain. Still, even with the enormous amounts of acids created by nature annually, normal rainfall is able to assimilate them to the point where they cause little, if any, known damage. But, it is the contributions of SO_x, NO_x, etc. from anthropogenic activities that disturb this acid balance and convert natural and mildly acidic rain into precipitation with far-reaching environmental consequences.

Acid rain represents one of the major consequences of air pollution, because of large SO_x and NO_x emissions from big industrial areas into the atmosphere. The longer the SO_x and NO_x remain in the atmosphere, the greater are the chances of their oxidation to H_2SO_4 and HNO_3 by the various photochemical and catalytic chemical reactions. Acid rains may cause extensive damage to materials and terrestrial ecosystems such as water, fish, vegetation, stone, steel, paint, soil and mankind.

The only practical approach to counter the problem of acid rain is to reduce SO_x and NO_x emissions. The following three general options are considered for this purpose :

1. *Energy conservation* resulting in reduced fuel consumption and hence slower emissions of SO_x and NO_x. Conservation via more efficient fuel use and through improved thermal insulation is also being studied.

2. *Desulphurisation and denitrification of fuels* of stack gases and increased use of fuels naturally low in sulphur content or use of technologies that reduce the SO_x and NO_x emissions. Desulphurisation and use of low-NO_x-producing technologies are the only viable control options today and will perhaps continue to be so for some more time.

3. *Substitutions for fossil fuels* by other alternative energy forms may offer future solutions to this problem.

Reduction of SO_x emissions can be accomplished by: (i) removing the sulphur content before the fuel is burnt with the help of techniques such as coal cleaning, coal gasification and desulphurisation of liquid fuels; (ii) removing the sulphur content during combustion, as in fluidised-bed combustion; and (iii) removal of sulphur emissions after combustion, as in stack or flue gas desulphurisation systems or scrubbers. The future of SO_x control from traditional fuel sources lies in the perfection of these techniques.

Reduction of NO_x emissions from stationary combustion sources can be achieved by modification of furnace and burner design and/or modification of operating conditions. The combustion modification techniques available now include using 2-stage combustion, precisely controlling air, injecting water during combustion, recirculating flue gases, and/or by altering design of firing chambers. Reductions in NO_x emissions from mobile combustion sources may be achieved by lowering the combustion temperatures in the engine and catalytic removal of NO_x from the exhaust gases using devices such as three-way system that reduces carbon monoxide, hydrocarbons and NO_x simultaneously.

WATER POLLUTION

Water is an essential ingredient for life. A man can survive few days without food, but starts struggling for life, if water is not made available even for one day. Water is also needed for the maintenance of plants and animals, for navigation, for the generation of hydroelectric power and for disposal of sewage. Standard of health also depends on the quality, quantity and purity of water. Supply of fresh water provides aquatic organisms with dissolved oxygen and some essential minerals and nutrients, making it most vital element for life.

Lack of availability of fresh water for drinking caused outbreak of water borne diseases which wiped out entire population of cities in the earlier days. At present, the menace of water borne diseases and epidemics is still large in all the developing countries. Polluted water is the main cause for such diseases. Water may get contaminated with a number of impurities like dust particles, dissolved gases, minerals and micro-organisms. Waterborne diseases are caused by bacterial, virus, protozoa and worms. Similarly, presence of chemical agents like fluorides and nitrates is responsible for causing human diseases.

Generally, water pollution is a state of deviation from the pure condition, whereby its normal functions and properties are affected. Water is said to be polluted when it is contaminated with :

1. Dissolved gases like H_2S, CO_2, NH_3 and N_2.
2. Dissolved minerals like sodium, calcium and magnesium salts.
3. Suspended impurities like clay, sand, mud and organic matter.
4. Micro-organisms like bacteria, viruses, protozoas and worms.
5. Contamination of isotopes (radiologically active substances).

Pollution of water is defined as the presence of some foreign organic, inorganic, biological, radiological, and physical substance or property in the water that tends to degrade its quality and either constitutes a health hazard or otherwise decreases the utility of water. Thus, the water pollution may be defined as the addition of any foreign material or any physical change in the natural water which may adversely affect living life directly or indirectly either in the short run or in the long run.

Though the problem of water pollution is age old but in the modern era, the problems like population increase, sewage disposal, industrial waste, etc. have polluted our water resources considerably. The prominent contributors to water pollution are sewage, oil, industry, agriculture and radioactive waste etc. Water pollution not only changes the physical properties of water like colour, odour, turbidity, taste and temperature but also makes it acidic or saline due to presence of dissolved or suspended chemical pollutants. (For details please refer chapter 4 Volume I Waste water and its treatment.

THERMAL POLLUTION

The atmospheric pollution raised in terms of changes in the temperature of the surrounding atmosphere due to the release of hot vapours from nuclear power plants, industrial effluents, nuclear reactors, coal-fired power plants, hydroelectric power, hot waste solids and liquids and domestic sewage, etc. is termed as 'Thermal pollution' (or) heat pollution. Due to the thermal pollution the average temperature of, cities will be 10°C greater than that of villages.

Domestic sewage is commonly discharged into rivers, lakes, canals or streams. Thereby the temperature is raised in such water than that of normal water temperature such discharged sewage creates numerous deleterious effects on quatic biota. The organic matter present in the sewage and other oxidisable matter utilise the DO present in the surface water for oxidation. With the result oxygen depletion takes place. Hence, the anaerobic conditions will set up resulting in the release of foul and offensive gases in water. Subsequently water quality is lowered causing health hazards to aquatic life.

Heat energy released from industrial operations, power generation stations, space heating and cooling in the atmosphere will upset the balance existing between solar energy input and absorption of solar energy at the earth's surface. This type of temperature (or) heat changes will affect the temperature in the global environment. Industries generating electricity, like coal power and nuclear power plants, require huge amounts of cooling water for heat removal. Other industries like textile, paper, pulp as well as sugar industries release heat in water. The hot water released from industries will enter into the rivers, canals and reservoirs. Thereby the temperature of such water regions will be raised. DO value is lowered. Due to lack of sufficient oxygen content, the rate of respiration in living aquatic organisms is reduced slowly and ultimately the organisms may die. For example the DO in water at 0°C is 14 ppm. As the temperature increases to 20°C, 35°C the DO concentration will be lowered to 9 ppm and 7.1 ppm respectively. The cold water fish which requires about 6 ppm to survive will not tolerate the high water temperature due to thermal pollution. If such type of fish remained in the area, they will die from oxygen starvation.

In oxygen rich water only bacteria, protozoas and microorganisms multiply rapidly and then they become food for more advanced aquatic creatures. Hence, DO plays an important role in food chain processes. This type of environment will be disturbed under thermal pollution.

A rise in temperature also changes the physical and chemical properties of water. As temperature increases, the vapour pressure of water increases, while the viscosity, density and solubility of gases in water decreases. With the result settling speed of suspended particles increases which badly affects the food supplies of aquatic organisms. In fishes several activities like nest building, spawning, hatching, migration and reproduction, etc. are carried smoothly at 8.9°C temperature. Other than this rise in temperature due to thermal pollution cause health hazards in terms of their biological activities including reproduction systems. All the major groups of algae like diatoms, green, blue-green algae have distinct tolerance ranges for water temperature. High water temperature due to thermal pollution promote blue-green algal booms which disrupt the aquatic food chain.

Coal fired power plants constitute the major sources of thermal pollutants. Their condenser coils are cooled with water from near by lake or river and discharge hot water back to the stream increasing the temperature of near by water to about 15°C. The heated effluents decrease the DO content of water. It results into killing of fish and other marine organisms.

By the combustion of coal and petroleum products the CO_2 concentration in the atmosphere is raised and there by earth's temperature increased, with the result green house effect takes place to produce floods.

Control of Thermal Pollution

The following points are to be observed to control high temperature caused by thermal discharges.

1. Cooling towers are to be employed to reduce the temperature. The hot water released from the industries should be passed through cooling towers to control the temperature.
2. The hot water released from industries should be cooled before sending into rivers, lakes, canals and reservoirs.
3. The efficiency of electrical energy productive machines should be increased.
4. By producing the protective modern engineering designs, the temperature controlling models are to be used.

RADIATION POLLUTION

One of the major sources of environmental concern now-a-days is the production of nuclear power which releases radioactive substances in the environment. Radioactivity is toxic in the sense that it causes harm to living organisms. Accidents at Three-Mile Island in the United States and Chernobyl in the Soviet Union, both of which released radioactive rays, have increased the public concern world wide about the safety of nuclear power. The management and disposal of the accumulating radioactive wastes remain the key long-term problem facing the nuclear power sector. Wastes related to nuclear power generation occur in a variety of physical and chemical forms and in different class levels based on the relative radioactivity of the radio nuclides that they contain.

Radioactive wastes can be divided into two categories: High level wastes and low level wastes. Both these wastes are temporarily stored at product sites awaiting disposal.

High Level Wastes

These wastes are predominantly used as fuel. Uranium oxide is fabricated into small pellets and sealed in metal tubes, which are then assembled in bundles. After about 18 months in the reactors, the fuel bundles need to be replaced. They are intensely radioactive, and some elements in them continue to emit radiation for tens of thousands of years (e.g., plutonium-239 has a half-life of 24,000 years). Extreme caution must be exercised to ensure that such high-level wastes do not contaminate the air, water and food materials.

High-level wastes are stored mainly in deep, water-filled pools at the reactor sites to cool them and shield their radiation, and a small fraction is stored dry in concrete containers. Although on-site storage is an adequate means of dealing with these wastes over the short term, eventually long-term solution is required. For the disposal of nuclear fuel wastes on long-term, deep geological formations are required. The buried radioactive material would have to be encased in non-corrosive (titanium or copper) containers to ensure safe and secure disposal so that radioactive wastes cannot enter ground water.

Low Levels Radioactive Wastes

These are related to the nuclear energy production and the wastes arising from uranium refineries and fuel fabrication plants. In low-level mines low-level radioactive wastes are generated from the extraction of uranium from ores. To check the low-level radioactive wastes, monitoring of temporary sites is necessary to ensure the safety for the surrounding public.

Radionuclides

Radiation from human activities has increased since the invention of atomic power and introduction of nuclear energy. Different sources produce radiation with different energies and have different biological effects. Ionising radiations occur when uncharged chemicals are changed into charged ion pairs. There are more than 1000 different nuclides in the atmosphere. Some of them are stable in nature while others can split up into parts and give off radiation, which are known as radionuclides. The *radionuclides* are the products of the natural decay of uranium and are not extracted during milling process. In addition to acid precursors and heavy metals, uranium tailings also contain a variety of radioactive forms of atoms, or radionuclides, including radium-226, thorium-230, lead 210 and polonium-210. Some of these radionuclides have long half lives. Thorium-230 has a half-life of about 80,000 years, which means that radioactivity from thorium will be present in the tailings for hundreds and thousands of years. The pH of the tailing plays a significant role in determining how soluble some of these radionuclides become. Thorium, in particular, becomes more available when pH drops to below 4. Radionuclides with short-half lives, i.e., upto a few days, may be very dangerous when produced, but the danger does not last.

The main exposure pathways for radioactivity from tailing are direct *gamma* radiation, inhalation of radioactive particulates, and ingestion of radionuclides through the food chain. The environmental contaminants in the tailings, such as metal and acid drainage are of great concern. The primary long-life fission products that enter the ecosystem in significant amounts, are due to the nuclear weapon tests. Table 24.8, indicates the radionuclides produced and released in the atmosphere during atomic explosions and their periods.

Table 24.8. Some important radionuclides and their half lives.

Radionuclide	Target tissue	Half-life
Calcium-45	Bone	165 days
Carbon-14	Whole body	5760 years
Caesium-137	Soft tissue, genital organs	27 years
Iodine-129	Thyroid	17 million years
Iodine-131	Thyroid	8 days
Plutonium-239	Bone, liver, spleen	24,400 years
Radium-226	Bone	1620 years
Strontium-90	Bone	28 years
Tritium (^{3}H)	Whole body	12.3 years

Other radioactive threats to the environment are the accidents connected with the activities of nuclear powered vessels and satellites, such as Cosmos 954, which crash-landed in 1978 near the Thelon River in the Northwest Territories.

SOIL OR LAND POLLUTION

The contamination of soil (or) Land with rain, excess of fertilisers, wrong fertilisers, insecticides and herbicides is termed as 'soil pollution'.

Sources of Soil Pollution

The following are some of the important sources which will pollute the soil :

1. Repeated use and also excess use of fertilisers and pesticides at random cause soil pollution. For example excess use of $(NH_4)_2 SO_4$ chemical fertiliser for several times, the SO_4^{-2} ion present in it get accumulated into the soil . The soil then becomes infertile to plant growth due to the acidity developed in the soil through the existence of SO_4^{-2} ions. Similarly due to repeated and excess use of fertilisers containing KNO_3 (or) $NaNO_3$, the plant growth will be retarded due to the accumulation of Na^+, K^+ ions in the soil which provides alkalinity to the soil.

2. Soil pollution arises due to the application of defective methods in the cultivation processes.

3. Soil gets polluted due to the release of their cysts from antameba, ascaris, pigwarm etc. into the soil, which enters into the human body through food chain.

4. Acid rain is another source for soil pollution. The air contents SO_2, NO_2 undergoes photolytic, catalytic oxidation followed by the interaction with rainy water or moisture to form H_2SO_4 and HNO_3.

$$2SO_2 + O_2 \rightarrow 2SO_3$$
$$SO_3 + H_2O \rightarrow H_2SO_4$$
$$4NO_2 + 2H_2O + O_2 \rightarrow 4HNO_3$$
$$2NO_2 + H_2O + O_3 \rightarrow 2HNO_3 + O_2$$

The H_2SO_4 and HNO_3 produced in the form of acid rain absorbed by the soil and the fertility of the soil for cultivation will be damaged, such soil is unfit for plant growth. Due to the acid rain the pH of the soil becomes 3.0 to 4.0. This soil causes health hazards to the living organisms on the earth. It reduces the agricultural yield. It leads to soil cracks due to its acidity. The toxic metallic compounds present in the soil will flow along with the acid rain and enter into the rivers, lakes, tanks and sea. Thereby the water bodies like fishes living in those places were affected with poison. Subsequently they enter into the human body through food chain causing health hazards.

5. Spraying the vegetables and fruit plants with insecticides and herbicides cause soil pollution. When the standing vegetable and fruit plants are sprayed with insecticides and herbicides to protect them from the harmful insects and herbs etc., the pesticides like insecticides; herbicides enter the living tissues of the growing plants and accumulate in them. The pesticides enter into human body by taking such contaminated grains, fruits and vegetables as food material causing health hazards that may damage even our heart. Hence, it is therefore advised that before eating food material like grains, fruits, vegetables, they should be properly washed with sufficient quantity of good water. Chemicals like $CaCl_2$ used for ripening of fruits like banana and mangoes can be eliminated by washing with water easily.

6. The heat falls over the land (or) earth surface from molecular reactors, industries also cause soil pollution.

7. The sewage is the dirty water which contains human and animal excretions. The discharge of large quantities of sewage into rivers, lakes, water tanks, discharge of industrial wastes on the

earth surface, sewage from house hold activities, dumping of synthetic, plastic waste material on the earth cause severe soil pollution.

8. Deforestation also leads to soil pollution. Due to the deforestation , the concentration of CO_2 in the atmosphere increases, there by the atmospheric (or) earth's temperature will be raised; with the result the water source in the soil is lowered.

9. Serious cadmium pollution of soil occurred in Japan. Over a 20 years period around 100 people (mostly older woman) living near the Jintsu river died of a syndrome called Itai-Itai (ouch-ouch) disease through bone damage.

Methods of Control of Soil Pollution

1. The utility of fertilisers and pesticides should be minimised as far as possible to control soil pollution.
2. The waste material should be covered in earth layers with soil.
3. The waste gases releasing from various automobiles and industries should be purified before they reach the earth surface.
4. Before dumping the sewage into rivers, lakes, tanks and sea, it should be treated chemically to remove harmful organic material present in it.
5. The polluted water should be purified by installing water purified machinery.
6. Deforestation (indiscriminate cutting of plants and trees) should be minimised and plantation is carried out as a universal standard project work under clean and green programme.
7. The usage of machines should be encouraged in which biogas is used as fuel; to get electrical production.
8. Various operations are to be carried out to neutralise the acidity of acid rain.
9. The digging of larger pits or grooves in view of granite mines, aqua culture, addition of various chemicals to grow water bodies etc. are to be avoided slowly to protect soil or land from pollution.

NOISE POLLUTION

The unwanted and undesirable sound plays at random quality which causes discomfort is termed as noise. The word noise is derived from the latin term "nausea" meaning a feeling of sickness at the stomach with an urge to vomit.

The environmental pollution caused due to hearing the unpleasant, undesirable sound at wrong places, at the wrong timings causing discomfort is termed as 'noise pollution'. The noise pollution is expressed in Decibel (DB) units. The magnitude of the smallest sound unit which can be audible reasonably without any harm to a healthy person having good reasonable hearing capacity is termed as 'one Decibel'. The unbearable bitter sound causing discomfort to the audience is considered as noise pollution. Crowded cities and towns, mechanised means of transport, new devices of recreation and entertainment are polluting the atmospheric environment with their continuous noise.

It has been observed that the hearing capacity of 75 years old man living peacefully at African forests will be equal to the hearing capacity of 25 years youngster living in cities or towns.

Sources of Noise Pollution

Noise is either natural such as 'thunder' or man made. The main sources of man made noise in developed urban areas are mechanised automobiles such as trucks, buses, motors, scooters, fire engines, police cars, ambulances etc. Factories, industries, trains, aeroplanes and accessory noise producers such as horns, sirens, loud speakers, musical instruments, TV, radio, transistors, taperecorders, shouting, barking of dogs, etc. Man made noise also includes social gatherings, use of airconditioners, washing machines, generators, marriage and birthday functions, etc. There has been a considerable increase in noise from man made sources during the last 100 years which is now doubling after every decade. An ever increasing number of common noise sources are being put into use daily. Hence, one can say that noise can be broadly classified into three types viz.: (i) industrial noise; (ii) transport noise; and (iii) neighbourhood noise.

The sound consists of wave motion in an elastic medium and caused by the vibrations of molecules. The quality of unpleasantness of sound wave depends on the frequency of sound waves, the intensity of sound waves, the time of exposure of sound waves and the intermittence of sound waves.

According to the World Health Organisation (WHO), a person can sleep conveniently under the environment having 35 Decibels of noise. If the sound exceeds 50 Decibels an individual cannot sleep. The intensity of sound is different in day times and at night times. WHO recommended 75 dB as explosive limit to industrial noise. Sound beyond 80 dB is harmful and damages hearing system. So 80 dB sound is safely regarded as pollutant.

At the places like temples, hospitals, living places and Ashrams noise level up to 65 dB has been fixed as tolerable limit as for international standards. The noise levels in the four metropolitan cities of India (New Delhi, Mumbai, Kolkatta and Chennai) are generally more than 90 dB.

Effects of Noise Pollution on Man

The following are some of the important effects of noise pollution on human health.

1. Noise pollution affects human health, comfort and efficiency. It causes contraction of blood vessels, makes the skin pale, helps in excessive secretion of adrenaline hormone into blood stream causing high blood pressure.
2. Excessive noise affects the digestive system.
3. Prolonged exposure to noise is harmful for central nervous systems and also affects memory power adversely.
4. Noise pollution causes muscles to contract leading to nervous breakdown, tension and even insanity.
5. Noise pollution increases the rate of heart beat and leads to the dilation of pupil of the eye.
6. It affects health efficiency and draws behavioural changes. It may cause damage to heart, brain, kidney, liver and also develop emotional disturbances.
7. The most immediate and acute effect of noise pollution is the impairment of hearing which reduces by the damage of some part of auditory system. Loud and sudden noise spoils the ear drum. Temporary deafness occurs at 4000-6000 Hz, and this effect is known as 'Temporary Threshold Shift' (TTS). Permanent loss of hearing occurs at 100 dB due to continuous noise exposure. This effect is called Permanent Threshold Shift (PTS).
8. Noise pollution decreases the efficiency of working.
9. A noise of high decibels i.e., 160 dB creates headache.
10. Noise pollution may increase violence and may cause a state of depression and tiredness.
11. Noise affects pregnant women and increases their heart beating and also has adverse effects on the newly-born child.
12. It disturbs sound sleep and proper rest.

13. Recently, USA scientists have discovered that high pitched noise leads to road accidents.
14. Noise pollution causes eosinophilia, hyperglycaemia, hypokalaemia and hypoglycaemia due to change in blood and other body fluids.

Control of Noise Pollution

The following methods should be followed to minimise noise pollution :

1. Noise pollution through vehicles can be minimised by fitting silencers to the vehicles.
2. The walls of the living houses, office buildings, cinema halls and factories should be covered with sound absorbers.
3. The educational institutions like schools, colleges and hospitals should be constructed at pleasant and decent places which are faraway from the living area of human society. Such places should be declared as noise prohibited areas.
4. The utility of speakers should be minimised in public meetings, temples and in churches etc.
5. The Government should impose certain rules and conditions to control noise
6. Noise creating machinery and equipment may be covered with insulating materials.
7. Workers working in industries should wear earplugs (or) ear muffs to minimise noise effects.
8. Certain codes should be enforced which require sound proofing in the construction of industries, buildings and apartments.
9. Noisy operations should be carried out in open spaces, far off from any residential colony (or) public place.
10. A big green belt of Ashok trees, Neem trees, Eucalyptus trees should be planted on either side of highway roads, industrial buildings, school and college buildings to absorb the noise as well as to purify air and also to produce food and fibre productions.
11. A provision should be arranged for checking hearing capacity of each industrial worker once in a year without fail so as to take precautionary measures.

CHAPTER 25

Environment and Energy

INTRODUCTION

Modern society cannot exist without the production and utilisation of energy. Every month we have to pay direct charges (fuel bills) for our use of electricity, oil and gas in our homes and for the petrol used in our cars. And there are also indirect charges that we pay for the energy used in manufacturing processes and for the transportation of the goods that we buy. In addition to these charges, we pay also in terms of the effects that energy production and energy utilisation have on our world in terms of environmental pollution. Energy is directly related to chemical phenomena in the formation and decomposition of compounds, the many important reactions that occur in electro chemistry, and in the release of energy in nuclear fission and fusion. Environmental pollution may be defined as the unfavourable alteration of our surroundings (air, water and land). It may not be possible to estimate monetary losses or many of the side effects associated with energy production and energy utilisation. What is the value of the health impairment, for example, caused by the car's exhaust fumes? What value do we place on the destruction of farmland and pollution of water caused by strip mining for coal? What value is associated with the loss of seaside beaches because of oilspills washing ashore? As a matter of fact, as long as we continue to produce and utilise energy, we will have to pay for these undesirable side effects.

THREATS FROM FOSSIL FUELS

Most of the energy that is generated throughout the world at present is derived from the burning of fossil fuels-coal, natural gas and petroleum products. There are numerous environmental problems associated with the extraction, transportation and utilisation of fossil fuels. The most plentiful fuel source in the world is coal. The highest quality coal (anthracite) generally occurs sufficiently far underground to require high-cost deep-mining techniques.

Further, anthracite generally contains a very high percentage of sulphur and it cannot be used as a fuel with out expensive treatment to remove sulphur. Unlike coal, the extraction of oil does not desecrate the land the way the strip-mining does. However, the most serious environmental problem associated with oil-well drilling occurs at offshore sites. Because of the many technical difficulties inherent in offshore drilling, if a rupture occurs or if the drilling opens a crack in the rock that contains the oil deposit, a major leakage of oil into the water can occur before the damage is repaired or the crack is sealed. The release of large amounts of oil into the water can be injurious to the marine life. When the oil spreads over water, the diffusion of oxygen into water is inhibited. This affects the respiration of fish and other marine life. Oil pollution of sea causes other problems too. Oil is pushed to the shore by the water currents and winds, thereby spoiling the beaches. Most of the world's oil is transported to refineries by sea. The huge size of modern tankers (some are capable of carrying more than 300,000-tonne of oil) makes them extremely slow to respond to commands. Consider, for example, a 300,000-tonne tanker travelling at 15 knots! Such a ship, because of its very large momentum requires at least 4 kilometres to come to a stop. This opens up a very strong possibility of collision with another ship or with reefs and rocks in narrow waters. When such an accident occurs, rupturing of the oil tanks can cause an enormous spill, endangering marine life and polluting even distant beaches and harbours.

Oil is also transported overland through pipelines. This mode of delivery poses serious problems in some cases. It is feared that if the pipeline passes through large areas of untouched land, it may upset the ecological balance of the region. Migratory animals may be forced to use new routes because of the obstruction of their usual migratory route by the pipeline. If the line were to leak, sizable areas could be soaked by oil before the flow could be stopped. Because the oil contains some wax, it must be maintained at a sufficiently high temperature while being pumped through the pipeline at high speeds. In cold countries near the arctic circle where permafrost is a perennial feature of the land scape, pipelines carrying heated oil may melt their way through the permafrost layer, then sag and rupture. Even if the rupture did not occur, the melting of the permafrost might cause irreversible changes in the local ecology. Like oil, natural gas too is transported through pipelines. Although this system is relatively trouble free, leaks and explosions have occurred in the past and several deaths result each year from these incidents.

COMBUSTION OF THE FUEL

The burning of fossil fuels releases a variety of noxious gases and particulate matter into the atmosphere. The major contributors to this atmospheric pollution are coal and oil; natural gas by far is the least offensive of the fossil fuels. One of the major problems with coal and oil is the presence of sulphur. Depending upon the source, the sulphur content can be up to several per cent and upon combustion, several oxides of sulphur (particularly sulphur dioxide) are produced. When sulphur dioxide is released into the atmosphere, it combines with water vapour and forms sulphuric acid. It is thus sulphuric acid which is injurious to plant and animal life. It has been found that atmospheric sulphuric acid is eating the limestone facings of many monuments and public buildings in urban areas. Excessive amounts of sulphur dioxide in the atmosphere have been directly linked to the high incidence of several types of respiratory ailments. Sulphur dioxide is believed to cause cough, shortness of breath and spasm of the larynx. It can cause acute irritation to the membranes of the eyes resulting in excessive flow of tears and redness. Sulphur dioxide has also a very adverse effect on plants. When absorbed by plants beyond a certain level the plant cells become inactive and are killed, resulting in tissue collapse and drying of leaves. Sulphur dioxide is also known to interfere with the respiration and photosynthesis in plants. The sulphur can be removed from coal and oil, but in most cases a major effort is needed to reduce the sulphur content to an acceptor level. Recently it has been discovered that mercury, an extremely poisonous element, is also let into the atmosphere by the burning of coal.

One of the ways to solve the pollution problem of coal and oil is to convert these fuels into cleaner burning gases and liquid hydrocarbons. With the sulphur removed and the particulate matter prevented form entering the atmosphere, a primary source of air pollution would be largely eliminated. Efforts are currently underway to perfect methods for coal gasification and liquefaction and it is hoped that in future we will no longer burn coal but will instead use coal gas or liquid fuel obtained from coal.

The burning of petrol in internal combustion engines is the major source of carbon monoxide, nitrogen oxides and hydrocarbons in the atmosphere. In addition, there are large quantities of lead which are released into the atmosphere from high octane petrol used in cars. All these pollutants and the products of the photochemical reactions they undergo in presence of sunlight contribute to the noxious mixture known as smog. There seems at present no escape from the health hazards of smog until some effective way is found to remove the pollutants from the vehicular exhaust gases or until some practical substitute for the internal combustion engine is developed.

EFFECTS OF CARBON DIOXIDE AND CARBON MONOXIDE

The consumption of oxygen and the formation of carbon dioxide are necessary consequences of every combustion process. One may think that this may deplete the world's supply of oxygen and thus upset the oxygen-carbon dioxide balance that is necessary for plant and animal life. Fortunately, this is not a problem associated with the burning of fossil fuels. All of the fossil fuels that have been burned have used up only seven out of every 100,000 oxygen molecules available to us.

Even the combustion of all the world's known reserves of fossil fuels would use less than 3 per cent of the available oxygen. Thus the use of fossil fuels does not present us with the spectre of exhausting our oxygen supply. On the other hand, the level of carbon dioxide in the atmosphere has increased considerably over the years and it is estimated that by the year 2025 the level will be almost double the value that prevailed in the early nineteenth century, before the large-scale use of fossil fuels began.

Carbon dioxide molecules strongly absorb heat radiations emitted from the surface of the earth heated by the sun. By holding back this energy in the earth's atmosphere, carbon dioxide reduces the heat lost by the earth to space. This is called 'greenhouse effect' and because of this, it is argued, the continued burning of fossil fuels will result in a steady increase in the earth's surface temperature. However, an increase in the temperature of the earth's surface and lower atmosphere has the compensating effect of increasing evaporation and cloudiness. Because clouds reflect some of the incident sunlight, increases in cloudiness tend to decreases the surface temperature. Further, the release of particulate matter into the atmosphere from fuel burning increases the number of condensation sites around which water droplets can form. The result is an increase in the amount of rain, hail and thunderstorms which lead to a lowering of the temperature.

Carbon monoxide is another pollutant produced by burning of fossil fuel. It is another pollutant produced when there is insufficient oxygen for burning. It is released into the atmosphere mainly from automobile exhaust gases. But it does not so far constitute a serious environmental problem.

Thermal Pollution

The term 'thermal pollution' basically refers to the detrimental effects of discharges of unwanted heat into the environment. All electricity generating plants (except hydroelectric plants) produce electricity by driving huge turbine generators with steam. The steam is condensed in a cooling system and is cycled back to the heating unit for reuse. The 'cooling system' can be water that is pumped from some nearby reservoir (a river, lake, or bay) and discharged back into it, or it can be a cooling tower in which the heat is dissipated into the atmosphere. Both cause thermal pollution. If the heated water is discharged into a static reservoir, such as a lake, the effect can be even more severe.

Concentration of dissolved oxygen decreases. Besides, when the temperature of the aquatic environment increases, the metabolic processes taking place in the body of a fish are speeded up and its need for oxygen and its rate of respiration therefore increase. Above a particular temperature, the fish dies due to failure in its nervous or respiratory system.

Thermal pollution is generated by the energy producer as well as the energy user. Almost all of the energy we use is eventually converted into heat. Most of this waste is dissipated into the air where it contributes to the general atmospheric heating.

EFFECTS OF NUCLEAR RADIATIONS

Nuclear reactors, unlike the other sources of power, offer a lot of advantage. Nuclear reactors generate electrical power without the smoke and fumes that are characteristic of fossil fuel-burning plants. Also the mining of uranium produces much less degradation of the countryside than the mining of fossil fuels, particularly coal. Nuclear reactors, therefore, offer the prospect of long-term relatively clean power. However, nuclear reactors have their own peculiar set of disadvantages, mainly associated with the production of radioactive materials. Some radioactive waste is released into the environment both as gases into the atmosphere and in the form of low activity waste such as tritium in cooling water.

All radioactive substances emit harmful radiations, some of which can cause cancer in man and animals and damage the genetic material of the cell, producing long-term harmful effects in living organisms.

ENERGY CONSERVATION - A PRAGMATIC APPROACH NEEDS APPROPRIATE IMPLEMENTATION

Conservation of energy is an important issue in National Forum. Today this issue is closely attached to our day to day affair at every moment. Though our country is having plenty of sources of energy, by and large we are usually depending on a few sources of energy like coal, electricity, oil etc. Among all, oil is the main source of energy. In the recent years unusual hike of energy price in all areas, common people specially in urban areas are in tight spot. Government of India has taken vigorous efforts to find out the alternative energy source to minimise expenses and make it cheaply available to the consumers. Recently in some specific areas solar energy has been tapped and consumers in different areas are using it. However, this has not gained sufficient popularity in urban and semi urban areas as it is expected. Today there is an acute crisis of energy in different areas due to some technical reasons and lack of awareness of consumers. Scientists and technologists are working day and night to over come this problem and advised the consumers to conserve energy by taking appropriate measures at different levels. Global energy profile; past and present is given in Table 25.1.

Table 25.1. Global energy profile: Past and present (Percentage of high intensity total energy).

Source	1850	1900	1950	2000
Wood and biomass	68–80	40–60	10–20	2–5
Coal and lignite	20–40	30–50	40–60	25–35
Petroleum	–	5–10	30–40	55–65
Natural gas	–	0–5	0–5	5–10
Others*	0–5	0–5	5–10	5–8

*Others include hydroelectric, nuclear, etc.

Concept of Energy Conservation

Energy efficiency improvements also bring about substantial reduction in environment impacts, both on a local and global level. Undoubtedly efficiency again would come from the implementation of institutional and regulatory changes, but also overwhelmingly from the introduction and use of energy efficient technologies. Market forces and pricing of energy also play an important role in efficient energy utilisation. The issue of energy security is an important concern, which needs to be taken into account in formulating India's future energy policy. However, most of the planning activities in India particularly as they relate to public investments in different sector are based on five year plans. However, it will be appropriate, in future, if technologies are to be developed in response to expected price change or change in security of supply and planning has to cover a reasonably long gestation period.

The role of the state in developing and promoting new technologies needs emphasis in areas such as energy production and consumption. There are enormous social benefits stemming from energy efficiency of capital investments and environmental benefits. Energy technologies available worldwide, provide substantial opportunities for improving the Indian economy. However, the fact is that these are not always used or adopted by the users in different areas.

India has to be conscious of global impacts of energy use particularly those that are based on the use of fossil fuels, and biomass. At present there is a need to give further momentum and reduce the gap between supply and demand at the rate of 90 per cent due to the increasing rate of industrialisation, urbanisation and mechanisation of agricultural activity. At present there is a peak demand shortage of about 8–9 per cent. To supplement the shortage, energy conservation is absolutely necessary and some of the following measures can be taken:

1. Removal of current inefficiencies in generation, transmission, distribution and utilisation.
2. Exploring the options of using cheaper forms of energy.
3. Practising and implementing energy efficiency schemes.

ENERGY CONSERVATION IN CHEMICAL INDUSTRIES IN INDIA

Energy plays a predominant role in the economic development of a country. Sustained supply of energy at reasonable cost is essential to ensure dynamism in economic growth. In India, the increased use of energy has been accompanied by depletion of resources and compounded with environmental degradation. The energy demand and supply gap in India is more than 400 billion units and continues to increase at a frantic pace.

Bridging the gap to increase the supply is an expensive option. The estimated funds requirement is in the order of around Rs. 100,000,00 million which is almost 3–4 times the national budget. The consumption of electricity in various sectors is shown in Table. 25.2.

Table 25.2. Consumption of electricity in various sectors.

Sector	Electricity (MTOE)	%
Industry	8.49	34.48
Agriculture	7.69	31.24
Residential	4.15	16.86
Transport	0.56	2.27
Others	3.73	15.15
Total	24.62	100

It is readily noticeable that the industrial sector consumes by far the largest proportion amounting to 34.50 per cent followed by agriculture, then domestic use and then the commercial sector.

Indian Chemical Industry

The Indian Chemical Industry has progressed significantly, from the setting up of small batch units in the 50's to the setting up of a world class integrated petrochemical complexes in the 90's. The size of chemical industry in 1997–98 was $22 billion. The individual sub-segment share of fertilizer, drugs and pharmaceutical, synthetic fibre, synthetic detergent, soaps and toiletries and organics, inorganics and agrochemicals etc. are 19 per cent, 19 per cent, 17 per cent, 8 per cent, 9 per cent and 28 per cent respectively. The chemical industries use the highest share of energy @35 per cent of the total industrial consumption. However, for many chemical manufacturing/processing the cost of energy is relatively low when compared with the cost of both the raw materials and the finished products, with the result that there has been little emphasis on cutting energy costs. The cost of utilities, i.e. fuel and power, ranges between five per cent to as high as 70 per cent in an industry such as caustic soda.

ENERGY CONSERVATION — THE NEED

Undoubtedly, in today's global economy, especially after the implementation of WTO recommendations, anything that increases profitability enhances our market position and competitive edge.

The enormous potential of energy conservation is evident in Japan, where energy per unit GDP got reduced by 34 per cent during the period 1973 to 1990. It was made possible by clear, long-term government policies involving financial incentives, regulations, standards, education etc. As a result, Japan is now one of the most energy efficient countries in the world.

The requirement for managing energy will be a continuous need for the following reasons:

1. Energy is by far the largest component of the total life cycle cost of most energy using systems. The significant portion of the operating cost of any manufacturing or service industries is in the form of energy. Hence, energy conservation is in an important activity within the organisation as it can reduce operating cost.
2. Energy Security will become even more important factor.
3. As we become more global, companies will look forward for a more competitive edge.
4. Future price shocks can not even be predicted because of the complex nature of energy.
5. The technology available for managing energy is changing so rapidly that state-of-the-art techniques have a half life of 10 years at the most.

Energy Audit

The energy audit is the route chosen in order to monitor the efficient and effective utilisation of energy. It plays a fundamental role in any organisation which wishes to control its energy cost. It is a system designed to monitor, record and analyse all key operating parameters, energy consumption patterns etc. The audit is the systematic gathering and evaluation of energy data about the plant and process. The audit determines the following:

1. Where energy is being used, in what form, in what quantities and at what cost.
2. It relates energy use to output of products or services.
3. It highlights areas of significant energy use and identifies areas of efficient and inefficient use.
4. It recommends actions to reduce energy consumption.
5. It provides estimates of potential energy savings, implementation costs and pay-back periods for each recommended action.
6. It establishes base from which future energy savings can be measured.

7. It provides base for an ongoing energy management programme in the organisation, including review and evaluation.

Measuring Instruments

A properly conducted energy audit is an essential blend of expertise, experience and 'state-of-the-art' energy measuring instrumentation. Essential instruments are tabulated hereunder:

1. Power analyser (for balanced and unbalanced load).
2. Harmonic analyser.
3. Portable clamp-on ultrasonic transit-time flow-computer.
4. Fluegas analyser.
5. Ultrasonic leak detector.
6. Hot wire annemometer.
7. I.R. thermometer (Heat spy).
8. Pressure transmitter.
9. Non contact tachometer.
10. Digital light meter.
11. Digital micro-ohm meter.
12. RTD thermometer (contact and immersion type).

Energy conservation—relevance to Chemical Process Industries (CPI)

Energy is consumed in the chemical process industries in the following ways:

1. As direct fuel in process heaters.
2. As indirect fuel for raising steam or generating power.
3. As power for drivers.

A typical chemical plant may have energy consumption pattern with utilities having the largest share. Although, existing plants might be considered stuck with their installed equipments, they should take steps to detect and evaluate potential and real energy losses.

Potential energy saving areas

Some of the potential energy saving areas are described as under:

1. In many instances, low pressure steam that has been vented can be recovered by condensation and the condensate pumped back for reuse. In other instances, the low pressure steam can be recompressed to a higher pressure for more effective-use. The flashing of pressurised condensate can also provide useful steam for a specific application.
2. A common source of steam losses involves the mis-application and poor operation of many steam traps. In closed systems, plant personnel cannot fully realise how much semi to continuous blowby exists. The answer to this problem is not to hastily replace the trap with a new one of the same size or design, but rather to carefully examine the operation of the equipment that the trap serves and to attempt to select a trap better rated for the system characteristics.
3. For those rather straight forward situations where high temperature flue gases or air or acceptable process vapours are being released to atmosphere, it may become only matter of adding a waste-heat recovery unit, which may be anything from a pre-heater for another stream to a wasteheat boiler for steam generation.

4. The power consumption of new refrigeration systems is being examined in greater detail to achieve the lowest energy level. This contrasts with the previous practice of selecting the lowest capital cost. Such refrigeration systems involve normal compression condensation with evaporation, plus interstage flashing and process heat exchange.

5. Precooling and dehumidification of air entering the compressor by refrigeration can save about 20 per cent of compressor power requirement and investment made on new refrigeration system can be fully justified.

6. Improving power factor nearer to load centres serves dual purpose of offering attractive energy savings by way of reduced cable losses in distribution network and eliminating penalty for low power factor at source. Need for capacitor relocation can be easily identified through a careful study of electrical distribution network.

7. Appropriate selection of cables may reduce the power losses in cables to the tune of 0.6–0.8 per cent.

8. Replacement of frequently rewound motors with high efficiency motors can offer energy saving potential in the range of 3–10 per cent.

9. Some processes have water requirements that vary over a wide range. When flow rates range between 50 and 100 per cent of design, and atleast 50 per cent of the pumping head consists of friction loss, variable speed drives can substantially reduce operating energy costs and even improve system reliability, which in turn hikes production.

In ABS granule manufacturing industry (Fig. 25.1) the process cooling water system was designed to circulate cooling water through three different streams, operating in batch cycles, resulting in variation of cooling water demand as per production schedule. After a detailed study of pumping system including pump efficiency evaluation, estimation of heat and flow requirements at various conditions, variable frequency drive was suggested for cooling water pump. The actual results indicate saving of Rs 12.36 lacs/annum. i.e., 56.92 per cent of power consumption in pumping system.

Existing Equipment Modification

Some existing equipments are very difficult to modify to reduce energy consumption. These type of equipments are usually designed to suit a set of, or range of, operating conditions, and if these are changed, the performance may suffer. Examples are shelf-type dryers, rotating kilns or drum dryers, certain scrubbers/separators. In order to achieve worth-while improvements for such equipments specific test and re-evaluation must be made, because no standard modification can be expected to produce fixed results. High-horse power equipment mixing liquids and suspensions can usually effect some power savings by replacing small high-speed units with one that is large and lower speed. However this requires careful evaluation to ensure equipment process performance. The gear drives on such units can often be replaced by units of improved efficiency. Rotating equipment such as fans or blowers can often have wheels or impellers replaced or modified to a more efficient configuration. Some gear drives for such equipment can be replaced by gears with more efficiency.

Efficient Process Technologies

Process engineers should constantly keep track record of emerging, more energy efficient technologies. Both retrofits and changes of process technology should be evaluated from the point of view of energy cost also. Some examples are as follows:

1. In chlor-alkali industry, use of dimensionally stable titanium anodes and membrane cells have reduced the specific energy consumption from 3300 kwh/tonne to 2848 kwh/tonne.
2. The specific energy consumption of producing ammonia has come down over a period of years from 16 million kcal/kg to 7 million kcal/kg due to constant improvement in design and engineering using conservation concepts.
3. In distillation column for oxygen manufacture, the use of compressed air pressure requirement is reduced from 150 bar to 45 bar, resulting in reducing the electricity consumption from 2.25 kwh/Nm^3 to 1.1 kwh/Nm^3. Operating conditions are give in Table 25.2.

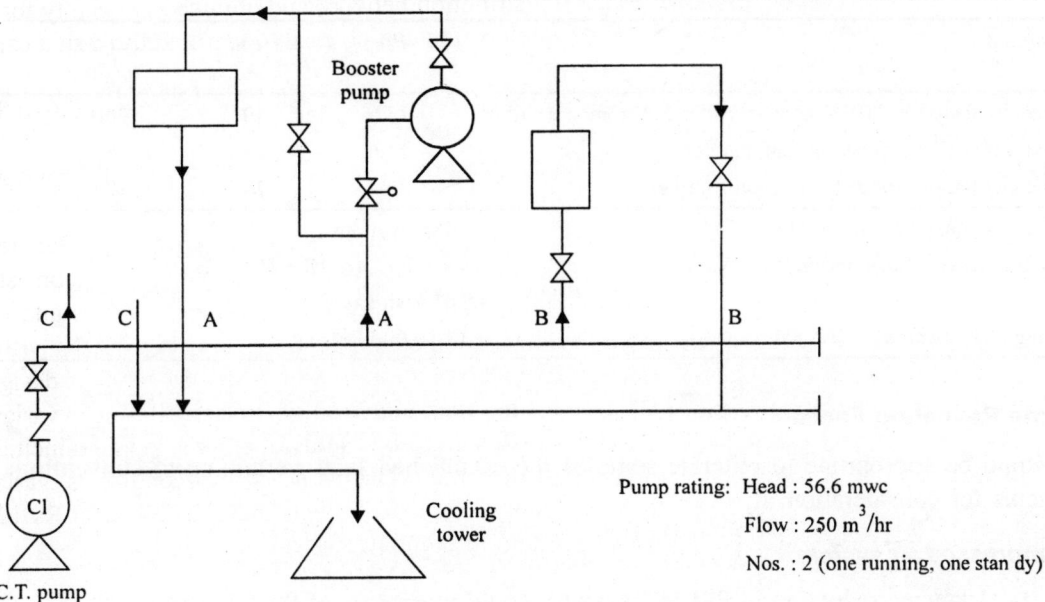

Fig. 25.1. Application of VFD in cooling water pumping system at Bayer ABS industries Ltd. Baroda (Phase-I Plant).

Stream	Location	Cooling water flow requirement m^3/hr
Stream-A	Acrylo-butadine styrene reactor	100
Stream-B	Poly-butadine reactor refrigeration condensor	100
Stream-C	Acrylo-butadine styrene powder extrusion	50
Total		250

Instruments deployed for study
Ultrasonic flow meter
Power analyser
Pressure transmitter

Considering the design flow requirement, the recurring saving of Rs. 12,36,000 with an attractive payback period of four months is achieved using the variable frequency drive. The improved operating parameters and energy consumption is given here. Ascertained results after installation of variable frequency drive are given in Table 25.3.

Table 25.2. Operating conditions.

Condition	Head m	Total flow m^3/hr	Input kw
Water circulation in ABS reactor, extruder and ammonia chiller	50	290	67.8
Water circulation stopped in ABS reactor	56	251	64.0
Water circulation stopped in ammonia chiller	58	220	59.9

Table 25.3. Ascertained results after installation of variable frequency drive.

Condition	Rpm	Head m	Total flow m^3/hr	Input kw
Water circulation in ABS reactor, extruder and ammonia chiller	1326	40	260	43.0
Water circulation stopped in ABS reactor	1106	34	165	24.0
Water circulation stopped in ammonia chiller	985	24	175	18.5
1. Present consumption = 62 kw × 24 hrs.	= 1488 kwh/day			
2. Actual power consumption using V.F.D	= 43 × 4 + 24 × 18 + 18.5 × 2			
	= 641 kwh/day			
Saving : 847 kwh/day × 365 × 4	= 12,36,000 Rs/year			

Some Revealing Facts

It would be appropriate to reiterate some of the established facts pertaining to conventional utility systems for consideration.

Compressed air system

1. 1 kg/cm² reduction in PSI (4 Rs./kwh) would mean loss of Rs. 5500/year.
2. Every 5°C temp. rise in intake air temp. increases the power consumption by 1 per cent.
3. Every 5°C rise in inlet air temp. of second stage to improper cooling, the specific power consumption increases by 2 per cent.
4. Pressure drop of 200 mm WC across the inlet air filter increases the power consumption by 1.5 per cent.

Refrigeration system

1. 5°C reduction in evaporator temp. decreases the specific power consumption by 8–10 per cent.
2. Vapour absorption system (VAHP) is viable against vapour compression system, if the electricity to fuel cost ratio is greater than 4.

Pumping system

1. Discrimination of low head and high head applications gives substantial savings.
2. Coating of pump internals improves pump efficiency by 3–4 per cent.
3. Elimination of undue throttling and overall flow balancing of close-loop system gives substantial savings.
4. Application of variable speed drive for varying load condition gives substantial savings.

Thermal system

1. 6°C rise in boiler feed water temp. would lead to 1 per cent improvement in thermal efficiency.
2. 20°C reduction in stack temp. would lead to 1 per cent improvement in thermal efficiency.
3. 1 kg/hr of flash steam loss due to improper condensate recovery (at 90 per cent boiler efficiency and 0.10 paise/kcal) leads to loss of Rs. 6000/year.
4. The economics of heat loss in insulation is changing rapidly. Thicker layers of insulting material or the use of high quality material is often justified.
5. 5 per cent failure of steam traps causes 15 per cent loss of generated steam.
6. 1 tonne of condensate recovery saves 6.75 lacs Rs./annum.

Thus, energy conservation is a continuous process. It can earn profits without blocking additional working capital, expanding plant capacity and employee strength and needs no effort to increase sales volume. With today's energy cost it does make sense to treat energy as a major raw material and not merely an overhead.

INDUSTRY INITIATIVES

In certain segments of industries, (such as textile, chemical, ceramic and engineering industries), considerable technological development has taken place to conserve energy and promote efficient utilisation of energy. For example:

1. Steam consumption has been reduced by redesigning boilers with innovative ideas.
2. Replacing mechanical parts by nylon parts in different machines and in the automobile sector has improved fuel efficiency and reduced consumption of petroleum products substantially.
3. Adoption of low liquor ratio dyeing machines in textile industry saves energy.
4. Adoption of foam technology in dyeing and printing will reduce consumption of petroleum products.
5. Developing new variety of ceramic products (Kiln) for ceramic and engineering industry reduces consumption of energy.
6. Use of appropriate chemicals reduces consumption of energy in different industries.
7. Introducing automatic control devices in different textile machines.
8. Introducing new design bearings, engineering parts also reduces energy consumption in household goods and other electrical and home appliances.
9. In certain areas of engineering industry the driving system has been remodelled by introducing profile cone drum to reduce energy consumption.
10. By modifying combustion techniques substantial quantity of energy is saved.

Technological Development in Boilers for an Efficient use of Energy

1. Incorporation of efficient burners with forced draft systems.
2. Reduced stack losses by minimising excess air and increasing the number of flue gas passages together with adequate water softening and consequent reduced scale formation.
3. Development of instrumentation systems etc. for proper control and efficient burning of fuel.
4. More heat recovery.
5. Advances in lagging material used for steam pipes, i.e., glass wool in place of magnesia and asbestos.

Technological development in textile machinery to minimise energy consumption

1. Vapour-loc system: A newly developed technology for a continuous washing and bleaching unit. It works under low pressure and takes a very little time to complete the job.
2. Combined hot mercerisation techniques substantially shortens the process and provides a good quality processed fabric and saves energy.
3. Solvent scouring units have been developed to carry out the processes of desizing and scouring simultaneously by using a suspension of enzyme and chlorinated solvent. About 80 per cent saving of steam consumption in normal process is reported.
4. High speed heat setting system, developed by different stenter manufacturers, are used to save energy. Super heated steam is used instead of dry steam to dry the fabric.
5. Transfer printing technique needs less energy while printing thermoplastic fabrics (specially polyester) as no after-treatments are required.

Thus with the advent of modern amenities our social structure has seen radical changes. Due to paucity of time, people in urban areas are more inclined to use mechanical devices to complete day-to-day domestic work in less time. Besides, in the post-liberalisation era, industrial activity has been spread out and it has received a boost in recent years. It is therefore vital to promote efficient use of energy conservation by taking appropriate control measures in different areas.

ENERGY CRISIS IN MODERN AGE: ALTERNATIVES

Energy crisis has hit and badly marred the power sources on our planet. Excessive, uneconomic and unwise usage of available non-renewable sources has ultimately resulted in their acute shortage. The scope of expansion of non-renewable sources is however an abstract concept. By and large, the entire 'blue' planet has been explored in search of reserves of conventional energy sources. Within a few years from today there could be no existence of conventional energy sources on the earth, if the present foolish trend of uneconomic energy usage continues. The time has come to have a wider outlook, newer experimentation on optional sources of energy and search for better alternative to the 'black-faced' energy crisis. We discuss and elaborate the scope, merits and demerits of several non-conventional sources, which are renewable and surpass the non-renewable energy sources in several regards. All have recently earned large acclaim.

Energy is the primary and most universal measure of all kinds of work by human beings and nature. Everything that happens in the world is the expression of flow of energy in one of its forms. Most people use the word energy for input to their bodies or to the machines and thus think about crude fuels and electric power. The energy sources available can be divided into two types:

1. Primary/conventional sources.
2. Secondary/non-conventional sources.

Coal, natural gas, oil and nuclear energy using breeder reactor are net energy yielders and are primary sources of energy, wind energy, water energy etc. Various sources of non-conventional/renewable sources of energy are discussed briefly.

Wind Energy

Wind energy is the utilisation of the energy due to moving air. The main reasons for wind are:
1. Uneven heating of earth's surface by the solar energy.
2. Climatic changes occurring by the climatic changes on the earth.

The wind's mechanical energy is usually captured by wind turbines, which convert it into electricity.

Merits of wind energy

1. Does not require any fuel.
2. Is one of the cleanest forms of alternative energy.
3. Does not face problems of off-time (night times).

Demerits involved in harnessing of wind energy

1. The initial cost for wind turbines is greater than that of conventional fossil fuel generators.
2. Noise pollution due to rotor blades.
3. Interference with television signals.
4. Wind resources might not be available near cities. So, the space might be used for other purposes that can generate larger profits.
5. In addition, wind does not blow consistently 24 hours a day and that could cause a problem when the demand for electricity peaks.
6. In India, the major drawback is that sufficient wind power density is not achieved, so the technology needs to be developed for deriving wind power from available density. Classification of wind power is given in Table 25.4.

Table 25.4. Classification of wind power.

Wind power class	Wind power density (Watts/m^2)	Speed
1.	· < 200	< 5.6
2.	200–300	5.6–6.4
3.	300–400	6.4–7.0
4.	400–500	7.0–7.5
5.	500–600	7.5–8.0
6.	600–800	8.0–8.8

Class 3 and above — good wind resources; class 7 — extremely strong; class 2 — mild breeze; class 4 — useful for electricity production.

Working of a wind-turbine

Three-bladed wind turbines are operated 'upward', (i.e., with the blades facing into the wind). The other common wind turbine is the two-bladed, downwind turbine. A wind turbine works in the opposite direction of a fan. The wind turns the blades, which spins a shaft, which is connected to a generator and makes electricity. Utility-scale turbines range in size from 50 to 750 kilowatts. Single small turbines, below 50 kilowatts, are used for homes, telecommunications dishes, or water pumping.

Tidal Energy

Tidal energy is the utilisation of the sun and moon's gravitational forces — as the tide is the result of their influences. To utilise tidal energy a barrage (barrier) has to be build with gates of some kind at the opening of a bay or a river system to create an estuary (a big basin). The gates create differences in the water levels between the estuary and the ocean, thereby enabling the generation of electricity. For example when the tide falls, the receding water retreats back to the ocean by passing through a turbine located in the barrage; generating electricity. Electricity can also be generated when the open gate lets water flow into the estuary during peak periods of high tide.

Merits of tidal energy

1. Does not require any fuel.
2. Tides rise and fall every day in a very consistent pattern thus it is non-intermittent.
3. Economics life of a tidal power plant is 75 to 100 years.
4. Clean and renewable, unlike fossil fuels.

Demerits of tidal energy

The altering of the ecosystem at the bay is the biggest draw-back to tidal power. Damages like reduced flushing; winter icing and erosion can change the vegetation of the area and disrupt the ecological balance. For a tidal power plant to produce electricity effectively (about 85 per cent efficiency), it requires a basin or a gulf that has a mean tidal amplitude (the differences between spring and neap tide) of seven meters or above. It is also desirable to have semi-diurnal tides where there are too high and low tides everyday.

Solar Energy

This energy comes from processes called solar heating, solar water heating, photovoltaic energy (converting sunlight directly into electricity), and solar thermal electric power (when the sun's energy is concentrated to heat water and produce steam, which is used to produce electricity).

Merits of solar energy

1. No requirements of fuel and it's recovery.
2. Pollution free.
3. Noise less
4. Cells are without moving parts.
5. Cells can be wired together to form a module (of about 40 cells), often enough to power a small light bulb.

Demerits of solar energy

1. The initial high cost is the main disadvantage of solar energy.
2. Requires large space to achieve average efficiency.
3. Location of the sun affects efficiency.
4. Air pollution and weather can reduce the productivity of the power plants.
5. High cost for generating electricity, as compared to conventional sources.
6. Nocturnal downtimes are another set back.
7. 'Intermittence', which means that if the sun is not shining, the system cannot make electricity.
8. Efficiency of PV cells is very less, i.e., maximum 18 per cent.
   Equipments components and uses of solar energy are given in Table 25.5.

Table 25.5. Equipments, components and uses of solar energy.

Equipments/uses	Basic components	Function
Focusing collectors	Heliostate	The moveable mirrors used to track the sun
(Solar furnace, electricity generation)	Parabolic collector	To collect and reflect the solar energy
	Receiver	Utilises the concentrated energy

(Contd ...)

Equipments/uses	Basic components	Function
Flat plate collectors	An absorber plate	Black coated metal (steel, Al, Cu) to trap solar radiation to turn it into heat
(Households or schools in such applications as swimming pool heating)	A glass or plastic plate An array of pipes Insulation	Above absorber plate to preserve the heat. Between the plates, cold water pass through them
Solar distillation	Similar to flat plate collectors except that, instead of heat, distilled water is the end product. Salt can also be 'produced' during this process	
Photovoltaic cell (electricity generation) (PV portable lanterns)	Photovoltaic plate	These cells are made up of thin pieces of semiconductors, such as crystalline silicone and gallium arsenide

Ocean Thermal Energy Conversion (OTEC)

OTEC or ocean thermal energy conversion, is an energy technology that converts solar radiation to electric power. OTEC systems use the ocean's natural thermal gradient to drive a power-producing cycle. As the temperature between the warm surface water and the cold deep water differs by about 20°C (36°F), an OTEC system can produce a significant amount of power. This potential is estimated to be about 10^{13} watts of base load power generation according to some experts.

Three approaches for OTEC

Closed cycle OTEC

Closed cycle OTEC employs a low-boiling-point liquid like propane or ammonia as an intermediate fluid. The OTEC plant first pumps in the warm seawater and boils the intermediate fluid. Then, the intermediate fluid vapour pushes the turbine to generate electricity. Finally, the vapour is cooled down by cold seawater (Fig. 25.2).

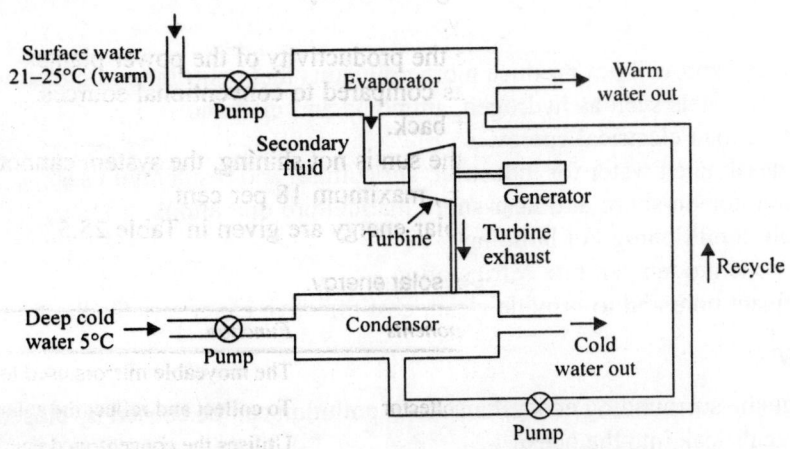

Fig. 25.2. Closed cycle OTEC.

Open cycle OTEC

It is very similar to the closed cycle one, except that it does not use intermediate fluid. The seawater is the fluid that pushes the turbine. The warm seawater on the ocean surface is turned into low-pressure vapour under a partly vacuumed environment. Vapour used to turn turbine can be condensed to obtain desalinated water by the colder water deeper into the ocean (Fig. 25.3).

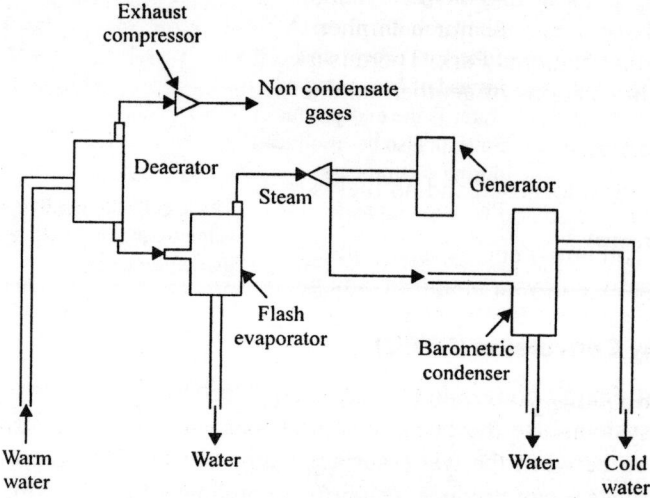

Fig. 25.3. Open cycle OTEC.

Hybrid cycle OTEC

It is a theoretical method of maximising the use of OTEC. There are two concepts:
1. Use of closed cycle OTEC to generate electricity to create the necessary low-pressure environment for the open cycle OTEC.
2. Integrate two open cycle OTEC (one is used to create the vacuumed environment) so that there will be twice the amount of the original desalinated water.

Merits of OTEC

1. OTEC is clean and will not produce more pollutants that contribute to global warming.
2. Also produces fuels such as hydrogen, ammonia and methanol.
3. Produce base load electrical energy.
4. Produces desalinated water for industrial, agricultural and residential uses.
5. Is a resource for on-shore and near-shore mariculture operations.
6. Provides air-conditioning for buildings.
7. Provides moderate-temperature refrigeration.
8. Has significant potential to provide clean, cost-effective electricity for the future.

Demerits of OTEC

OTEC may damage the surrounding ecosystem. Also, pollution can be caused by closed cycle OTEC if intermediate chemicals leak into the ocean.

Geothermal Energy

The temperature in the core of the earth is at least 6650°C, yet the cooling rate of earth is 300 to 350°C per three billion years. There are 42 x 10^{12} W of heat in the earth, 2 per cent is in the crust and 98 per cent in the mantle and core. Energy residing too deep within the earth is inaccessible, but the 840,000,000,000 W (2 per cent) of geothermal energy that is accessible is sufficient for humans to use for a long period of time. Areas around the plate margins are the best geothermal energy extraction sites, because the crusts in those areas are much thinner. A lively example of geothermal energy is 'Old faithful' in the yellow stone National Park. There is a continual flow of heat energy out wards towards the surface. Surface manifestations of geothermal energy include: volcanoes, hot springs and geysers.

Merits of geothermal energy

1. Geothermal energy is localised and no fuel is required.
2. Economical
3. The supply of geothermal energy is continuous
4. Geo-thermal energy enjoys a major advantage over other sources of regenerative energy, such as wind and solar energy. It is not dependent on meteorological fluctuations and is available on a regular and steady basis, day and night all the year round.

Demerits of geothermal energy

1. Environmental impacts may take place.
2. The increased temperature of the area can kill life forms in the water.
3. The geothermal fluid usually contains gases and dissolved substances (CO_2, H_2S, CH_4, NaCI, B, As and Hg) that can pollute the environment.
4. It is available only in a few places.
5. Production is phenomenally expensive.

Fuel Cells

A fuel cell has four basic components, two electrodes (anode and cathode), one electrolyte and a catalyst (enzyme) between the electrodes and the electrolyte in centre (Table 25.6). These components are made of different kinds of materials in different types of fuel cells.

Table 25.6. Types of Hydrogen fuel cells.

Type	Characteristics
Phosphoric acid	It operates at about 204°C. The normal energy efficiency of this type of fuel cell is around 40–85 per cent.
Proton exchange membrane (PEM) or solid polymer	This type of fuel cell is named after the material used for the electrolyte, polymer membrane. The catalyst of a PEM cell is made of platinum. This type of cell has the most potential on small applications because this type of cell can vary its output. This type of fuel cell also operates at a relatively low temperature, 85°C. The Vancouver-based Ballard Power Systems are currently using PEM fuel cells.
Molten carbonate	It operates at about 1,200°F. The ability of consuming coal-based fuel is the unique characteristic of the molten carbonate fuel cell.
Solid oxide	This type of fuel cell got its name from its electrolyte material. The solid oxide fuel cell uses hard ceramic material for the electrolyte. This type of fuel cell can generate large amounts of energy at the rate of 60 per cent, suitable for power-plant type of operations. The cell operates at a temperature of 1,800°C.

(Contd...)

Type	Characteristics
Alkaline	NASA uses alkaline fuel cells for space missions as it can achieve up to 70 per cent fuel to electricity efficiency. This cell has a big disadvantage, of being extremely costly.
Direct Methanol (DMFC)	It uses polymer membrane as its electrolyte. The main difference between a DMFC and PEM fuel cell is that DMFC does not require pure hydrogen as fuel, and it can consume liquid methanol directly without a 'fuel reformer'. This cell has an efficiency around 40 per cent at 120–190°F.
Regenerative	This technology combines solar electrolysis with fuel cells. Solar electrolyers separate H_2 from O_2 in water. The O_2 and H_2 are combined at the fuel cell, turned back into water and go back to the electrolyser.
Aluminium oxygen	This type of fuel cell uses Al anodes as fuel, and with the combination of oxidant, the cell is 'air-independent'. This cell is best suited for underwater operations. The energy production of Al=O. Semi-Fuel Cells are three times, 10 times and 6 to 7 times greater than silver Zn battery, Pb acid battery and Ni-Cd battery respectively.

The process of generating electricity from a fuel cell requires four cells. The process of generating electricity from a fuel cell requires four basic steps. First, hydrogen enters the fuel cell through the anode electrode. Then the catalyst splits the electron and proton apart in the hydrogen and produces a hydrogen ion. ($2H_2 \longrightarrow 4H + 4e$). After that, the proton travels to the cathode through the electrolyte while the hydrogen ion creates a flow of electricity.

After the energy has been utilised, the electron is sent back to the cathode and it is combined with the proton and oxidises to form a water molecule ($4e^- + 4H^+ + O_2 \longrightarrow 2H_2O$).

Hydro Energy

Hydro electricity is the generation of electricity through water pressure. Embankments usually are built to reserve water and create differences in water levels. Lakes in high altitudes are also used for the same purposes. Power stations that contain turbines and generators are built near the downstream side of the dam. Pipes or channels are used to direct water from the storage to the stations. Within the station, water pushes the turbine that generates electricity and then exits through the tailrace.

Types of dams

Embankment dams

Concrete faced rockfill dam
Clay or soil cored rockfill dam

Concrete dams

Concrete gravity dam
Concrete arch dam

Advantages of hydroelectricity

1. The 'fuel' for hydro electricity is renewable and cheap.
2. The only cost of hydro electricity is the expenses for building and maintaining the power stations and the dams.
3. There are no costs for fuel or the transportation of such.

4. It does not create any air, chemical, water or thermal pollution.
5. The higher level of water gives a better habitat for fish to grow and live.

Biomass Energy

Biomass energy is the utilisation of energy stored in organic matter. Wood, crops, animal waste, bones, leaves and scales are examples of 'Biomass'. These organic matters are sometimes burned directly to produce heat; sometimes they are refined to produce fuel like ethanol or other alcoholic fuels. In fact, using biomass energy is actually an indirect way of using energy from the sun.

Forms of biomass energy

Agriculture industry

Residuals like bagasse (fibres) from sugarcane, straw from rice and wheat, hulls and nutshells.

Animal wastes

Manure lagoons from cattle, poultry and hog farms.

Wood wastes

Sawdust, timber slash and mill scraps are considered organic materials.

Even in cities, paper and yard wastes are usable as biomass.

Merits of biomass energy

Fully utilised biomass reduces pollution in underground water bodies by offsetting the amount of waste in landfills. Methane and other poisonous gases formed from dead organic matters can be captured and converted into fuels suitable for generating electricity. Biomass energy reduces or may even eliminate some of the pollution found in the atmosphere, land and water.

Demerits of biomass energy

The burning method of biomass is not clean. Similar to burning fossil fuels, it produces large amounts of carbon dioxide. However, it produces much less harmful pollutants (e.g. sulphur), as the main elements found in organic materials are hydrogen, carbon, oxygen and nitrogen.

Technique to use biomass energy

There are three ways to use biomass.
1. It can be burned to produce heat and electricity.
2. It can be changed to a gas-like fuel such as methane.
3. It can be changed to a liquid fuel. (Biofuels: ethanol and methanol). Diesel fuel can also be replaced by biodiesel made from vegetable oils, soyabean oil, corn, cottonseed, peanut, sunflower, or canola, algae.

Biogas production

Biogas is a mixture of gases produced by anaerobic fermentation at 65°C. Methane is it's major constituents. Equipment used is digester (underground well). Production of biogas takes place in three steps:
1. Enzymatic hydrolysis.
2. Formation of acid.
3. Formation of methane.

Biogas plant can be set up for both small and large family by varying its size. Biogas can be produced from variety of waste in a large plant. India being an agricultural country has lot of biomass energy, which can be utilised, if proper techniques are developed. Lot of land is available over which plantation of such trees, which required less water and have faster rate of growth and provide a good amount of biomass, should be carried out. One company near Nasik is carrying on such a project where eucalyptus trees are grown, which are then supplied for pulp industry. Tissue Culture Technique increases production four folds. Government should finance biogas plants. Such biogas plants can be set up in urban areas providing domestic fuel. Biomass supplies for such plants can be given adjoining rural areas.

Sonoluminescence

Sonoluminescence is the emission of light by bubbles in a liquid excited by sound. The first process of sonoluminescence is to create a bubble with one per cent argon impurity in a container filled with liquid. The size of this bubble should be about 4 microns in diameter. Next, sound waves with a frequency at around 110 decibels will bombard the bubble, causing sonoluminescence to be initiated. At first, the tiny bubble grows to at least one hundred microns wide. Then, the bubble collapses to 1 micron. During this process, the temperature of the bubble can rise as high as 72,000 K.

Merits of sonoluminescence

1. Sonoluminescence is clean and renewable.
2. Produces large amounts of energy.
3. It is possible with sonoluminescence to break down materials at the subatomic level.
4. It can help to recycle different types of materials.
5. Sonoluminescence can also be used to create fusion. As the pressure and the temperature are very high, several industrial laboratories have used the sonoluminescence process to fuse hydrogen into metal.

TIPS FOR THERMAL ENERGY CONSERVATION

General

1. Undertake regular energy audits.
2. Plug all oil leakage. Leakage of one drop of oil per second amounts to a loss of over 2000 litres/year.
3. Filter oil in stages. Impurities in oil affect combustion.
4. Pre-heat oil. For proper combustion, oil should be at right viscosity at the burner tip. Provide adequate pre-heat capacity.
5. Incomplete combustion leads to wastage of fuel. Observe the colour of smoke emitted from chimney. Black smoke indicates improper combustion and fuel wastage. White smoke indicates excess air and hence loss of heat. Hazy brown smoke indicates proper combustion.
6. Use of Low air pressure 'film burners' helps save oil upto 15 per cent in furnaces.

Furnace

1. Recover and utilise waste heat from furnace flue gases for pre-heating of combustion air. Every 21°C rise in combustion air temperature results in 1 per cent fuel oil savings.

2. Control excess air in furnaces. A 10 per cent drop in excess air amounts to 1 per cent saving of fuel in furnaces. For an annual consumption of 3000 kl of furnace oil, this means a saving of Rs 3 Lakhs. (Cost of furnace oil-Rs. 10 per litre).

3. Reduce heat losses through furnace openings. Observations show that a furnace operating at a temperature of 1000°C having an open door (1500 mm × 750 mm) results in a fuel loss of 10 lit/hr. For a 4000 hrs furnace operation this translates into a loss of approx. Rs. 4 lakhs per year.

4. Improve insulation if the surface temperature exceeds 20°C above ambient. Studies reveal that heat loss from a furnace wall 115 mm thick at 650°C amounting to 2650 Kcal/m^2/hr can be cut down to 850 kcal/m^2/hr by using 65 mm thick insulation on the 115 mm wall.

5. Proper design of lids of melting furnaces and training of operators to close lids reduce losses by 10–20 per cent in foundries.

Boiler

1. Remove soot deposits when flue gas temperature rises 40°C above the normal. A coating of 3 mm thick soot on the heat transfer surface causes an increase in fuel consumption up to 2.5 per cent.

2. Recover heat from steam condensate. For every 6°C rise in boiler feed water temperature through condensates return, there is 1 per cent saving in fuel.

3. Improve boiler efficiency. Boilers should be monitored for flue gas losses, radiation losses, incomplete combustion, blow and losses, excess air etc. Proper control, can decrease the consumption upto 20 per cent.

4. Use only treated water in boilers. A scale formation of 1 mm thickness on the waterside increases fuel consumption by 5–8 per cent.

5. Stop steam leakage. Steam leakage from a 3 mm-diameter hole on a pipeline carrying steam at 7 kg/cm^2 wastes 32 kl of fuel oil per year amounting to a loss of Rs 3 lakhs.

6. Maintain steam pipe insulation. It has been estimated that a bare steam pipe, 150 mm in diameter and 100 m in length, carrying saturated steam at 8 kg/cm^2 wastes 25 kl of furnace oil in a year amounting to an annual loss of Rs. 2.5 lakhs.

DG sets

1. Maintain diesel engines regularly.
2. A poorly maintained injection pump increases fuel consumption by 4 gm/kWh.
3. A faulty nozzle increases fuel consumption by 2 gm/kWh.
4. Blocked filters increase fuel consumption by 2 gm/kWh.
5. A continuously running DG set can generate 0.5 Tonne/hr of steam at 10 to 12 bar from the residual heat of the engine exhaust per MW of the generator capacity.
6. Measure fuel consumption per Kwh of electricity generated regularly. Take corrective action in case this shows a rising trend.

TIPS FOR ELECTRICAL ENERGY CONSERVATION

General

1. Improve power factor by installing capacitors to reduce KVA demand charges and also line losses within plant.

2. Improvement of power factor from 0.85 to 0.96 will give 11.5 per cent reduction of peak KVA and 21.6 per cent reduction in peak losses. This corresponds to 14.5 per cent reduction in average losses for a load factor of 0.8.

3. Avoid repeated rewinding of motors. Observations show that rewound motors practically have an efficiency loss of upto 5 per cent. This is mainly due to increase in no load losses. Hence use such rewound motors on low duty cycle applications only.

4. Use of variable frequency drives, slip power recovery systems and fluid couplings for variable speed applications such as fans, pumps etc. helps in minimising consumption.

Illumination

1. Use of electronic ballast in place of conventional choke and save energy upto 20 per cent.
2. Use of CFL lamp in place of GLS lamp and save energy upto 70 per cent.
3. Clean the lamps and fixtures regularly. Illumination levels falls by 20–30 per cent due to collection of dust.
4. Use of 36 W tubelight instead of 40 W tubelight saves electricity by 8 to 10 per cent.
5. Use of sodium vapour lamps for area lighting in place of mercury vapour lamps and save electricity upto 40 per cent.

Compressed Air

1. Compressed air is very energy intensive. Only 5 per cent of electrical energy is converted to useful energy. Use of compressed air for cleaning is rarely justified.
2. Ensure low temperature of inlet air. Increase in inlet air temperature by 3°C increases power consumption by 1 per cent.
3. It should be examined whether air at lower pressure can be used in the process. Reduction in discharge pressure by 10 per cent saves energy consumption upto 5 per cent.
4. A leakage from a 1/2″ dia hole from a compressed air line working at a pressure of 7 kg/cm^2 can drain almost Rs. 2500 per day.
5. Air output of compressors per unit of electricity input must be measured at regular intervals. Efficiency of compressors tends to deteriorate with time.

Refrigeration and Air Conditioning

1. Use of double doors, automatic door closers,. air curtains, double glazed windows, polyester sun films etc. reduces heat ingress and air-conditioning load of buildings.
2. Maintain condensers for proper heat exchange. A 5°C decrease in evaporator temperature increases the specific power consumption by 15 per cent.
3. Utilisation of air conditioned/refrigerated space should be examined and efforts made to reduce cooling load as far as possible.
4. Utilise waste heat of excess steam or flue gases to change over from gas compression systems to absorption chilling systems and save energy costs in the range of 50–70 per cent.
5. Specific power consumption of compressors should be measured at regular intervals. The most efficient compressors to be used and for continuous duty and others on standby.

Cooling Towers

1. Replacement of inefficient aluminium or fabricated steel fans by moulded FRP fans with aerofoil designs results in electricity savings in the range of 15–40 per cent.

2. A study on a typical 20 ft. dia fan revealed that replacing wooden blade drift eliminators with newly developed cellular PVC drift eliminators reduces the drift losses from 0.01–0.02 per cent with a fan power energy saving of 10 per cent.
3. Install automotive 'on'–'off' switching of cooling tower fans and save upto 40 per cent on electricity costs.
4. Use of PVC fills in place of wooden bars results in a saving in pumping power of upto 20 per cent.

Pumps

1. Improper selection of pumps can lead to large wastage of energy. A pump with 85 per cent efficiency at rated flow may have only 65 per cent efficiency at half the flow.
2. Use of throttling valves instead of variable speed drives to change flow of fluids is a wasteful practice. Throttling can cause wastage of power to the tune of 50 to 60 per cent.
3. It is advisable to use a number of pumps in series and parallel to cope with variations in operating conditions by switching 'on' or 'off' pumps rather than running one large pump with partial load.
4. Transmission drive between pumps and motors is very important. Loose belts can cause energy loss upto 15–20 per cent.
5. Modern synthetic flat belts in place of conventional V-belts can save 5 per cent to 10 per cent of energy.
6. Properly organised maintenance is very important. Efficiency of worn out pumps can drop by 10–15 per cent unless maintained properly.

ENERGY AND THE FUTURE

The worldwide demand for energy is increasing day by day. The increasing use of modern means of transport—buses, trucks, trains, aeroplanes, ships, etc.; the rapid rise in the overall industrialisation; the tremendous growth in population, are some of the factors that have lead to a tremendous spurt in mankind's energy requirements. Since all this energy has to come from the energy sources available on this planet. Scientists have calculated (taking into account our present rate of consumption and the likely increase in this rate of consumption per year) that the world's present known stocks of fossil fuels may not last fore more than a 100 years or so.

Need for Judicious Use of Energy

It follows therefore that mankind has to adopt a judicious approach towards consumption of energy sources to ensure that these are not depleted too fast. This approach needs to be supplemented by optimum utilisation of our natural sources. We have, for example, reserves of billions of tonnes of coal spread across—Bihar, West Bengal and Orissa region. This coal may not be of the best quality but coal mining in this area can always be stepped up to meet our energy requirements. In India, technology used in coal mining and handling after it is mined is still primitive as compared to, say the German coal mining industry where mechanical wheels are used in open pit mining. Any improvement in material handling system can lead to a saving of a lot of coal which is otherwise lost. One source of energy which has remained under utilised is the hydroelectric energy. The Indian subcontinent has many large rivers with substantial hydroelectric potential, much of which still remains unutilised. These can be tapped to provide energy which is clean, renewable and cheap. Large numbers of small hydroelectric power projects across the country over small rivers could also yield a fair amount of energy.

Wind energy has a tremendous scope as an alternative source of energy not only in India but the entire region stretching from Afghanistan to Vietnam. This is also the region where oil is not found in sufficient quantities and people are extremely poor.

Wind electric generators are at present operating successfully in many parts of India. Windmills are also being used for pumping water and this use of windmills should be encouraged. If India develops a system whereby windmills and generators could be manufactured on a large scale, it will really be a tremendous boon to the rural economy of this vast region. Wind energy is a non-polluting, cheap, renewable source of energy.

A substantial portion of our energy requirement is met by firewood. It necessitates felling of trees, resulting in deforestation, soil erosion, and floods. To prevent this and to maintain the stability of forest reserves a massive afforestation programme is necessary. The use of firewood as fuel must be avoided as far as possible by encouraging the use of biogas plants. Benefits accruing from biogas plants are immense and manifold.

As India is dependent on imported oil for meeting its energy requirements, it would be prudent to reduce the consumption of petroleum products.

Minimising Wastages

Not only have we to adopt a judicious approach to using our energy sources, we have also to lay a great stress on prevention of wastage. Even a casual look at our day-to-day activities reveals that energy is wasted in many ways. Careless habits, like leaving the lights and fans on when no one is around, keeping the car or scooter engine on while gossiping with a friend on the road, etc. contribute to wastage of energy. We have to know about the various ways in which energy is wasted at home and in industries, and then develop—and encourage others to develop—proper habits to prevent such wastage. Industries must use devices of proper design and also ensure that all machinery is kept well maintained and in proper running condition. This helps save a lot of energy. With the impending energy crisis facing mankind, saving 'every bit of energy' is of great importance. This saved energy can then be put to some useful 'use' in future. We must remember 'energy saved is energy produced'.

Thus of all the dizzying visions of the future of energy, the one most remarkable uses is hydrogen—one of the most abundant elements. The hydrogen could come from nuclear energy (if we ever figure out what to do with the waste), from hydrocarbons (if we manage to find a technical solution to sequester the carbon away), or even from renewables such as solar energy (if we make the process economically viable). With the simplest of chemistries it could provide an abundance of clean energy, with water as the only by-product. Will it ever happen? Quite probably. When? In the next 25–50 years. Or make it a hundred. However long it takes – it will make the world quite unrecognisable from what it is today.

Thus we have seen several non-conventional sources so far. We have learned the basic principles, pros and cons of these energy sources, which have been practically employed.

The basic problem is that these energy sources have not been successfully harnessed on a large scale. It is still longing for a better and economic technology. There is a wide scope for the engineers of today to extend their interest and alacrity to this burgeoning arena of sciences, which is still in its infancy. There is unlimited scope to it.

CHAPTER 26

Safety Considerations in Chemical Process Industries

INTRODUCTION

In view of the hazardous nature of the processes involved, chemical and petrochemical units need to ensure the highest standards of safety. Such rigorous safety standards require association of technical expertise of the best quality. One effective method of conceptualising some of the technical perspectives involved is to evolve certain guidelines to act as a reference for the chemical industry. At the same time, such guidelines can also assist the statutory inspection authorities to bring about qualitative improvements in inspections. The Indian chemical industry is about one hundred years old. From very modest beginnings the industry registered significant post-independence landmarks in the areas of inorganic fertilisers, synthetic dyestuffs, drugs and pharmaceuticals, pesticides, man-made fibres, petrochemicals, thermoplastics and thermosets. All basic needs of society such as food, clothing, housing and health depend on the chemical industry. This industry is poised for impressive growth and performance in the coming years. This industry is now increasingly responsive to the hazards of pollution and is willing to bestow greater care in the selection of process technologies and plant sites.

CONCERN FOR CHEMICAL SAFETY

Recent chemical disasters particularly Seveso toxic chemical leak, Mexico LPG disaster and Bhopal gas leak holocaust have increased public awareness of potential hazards to the community from chemical and petrochemical manufacturing activities. These disasters have given a blow to the public image of the chemical industry in general. Nonetheless, it remains true that the chemical industry is one of the safest of industries. However, in India we have to pay more attention and put in more efforts to achieve the same or better standards in HSE (Health, Safety and Environment) Management as in the developed countries. Disasters could arise due to runaway reactions, design faults, storage of unacceptable large quantities of hazardous substances, contamination, accidents during transportation, improper waste disposal practices or human failures. Crucial for the safety of such plants are: Hazard identification, location in relation to surroundings and assessment of likely damage including those in the event of expansions or additions. By using the concept of inherent and intrinsic safety, new plants can be designed such that they use relatively safer raw materials and intermediates and milder conditions together with careful assessment of risks and safety.

Observations of the Expert Panels

On the basis of their visits to the industrial units and analysis of gathered data, the expert groups made important observations which are generally applicable to the broad spectrum of chemical industry. They noted with concern the poor safety level in those public and private sector units which do not have the necessary financial resources to implement even basic safety features. The experts observations relate to: (i) management's a concern for safety; (ii) plant design; (iii) hazard identification and safety audit; (iv) systems and procedures; (v) maintenance; (vi) technical services; (vii) safety organisation; (viii) fire-fighting facilities; (ix) emergency preparedness; (x) human inhabitation in the vicinity of chemical plants; (xi) factory act and; (xii) statutory bodies.

Specific Recommendations for Hazard Control and Improved Plant and Process Safety

After consulting the various acts/regulations in vogue in India and abroad for the protection and improvement of industrial safety and in particular the recently enacted Environment (protection) Act 1986, the expert groups have formulated the following recommendations:
1. Hazardous substances and installations must be identified and notified.
2. Safety checks must be strengthened for plants handling hazardous chemicals.
3. A multi-disciplinary in-house safety unit is needed in every industry for effective implementation to safety policies and practices.
4. Safety should be an integral part to chemical plant design engineering.
5. Potential disaster areas in the country need to be identified and co-ordinated master plans prepared for design with emergencies in chemical plants.
6. Human habitation must be restricted in the vicinity of hazard-prone industries.
7. Safety training programmes must be encouraged, improved and strengthened.
8. Transportation and disposal of chemicals must conform to specified guidelines.
9. Technical capabilities and infrastructure of Factory Inspectorates need augmentation urgently.
10. Diverse regulatory functions currently operating under different statutes need streamlining and co-ordination.
11. A 'National Board on Industrial Safety and Hazards' may be constituted with Jurisdiction over all process industries.
12. Fiscal benefits are needed to encourage investment in safety.

Checklists for Safety and Risk Assessment for Chemical and Petrochemical Plants

Chemical and petrochemical plants help in evaluating the organisational and operational safety practices and policies and their effectiveness against identified hazards. The checklists for the assessment of safety and risk is applicable in the following stages: (i) design; (ii) pre-commissioning; (iii) periodic auditing of existing installations; (iv) routine checking of safety during normal plant operation; and (v) emergency preparedness.

HAZARDS AND THEIR CONTROL IN PETROCHEMICAL INDUSTRIES

Petrochemical have become part of out daily life. These chemicals provide us clothing, health and hygiene products, cosmetics, etc. Many of these chemicals have even replaced the naturally occuring materials. To cite an example, plastics have replaced wood and metals in many applications. They have contributed in improving the quality of our lives and have raised our standard of living. We cannot do without these chemicals.

Major Hazards

Fire and explosion are the major hazards in a petrochemical plant, though some chemicals may also be toxic. Petrochemicals plants have many high value equipment (which are usually installed close to each other), handling large quantities of flammable hydrocarbons with high energy content. Hence there is always a danger of fire and explosion hazard which may also have a cascading effect.

Management Commitment to Health, Safety and Environment Protection

Top management's genuine commitment to health, safety and environmental (HSE) protection is a prerequisite for the safety in plant operation and chemical handling. The management should demonstrate this commitment in practice.

The first step in this direction is to have a written HSE policy signed by the top executive of the company and issued to every employee. The policy should be discussed with all employees to ensure that they know in clear terms what are the HSE goals, what the management expects of them and what are the individual employee's responsibilities.

It is important that there should be a personal involvement of management staff at all levels, setting examples to their subordinates. This is essential for an effective safety programme, motivating employees and achieving high HSE standards. Also, it is necessary to have regular liaison and communication with the factory inspectorate, the nearby fire brigade, the police, the local hospitals and doctors, etc. to apprise them of the hazards associated with the process and the chemicals handled. The management should have a suitable HSE department to act as catalyst in the implementation of it's announced HSE policy and to be a watchdog for the company's overall HSE performance. There should also be an department as part of the onsite and offsite emergency management plan. Of course, the management should allocate adequate budget to enable implementation of the HSE policy.

Assessment of Hazards and Risk Management

All foreseeable hazards and risks and the magnitude of their consequences should be assessed at the time of selecting a process. This will be helpful either in eliminating the hazards or in minimising it to a acceptable level, thus enabling the designing of the control measures, preparing for any mishap in an orderly manner and in mitigating its effects. Various techniques are now available for assessing the hazards, some of which are:

1. HAZOP (Hazard and Operability Studies)
2. FTA (Fault-Tree Analysis)
3. ETA (Event-Tree Analysis)
4. CCA (Cause-Consequence Analysis)
5. MPPD (Maximum Probable Property Damage), using Dow Fire and Explosion Index.
6. Dispersion Models (for unintended gas releases)

Hazard analysis of the critical sections of each manufacturing unit, including storage and handling should be carried out and appropriate measures taken. The hazard must be notified to the concerned authorities. The population in the neighbourhood should also be made aware of the hazards with the help of authorities.

Plot plans or maps showing areas which will be affected in case of an accidental release of any hazardous chemical for various meteorological conditions should be prepared. Computers can also be employed for this purpose. It is important (and now mandatory) to prepare the material safety data sheets (MSDS) for all the chemicals (raw materials, intermediates and final products). Tremcards

(Transport Emergency Cards) for all the hazardous products to be transported by road must also be prepared in English, Hindi and in regional languages as necessary. Fig 26.1 highlights the various steps involved in risk management. A brief description of the important steps and some other suggestions for achieving a high standard of HSE are given below.

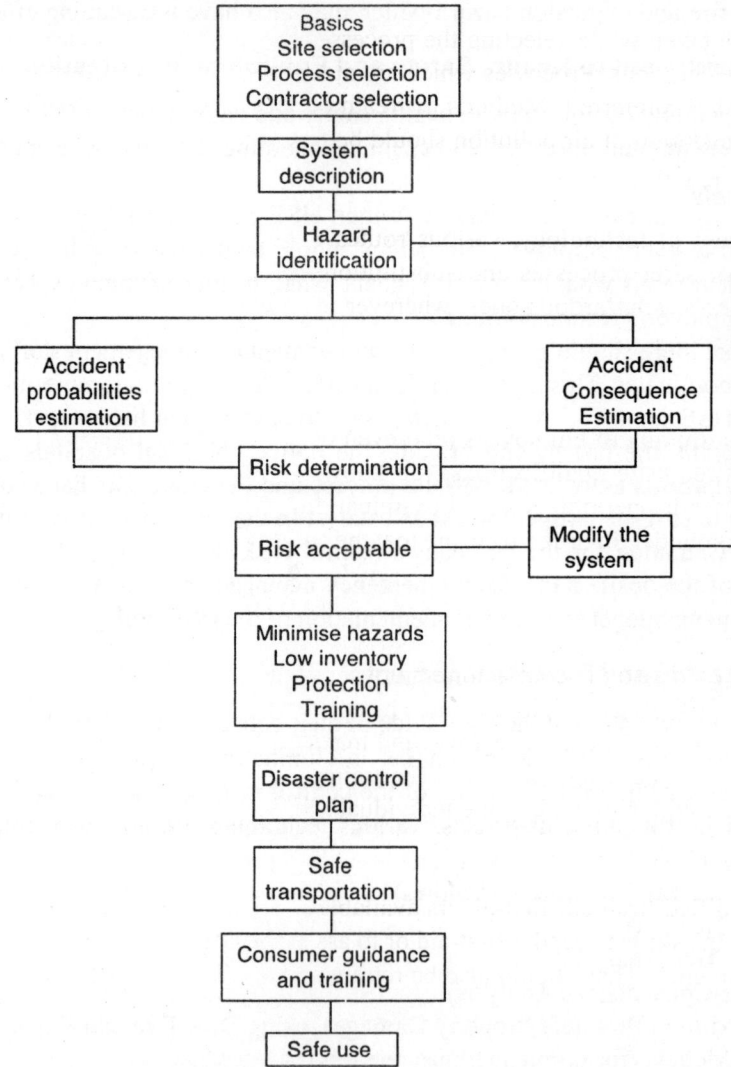

Fig. 26.1. Various steps involved in risk management.

Site selection

Selection of plant site plays vital role from hazard point of view. As far as possible, the site should be away from thickly populated areas. Other factors such as effluent disposal, means of communication, availability of certain services and facilities like fire brigades, hospitals, police station, etc., should also be taken into account.

Green belt

A fenced green belt around the plant is very useful. It not only adds to the aesthetics, but also helps in noise abatement, dust reduction and air quality improvement.

Process selection

Great care should be taken while selecting the processes. Emphasis should be as much on HSE as on other important considerations. Processes which are at lower pressures and temperatures and use safer raw materials should be preferred. Methods of treatment of effluent and disposal of solid and liquid wastes as well as prevention of air pollution should be considered at the design stage itself.

Use of safe chemicals

With the advancement of technology, various routes for manufacturing the products are available. Choice should be for safer processes and substitution of dangerous raw materials and intermediate products with safer or less hazardous ones, wherever feasible. Also, the inventory of chemical should be as low as possible.

Personnel

Proper selection and training of employees (and even of contractors) is most important. They must be healthy and possess requisite qualifications, job knowledge and experience. There should be a pre-employment and thereafter periodic medical examinations, in line with the provisions of the Factories Act and the Rules thereunder, in order to ensure that the employees are physically and mentally fit for the job.

Training

All fresh employees (and also the contractor employees) must be trained to suit the requirements of the job, including the HSE aspects. Employees must be apprised of the hazards associated with the job, the safety precautions to be taken, the correct use and maintenance of personal protective equipment, the correct way of choosing and using tools, etc. The training should also cover fire fighting, first-aid and emergency management plan, including the individual's role in an emergency.

Trained first-aiders

There should be qualified (trained) first-aiders available in adequate numbers round the clock amongst the employees in all the shifts to render first-aid or to assist the doctor or other medical staff in handling casualties in an emergency. They should also be retrained periodically. A first-aid rendered in time may save a life.

Mutual aid

Mutual aid arrangement with similar factories in the vicinity can be very useful in case of a major disaster. The mutual aid arrangement can be in the areas of fire fighting, medical aid (including ambulance) and rescue, since in a major disaster, it might be beyond the capacity of the individual factory to handle it efficiently or the available facilities may become in operative as a result of the disaster. Moreover, the mutual aid arrangement is appreciated by the insurance companies and they offer rebate in the premium to the members of the mutual aid scheme.

It is desirable to have a similar mutual aid scheme also for handling emergencies during product transportation.

Plant changes

In order to retain the technical and mechanical integrity of plants, it is essential to have a foolproof procedure for plant changes. All plant changes, however minor, must be scrutinised to the same standards as the original design (or even better, since the standards are continually being updated) in order to avoid risks to the plant, people and environment from the unsound or badly executed plant changes.

Any change in the plant involving process alteration or any change in the process hardware like alteration to equipment or piping resulting in change in capacity to equipment or specification, changes in monitoring systems and instrumentation (including, but not limited to, alarms, trip settings, sequence programming, computer software), electrical changes (including but not limited to changes in capacity, electrical area classification, relocation of equipment, etc.), changes to buildings or structures, changes in process parameters (like flow, level, pressure or temperature, etc.) any temporary facility, etc., should be regarded as a plant change.

The written plant change request with appropriate drawing should be routed through the competent persons (say, managers) of each concerned discipline like Operations, Technology, Office engineering (Design), Instrumentation, Electrical, Civil, Inspection, HSE, who should put down their comments and sign and then send the plant change request to the Chief Executive for final approval.

Monitoring environment

Monitoring of environment is essential for providing a healthy workplace, reducing losses and averting a mishap. Continuous online monitoring of atmosphere for hydrocarbons and certain other specific substances like carbon monoxide, benzene, etc., with alarms and/or trips are used in industry quite often. Periodic air sampling at different locations inside and outside the factory premises as well as the source monitoring of stacks should be done regularly to serve as guide to plant operations and to ensure compliance in accordance with the statutory requirements in respect of air quality.

Similarly, the liquid effluent must be suitably treated and monitored to ensure that it meets the statutory regulations. The treated water should be re-used within the factory to the extent possible, thus conserving the scarce fresh water. The ultimate aim should be to achieve 'zero effluent'. Treated effluent water, for example, can be used in the scrubbers and sample-coolers as well as for the hydrostatic testing of equipment, flushing of floors, watering of lawns and planting of trees, etc.

Managing safety in trasportation of products

Special attention should be given to the following factors for improving safety in transportation of hazardous chemicals:
1. Selection of hauler (transport contractor).
2. Hauler audit.
3. Periodic vehicle audit.
4. Vehicle fitness certification by the hauler.
5. Checking of vehicle prior to loading.
6. Crew recruitment.
7. Ability of crew to read and write.

8. Ability of crew to communicate.
9. Pre-employment and periodic medical examination of the crew.
10. Discouraging alcoholism and drug abuse amongst the crew.
11. Training of drivers in defensive driving and handling of emergencies.
12. No-accident bonus (instead of tonne-kilometre bonus) to the crew.
13. Emergency kit.
14. Emergency procedures (for managing any emergency onsite while loading/unloading as well as during trasportation).
15. Tremcards (Transport Emergency Card) which is to be prepared specifically for each hazardous substance to be transported by road.
16. Written instructions to drivers.
17. Facilities for the crew (like safe and decent clothing, resting places on product routes, etc.).
18. Survey and approval of product routes (and insisting upon the crew not to deviate from approved routes).

Inspection and preventive maintenance

Management must establish a programme for regular inspection of all equipment, piping, etc. Various non-destructive testing (NDT) methods are available for inspection of running plants for static and rotating equipment. All inspections or tests should be documented, written inspection reports prepared and copies of the same sent to the concerned departments for action. An equipment history sheet for each major equipment like compressors, pumps, tanks, vessels, columns, etc., should also be maintained.

There should be a maintenance planning cell to compile all inspection reports, plant change requests and other recommendations in order to plan for the preventive maintenance as well as for implementation of approved plant-changes and various recommendations of the inspection reports, etc.

A good preventive maintenance programme will minimise chances of accidents or disaster due to equipment failure and also of the break-down maintenance.

Care should be taken to update all the concerned engineering flow diagrams after implementing any plant-change. There may also be a need for revision of procedures, etc., due to plant-change to ensure safety of personnel and equipment.

Emergency management plan

The management should make a comprehensive onsite emergency (disaster) management plan to meet any residual risks after assessing the hazards. The plan must be submitted to the local factory inspectorate for approval. Offsite emergency management plan must also be prepared by the management to dove-tail with the overall district offsite disaster management plan of the local authorities to take care of any unforeseen eventuality. The management must render all necessary assistance to the authorities in the preparation of the district offsite disaster management plan.

It is a good practice to involve the local authorities, neighbouring factories, nearby fire brigades, local police, etc., in the rehearsals of the emergency management plans (both onsite and offsite) to assess the degree of preparedness of various agencies and their response time. After rehearsal, the plan should be critically reviewed and updated, if necessary. This should be an ongoing exercise to continually improve the plans.

The nearby hospitals and the doctors must also be informed of the toxic effects of the chemicals handled by the factory and the method of treatment of persons exposed to these chemicals. Similarly, the management should liaise with the nearby fire brigades and the police stations. Holding of regular conferences with the local doctors, fire bridge personnel and the police is, therefore, essential.

HSE audits

There should be a system of periodic HSE audits of the plants, facilities and procedures. The audits should include weekly micro-audits, monthly mini-audits and a major audit of the entire complex, say, once a year. There should also be a comprehensive external HSE audit by outside experts, say, once in three years or so. The audits should emphasise more on the unsafe acts rather than on unsafe conditions since over 80 per cent of the accidents are due to human failure. HSE audits enable the management to review and update the onsite emergency management plans, to design suitable training programmes, to carry out modifications in the process and/or procedures and to prevent possible accidents and disasters.

Recently a new concept of people's audit or 'safe attitude encouragement' (SAE) has been introduced by some enlightened managements. In this, a senior executive of the plant, along with a manager or supervisor goes round the works for the people's audit. The SAE is done exclusively for people's audit and is not combined with any other job. Both these persons together go round certain areas within the plant premises with a critical eye for any unsafe act or unsafe condition. On spotting someone working unsafely, or not following procedures, they engage the person in a sort of informal talk to put him at ease and then gradually come to the topic of the work being done by the person. They do not admonish the person or enter into any argument with him or do anything to hurt his ego in any way. Also, they do not take down any notes. They simply ask the person about the job he was doing, the hazards associated with the job and precautions to be taken, etc. Ultimately, as a result of this talk the person realises the violation of the safety norms by him. The executive and the manager/supervisor then leave the place and go to their office cabin, fill up the prescribed SAE form (which is a simple single-sheet form) to record their observations without mentioning the names of the person interviewed. They also mention there any further action to be taken. Copies of the form are then sent to the concerned managers and also to the safety department for necessary action and follow-up.

HAZARDS AND THEIR CONTROL IN PETROLEUM REFINERIES AND LPG BOTTLING PLANTS

Like most of enterprises chemical industry started as a cottage industry and gradually expanded. The use of chemicals in the products made by other industries contributed to chemical industry's rapid growth. This rapid growth was not without problems however; it quickly became apparent that handling the various chemicals and their reactions could be a dangerous proposition. As early as 1818 an explosion wrecked Elevthere Irenee Du Pont's first powder mill and killed 40 workers. As plants grew larger and processes became more complex, the need for good maintenance also came in picture. In the year 1974, in Fixborough's Caprolactam plant, due to a fracture in a provisional by-pass pipe line connecting two reactors, gas cloud leaked out which detonated and destroyed the plant section covering approximately 300,000 square metres resulting in 28 fatal injuries and 100 minor burn cases. The plant was destroyed and 2000 residential and office buildings were damaged due to this.

It must be recognised that, all process and operations have hazards associated with it. Today technology has produced and continuing to produce new sets of complex hazards which can threaten safety and health of the employees, the general public and the consumer.

The role of government agencies in the field of safety and health is also tending to increase in a number of directions for instance—New Factory Act amendments, formations of O.I.S.D. under Ministry of Petroleum, formation of crisis management cell under Home Ministry, to the extent that good concern in the direction of enforcement of rules strictly, to ensure public safety at large.

Fire ranks as the foremost major hazards in petroleum refineries and LPG bottling plants. Fire in process plant or in LPG bottling plants causes serious damage to plant, property and human lives not only inside the refinery but outside the plant also. The loss can disrupt production and have adverse effect on the economy of the country as well.

Fire and explosion risks are given top priority in petroleum refineries and in LPG Bottling plants because all the three basic requirements for fire to break out exist in the plants, i.e. heat/ignition source, oxygen/air and fuel vapours. It is an established fact that we cannot run the refinery without coming across these components. To make sure that these three components which cause fire, do not come together at one place at one time. Important steps are to be taken to identify such components and evaluate them to control them before they surface in such proportions to cause damage to plant. Some important sources of fire hazards and their elimination are as follows:

Heat and Ignition Sources

Electricity

Electrical installations should be made in accordance with relevant codes/standards. Standards such as ISI, National Electrical Code of NFPA, USA, etc. have to be followed carefully in provision of electrical equipment in hazardous areas. Provision of properly designed electrical equipment alone does not solve the problem of hazards. All electrical equipment should be regularly inspected and properly maintained.

In LPG bottling plants, more care is exercised to ensure that all electrical equipments are kept in good condition and all fixtures are explosion-proof equipment, i.e. flame-proof fittings. Even innocent looking equipments like telephones, fans and lights should be explosion-proof. Strict preventive electrical maintanance schedule ensures that no ignition source is present in these equipment.

Hot work

Mechanical work, particularly hot work has been a source of fires/explosions in petroleum refineries. In an effort to establish control over operations using open flames or producing sparks or entry into closed vessels, it is necessary to institute hot work permit system. The system, by definition, requires that authorisation be secured before the equipment is capable of igniting combustible materials is handled outside areas normally specified for its use. Work permit system has to be followed to ensure that an equipment or area has been made safe for employees who are required to work in or around the equipment or area, and to fix the responsibility for the person who is accountable for authorising the work and for the person who is carrying out the work, so that the work will be carried out safely.

In LPG bottling plant, hot work is normally avoided and is carried out when the unit is shut down and made gas free. The permit is issued by Manager—LPG bottling plant only after clearance from higher authority. This is the importance it deserves and is given. Even for jobs in vicinity of LPG bottling plant all above precautions are to be followed.

Static electricity

The generation of static electricity cannot be prevented absolutely, because its intrinsic origins are present at every interface. Static becomes a source of ignition, under the following four conditions and hence these should not be allowed to occur:

1. An effective means of static generation like high velocity flow.
2. A means of accumulating the separate charges and maintaining a suitable difference of electrical potential.

3. The spark discharge of adequate energy.

4. The spark must occur in an ignitable mixture.

Some of the safe methods adopted to prevent ignition from static electricity are: grounding, bonding, avoidance of splash filling or agitation, high liquid flow velocities and water contamination. Provision of a 30 second relaxation time down stream of filters before liquids enter the tanks is also necessary. In LPG bottling plants also all these precautions are strictly adhered to.

Sparks from vehicles

It has been a proven fact that, in case of hydrocarbon gas leaks, even a spark from the nearby passing vehicle can cause and start a big fire. To ensure that this cause is eliminated spark arrestors are provided in exhaust of vehicles. In LPG Bottling Plants no trucks are allowed without spark arrestors.

Ignition due to hot lines/furnaces

Hot uninsulated lines and furnaces become sources of ignition when hydrocarbon liquid is spilled near them or during hydrocarbon gas release. These fires can be kept well in control by remote cut-off systems and by proper maintenance care. If a hot line is exposed it should be properly covered hermetically on warfooting basis. Controls on furnace must be maintained in good condition so that they can be put off from remote locations.

Air/Oxygen

Most of the refined petroleum products are used as sources of heat energy by burning with air in engines or burners. The same type of combustion can occur on a larger scale in refining units with disastrous results if air (oxygen) mixes with petroleum fractions in the wrong places or in wrong proportions. It is difficult in refinery operations to eliminate all sources of ignition and as large quantities of fuel will be present in the refinery, our effort must be directed to eliminate air and oxygen before admitting hydrocarbon (before start up). This is achieved by following methods:

1. By purging of air from the vessels with steam followed by fuel gas.

2. By purging of air from the vessels with inert gas where catalyst beds are involved.

3. In LPG vessels, this is done by filling the vessel with water to overflow and followed by introduction of gas to remove water.

4. When the process requires the mixing of air or oxygen with fuel, this is done in the refinery under rigorously controlled conditions such as in asphalt blowing.

Hydrocarbons/Fuels

It is stated lightly that in a petroleum refinery everything but water burns, so, we treat all fuels with due respect and caution. These fuels or hydrocarbons are alike in many respects but yet differ. Some vapourise at room temperature, others do not. Hydrocarbon gases mix with air in all proportions but the proportion must be within certain limits to catch fire. These limiting proportions are called as 'Flammable limits' or 'explosive limits'. These are expressed as volume of the fuel in the fuel-air mixture. For hydrocarbons, this range is between 1.5 to 9.5 per cent, so all efforts are made in the refinery and LPG bottling plants to make sure that these proportions are not reached to create a fire hazard by following methods.

Design layout

By process hazard analysis, basic planning for safety of process units is done at design stage itself.

In LPG bottling plant, layout of plant is also given equal importance to ensure that, even in event of minor leaks, a hazardous situation does not arise, by proper ventilation, lighting etc.

Safe operations

Normal start-up and operating procedures and emergency shutdown procedures are periodically updated and personnel trained accordingly. Extra care is taken during start-up and shutdown of process units, to avoid following hazards:

1. Mixing of air and hydrocarbons.
2. Contacting hot oil and water.
3. Water shots and water freezing.
4. Over and under pressuring of equipment.
5. Thermal and mechanical shocks.
6. Corrosive and poisonous fluids.
7. Pyrophoric iron sulphide.

In LPG bottling plant for normal operations also, safety check lists are maintained. These are checked by safety technician.

Further safe operations of plant can be achieved by intensive training of all employees in the plant.

Maintenance

Special attention is given to maintenance standards, as proper maintenance plays an important factor in elimination of hazards. Maintenance jobs if carried out as scheduled at proper time will eliminate the major fire hazards which occur due to leaks and spillages of hydrocarbons, equipment failures, etc.

In LPG bottling plant preventive maintenance is a must and it is carried out to ensure all equipment like flexible hoses, (testing to be done once in a year), electricity fittings and fixtures, pipelines, valves, etc., are in good condition.

Lastly in petroleum refineries and LPG bottling plants, while every effort is made towards fire prevention, as a basic necessity, fire fighting facilities are also provided so that any fire can be efficiently fought and extinguished in the shortest time. Following are some important aspects that are given consideration in the refinery:

1. Adequate fire protection facilities i.e., Hydrants, Monitors, Deluge, Sprinklers, etc.
2. Provision is made for adequate fire crew manpower.
3. Training of personnel in fire fighting from operations, and maintenance who are first line fire fighters in the company before fire crew takes over.
4. Maintenance of all fire fighting equipment in top condition.
5. Ready emergency and disaster preparedness plan manuals.
6. Periodic mock drills and dry runs.

As fire at LPG bottling plant may lead to a disaster, effective detection alarm systems for gas leaks are provided, so that explosive range of gas is not reached at any time.

As fire generally starts in small way an additional detection system is also provided in LPG bottling plants which activates fire water system immediately and also effectively isolates LPG source to plant in case of a fire.

'Prevention is better than cure' is an old adage in respect of fire prevention. It is very appropriate and can a long way to eliminate hazards in petroleum refining and LPG bottling plant Operations.

HAZARDS IN STORAGE, HANDLING AND USE OF CHEMICALS

Most chemicals are hazardous, though the degree of hazard varies considerably. About 60,000 chemicals are now in common use. The extent of risk depends on the properties of the materials stored and processed, their inventory and the processes to which they are subjected, like high pressure, oxidation, hydrogenation etc. Hazardous chemicals have been classified into following nine groups, in decreasing order of degree of hazard.

Hazard Class

1. Explosive e.g., black powder, nitroglycerine, trinitrotoluene.
2. Non-flammable and flammable gases e.g., chlorine, ammonia, ethylene oxide etc.
3. Flammable liquids; (i) liquids with flash point below 23°C e.g., petrol acetone (petroleum class A); (ii) liquids with flash point between above 23°C but below 65°C e.g. kerosene (petroleum class B); and (iii) liquids with flash point above 65°C but below 93°C e.g., L.D.O., furnace oil (petroleum class C).
4. Combustible solids e.g., naphthalene, sulphur, aluminium powder, sodium nitrite.
5. Oxidising materials e.g., sodium peroxide, sodium nitrate, potassium chlorate.
6. (i) Poisonous liquids/solids/gases; (ii) toxic substances, aniline, phenol, pesticides, phosgene etc.; (iii) infectious substance-containing disease producing micro-organisms.
7. Radioactive materials e.g., radioactive ores, monazite.
8. Corrosive liquids e.g., acids, alkalies, oleum.
9. Other restricted articles e.g., cinematographic films.
   From the above list, following are the major hazards: (i) explosion; (ii) fire and combustion; (iii) toxicity and poisons; (iv) radioactivity; and (vi) corrosion.

There can be a combination of two or more hazards e.g. explosion often results, when a chemical catches fire. A fire may be accompanied by toxic release. Benzene is both toxic and flammable and when vapourised, it can even form an explosive mixture with air. Explosion hazard is connected basically with substances, which are unstable, while the hazard of fire is more widespread, because a large majority of the chemicals being organic, can burn in the presence of air. Toxicity depends on the biological properties and the specific concentration of the chemical and the duration during which human beings may be exposed to it. Radioactivity is limited to a small group of elements and their compounds, while corrosion is mostly caused by acids and alkalies. This hazard plays a major role, in selecting the proper container for holding the chemical.

Explosions

Chemical explosions can be of two types, deflagration and detonation, depending on the speed at which the products of combustion are propagated. Explosion can take place in an exothermic reaction, when considerable energy is evolved. Any flammable vapour or dust with air, can form an explosive mixture, but within certain concentrations. For each chemical or gas, there are two concentrations, called the upper and lower limits of explosion. Beyond the upper limit, it will not ignite because the mixture is 'too rich' in fuel, while below the lower limit, again it will not ignite, because it is 'too lean'. Hence, the hazard exists, only if the chemical vapour is present in the air, within these two limits.

Flame/combustion

The two words flammable and inflammable are synonyms and are used for gases, vapours and liquids, while the expression combustible is mainly used for solids and powders. For each flammable material, there is a flash point and ignition temperature. The flash point of a liquid is the lowest temperature at which its vapour will form a mixture with air, which will ignite near the surface of the liquid, when a spark is applied to it. There are two methods of determining it, the 'closed cup' or the 'open cup' method, the former giving a slightly lower figure. The lower the flash point, greater is the danger, especially if the ambient temperature is higher than the flash point. Ignition temperature, also referred as self-ignition temperature, is the lowest temperature, at which the chemical or its vapour will spontaneously ignite in a given atmosphere and burn without further heat input. Thus it is also a useful guide to the maximum safe temperature of any surface, in contact with the vapour mixture above the flammable liquid. Flammable mixtures may be formed and exist harmlessly so long as no source of ignition is present within the system. The combustion process must be initiated by the introduction of a finite amount of energy.

All combustible materials can form explosive dusts. These include metals like aluminium and magnesium, plastics, resins, sulphur, coal etc. In general, the dust explosions are equally or more intensive than the gas explosions and therefore, more hazardous. A mixture of combustible dust and a flammable gas with air is called a hybrid mixture, an example of which is found in coal mines. Here coal dust, methane gas and air can be present as a mixture. In chemical industry, such mixtures occur in fluid-bed dryers or when combustible powders are charged into kettles containing flammable solvents.

The two most important safety measures are: 'insertation' and 'pressure relief devices'. Insertation means removal of air or oxygen by an inert gas, which is normally nitrogen. If the oxygen concentration in the air is reduced below 8 per cent, then the vapour will not burn or detonate. The other important precaution is to provide a pressure relieving device to any vessel holding flammable or explosive material. Such a device is, in fact, a suitable aperture, which by opening at the right time, will allow the unburnt mixture to be expelled into the atmosphere. Such devices include safety valves, relief valves or rapture discs. Other precautions include, ensuring absence of any ignition source. Flame proof light fittings and fixtures are necessary for areas, where such materials are handled. Similarly, precautions have to be taken against static electricity.

Toxicity and poisons

Almost every chemical will produce injurious effects in a living body to some degree. We have to know, how much is needed to produce a toxic effect and what are its chances of entering the body. Chemicals differ widely in their relative toxicity, in their ability to enter the body and the effects they produce. Their effect also varies according to the route of entry. It can be by skin contact (cutaneous), by breathing (inhalation) or by mouth (ingestion or oral).

Many gases and vapours can be absorbed through intact skin. The chemical may cause irritation, dissolve the lipid and fatty matter, combine with the tissue protein or enter the blood stream, in which case it becomes a systemic poison. Examples are nitrobenzene, aniline and chlorinated hydrocarbons.

A toxic substance may result in deficiency of oxygen and high carbon dioxide in the blood. These are called asphyxiants, when the affected person grows pale and starts perspiring. Examples are carbon monoxide, hydrogen cyanide, hydrogen sulphide and aniline. Certain chemicals act as depressants, such as ethyl alcohol, while certain others act as hypnotics, like paraldehyde, chloral hydrate and barbiturates.

Recently it has been found that several chemicals will cause or are suspected of causing cancer to internal organs or systems of the body. Vinyl chloride, benzidine, chloromethyl ether, β-naphthyl amine are examples of such chemicals, termed carcinogens. Organophosphorus pesticides depress the cholinestrase enzyme activity, which controls the hydrolysis of acetylcholine in the body. Excess of acetylcholine prevents transmission of nerve impulses, resulting in convulsions and death. Some chemicals, after absorption in the body, release chemicals, which are more harmful than the parent material. One such example is the oxidation of methanol to formaldehyde, by an oxidative enzyme. Formaldehyde is an optic nerve poison. Nitrobenzene is reduced to *p*-aminophenol, which is 50–80 times more toxic. Poisoning by inhalation is much faster than by ingestion or by skin absorption.

Certain gases like ammonia, chlorine, formaldehyde or hydrogen chloride are easily detected by their irritating odour but in other cases like hydrogen cyanide, nitric oxide or phosphine, the concentration at which their odour can be detected is above the permissible limit of exposure. Threshold limit value (TLV) is primarily for a toxic chemical, which enters the body through the respiratory system. It indicates the average concentration that can be tolerated during exposure for a 40-hour week continuously. Toxicity tests are normally performed on animals like mice and LD50 is the median dose that will be lethal to 50 per cent of the group under test.

To determine whether or not a hazard exists in an environment, a survey has to be made by taking samples of the air and analysing them. This technique is called air monitoring. At the same time, the body response of the workers exposed to toxic chemicals has to be determined by clinical tests, which is called biological or man-monitoring. Industrial hygiene survey covers both these aspect, since both are interrelated.

Once a hazard is judged to exist, appropriate measures are required to be taken. These include, process modifications, substitution of materials, better personal protection, improved ventilation and other controls. The use of respiratory protective devices like respirators, filter masks etc. should be strictly enforced. Chemical cartridge respirators will protect only against specific gases and vapours which carry some odour, since this would provide the only warning of failure or exhaustion of the canister. Methyl chloride is an example, where odour indication is inadequate. If the identity or the concentration of the gas is in doubt or a mixture of several gases is present, it is better to use a self-contained breathing apparatus. Air-hoods and air-supplied suits for protection of whole body also serve the same purpose.

Industries linked with cancer, include asbestos manufacture and use, certain dyestuffs, some metals, coal-tar, mineral oils and radioactive chemicals. All technical measures are required to be taken to reduce the exposure of workers to any carcinogen to the absolute minimum.

Radioactive chemicals

Such chemicals can be hazardous, when they emit ionising radiation. Exposure to such radiation may cause harm to the human tissue. Individual cells are damaged, following ionisation of their water content. Excessive background radiation may result from spillages, leaks and general contamination. Areas have to be reserved for radioactive work with a high standard of ventilation. Work has to be carried out in closed systems, e.g. in glove boxes under slight negative pressure, with specially designed shields.

Corrosive chemicals

These include strong acids and alkalies, and such other materials, which will cause burns or irritation of the skin, mucous membrane or eyes. The hazard increases, because the chemical will corrode the container and may leak into the atmosphere. Acid fumes will also corrode structural material and equipment. Storage area for such chemicals should be isolated from the rest of the plant, and should be lined with acid proof tiles, with provision for safe disposal or spillage. Emergency showers and eye wash fountains should be provided nearby. Acids can react with mild steel to generate hydrogen, causing pressure build-up within the drum. Material of construction of a storage tank has to be selected, with due consideration of the chemical properties of the materials. Several chemicals in pure condition are not corrosive, but if moisture enters, corrosion can start rapidly, e.g. sulphuric acid or phosphorus trichloride. Wall thickness of the tank will depend on the corrosion rate.

Storage of chemicals

Storage of chemicals in warehouses or in tank farms is governed by specific guidelines. In case of hazardous chemicals, inventory should be kept to the minimum. Drums or carboys can be stacked in stacks or racks, but must remain off the floor (by using pallets), particularly on open ground. This would enable a speedy identification of a leaking container. Normally, drums should be stored vertical with their plugs up.

There should be sufficient space for a person to move between bays and along the walls. Similarly, in a tank farm, proper distances are to be maintained between two tanks as well as from other installations. When the tanks contain hazardous material, it is necessary to provide a dyke wall around them with an outlet valve leading to a chamber. The volume of this bund should be at least 1.5 times the volume of the largest tank inside. The pump connected to the tank should be outside the bund. Use of the same pump for different liquids should be avoided, tanks holding flammable materials should be provided with flame arrestors.

Two other aspects may be briefly touched, while discussing safety in handling chemicals. One is the effect of impurities and the other is compatibility with other chemicals in the same area. Often presence of impurities can increase the hazard. It has already been mentioned that moisture present as an impurity in certain chemicals increases the corrosion hazard. Traces of inorganic salts like copper sulphate, lead nitrate and zinc chloride also accelerate corrosion. Toxicity of xylene is increased if it contains traces of benzene. While pure ethylene oxide, polymerises very slowly, presence of certain impurities can initiate a rapid reaction with explosive consequences.

The other aspect to be borne in mind, during storage, is compatibility among the two or more chemicals. As such a chemical may not be so hazardous but on reacting with another chemical, it may create a highly hazardous situation e.g. ammonia and chlorine, aniline and nitric acid, sulphuric acid and sodium cyanide. Thus safety in storage, handling and use of chemicals will depend on several factors, each to be considered separately as well as in conjunction with others. The safety measures will need technical, managerial and financial resources for their implementation.

CHEMICAL STORAGE—SAFETY ISSUES

With the growth of the chemical industry from the early fifties storage of chemicals assumes great importance. The trend towards capacity upgradation of chemical plants to achieve economies of scale has led to provision of high capacity storage facilities. Today a single tank of more than 5,00,000 barrels capacity is quite common.

The hazards in storage depend on the material and the type of storage. The magnitude of loss in case of untoward incident varies directly with the quantum of material stored. This factor is critical for large inventories of chemicals as the mechanism to handle the emergency situations also needs to be upgraded. Chemical plants use chemicals mainly in the liquid form and to some extent in the gaseous form; as solid their use is marginal. The storage of liquid chemicals is of prime concern. The principal hazards of liquid or gaseous chemicals are illustrated in Tables 26.1 and 26.2 respectively.

Table 26.1. Hazardous events.

Material	State	Storage	Hazardous events
Flammable	Liquid	Atmospheric	Liquid release, then
			Tank or bund fire
			Tank explosion, then
			Tank or bund fire
	Liquefied gas	Pressure	Flashing liquid release — flammable vapour cloud, liquid pool, then
			Pool fire
			Running liquid fire
			Jet fire
			Vapour cloud fire
			Vapour cloud explosion
			Fire engulfed vessel, then
			Jet fire
			BLEVE
	Liquefied gas	Refrigerated	Flashing liquid release — flammable vapour cloud, liquid pool, then
			Tank or bund pool fire
			Running liquid fire
			Vapour cloud fire
			Vapour cloud explosion
			Fire engulfed tank, then
			Tank or bund pool fire
			Running fire
	Liquid	Atmospheric	Liquid release, then
			Toxic gas cloud
			Tank explosion, then
			Toxic gas cloud
	Liquefied gas	Pressure	Flashing liquid release — flammable vapour cloud, cloud pool, then
			Toxic gas cloud
			Fire engulfed vessel, then
			BLEVE
			Toxic gas cloud
	Liquefied gas	Refrigerated	Flashing liquid release — flammable vapour cloud, liquid pool, then
			Toxic gas cloud
			Fire engulfed tank, then
			Toxic gas cloud

Table 26.2. Initiating events.

Catastrophic failure of vessel or tank

Failure of or leak from other equipment, (pipe work or fittings)

Explosion in vessel or tank

Fire engulfing vessel or tank

Jet flame playing on vessel or tank

Overfilling of vessel or tank

Release occasioned by operations

Release occasioned by maintenance

Impact event

Natural event

Arson, Sabotage

Irrespective of the size of storage the first safety requirement for storage of hazardous chemicals—liquid or gaseous, is a properly designed container. The objective is to minimise the chance of leakage. Leakage of toxic chemicals can have particularly serious consequences, the Bhopal incident being one such example.

In large chemical storage areas fires have occurred mostly from leakages which have resulted in financial loss rather than loss of lives. Thus statistically and historically it would seem that fires in bulk storage installation have relatively little impact on public safety. There are exceptions such as disasters at Feyzin (France) in 1966, Mexico city in 1984, etc.

Past Experience

The record of accidents shows that majority of the losses involved storage installations. In the process plants the quantity of hazardous material in storage is usually much higher than that in process. In his book 'Loss Prevention in Process Industry' *Lees* has considered accidents in the process industry world wide between 1926 and 1933.

The study shows that 39 per cent of incidents involved storage of chemicals and 72 per cent of the financial losses was attributed to storage. Based on this study *Lees* commented in this book, 'thus the idea that material in storage presents a much lower risk than that in process is half the truth and cannot be accepted without qualification'.

Layout

Large storage tanks sited in or near urban areas have become the cause of some public concern. This in spite of a good deal of legislation from the angle of public safety in most of the industrialised countries. An important feature of the legislation is siting of large storage tanks away from habitated areas. Sites chosen for industrial installations of bulk storage facilities range from rural to urban with population density varying from virtually negligible to very high. The selection of the site however, is not a substitute for high standard of design and operation of the plant. Therefore, it should not be forgotten that the people most at risk are the people on site and standards should be such as to safeguard this work-force.

Topography is another relevant feature. It is desirable to avoid terrain where hazardous chemicals—be it a liquid or gas—can flow down into populated areas. Another consideration to be taken into account is contamination of water bodies by liquid spills. In selecting sites, due importance should be given to infrastructural facilities available for handling emergencies in relation to the large inventories involved. The location of large storage areas in relation to the process areas is important. It is prudent not to expose storage areas to risks originating from process areas. It is therefore necessary to segregate the two. This will reduce the risk of process incidents affecting the inventories in storage and vice versa.

Safety Standards

There are numerous standards and codes of practice available for safety of large chemical storages. Over the decades, much has been learnt about construction, inspection, design, storage facilities, etc. The current storage tank safety standards are based on years of operating experience and have translated lessons learnt from past accidents into reliable engineering standards, safe operating practices and training programmes. Although there may be some difference between the various codes, their basic objective remains same, i.e. safety of plant and personnel. Codes may not be available for all the chemicals; what is important is the prudent and judicious use of codes. The guidance available relates mainly to separating distances for storage, either of petroleum products or flammable liquids. The prominent codes available in connection with hazardous chemical storage include those of the National Fire Protection Association, USA; American Petroleum Institute, USA; Industrial Risk Insurers, USA; Institute of Petroleum, USA; Health and Safety Executive, UK, and the Oil Industry Safety Directorate, India.

However, good operation and maintenance depend on an effective management system. While it should be possible to assume with some justification that such a system exists in large chemical plants, there is a potential problem in situations where products such as liquefied petroleum gas (LPG) or chlorine are stored and used in essentially non-chemical industries like glass making, textiles, engineering. In these cases the supplier of the chemicals must provide safety guidelines on safe handling of emergencies, protection measures in the form of instruction manuals, chemical data sheets, etc.

Safety Feature

The codes and standards and good engineering practices for chemical storage take into account minimum safety requirements. Modern chemical storage facilities are generally built to these standards. With respect to large chemical storage, the approach is to reduce the effect of one event in a storage tank on the other and avoiding loss of containment through a number of passive and active protection systems. The passive restrainment measures are: segregation and adequate separating distance.

Apart from features like properly designed venting systems, earthing and safe drainage, the active restrainment measures include: (i) leak detection system for early warning; (ii) over spill protection; and (iii) fire protection.

Although the restrainment methods applicable to individual situations vary, over the years, from practical experience most protection systems have been standardised. Major emergencies arise when these systems are rendered ineffective or inoperable by inadequate maintenance or improper operating procedures.

Operational Hazards

Loss of containment of chemicals is a major loss producing event in case of large chemical storages. There were occasions, though rare, of catastrophic failure of the vessel or tank. The cause may be a mechanical or metallurgical defect. The most frequent cause of loss of containment is over filling leading to over pressure of tank or vessel. A tank may be subjected to over pressure by too rapid filling and too negative pressure by rapid emptying.

The release of chemicals from the storage tank also may occur due to explosion in the tank. There are various ways in which this can happen; one is the physical over pressure referred to earlier which causes vessels or tanks to burst and the other is ignition of flammable mixtures. There could be the evolution of gas or a runaway reaction within the vessel or tank due to reaction of an impurity in the product being stored with the material of construction of tank or vessel. Natural phenomenon like high wind, rain, storm, flooding, earthquake, etc. can also cause loss of containment.

Packaged Chemicals Storage

Packaged chemicals in large quantities are normally stored in warehouses. Chemical warehouses require systematic, scientific method of storage based on the hazardous properties of the chemicals. Materials may be simply stored in a stock pile on the warehouse floor or in racks; where space is a constraint, high pile storage system is used.

In a large chemical warehouse, a wide range of chemicals are stored. The chemicals to be stored vary widely in their characteristics; they could be inert, toxic, reactive and sometimes their chemical composition may not be known. The hazards inherent in the chemicals received and stored in the warehouses are: (i) fire; (ii) explosion; (iii) emission of toxic gases, vapour and dusts; and (iv) various combinations of these effects.

Safety Measures

The first and foremost objective of running a chemical warehouse from the safety angle is to create an awareness about the dangerous properties of the chemicals amongst the persons handling the chemicals and the general public at large in and around the area. Therefore the points considered for siting liquid and gaseous chemical tanks should also be applicable for large chemical warehouses.

There should be a formal system for the identification and tracking of dangerous substances. Substances arriving on site should be marked with labels as per specified codes and standards. The most important factor in warehouse operations is the segregation of packed chemicals as per their compatibility. Many guidelines and codes are available for this purpose, the most important is the UN recommendation for transport of hazardous goods. Serious accidents occur in warehouses, when a small event escalates and involves large quantities. The main cause of this is fire. The fire potential in a large warehouse is quite high and the conditions for the fire spread are nearly ideal; large quantities of combustibles are stored close together with good air circulation. In fact a large proportion of fire loss in storage is due to warehouse fire. The fire prevention in chemical warehouses is therefore extremely important.

Fire Protection and Loss Prevention

The main loss prevention and fire protection measures for warehouses are as follows: (i) Building construction; (ii) operational measures; (iii) control of ignition sources; (iv) fire detection system; and (v) fire protection system.

The warehouse building should be a non combustible one. As most of the chemical warehouses remain unoccupied for a considerable length of time, fire detection systems assume great importance. For better fire protection, detection systems should be backed by properly designed fire protection systems. Sprinklers are useful fire protection devices and are effective if designed and maintained properly. For warehouses with high rack storage, the sprinkler system needs special designing to be effective. There should be work discipline, essentially, similar to that required in the process plant. This includes good house-keeping, avoidance of obstruction in corridors and near fire doors, and elimination of ignition sources. Large warehouses storing packed chemicals are inherently hazardous by virtue of their contents being combustible, flammable, explosive, toxic, corrosive, unstable, reactive with air and/ or water, pyrophoric or liable to spontaneous combustion or oxidising. There should be a systematic approach to hazard identification especially from a liability point of view. Hence it is very important to provide safe drainage facility for the warehouse capable of handling both storm and fire water. A detailed consequence analysis for fire and explosion should be done. There should be an emergency plan for the warehouse suited to the level of assessed risk and should be developed in consultation with the local authorities. Personnel should be trained in handling emergencies.

OBSERVATIONS RELATED TO SAFETY ASPECTS

Major observations which are generally applicable to the broad spectrum of chemical industry are highlighted here.

Management's Concern for Safety

The attitude and awareness of management are very important factors in ensuring plant safety. These will manifest in safety policy, organisation, system and procedures adopted by the company. It is the general impression that, though a sense of concern to safety is evident, the managements are not clear as to the suitable course of actions to be taken. Well-defined management policy on matters of safety is lacking in our country. No emphasis is also placed on delegating the responsibility at various levels of safety management. There are, however, a few exceptions.

Plant Design

There are various facets to plant design viz., process selection, process engineering, detailed engineering, plant installation and commissioning. It is generally observed that during process selection, adequate importance is not given for safety aspects especially in medium scale plants. In a number of plants, the quality of engineering construction is rather poor. Internationally accepted engineering practices, codes and norms are not followed. This is particularly evident in the case of medium scale factories owned by private entrepreneurs. In many plants it was observed that adequate instrumentation is lacking for plant control and protection.

Hazard Identification and Safety Audit

Hazard analysis provides a systematic approach to the identification of risks, the probability estimates of an unwanted event which together with the frequency of incidence provides a quantification of risk involved. Safety assessment by HAZAN and HAZOP techniques have not taken root in the visited companies except one or two. Companies in medium scale sector do not have specialists in hazard analysis. These plants have, therefore, not undergone any safety audit or other conventional means of hazard identification ever since the plants were commissioned.

Systematic hazard analysis has to be carried out by multi-disciplinary teams of qualified and experienced specialists. There is a need to build up such expertise in an organised manner. Large chemical and petrochemical plants can take a lead by lending their experts to form hazard study teams which should draw some members from the other organisations lacking in hazard analysis expertise. This is one of the ways to spread the safety audit culture.

System and Procedures

The nature of systems and procedures adopted by the companies varied widely. These relate to operation manuals, permit system for maintenance jobs, measures for preventive maintenance, plant modifications, accident reporting and subsequent investigations, implementation of general plant safety rules etc. In the petrochemical and large private units, systems and procedures are well laid out to minimise accidents due to human lapses. Operating manuals are updated and reviewed. These areas deserve the attention of senior management in all other plants. The operating manuals must also contain emergency procedures to cope with failures, such as power, water, air etc. It is during such emergencies (if proper steps to shut down the plant are not taken) serious problems could arise. It is not only enough to have the procedures recorded but regular drills must be conducted to maintain the required vigil amongst the operating personnel.

Quality and nature of safety training programmes also varied widely from plant to plant in terms of their course content and frequency of organisation. There is a need to bring more uniformity in the course content to augment operator's skill and alertness in all chemical plants irrespective of their size and organisational structure.

Maintenance

Mechanical integrity of process equipments and associated systems is of prime importance in achieving plant safety. In more than 50 per cent of the plants, quality of maintenance work and housekeeping is poor. Marks of external corrosion, unchecked safety instruments and accessories and non-compliance of piping support standards are particularly noted in the case of small and medium scale industries, with a few exceptions.

To minimise breakdowns and unexpected release of hazardous chemicals, it is very essential that appropriate preventive maintenance procedures are adopted by these companies. Permit system for maintenance is to be strictly enforced to avoid accidents.

Comprehensive preventive maintenance programmes have to be drawn up for achieving consistent performance of process control instrumentation.

Technical Services

Technical services group has many important functions to perform in plant safety. The foremost is the understanding of process technology and interaction of process parameters. The impact of plant modifications on process technology is another important study area for the group. The expert groups noted with concern that lack of such knowledge normally results in unsafe plant operations.

Safety Organisation

Except in large chemical/petrochemical companies in public as well as in private sectors, the role of safety officer is not well-defined. In the organisational hierarchy, safety officer's position does not seem to carry much authority. In smaller companies, the position is merely created to meet the statutory requirements.

Firefighting Facilities

Except in large petrochemical and private chemical companies, the firefighting facilities are generally inadequate. Factories in medium scale sector do not even meet Tariff Advisory Committee (TAC) norms—which by themselves are considered as minimum requirements. The deficiencies in the plants pertain to pumping capacity, type and number of pumps, water storage capacity, ring main layout etc. The other point of relevance is the expansion of main plant without enhancing the capability of fire-water system.

Emergency Preparedness

Emergency preparedness measures have not attained the required maturity and perspective in more than 70 per cent of the companies. There are multiple reasons for this state of affairs; the most important being the non-availability of basic information. Another reason appears to be poor communication with the local authorities and public.

It is suggested that on-site contingency planning is the basic responsibility of the company management. On the other hand, off-site emergency planning falls in the domain of local authorities who are required to draw up the plans in consultation with the companies. Mutual aid scheme during emergencies is not widely practised in the companies. More attention is to be paid to this aspect in view of its advantages in terms of sharing available resources and expertise. Factory inspectorates can play an useful role in bringing groups of industries on a common front in preparing for the emergencies.

Human Inhabitation in the Vicinity of Chemical Plants

Thickly populated areas in the vicinity of the chemical plants are causing concern to the safety experts of the majority of plants. In a few cases managements claimed to have drawn the attention of local civil authorities but no serious action resulted. One of the general features of the problem is the development of human inhabitations in the close vicinity of the plant in an area originally declared as industrial zone. Lack of coordinated and effective action is reported to be one of the major causes for this situation.

Factory Act and Statutory Bodies

It is generally felt that mere compliance with factory acts and rules is not adequate to ensure a safe working environment—to the employees and the surrounding population. The Factory Act, for example, mainly concerns itself with the safety, health and welfare of employees within the factory premises.

Long drawn legal procedures are involved in penalising the defaulters; this consumes a lot of valuable time and resources of factory inspectors. Moreover, the penalty provided for in the act is not deterrent enough. Factory inspectors would need more technical support and need to be better equipped to identify the special hazards of a chemical plant. Considering the number of factories falling under the jurisdiction of a factory inspector, the staff strength of inspectors is woefully inadequate. Consequently, their ability to effectively monitor the safety hazard levels in the individual units is rather limited. It must be noted that understanding hazard identification techniques calls for very special skills, process knowledge, plant operation and design experience.

SPECIFIC RECOMMENDATIONS FOR HAZARD CONTROL AND IMPROVED PLANT SAFETY

As per the various Acts/Regulations in vogue in India and abroad an appropriate steps should be taken for the protection and improvement of industrial safety and living environment in the country.

Some of the documents consulted are:

Indian Regulations/Guidelines/Acts

1. The Indian Explosives Act, 1884 and Rules thereunder.
2. The Petroleum Act, 1934 and Rules, 1976.
3. The Factories Act, 1948.
4. Indian Boilers Act, 1923 and Rules thereunder.
5. The Indian Insecticides Act, 1968.
6. The Water (Prevention and Control of Pollution) Act, 1974.
7. The Air (Prevention and Control of Pollution) Act, 1981.
8. Smoke Nuisance Control Act (Gujarat, Maharashtra, West Bengal).
9. Environment (Protection) Act, 1986.
10. Safety and Health Accident Reduction Plan (SAHARA), 1985.

Regulations/Guidelines/Acts from Developed Countries

1. Occupational Safety & Health Act (OSHA), 1970, USA.
2. Health and Safety at Work etc., Act (HSWA), 1974, UK.
3. EEC Directive on Hazardous Installations.
4. Control of Industrial Major Accidents Hazards Regulations (CIMAH), 1984, UK.
5. World Bank and IFC Guidelines for identifying analysing and controlling major hazard installations in developing countries, 1985.
6. Federal list of 403 very toxic chemicals, EPA, USA.
7. Transportation Emergency Assistance Plan (TEAP), Canadian Chemical Producers Association.

Environment Protection Act, 1986

The following provisions of Environment (Protection) Act, 1986, are noted as important for chemical and petrochemical industries as well:

1. Coordination of actions by all concerned authorities.
2. Planning and execution of a nationwide programme for the prevention, control and abatement of environmental pollution.
3. Laying down: (i) environment standards; (ii) procedures and safeguards for the prevention of accidents; and (iii) procedures for handling of hazardous chemicals.
4. Inspection of hazardous manufacturing processes.
5. Establishment or recognition of environment laboratories/institutes.
6. Preparation of manuals, codes or guides relating to the prevention of environmental pollution.
7. Constitution of authority/authorities for exercising/performing such powers/functions as necessary for environmental protection.

Hazards and Safety of Chemical Plants

The following points are important for hazards and safety of chemical plants:

1. National programme for coordinated action plan for control of hazards and protection of occupational health and safety of the workers, 1985.
2. To examine involvement of different inspecting agencies with a view to rationalising various functions of the Directorate of Explosives, 1986.
3. The specifications of services for risk, reliability and safety audit studies of chemical process, petrochemicals and fertiliser plants by Lloyd's Register, 1986.

4. The Lessons of Bhopal–a community resource manual on hazardous technologies, IOCU, Malaysia, 1985.
5. Report of the Committee on Safety Standards in small chemical units in India, Ministry of Labour, February, 1986.
6. Report on General and Comprehensive Legislation on Occupational Safety and Health at Work Place, ILO, 1986.

RECOMMENDATIONS

The following twelve recommendations have been made which are given below:
1. Hazardous substances and installations must be identified and notified.
2. Safety checks must be strengthened for plants handling hazardous chemicals.
3. A multidisciplinary in-house safety unit is needed in every industry for effective implementation of safety policies and practices.
4. Safety should be an integral part of chemical plant design engineering.
5. Potential disaster areas in the country need to be identified, and coordinated master plans prepared for dealing with emergencies in chemical plants.
6. Human habitation must be restricted in the vicinity of hazard-prone industries.
7. Safety training programmes must be encouraged, improved and strengthened.
8. Transportation and disposal of chemicals must conform to specified guidelines.
9. Technical capabilities and infrastructure of factory inspectorates need augmentation urgently.
10. Diverse regulatory functions currently operating under different statutes need streamlining and coordination.
11. A National Board on Industrial Safety and Hazards may be constituted with jurisdiction over all process industries.
12. Fiscal benefits are needed to encourage investment in safety.

PACKAGING OF CHEMICALS AND DANGEROUS GOODS

Many chemicals have recognised obnoxious or dangerous properties. In choosing their packaging, it is necessary that these properties are accurately known. A basic obligation is to take 'reasonable care' to ensure that other persons or property are not injured or damaged by the product during its transport, storage or handling. Hazardous chemicals can be put into the following categories:
1. Explosive
2. Inflammable
3. Corrosive
4. Liable to spontaneous combustion
5. Strongly oxidising
6. Reacting with air or water
7. Radioactive
8. Poisonous—by inhalation, by oral ingestion or by skin absorption
9. Skin-irritant or dermatically toxic
10. Gaseous—requiring special considerations.

However, although the question of safety must be ever present, it must be recognised that not all chemicals are dangerous. For packaging purposes, the products should be divided into the physical forms, i.e. solids, liquids, semi-solids and gaseous. Further sub-division of these groups is as given in Table 26.3.

Table. 26.3. Physical form.

Solids	
Powders, granules	Easy to fill
Tablets, bars, blocks	Not easy to fill
Low melting point	Flow on heating
High melting point	Does not flow on heating
Viscosity	High
Viscosity	Medium
Liquefiable under pressure	Medium
Not liquefiable under pressure	Low
Dissolved in a liquid	High

Basic Requirements

A number of factors has to be taken into account in selecting packages for chemical products: safety; security; customer convenience; cleanliness of the product; shelf-life; cost; ease of filling, marking and storage; appearance—information, sales appeal; disposability; re-use value; availability.

The above points must be considered before a decision can be made on the suitability of a package for a specific application and are explained below.

Safety

For each category of chemical, there may be prescribed forms of packaging. These regulations are found in handbooks of the specific road, rail, sea and air transport authority. It is essential that all packaging used should comply with the authorised requirements and other requirements.

Amongst the main requirements are:

1. International Air Transport Association Regulations relating to the carriage of restricted articles by air.
2. International convention concerning the Carriage of Goods by Rail (C.I.M).
3. The Poisons Regulation (Great Britain).
4. List of dangerous goods and conditions of acceptance for merchandise train.
5. The carriage of dangerous goods and explosives in ships (Ministry of Transport and Civil Aviation).

It is necessary to comply with the regulations of the country of destination for transport of goods abroad. The local Chambers of Commerce would be reliable sources for up-to-date requirements. Labelling requirements to indicate and warn of the hazardous nature of the packed product are very detailed and must be followed accurately. Labels generally cover the nature of the main hazard, precautions necessary during handling and, in the event of an accident, the attention and action to be taken.

The manufacturers of hazardous chemicals are expected to take extra care to pack the product adequately for transport and acceptable for use by the customer. Knowledge of the properties of the product, the functioning of the packaging materials and closures, and of the potential hazards which can be encountered during transport, storage and handling of this class of product, is particularly important. The user should not be at risk due to faulty performance of the pack or closure.

Security

The degree of security which is required must be judged against the value of the product. Cost becomes an important factor where maximum security is required. Containers and closures must be reliable, and the packaging and closing operations must be capable of being performed without difficulty or doubt. It is not usually economical to test each individual pack for optimal performance (closure) after filling.

Convenience

The handling by the customer of the package, for a chemical product, must be considered in the package design. Very large containers are often more economical but if there are no easy methods of handling it, particularly with the safety aspect in mind, the customer may turn to an alternative supplier. Packages which are too small, thus too numerous in use, can also be a disadvantage.

The size and shape of the package must suit the customers' needs and present the product in the manner required. Removal of closure to gain access to the product must be straight forward and safe; clear instruction should be provided. The closure should be simple and secure.

Cleanliness of the product

Inner cleanliness, to prevent reaction, contamination or adulteration of the product is the first priority. However, where necessary, steps should be taken to ensure that the outside of the pack is not affected by the product. A product may be effectively packed and not affected by contact with the inside of the pack, but if handling damages removes the label during normal use this is highly undesirable.

Shelf-life

When export marketing, it is difficult to estimate accurately how long the packed product will be stored at the various stages of distribution. Prolonged storage before use may also occur at the destination. The main consideration (particularly corrosive materials) is that the product could 'eat' its way out of the container, or that the container could be badly corroded by atmospheric conditions.

Hygroscope products could take up excessive moisture from the atmosphere by prolonged storage, and become out of specification or unsuitable for use. The manufacturer must be able to give a guarantee of the shelf-life and the role of the packaging is vital in maintaining this. Storage trials with the packed product, if possible simulating the conditions which are likely to be encountered, should be performed. All the factors which could detrimentally affect the packed product during its journey from manufacturer to customer should be taken into account in the hope of anticipating problems. Any factor which will affect the security of the pack (corrosion of metal, softening of paperboard cartons, etc.) is vital.

Cost

Review and assessment of the packaging cost need to taken into account:
1. Cost of package or packaging material.
2. Cost of warehousing and handling.
3. Cost of weighing, filling and closing the container.
4. Freight cost (of particular importance)
   (a) actual gross weight.
   (b) actual gross weight or overall volume.
   In order to compare the economics of two types of containers, the initial cost of the container as well as the transport cost must be considered.

5. Returnable packages may be used for economic reasons for local trade, but are unlikely to be interesting for the export market. The cost should always be considered in relation to the value of the packed product.

Convenience for filling and storing

Packs should be designed to take advantage of the situation where the products can be automatically filled and weighed or metered. Some products are too objectionable to be filled in a normal building; a container may have to be designed for automatic enclosed filling and its design may have to be integrated with the filling process. Alternative containers may be easier to fill, but the choice for easier handling at the customer may outweigh this. Marking is invariably performed by the filling factory and, where necessary, modifications to the package may be needed to facilitate this.

A large number of containers, both filled and empty, may have to be stored at the manufacturer of the product. Drums, bottles and barrels are voluminous and difficult to store. Containers which can be stacked easily are an advantage in reducing storage space. Brick-shaped packages stack easily, whereas cube-shaped packs are more difficult. Wooden cases can be stacked to very great height, but fibreboard cases only to limited heights. Drums designed to nest on each other will be safer than those which are not so designed.

Appearance and sales appeal

When exporting, critical attention should always be paid to the appearance of the package; untidy or badly labelled packages are inexcusable. However, for the packaging of chemical or dangerous goods there is less emphasis on sales appeal than in the normal retail trades.

Re-use value

A pack may be technically suitable, of the right price and attractive. The next point which should be considered is the possibility of re-use after the consumer has used its contents. Re-use by the producer of the product is not an attractive option in export, because of the freight cost and the risk of damage on both outward and return journey. As a rule non-returnable containers will be cheap enough to be disposed off after initial use, but it can be an advantage if the packs have a further use to the customer or have a resale value.

Availability

Continuity of supply of the specific package is very important. Standard containers are usually cheaper, more readily obtained (thus reducing stock holding), of reliable quality, and should therefore be used whenever possible. When non-standard packs are necessary, or a special closing feature is required, more than one supplier should be engaged.

Non-standard packs should be specified in such a way so that they can be made with equal ease by different suppliers. This ensures the reliability of the suppliers, as the aim is to make the availability more a function of administration than a technical problem.

Types of Container

The packaging of explosives and gases will not be treated in this guide. An outline is given of the more common types of packages which can be used for chemicals and dangerous products.

Metal drums

This is one of the most important forms of container in the chemical industry; and three main types with a number of variations are used:

Closed-end drums

Heavy-duty drums—constructed of mild steel, with body and end seams all welded and roller hoops of I-section rolled steel. These are generally galvanised and are suitable for liquids which are highly corrosive, or poisonous or which develop high vapour pressure. The gauge of metal used depends on the expected journey. Products which attack or are affected by mild steel can be packed in similarly designed drums made of pure aluminium. These are, however, much more expensive. Aluminium alloy drums are slightly more expensive than steel and are very useful. Light duty drums—constructed basically from mild steel with welded side seams, double-seamed and solutioned end with pressed out rolling hoops.

Drums may be made from tin-plate or galvanised sheets, but the use of two or more internal coatings or lacquer (polyurethane resins; phenolic modified epoxy lacquers; vinyl-based lacquers) is increasing. Trials should always be performed to determine if a lacquer is suitable for the intended product.

Light-duty drums are more susceptible to damage but, when handled properly, are safe and capable of a number of return journeys on the local market. The 45 gallon drum is the most in common size but the 5 gallon of similar construction, when suitable palletised is particularly satisfactory for export.

Full-opening drums

These have totally removable lids, sizes range from 45 gallon with sealed gaskets and latch, lever or bolt-type closing hoops to 5-gallon pails with flowed-in gaskets in the multilugged covers. They are usually constructed from mild steel or lacquered steel and flexible liners can be fitted. The side seams are usually welded and a double seamed end solutioned bottom seam, but for hazardous products fully welded seams may be used.

Small-aperture drums

These have a fixed head and a filling opening of up to 9" (23 cm) in diameter. Access to the product is not very convenient, but the price is lower than full-opening drums.

Fibreboard drums

The construction is generally of a spirally wound kraft body with a metal or paperboard base and a metal lid which may be secured by a latch or lever-type securing ring. The side walls may incorporate a wide range of barrier materials to give special performance to the drums, e.g. aluminium foil, polyethylene film, grease-proof paper for moisture resistance. Provided there is a suitable lining, semi-solids and pasters may be packed. The outer surface must be treated to provide adequate weather-proofing for general shipment to overseas markets. Where transport bulk containers are used, this extra treatment might not be necessary.

Sacks

These may be of textile, textile laminate, aluminium laminates or of paper. Two main point must be remembered when sacks are being considered:

1. Sacks do not provide support for the product, thus the product should not be sensitive to damage by compression.
2. Sacks are very susceptible to mechanical damage, i.e. puncture, tears and snags, thus there is a risk of product loss.

Multi-walled sacks

The basic material is plain kraft paper and a sack may consist of two, three or more plies. These can incorporate plies of other materials to give the required properties for packaging particular products.

Bitumen-laminated kraft paper can provide adequate moisture protection for a wide range of chemicals. Polyethylene waxed kraft paper can be used for still higher moisture barrier. An inner layer of polyethylene coated kraft can be used to give protection of the sack against the product. A wide range of opening features, for facilitating filling and closing with various types of products, is available.

Plastic sacks

These are available in a wide range of sizes and thickness and are generally produced from plain polyethylene or polyethylene coated with one or more types of plastic films to give the sack special properties, e.g., strength, tear strength, gas barrier, scratch resistance.

Glass Carboys

The inert property of glass makes it ideal for containing many corrosive and dangerous products. However, fragility of glass means that extra protection to prevent damage during transport is required. In general, carboys are wasteful of space and unattractive, but the re-usable quality, long life and possibility of re-sale can compensate for this.

Carboys are usually used for chemicals which are not suited to other types of containers. The quality and reliability of the closure is important.

Plastic Containers

Rigid and semi-rigid plastic containers are available in a very wide range of sizes, shapes, special design features and closure. Plastic can be produced with excellent chemical resistance to a large number of products. In addition, when necessary, materials for special application can be manufactured. Tests to ensure acceptance of the intended product are essential before a definite decision can be made. Plastic containers are lighter than glass and the possibilities for economical functional designs make them very competitive for this class of products.

Intermediate-bulk Containers

Containers made of metal, rubber, fibreboard, crates or box pallets with disposable liners form a wide range of available intermediate-bulk containers. Basically they are returnable containers and fall in size between a bulk-delivery vehicle and the larger shipping container or drum.

CHEMICAL PLANT SAFETY—FROM CONCEPT TO DECOMMISSIONING

In today's competitive chemical industry, it is extremely important to operate chemical plants safely and eliminate accidents, which result in the loss of material, man-hours, property and invaluable human life. Accidents in turn lead to litigation, causing payment of heavy compensation and project poor image of the company. Safety considerations play a very vital role from the technology selection stage or laboratory process development and maintain safety on a day to day basis to avoid accidents. This section describes in details various safety considerations from initial stage to all the stages throughout the plants life.

In today's world of growing economic competition and environmental awareness, it is necessary for the survival of chemical process industry to have safe, accident free and eco-friendly plants. Safety and environmental aspects are of as much importance as the economic viability of a project.

Safety aspect must be given the prime importance during each of the following stage of the process plant's life for a near zero or accident free process plant:

1. Technology selection or development.
2. Process design.
3. Instrument design.
4. Equipment design.
5. Plant layout.
6. Plant erection.
7. Training.
8. Start up and commissioning.
9. Regular plant operation maintenance and modifications.
10. Decommissioning.

Technology—Selection and Development

Managing the process plant safety begins from the conceptual, process development and technology selection stage itself. To select or develop an inherently safer technology is of utmost importance for the plant safety. During this stage, following aspects are to be considered:

Inherently safer process

The process, inherently safer, using less hazardous raw materials and less severe operation conditions, is to be preferred over a process economically more attractive buy not inherently safe. So, properties like toxicity, volatility, flammability and explosibility of each raw material must be evaluated right at the conceptual stage.

Inventory of hazardous material

Wherever hazardous materials are to be used inventory of hazardous materials should be minimum. Highly toxic, volatile and explosive materials preferably should be generated and consumed *in situ*, e.g., highly explosive catalyst copper acetylide is generated *in situ* in the manufacture of butanediol from copper oxide. It is also destroyed *in-situ* before discharging.

Volatile liquids

Highly volatile liquids should be stored at low temperature and moderate pressures instead of at ambient temperature and high pressure particularly when the reactivity is lower at lower temperature.

One of the important factors for MIC leakage at Union Carbide plant of Bhopal, India was that it was stored at ambient temperature because of the failure of refrigeration system instead of the recommended lower temperature.

Less hazardous form

Possibility of using hazardous materials in a less hazardous form should be given the first priority and a serious consideration during the process design stage.

Continuous processes

For high capacity plants (more than 3 to 5 tonnes per day) continuous processes should always be preferred over batch particularly if the operating conditions like temperature and pressure are severe in a batch process. Batch processes are more prone to human error and have large reactor inventory, which is potentially dangerous.

Process Design

During process design stage following need to be considered.

Broad range

The process should be developed and designed to operate safely within a broad range of operating parameters and thus should be more forgiving to human errors.

Containment systems

The process should have adequately designed secondary containment systems like scrubbers, flares etc. to minimise the damage in the case of accidental release of chemicals. It is equally important to keep these systems running during the normal operation as well as during the shutdown. There must be alternate redundant system, so that even if one safety system is shutdowns at least there is another in operation.

HAZOP analysis

HAZOP study, safety review and evaluation of alternatives should be carried out at an early stage of the process design. This will allow early design changes in the equipment, help in considering installation of additional equipment, instruments, if required and design interlocks and safety systems.

Hands on experience

Involvement of a person with real hands on experience in the plant operation from the early stages of the process design and throughout the project upto commissioning is essential. This would ensure that key operating and safety issues are taken care of at the design stage itself and do not come at the last stage of the project, when it will be difficult to make any changes.

Batch processes

Process design for the batch plants need to have special consideration, which are explained in the next section.

Special safety consideration for the batch process during the process design: Considerable number of high value products, particularly upto the capacities of 1 to 2 tonnes per day are produced by batch processes because of the following advantages:

1. Manufacturer can respond quickly to changes in market demand.
2. Batch process can be carried out in a multipurpose facility using relatively simple production facilities.
3. Expensive speciality chemicals can be made in such plant.

However, special safety considerations are to be given for the batch process in the laboratory development stage, in design stage and during operations. Continuous plants by nature are safer plants, because once the plant conditions are stabilised, plant can run smoothly without operator intervention for a number of days. Operator's intervention is required only to make minor adjustments in the operating parameters to maintain normal stage. However, batch plants are more prone to accidents caused by human error because of repetitive operations, which need constant attention of the operator. Similarly batch plants are more prone to accidents because of normally poor understanding of process chemistry, specially effect of accumulated impurities and varying process conditions for different products. Frequent product charges leading to contamination and lack of special training to operate batch plants also lead to accidents. Many times batch processes are examined less thoroughly, because the operations are on small scale and scale up ratios are kept as high as 10,000: 1. Thermal explosions are of particular concern in multipurpose batch plants.

A statistical review of incidents of thermal explosion between 1962 to 1987 in chemical reaction of type $A + B \longrightarrow$ products revealed following causes:

1. Insufficient knowledge of process and thermochemistry.
2. Inadequate or wrongly designed cooling systems.
3. Wrongly designed monitoring and safety system.
4. Incorrect process management and poorly trained personnel.

Hence, for designing safe batch plants, thorough understanding of the process chemistry and thermochemistry is essential as given below:

1. Process chemistry to know desirable and undesirable reactions taking place.
2. Heat of reactions and specific heats.
3. Heat generation rate and maximum heat generated under adiabatic conditions.
4. Effect of feed rate, temperature, mixing, accumulation of reactants, effect of contamination and reaction medium on the rate of reaction.
5. Heat transfer rate i.e., cooling rates and heat transfer area required.
6. Temperature and quantity of accumulated unreacted component at which runaway conditions can occur.

All this information can be obtained on laboratory scale by using the following techniques.

1. Reaction calorimetry for the measurement of thermodynamics and kinetics.
2. Differential Scanning Calorimetry (DSC) and Differential Thermal Analysis (DTA) for the determination of heat of decomposition.
3. ARC (Accelerating Rate Calorimetry) for the measurement of time vs. temperature/pressure rise for the runaway reactions.

The above information and data should then be used to ascertain whether the reaction is inherently safe or not, identify hazardous conditions, and generate runaway scenarios. Consequently one can define procedure for safe plant operation define limits of operating conditions, prepare operating manual, design reactor and adequate cooling system, design alarm setting, interlocks and train the operators. The information can also be used to design adequate protective system to contain the effects of any release in case of runaway reaction, i.e., minimise the damage to men, material and environment.

Instrument Design

It is recommended that during the safety review/HAZOP set point for alarms should be finalised based upon the operating conditions. The logic followed for deciding the control philosophy should also be documented. This document will be useful during the training of operating personnel. Once operating personnel understand the logic, they will be able to take corrective action during the abnormal conditions.

Equipment Design

Normally the equipment and piping are designed, fabricated and inspected as per the standard design and fabrication codes. However, for the equipment and piping handling hazardous chemicals special considerations are to be given. The containment should be the first line of defence. Accordingly equipment/piping wherever possible can be designed to withstand the maximum pressure/temperature predictable under the runaway conditions. For example piping for acetylene is designed for 10 times the operating pressure i.e., piping design pressure is 10 atm for an operating pressure of 1 atm when handling acetylene.

Plant Layout

Safety considerations during this stage are given below:

Open structure

Use of open structure of the plants using flammable raw materials is recommended as even small quantity of flammable material igniting inside a building will cause serious damage to the building. Open structures also provide effective ventilation, reducing the chances of vapour accumulation due to small leakage.

Improved diking method

Insted of using traditional diking method for the storage tanks, which accumulates leakages, it is recommended to use improved diking method. In this slope is provided for the floor and leakages are accumulated outside the dike wall in a pit from which it can be pumped out.

Legal provisions

Maintaining safe distance between the tanks, equipment, providing adequate number of staircases, escape ladders has to be as per the legal provisions.

Adequate/safe distance

It is important to keep adequate distance between the equipment, various process areas and various plants to minimise the effect of fire or explosions. Many times, the reactors handling hazardous materials and likely to explode are isolated by concrete or brick walls from three sides. Relief flares are located far away from the main plant.

Plant Erection

After the completion of about 70 per cent mechanical erection, all the other contractors for instrumentation, electrical, insulation and painting along with the mechanical and civil contractors are at work simultaneously. Hence, at this stage maximum co-ordination is necessary for the smooth working as well as for the plant safety.

Power connections for equipment used by various contractors need to be planned properly. Power connection cables, though temporary must be routed at height. Cables lying on the ground are potential safety hazard.

Use of safety belts for the person working at height must be insisted.

During the construction period, it is necessary to have work permit system for vessel entry to avoid accidents. Before the entry, vessel should be inspected by a competent authority and certified that it has been isolated, and that proper ventilation has been provided. Supervisors of the contractor should ensure that the persons who have been working inside the vessel have come out safely at the end of the day or before boxing up of the equipment.

Training

Training of the operators via classes before start-up and in the plant during the water runs, start up, commissioning and refreshing the same during the regular plant operation is most important of the plant safety.

Generally, the operators are expected to keep the plant in normal conditions by maintaining the operating parameters within the prescribed limits. In case of disturbance they adjust conditions to return to normal state. In case, the situation worsens, operator is supposed to take a safe shutdown. In the worst case, he is expected to take the actions to minimise the damage/impact due to the accident inside as well as outside the plant.

Operating and safety manuals, a parts of the process know-how supply, form an important part of the document of reference and safety training. These should contain the following information in addition to the regular plant operating instructions:

1. Material Safety Data Sheets (MSDS) for all the raw materials, products, by-products being handled.
2. Process description, chemical reactions involved, thermodynamic data for all the reactions, design criterion for important equipment like reactor, heat exchangers etc.
3. General control philosoply, which explains the logic for alarm settings, mode of control valve actions in case of air failure, trips and interlocks.
4. Emergency actions to be taken in case of power and/or air failure.
5. Process boundary conditions, upper and lower limits of operations, consequence of operating beyond the limits and instructions to return to the normal operating parameters preferably in the format as given in Table 26.4.

Table 26.4. Format process boundary conditions.

Normal parameter or range	Abnormal condition	Consequence	Probable Reasons	Action to be taken

This information can be collected during the process design, from the operating experience of the similar facility, HAZOP study and needs to be updated is based upon the problems faced, analysed and solutions implemented during the plant run.

Operator should be familiar with controls, interlocks, logic behind the same and particularly the logic behind sequencing of operations in a particular order especially for the batch process. Though, the strict rules should be there for not changing the sequence of operations, (particularly for the batch process, where the operations have to be repeated for each batch) to override alarms, trips and interlocks, proper understanding of logic behind these will refrain the operations from making such changes.

Operators should be highly involved, and require high degree of awareness and should be trained not to ignore any abnormality including abnormal sounds, vibration or smell. They need to investigate the reasons for abnormally and take corrective action.

Plant Start-up and Commissioning

Water flushing

During the water runs, the equipment and piping are thoroughly flushed with water. During this time, one has to be careful to avoid the tank failures, which occur mainly due to the following reasons:

Overfilling

If the rate of the water filling and pumping in rate and pressure exceed the vent capacity of the tank or if the vents are closed, tank can fail due to over pressurisation.

Sucking in

This can occur when the tank is emptied at a rate higher than the rate at which air can enter the tank through vents, or if the vents are closed/choked. It can also occur due to sudden cooling of the tank without proper venting.

Nitrogen purging

During the start-up/commissioning the equipment handling flammable gases are purged with nitrogen. One has to be careful during the nitrogen purging and particularly if the vessel entry is required after the nitrogen purging. Since nitrogen is not 'Inert' as normally understood, if a person enters the atmosphere of nitrogen (or the atmosphere rich in nitrogen), he can loose, consciousness without any warning symptoms or distress in as little as 20 seconds.

Death can flow within three to four minutes. Accidents have occurred due to asphyxication because of entry into the vessel purged with nitrogen. Vessels have to be purged with air again after the nitrogen purging for the complete replacement of nitrogen and should be tested for oxygen before allowing the vessel entry.

Dummy runs

It is recommended that plant trial run with water and nitrogen/air (in case of gas) should be carried out for 72 hours, under the simulated conditions with all the instrumentation in line, and all the alarm trips and interlocks can be tested during the run. This also helps the operating staff to get the feel and hands on operating experience.

Introduction of reactants

As a checklist before introducing the hydrocarbon into the system, one must ensure that the following have been carried out:
1. Evacuation procedure is in place and a drill has been carried out with all the personnel involved in commissioning.
2. Company's emergency response procedure is briefed to all the personnel.
3. All the fighting equipments are in place, fire hydrants are charged and are accessible, fire hoses are laid out etc.
4. All the construction activities are to be stopped and erection contractor's personnel should have been evacuated. Entry procedure to the plant should be finalised with preferably single point entry.

Regular Plant Operation and Maintenance

Good operating practices

Safety during the regular operation is achieved by preventing abnormal situation by operating the plant within the prescribed parameters. Teamwork is very important. Communication between the operating and maintenance persons and proper handing over and taking over during the shift change is equally important. A good operator relies on hearing, observation and often intuition to spot tell tale signs of trouble and abnormality. By reporting such signs and analysis the same with the help of other members of the plant, potential problems can be prevented.

Near misses along with the potential situations spotted as above should be thoroughly reviewed, documented and the action should be taken to prevent the occurrence in future.

After considerable operating experience, operators tend to take short cuts. Accidents invariably occur as a result of cumulative effect of these short cuts and other mistakes and not due to a single mistake. Hence this should be strictly avoided.

Modifications

Modifications carried out in the equipment, piping, or operating procedure without thorough analysis can results in an accident. Modifications producing such accident or unforeseen and undesirable side effects can be avoided as follows.

1. All the modifications small, inexpensive or temporary and modification to the operating procedures should be analysed by carrying out HAZOP, safety audit and should be properly examined and approved by at-least 2 levels of hierarchy.
2. After completion of the modification, it should be inspected to make sure that it has been carried out as designed and it looks right. What does not look right is usually not right and should be checked.
3. Drawings must also be updated and replace the earlier drawing in the file.

Maintenance

During maintenance the following considerations are important:

1. All the process material, hazardous or otherwise should be completely removed from the equipment or system and the equipment should be purged properly. Gas analysis after purging should be carried out to ensure safe conditions.
2. Lines carrying hazardous materials or material under pressure should be provided with double block and bleed valves.
3. For any change in scope of work, the old work permit should be withdrawn and a new one should be issued.
4. Isolation should not be removed until the maintenance and testing is complete.
5. Electrical equipment should be positively isolated by removing fuses.

Work permit: Maintenance work permit should setout what is to be done, how the equipment has been isolated and identified and what precautions should be taken.

Plant Decommissioning or Demolition

Process safety continues to play an important role in the final life cycles of a process plant, i.e., in decommissioning and demolition. Due to economic considerations, it may be decided to shut down a plant for a long time, to decommission it or even demolish it.

In this case all the material should be removed from the system. Proper procedure should be developed to consume the process materials from all the equipment to the maximum extent, leftovers need to be drained from the equipment and piping in drums for proper disposal. The equipment and piping can be washed with water or a suitable solvent to make the system free from all the process chemicals.

Hazard gases should be purged first with nitrogen and then with air. After purging, the samples from these equipment may be analysed for the original gas and oxygen content. Equipment and piping may be kept open to atmosphere by proper vents to avoid pressurisation. The plant then should be positively isolated from the rest of the system by the removal of spool pieces from all the incoming and outgoing process and utility lines and blinding (inserting blind at both the ends).

All the electrical supplies should be positively isolated by removing fuses.

Procedure for work permits for vessel entry, hot work should be strictly followed during dismantling and removal or the equipment to avoid accidents.

To sum up the plant management starts right from the conceptual, technology selection/development stages. Selecting inherently safe process and designing inherently safe plant is of prime importance. Simultaneously safety aspects are to be considered during each stage of the project implementation. Teamwork, vigilance and training play vital role in achieving accident free safety record during the regular plant operation.

The petrochemicals form part of our daily life and we cannot do without them. Though chemicals are extremely useful to us, if handled improperly, they can pose danger. In view of the large potential for causing damage to life, property and environment, it is important that the above mentioned steps be put into practice.

Proper HSE management is required right from selecting a process and site till the final product is used or consumed by the last user.

References

Adams, J.A., and **Rogers, D.F.**, *Industrial Chemistry*, Mcgraw-Hill, New York.

Alliger, G., *Industrial Enzymes and Catalysts*, Interscience, New York.

Baum, B., and **Parker, C.H.**, *Handbook of Natural and Synthetic Resins*, Reinhold, New York.

Berenson, C., *Inorganic Chemistry*, Wiley-Interscience, New York.

Billmeyer, W.F., *Text Book of Polymer Science*, John Willey and Sons, New York.

Blake, W.P., *Modern Plastics Encyclopedia*, Mcgraw-Hill, New York.

Boynton, R.S., *Chemistry and Technology of Paints and Synthetic Resins*, John-Willey, USA.

Brennan, J.J., *Toxicity and Safe Handling of Hazardous Chemicals*, Academic Press, New York.

Brydson, J.A., *Corrosion and Its Control*, Elsevier Applied Science, London.

Brydson, J.A., *Handbook of Electroplating*, Elsevier Applied Science, London.

Budnikov, P.O., *Green Chemistry*, Prentice-Hall, UK.

Burke, J.E., *Chemistry and Technology of Dyes*, Elsevier Applied Science, London.

Camp, T.R., *Water and Its Impurities*, Reinhold, New York.

Chanlett, E.T., *Surfactants and their Applications*, Mcgraw-Hill, New York.

Clauser, H.R., *Organic Chemistry*, Reinhold, USA.

Cooper, H.B.H., and **Rossano, A.T.**, *Industrial Fermentation*, Mcgraw-Hill, New York.

Conover, M.S., *Pesticides and Insecticides*, Wiley-Interscience, New York.

Craig, A.S., *Industrial Chemicals*, Philosophical Library, New York.

DeGarmo, E.P., *Industrial Chemistry*, John Willey-Interscience, New York.

Dian, K.L., *Organic Chemistry of Macromolecules*, Prentice-Hall, Englewood Chiffs, New Jersey.

Donnet, J.B., *Explosives and propellants*, Marcel Dekker, New York.

Doremus, R.H., *Industrial Chemistry of Carbon*, Wiley-Interscience, New York.

Eckenfelder, W.W., and **O'Connor, D.J.**, *Petroleum Refinery and Petrochemicals*, Pergamon, UK.

Ewing, G.W., *Handbook of Chemistry*, Mcgraw-Hill, New York.

Faulk, B.F., *Technology of Carbon and Graphite*, Van Nostrand Reinhold, New York.

Gardon, J.L., *Encyclopaedia of Polymer Technology*, Interscience, New York.

Hampel, B.H., *Electrochemical Encyclopaedia*, Reinhold, New York.

Harvey, B.G., *Nuclear Physics and Chemistry*, Prentice-Hall, UK.

Heckert, W.W., *Technology of Synthetic Fibres*, Willey-Interscience, New York.

Jacobs, M.B., *Ceramic and Glass*, Wiley-Interscience, New York.

James, A.N., *Industrial Chemistry*, Noyes, UK.

Jolles, Z.E., *Chemistry of Polymerisation*, Noyes, UK.

Jurah, J.M., *Chemistry of Petroleum*, Mcgraw-Hill, New York.

Kent, J.A., *Riegel's Industrial Chemistry*, Reinhold, USA.

Kingzett, M.J., *Chemical Encyclopaedia*, Van Nostrand, USA.

Kirk-Othmer., *Encyclopaedia of Chemical Technology*, Wiley-Interscience, New York.

Kit, B., and Evered, D.S., *Chemistry and Technology of Plastic and Rubber*, Mcgraw-Hill, New York.

Laitinen, H.A., and W.E., Harris, *Physical and Inorganic Chemistry*, Mcgraw-Hill, New York.

Lenz, M.Z., *Mechanism of Catalysts*, Pergamon Press, New York.

Lewis, F.W., *Chemical Synthesis*, John Wiley and Sons, New York.

McKelvey, J.M., *Polymer Processing*, Wiley Interscience, New York.

Morton, M., *Rubber Technology*, Van Nostrand, Reinhold, USA.

Noll, W., *Chemistry and Technology of the Silicones*, Academic Press, New York.

Ostwald, W.J., *Heavy Inorganic Chemicals*, Academic Press, New York.

Patton, T.C., *Alkyd Resin Technology*, Wiley Interscience, New York.

Pauling, L.J., *Chemistry of Paints, Pigments and Varnishes*, Cornell University Press, Ithaca, New York.

Penn, W.S., *Insulating Polymers*, Maclaren, London.

Perry, R.H., *Chemical Engineer's Handbook*, Mcgraw-Hill, New York.

R. Norris, Shreve., and Joseph, A. Brink., *Chemical Process Industries*, Mcgraw-Hill, New York.

Rhodes, T., and Carroll, G., *Organic Chemistry*, Mcgraw-Hill, New York.

Roger, S.W., *Chromatographic Fractionation*, Academic Press, New York.

Singer, F.S., *Agro-Chemical Industries*, Chemical Publishing, UK.

Siting, M., *Chemistry of Fermentation*, Noyes, UK.

Stephenson, R.N., *Introduction to Chemical Process Industries*, Reinhold, USA.

Swern, W., *Environmental Engineering*, Reinhold Publishing Company, USA.

Thompson, D.C., *Chemistry of organo metallic compounds*, John Wiley and Sons, New York.

Urbanski, T., *Chemistry and Technology of Explosives*, Pergamon, UK.

Van Wazer, J.R., *Industrial Chemistry*, Wiley-Interscience, New York.

Walker, E.E., *Separation and Crystallisation*, Wiley interscience, New York.

William, A.Y., *Introduction to Science and Technology of Rubber*, Van Nostrand Reinhold, New York.

Wright, P., *Purification of organic compounds*, Maclaren, London.

Yaffe, L., *Nuclear Chemistry*, Mcgraw-Hill, New York.

Index